BASIC AND APPLIED ASPECTS OF SEED BIOLOGY

Current Plant Science and Biotechnology in Agriculture

VOLUME 30

Aims and Scope
The book series is intended for readers ranging from advanced students to senior research scientists and corporate directors interested in acquiring in-depth, state-of-the-art knowledge about research findings and techniques related to all aspects of agricultural biotechnology. Although the previous volumes in the series dealt with plant science and biotechnology, the aim is now to also include volumes dealing with animals science, food science and microbiology. While the subject matter will relate more particularly to agricultural applications, timely topics in basic science and biotechnology will also be explored. Some volumes will report progress in rapidly advancing disciplines through proceedings of symposia and workshops while others will detail fundamental information of an enduring nature that will be referenced repeatedly.

The titles published in this series are listed at the end of this volume.

Basic and Applied Aspects of Seed Biology

Proceedings of the Fifth International Workshop on Seeds, Reading, 1995

Editors

R.H. ELLIS[1], M. BLACK[2], A.J. MURDOCH[1], T.D. HONG[1]
[1]Department of Agriculture, The University of Reading, Earley Gate, P.O. Box 236, Reading RG6 6AT, UK
[2]Division of Biosphere Science, King's College London, Campden Hill Road, London W8 7AH, UK

KLUWER ACADEMIC PUBLISHERS
DORDRECHT / BOSTON / LONDON

Library of Congress Cataloging-in-Publication Data is available.

ISBN 0-7923-4363-8

Published by Kluwer Academic Publishers BV,
PO Box 17, 3300 AA Dordrecht, The Netherlands

Kluwer Academic Publishers BV incorporates
the publishing programmes of
D. Reidel, Martinus Nijhoff, Dr W. Junk and MTP Press.

Sold and distributed in the United States and Canada
by Kluwer Academic Publishers, PO Box 358,
Accord Station, Hingham, MA 02018-0358, USA

In all other countries, sold and distributed
by Kluwer Academic Publishers Group, Distribution Center,
PO Box 322, 3300 AH Dordrecht, The Netherlands

Printed on acid-free paper

Contents

SESSION II
DORMANCY

SESSION III
GERMINATION AND VIGOUR

Preface

The Fifth International Workshop On Seeds was held at The University of Reading, UK, from Sunday 10 to Friday 15 September, 1995. Some 230 seed scientists, from a wide range of disciplines (botanists, biochemists, ecologists, agriculturalists, foresters, and commercial seedsmen), from 31 countries (from Europe, the Americas, and Asia) participated in the workshop. A large number of oral and poster presentations were made during the workshop and we are pleased to publish so many of them in these Proceedings. The papers herein are listed by the sessions in which they were presented but, as is often the case, many papers cover a broader range of topics than the session titles imply. For seed physiologists, ecologists, and technologists, this book collates much of the current research on seeds.

The Workshop continued the fine tradition begun in 1979 by Professor Mayer in Jerusalem, Israel, and continued in Wageningen, The Netherlands (1985), in Williamsburg, USA, (1989), and in Angers, France (1992). The Fifth Workshop was held at Reading to mark the retirement (at least from full-time teaching and research!) from The University of Reading of Professor E.H. Roberts. In the first paper some of his significant contributions are highlighted. But, as ever with research, some of it he regards as 'unfinished business' and so he outlines a few tasks for us in the coming years. The next workshop will be held in Mexico in late 1998/early 1999, though it is too early to say just how much of the unfinished business described by Eric Roberts herein will be completed by then!

We are of course indebted to the many people and organizations who helped make this Workshop possible and successful.

Financial support was provided by the following organizations whom we thank most sincerely:

Annals of Botany Company
CABI
Department of Agriculture, The University of Reading
Elsoms Seeds
The Gatsby Foundation
Germains
Hilleshög
International Science Foundation
Kluwer Academic Publishers
Marcel Dekker
Maribo Seed
PBI

R.H. Ellis, M. Black, A.J. Murdoch, T.D. Hong (eds.), Basic and Applied Aspects of Seed Biology, pp. xiii–xvi.
© *1997 Kluwer Academic Publishers, Dordrecht. Printed in Great Britain.*

The Royal Society
Sharpes International Seeds
Zeneca Seeds

Scientific Committee
J.D. Bewley
M. Black
D. Côme
A. Mayer
C.M. Karssen

Organising Committee
M. Black
R.H. Ellis
P. Gosling
D. Gray
P. Halmer
R.D. Smith

We are most grateful to Horticulture Research International, Wellesbourne, Warwickshire, Forest Research Station, Farnham, Surrey, and Royal Botanic Gardens Kew, Wakehust Place, Sussex, for receiving visitors from the Workshop so generously, and to the Vice-Chancellor of The University of Reading for opening the Workshop and providing such an enjoyable reception.

Finally, we thank Mrs Dorothy Roberts and Mrs Sue Redman for their unfailing hardwork and support in enabling the Workshop to proceed.

R.H. Ellis
M. Black
A.J. Murdoch
T.D. Hong

September 1996

The Sixth International Workshop on Seeds will be held in late 1998/ early 1999 in Mexico, and will be organized by Dr J. Vasquez-Ramos, Departamento de Bioquimica, Facultad de Quimica, UNAM, Mexico DF, Mexico.

Plate 1. Participants, Fifth International Workshop on Seeds, The University of Reading, 1995

Plate 2. *(left to right)* Professor M. Black, Professor E.H. Roberts, Professor R.H. Ellis and Dr A.J. Murdoch

Unfinished Business

E.H. ROBERTS

Department of Agriculture, University of Reading, Earley Gate, PO Box 236, Reading, RG6 6AT, UK

Abstract

The relevance of seed science – especially that concerned with viability and dormancy – to solving practical problems concerned with plant breeding and genetic conservation is discussed. It is unfortunate that, in spite of considerable effort and voluminous literature, we are still uncertain about the physiological and molecular mechanisms involved in these processes. But take heart, for sometimes such knowledge is not necessary, and may not even be helpful for predicting what will happen under different circumstances to populations or organisms such as seeds. Solutions to practical problems, however, often demand quantitative predictions which are valid in a wide range of circumstances; and such solutions are sometimes achievable. Nevertheless, the possibilities of productive interactions between those concerned with fundamental problems and those concerned with practical problems are increasing, and this series of International Workshops provides important opportunities to encourage this symbiosis.

Introduction

In his introduction to the Fourth International Workshop on Seeds, Cees Karssen (1993) pointed out that at least two categories of seed scientists can be distinguished: those whose primary interest is in plant physiology, biochemistry or molecular biology and who use seeds as convenient material to study the phenomena which interest them; and those who aim to learn more about seeds as such. I suggest the latter category can be sub-divided into those who are driven essentially by scientific curiosity, and those whose research arises out of attempts to solve practical problems in agriculture, horticulture and forestry.

Whichever category one belongs to it has become increasingly relevant to be aware of what is going on in the other camps. For attempts to solve practical problems often raise issues or suggest approaches to more fundamental problems; while on the other hand many fundamental studies are beginning to advance sufficiently to suggest approaches towards solving practical problems. It is for these reasons that this series of International Workshops on Seeds is becoming increasingly valuable: they provide a good opportunity for all categories of seed scientists, who might otherwise not meet, to exchange information and ideas and learn from each other.

When invited to open the 1995 Workshop I was told I was allowed a certain degree of self-indulgence, and so I propose to begin by explaining how my own interest in seed science arose out of trying to solve practical problems. In some

R.H. Ellis, M. Black, A.J. Murdoch, T.D. Hong (eds.), Basic and Applied Aspects of Seed Biology, pp. xvii–xxv.
© *1997 Kluwer Academic Publishers, Dordrecht. Printed in Great Britain.*

cases the research suggested some approaches to more fundamental aspects of the subject. But I came to realise that fundamental studies by themselves seldom offered practical solutions.

One analogy would be that although it is important for an engineer who designs a bridge to know the physical properties of the metal he or she uses – tensile strength, coefficient of expansion, susceptibility to corrosion, etc. – a knowledge of inter-atomic forces which underlie many of these properties is of little help. Furthermore, it is not always possible adequately to predict the relevant properties from such knowledge. In any case, knowledge of inter-atomic forces does not suggest the design of an appropriate structure, because the problems facing the engineer are of a different hierarchical level of organization, and different concepts are needed. One can also draw relevant examples from plant physiology. After extensive work on phytohormones over the last 60 years, if we are honest we have to admit we still do not know how they work, and current applications depend little on such speculations concerning mechanisms that we do have. Then again the detailed biochemical knowledge of photosynthesis is now impressive, but none of this has led to improved crop production where improvements have depended on research at a different hierarchical level and more empirical approaches.

The paper is entitled 'Unfinished Business' because, in all the projects in which I have been involved, I am fully aware that any solutions which have been arrived at are based on approximations which are capable of further improvement. Furthermore, in no case which I mention do we yet understand the underlying mechanisms.

Seed Science in Plant Breeding and Genetic Conservation

My first job as a scientist was as a plant breeder and crop physiologist in the West African Rice Research Station – a very small organization in Sierra Leone. It is a country with a difficult climate which was originally known as the 'White Man's Grave' because many early visitors from Europe died from yellow fever and malaria after only brief sojourns there. By the time I arrived, such problems had been solved by reliable inoculation for the first disease and reliable prophylactics for the second. But this did not alter the fact that the climate, hot and humid, is bad for seed survival: conventionally stored sun-dried rice seeds survive adequately for sowings at the beginning of the next season, but viability is soon completely lost after that. As a consequence the germplasm collection had to be re-planted every year – a considerable undertaking, especially in a transplanted crop, because stringent precautions were needed to prevent mechanical mixture or mistakes at sowing, transplanting, harvest, threshing and drying. It was therefore time-consuming and costly. Hence my interest in seed viability and longevity.

Attempts to discern patterns of survival behaviour in the published literature on stored wheat seed led to the development of three seed viability equations which predicted percentage viability after any period in a wide range of seed moisture contents and temperatures (Roberts, 1960). These equations were then shown experimentally to be valid also for rice (Roberts, 1961), and subsequently, after leaving Africa, for other cereal and legume crops (Roberts and Abdalla, 1968). While these equations were satisfactory enough to contribute to the rational design of the first National Seed Storage Laboratory for Genetic Resources in Japan (Ito, 1972), their predictive value was limited for two reasons: (1) they became inaccurate if extrapolated much beyond ambient or near-ambient conditions, and (2) they took no account of initial seed quality which can have a profound effect on subsequent longevity; they assumed that the seed was of good but undefined quality.

These problems were later resolved by research in which Richard Ellis took the lead, as a research student and then post-doctoral fellow (Ellis and Roberts, 1980a,b). The extrapolation problem was solved by extensive experimentation over a wide range of conditions which resulted in modification of the temperature and moisture-content terms. The problem concerning the variation in initial seed quality was overcome by recognizing that the slope of the linear probit seed survival curve is a function of moisture content and temperature, but not initial seed quality; whereas initial seed quality affects the intercept but has no effect on the slope. Accordingly the intercept of a probit survival curve (K_i) is a measure of seed quality and potential longevity. The survival curve originates from this intercept but actual longevity is then modified by the slope ($1/\sigma$) of the survival curve which is determined by the conditions the seeds are subjected to during storage. This measure of seed-lot quality K_i also turned out to be a good indicator of seed vigour or field performance (e.g. Ellis and Roberts, 1980c; Roberts, 1986; Khah *et al.*, 1989; Pieta Filho and Ellis, 1991; Demir and Ellis, 1993).

Thus was established a single viability equation which applies to very many, if not all, species which produce orthodox seeds (Dickie *et al.*, 1990; Zewdie and Ellis, 1991b). Furthermore there is now evidence that the temperature coefficient in the viability equation is identical for widely different species (Dickie *et al.*, 1990) and that probably the response to moisture would also be identical if it were expressed in terms of water potential or equilibrium relative humidity (Roberts and Ellis, 1989; Zewdie and Ellis, 1991b). Thus here is some unfinished business – to simplify further the viability equation in which the temperature and water potential coefficients are identical for most if not all species.

The original empirical viability equations – taken in conjunction with considerations of an association between viability and the accumulation of mutations in the surviving population (see below), engineering and economic factors – had been used to establish the 'preferred conditions' for long-term seed storage for genetic conservation adopted by the FAO Panel of Experts in 1974 (FAO, 1975), *viz.* storage in sealed containers at $5\pm1\%$ moisture content (wet weight basis) at –18°C or less. The development of the improved viability equation did not affect these recommendations and so they were subsequently adopted by the International Board for Plant Genetic Resources (Cromarty *et al.*, 1982), and essentially have recently been reaffirmed by the organization into which it has evolved, the International Plant Genetic Resources Institute, and also by FAO (FAO/IPGRI, 1994).

It has always been recognized that there are limits to the application of the viability equation. At high moisture contents there is a halt to the trend of decreasing longevity with increase in moisture content, and indeed a reversal of the trend under aerobic conditions (Ibrahim and Roberts, 1983; Ibrahim *et al.*, 1983). Otherwise the well-known records of long periods of survival of wild orthodox seeds in the soil would be impossible. The moisture content at which this changed relationship occurs (the upper limit for application of the viability equation) varies from about 15 to 26% moisture content depending on the species, but at a more-or-less common water potential, *viz*, -10 to -20 MPa or in equilibrium with about 90% relative humidity at 20°C (Zewdie and Ellis, 1991a). This is the level at which some metabolism including respiration becomes possible and the coincident change in longevity relationships in response to water seems to support the original idea of Villiers (1975) that the prolonged survival of moist orthodox seeds depends on repair mechanisms.

There is also a limit at very low seed moisture contents to the application of the viability equation. There is agreement that the water potential at which this happens is probably common to all orthodox species but the precise value is still unresolved. We believe the evidence suggests a value in equilibrium with about 10 to 13% relative humidity at 20°C (Ellis *et al.*, 1995) whereas Vertucci and her colleagues argue a greater value, about 20–25% relative humidity at 25°C (Vertucci *et al.*, 1994).

Here, then, is more unfinished business which is vitally important from the practical point of view since it affects the advice given to genebanks and the security of millions of accessions (600 000 of which are held within the seed banks of the International Agricultural Research Centres of the Consultative Group on International Agricultural Research). Hence, there is some urgency for doing further business in this area.

When storing seeds for genetic conservation, there is a concern not only for maintaining viable seeds but also for maintaining the genetic intregity of each accession. Genetic integrity can be degraded in two ways: (1) by differential loss of viability amongst different genetic components of genetically heterogeneous accessions – especially important in the case of landraces, and (2) the accumulation of mutations during storage – a potential problem with both heterogeneous and homogeneous accessions. Here I will refer only briefly to the mutation problem.

The amount of mutation in the surviving seeds of an accession is related to the percentage loss of viability. Originally it was thought there was a simple relationship between percentage viability and chromosome damage irrespective of storage conditions, and therefore irrespective of how rapidly viability was lost (Roberts *et al.*, 1967; Abdalla and Roberts, 1969). Experiments over a much wider range of conditions, however, have shown that in very dry seeds there is far more chromosome damage in surviving seeds for a given loss of viability than in more moist seeds (Rao *et al.*, 1987). This is because rates of loss of viability and rates of accumulation of chromosome damage have different moisture-content coefficients (Rao *et al.*, 1988). It nevertheless remains true that it is still best to store seeds very dry because the rate of accumulation of chromosome damage is less than at higher moisture contents. However it does mean that when seeds are stored under such conditions for genetic conservation, it is even more important than we originally thought not to allow the percentage viability of accessions to fall very far before a new stock of the accession is regenerated. The current preferred standards are 85% viability for most crop species but 75% for some vegetables and even lower for some wild and forest species (FAO/IPGRI, 1994).

The reasons for the accumulation of lesions which lead to chromosome breakage and gene mutations are not clear (Roberts, 1988); hence more unfinished business.

Fortunately all the major annual crops, many perennial crops and most weed species produce seeds which obey the viability equation and are therefore easy to store and handle and were therefore defined as orthodox (Roberts, 1973a). In contrast a second category of seeds was recognized at that time which did not obey the simple laws defined by the viability equation, which cannot be dried much without damage, and are difficult to store for more than short periods. They were therefore described as recalcitrant. Despite many efforts, little progress has been made in significantly improving the storage of recalcitrant seeds, although cryogenic procedures as used on other moist tissues still hold some hope; but the problems are formidable. More speculative business is needed. I say speculative because the probability of a real return on research investment is small, but, if successful, the rewards would be tremendous.

We now know that there is at least one other category of seeds which can be dried and cooled to some extent with benefit, but even when dry enough to prevent metabolism as we know it – say about 10% moisture content – they are damaged if cooled much below 10°C; hence conventional long-term storage is impossible (Ellis *et al.*, 1990; 1991a,b). For want of a better term, seeds in this category have been termed 'intermediate'. It is not clear how numerous such species are. A recent critical survey of the literature has concluded that out of 6866 species on which there is some relevant information, 6091 are orthodox or probably orthodox, 502 are recalcitrant or probably recalcitrant and 128 are intermediate or probably intermediate; information on the remainder was equivocal (Hong *et al.*, 1996). This is a very small sample out of the 250 000 species of flowering plants which currently inhabit this planet, but nevertheless it shows that the intermediate category is not rare. There is a lot of business to be done on such species.

I now return to rice breeding in Sierra Leone. Plant breeding is a slow process. For example, if genetic variation is created by hybridization in a self-pollinating crop such as rice, several generations of selections are needed before segregation is reduced sufficiently to be ignored and the progeny breeds 'true'. No selection can occur in the F_1 generation, and vigorous selection is not advantageous until

F_4 and F_5. After a few years of selection, about three years of testing at several sites is needed before a new cultivar can be released. Anything that can be done to speed up this process, which otherwise may take ten years or more, will be an advantage.

One approach is to get through the first few generations, when selection in the intended growing environment is not important, as rapidly as possible. To do this one needs to minimize the generation time from sowing to harvest and minimize the inter-generation time from harvest to sowing. The former was achived by subjecting the plants to short-days (Roberts and Carpenter, 1962) and the latter by removing seed dormancy. It would, of course, be possible to remove seed dormancy by genetic selection, but this would be a mistake because reduced innate dormancy results in a tendency to sprout in the panicle. A physiological method for temporary removal was required. As a result of a wide range of physiological investigations, it was found that the rate of loss of dormancy is linearly related to temperature with a Q_{10} of about 3.1–3.4, and that this Q_{10} does not vary amongst genotypes (Roberts, 1962; 1965). From this relationship a simple procedure was developed in which seeds immediately after harvest were incubated at 47°C so that dormancy was completely lost within a week and the next generation could be sown. This particular temperature was chosen by combining a knowledge of viability and dormancy studies so that rapid dormancy loss could be achieved with no detectable loss of viability. Thus a combination of photoperiodic treatment of plants and high temperature treatment of seeds resulted in a technology for accelerating breeding pro-grammes (Carpenter and Roberts, 1962).

Much later an alternative dormancy-removal method was developed, based on a combination of physical and chemical methods which exploits the synergism of both physical and chemical treatments and which is sufficiently powerful to deal with the even more dormant cultivars of the African species of rice, *Oryza glaberrima* (Ellis *et al.*, 1983). The main motive here was to provide a technique for dormancy removal to facilitate efficient viability testing in genetic resources centres.

The research on rice seed dormancy serendipitously led to the discovery that terminal oxidase inhibitors stimulate loss of dormancy (Roberts 1964a,b). This and other evidence led to further investigations which suggested the pentose phosphate respiratory pathway is important in the early stages of germination and needs to be stimulated for dormancy to be lost (Roberts, 1969; 1973b; Roberts and Smith, 1977). However others have raised criticisms and so the role of this highly active pathway in the early stages of seed germination remains obscure. Since there is still no satisfactory dormancy hypothesis and no satisfactory explanation of photoperiodism, it is by no means clear why the techniques for speeding up breeding programmes are effective – more unfinished business.

Seed Science and Seed Ecology

When I left Africa and accidentally became an academic, I wondered whether a knowledge of factors which affect seed dormancy and viability could be used to develop more rational methods of weed control. The premise was this: most weed species which depend on seed propagation are present for the most part as seeds in the 'soil bank' in very large numbers. They are ephemeral opportunists that rely on soil disturbance to bring them to the surface of bare soil to enable them to germinate and produce another generation to replenish the seed bank. For this life style they need to be capable of long periods of survival in the soil which, in turn, depends on two properties – intrinsic longevity under these conditions and an effective dormancy mechanism which prevents germination until conditions are such that the germinated seedling has a reasonable probability of achieving reproductive maturity. Originally the long viability periods were a puzzle until Villiers (1975) suggested the role of repair mechanisms in maintaining viability under moist conditions. So the major concern became dormancy.

In examining the environmental factors which control dormancy it became clear that positive interactions between several factors – especially light, alternating temperatures and nitrate ions, and sometimes stratification – are very important, even second and third order interactions (Popay and Roberts, 1970a,b; Vincent and Roberts, 1977; Roberts and Benjamin, 1979; Totterdell and Roberts, 1980; Roberts and Totterdell, 1981; Goedert and Roberts, 1986). Thus if some seeds were tested for light sensitivity under typical laboratory conditions at constant temperature an experimenter could easily be misled into concluding the phytochrome response was absent. Orthogonal multi-factorial experiments are crucial in research in this area.

We have argued that the powerful positive interactions which are so common in the responses of these small-seeded weed species to environmental factors have evolved in response to a need for a reliable set of signals which inform the seed whether conditions are suitable to embark on the next generation – not only signalling whether the seed is at or near the soil surface but also whether there is competitive vegetation above and whether it is a suitable time of the year (Roberts, 1973b; Roberts, 1981).

However, although we now know much more about these responses and have achieved partial success in modelling the complex alternating-temperature responses (Murdoch *et al.*, 1989) and can understand more about field behaviour, we are still a long way from being able to predict the consequences of agricultural operations in sufficient detail. More business is needed here. Nevertheless, it has been possible to use such knowledge as we have to decide rationally on the feasibility of either eradication or containment policies in different species (Murdoch and Roberts, 1982; Murdoch, 1988).

I end with a quotation from Thornley (1987) which I think encapsulates what I have been talking about:

"The task of the scientist is to describe, predict, understand and apply – not necessarily in that order. Prediction and application may precede understanding, and much understanding may be inapplicable to practical problems."

Acknowledgements

I would like to thank all those research students who have worked on the topics touched on here and who have contributed a great deal to the ideas expressed, notably (in alphabetical order): Farouk Abdalla, Sampath Benjamin, Wendy Bridle (neé Major), Anna Dourado, Richard Ellis, Sala Gaber, Mahteme Giorgis, Clara Goedert, Tran Dang Hong, Ahmed Ibrahim, Massoud Khah, Philippe Koole, Alistair Murdoch, Sam Olosuyi, Kwabena Osei-Bonsu, Ian Popay, Nanduri Kamaswara Rao, Luigi Russi, Roger Smith, Usep Soetisna, Sue Totterdell, Elizabeth Vincent, and Mehari Zewdie.

References

Abdalla, F.H. and Roberts, E.H. 1969. *Annals of Botany* 32: 119–136.

Carpenter, A.J. and Roberts, E.H. 1962. *Empire Journal of Experimental Agriculture* 30: 117–131.

Cromarty, A.S., Ellis, R.H. and Roberts, E.H. 1982. *The Design of Seed Storage Facilities for Genetic Conservation*, pp. 100. Rome: International Board for Plant Genetic Resources.

Demir, I. and Ellis, R.H. 1993. *Seed Science Research* 3: 247–257.

Dickie, J.B., Ellis, R.H., Kraak, H.L., Ryder, K. and Tompsett, P.B. 1990. *Annals of Botany* 65: 197–204.

Ellis, R.H., Hong, T.D. and Roberts, E.H. 1983. *Seed Science and Technology* 11: 77–112.

Ellis, R.H., Hong, T.D. and Roberts, E.H. 1990. *Journal of Experimental Biology* 41: 1167–1174.

Ellis, R.H., Hong, T.D. and Roberts, E.H. 1991a. *Journal of Experimental Biology* 42: 653–657.

Ellis, R.H., Hong, T.D. and Roberts, E.H. 1991b. *Seed Science Research* 1: 69–72.

Ellis, R.H., Hong, T.D. and Roberts, E.H. 1995. *Annals of Botany* 76: 521–534.

Ellis, R.H. and Roberts, E.H. 1980a. *Annals of Botany* 45: 13–30.

Ellis, R.H. and Roberts, E.H. 1980b. *Annals of Botany* 45: 31–37.

Ellis, R.H. and Roberts, E.H. 1980c. In: *Seed Production*, pp. 605–635 (ed. P.D. Hebblethwaite). London: Butterworths.

FAO (Food and Agricultural Organization of the United Nations). 1975. *Report of the Sixth Session of the FAO Panel of Experts on Plant Exploration and Introduction*, pp. 37. Rome: FAO.

FAO/IPGRI. 1994. *Genebank Standards*, pp. 13. Rome: FAO/International Plant Genetics Resources Institute.

Goedert, C.O. and Roberts, E.H. 1986. *Plant, Cell and Environment* 9: 521–525.

Hong, T.D., Linington, S. and Ellis, R.H. 1996. *Seed Storage Behaviour: a Compendium, Handbook for Genebanks No. 4*, pp. 656. Rome: International Plant Genetic Resources Institute.

Ibrahim, A.E. and Roberts, E.H. 1993. *Journal of Experimental Botany* 34: 620–630.

Ibrahim, A.E., Roberts, E.H. and Murdoch, A.J. 1983. *Journal of Experimental Botany* 34: 631–640.

Ito, H. 1972. In: *Seed Viability*, pp. 405–416 (ed. E.H. Roberts). London: Chapman and Hall.

Karssen, C.M. 1993. In: *Proceedings of the Fourth International Workshop on Seeds*, Vol.1. pp. 3–9 (eds. D. Côme and F. Corbineau). Paris: ASFIS.

Khah, E.M., Roberts, E.H. and Ellis, R.H. 1989. *Field Crops Research* 20: 175–190.

Murdoch, A.J. 1988. *Aspects of Applied Biology* 18: 91–98.

Murdoch, A.J. and Roberts, E.H. 1982. *Proceedings of the 1982 British Crop Protection Conference – Weeds*, 741–748.
Murdoch, A.J., Roberts, E.H. and Goedert, C.O. 1989. *Annals of Biology* 63: 97–111.
Pieta Filho, C. and Ellis, R.H. 1991. *Seed Science Research* 1: 179–185.
Popay, A.I. and Roberts, E.H. 1970a. *Journal of Ecology* 58: 103–122.
Popay, A.I. and Roberts, E.H. 1970b. *Journal of Ecology* 58: 123–139.
Rao, N.K., Roberts, E.H. and Ellis, R.H. 1987. *Annals of Botany* 60: 85–96.
Rao, N.K., Roberts, E.H. and Ellis, R.H. 1988. *Annals of Botany* 62: 245–248.
Roberts, E.H. 1960. *Annals of Botany* 24: 12–23.
Roberts, E.H. 1961. *Annals of Botany* 25: 381–390.
Roberts, E.H. 1962. *Journal of Experimental Botany* 13: 75–94.
Roberts, E.H. 1964a. *Physiologia Plantarum* 11: 14–29.
Roberts, E.H. 1964b. *Physiologia Plantarum* 17: 30–43.
Roberts, E.H. 1965. *Journal of Experimental Botany* 16: 341–349.
Roberts, E.H. 1969. *Symposium of the Society of Experimental Biology* 23: 181–192.
Roberts, E.H. 1973a. *Seed Science and Technology* 1: 499–514.
Roberts, E.H. 1973b. In: *Seed Ecology*, pp. 189–218 (ed. W. Heydecker). London: Butterworths.
Roberts, E.H. 1981. *Annals of Applied Biology* 98: 522–555.
Roberts, E.H. 1986. In: *Physiology of Seed Deterioration*, pp. 101–123 (eds. M.B. McDonald, Jr. and C.J. Nelson). Madison: Crop Science of America.
Roberts, E.H. 1988. In: *Senescence and Aging in Plants*, pp. 465–498 (eds. L.D. Noodén and A.C. Leopold). New York: Academic Press.
Roberts, E.H. and Abdalla, F.H. 1968. *Annals of Botany* 32: 97–117.
Roberts, E.H., Abdalla, F.H. and Owen, R.J. 1967. *Symposium of the Society of Experimental Biology* 21: 65–100.
Roberts, E.H. and Benjamin, S.K. 1979. *Seed Science and Technology* 7: 379–392.
Roberts, E.H. and Carpenter, A.J. 1962. *Nature* 196: 1077–1078.
Roberts, E.H. and Ellis, R.H. 1989. *Annals of Botany* 63: 39–52.
Roberts, E.H. and Smith, R.D. 1977. In: *The Physiology and Biochemistry of Seed Dormancy and Germination*, pp. 385–411 (ed. A.A. Khan). Amsterdam: Elsevier, North Holland.
Roberts, E.H. and Totterdell, S. 1981. *Plant, Cell and Environment* 4: 97–106.
Thornley, J.H.M. 1987. *Manipulation of Flowering* pp. 67–79 (ed. J.G. Atherton). London: Butterworths.
Totterdell, S. and Roberts, E.H. 1980. *Plant, Cell and Environment* 3: 3–12.
Vertucci, C.W., Roos, E.E. and Crane, J. 1994. *Annals of Botany* 74: 531–540.
Villiers, T.A. 1975. In: *Crop Genetic Resources for Today and Tomorrow* pp. 297–316 (eds. O.H. Frankel and J.G. Hawkes). London: Cambridge University Press.
Vincent, E.M. and Roberts, E.H. 1977. *Seed Science and Technology* 5: 659–670.
Zewdie, M. and Ellis, R.H. 1991a. *Seed Science and Technology* 19: 295–302.
Zewdie, M. and Ellis, R.H. 1991b. *Seed Science and Technology* 19: 319–329.

1. Desiccation Tolerance and Long Term Structural Stability

F.A. HOEKSTRA, W.F. WOLKERS, J. BUITINK and E.A. GOLOVINA

Department of Plant Physiology, Wageningen Agricultural University, Arboretumlaan 4, 6703 BD Wageningen, The Netherlands

Abstract

Suspended life in completely desiccated organisms requires the protection of proteins and phospholipids in the cellular membranes. Desiccation tolerant organisms usually contain large amounts of sugars. Di-, tri- and tetra-saccharides in particular, play a major role in preventing fusion, phase transitions and most likely phase separations in dry membranes. The ability of the sugars to form a glass and to directly interact with the polar headgroups of the phospholipids are the hypothesized mechanisms involved in this protection. Sugars also protect labile proteins from major structural rearrangements during dehydration.

Dry membranes *in situ* often have elevated phase transition temperatures (T*m*) which may be explained by insufficient interaction of the sugars with the polar headgroups. Thus, gel phase may occur in dry desiccation tolerant organisms at room temperature. It seems necessary to melt this gel phase before imbibition, or leakage and damage will ensue. The presence of a glassy state at imbibition is not involved in imbibitional injury. Alternatives for the role of sugars in the suppression of T*m* with drying are discussed.

Drying of desiccation sensitive organisms and ageing of desiccation tolerant organisms lead to phospholipid breakdown, free fatty acid accumulation and leakage of endogenous solutes, all of which are indicative of membrane deterioration. In contrast, protein secondary structure is very stable during ageing.

Introduction

In higher plant organs such as seed and pollen, tolerance to desiccation is widespread. Periods of suspended life in open storage at 20°C may last several years or even decades in the case of seeds (Priestley, 1986), but only a few months or less in the case of pollen (Hoekstra, 1995). The mechanism of tolerance when H_2O is almost completely removed (5% moisture content) has long been a focus of research (reviewed in Crowe *et al.*, 1984; 1992; 1995; Hoekstra *et al.*, 1989a; Vertucci and Farrant, 1995). What emerges from these studies is that for the desiccated organs to survive, protection of biomolecules such as proteins and phospholipids in the membranes is required.

When non-tolerant pollens or embryos are air-dried, they leak all of their endogenous solutes upon imbibition, in contrast to tolerant ones (Senaratna and McKersie, 1983; Hoekstra and Van Roekel, 1988; Hoekstra *et al.*, 1989b; Tetteroo *et al.*, submitted). This excessive leakage caused by drying indicates

R.H. Ellis, M. Black, A.J. Murdoch, T.D. Hong (eds.), Basic and Applied Aspects of Seed Biology, pp. 1–12.
© *1997 Kluwer Academic Publishers, Dordrecht. Printed in Great Britain.*

massive loss of membrane integrity in the non-tolerant organisms.

Desiccation tolerant cells are characterized by large amounts of sugars (up to 30% of dry weight) generally accumulating during the acquisition of desiccation tolerance (Madin and Crowe, 1975; Crowe *et al.*, 1984; Fischer *et al.*, 1988; Hoekstra and Van Roekel, 1988; Leprince *et al.*, 1990). This led to the suggestion that sugars may play a role in desiccation tolerance (Crowe *et al.*, 1984). Experiments on dehydrating model membranes in the presence of sugars have confirmed that certain sugars can protect membranes (Crowe *et al.*, 1986; 1987; Crowe *et al.*, 1988). Similarly, when labile proteins are dehydrated in the presence of sugars, enzymic activity and secondary structure are maintained (Carpenter *et al.*, 1987a; 1987b; Carpenter and Crowe, 1989; Prestrelski *et al.*, 1993).

Mechanisms of Membrane Protection by Sugars

Several soluble carbohydrates, the disaccharides, trehalose and sucrose in particular, enable the retention of entrapped solutes inside liposomes during freeze-drying or air-drying (Crowe *et al.*, 1986; Fig. 1). After rehydration these protected liposomes assume their original size. Without sugars dehydration causes fusion and the entrapped solutes are completely lost. The sugar interacts with the phosphate of the phospholipid polar headgroups during the loss of the hydration shell below 0.3 g $H_2O.g$ dm^{-1}, which keeps the individual phospho-

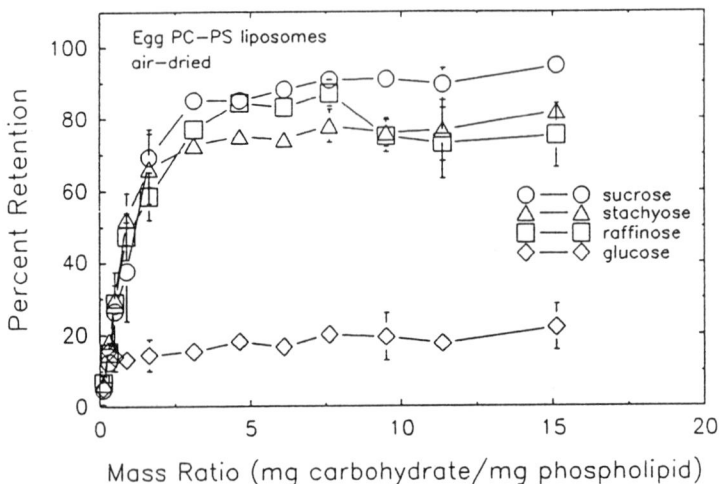

Figure 1. Retention of carboxyfluorescein (CF) by egg-phosphatidyl-choline/phosphatidylserine (9:1) liposomes dried in the presence of glucose, sucrose, raffinose and stachyose. Controls were dried in distilled water

lipid molecules at a distance from one another, a proposal called the "water replacement hypothesis" (reviewed in Crowe and Crowe, 1992; Crowe *et al.*, 1995). Thus, the acyl chains cannot tightly pack, which results in depression of the gel-to-liquid crystalline phase transition temperature (T*m*). Without sugar T*m* would have increased by about 60–70°C. There is considerable evidence that this sugar effect is involved in the stabilization of liposomes (Crowe and Crowe, 1992), intact membranes (Mouradian *et al.*, 1984), and whole cells (Leslie *et al.*, 1994; Hoekstra *et al.*, 1991; 1992c) during drying. Evidence for similar direct interactions between sugars and dry proteins has also been provided (reviewed in Carpenter *et al.*, 1992).

The temporary co-existence of solid and liquid phases during a phase transition causes a transient loss of membrane integrity leading to leakage during rehydration (Crowe *et al.*, 1989a). The importance of suppression of T*m* by sugars during drying lies in the fact that phase transitions can thus be prevented and leakage minimized.

Water Replacement Hypothesis versus Glass Formation Hypothesis

Koster *et al.* (1994) have recently suggested an alternative mechanism to direct interaction. They proposed, based partly on previous work of Green and Angell (1989), that vitrification of the sugar is in itself sufficient to reduce T*m* and protect liposomes. However, it was recently shown that certain large soluble carbohydrates that vitrify excellently during drying do not depress T*m* and are not able to protect liposomes (Crowe *et al.*, submitted). The following example shows that both direct interaction and vitrification (glass forming) are required for membrane preservation.

The monosaccharide glucose is not effective at protecting egg-phosphatidyl-choline liposomes during air-drying at room temperature (see also Fig. 1), which is attributed to fusion. However, when this drying is carried out at 4°C glucose fully protects the liposomes. This protection is lost within a few hours during heating of this dehydrated sample at 35°C, again due to fusion. The difference between the two samples is that dry glucose is in the glassy state at 4°C, but in the liquid state at 35°C. Apparently, the sample will be in a rubbery or glassy state of high viscosity during drying at 4°C fast enough to avoid fusion, but at elevated temperatures fusion ensues. This result would suggest that vitrification is the only factor of importance for protection. However, if excellent glass formers such as hydroxy-ethyl starch or dextran are tested, the glassy state is present, but the liposomes leak nevertheless. Fourier transform infrared spectroscopy (FTIR) has provided evidence that glucose interacts with the polar headgroups, whereas hydroxy-ethyl starch or dextran do not, probably due to their large size (Crowe *et al.*, submitted). This example shows that for protection of liposomes against the deleterious effects of drying, both the interaction of the sugar with the polar headgroup and the ability of the sugar to form a stable glass are crucial for depressing T*m* and preventing fusion, respectively. The disaccharides trehalose and sucrose combine both requirements within one compound. Mixtures of a good interactant such as glucose and an excellent glass former such as hydroxy-ethyl starch, are able to protect the liposomes. But neither compound is effective alone. The different hypotheses for the mechanisms of liposome protection of desiccation, *i.e.* the "water replacement hypothesis" and the "glass formation hypothesis" are thus combined.

It can be envisaged that *in situ* also sugars function to prevent membrane fusion by their glass forming capabilities. It is remarkable that sugars that are bad glass formers at room temperature, such as glucose and fructose, are generally absent in desiccation tolerant seeds and pollens (Amuti and Pollard, 1977; Hoekstra and Van Roekel, 1988; Koster and Leopold, 1988; Hoekstra *et al.*, 1992c). The presence of a glassy state in dry seed embryos has been reported by Williams and Leopold (1989), Koster (1991) and Sun *et al.* (1994). We propose that in relation to long term survival, the viscous glasses not only have a role in slowing deteriorative free-radical-mediated processes, but also in maintaining desiccation tolerance by preventing membrane fusion.

Sugar Concentration and Mass Ratio

Air-drying enables us to monitor at which water content (control) liposomes without sugar begin to leak. Surprisingly, leakage due to fusion commences already below 1.5 g $H_2O.g$ dm^{-1} (Hoekstra *et al.*, submitted). Fusion at this comparatively high water content can be readily stopped by any soluble carbohydrate. However, below 0.3 g $H_2O.g$ dm^{-1} leakage will occur unless the proper type of carbohydrate is present. As discussed above, monosaccharides and large oligosaccharides are not suitable, because they are poor glass formers and poor interactants, respectively. Disaccharides, trisaccharides and tetrasaccharides are good at protecting liposomes, because they are excellent glass formers and in addition also interact sufficiently with the polar headgroups.

It has always been assumed that in anhydrobiotes, disaccharides prevent fusion between membrane systems and subsequent leakage by mechanisms similar to those mentioned above for the model membrane systems (Crowe *et al.*, 1992; 1995). The recently developed technique of air-drying liposomes (Hoekstra *et al.*, submitted) could mimic the situation during *in vivo* drying, because it lacks the freezing step of freeze-drying. The different roles of the sugar implicate that its concentration in the cytoplasm is important, as well as the molecular ratio of sugar to phospholipid. In the case of sucrose, it has been found that for full protection of egg-phosphatidylcholine liposomes the concentration has to be above 20 mg.mL^{-1} and the amount of sucrose per phospholipid (mass ratio) approximately 5:1 or higher. Insufficient concentration causes fusion, and too low a molecular ratio causes leakage.

The total sugar concentration in a number of desiccation tolerant seeds and pollens is usually more than the required 20 mg.mL^{-1} (Hoekstra *et al.*, 1992c; Hoekstra *et al.*, 1996). However, the sugar to phospholipid ratio is often much less than what is marginally required. This means that either full depression of Tm is not necessary or that also other compounds interact with the phospholipids, with depression of Tm as the result.

In Situ Measurement of Tm

FTIR has enabled us to examine membrane phase behaviour in desiccation tolerant pollen *in situ*, i.e. in the presence of the endogenous sugars (Crowe *et al.*, 1989b; Hoekstra *et al.*, 1992a). When the band positions of the symmetric CH_2 stretching vibration in FTIR spectra are plotted against temperature, the average phase transition temperature (Tm) of membranes in intact pollen can be determined. A number of very dry anhydrobiotic pollen species have thus been analysed and were found to have Tms that are approximately 20–30°C higher than in the hydrated state (Hoekstra *et al.*, 1992b, 1992c), which means that Tm is not fully suppressed. However, this increase of Tm with drying is much less than in isolated membranes (Fig. 2), which is attributed to the effect of the endogenous sucrose (Hoekstra *et al.*, 1991).

The elevated Tms in very dry pollen indicate that gel phase domains may occur at room temperature. After humid air treatment this pollen can germinate, which implies that the fore-mentioned gel phase is fully reversible. This is also true with regard to cold storage at –20°C, a condition that promotes the presence of gel phase domains in the membranes of dry orthodox pollen and seed. As storage at –20°C can be endured without loss of viability, this sort of gel phase is also reversible. It is not expected that these gel phase domains lead to extensive phase separations, since such events are much more disrupting and not entirely reversible. In desiccation tolerant plant organs, phase separations may be effectively prevented due to the direct interaction between the sugars and phosphate of the phospholipids, decreasing the lateral mobility of the lipids (Crowe and Crowe, 1988; Crowe *et al.*, 1995). However, there is no direct evidence that this is so.

Data from FTIR analyses of membrane behaviour in seeds and somatic embryos are much more difficult to interpret than those from pollen (Hoekstra *et al.*, 1993). This is generally due to the large amount of oils interfering with the phospholipid signal. Yet it was found that Tm in some seeds is fully suppressed, but in others Tm was only partially suppressed as in pollen. Attempts to demonstrate interactions with the phosphate by *in situ* FTIR spectroscopy have, in general, been positive, but the identity of the phosphate and the interacting substance was uncertain in some cases (Hoekstra *et al.*, 1996).

In conclusion, it has become clear that the amount of sugars may not be sufficient for a full interaction with the available lipid phosphates. The efficiency of interaction *in situ* may even be much less than in model systems, since there are many other cell constituents with which the sugar can interact. This incomplete interaction with the lipid phosphate does not appear to harm viability as long as imbibitional damage is carefully avoided.

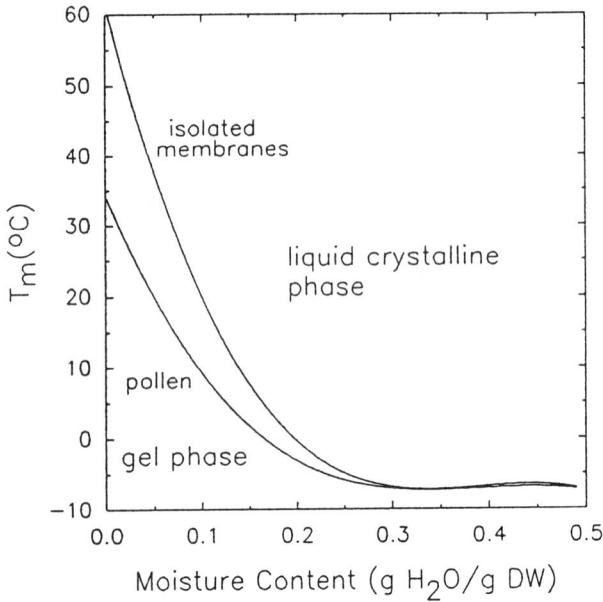

Figure 2. Effect of moisture content on transition temperatures (T*m*) of pollen membranes *in situ* and of isolated membranes (*T. latifolia* pollen; determined by FTIR)

Imbibitional Stress

The presence of domains with densely packed phospholipids in the membranes of dry anhydrobiotes implies that upon re-hydration these phospholipids melt, which may lead to a transient loss of membrane integrity. Dry viable seeds and pollen often suffer from excessive leakage of endogenous solutes and loss of germination capacity when they are imbibed rapidly, particularly at low temperatures (Hoekstra, 1984; Hoekstra and Van der Wal, 1988; Crowe *et al.*, 1989b; Hoekstra *et al.*, 1992a). This damage can be prevented by pre-hydration from the vapour phase or by pre-heating. These treatments melt gel phase phospholipids prior to the uptake of liquid water. Thus, a membrane phase transition, which is considered to be the cause of the leakage during imbibition is avoided. Figure 2 illustrates this in the phase diagram of membrane phospholipids in pollen. Both water vapour absorption and heating lead to a position above the curve, which represents the liquid crystalline phase and is associated with improved germination. The common occurrence of imbibitional damage in anhydrobiotes even during imbibition at room temperature, suggests that some amount of gel phase lipid had occurred in the membrane, which may be the result of a marginally suppressed T*m*, as we have discussed above. Seed is relatively insensitive to imbibitional stress, because it has a testa or/and an enveloping endosperm, which reduces the rate of water uptake and prevents the loss of endogenous solutes. The intactness of the seed coat is very important in this respect (Tully *et al.*, 1981; Herner, 1986).

State diagrams of glasses (melting temperature of the [sugar] glass as a function of the water content) have a similar shape as phase diagrams of membranes in Fig. 2 (Levine and Slade, 1988). This has drawn our attention to the possibility that the mechanism of imbibitional injury is linked to a glassy state change rather than to a membrane phase change. Figure 3 shows the state diagram Tg and the phase diagram T*m* of *Typha latifolia* pollen as a function of moisture content. Although both curves increase with reduction in water content, their slopes are considerably different from one another. Imbibitional injury is less above the T*m* curve, in the presence of either glassy state or liquid state at the onset of imbibition. This limits the likelihood of glassy state involvement in the imbibitional injury.

Depression of T*m* by Compounds other than Sugars

Sugars may interact with the polar headgroups as they happen to be located in close proximity to membranes. However, such a low efficiency of interaction may be compensated by the relatively large amount of sugars in desiccation tolerant cells. A much better efficiency of interaction may come from amphiphilic compounds when they partition into the membrane after becoming undissociated at low water content (Hoekstra *et al.*, 1996). This possibility is supported by results of experiments with the amphiphilic spin probe, tempamine (4-amino-TEMPO). After penetration in hydrated wheat embryos, this spin probe occurs exclusively in the cytoplasm, but after drying it is redistributed in the hydrophobic surroundings (Golovina *et al.*, submitted). Pollen and seed contain sufficient quantities of endogenous amphiphilic compounds (Liao *et al.*, 1989; Rizk *et al.*, 1992; Ylstra *et al.*, 1992) to allow them to play a role in fluidizing membranes. In the case of the flavonol, quercetine, the interaction is at its maximum at the incredibly low mass ratio of 1:20 (quercetine:phospholipid). Quercetine depresses T*m* of the dry phospholipid by about 30°C, probably by insertion in the more a-polar part of the membrane and not by interaction with the polar headgroups. As flavonols cannot protect dry liposomes by themselves, the role of sugars remains unaffected, particularly in providing high viscosity conditions and glassy state. The advantage of flavonols as compounds involved in desiccation tolerance may be great. Flavonols usually have strong antioxidant properties (Wang and Zheng, 1992; Terao *et al.*, 1994). In the undissociated form they readily partition in membranes, a condition prevailing in dry organisms. Thus, we have suggested that protection of membranes in dry anhydrobiotic organisms results from a multi-compound interaction with the phospholipids to keep T*m* suppressed on the one hand (Hoekstra *et al.*, 1996), and from sugar–sugar interaction to

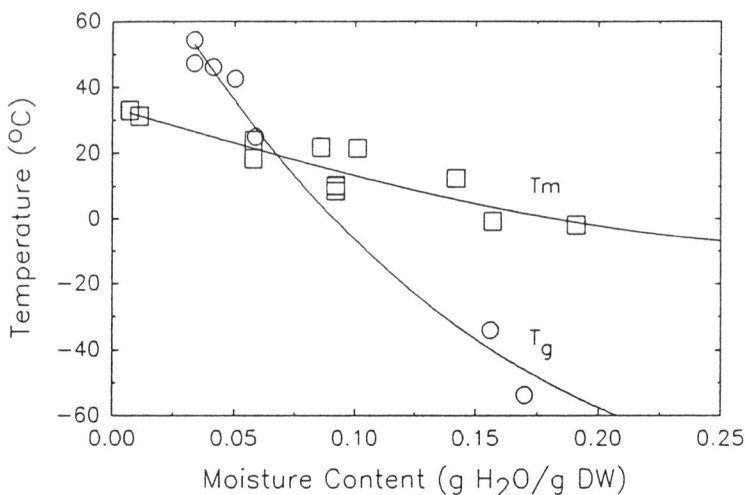

Figure 3. Plot representing the state diagram T*g* (glass-to-liquid transition curve; data from Buitink et al., submitted) and the phase diagram T*m* (replotted from Hoekstra et al. [1992]), of *Typha latifolia* pollen. The T*m* curve resembles the curve representing combinations of imbibition temperature and initial moisture content at which half of the initially viable grains germinated. Above the T*m* curve liquid crystalline phase prevails and germination improves, whereas below the T*m* curve gel phase prevails and germination becomes poor

provide high viscosity and glassy state for preventing fusion, on the other hand (Crowe *et al.*, submitted). In addition, interaction with the polar headgroups may be required to prevent lateral phase separations below T*m*.

Membrane Stability and Long-Term Survival

Loss of viability during the storage of desiccation-tolerant seeds and pollen coincides with an increased leakage of endogenous solutes upon rehydration (Senaratna *et al.*, 1988; Van Bilsen and Hoekstra, 1993; Van Bilsen *et al.*, 1994b), indicative of reduced membrane integrity. In that respect the similarity in membrane-associated phenomena between dried desiccation-sensitive cells and aged desiccation-tolerant cells is striking. For example, fast drying of carrot somatic embryos prevents acquisition of desiccation tolerance and leads to a consistent 20–30% decrease in phospholipid content (de-esterification) and an increase in free fatty acid (FFA) content compared to slow-dried, desiccation tolerant control embryoids (Tetteroo *et al.*, submitted). Here, lipid peroxidation appears to play only a minor role. Membrane deterioration apparently has begun, even before the embryoids are completely dry. After rehydration these embryoids contain plasma membranes having clustered intra-membraneous particles and particle-less areas, which is typical of demixing of the membrane components. Furthermore, isolated membranes are in the gel phase and the membrane proteins are aggregated.

Also during ageing of dry desiccation-tolerant pollen and seeds phospholipids are de-esterified, and FFAs and lysophospholipids (LPLs) accumulate, which occurs more rapidly at elevated relative humidity (Senaratna *et al.*, 1988; Van Bilsen and Hoekstra, 1993; Van Bilsen *et al.*, 1994b). Both, FFAs and LPLs were indentified in liposome studies as being reponsible for membrane leakage (Van Bilsen and Hoekstra, 1993). The fatty acid composition of the remaining phospholipids hardly changes during ageing, which was interpreted to mean that during ageing lipid peroxidation plays a minor role in the degradation process. Model lipid work has established that FFAs and LPLs can cause phase separation in dry membranes, even in minute amounts (Crowe *et al.*, 1989c; Van Bilsen *et al.*, 1994a). Furthermore, ageing of pollen and seed leads to a permanent increase in the gel-to-liquid crystalline phase transition temperature (Tm) of isolated membranes, which in itself can explain the leakage from the aged organism (McKersie *et al.*, 1988; Van Bilsen *et al.*, 1994a).

On account of results of *in vitro* membrane studies with free radicals it has been proposed that during natural seed ageing free radicals are involved in the de-esterification of fatty acids from the glycerol backbone of the phospholipids (McKersie *et al.*, 1988). This may also apply to desiccation-sensitive cells. It is to be expected that for desiccation tolerance effective defence mechanisms against free radicals are required.

The similarity between what happens during drying of desiccation sensitive cells and during ageing of desiccation tolerant cells is further supported by observations with *Arabidopsis thaliana* mutants which are insensitive to abscisic acid (Ooms *et al.*, 1993). Some of the mutant seeds can be dried to low moisture contents without losing viability. However, the life span of such seeds is very limited.

We may thus consider life span of an organism in the dry state as the result of the extremely decelerated accumulation of damages associated with desiccation, with the ultimate example of longevity of zero days for the desiccation sensitive type. The cause of the extreme difference in life span between pollen and seed for example, is far from clear, however. From the above it can be learned that membranes are a primary target associated with loss of viability. The next paragraph shows that protein secondary structure is much more stable.

Protein Secondary Structure and Long-Term Stability

Loss of viability with seed ageing is generally linked with loss of membrane integrity (see Priestley, 1986; Bewley and Black, 1994). However, highly ordered lipid domains indirectly cause the formation of protein domains, since membrane proteins will be excluded from these ordered lipid domains and move to the remaining fluid parts of the membrane, with the possibility for irreversible protein aggregation. This aggregation may in itself cause the loss of membrane integrity. Until recently little was known about possible changes in protein secondary structure with ageing, although it has been suggested that protein denaturation might play a role as early as 1915 by Crocker and Groves. Only the extractability of proteins from aged seeds has been shown to decrease (Priestley, 1986 and references therein), which can be caused by denaturation or disulphide bridge formation. Dry seed and pollen have considerable heat tolerance, which gradually declines with increasing moisture content (Barton, 1961; Ellis and Roberts, 1980; Marcucci *et al.*, 1982; Wolkers and Hoekstra, submitted). From this heat tolerance it could also be inferred that the dry proteins inside are heat stable, thereby providing long term stability.

Some enzymes retain activity for many decades, even long after the seed has lost its viability (Priestley, 1986 and references therein). The cytosolic proteins in particular retain enzymic activity, even when they are stored dry after isolation from the seed (Priestley and Bruinsma, 1982). The works of Carpenter *et al.* (1987a, 1987b, 1989) have demonstrated that stability and activity of dehydrated enzymes improves considerably when drying or freeze-drying is carried out in the presence of sugars. The probable mechanism of this protection is hydrogen bonding between the sugar and polar residues in the protein (Carpenter *et al.*, 1992; Prestrelski *et al.*, 1993). Labile proteins retain their secondary structure in the presence of sugars.

With the introduction of *in situ* FTIR microspectroscopy, studies on protein secondary structure in its native state in dry biological materials has become possible. In our previous work using FTIR spectroscopy, we have shown the extreme stability of proteins in *Typha latifolia* pollen during accelerated ageing (Wolkers and Hoekstra, 1995). Protein secondary structure remains unchanged even after the complete loss of viability, and aggregation of the membrane proteins does not occur. Apart from intrinsic properties of these proteins, the excellent stability might stem from the glassy environment in which the proteins are embedded. Glasses occur in this pollen at low water content (Buitink *et al.*, 1996), which is likely considering the high endogenous sucrose content (23%, dw basis) (Hoekstra *et al.*, 1992c). The heat stability of the proteins in the dry pollen is also very good since signs of denaturing appear only above 90°C (Wolkers and Hoekstra, submitted).

Naturally aged seeds of different plant species were also studied with respect to their protein secondary structure. Even after several decades of open storage and long after they became nonviable, protein secondary structure was maintained (Golovina *et al.*, submitted; Golovina *et al.*, these proceedings). In conclusion, the loss of membrane integrity with natural or accelerated ageing seems to precede the loss of protein secondary structure. In the case of orthodox seeds, protein secondary structure is maintained during several decades of open storage.

References

Amuti, K.S. and Pollard, C.J. 1977. *Phytochemistry* 16: 529–532.

Barton, L.V. 1961. *Seed Preservation and Longevity*. New York: Interscience Publishers.

Bewley, J.D. and Black, M. 1994. *Seeds, Physiology of Development and Germination*. Second Edition, pp. 445. New York and London: Plenum Press.

Buitink, J., Vertucci, C.W., Hoekstra, F.A. and Leprince, O. 1996. *Plant Physiology* (in press).

Carpenter, J.F., Arakawa, T. and Crowe, J.H. 1992. *Developmental Biology Standard* 74: 225–240.

Carpenter, J.F. and Crowe, J.H. 1989. *Biochemistry* 28: 3916–3922.

Carpenter, J.F., Crowe, L.M. and Crowe, J.H. 1987a. *Biochimica et Biophysica Acta* 923: 109–115.

Carpenter, J.F., Martin, B., Crowe, L.M. and Crowe, J.H. 1987b. *Cryobiology* 24: 455–464.

Crocker, W. and Groves, J.F. 1915. *Proceedings of the National Academy of Sciences U.S.A.* 1: 152–155.

Crowe, J.H. and Crowe, L.M. 1992. In: *Liposome Technology*, 2nd Edition, (ed. G. Gregoriadis). Boca Raton, FL: CRC Press.

Crowe, J.H., Crowe, L.M., Carpenter, J.F. and Aurell Wistrom, C. 1987. *Biochemical Journal* 242: 1–10.

Crowe, J.H., Crowe, L.M., Carpenter, J.F., Prestrelski, S.J. and Hoekstra, F.A. 1995. In: *Handbook of Physiology*, (in press) (ed. W. Dantzler). Oxford: Oxford University Press.

Crowe, J.H., Crowe, L.M., Carpenter, J.F., Rudolph, A.S., Aurell Wistrom, C., Spargo, B.J. and Anchordoguy, T.J. 1988. *Biochimica et Biophysica Acta* 947: 367–384.

Crowe, J.H., Crowe, L.M. and Chapman, D. 1984. *Science* 223: 701–703.

Crowe, J.H., Crowe, L.M. and Hoekstra, F.A. 1989a. *Journal of Bioenergetics and Biomembranes* 21: 77–91.

Crowe, J.H., Hoekstra, F.A. and Crowe, L.M. 1989b. *Proceedings of the National Academy of Sciences U.S.A.* 86: 520–523.

Crowe, J.H., Hoekstra, F.A. and Crowe, L.M. 1992. *Annual Review of Physiology* 54: 570–599.

Crowe, J.H., Hoekstra, F.A. and Crowe, L.M. *Cryobiology* (submitted).

Crowe, J.H., McKersie, B.D. and Crowe, L.M. 1989c. *Biochimica et Biophysica Acta* 979: 7–10.

Crowe, L.M. and Crowe, J.H. 1988. In: *Physiological Regulation of Membrane Fluidity*, pp. 75–99 (ed. R.C. Aloia). New York: Alan R. Liss, Inc.

Crowe, L.M., Womersley, C., Crowe, J.H., Reid, D., Appel, L. and Rudolph, A. 1986. *Biochimica et Biophysica Acta* 861: 131–140.

Ellis, R.H. and Roberts, E.H. 1980. *Annals of Botany* 45: 31–37.

Fischer, W., Bergfeld, R., Plachy, C., Schäfer, R. and Schopfer, P. 1988. *Botanica Acta* 101: 344–354.

Golovina, E.A., Wolkers, W.F. and Hoekstra, F.A. *Comparative Biochemistry and Physiology* (submitted).

Green, J.L. and Angell, C.A. 1989. *The Journal of Physical Chemistry* 93: 2880–2882.

Herner, R.C. 1986. *HortScience* 21: 1118–1122.

Hoekstra, F.A. 1984. *Plant Physiology* 74: 815–821.

Hoekstra, F.A. 1995. In: *Collecting Plant Genetic Diversity*, pp. 527–550 (eds. L. Guarino, V. Ramanatha Rao and R. Reid). Oxon: CAB International.

Hoekstra, F.A., Bomal, C. and Tetteroo, F.A.A. 1993. In: *Proceedings of the Fourth International Workshop on Seeds*, pp. 755–762 (eds. D. Come and F. Corbineau). Paris: ASFIS, 44 Rue du Louvre.

Hoekstra, F.A., Crowe, J.H. and Crowe, L.M. 1989a. In: *Recent Advances in the Development and Germination of Seeds*, pp. 77–88 (ed. R.B. Taylorson). New York: Plenum Press.

Hoekstra, F.A., Crowe, J.H. and Crowe, L.M. 1991. *Plant Physiology* 97: 1073–1079.

Hoekstra, F.A., Crowe, J.H. and Crowe, L.M. 1992a. *Physiologia Plantarum* 84: 29–34.

Hoekstra, F.A., Crowe, J.H. and Crowe, L.M. *Plant Physiology* (submitted).

Hoekstra, F.A., Crowe, J.H., Crowe, L.M. and Van Bilsen, D.G.J.L. 1992b. In: *Angiosperm Pollen and Ovules*, pp. 177–186 (eds. E. Ottaviano, D.L. Mulcahy, M. Sari-Gorla and G. Bergamini-Mulcahy). New York: Springer-Verlag.

Hoekstra, F.A., Crowe, J.H., Crowe, L.M., Van Roekel, T. and Vermeer, E. 1992c. *Plant Cell and Environment* 15: 601–606.

Hoekstra, F.A., Crowe, L.M. and Crowe, J.H. 1989b. *Plant Cell and Environment* 12: 83–91.

Hoekstra, F.A. and Van der Wal, E.W. 1988. *Journal of Plant Physiology* 133: 257–262.

Hoekstra, F.A. and Van Roekel, T. 1988. *Plant Physiology* 88: 626–632.

Hoekstra, F.A., Wolkers, W.F., Buitink, J., Golovina, E.A., Crowe, J.H. and Crowe, L.M. 1996. *Comparative Biochemistry and Physiology* (in press).

Koster, K.L. 1991. *Plant Physiology* 96: 302–304.

Koster, K.L. and Leopold, A.C. 1988. *Plant Physiology* 88: 829–832.

Koster, K.L., Webb, M.S., Bryant, G. and Lynch, D.V. 1994. *Biochimica et Biophysica Acta* 1193: 143–150.

Leprince, O., Bronchart, R. and Deltour, R. 1990. *Plant Cell and Environment* 13: 539–546.

Leslie, S.B., Teter, S.A., Crowe, L.M. and Crowe, J.H. 1994. *Biochimica et Biophysica Acta* 1192: 7–13.

Levine, H. and Slade, L. 1988. In: *Water Science Reviews Vol 3*, pp. 79–185 (ed. F. Franks). Cambridge: Cambridge University Press.

Liao, M.C., Liu, Y.L. and Xiao, P.G. 1989. *Acta Botanica Sinica* 31: 939–947.

Madin, K.A.C. and Crowe, J.H. 1975. *Journal of Experimental Zoology* 193: 335–337.

Marcucci, M.C., Visser, T. and Van Tuyl, J.M. 1982. *Euphytica* 31: 287–290.

McKersie, B.D., Senaratna, T., Walker, M.A., Kendall, E.J. and Hetherington, R.R. 1988. In: *Senescence and Aging in Plants*, pp. 412–465 (eds. L.D. Nooden and A.C. Leopold). New York: Academic Press.

Mouradian, R., Crowe, L.M. and Crowe, J.H. 1984. *Biochimica et Biophysica Acta* 778: 615–617.

Ooms, J.J.J., Leon, K.K.M., Bartels, D., Koornneef, M. and Karssen, C.M. 1993. *Plant Physiology* 102: 1185–1191.

Prestrelski, S.J., Tedeschi, N., Arakawa, T. and Carpenter, J.F. 1993. *Biophysical Journal* 65: 661–671.

Priestley, D.A. 1986. *Seed Aging: Implications for Seed Storage and Persistence in Soil*. Ithaca and London: Comstock Publishing Associates.

Priestley, D.A. and Bruinsma, J. 1982. *Physiologia Plantarum* 56:303–311.

Rizk, A.M., Ismail, S.I., Azzam, S.A. and Wood, G. 1992. *Qatar University Science Journal* 12: 69–72.

Senaratna, T., Gusse, J.F. and McKersie, B.D. 1988. *Physiologia Plantarum* 73: 85–91.

Senaratna, T. and McKersie, B.D. 1983. *Plant Physiology* 72: 911–914.

Sun, W.Q., Irving, T.C. and Leopold, A.C. 1994. *Physiologia Plantarum* 90: 621–628.

Terao, J., Piskula, M. and Yao, Q. 1994. *Archives of Biochemistry and Biophysics* 308: 278–284.

Tetteroo, F.A.A., de Bruijn, A.Y., Henselmans, R.N.M., Wolkers, W.F., van Aelst, A.C. and Hoekstra, F.A. 1996. *Plant Physiology* (submitted).

Tully, R.E., Musgrave, M.E. and Leopold, A.C. 1981. *Crop Science* 21: 312–317.

Van Bilsen, D.G.J.L. and Hoekstra, F.A. 1993. *Plant Physiology* 101: 675–682.

Van Bilsen, D.G.J.L., Hoekstra, F.A., Crowe, L.M. and Crowe, J.H. 1994a. *Plant Physiology* 104: 1193–1199.

Van Bilsen, D.G.J.L., Van Roekel, T. and Hoekstra, F.A. 1994b. *Sexual Plant Reproduction* 7: 303–310.

Vertucci, C.W. and Farrant, J.M. 1995. In: *Seed Development and Germination*, pp. 237–271 (eds. J. Kigel and G. Galili). New York: Marcel Dekker.

Wang, P.F. and Zheng, R.L. 1992. *Chemistry and Physics of Lipids* 63: 37–40.

Williams, R.J. and Leopold, A.C. 1989. *Plant Physiology* 89: 977–981.

Wolkers, W.F. and Hoekstra, F.A. 1995. *Plant Physiology* (in press).

Wolkers, W.F. and Hoekstra, F.A. *Comparative Biochemistry and Physiology* (submitted).

Ylstra, B., Touraev, A., Benito Moreno, R.M., Stöger, E., Van Tunen, A.J., Vicente, O., Mol, J.N.M. and Heberle-Bors, E. 1992. *Plant Physiology* 100: 902–907.

2. Desiccation Tolerance in Immature Embryos of Maize: Sucrose, Raffinose and the ABA–Sucrose Relation

A. BOCHICCHIO[1], P. VERNIERI[2], S. PULIGA[1], C. MURELLI[3] and C. VAZZANA[1]

[1]*Dipartimento di Agronomia e Produzioni Erbacee, Università di Firenze, P. le Cascine 18, 50144 Firenze;* [2]*Dipartimento di Biologia delle Piante Agrarie, Sez. di Orticoltura e Floricoltura, Università di Pisa, Viale delle Piagge 23, 56124 Pisa;* [3]*Dipartimento di Chimica Organica, Università di Pavia, Viale Taramelli 10, 27100 Pavia, Italy*

Abstract

Our previous experiments on desiccation tolerance in immature embryos of maize cast doubt on a crucial role of high concentrations of sucrose and high raffinose to sucrose mass ratio. In the present work we report experiments aimed at gaining a further insight into the above mentioned subject and to ascertain whether endogenous ABA may stimulate the accumulation of sucrose. Immature and desiccation sensitive embryos of maize were isolated and exposed to different drying conditions either leading or not leading to desiccation tolerance acquisition and allowing for different ABA and sugar responses. ABA concentrations as well as sugar concentrations were compared in embryos undergoing such different dehydration treatments. We conclude that in maize embryos: 1) sucrose, even at high concentrations, is not the sole factor responsible for desiccation tolerance; 2) high sucrose concentrations do not appear to be an absolute requirement of desiccation tolerance; 3) a high (>0.05) raffinose to sucrose mass ratio is not a prerequisite of desiccation tolerance; 4) ABA doesn't seem to stimulate sucrose accumulation. We suggest that desiccation tolerance can occur even in the absence of raffinose.

Introduction

Recent literature provides results that are not always fully consistent with the hypothesis that sucrose and oligosaccharides of the raffinose family have a key role in desiccation tolerance of orthodox seeds. Oligosaccharides were not a requirement of desiccation tolerance in embryos of cauliflower, while high levels of sucrose (8–10% on a dry weight (DW) basis) were associated with both the tolerant and the intolerant state (Hoekstra *et al.*, 1994). Accumulation of raffinose and stachyose were not temporally linked to acquisition of desiccation tolerance in seeds of wild type or abi3-1 mutant of *Arabidopsis thaliana* (Ooms *et al.*, 1993). Moreover, carbohydrates were found unlikely to be the sole factor determining desiccation tolerance in seeds of *Arabidopsis* (Ooms *et al.*, 1994). Sensitivity of wild rice seeds could not be ascribed to their inability to synthesize sucrose or olgosaccharides and embryonic axes of wild rice were found unable

R.H. Ellis, M. Black, A.J. Murdoch, T.D. Hong (eds.), Basic and Applied Aspects of Seed Biology, pp. 13–22.
© *1997 Kluwer Academic Publishers, Dordrecht. Printed in Great Britain.*

to survive very low water contents ($< 7\%$) notwithstanding a sucrose concentration of 6.9% on a DW basis (Still et al., 1994). Moreover, sucrose was not the sole factor re-inducing desiccation tolerance in germinated seeds (Bruggink and van der Toorn, 1995) and a comparison between recalcitrant seeds of Avicennia marina and desiccation tolerant types suggested that sugars alone do not provide an explanation for desiccation tolerance (Farrant et al., 1993).

On the other hand the literature also provides results consistent with the hypothesis of an essential involvement of raffinose and/or sucrose in desiccation tolerance of orthodox seeds. Stachyose accumulated in immature soybean axes during slow drying, a treatment leading to desiccation tolerance (Blackman et al., 1992); appearance of stachyose and an increase of sucrose were found in radicles of developing embryos of Brassica campestris coincidently with the onset of desiccation tolerance (Leprince et al., 1990). Studies performed on the acquisition or loss of desiccation tolerance by maize kernels have given results consistent with the hypothesis of a key role of raffinose in the presence of sucrose in the mechanism of tolerance. In fact the percentage of embryos tolerating high temperature (50°C) drying was correlated with the raffinose to sucrose mass ratio in maturing embryos preconditioned at 35°C for different lengths of time and the percentage of sucrose on total soluble sugars increased during the same period. The authors concluded that the presence of raffinose at certain levels may be a key factor in protecting maturing seeds from high temperature drying damage (Chen and Burris, 1990). Sucrose and raffinose were found in mature axes of maize and loss of raffinose was found coincident with loss of desiccation tolerance during their germination (Koster and Leopold, 1988).

Our preliminary experiments concerning acquisition of desiccation tolerance by immature embryos of maize gave contradictory results (Bochicchio et al., 1994a). In fact this depended on the type of drying treatment to which embryos were exposed. Consequently the first aim of the present work was to further investigate the relation between sucrose, raffinose to sucrose ratio and desiccation tolerance in maize embryos. Experiments with different drying treatments were repeated and extended to less immature embryos. Moreover, since sugars are believed to fulfill their role in the mechanism of desiccation tolerance when cells are almost dry, we have always related desiccation tolerance to sucrose or raffinose concentrations of the dried embryos.

Recently studies have been performed on both ABA and sugars; relations have been found between exogenous ABA treatments and a modification of sugar concentrations (Ooms et al., 1994 for embryos of aba-1 and abi3-1 double mutant of Arabidopsis thaliana and Tetteroo et al., 1994 for somatic embryos of carrot). It is suggested that different patterns of carbohydrate accumulation in the ABA mutants of Arabidopsis thaliana are indicative of a role of endogenous ABA in the regulation of carbohydrate metabolism. Moreover, a protein with homology to an ABA-dependent aldose reductase found in tolerant barley embryos (Bartels et al., 1991) has been detected in tolerant seeds of Arabidopsis thaliana (Ooms et al., 1993). Dehydration and ABA were found to increase levels of transcripts of the cytosolic glyceraldehyde-3-phosphate dehydrogenase

(Velasco *et al.*, 1994; Kleines *et al.*, 1995). Our research was also aimed at gaining some preliminary knowledge on an eventual relationship between endogenous ABA and sucrose accumulation in immature embryos of maize. We have focused our attention on possible relationships between changes in ABA concentrations and changes in sucrose concentrations during development of the embryo on the ear and during the artificial drying treatments.

Material and Methods

Plant Material and Drying Treatments

Immature embryos of the inbred line Lo904 were isolated from the kernels at fixed days after pollination (DAP) in 1993, 1994 and 1995. Isolated embryos were dried by rapid or slow drying. For rapid drying, embryos, laid between two discs of filter paper, were kept for 48 h in an incubator at 35°C. For the slow drying treatment, embryos were kept for 120 h in sealed 500 ml jars over a saturated solution of K_2SO_4 (RH 98%) and for the following 24 h in the same type of jars over a sludge of LiCl (RH 13%). Embryos were exposed to slow drying either upon excision, or after 12 h incubation on a solution of fluridone (150 μM) in water or after a double incubation on the fluridone solution (1995 harvest); double incubation consisted of the usual 12 h incubation followed by 6 h drying at 98% RH and a subsequent, second 3 h incubation on the fluridone solution. Fluridone was added to inhibit the biosynthesis of ABA as required by the second part of the present work. Slow drying and incubation were carried out at 28°C in a growth chamber in dim light. Similar to our previous work (Bochicchio *et al.*, 1994a and b), rapid drying was a quite severe treatment; in fact embryos lost more than 90% of their water content within the initial 3 h of the treatment. Slow drying was a much milder treatment and when embryos were incubated before being exposed to slow drying, they absorbed water; in fact at the start of the drying treatment their water content was higher than at excision. While different, all the drying treatments resulted in a final moisture content always lower than 7% on a DW basis (data not shown).

Tolerance Test

From at least 14 DAP onward isolated fresh embryos are able to germinate *in vitro* on a solid culture medium. Desiccation tolerance of dried embryos was assessed by their capacity to germinate on a solid culture medium. Germination was carried out in a growth chamber at 25°C for up to 15 days, embryos were scored as germinated if their radicle or coleoptile were visibly grown (longer than 0.2 cm). Tolerance percentage is the percentage of embryos capable of germinating after a drying treatment.

Sucrose and Raffinose Analysis

Sucrose and raffinose were quantified by HPLC using the same procedure, with minor modifications, reported by Bochicchio *et al.* (1994a).

ABA Analysis

Quantitative analysis of ABA was performed on the crude aqueous extracts obtained for sugar extraction; aliquots were taken from the extracts before their filtration through the ion exchange columns. Quantitative analysis of ABA was performed by a solid-phase radioimmuno-assay (RIA) based on the use of a monoclonal antibody (DBPA1) raised against free (S)-ABA (Vernieri *et al.*, 1989). Validation of RIA results is reported in Bochicchio *et al.* (1994b).

Neutral Lipid Quantitation

Embryo samples (50–80 mg) were suspended in $CHCl_3$ (10 ml, room temperature), shaken for 5 min and filtered. The process was repeated a second time in the same volume of $CHCl_3$ for 30 min. The two $CHCl_3$ extracts were pooled and evaporated under reduced pressure at 40°C. In order to verify that the only lipids present in the extracts were the neutral ones, the extracts were subjected to TLC analysis. TLC silica gel plates were developed in $CHCl_3$. Lipids were identified by using authentic samples as reference and by spraying with sulphocromic mixture followed by heating at 120°C. Total neutral lipids were estimated by gravimetry (Bianchi *et al.*, 1990).

Results

Sucrose, Raffinose and Desiccation Tolerance

Sucrose concentration declined during the developmental period considered (18–27 DAP); in rapidly dried embryos sucrose concentrations were slightly higher than in fresh ones (Fig. 1A); 25 and 27 DAP embryos, the ones with the lower sucrose concentrations were also the ones able to survive the rapid drying treatment (Fig. 1B).

Experiments were performed also with embryos of 1995 harvest, but they were limited to tolerance of the rapidly dried embryos and to sugar concentrations of fresh embryos: results (not reported) showed the very same trend as the ones of Fig. 1. As it is well known, maize embryos are rich in neutral lipids. We wanted to ascertain whether neutral lipid concentrations increased during the developmental period considered. In fact, if this was the case, then the decrease of sucrose concentration found during development might not be relevant to desiccation tolerance. Neutral lipid concentration did not show any variation during development (Table 1). Since in sensitive embryos sucrose concentration

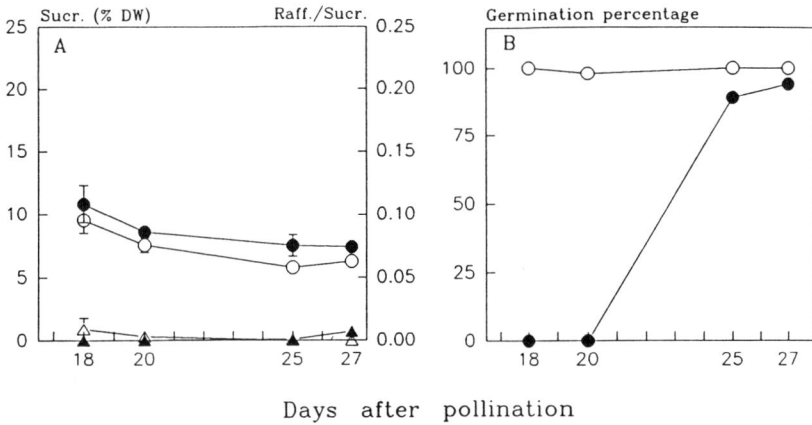

Figure 1. Embryos of 1994 harvest. (A) Sucrose concentrations (mg/100 mg DW) (O●) and raffinose to sucrose ratio (△▲) in fresh (O△) or rapidly dried embryos (●▲) as a function of days after pollination. Data points are means±s.e. of 3 samples. When larger than symbols error bars are shown. (B) Percentage of embryos able to germinate fresh (O) or after the rapid drying treatment (tolerance percentage) (●) as a function of days after pollination. Germination percentage values are means of at least 2 samples of, at least, 20 embryos each

was higher than in tolerant ones, sucrose couldn't be the sole factor determining desiccation tolerance in immature embryos of maize. Desiccation tolerance was found associated with the appearance of raffinose, but raffinose concentration was variable and barely detectable at 25 DAP and extremely low at 27 DAP (0.05% at 27 DAP giving a raffinose to sucrose ratio of 0.007) (Fig. 1A).

Experiments were performed to verify whether desiccation tolerance is always associated with the appearance of raffinose and to ascertain whether a high sucrose concentration, as shown by 25 and 27 DAP rapidly dried embryos, is an absolute requirement of desiccation tolerance in embryos of maize. Embryos, isolated at 18 and 20 DAP, were exposed to different pre-treatments capable of reducing sucrose concentration. Once pretreated they were exposed to slow drying, which, normally, leads 18 and 20 DAP embryos to tolerance acquisition; embryos were also exposed to slow drying upon excision, without any previous

Table 1. Neutral lipids (NL) concentrations (mg/100 mg DW) of embryos rapidly dried at three different days after pollination

DAP	18	25	27
NL	22.8	23.9	24

pre-treatment. All the treatments resulted in the appearance or increasing of raffinose; embryos were able to germinate independently of the value of their raffinose to sucrose mass ratio which was found also lower than 0.03 (Fig. 2A). Sucrose concentrations increased slightly when embryos were exposed to slow drying upon excision (Fig. 2B, △). Pretreatments by one or two incubations resulted in an evident decrease of sucrose concentrations (compare sucrose concentrations at time zero in Fig. 2B with the ones of 18 and 20 DAP fresh embryos in Fig. 1A), but during the subsequent slow drying sucrose concentrations increased to levels higher than 10% except for embryos of the 1993 harvest: in this case sucrose concentrations remained very low (<2%). The percentage of embryos able to survive drying, in this case, was lower but still 55.3% of the embryos withstood the treatment; even if it is claimed that in the samples used for sugar analysis sucrose was present only in the 55.3% of the embryos, which is only in the ones which acquired desiccation tolerance, still their sucrose concentration would be very low (about 3.2%); hence a high sucrose concentration and a high raffinose to sucrose ratio do not appear an absolute requirement of desiccation tolerance, and anyway the desiccation tolerance state was always found to be associated with the presence of raffinose.

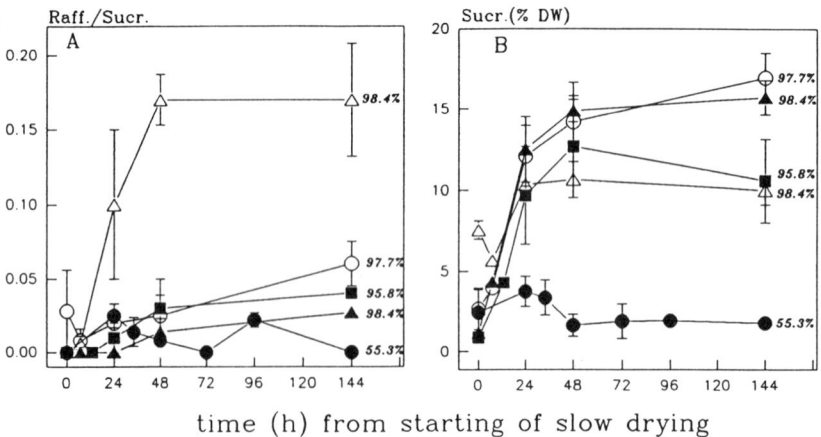

time (h) from starting of slow drying

Figure 2. Time course of (A) raffinose to sucrose ratio and (B) sucrose concentrations (mg/100 mg DW) during slow drying. Embryos were exposed to slow drying either upon excision (△) 20 DAP-1994 harvest, or after a 12 h incubation on a solution of fluridone (150 µM) in water ((●) 18 DAP-1993 harvest, (▲) 20 and (○) 18 DAP 1994 harvest), or after a double incubation ((■) 18 DAP-1995 harvest). Data points are either means±s.e. of 3 samples (△○), or means of two samples (▲■●); when larger than symbols standard errors or difference between single values are shown with vertical bars from the mean. The number at the right end of each curve shows the percentage of embryos tolerating the drying treatment. Data relative to 1993 harvest have been re-arranged from Bochicchio *et al.* (1994)

Relationship between ABA and Sucrose Accumulation

As already reported (Bochicchio *et al.*, 1994b), ABA concentrations increased in the period of development considered and showed a peak at about 25 DAP (coincident with acquisition of tolerance to the rapid drying treatment); while ABA increased, sucrose decreased steadily (Fig. 3) . No evidence for a relation between ABA and sucrose accumulation could be gained from the latter results: increasing ABA levels were not accompanied by sucrose accumulation. No evidence for an ABA and sucrose relation could be gained either from experiments during which sucrose was found to accumulate (Fig. 4). In fact sucrose concentration increased during slow drying subsequent to an ABA clear peak (Fig. 4A) as well as when no ABA peak occurred as happened when embryos were treated with an inhibitor of ABA biosynthesis before slow drying (Fig. 4B). Experiments were repeated with 18 DAP embryos and results (not reported) showed the same trend as those in Figure 4.

Discussion and Conclusions

Sucrose, even at high ($>10\%$) concentrations, is not the sole factor responsible for desiccation tolerance in maize embryos, and in fact in sensitive embryos sucrose concentrations were higher than in tolerant ones (Fig. 1A). In one of the harvests (1993) desiccation tolerance was associated with a low sucrose concentration. Consequently it is unlikely that high sucrose concentrations are an absolute requirement of desiccation tolerance. Tolerance to desiccation occurred in embryos showing high (even >0.1) or very low (0.007) raffinose to sucrose ratio (Figs. 1 and 2A). Hence a high raffinose to sucrose ratio does not appear a requirement of desiccation tolerance in maize embryos. In our experiments tolerance to desiccation was always associated with the presence of raffinose, but its concentration and its ratio to sucrose could be so low and so far from the value of 0.3 found as the effective one for an optimum preservation of dry model membranes and prevention of sucrose crystallization (Caffrey *et al.*, 1988) that they were likely not to be sufficient to fulfill the latter proposed functions.

Our results are not easily comparable to those of Koster and Leopold (1988) and Chen and Burris (1990) (see Introduction) for maize embryos. Koster and Leopold analyzed sugars in axes from imbibed seeds not yet exposed to the drying treatment and related sugar concentrations to tolerance of the dried axes. On the contrary we have related the sugar concentrations found at the end of the drying treatments to the tolerance of isolated embryos exposed to the same treatments; we have chosen this method because sugar concentrations and composition may change even during a very rapid dehydration.

Chen and Burris (1990), using less immature embryos, studied tolerance to high temperature drying, a stress condition more severe than the one studied in this work and it is likely for the seed to have different requirements to withstand

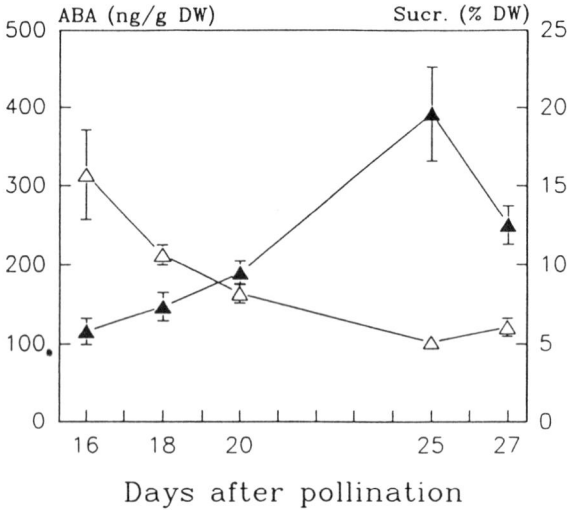

Figure 3. ABA (▲) and sucrose (△) concentrations as a function of days after pollination. ABA (ng/g DW): data points are means±s.e. of at least 4 samples relative to 1993, 1994 and 1995 harvest. Sucrose (mg/100mg DW): data points are means±s.e. of at least 5 samples relative to 1993,1994 and 1995 harvest. When larger than symbols error bars are shown. Data relative to 16 DAP are means of only 2 samples (1993 harvest), in this case difference between single values is reported

different stresses. The authors assign tolerance acquisition during preconditioning of the kernel at 35°C to a complex of changes concerning soluble sugars, raffinose to sucrose mass ratio included; the increase of raffinose concentrations found by Chen and Burris during preconditioning is in agreement with our data, in fact preconditioning is actually a kind of slow drying, a condition during which raffinose increases also in more immature isolated embryos.

Our conclusions are obviously limited to the criterion we have chosen to evaluate, desiccation tolerance, and our results do not exclude the possibility that sucrose and raffinose may be components of the complex mechanism of tolerance; they may lead to hypothesize that the role normally assigned to sucrose and raffinose may be also fulfilled by other compounds, the latter hypothesis would be supported by results obtained by Koster *et al.* (1994) and by Strauss and Hauser (1986) on model systems. Our results cannot give any evidence for this hypothesis since we have limited our attention to sucrose and raffinose. Accumulation of raffinose during development and maturation of maize embryo may be more important to its longevity than to tolerance to desiccation: in mature seeds from four cultivars of maize, dry storage stability was found correlated to their raffinose concentration and to their raffinose to sucrose mass ratio (Bernal-Lugo and Leopold, 1995), and vitrification of

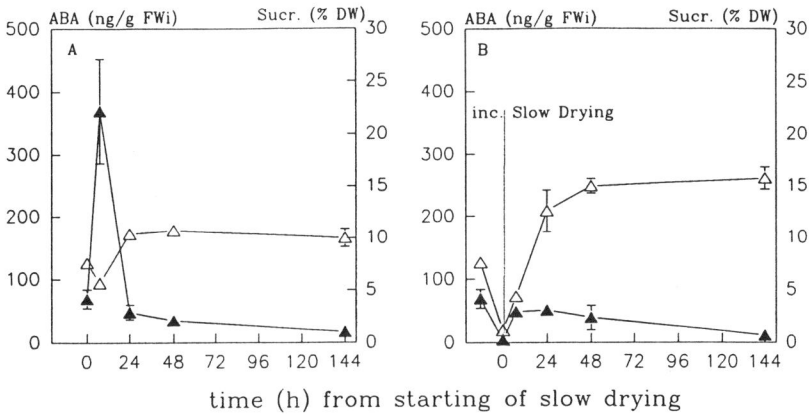

time (h) from starting of slow drying

Figure 4. Time course of ABA (▲) concentrations (ng/g of the fresh weight upon excision) and sucrose (△) concentrations (mg/100 mg DW) in 20 DAP embryos of the 1994 harvest undergoing slow dehydration either upon excision (A) or subsequently to 12 h incubation on a fluridone solution (150 µM) in water (B). (A): data points are means ± s.e. of 3 samples, when larger than symbols error bars are shown. (B) data points are means of 2 samples, when larger than symbols difference between single values is shown with vertical bars from the mean

cytoplasm to which oligosaccharides of the raffinose family are relevant, was not found sufficient to account for desiccation tolerance (Sun *et al.*, 1994).

Our results do not give any evidence of a relation between ABA and sucrose accumulation. Moreover in our past research we have never found an ABA increase during rapid drying (Bochicchio *et al.*, 1994b) while results here reported show a small, but repeatable, increase in sucrose during the same treatment.

Acknowledgements

The authors thank the Istituto Sperimentale per la Cerealicoltura, Sez. Maiscoltura, of Bergamo, Italy, for growing and hand pollinating the inbred line Lo904, and Mr L. Massaini and Mrs G. Casella for their skillful technical assistance in embryo isolation and culture.

References

Bartels, D., Engelhardt, K., Roncarati, R., Schneider, K., Rotter, M., Salamini, F. 1991. *The EMBO Journal* 10: 1037–1043.
Bernal-Lugo, I. and Leopold, A.C. 1995. *Seed Science Research* 5: 75–80.
Bianchi, G., Murelli, C., Ottaviano, E. 1990. *Phytochemistry* 29: 739–744.

Blackman, S.A., Obendorf, R.L., Leopold, A.C. 1992. *Plant Physiology* 100: 225–230.
Bochicchio, A., Rizzi, E., Balconi, C., Vernieri, P., Vazzana, C. 1994a. *Seed Science Research* 4: 123–127.
Bochicchio, A., Vernieri, P., Puliga ,S., Balducci, F., Vazzana, C. 1994b. *Physiologia Plantarum* 91: 615–622.
Bruggink, T. and Van der Toorn, P. 1995. *Seed Science Research* 5: 1–4.
Caffrey, M., Fonseca, V., Leopold, A.C. 1988. *Plant Physiology* 86: 754–758.
Chen, Y. and Burris, J.S. 1990. *Crop Science* 30: 971–975.
Farrant, J.M., Pammenter, N.W., Berjak, P. 1993. *Seed Science Research* 3: 1–13.
Hoekstra, F.A., Haigh, A.M., Tetteroo, F.A.A., Van Roekel, T. 1994. *Seed Science Research* 4: 143–147.
Kleines, M., Elster, R., Ingram, J., Bernacchia, G., Schwall, G., Salamini, F., Bartels, D. 1995. In: *Proceeding of the Second Stressnet Conference. Salsomaggiore, Italy 21–23 september 1995*, pp. 131–137 (eds. R.A. Leigh and M.M.A. Blake-Kalff). European Commission, Directorate General VI. VI/666C/95-EN.
Koster, K.L. and Leopold, A.C. 1988. *Plant Physiology* 88: 829–832.
Koster, K.L., Webb, M.S., Bryant, G., Lynch, D.V. 1994. *Biochimica et Biophysica Acta* 1193: 143–150.
Leprince, O., Bronchart, R., Deltour, R. 1990. *Plant, Cell and Environment* 13: 539–546.
Ooms, J.J.J., Léon-Kloosterzie, K.M., Bartels, D., Koornneef, M., Karssen, C.M. 1993. *Plant Physiology* 102: 1185–1191.
Ooms, J.J.J., Wilmer, J.A., Karssen, C.M. 1994. *Physiologia Plantarum* 90: 431–436.
Still, D.W., Kovach, D.A., Bradford, K.J. 1994. *Plant Physiology* 104: 431–438.
Strass, G.and Hauser, H. 1986. In: *Membranes, Metabolism and Dry Organisms*, pp. 318–326 (ed. A. C. Leopold). Ithaca, London: Comstock Publishing Associates.
Sun, W.Q., Irving, T.C., Leopold, A.C. 1994. *Physiologia Plantarum* 90: 621–628.
Tetteroo, F.A.A., Bomal, C., Hoekstra, F.A., Karssen, C.M. 1994. *Seed Science Research* 4: 203–210.
Velasco, R., Salamini, F., Bartels, D. 1994. *Plant Molecular Biology* 26: 541–546.
Vernieri, P., Perata, P., Armellini, D., Bugnoli, M., Presentini, R., Lorenzi, R., Ceccarelli, N., Alpi, A., Tognoni, F. 1989. *Journal of Plant Physiology* 134: 441–446.

3. DNA Status, Replication and Repair in Desiccation Tolerance and Germination

D.J. OSBORNE and I.I. BOUBRIAK[1]

Oxford Research Unit, Open University, Foxcombe Hall, Boars Hill, Oxford, OX1 5HR, UK
[1] *Permanent address: Institute of Cell Biology and Genetic Engineering, Ukrainian Academy of Sciences, Kiev, Ukraine*

Abstract

We have reported previously on the stability of DNA and the tolerance to desiccation in cells of imbibed embryos and in wind dispersed pollen at a stage when nuclear DNA is not replicating. In embryos of rye and wild oat, the tolerance is lost at the onset of S-phase coincident with radicle extension but in birch pollen tolerance is lost on pollen tube extension in the absence of DNA replication. The current experiments were aimed to determine whether a changed DNA conformation alone, or an alteration in DNA repair capability could account for the conversion from the desiccation-tolerant to the desiccation-intolerant condition. The results indicate that a competent DNA repair capacity is an essential component for the maintenance of desiccation tolerance.

Introduction

The acquisition of desiccation tolerance permits a cell to survive for long periods in the dry state despite the loss of a major percentage of its water composition. In the dry state metabolic activity is arrested, but when the cell again takes up water, growth is resumed and both synthetic processes and new gene expressions are revealed. Examples of the desiccation-tolerant condition occur throughout the plant and animal kingdom though the mechanisms by which it is initiated and achieved may differ and the limits to which the tolerance may extend can vary (Crowe *et al.*, 1992). In plants, seeds, spores and wind dispersed pollen, the leafy parts of mosses and other lower plants and the aerial parts of the specialized 'resurrection' species all survive dehydration without irretrievable damage to nuclear, mitochondrial or plastid genomes. (For review, see Leopold, 1986).

A cell in the dry state (less than 14% moisture content) is not held without change. The denaturation of proteins, loss of enzyme function, destabilization of membranes and loss of integrity of nucleic acids all take place at rates that are determined both by the external environment and the genetic constitution of the species. If the cell remains dry, loss of viability and death are eventually inevitable.

The performance of a dry cell on the addition of water therefore depends upon its past history. Namely, upon the extent to which that cell survived the

R.H. Ellis, M. Black, A.J. Murdoch, T.D. Hong (eds.), Basic and Applied Aspects of Seed Biology, pp. 23–32.
© *1997 Kluwer Academic Publishers, Dordrecht. Printed in Great Britain.*

original desiccation stress and thereafter conserved integrity of macromolecular organization during the dehydrated state. Whereas the removal of damage and the revival of full cytoplasmic performance rests upon the ability of the cell for renewed and continued synthesis of proteins, lipids and all classes of RNA, this is achieved only if the integrity of the genetic information has also been assured.

Damage to DNA occurs in cells if they lose water when chromatin or the conformation of DNA is not organized to withstand dehydration (Osborne and Boubriak, 1994). Desiccation tolerance of bacterial spores, for example, depends upon the conversion of B-form DNA to the more condensed A-form (Setlow, 1992; Griffith *et al.*, 1994). Loss of DNA integrity with an increase in both single and double strand breaks progresses during the dry state (Cheah and Osborne, 1978) but if function of the dark excision repair enzymes is retained, then re-ligation of breaks and major restoration of genomic integrity can occur within 30–60 min of rehydration (Elder *et al.*, 1987).

In plants, with seeds called orthodox and in wind-dispersed pollen, acquisition of the desiccation-resistant state occurs before shedding and this state is retained for a limited period during and after rehydration. It is lost again with cell growth or division at germination. In the embryos of cereal seeds loss of desiccation tolerance appears linked to the first round of cell divisions following S-phase replication. Furthermore, in imbibed but dormant embryos of *Avena fatua* (in which no DNA replication occurs), desiccation tolerance can be retained for long periods until dormancy is broken and cell cycling commences (Elder and Osborne, 1993).

Knowing that DNA dark-excision repair takes place within minutes upon imbibition of viable embryos or pollen and that the integrity of the repaired DNA is retained upon a subsequent dehydration and rehydration (Elder and Osborne, 1993; Osborne and Boubriak, 1994), we now ask the question is the operation of a functional DNA repair mechanism an essential component of desiccation tolerance or could the conformation of DNA in a desiccation-tolerant condition be sufficient to maintain a stable genome through cycles of hydration and rehydration?

We have approached the question in two different desiccation tolerant systems: the pre-DNA replicative stage (pre-S-phase) of rye embryo imbibition; and during periods of atmospheric hydration prior to germination in the G_2 pollen of birch.

Our results to date indicate that for both systems DNA is not stable to desiccation when DNA repair is blocked or incomplete.

Materials and Methods

The plant materials used in these studies were seeds of the 1994 harvest of rye, *Secale cereale*, cv Admiral from J. Picard and Co., Wirral, Cheshire, UK and the 1995 pollen of birch, *Betula verrucosa* from Kiev, Ukraine.

Whole embryos of rye were isolated by hand following imbibition of the seed for 2–3 h. They were then air dried to constant weight (1 h) and stored dry (1–2 days) before use. Embryos were imbibed either in 1 or 2% sucrose including $10 \mu g \ ml^{-1}$ chloramphenicol or, where inhibitors were used, all material was imbibed in $2.5 \mu l \ ml^{-1}$ DMSO with or without aphidicolin (Ap) or dideoxythymidine triphosphate (dTTP) (Sigma-Aldrich, UK).

Dry pollen was maintained desiccated at $-20°C$. For hydration, the weighed pollen was exposed in open petri-dishes in desiccators at 100% humidity at 30°C for 2.5 h. The pollen was then either dehydrated back in air to the original dry weight (2 h) or set to germinate (3 h) in a shaker bath in 20% sucrose solution containing 0.01% boric acid at 26°C. Germination was assessed by viewing samples under the microscope at $\times 10$ magnification.

Irradiation

Both embryos and pollen were γ-irradiated dry (500–1000 Gy) from a ^{137}Cs source (Gravatom, UK) and used immediately in experiments.

Incorporation Studies

Unscheduled DNA synthesis was measured by the incorporation of ^3H-methyl-thymidine (Amersham International, UK) into TCA-insoluble or PCA-soluble material. Radioactivity in the samples was determined by scintillation counting.

DNA Extraction

Several methods have been employed including proteinase K, with or without phenol-chloroform extraction or a rapid genomic DNA kit (Invisorb, Bioline UK). The details are given in the respective legends. For all embryo DNA extractions the material was first frozen and homogenized in liquid nitrogen. For pollen, the material was frozen with solid CO_2 before homogenization.

DNA Synthesis in Embryos on Imbibition

In embryos of rye and in many other species, the cells are arrested in the G_1 phase of the cell cycle on maturation drying. On rehydration, RNA and protein synthesis commence within minutes (Sen *et al.*, 1975) and continue unabated through germination (Fig. 1).

In contrast, DNA synthesis, which also starts within minutes of hydration, is initially restricted to DNA repair until the onset of first S-phase. Then replicative DNA synthesis starts as the prelude to the G_2 phase and eventual cytokinesis (G_2M). The S-phase interface with G_2M is very close, and at this stage the tolerance to desiccation exhibited throughout G_1 is lost. Until then, the repair of single-stranded breaks induced in DNA during storage or by 1000 Gy γ-irradiation of embryos in the dry state is fully repaired during 0–2 h imbibition (Fig. 2, lane 2 compared with lane 1) and the restoration of the DNA to high

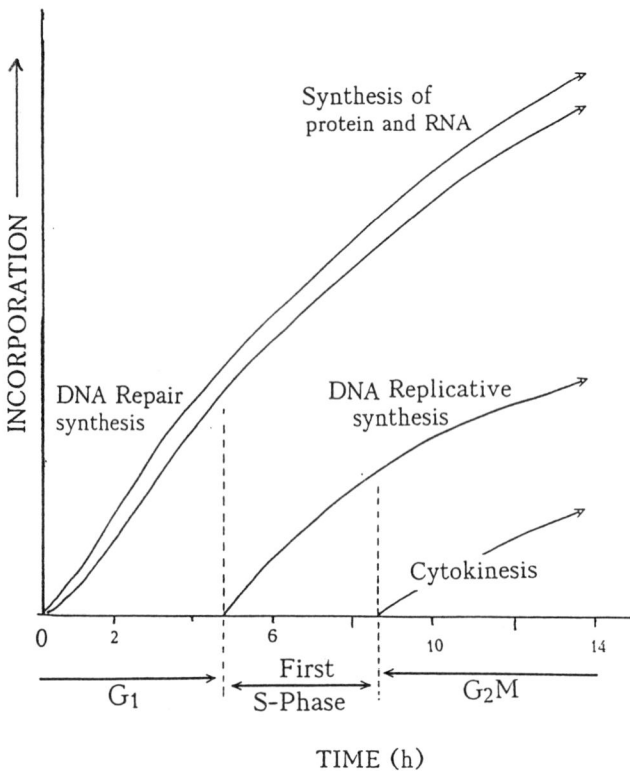

Figure 1. The onset of synthetic events in the embryos of dry seeds on rehydration

molecular weight is stably maintained when these embryos are then dehydrated back to their original dry weight (Fig. 2, lane 3).

DNA Polymerases Involved in DNA Repair

In most eukaryotic systems a β-polymerase is the major polymerase involved in dark excision DNA repair and the activity of this enzyme is inhibited by the presence of dTTP. Following irradiation damage however, additionally polymerase α and possibly also δ polymerase operate to complete the repair and these are subject to inhibition by aphidicolin (Popanda and Thielmann, 1992). We have determined the extent of inhibition of DNA repair (as measured by incorporation of ^{3}H-methyl-thymidine) during the first 3 h of imbibition in untreated embryos and in embryos subjected to γ-irradiation (1000 Gy) in the dry state. In duplicated experiments (Table 1), the β-polymerase inhibitor dTTP reduces control DNA repair by more than 40% but Ap has no inhibitory effect. In the γ-irradiated embryos, both Ap and dTTP reduce the thymidine

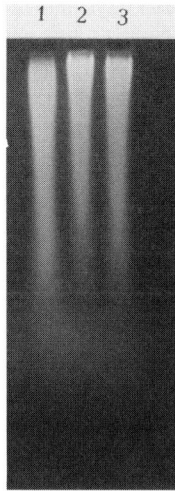

Figure 2. Desiccation stability of repaired DNA in G_1. Molecular weight profiles of DNA extracted from (lane 1) weighed dry embryos of rye γ-irradiated (1000 Gy) in the dry state, (lane 2) γ-irradiated embryos imbibed for 2 h in 2% sucrose containing 10 μg ml^{-1} chloramphenicol and (lane 3) then dried back to their original weight. DNA was extracted by the proteinase K method as described in Elder and Osborne (1993) and fractionated by electrophoresis on 0.8% alkaline denaturing gels

incorporation of unscheduled DNA repair synthesis up to 39%.

With this information, we have now used Ap and dTTP together to determine if inhibition of DNA repair leads to loss of DNA integrity in the desiccation-tolerant G_1 stage of embryo imbibition.

Embryo DNA Integrity on Desiccation following Inhibition of DNA Repair

For these experiments, weighed samples of embryos of rye (controls or γ-irradiated in the dry state) were imbibed for 0–2.5 h either with or without Ap and dTTP. Other embryos, similarly treated, were dried back to their original fresh weight. DNA was extracted from each sample and the molecular weight profiles compared after electrophoretic fractionation (Fig. 3). The absorbance scans (a), (c) and (e) show that a desiccation-stable restoration of high molecular weight DNA follows upon the DNA repair synthesis during the 0–2.5 h of imbibition. Where polymerase activities were inhibited by the presence of Ap and dTTP however, DNA integrity is not restored (d), and on desiccation of these samples (f) further DNA fragmentation ensues.

If these DNA isolates are first incubated with S1 nuclease (so that single stranded regions of the DNA are cleaved) before fractionation on the neutral agarose gels then the extent of damage in DNA is amplified. The molecular

Table 1. Effect of aphidicolin (Ap) and dideoxythymidine (dTTP) on DNA repair synthesis after γ-irradiation (1000 Gy) of dry embryos. Embryos were imbibed in 2% sucrose with 2.5 $\mu l\,ml^{-1}$ DMSO, 50 μCi ^3H-methyl-thymidine (specific activity 50 Ci $mmol^{-1}$) and either Ap or dTTP

Gy	DNA		Inhibition (%)
	Control	Ap 1.5×10^{-4} M	
0	2190 ± 132	2002 ± 326	0
	2155 ± 87	2050 ± 140	0
1000	3762 ± 256	2700 ± 108	28
	5554 ± 246	3616 ± 178	35
	Control	dTTp 1×10^{-3} M	
0	2098 ± 187	1062 ± 101	49
	1410 ± 111	823 ± 67	42
1000	1388 ± 62	839 ± 12	39
	706 ± 45	571 ± 5	19

weight profiles (not shown) confirm the integrity and stability to desiccation of the DNA following normal repair (Figs. 3a,e) and the DNA instability that follows the failure to repair (Fig. 3f). These results indicate to us that irrespective of the conformational state of the DNA, desiccation tolerance during the pre-S-phase of embryo imbibition requires a continuous and functional DNA repair to ensure integrity of the genome.

DNA Integrity in Pollen following Hydration and Germination

We have shown previously that dry birch pollen (14% moisture content) can be fully hydrated in moist air (to 28% moisture content) and dehydrated back to the original dry weight without loss of ability to germinate and without fragmentation damage detectable in the extracted DNA (Osborne and Boubriak, 1994). At shedding, birch pollen is haploid and bi-nucleate and arrested at the G_2 stage of the cell cycle. Whereas hydration alone does not result in germination or a loss of desiccation tolerance, access to free water does, and the emergence of the pollen tube signals loss of desiccation tolerance and a related loss of DNA integrity. This can occur at the limited percentage water loss of 16%, even though there is no DNA replication synthesis in this system (Osborne and Boubriak, 1994).

We have therefore questioned whether there is evidence for DNA repair during the air hydration of this wind-dispersed pollen. If repair operates during

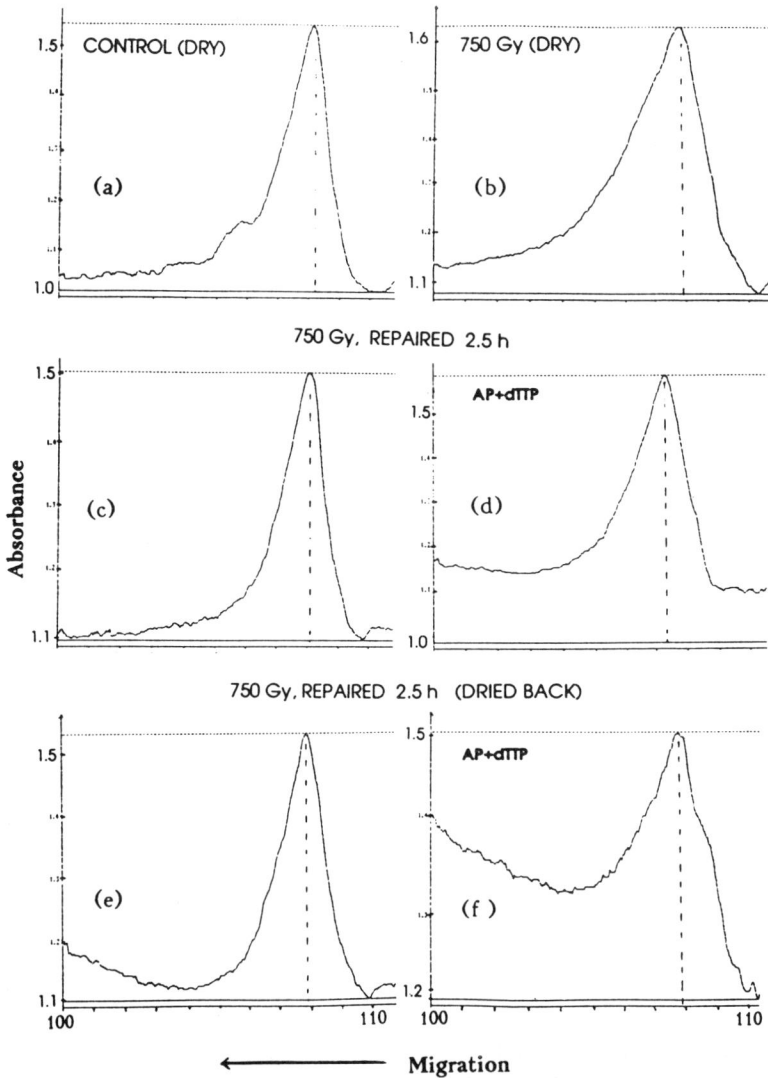

Figure 3. Comparison of molecular weight profiles of DNA from rye embryos. Control (a); γ-irradiated (b); repaired and dried back in the absence (c,e) or presence (d,f) of DNA repair inhibitors. For (a,b) DNA extracted from weighed dry embryos; for (c,d,e,f) embryos imbibed for 2.5 h in 1% sucrose solution with 2.5 μl ml⁻¹ DMSO (c,e), or DMSO with 1.5×10^{-4} M Ap and 1.0×10^{-3} M dTTP (d,f). Before DNA extraction, embryo samples (e,f) were blotted and air-dried back to their original weight. Rapid genomic DNA kit (InViSorb, Bioline, UK) was used for DNA isolation with some modification; i.e. ground embryos were lysed for 1 h at 24°C, before the addition of 15 μl of matrix carrier per sample. Bound DNA was washed 3×1 ml with washing solution and eluted in two stages to a final volume of 55 μl. DNA was fractionated on neutral 0.8% agarose gels, stained by ethidium bromide and photographed. Negatives from Polaroid 665 film were scanned on an LKB (ULTROSCAN XL) Laser Densitometer

Table 2. DNA repair synthesis on the germination of birch pollen in 20% sucrose with 0.01% H_3BO_3 and 30 μCi ^{3}H-methyl-thymidine (specific activity 45 Ci $mmol^{-1}$) following increasing periods of hydration in moist air. Values expressed as dpm $μg^{-1}$ DNA

Hydration (min)	Germination (%)	Incorporation (dpm)
0	32.7 ± 3.8	3300 ± 400
20	76.4 ± 4.6	5100 ± 700
40	78.3 ± 3.9	5800 ± 600
60	77.3 ± 6.1	6100 ± 300
80	75.8 ± 4.7	5500 ± 400

this period then it might be expected that the extent of repair when liquid water is available would be diminished. This does not appear to be so (Table 2).

Pollen that has been exposed to increasing durations of hydration in moist air before transfer to liquid medium shows essentially similar levels of incorporation of ^{3}H-methyl-thymidine (on a percentage germination basis) irrespective of the period of previous air hydration. This suggests that a normal DNA excision repair does not operate until the pollen encounters a liquid medium. However, extraction and electrophoretic fractionation of DNA from air hydrated pollen that is first irradiated in the dry state shows that an overall higher molecular weight DNA is achieved over a 2.5 h exposure period (Fig. 4, lane 3 compared with lane 2). But when this air-hydrated pollen is dried back to the original weight (lane 4) the level of high molecular weight DNA returns to that of the original γ-irradiated sample (lane 2) indicating desiccation instability. When these DNA samples are exposed to S1 nuclease, to reveal single stranded breaks in the DNA, the shift to a higher molecular weight on hydration is confirmed (lane 9 compared with lane 10) and the loss of this high molecular weight DNA component on desiccation (lane 11) becomes even more evident.

Because the incorporation data of Table 2 suggest that an excision repair measured as unscheduled DNA synthesis may not occur during the relatively low degree of hydration possible in moist air, the molecular weight shift in DNA must presumably result from a direct ligation of single strand breaks. It then follows that the DNA conformation that is acquired as a result of this ligation is that of a desiccation-intolerant form as normally found in germinating pollen (lanes 5 and 6; lanes 12 and 13).

Interpretation

Firstly, it would seem that the integrity of what is primarily nuclear DNA is not determined by the stage of the cell cycle or by the genome copy number. Embryos whose diploid cells are predominantly in G_1, or pollen whose two

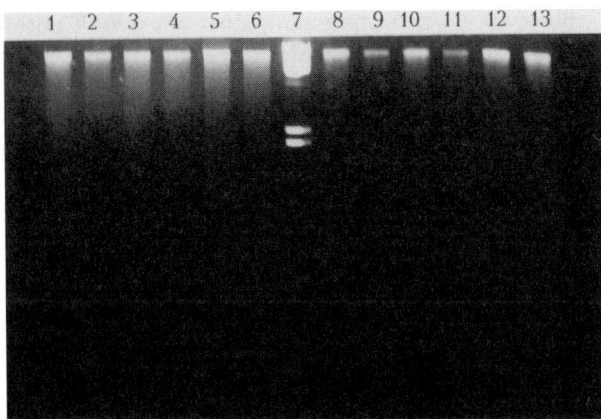

Figure 4. Neutral agarose gels of DNA from birch pollen extracts prior to (1–6) and after (8–13) S1 nuclease treatment (37 units per sample for 2 h at 37°C). DNA isolation, fractionation and photography as described in Fig. 3. Lanes 1, 8 dry pollen, lanes 2, 9 dry γ-irradiated (500 Gy) pollen; lanes 3, 10 γ-irradiated pollen hydrated in a humidity chamber for 2.5 h and then (lanes 4, 11) dried back in air to the original dry weight; lanes 5, 12 γ-irradiated pollen after a hydration–dehydration cycle (as before) then 3 h in germination solution; lanes 6, 13 germinating pollen dehydrated back to 50% moisture loss; lane 7: Hind III digest of λ DNA

haploid nuclei are in G_2 are both stable to hydration and dehydration provided germination has not taken place. Whereas embryos undergo a first round of replicative DNA synthesis and approach G_2M before losing desiccation tolerance, neither of the two nuclei of the wind-dispersed birch pollen that we have studied will replicate DNA on pollen tube emergence; the generative nucleus undergoes division to form the two G_1 sperm cells only late in germination (after 8 h), well beyond the time when desiccation tolerance is lost.

With the knowledge that certain conformational forms of DNA and the binding of specific low molecular weight proteins can confer protection against DNA damage in bacterial spores (Setlow, 1992; Fairhead *et al.*, 1993) it is relevant to consider whether the loss of desiccation-tolerance in both embryos and pollen could result from alterations in DNA conformation or if an aberrant DNA repair itself could contribute to DNA instability.

In embryos, it seems clear that γ-irradiation damage to DNA is fully repaired during the pre-replicative stages of imbibition and this DNA remains stable to desiccation if full excision DNA repair processes operate (Fig. 2 and Fig. 3c). However, if repair enzymes are inhibited during this desiccation tolerant phase, then overall, DNA integrity and desiccation tolerance are lost. We suggest that the DNA excision repair that is normally achieved during the G_1 stages of germination (Elder *et al.*, 1987) generates sites on the DNA strands which, like the DNA of the undamaged regions, are in a conformational form that is not disrupted when water is removed.

With pollen, however, air hydration alone does not permit normal excision patch repair. Instead, ligation events which appear to restore high molecular weight to DNA after irradiation damage are, we suggest, then held in a conformation that does not resist degradation on water loss.

As in bacterial spores, the holding of DNA in particular conformational forms may depend upon a concurrent synthesis of specific DNA-binding proteins. These could be synthesized by embryos in G_1, since a fully functional transcriptional and translational machinery operates as soon as water is available (Sen *et al.*, 1975) but in pollen, which air hydrates to less than 30% moisture content such synthesis may not be possible. This could, in part, explain the instability of the apparent repair of damaged DNA in pollen on dehydration (Fig. 4 lanes 4, 11).

Although speculative, the knowledge we have to date is not at variance with the interpretations of DNA status, replication and repair with respect to desiccation tolerance that we have proposed. Further, it serves to emphasize the central role that DNA repair can play in determining the integrity of the plant genome.

Acknowledgements

We are indebted to Mrs Vivian Reynolds for preparing the manuscript and one of us (I.I. Boubriak) acknowledges financial support of the Royal Society, London.

References

Cheah, K.S.E. and Osborne, D.J. 1978. *Nature* (London) 272: 593–599.

Crowe, J.H., Hoekstra, F.A. and Crowe, L.M. 1992. *Annual Reviews of Physiology* 54: 579–599.

Elder, R.H. and Osborne, D.J. 1993. *Seed Science Research* 3: 53–63.

Elder, R.H., Dell'Aquila, A., Mezzina, M., Sarasin, A. and Osborne, D.J. 1987. *Mutation Research* 181: 61–71.

Fairhead, H., Setlow, B. and Setlow, P. 1993. *Journal of Bacteriology.* 175: 1367–1374.

Griffith, J., Makhov, A., Santiago-Lara, L. and Setlow, P. 1994. *Proceedings of the National Academy of Sciences* (USA) 91: 8224–8228.

Leopold, A.C. 1986. *Membranes, Metabolism and Dry Organisms.* Ithaca: Cornell University Press.

Osborne, D.J. and Boubriak, I.I. 1994. *Seed Science Research* 4: 175–185.

Popanda, O. and Thielmann, H.W. 1992. *Biochimica et Biophysica Acta. Gene Structure and Expression* 1129: 155–160.

Sen, S., Payne, P.I. and Osborne, D.J. 1975. *Biochemical Journal* 148: 381–387.

Setlow, P. 1992. *Molecular Microbiology* 6: 563–567.

4. The Role of Protein Biotinylation in the Development and Germination of Seeds

M. DUVAL[1], L. DEHAYE[1], C. ALBAN[1], R. DEROSE[2], R. DOUCE[1], C. JOB[1] and D. JOB[1]

[1]Unité Mixte CNRS/Rhône-Poulenc (UMR41); [2]Département des Biotechnologies, Rhône-Poulenc Agrochimie, 14–20 rue Pierre Baizet, 69263 Lyon Cedex 9, France

Abstract

Biotin levels of pea seeds were investigated in order to characterize the temporal expression of the vitamin and the pattern of protein biotinylation that can be used to define developmental stages of seed formation and germination. These studies disclosed that metabolically active and quiescent tissues can be distinguished by their biotin content. In young developing seeds, as well as in germinating seeds, there is an excess of free versus bound biotin. In young seeds biotin is bound to the housekeeping biotin enzymes. A different pattern is observed in late stages of seed formation and in mature dry seeds. Here, protein-bound biotin is in excess with respect to free biotin. This is accounted for by the accumulation of a biotinylated polypeptide called SBP65 that behaves as a putative sink for the free vitamin, representing more than 90% of the total protein-bound biotin in mature seeds. Because the biotinylation domain of SBP65 differs markedly from that of presently known biotin enzymes, this protein defines a previously unrecognized class of biotinylated proteins. Furthermore, SBP65 shares many features in common with the LEA (Late Embryogenic Abundant) group of proteins. These results suggest that metabolic control of seed maturation and germination may occur through modulation in the level of an essential cofactor such as biotin. They also document an as yet undescribed role for some LEA proteins.

Introduction

Although the existence of biotin is well recognized in plants, little is known, however, regarding its biosynthesis and function. In plants, the most extensively studied biotin-dependent carboxylase is acetyl-CoA carboxylase, because of its obvious function in membrane biogenesis, and because the enzyme isolated from vegetative tissues and developing seeds of monocot plants is the target of the very potent cyclohexanedione and aryloxyphenoxypropionate herbicides (Harwood, 1988). This enzyme is involved in both growth of vegetative tissues and synthesis of storage lipids in developing seeds (Stumpf, 1980; Harwood, 1988). The role of biotin in plants has been also particularly well illustrated by the discovery of a mutation that causes defective embryo development in *Arabidopsis thaliana* and requires the vitamin at a critical stage of embryogenesis. Thus, no biotin was detectable in the arrested embryos of the mutant, and mutant embryos were specifically rescued when grown in the presence of biotin,

R.H. Ellis, M. Black, A.J. Murdoch, T.D. Hong (eds.), Basic and Applied Aspects of Seed Biology, pp. 33–43.
© *1997 Kluwer Academic Publishers, Dordrecht. Printed in Great Britain.*

thus demonstrating a specific housekeeping function for biotin during plant embryo development and initial plant growth (Schneider *et al.*, 1989; Shellhammer and Meinke, 1990).

We have analysed the content of free and protein-bound biotin of pea seeds at various developmental stages with the belief that an understanding of protein biotinylation in seeds would provide some information on the regulation of metabolic activity within embryonic cells during maturation and germination. There are specific features that appear typical of the physiology of seeds. These include: (i) the level of free biotin in vegetative and reproductive organs, and (ii) the discovery of a novel, seed-specific, biotinylated polypeptide of 65 kDa that we called SBP65. This paper will discuss the results of these investigations.

Biological Role of Biotin

In all organisms biotin is an essential cofactor for a small number of enzymes involved in the transfer of CO_2 during carboxylation and decarboxylation reactions (reviewed in Dakshinamurti and Chauhan, 1989; Knowles, 1989). Only very few proteins are naturally biotinylated, only one in *Escherichia coli*, three in *Saccharomyces cerevisiae*, and four in mammalian cells. The four biotin-dependent carboxylases found in mammals play essential cellular housekeeping functions. Thus, acetyl-CoA carboxylase (ACCase; EC 6.4.1.2) that catalyses the ATP-dependent carboxylation of acetyl-CoA is recognized as the regulatory enzyme of lipogenesis; methylcrotonoyl-CoA carboxylase (MCCase; EC 6.4.1.4) catalyses the conversion of methylcrotonoyl-CoA to methylglutaconoyl-CoA, a key reaction in the degradation pathway of leucine; propionyl-CoA carboxylase (EC 6.4.1.3) is a key enzyme in the catabolic pathway of the amino acids, isoleucine, threonine, methionine and valine; pyruvate carboxylase (EC 6.4.1.1) has an anapleurotic role in the formation of oxaloacetate. Because mammals lack the biosynthetic pathway for biotin, the vitamin appears to be recycled into new holocarboxylases by the combined action of proteases and of biotinidase (EC 3.5.1.12). The latter enzyme catalyses the hydrolysis of the lysyl-biotin bond (reviewed in Dakshinamurti and Chauhan, 1989). The family of biotin enzymes also includes the ATP-independent decarboxylases (oxalacetate, methylmalonyl-CoA, and glutaconyl-CoA decarboxylases) that are involved in sodium transport in anaerobic procaryotes (Dimroth, 1985), and transcarboxylase (EC 2.13.1) that participates in propionic acid fermentation in *Propionibacterium shermanii* (Wood and Kumar, 1985). In all biotin enzymes, the biotin is covalently linked to the epsilon amino group of a specific Lys residue located within a highly conserved Met-Lys-Met motif by the action of biotin holocarboxylase synthetase (also called biotin ligase) (reviewed in Samols *et al.*, 1988). Mutation experiments on the 1.3-S subunit gene of *Propionibacterium shermanii* transcarboxylase have established that -1 Met is dispensable for biotinylation *in vivo* by biotin holocarboxylase synthetase of *E. coli*, but has a critical role in the carboxyl transfer reaction (Shenoy *et al.*, 1988). A similar functional role has been demonstrated for the corresponding -1 Met of the biotinylated subunit of *E. coli* ACCase (Kondo *et al.,* 1984). Consistent with this finding Schatz (1993) demonstrated that synthetic peptides containing a central Lys residue that lack the two flanking Met residues can act as substrates for biotinylation *in vivo* by *E. coli* biotin holocarboxylase synthetase.

Biotin Synthesis and Biotin Levels in Vegetative and Reproductive Organs

Although it is recognized that plants, like microorganisms, have the capability to synthesize most vitamins, only two recent studies have addressed the question of biotin synthesis in plants. This is because plants, as all living organisms, only need trace amounts for general metabolism (Dakshinamurti and Chauhan, 1989; Shellhammer and Meinke, 1990). Thus, highly sensitive assays are needed to measure biotin levels in plants. These include streptavidin binding assays and assays based on the use of biotin-requiring bacteria (see Shellhammer and Meinke, 1990; Baldet *et al.*, 1993; Duval *et al.*, 1994a). Furthermore, much care should be taken to avoid contamination by trace amounts of biotin. For example, all solutions used for biotin extraction from plant materials must be rendered 'biotin-free' by passing through an avidin-affinity column. By these techniques plants were shown, indeed, to contain only very low amounts of biotin, in the range of 10 to 140 pg of total biotin per mg fresh weight in the case of various tissues of *A. thaliana* (Shellhammer and Meinke, 1990), and in the order of 20 pg free biotin per mg fresh weight in the case of mature pea seeds (Duval *et al.,* 1994a).

Initial information on biotin synthesis and transport in higher plants came from analysis of the *bio1* biotin auxotroph of *A. thaliana* discovered by Schneider *et al.* (1989). The mutant *bio1* embryos show a block in development and contain virtually no detectable biotin. Furthermore, these mutant embryos can produce phenotypically normal plants when cultured in the presence of either biotin, desthiobiotin or diaminopelargonic acid, the two latter compounds being the final intermediates of biotin synthesis in microorganisms (Eisenberg, 1985). Interestingly, Shellhammer and Meinke (1990) also observed that maternal sources of biotin are insufficient to rescue mutant embryos produced by heterozygous plants grown in the absence of supplemental biotin. This means that embryo development cannot be achieved without some biotin synthesis occurring within the embryo itself. Baldet *et al.* (1993) showed that all the intermediates of biotin synthesis established in bacteria, including pimelyl-CoA, 8-amino 7-oxo pelargonic acid, 7,8-diaminopelargonic acid, desthiobiotin, 9-mercaptodesthiobiotin and desthiobiotin are also present in plants. Altogether, the results of Shellhammer and Meinke (1990) and of Baldet *et al.* (1993) demonstrated that the pathway of biotin synthesis in bacteria is therefore conserved in higher plants.

Baldet *et al.* (1993) also showed that in plant cells (i) biotin synthesis occurs exclusively in the cytosolic compartment, (ii) free biotin is detected exclusively in this compartment, although biotin-containing enzymes are found both in plastids and mitochondria (see later), (iii) there exists a large excess of free over protein-bound biotin. For example, in cells from pea leaves, the free biotin pool accounts for about 90% of the total (free plus protein-bound) biotin. The precise role of this free-biotin pool is not known. One possibility would be that, as in bacteria, the level of free biotin might control the expression of genes encoding the biotin-containing enzymes and/or the enzymes involved in biotin synthesis. Duval *et al.* (1994a) confirmed that the amount of protein-bound biotin was almost negligible compared with that of free biotin in all vegetative tissues of pea plants examined, including leaves, roots, stems, flowers, pods and young developing tissues. Although free- and bound-biotin levels varied substantially during development of leaves and roots their relative levels remained almost unaffected during development, indicating therefore that an excess of free over bound-biotin is a characteristic feature of all metabolically active plants tissues. Mature pea seeds were a remarkable exception, since in this case there was a 5-fold higher amount of protein-bound biotin than of free biotin. On a per mg protein basis the level of free biotin was high in the young seeds, and then decreased during development. In contrast, the protein-bound biotin level was low in the young seeds, and then rose sharply during late stages of seed development and onset of desiccation, being maximal in the mature, dry seeds. That is, in contrast to the vegetative tissues, an excess of bound over free-biotin is a characteristic feature of quiescent tissues. This led us to examine in more detail the nature of protein biotinylation in the various organs of pea plants.

Biotinylated Proteins in Pea

Vegetative Tissues

Wurtele and Nikolau (1990) first demonstrated that in addition to ACCase, carrot cells contain the three other biotin-dependent carboxylase activities (*e.g.* MCCase, propionyl-CoA carboxylase and pyruvate carboxylase) found in mammals. Pea leaves only contain three biotinylated proteins (Baldet *et al.*, 1992; Alban *et al.*, 1993; 1994; Dehaye *et al.*, 1994; Duval *et al.*, 1994a). One of these, of 76 kDa, corresponds to the biotinylated subunit of MCCase. This enzyme, which is localized in mitochondria, is constitutively expressed, but its activity increases strongly during leaf senescence, where intense degradation of proteins occurs (Alban *et al.*, 1993). This feature is consistent with the idea that the plant enzyme, as for its animal counterpart, is involved in the degradation of leucine derived from protein mobilization. The enzyme purified from other plants exhibits similar properties (Chen *et al.*, 1993; Diez *et al.*, 1994).

The two other biotinylated proteins in pea leaves are of 220 kDa and 38 kDa, both of them being associated with ACCase activity. Chloroplasts of mesophyll cells contains an ACCase of the 'prokaryotic' type, consisting of several proteins (Alban *et al.*, 1994; Sasaki *et al.*, 1995) one of which, of 38 kDa, is biotinylated (for this reason we refer to this enzyme complex as ACC38). The enzyme subunits are organized in two functional domains that are readily dissociable and reassociable. The biotin carboxylase domain containing a 32-kDa biotin-free polypeptide tightly associated with the 38-kDa biotinylated subunit was isolated free from transcarboxylase activity. This functional domain is able to carboxylate exogeneous biotin, thus, catalysing the first half reaction of the overall ACCase activity (Alban *et al.*, 1995). Owing to its chloroplastic localization, ACC38 corresponds, most presumably, to the enzyme form involved in fatty acid synthesis. The level of ACC38 shows considerable variations during development, being maximum in young pea leaves, presumably reflecting a strong demand in fatty acids required for the biosynthesis of thylakoid membranes. A different form of ACCase is found in epidermal cells (Alban *et al.*, 1994), most probably in an extraplastidial compartment (Konishi and Sasaki, 1994). In contrast to ACC38 this enzyme is a homodimer of a single biotinylated polypeptide of 220 kDa (for this reason, we refer to this enzyme as ACC220), and therefore resembles typical eukaryotic ACCase described in mammals and yeast. Although it is recognized that malonyl-CoA is required, in addition to fatty acid synthesis, in a number of biosynthetic pathways, the physiological role of this second ACC is not yet established, and it can be involved in the biosynthesis of very-long-chain fatty acids required for cuticular waxes (Kolattukudy *et al.*, 1976), or the biosynthesis of flavonoids via chalcone synthase (Hrazdina *et al.*, 1982). In this context, it is interesting that, in pea leaves, both processes occur in epidermal tissue, thus matching the tissue localization of pea ACC220 (Alban *et al.*, 1994).

ACC220 exhibits a dual specificity, since it is also able to carboxylate propionyl-CoA, although with a lesser efficiency than acetyl-CoA (Dehaye *et al.*, 1994). These two reactions occur at separate sites on the enzyme. One site is not substrate specific, accepting either acetyl-CoA or propionyl-CoA and is strongly inhibited by quizalofop, an aryloxyphenoxypropionate herbicide. The other is specific for acetyl-CoA and is much less inhibited by quizalofop. In marked contrast, ACC38 which is not able to carboxylate propionyl-CoA is totally insensitive to quizalofop. From these results Dehaye *et al.* (1994) suggested that the sensitivity of plant ACCase towards aryloxyphenoxypropio-nates is somehow linked to the substrate specificity of the enzymes. That is, the active site of the highly selective ACC38 could have a more closed structure than that of the less selective ACC220, and thus quizalofop cannot react.

Maize leaves also contain two ACC isoforms, one in the chloroplasts of mesophyll cells, the other in an extraplastidial compartment of the mesophyll cells or in other cell types, but, in contrast to pea, both isoforms are composed of a single type of high-molecular-mass biotin-containing polypeptide of 227 kDa (ACC I, mesophyll chloroplast enzyme) or 219 kDa (ACC II, non-mesophyll-chloroplast enzyme) (Egli *et al.*, 1993; Ashton *et al.*, 1994). Both enzymes are inhibited by aryloxyphenoxypropionates, although ACC I is more sensitive than ACC II (Egli *et al.*, 1993). Thus, a comparison of our results with those of Egli *et al.* (1993) provides an explanation to why aryloxyphenoxypropionate herbicides will not kill all plants, since some Gramineae (*e.g.* maize) are susceptible, while other plants, for example of the families of the Dicotyledonae (*e.g.* pea), are resistant.

No other propionyl-CoA carboxylase activity, different from that catalysed by ACC220, could be detected from either reproductive or vegetative organs of pea plants at any stage of development (Dehaye *et al.*, 1994). Despite all our efforts, pyruvate carboxylase activity that has been shown to exist in carrot (Wurtele and Nikolau, 1990) cannot, as yet, be demonstrated in pea.

Seeds

Duval *et al.* (1993, 1994a) observed that the pattern of biotinylated polypeptides of mature pea seeds differs markedly from that seen in vegetative tissues, e.g. leaves. Thus, protein extracts from mature pea seed exhibit three major biotinylated polypeptides of 220 kDa, 76 kDa and 65 kDa. Activity measurements disclosed that the biotinylated form of 76 kDa corresponds to the biotinylated subunit of MCCase, and that of 220 kDa corresponds to the ACC220, the ACCase isoform found in epidermal tissues of leaves. Both activities were purified to near homogeneity from mature pea seeds (Dehaye *et al.*, 1994; Duval *et al.*, 1994a). These purifications revealed:

i. In pea seeds, it is the leaf-epidermis ACCase equivalent (ACC220) that contributes the major portion of ACCase activity. Thus, only ACC220 is detected, while ACC38 is absent. Similar observations have been reported by Bettey *et al.* (1992). Varying proportions of the two ACCase isoforms have also been described in different tissues of maize plants. Here, the major portion of ACCase activity detected in developing embryos and endosperm tissues of maize kernels is accounted for by the ACC I mesophyll-chloroplast equivalent, suggesting a role for this enzyme in synthesis of storage lipids (Somers *et al.*, 1993). Only leaves contain significant amount of the second isoform, ACC II (Somers *et al.*, 1993). This difference between maize and pea raises the question of the mechanisms involved in storage lipid synthesis in reproductive organs of plants.

ii. A biotinylated polypeptide of about 65 kDa accounts for the major part of total biotin-containing polypeptides in the mature pea seed extract. Duval *et al.* (1993, 1994a,b) have purified this protein to apparent homogeneity in order to prepare specific antibodies. It was immediately apparent that this protein was devoid of biotin-dependent carboxylase activity when using either acetyl-CoA, propionyl-CoA, methylcrotonoyl-CoA or pyruvate as a substrate. Furthermore, all the ACCase and MCCase activities in the mature pea seed extract were found to be associated with the 220-kDa and the 76-kDa biotinylated polypeptides, respectively. Duval *et al.* (1993) called this protein SBP65 for 'Seed Biotinylated Protein of 65 kDa'. The protein covalently binds biotin, with a stoichiometry of 1 mol of biotin per 1 mol of 65-kDa polypeptide. By using the specific anti-SBP65 antibodies Duval *et al.* (1993) showed the existence of biotinylated proteins immunologically related to pea SBP65 in other seed species, suggesting that they are ubiquitous in the seeds of legume plants. A cDNA encoding SBP65 has been cloned and sequenced (Duval *et al.*, 1994b). The deduced primary structure of the protein was confirmed by protein sequencing. Peptide sequencing also indicated that the biotinylation domain of SBP65 differs markedly from that of presently known biotin enzymes. In particular the two Met residues flanking the biotinylated biotin in most biotin enzymes are absent in the pea seed protein. The absence of these residues in SBP65 thus accounts for the lack of known biotin-dependent carboxylase activity associated with this protein. Since such an atypical biotinylation domain has never been described in naturally occurring proteins, this led us to conclude that SBP65 covalently binds biotin at a novel site (Duval *et al.*, 1994b). Interestingly, the cDNA clone that codes for SBP65, when expressed in *E. coli* yields a protein product that is specifically recognized by the anti-SBP65 antibodies but is not biotinylated (Duval, DeRose and Job, unpublished results). This suggests that the requirements for protein biotinylation in plants are not the same as those in bacteria, and/or that a particular biotin ligase exists in the developing pea seeds, so that covalent binding of biotin to the apo-SBP65 can proceed. (Enzymes catalyzing covalent attachment of biotin to specific apoproteins have not yet been described in plants.) The cDNA probe has further been used to clone the *SBP65* gene. Southern blot analyses showed that it is a single copy gene (Dehaye, Duval and Job, unpublished results). Also, by using the cDNA probe Dr Ellis (John Innes Institute, Norwich) localized the *SBP65* gene on chromosome IV (Ellis, Duval and Job unpublished results).

Further analyses disclosed that SBP65 exhibits a combination of several characteristics which renders it unique. Thus, it shares many features, both at the structural and developmental levels, in common with the LEA (Late Embryogenesis Abundant) group of seed proteins. That is, (i) SBP65 is synthesized in the embryo during the final stages of development, reaching a maximum level in the mature dry seed; (ii) its level rapidly decreases during seed germination (about 70% of the protein is degraded prior to radicle emergence); (iii) its synthesis is undetectable in all vegetative parts of the plant; (iv) its protein sequence reveals an extremely hydrophilic protein with many repeated motifs, some of which are similar to the 11-mer repeat unit present in LEA proteins such as carrot DC8 (Franz *et al.*, 1989) and birch BP8 (Puupponen-Pimiä *et al.*, 1993) proteins. In addition, SBP65 is localized mainly in the cytosol of cotyledon cells (Duval *et al.*, 1995), like two other LEA proteins from cotton seeds, D7 and D113 (Roberts *et al.*, 1993). SBP65 represents the first LEA protein known to bind with a small cofactor, such as biotin, thus making it the major biotinylated polypeptide in dry pea seeds. For this reason, SBP65 behaves as a putative sink for free biotin during the last stages of seed development, reflecting the fact that during seed development, the relative content of free over bound biotin evolves during seed development, so that in mature seeds, protein-bound biotin (*i.e.* that bound to SBP65) is in excess over free biotin (Duval *et al.*, 1994a). This feature is important in that, as outlined above, biotin synthesis occurs in the cytosolic compartment and because an excess of free over protein-bound biotin characterizes metabolically active tissues. Hence, the subcellular localization of SBP65 in the cytosol is compatible with its efficient trapping of this essential cofactor. Put together, these observations question the role of SBP65 as a LEA protein. It is generally admitted that the LEA proteins play a role in seed desiccation, that is, they are assumed to prevent the desiccating seed from losing all its water (Dure *et al.*, 1989). The fact, however, that SBP65 represents less than 0.01% of total soluble proteins rules out *a priori* the possibility that this biotinylated protein functions as a desiccation protectant.

A comparative study has been undertaken using pea zygotic embryos cultured *in vivo*, pea somatic (Loiseau *et al.*, 1995) and immature pea zygotic embryos cultured *in vitro* (Cook *et al.*, 1988) in order to address whether SBP65 synthesis is solely under the control of the developmental program imposed by the embryo itself, or whether this synthesis depends on specific interactions between the growing embryo and maternal tissue (Duval, Loiseau, Dehaye, LeDeunff, Wang and Job, unpublished results). For the zygotic embryos grown *in vivo* SBP65 was synthesized exclusively from the cotyledonary stage onwards, in the later stages of embryogenesis, i.e. after legumin synthesis had ceased. Cotyledonary zygotic embryos cultured *in vitro* kept the same pattern of synthesis for this protein, showing that once the cotyledonary stage is reached the informational context is established within the embryo so that the correct production of SBP65 can proceed independently of subsequent signal(s) from the mother plant. During somatic embryogenesis, SBP65 was also expressed, demonstrating that the induction of synthesis of this protein was under the control of the embryogenic program. In this instance, however, its synthesis occurred as early as the globular stage and before legumin synthesis started. This suggests the existence of a maternally derived signal that represses synthesis of SBP65. Nevertheless, as soon as somatic embryos rooted, the SBP65 level decreased, this being equivalent to the temporal expression pattern of SBP65 during germination of zygotic seeds.

Conclusion

We identified SBP65 as a biotinylated protein that accumulated specifically in developing seed of pea, being absent in all other vegetative parts of the plant, and decreased very rapidly during germination. Also, we observed that the synthesis of SBP65 occurs in embryonic tissue during normal embryo development at a time when storage proteins have accumulated and when the seed approaches maturity. This protein cannot be detected either in embryonic tissue before the cotyledonary stage or in embryos at the beginning of the cotyledonary stage by which time they have already gained the capacity to germinate (Cook *et al.,* 1988; our unpublished results). This indicates that SBP65 is not required for germination *sensu stricto* but rather is involved either in the process of seed maturation or in the growth of the seedling after germination. Consistent with this finding, immature pea embryos regenerate normal plants when cultured under appropriate conditions (*e.g.* low-sucrose-containing medium, see Cook *et al.*, 1988), but failed to do so in the presence of actithiazic acid, a potent inhibitor of biotin synthesis in bacteria (Ogata *et al.,* 1973) and plants (Baldet *et al.,* 1993). Such immature embryos can produce phenotypically normal plants when cultured both in the presence of actithiazic acid and minute amounts of biotin (Duval, Dehaye and Job, unpublished results). Furthermore, SBP65 synthesis was induced when the endosperm had nearly disappeared, the epidermis of the cotyledons being close to the inner surface of the testa. Thus, the accumulation of SBP65 was observed when the embryonic cells became metabolically quiescent. Since SBP65 is the major biotinylated protein in mature embryos (Fig. 1) and behaves as a putative sink for free biotin, it may dramatically alter the metabolic state of cells, for example, by depriving the cells of the biotin needed by the housekeeping, biotin-dependent, enzymes. Such a role for SBP65 would make it detrimental to cells requiring biotin and could provide an explanation as to why this protein is never detected in vegetative tissues, including young developing embryos, at any time of the plant life cycle. In this context, it is worth noting the observation of Shellhammer and Meinke (1990) that arrested embryos from the *bio1* mutant of *A. thaliana* contain reduced levels of biotin. Since SBP65 disappears at a high rate during germination (Duval *et al.,* 1993; 1994a), an alternative but not mutually exclusive possibility is that SBP65 constitutes a storage form of this vitamin that is required by the embryo to initiate its growth during the germination process. At present, it is not possible to rule out one further explanation, in that SBP65 may represent a new type of biotin-dependent enzymatic activity for which a role remains to be established. Yet, the fact that this protein is specifically expressed during embryogenesis indicates that its function may somehow be related to the physiology of the embryo and that the use of an unusual biotinylation site may be functionally advantageous in seed developmental programmes. This system may provide tools to answer basic questions concerned with regulation of metabolism during plant embryo development and initial plant growth.

Acknowledgements

We thank Trevor Wang and Rod Casey (John Innes Institute, Norwich) for help in the pea embryo cultures, as well as Jacques Loiseau and Yvon LeDeunff of the École Nationale d'Horticulture (Versailles) for providing samples of pea somatic embryos. This study was conducted under the Bio Avenir program funded by Rhône-Poulenc, the Ministère de la Recherche et de l'Espace and the Ministère de l'Industrie et du Commerce Extérieur.

References

Alban, C., Baldet, P., Axiotis, S. and Douce, R. 1993. *Plant Physiology* 102: 957–965.

Alban, C., Baldet, P. and Douce, R. 1994. *Biochemical Journal* 300: 557–565.

Alban, C., Julien, J., Job, D. and Douce, R. 1995. *Plant Physiology* 109: 927–935.

Ashton, A. R., Jenkins, C. L. D. and Whitfeld, P. R. 1994. *Plant Molecular Biology* 24: 35–49.

Baldet, P., Alban, C., Axiotis, S. and Douce, R. 1992. *Plant Physiology* 99: 450–455.

Baldet, P., Gerbling, H., Axiotis, S. and Douce, R. 1993. *European Journal of Biochemistry* 217: 479–485.

Bettey, M., Ireland, R. J. and Smith, A. M. 1992. *Journal of Plant Physiology* 140: 513–520.

Chen, Y., Wurtele, E. S., Wang, X. and Nikolau, B. J. 1993. *Archives of Biochemistry and Biophysics* 305: 103–109.

Cook, S. K., Adams, H., Hedley, C. L., Ambrose, M. J. and Wang, T. L. 1988. *Plant Cell Tissue and Organ Culture* 14: 89–101.

Dakshinamurti, K. and Chauhan, J. 1989. *Vitamins and Hormones* 45: 337–384.

Dehaye, L., Alban, C., Job, C., Douce, R. and Job, D. 1994. *European Journal of Biochemistry* 225: 1113–1123.

Diez, T. A., Wurtele, E. S and Nikolau, B.J. 1994. *Archives of Biochemistry and Biophysics* 310: 64–75.

Dimroth, P. 1985. *Annals of the New York Academy of Sciences* 447: 72–85.

Dure, L., Crouch, M., Harada, J., Ho, T. H. D., Mundy, J., Quatrano, R., Thomas, T. and Sung, Z. R. 1989. *Plant Molecular Biology* 2: 475–486.

Duval, M., Job, C., Alban, C., Sparace, S., Douce, R. and Job D. 1993. *Comptes Rendus de l'Académie des Sciences (Paris) Série III, Sciences de la Vie* 316: 1463–1470.

Duval, M., Job, C., Alban, C., Douce, R. and Job, D. 1994a. *Biochemical Journal* 299: 141–150.

Duval, M., DeRose, R. T,. Job, C., Faucher, D., Douce, R. and Job, D. 1994b. *Plant Molecular Biology* **26**: 265–273.

Duval, M., Pépin, R., Job, C., Derpierre, C., Douce, R. and Job, D. 1995. *Journal of Experimental Botany* 46:1783–1786.

Egli, M. A., Gengenbach, B. G., Gronwald, J. W., Somers, D. A. and Wyse, D. L. 1993. *Plant Physiology* 101: 499–506.

Eisenberg, M. A. 1985. *Annals of the New York Academy of Sciences* 447: 335–349.

Franz, G., Hatzopoulos, P., Jones, T., Krauss, M. and Sung, Z. R. 1989. *Molecular and General Genetics* 218: 143–151.

Harwood, J. L. 1988. *Trends in Biochemical Sciences* 13: 330–331.

Hrazdina, G., Marx, F. A. and Hoch, H. C. 1982. *Plant Physiology* 70: 745–748.

Knowles, J. R. 1989. *Annual Review of Biochemistry* 58: 195–221.

Kolattukudy, P. E., Croteau, R. and Buckne, J. S. 1976. In: *Chemistry and Biochemistry of Natural Waxes*, pp. 289–347 (ed. P. E. Kolattukudy). New York: Elsevier.

Kondo, H., Uno, S., Komizo, Y. and Sunamoto, J. 1984. *International Journal of Peptide and Protein Research* 23: 559–564.

Konishi, T. and Sasaki, Y. 1994. *Proceedings of the National Academy of Sciences, USA* 91: 3598–3601.

Loiseau, J., Marche, C. and LeDeunff, Y. 1995. *Plant Cell Tissue and Organ Culture*, in press.

Ogata, K., Izumi, Y. and Tani, Y. 1973. *Agricultural and Biological Chemistry* 37: 1079–1085.

Puupponen-Pimiä, R., Saloheimo, M., Vasara, T., Ra, R., Gaugecz, J., Kurten, U., Knowles, J. K. C., Keränen, S. and Kauppinen V. 1993. *Plant Molecular Biology* 23: 423–428.

Roberts, J. K., DeSimone, N. A., Lingle, W. L. and Dure, L. 1993. *Plant Cell* 5: 769–780.

Samols, D., Thornton, C. G., Murtif, V. L., Kumar, G. K., Haase, F. C. and Wood, H. G. 1988. *Journal of Biological Chemistry* 263: 6461–6464.

Sasaki, Y., Konishi, T. and Nagano, Y. 1995. *Plant Physiology* 108: 445–449.

Schatz, P. J. 1993. *Bio/Technology* 11: 1138–1143.

Schneider, T., Dinkins, R., Robinson, K., Shellhammer, J. and Meinke, D. W. 1989. *Developmental Biology* 131: 161–167.

Shellhammer, J. and Meinke, D. 1990. *Plant Physiology* 93: 1162–1167.

Shenoy, B. C., Paranjape, S., Murtif, V. I., Kumar, G. K., Samols, D. and Wood, H. G. 1988. *FASEB Journal* 2: 2505–2511.

Somers, D. A., Keith, R. A., Egli, M. A., Marshall, L. C., Gengenbach, B. G., Gronwald, J. W. and Wyse, D. L. 1993. *Plant Physiology* 101: 1097–1101.

Stumpf, P. K. 1980. In: *The Biochemistry of Plants*, Vol. 4, pp. 177–204 (eds P. K. Stumpf and E. E. Conn). New York: Academic Press.

Wood, H. G. and Kumar, G.K. 1985. *Annals of the New York Academy of Sciences* 447: 1–21.

Wurtele, E. S. and Nikolau, B. J. 1990. *Archives of Biochemistry and Biophysics* 27: 179–186.

5. The Effect of Harvest Time and Drying on Dormancy and Storability in Beechnuts

K.A. THOMSEN

Department of Ornamentals, Research Centre Aarslev, Kirstinebjergvej 10, 5792 Aarslev, Denmark

Abstract

The effect of harvest time and drying on dormancy and storability was studied in two seed sources of beechnuts (*Fagus sylvatica* L.). Seeds harvested before, around and after shedding, were dried at 15% relative humidity by three different methods; A: at 15°C to 8% moisture content; B: at 20°C four hours per day to 8% moisture content; and C: at 20°C eight hours per day to 20, 16, 12 and 8% moisture content. Intact seeds or seeds after removal of pericarp and seed coat were tested for germination at 5°C for 10 and 16 months. Dormancy measured as mean germination time (MGT), decreased in mature seeds. Drying and removal of pericarps also reduced MGT. The difference of MGT between fresh and dry seeds was approx. 3–4 weeks. Removal of pericarp resulted in a further reduction in MGT by two weeks. The two seed sources were not synchronous with regard to optimum time of harvest, collection of seeds long before or after shedding had a negative effect on the seed storability.

Introduction

Beechnuts are deeply dormant seeds which demand two to three months of cold treatment before they can germinate (Nyholm, 1986). The seeds are probably orthodox and can be dried to a moisture content of 7% if drying temperatures are below 20°C (Poulsen, 1993). Suszka (1975) found that drying reduced the germination period by 28 days.

Drying or reduced water content in seeds stimulates germination in both immature and mature seeds of many species (Evans *et al.*, 1975; Sawhney and Naylor, 1982; Welbaum and Bradford, 1989; Rasyad *et al.*, 1990; Nerson, 1991; Demir and Ellis, 1992a, b; Gray *et al.*, 1992) including dormant tree seeds e.g. *Aesculus hippocastanum* L. (Suszka, 1966) and *Fagus sylvatica* L. (Suszka, 1975).

In barley (*Hordeum distichum* cv Julia) grains (Evans *et al.*, 1975) and sunflower (*Helianthus annuus* L.) embryos (Bianco *et al.*, 1994), drying has been demonstrated to change the sensitivity towards gibberellin, which is believed to initiate the synthesis and secretion of hydrolases during the reserve mobilization (Jensen, 1994).

This study was made to investigate the effect of harvest time and drying on dormancy and seed quality in beechnuts.

R.H. Ellis, M. Black, A.J. Murdoch, T.D. Hong (eds.), Basic and Applied Aspects of Seed Biology, pp. 45–51.
© *1997 Kluwer Academic Publishers, Dordrecht. Printed in Great Britain.*

Materials and Methods

Beechnuts from two different Danish seed sources, Graasten and Soroe, were harvested at weeks 35, 38 (where the seeds started to shed) and 44, shedding in 1993. The two first collections were picked from the trees, the last was collected on the ground.

After manual extraction from the fruits, the nuts were dried at three different rates: A: at 15°C and 15% RH to approx. 8% moisture content, B: at 20°C and 15% RH, 4 hours a day to 8% moisture content and C: at 20°C and 15% RH, 8 hours a day to 20, 16, 12 and 8% moisture content.

Germination tests were performed before and after drying, and after 10 and 16 months' storage at 5°C. Four replicates of 50 whole seeds were germinated in vermiculite at 5°C. Two replicates of 50 embryos (without both pericarp and testa) and seeds without the pericarp were germinated on top of filter papers, also at 5°C, but only after the drying treatments. The germination was counted weekly and the test was not ended until all seeds were either germinated or dead.

Moisture content on a fresh weight basis was determined according to ISTA regulation (ISTA, 1985).

Results

Both seed sources were fully germinable at the first harvest, but the moisture content indicated that the Graasten nuts (44%) were less mature than the Soroe nuts (40%). The germination period decreased with harvest time and drying (Fig. 1). The mean germination time (MGT) for fresh seeds of both seed sources was 22 weeks, decreasing to around 18 weeks at the second and 13–14 weeks at the last harvest. Drying reduced the germination period further by an average of 3 weeks, when the seeds were dried to 8% moisture content. Removal of the pericarp (after the first two collections) advanced the germination another 2 weeks, but removal of the seed coat did not have any effect on the speed of germination.

In Figure 2, the MGT is plotted against moisture content after the different drying treatments. The results show that there was no effect of drying rates on the speed of germination. It seems that there is a relationship between the moisture content and the MGT: drier seeds germinated more rapidly than wet seeds. The main effect of drying was found in the interval between 8 and 20% moisture content.

There were no significant differences in survival following drying to 8% moisture content under the different drying rates except in the first harvest of Graasten-beechnuts at which harvest time the treatments B and C reduced viability, from 73% (of fresh seeds) to 51 and 39%, respectively, whereas viability was slightly increased by treatment A (Fig. 3, week 35). The faster the rate the more damage.

The storability of the beechnut seeds at 8% moisture content was good. As

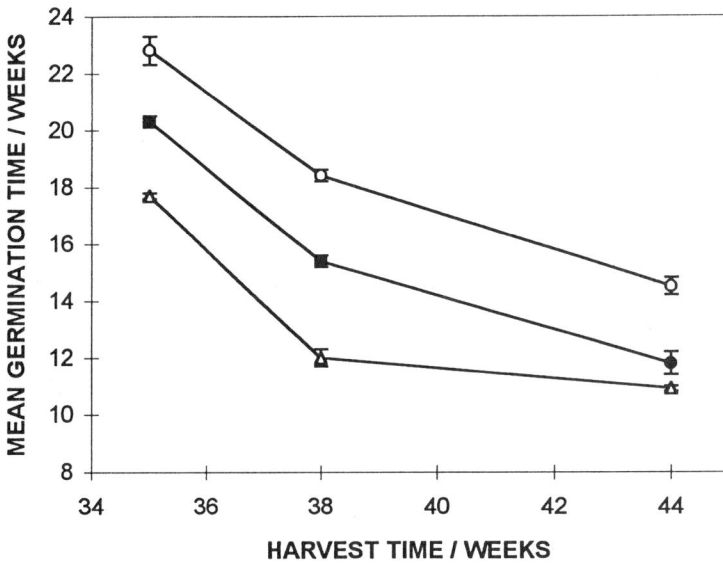

Figure 1. Mean germination time (at 5°C) for fresh (circles), dried (squares), and dried seeds without pericarp (triangles). Seed source: Soroe. Seeds were dried under treatment C to 8% moisture content. Vertical bars represent ±s.e.

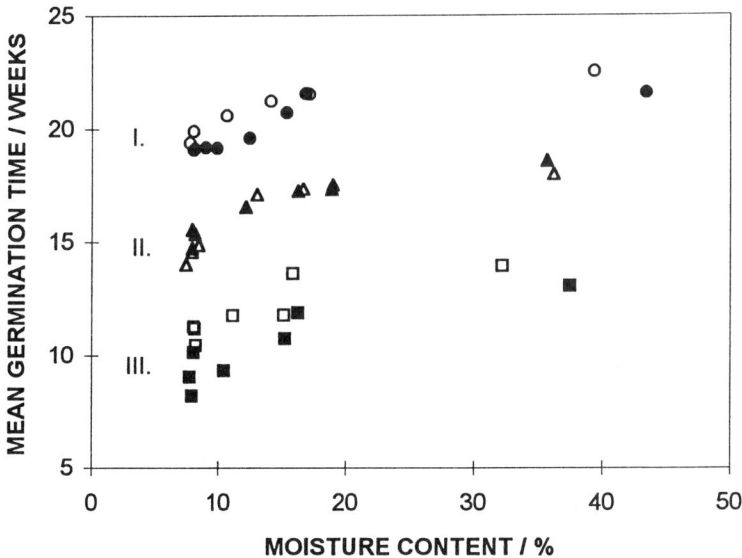

Figure 2. The relationship between seed moisture content and mean germination time (at 5°C) for two seed sources: Graasten (closed symbols) and Soroe (open symbols) harvested at week 35 (I.), 38 (II.) and 44 (III.)

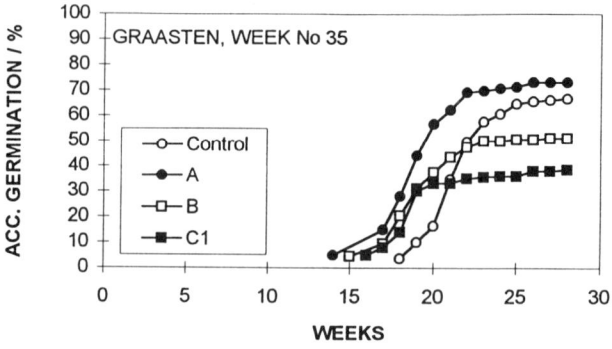

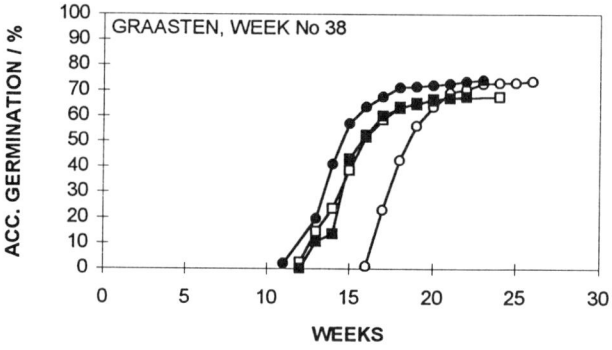

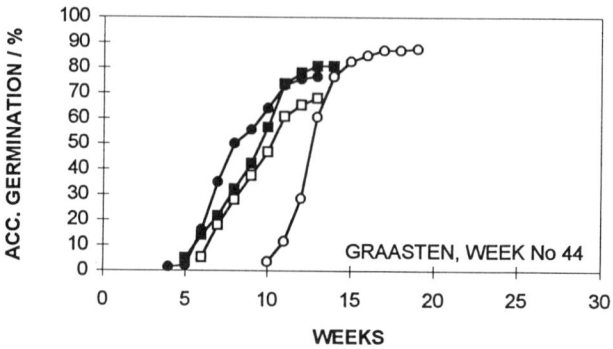

Figure 3. Accumulated germination of beechnuts harvested at weeks 35, 38 and 44 for fresh seeds (control) or seeds dried to 8% moisture content under different rates A, B and C. Seed source: Graasten

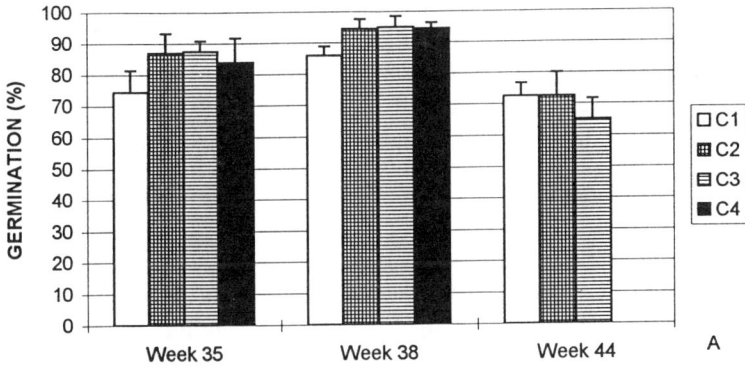

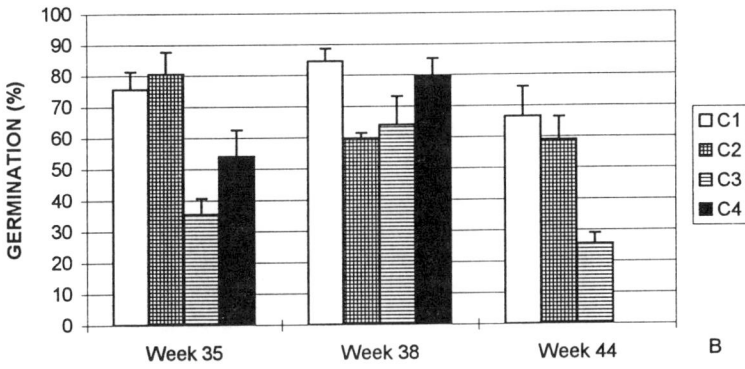

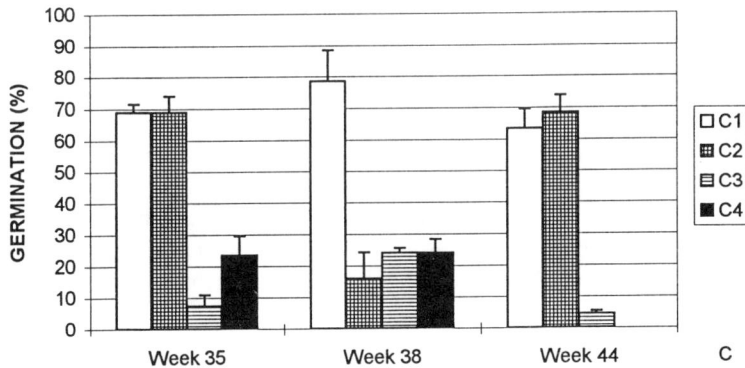

Figure 4. Survival of beechnuts of Soroe provenance following 0 (A), 10 (B) and 16 (C) months' hermetic storage at 5°C with 8 (C1), 12 (C2), 16 (C3) and 20% moisture content (C4)

expected, the seeds stored at higher moisture contents deteriorated more rapidly (Fig. 4). The MGT did not change with storage. For the Soroe nuts the optimal time of harvest nuts was week 38, since variability was reduced following 16 months' storage at 5°C with seeds which were harvested at week 44 (Fig. 4C). In contrast, the Graasten nuts had a better storability if collected at week 44 compared to week 38, even though the immediate viability was the same.

Discussion

The stimulation of germination by drying appears to be universal, since even in recalcitrant seeds a small reduction in moisture content has this effect (Suszka, 1966, Probert and Brierly, 1989). In beechnuts, drying can therefore substitute part of the cold stratification. Whatever the mechanism is, drying the beechnuts to 8% moisture content gave the same results as 3–4 weeks cold stratification. The apparent relation between moisture content and MGT suggested that the effect of drying was due to a reduction of the moisture content and not the drying process in itself, particularly since the rate of drying did not influence the MGT. It corresponds well with the decrease in MGT that has also been found with maturation (drying) of both non-dormant (e.g. Finch-Savage, 1992; Gray *et al.*, 1992) and dormant seeds (Hong and Ellis, 1992; Tompsett and Pritchard, 1993).

The best short and long term survival was obtained by drying the seeds at 15°C and 15% RH. When comparing the two drying treatments at 20°C it was found that the faster rate was more damaging than the slower.

Beechnuts are usually collected under the tree, after shedding. These results show that there is no point in an earlier harvest, but that if left on the ground too long the seeds may deteriorate. The optimal time of collection is therefore as soon after seed shed as possible.

The relatively new method, where seeds are pretreated at a controlled moisture content which prevents germination during the treatment (Muller and Bonnet-Masimbert, 1989), has facilitated the handling of beechnuts. There is however still no way to know the length of the pretreatment before the seeds have been germinated.

It is therefore useful to know the difference in cold requirement between fresh and dry seeds. As drying reduces the pretreatment period by three to four weeks, with lasting effect during storage, it may be worthwhile to dry the seeds.

Acknowledgements

The Tree Improvement Station, Humlebaek, carried out a major part of the experiment; thanks are especially due to Henrik Knudsen for technical assistance. Also thanks to Adriana Caznoch Kurten for technical assistance and the EU for financial support.

References

Bianco, J., Garello, G. and Le Page-Degivry, M.T. 1994. *Seeds Science Research* 4: 57–62.

Demir, I. and Ellis, R.H. 1992a. *Annals of Applied Biology* 121: 385–399.

Demir, I. and Ellis, R.H. 1992b. *Seed Science Research* 2: 81–87.

Evans, M., Black, M. and Chapman, J. 1975. *Nature* 258: 144–145.

Fenner, M. 1991. *Seed Science Research* 1: 75–84.

Finch-Savage, W.E. 1992. *Seed Science Research* 2: 17–22.

Gray, D., Steckel, J.R.A. and Hands, L.J. 1992. *Seed Science Research* 2: 89–95.

Hong, T.D. and Ellis, R.H. 1992. *Seed Science Research* 2: 169–172.

ISTA, 1985. International Seed Testing Association. *Seed Science & Technology* 13: 299–355.

Jensen, L.G. 1994. *Hereditas* 121: 53–72.

Muller, C. and Bonnet-Masimbert, M. 1989. *Seed Science and Technology* 14: 117–125.

Nerson, H. 1991. *Seed Science and Technology* 19: 185–195.

Nyholm, I. 1986. Håndbog i frøbehandling. Dansk Planteskoleejerforening.

Poulsen, K.M. 1993. *Seed Science and Technology* 21: 327–337.

Probert, R.J. and Brierly, E.R. 1989. *Annals of Botany* 64: 669–674.

Rasyad, A., Van Sanford, D.A. and TeKrony, D.M. 1990. *Seed Science and Technology.* 18: 259–267.

Sawhney, R. and Naylor, J.M. 1982. *Canadian Journal of Botany* 60: 1016–1020.

Suszka, B. 1966. *Arboretum Kornickie*, 11: 203–220.

Suszka, B. 1975. *Arboretum Kornickie*, 20: 299–313.

Tompsett, P.B. and Pritchard, H.W. 1993. *Annals of Botany* 71: 107–116.

Welbaum, G.E. and Bradford, K.J. 1989. *Journal of Experimental Botany* 40(221): 1355–1362.

6. Post-transcriptional Regulation of Storage Protein Synthesis during Early Embryogenesis

J.D. BEWLEY and S.K. PRAMANIK

Department of Botany, University of Guelph, Guelph, Ontario N1G 2W1, Canada

Abstract

Messenger RNA synthesis during somatic embryogenesis of alfalfa is controlled at the post-transcriptional level during the globular to torpedo stages of development, and transcriptionally during the cotyledonary stages. Storage protein messages are sequestered in messenger ribonucleoprotein (mRNP) complexes during the early stages of embryogenesis, in association with specific binding proteins. Reconstitution of the messenger-protein complex can be achieved *in vitro* using medicagin (11S) mRNA; cell-free extracts from pre-cotyledonary embryos contain different binding proteins from those at the cotyledonary stages. Those associated with mRNA at the early stages may prevent translation, whereas those at the late stages may enhance large polysome formation. The mRNA-bound proteins from pre-cotyledonary-stage embryos exhibit 50-fold more protein kinase activity than those from cotyledonary embryos, and several proteins in the mRNP fraction can be phosphorylated. While the physiological function of the phosphorylated proteins is unknown, it is possible that selective phosphorylation and dephosphorylation of proteins determines recruitment of individual mRNAs into polysomes, thus regulating translation during embryogenesis.

Introduction

Extensive characterization of the storage proteins of alfalfa (*Medicago sativa* L.) has shown that there is a low-salt-soluble component comprised of two major protein classes, the low weight (LMW) proteins and the vicilin-like proteins, and a high-salt-soluble class of proteins, the legumins (Krochko and Bewley, 1990; Coulter and Bewley, 1990). The LMW (2S) proteins are S-rich, disulphide-bonded polypeptides ranging in size from 11–13 kDa, while the 7S vicilin-like proteins, the alfins, are a heterogeneous group of 14–50 kDa polypeptides. The high molecular mass 11S legumin-like protein, medicagin, is made up of about 10 acidic (39–49 kDa) and 3 basic (20–24 kDa) polypeptides. In the Proceedings of the previous International Workshop on Seeds we reported that developing somatic embryos of alfalfa synthesize less of these major storage proteins than their zygotic counterparts, and that the resultant proteins are also qualitatively different (Bewley *et al.*, 1993). The amount of protein present on a dry weight basis in mature somatic embryos is only 18% of that in mature zygotics, and whereas the 11S and 2S proteins are predominant in zygotic embryos, in the somatics they are proportionately much less in relation to the 7S protein. Even

R.H. Ellis, M. Black, A.J. Murdoch, T.D. Hong (eds.), Basic and Applied Aspects of Seed Biology, pp. 53–61.
© *1997 Kluwer Academic Publishers, Dordrecht. Printed in Great Britain.*

the ratios of the component subunits of medicagin are different in somatics compared with zygotics (Krochko *et al.,* 1992).

During zygotic embryogenesis there is a generally good correlation between the amount of synthesis of storage proteins taking place and the presence of their mRNAs (Pramanik *et al.,* 1992). Hence it has been concluded that transcriptional control operates during the period when storage proteins are deposited within the embryos; this is the case during development of most zygotic embryos (Bewley and Marcus, 1990). Likewise, during the equivalent (cotyledonary) stages of somatic embryo development, the mRNAs for 2S, 7S and 11S storage proteins are abundantly present; again, this is indicative of transcriptional control. Surprisingly, mRNAs for these three major storage proteins are also present during the early stages of somatic embryogenesis (globular, heart and torpedo stages), even though there is no storage protein synthesis at this time (Pramanik *et al.,* 1992). This means that some post-transcriptional block occurs within these early-stage embryos which prevents the storage protein messages being translated.

The occurrence of untranslated mRNAs in eukaryotic cells has been reported, especially in animal cells undergoing embryogenic and morphological changes (Rosenthal *et al.,* 1980). These mRNAs are associated with proteins in the cytoplasm of the cell as messenger-ribonucleoprotein complexes (mRNPs) (Prebrazhensky and Spirin, 1978). Messages can be recruited from this fraction (which during animal embryogenesis can contain up to 80% of the total mRNA complement of a cell) into the active protein synthesizing complex (polysomes). A similar situation occurs during the early (pre-cotyledonary) stages of somatic embryo development (Krochko *et al.,* 1992; Pramanik *et al.,* 1992). For while there is no storage protein synthesis during these stages, mRNAs are abundantly present, but mostly in the non-polysomal, mRNP fraction. Later in development, as storage proteins accumulate during the cotyledonary stages, the distribution of mRNAs is reversed, with most being present in the polysomal fraction. We surmised that several factors could be in operation during early somatic embryogenesis to repress the translation of the storage protein messages (Bewley *et al.,* 1993). Here we report on our progress over the past three years to understand the nature of post-transcriptional regulation, with particular emphasis on the repression of the medicagin (11S) message.

Storage Protein Messages in the mRNP Fraction are Potentially Active, but their Translation is Repressed

When mRNAs are isolated from both the polysomal and non-polysomal (mRNP) fractions of zygotic and somatic embryos, and deproteinized, they can be translated *in vitro* in a wheat germ extract or a rabbit reticulocyte system (Pramanik *et al.,* 1992). The polypeptides synthesized by the polysomal mRNAs and those in the mRNP fraction are quite similar in both embryo types, although there are notable differences in the distribution of certain proteins synthesized by the messages from the two fractions. Of the storage proteins in somatic embryos, the amount of 11S message present in the mRNP fraction during the (pre-cotyledonary) torpedo stage is much greater than in the polysomal fraction, whereas later, in the early cotyledon stage, this distribution is reversed. In the cotyledonary-stage zygotic embryos the amount of mRNA for all three major storage proteins is greater in the polysomal fraction than in the mRNP fraction (Pramanik *et al.*, 1992).

We determined if the proteins associated with mRNAs in the messenger-ribonucleoprotein complex influence their capacity to be translated *in vitro*. An mRNA fraction was isolated from developing zygotic embryos at the cotyledonary stage (VI-VII, Xu *et al.*, 1991), when storage protein synthesis is proceeding. When translated *in vitro*, using a cell-free wheat-germ extract, several polypeptides were synthesized (Fig. 1). The addition to the *in vitro* system of proteins isolated from the cytoplasmic mRNP fraction of somatic pre-cotyledonary embryos resulted in an overall suppression of protein synthesis, whereas the addition of proteins isolated from cotyledonary-stage zygotic (Fig. 1) and somatic (not shown) embryos enhanced protein synthesis. The suppression of *in vitro* translation due to mRNP-fraction proteins from pre-cotyledonary-stage somatic embryos could be largely overcome by the addition of polysomal-fraction proteins from cotyledonary-stage zygotic embryos. Thus the proteins associated with the mRNP fraction in developing somatic embryos, at a stage when *in situ* storage protein synthesis is repressed, have the capacity to suppress protein synthesis *in vitro*; those associated with the polysomal fraction enhance synthesis.

Proteins Associated with mRNAs in the Cytoplasmic mRNP and Polysome Fractions

What, then, is the nature of the proteins that are associated with the mRNP fraction and that repress protein synthesis during early stages of somatic embryogenesis? This has been determined using the technique of UV-cross-linking, which involves the exposure of embryos or their homogenates to short-wavelength (254 nm) ultraviolet light, which covalently cross-links mRNA with proteins that lie within one band length of each other (Dreyfuss, 1986). Both purines and pyrimidines, in particular uracil, can form a covalent link with as many as 11 different amino acids (Smith, 1976). As shown previously (Bewley *et al.*, 1993), messages in both the polysomal and mRNP fractions become cross-linked with a common set of proteins but, in addition, the mRNA in the mRNP fraction of early-stage somatic embryos is associated with at least 4 unique proteins (30, 43, 55 and 65 kDa), and two which are much more abundant than in the polysomal fraction (80 and 92 kDa).

The type of proteins which became associated with a specific message, that for the 11S storage protein, during embryogenesis have been determined in an experiment involving *in vitro* reconstitution of mRNPs. First, legumin message

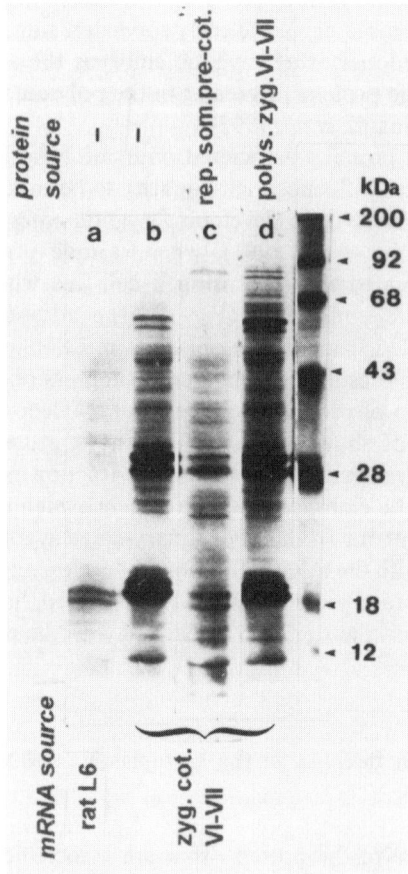

Figure 1. The effect of native mRNP-complex proteins from pre-cotyledonary-stage alfalfa somatic embryos and polysomal proteins from cotyledonary-stage zygotic embryos on *in vitro* translation in a wheat germ extract

Both polysomal and non-polysomal mRNP fractions were isolated from embryos (Pramanik *et al.*, 1992) and treated with micrococcal nuclease in the presence of Ca^{2+} to remove endogenous mRNAs prior to addition to the *in vitro* system

Lanes: (a) rat L6 myoblast mRNA was added to test the translational activity of the wheat; (b) poly-A^+ mRNA fraction from stage VI-VII zygotic (cotyledonary-stage) embryos; (c) as b, but with addition of 20 µg/ml messenger ribonucleoprotein fraction (repressed fraction) from pre-cotyledonary-stage somatic embryos; (d) as b, but with addition of 20 µg/ml protein from the polysomal fraction of cotyledonary-stage (VI-VII) zygotic embryos. Mol mass markers to the right

labelled with ^{32}P was transcribed by placing a pea legumin cDNA (pCD43, courtesy of Dr. R. Casey, John Innes Centre, Norwich, U. K.) insert under the control of T7 promoter, to make a sense strand, in the plasmid vector pSPT18 (Pharmacia). The resultant legumin mRNAs were then incubated separately with cell-free translation extracts prepared from pre-cotyledonary- and cotyledonary-stage somatic embryos and cotyledonary-stage zygotic embryos. When translation was underway and polysomes were formed, the reaction was stopped and intimately-associated polysomal and non-polysomal mRNAs were cross-linked by UV irradiation. The legumin mRNAs became UV-cross-linked to several proteins in the cell-free translational extracts prepared from the three embryo types (Fig. 2). The greatest difference in the types of proteins that cross-linked was between the two stages of somatic embryos. The proteins in the translational extracts of cotyledonary-stage zygotic embryos which became cross-linked to legumin mRNA included the most predominant ones in the cotyledonary-stage somatic embryos (36–38 kDa), but not those most evident in the pre-cotyledonary-stage embryos (Fig. 2).

To determine if there is any relationship between the ability of legumin mRNA to be translated and the activity of the cell-free translation extracts from the three embryo types, legumin mRNA was incubated in their presence and incorporation of ^{35}S-methionine measured. After a 30-min incubation period, cell-free extracts from pre-cotyledonary stage embryos promoted incorporation of 909 cpm/µl extract, compared to 2523 cpm/µl extract and 2706 cpm/µl extract by cotyledonary-stage somatic and zygotic embryos, respectively. The differences in incorporation capacity did not change with larger periods of incubation of the cell-free extracts with the legumin message.

It is possible that the defective translation of legumin mRNA during the pre-cotyledonary stages of somatic embryogenesis is due to some deficiency in the translationally-active legumin-message-binding complex. Perhaps the 36–38 kDa proteins (Fig. 2) are required, but are present only later in the cotyledonary stages, when legumin is synthesized. The presence of other mRNA binding proteins (21–23, 28, 50, 55, 62 kDa) might result in the masking of the legumin message, thus forming free mRNA complexes and thereby preventing initiation of translation in the pre-cotyledonary stage of development. As development proceeds, the masking proteins are either turned over and are no longer present, or they cannot compete with newly-synthesized, translationally-active legumin-mRNA-binding proteins, which positively regulate translation.

Phosphorylation of Proteins in the mRNP and Protein-synthesizing Complex

The biochemical and physiological significance of these mRNA-associated proteins remains to be elucidated but, as discussed, they could act as controllers (repressers or enhancers) of protein synthesis. For example, in animal tissues it is known that covalent modification of components associated with translation

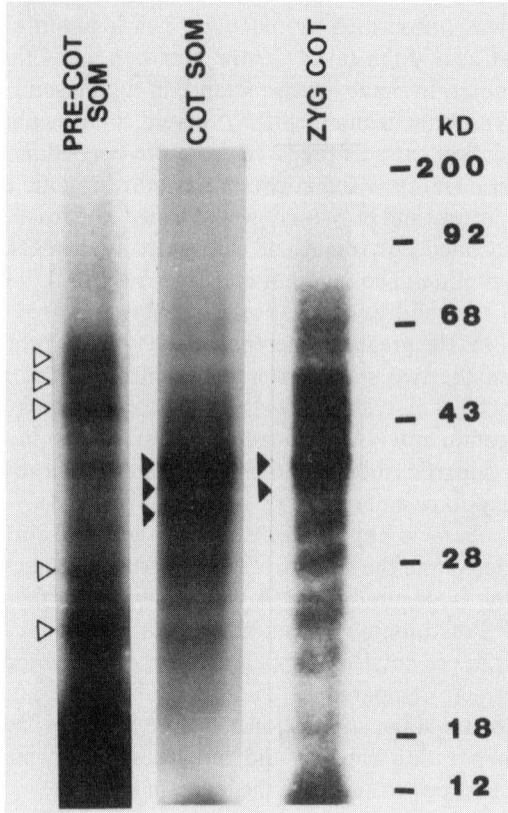

Figure 2. The association of proteins and legumin mRNA into an mRNP-complex using cell-free translational extracts of developing somatic and zygotic embryos of alfalfa

Sense-strand ^{32}P-labelled legumin mRNA, transcribed from pea legumin cDNA in plasmid vector pSPT18, was incubated with cell-free translational extracts from developing zygotic and somatic embryos. When protein synthesis had commenced the translation extracts were exposed to UV to forge covalent links between the mRNA and intimately-associated proteins. The resultant complex was digested with a nuclease cocktail and the proteins remaining associated with the fragment of ^{32}P-mRNA were resolved by SDS-polyacrylamide gel electrophoresis. UV-cross-linked proteins from pre-cotyledonary- and cotyledonary-stage somatic embryos and cotyledonary-stage zygotic embryos are shown. Proteins from the extracts of cotyledonary-stage embryos which cross-link with the mRNA are indicated by dark arrows, and those from the pre-cotyledonary stage by open arrows

of mRNAs, *viz.* initiation or elongation factors, ribosomal proteins, etc., regulates protein synthesis, and particularly effective changes are brought about by protein-kinase-mediated phosphorylation of proteins which are integral to the protein-synthesizing machinery (Hershey, 1990). Many well-characterized events in animal cells, e.g. mitosis, meiosis, wound-healing, starvation and activation of cell growth by oncogenes and growth factors are controlled at the

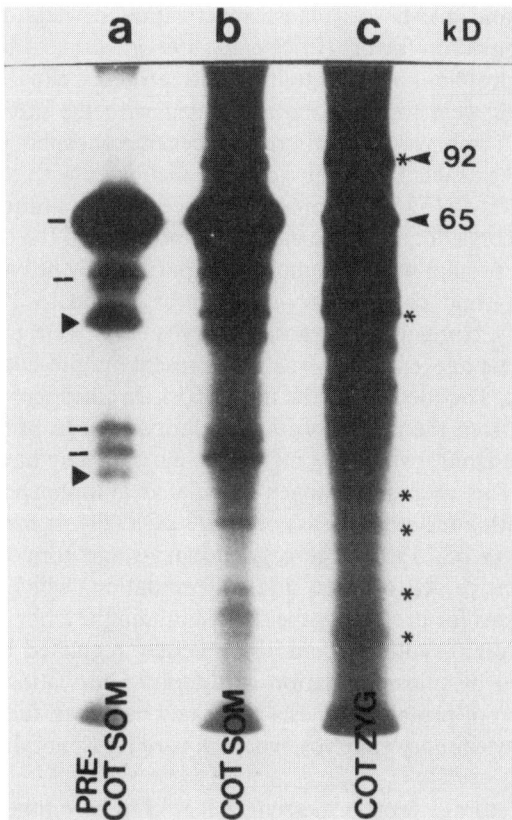

Figure 3. Sodium dodecyl sulphate-polyacrylamide gel electrophoresis (SDS-PAGE) analysis of phosphorylated proteins in cell-free extracts prepared from developing somatic and zygotic alfalfa embryos

mRNP fraction was used as the source of kinase enzymes and protein substrate and incubated with ^{32}P-α-ATP. The proteins were precipitated and then analyzed by SDS-PAGE. (a) Pre-cotyledonary-stage somatic embryos; (b) cotyledonary-stage somatic embryos; (c) cotyledonary-stage zygotic embryos. Proteins which occur predominantly in cotyledonary-stage embryos are marked with an asterisk; arrowed (left) are those unique to pre-cotyledonary-stage somatic embryos; lines indicate those most prevalent in somatic embryos. The amounts of protein used in the kinase assay were 1:10:20, pre-cotyledonary-stage somatic : cotyledonary-stage somatic : cotyledonary-stage zygotic embryos

level of translation by covalent modifications of the translational apparatus, a process which is reversible. The control is effected by the presence of intracellular protein kinases, of which three main classes have been reported: (i) tyrosine kinase; (ii) cyclic AMP-dependent protein kinases; (iii) the protein kinase C family (Proud, 1992). Protein kinase C regulates translation in several eukaryotic systems by phosphorylating several initiation factors (Morley and

Traugh, 1989), and cAMP can cause inactivation or dephosphorylation of initiation and elongation factors (Price *et al.*, 1991).

Endogenous protein kinase activities and proteins capable of accepting phosphate groups were found to be associated with the mRNP fractions of developing alfalfa embryos. Several proteins became phosphorylated when cell-free translational extracts from embryos were incubated with ^{32}P-α-ATP to test for kinase activity (Fig. 3). Some appeared to be unique to, or more predominant in cotyledonary-stage embryos, and some in the pre-cotyledonary stage. A 65-kDa protein was phosphorylated particularly heavily in developing pre-cotyledonary- and cotyledonary-stage embryos, up to 100- and 50-fold more, respectively, compared to zygotic embryos. A 92 kDa protein present in the cotyledonary-stage embryos was not present in pre-cotyledonary-stage somatics (Fig. 3). The identity of the major 65 kDa phosphorylated protein is not known, but from the nature of its phosphorylation (a process sensitive to heat-shock in the embryos) and its molecular mass, it may have equivalents in animal tissues. This includes mitogen-stimulated autophosphorylated S6 kinases I and II with a molecular mass of 65–70 kDa (Blenis *et al.*, 1987; Jeno *et al.*, 1988; Price *et al.*, 1989). Pre-cotyledonary-stage somatic embryos are undergoing extensive cell division and differentiation, which require precise control mechanisms for protein synthesis to maintain the embryogenic pathway. At this stage, protein synthesis could be precisely regulated by mitogen, and mainly controlled by phosphorylation and dephosphorylation reactions. This type of regulation of protein synthesis will likely be less in fully expanded and differentiated cotyledonary embryos, where *in vitro* phosphorylated proteins are fewer.

The failure of storage protein messenger RNAs to be synthesized during the early (pre-cotyledonary) stages of somatic embryo development is due to post-transcriptional regulation, with the messages being sequestered outside of the polysomes, in a messenger ribonucleoprotein fraction. These are recruited at later (cotyledonary) stages of development for storage protein synthesis in somatic embryos, as in zygotic embryos, when regulation is transcriptionally effected. The failure of certain messages, including those for storage proteins, to be recruited into polysomes in early-stage somatic embryos could be related to the nature of the proteins with which they are associated in the mRNP-complex. Proteins which are inhibitory to protein synthesis may be associated with storage protein messages, or those necessary for recruitment into polysomes may be absent. The mRNP fraction contains both kinases and proteins that can be phosphorylated; phosphorylation and dephosphorylation reactions may be a means whereby control is exerted over the types of mRNA that are recruited into polysomes.

References

Bewley, J. D., Pramanik, S. K. and Krochko, J. E. 1993. In: *Basic and Applied Aspects of Seed Biology. Proceedings of the Fourth International Workshop on Seeds*, pp. 29–38 (eds. D. Côme and F. Corbineau). Paris: ASFIS.

Bewley, J. D. and Marcus, A. 1990. *Progress in Nucleic Acids Research and Molecular Biology* 38: 165–193.

Blenis, J., Kuo, C. J. and Erikson, R. L. 1987. *Journal of Biological Chemistry* 262: 14373–14376.

Coulter, K. M. and Bewley, J. D. 1990. *Journal of Experimental Botany* 41: 1541–1547.

Dreyfuss, G. 1986. *Annual Reviews of Cell Biology* 2: 459–498.

Hershey, J. W. B. 1990. *Annual Reviews of Biochemistry* 60: 717–755.

Jeno, P., Ballou, L. M., Novak-Hofer, I. and Thomas, G. 1988. *Proceedings of the National Academy of Science U.S.A.* 85: 406–410.

Krochko, J. E. and Bewley, J. D. 1990. *Journal of Experimental Botany* 41: 505–514.

Krochko, J. E., Pramanik, S. K. and Bewley, J. D. 1992. *Plant Physiology* 99: 46–53.

Morley, S. J. and Traugh, J. A. 1989. *Journal of Biological Chemistry* 264: 2401–2404.

Pramanik, S. K., Krochko, J. E. and Bewley, J. D. 1992. *Plant Physiology* 99: 1590–1596.

Prebrazhensky, A. A. and Spirin, A. S. 1978. *Progress in Nucleic Acids Research and Molecular Biology* 21: 1–38.

Price, D. J., Nemeoff, R. A. and Auruch, J. 1989. *Journal of Biological Chemistry* 264: 13825–13833.

Proud, C. G. 1992. *Current Topics in Cellular Regulation* 32: 243–269.

Rosenthal, E. T., Hunt, T. and Rudeman, J. V. 1980. *Cell* 20: 487–494.

Smith, K. C. 1976. *Photochemistry and Photobiology of Nucleic Acids* 2: 187–212.

Xu, N., Coulter, K. M., Krochko, J. E. and Bewley, J. D. 1991. *Seed Science Research* 1: 119–125.

7. Choice of the Matrix for Encapsulating and Dehydrating Carrot Somatic Embryos: Importance of the Rheology, the Desorption and the Diffusion Parameters

R. TIMBERT, J.-N. BARBOTIN and D. THOMAS

Laboratoire de Technologie Enzymatique, URA 1442 du CNRS, Université de Technologie de Compiègne, B.P. 649, F-60206 Compiègne Cédex, FRANCE

Abstract

Carrot somatic embryos were encapsulated in gel beads prior to slow dehydration. To improve the survival of future "synthetic seeds", the rheology, desorption and diffusion properties of different coatings were investigated. By increasing alginate and $CaCl_2$ concentrations, and by adding mineral elements to alginate, the resistance of the matrix to rupture was increased, but the germination of somatic embryos was depressed. Alginate-kaolin or alginate-gellan gum beads, and alginate-starch beads, could take up and retain respectively about 29% and 88% of sucrose. A polysaccharide addition was found to slow the alginate matrix dehydration: gellanate and alginate-gellan gum beads had the best capacity to rehydrate, while alginate-kaolin beads were more difficult to rehydrate. Alginate-gellan gum and alginate-kaolin could preserve the viability of somatic embryos to a greater extent than other matrices. Desiccation tolerance could be enhanced by slowing the dehydration process.

Introduction

The synthetic seed concept was initiated by Murashige *et al.* (1977). They suggested to encapsulate somatic embryos in a matrix that would play the role of an endosperm and of a tegument. The aim of this concept was to stabilize the somatic embryos and to enhance their vigour by adding nutrients to the capsule.

Many authors have worked on embryo physical protection using encapsulation (Kitto and Janick, 1985a; Redenbaugh *et al.*, 1986; Kersulec *et al.*, 1993). Some others have tried to harden and dehydrate naked somatic embryos (Kitto and Janick, 1985b; Anandarajah and McKersie, 1990; Lecouteux *et al.*, 1992; Tetteroo *et al.*, 1995).

Encapsulated somatic embryos are generally germinated just after encapsulation, while desiccated somatic embryos cannot be handled and planted easily. To our knowledge, the Janick group is the only one that tried to dehydrate encapsulated somatic embryos (Kitto and Janick, 1985b; Janick *et al.*, 1989; Kim and Janick, 1990; 1991), but they only tested one encapsulation matrix (i.e. polyox). Consequently, the aim of this work was to define the physical environment of carrot somatic embryos encapsulated in different matrices. The

R.H. Ellis, M. Black, A.J. Murdoch, T.D. Hong (eds.), Basic and Applied Aspects of Seed Biology, pp. 63–73.
© *1997 Kluwer Academic Publishers, Dordrecht. Printed in Great Britain.*

ability of such matrices to protect them during dehydration was also investigated.

Experimental Methods

Experiments involving living material were carried out in sterile conditions.
Somatic embryos were produced according to Timbert *et al.* (1995), as for beads
production, somatic embryos encapsulation and compression measurements on
beads.

Germination of Somatic Embryos

Encapsulated and non-encapsulated somatic embryos were cultured on half-
strength MS with 2% sucrose (Murashige and Skoog, 1962). Germination was
noted when the radicle presented a suitable growth or when it was seen to be
emerging from the bead. Results were expressed as "germination frequency"
(%).

Diffusion of Sucrose

A thousand beads of each composition were dropped at time zero in 100 ml of a
well-stirred 2% sucrose solution, at 25°C. The sucrose remaining in the solution
was determined with the Boehringer-Mannheim UV method. In the same
conditions, 1000 beads of each composition containing 10% sucrose were
dropped at time zero in 100 ml of water. Appearance of sucrose in water was
also determined. The beads volume, for each composition, was measured by
water displacement before and after the experiments.

Calculation of Diffusion Coefficient

The theoretical background is given by Crank (1975). Initially, beads do not
contain sucrose. The sucrose amount in the beads (M_t) is given by:

In Eq. (1), q_n represents the non-zero positive roots of:

$$\frac{M_t}{M_\infty} = 1 - \sum_{n=1}^{\infty} \frac{6 . \alpha . (\alpha + 1) . e^{\left[-\frac{D . q_n^2 . t}{a^2} \right]}}{9 + 9 . \alpha + q_n^2 . \alpha^2} \tag{1}$$

$$\tan qn = \frac{3 \cdot qn}{3 + \alpha \cdot qn^2} \qquad (2)$$

With the total numbers of beads and their volume, the bead radius a and the volume ratio of liquid and beads, α, were calculated. When the sucrose concentration M_t is recorded as a function of time, D can be estimated. In the symmetric case, the beads contain 10% sucrose at time zero. M_t is replaced by $(M_1)_t = M_0 - M_t$ and M_∞ by $(M_1)_\infty = M_0 - M_\infty$.

Slow Dehydration

Beads with or without somatic embryos were dehydrated in a thermostated chamber with controlled relative humidity, as described previously (Timbert *et al.,* 1995).

Protocol I is the following: a decrease of 5% relative humidity (RH) occurred each 18 h, starting from 95% RH until 25% RH. Thus, it took 11.5 days to reach 25% RH. Protocol II is the next: a decrease of 5% RH occurred each 10 h between 95% RH and 70% RH, each 24 h between 70% RH and 45% RH, and each 10 h until 15% RH was reached (after 12 days).

Weighing was performed at humidity equilibrium, so the results were expressed as water activity (a_w). Samples of encapsulated somatic embryos were laid in the chamber, removed at different humidities and tested for viability after partial dehydration.

Results and Discussion

Compression Measurements on Beads

The results of the compression measurements describe two characteristics of the beads:

– the mechanical protection offered to the somatic embryo;

– the resistance the embryo will face in emerging from them.

Consequently, we have correlated the maximum compression strength with the germination frequency.

Increasing alginate and calcium chloride concentrations tend to increase the maximum compression strength except in the case of high-viscosity alginate. An enhancement of the germination frequency is not clearly related to a higher resistance of the beads to rupture (Figs. 1A,B,C). Increasing calcium chloride concentrations were found to decrease the germination frequency for beads made with less than 2% alginate. This behaviour not only depends on mechanical factors, but also on diffusional or chemical ones: in experiments performed with high-viscosity alginate, higher resistances to rupture were linked to lower germination frequencies, the best ones being obtained at 0.5 mJ/bead.

Gels composed of a mixture of alginate and other polysaccharides or mineral elements can be obtained. Addition of mineral elements (silica, kaolin) to alginate solutions along with high calcium chloride concentration slightly increased the resistance to rupture (Fig. 1D). As the beads become harder, the

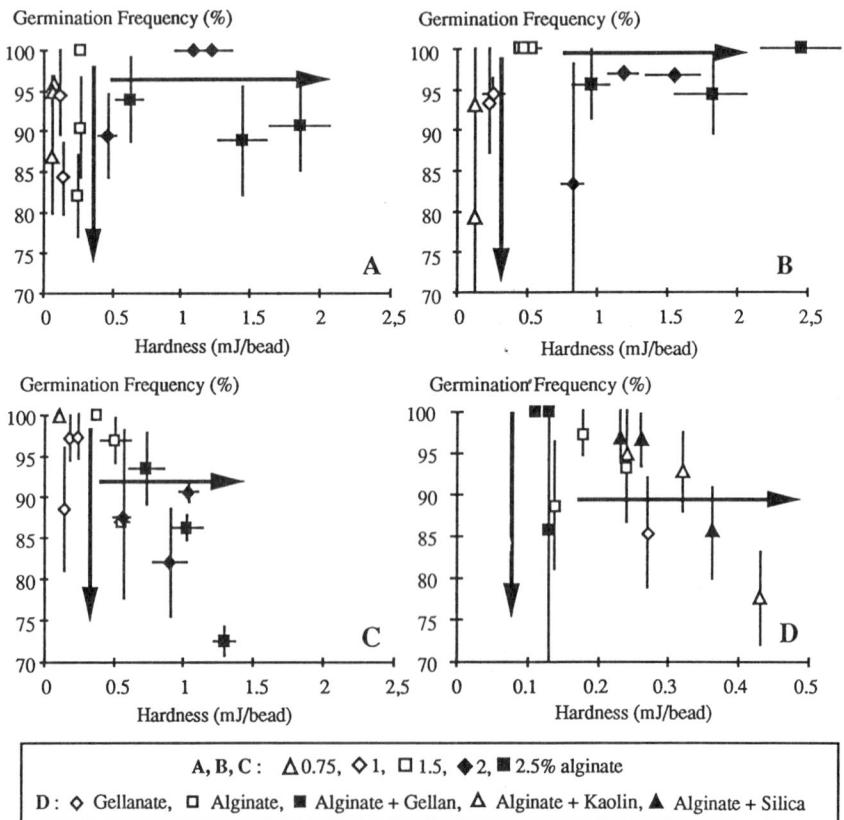

A,B,C : △0.75, ◇1, □1.5, ◆2, ■2.5% alginate

D : ◇ Gellanate, □ Alginate, ■ Alginate + Gellan, △ Alginate + Kaolin, ▲ Alginate + Silica

Figure 1. Germination frequency (%) of encapsulated somatic embryos, and hardness (mJ/bead) of their encapsulation matrices: low (A), medium (B) and high (C) viscosity alginate, or 1% gellanate, and 1% high-viscosity alginate plus 0.6% of different additives (D, different scale for hardness). Arrows symbolize increasing [CaCl$_2$] (25 to 100 mM). Bars=standard deviation

somatic embryos cannot emerge easily from the capsule. Using gellanate or adding gellan gum to alginate resulted in a similar germination frequency and hardness as that with alginate alone.

Similar to Cheetham *et al.* (1979), we showed that the mechanical resistance of alginate beads increased with calcium chloride and alginate concentrations. The effect of alginate-encapsulation on somatic embryos germination has been examined by Redenbaugh *et al.* (1986) and Shigeta *et al.* (1990). Redenbaugh *et al.* (1986) showed that this frequency decreased when the alginate capsule hardness exceeded 2 kg/bead. We demonstrated that this limit was not absolute, and that it was different for each matrix tested. This can be explained by a given physico-chemical behaviour of the matrix which is a function of the nature of the additive and of alginate viscosity. As said above, a compromise must be obtained between a good resistance of the matrix, which allows protection of the somatic embryo, and a limited internal resistance, in order to allow germination. The overall resistance of beads could be enhanced by increasing the alginate and calcium chloride concentrations, and slightly enhanced by adding minerals in the matrix. The germination of somatic embryos was depressed by increasing calcium chloride concentration, and in the case of high-viscosity alginate, by alginate concentrations over 1.5%. Consequently, the compromise is difficult to obtain as far as high viscosity alginate and calcium chloride concentrations are concerned.

Diffusion of Sucrose

Among many elements, sucrose may be the first to be included in an "artificial seed" because of the somatic embryo heterotrophy, and of the osmotic protection sucrose may offer to the somatic embryo during dehydration. At the nutritional level, the problem is whether to put an artificial seed without sucrose on a rich medium or the contrary. We addressed this problem by testing different compositions of the encapsulation matrix. Different solutions were found for alginate beads and for mixed beads. We must distinguish final uptake and release percentages from apparent diffusion coefficients.

Alginate, alone or with silica, has a small uptake percentage of sucrose (Fig. 2A). These kinds of beads may not be adapted for an *in vitro* conversion of plants, because it would not allow sufficient nutrient uptake for the somatic embryo before it becomes autotrophic. On the contrary, beads made of gellanate, or made of alginate with starch, gellan gum or kaolin, can take up more than 25% of the solution sucrose (Fig. 2A). They may be particularly adapted for encapsulation and *in vitro* culture of somatic embryos. Beads made of alginate and starch or silica show little release percentages of sucrose: they could be used for *in vivo* regeneration, as far as they permit the retention of the nutrients in a poor environment (Fig. 2A).

The sucrose diffusion coefficient varies with bead composition. Heterogenous matrices were slower to equilibrate than pure ones. Desorption and absorption coefficients were the same for pure alginate and k-gellanate, and for their most

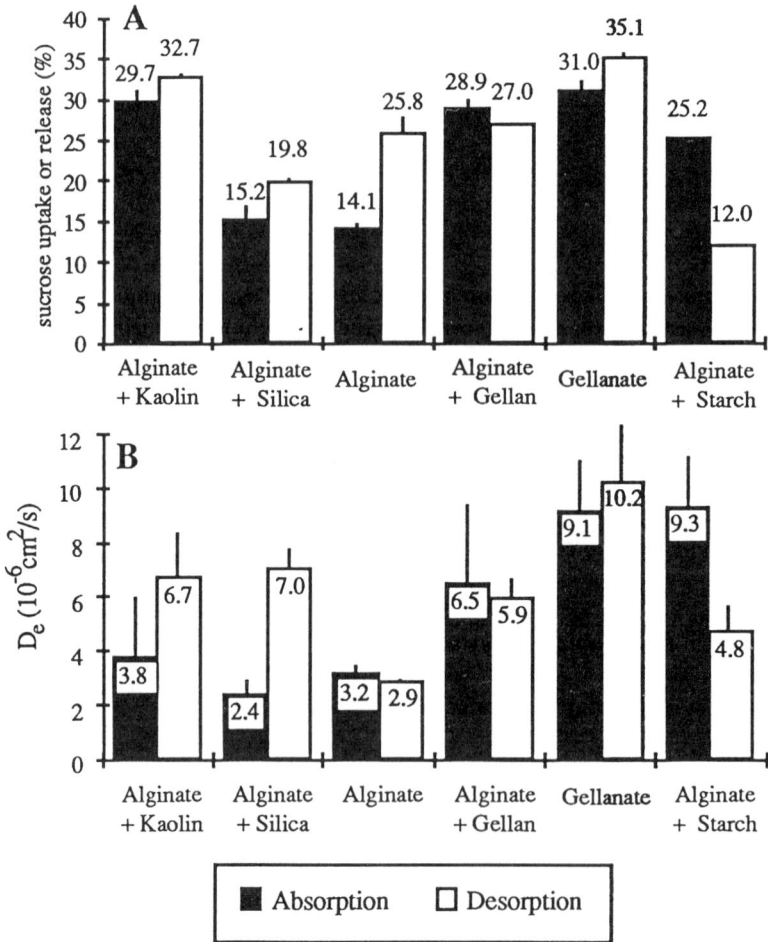

Figure 2. Sucrose diffusivity of different matrices, made with 100 mM $CaCl_2$. A: final uptake and release percentages of sucrose. B: apparent diffusion coefficients. Bars = standard deviation

related matrix, alginate–gellan gum (Fig. 2B). Addition of minerals to alginate increased the desorption coefficient, while starch increased the alginate absorption coefficient of sucrose (Fig. 2B). The matrix made of alginate and starch gave low germination percentages *in vitro*, with rich medium, but it would be of interest to try it for *in vivo* regeneration, with poor or no medium.

Sucrose is assumed to diffuse freely in calcium alginate (Tanaka *et al.*, 1984). Nevertheless, Pu and Yang (1988) found sucrose diffusion coefficients varying from 4.7 to 4.98×10^{-6} cm^2/s for calcium alginate, i.e. 87% of water. We did not calculate our water reference, but Tanaka *et al.* (1984) also showed that glucose absorption and desorption coefficients in calcium alginate were comparable. Adding polysaccharides (starch or gellan gum) to alginate was found to enhance its absorption diffusivity, but no additive was able to decrease its desorption diffusivity. The modification of diffusivity is probably correlated with structural changes of the gel (i.e. porosity, homogeneity and hydration). For synthetic seeds, matrices allowing a high uptake percentage and a high absorption diffusion coefficient, like alginate–starch, may be of interest.

Slow Dehydration

Encapsulated carrot somatic embryos have a short storage duration: a desiccation procedure may be helpful to overcome this limitation. As sucrose can osmotically protect the somatic embryo against dehydration (Janick *et al.*, 1989; Anandarajah and McKersie, 1990; Lecouteux *et al.*, 1992; Etienne *et al.*, 1993), beads have been produced in the presence of 10% sucrose, before dehydration with the first protocol. The effect of the addition of 10% sucrose on the a_w of beads was measured before dehydration and a significant decrease of this parameter is observed (Timbert *et al.*, 1995). During Protocol I (18 h/plateau), the presence of gellan gum or starch delayed significantly the desorption (Fig. 3A). On the contrary, the desorption curves obtained for beads with mineral additives were not significantly different of those obtained for the alginate ones (Fig. 3A). Desorption of alginate beads containing somatic embryos was slower than desorption of somatic embryos alone (Fig. 3B). Thus, encapsulation succeeded in slowing the desorption of somatic embryos, whatever the matrix composition.

Germination of the encapsulated somatic embryos was not affected by a dehydration above 80% relative humidity (RH) (corresponding to a 3-day experiment). For a 50% RH (7 days), the best results were obtained with somatic embryos encapsulated in alginate–gellan gum (30.3% germination) and alginate–kaolin beads (34.0% germination, Fig. 4). For somatic embryos encapsulated in alginate–gellan gum and alginate-starch, the loss of germination capacity is much more regular than for the others (Fig. 4). In all the other cases, this loss was dramatic between 80% and 60% RH (3 to 6 days), especially for alginate alone, and resulted in very low germination frequencies under 60% RH. It must be noted that, since the desorption curves of the different matrices were not exactly the same, the a_w obtained at the same RH may be different; a_w was correlated with germination frequency of the embryos (results not shown). Mineral-containing alginate could preserve viability at low a_w to a greater extent than other matrices, while polysaccharides could preserve higher a_w, and therefore greater viability. Polysaccharides, pure or mixed (i.e. k-gellanate, alginate–starch and alginate–gellan gum), could also rehydrate more easily than other matrices (results not shown).

Changing the dehydration protocol could enhance the viability of encapsulated/dehydrated somatic embryos (Fig. 5). Protocol II slowed the dramatic water loss occurring between 70 and 45% R.H., and permitted maintenance of a

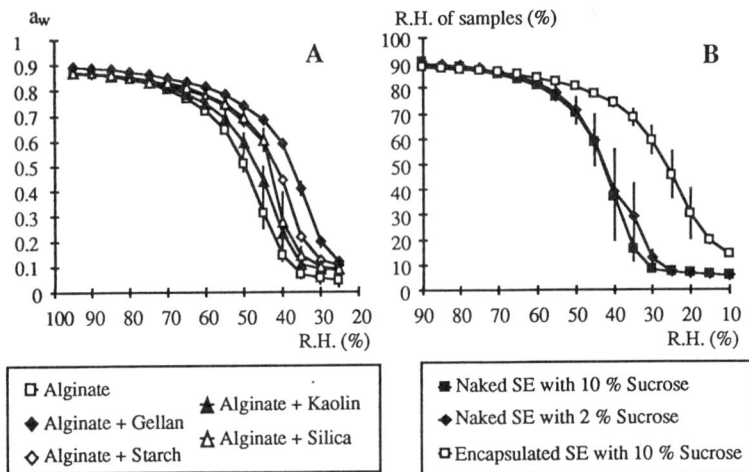

Figure 3. Desorption isotherms of 1% high-viscosity alginate plus 0.6% of different additives (A), and of naked, or alginate-encapsulated, somatic embryos (SE) (B), beads made with 100 mM $CaCl_2$. First dehydration protocol. Bars = standard deviation

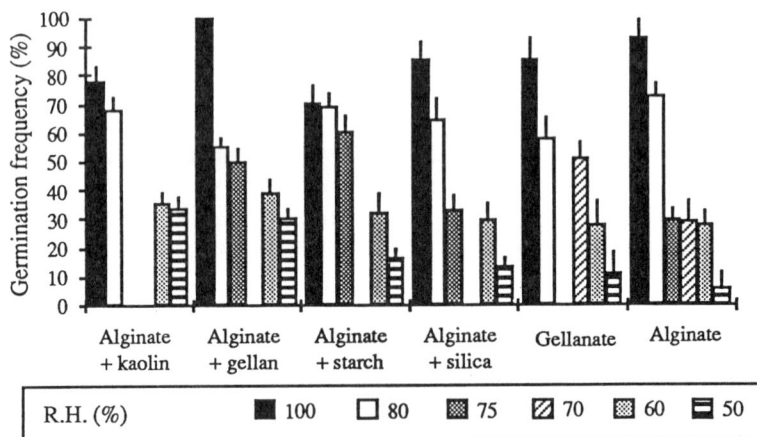

Figure 4. Germination frequency (%) of encapsulated and dehydrated somatic embryos. Encapsulation in 1% gellanate and 1% alginate plus 0.6% of different additives, beads made with 100 mM $CaCl_2$. First dehydration protocol. Bars = standard deviation

Germination frequency (%)

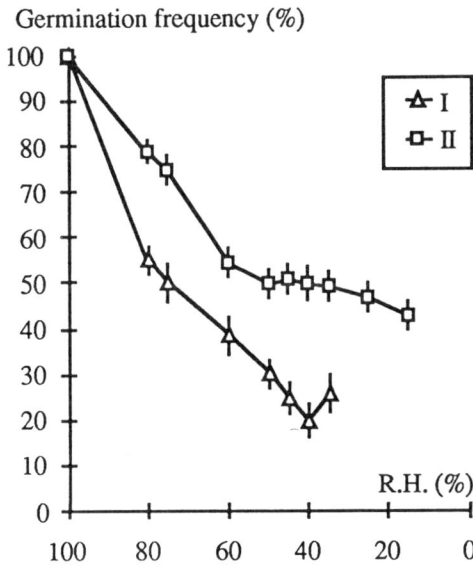

Figure 5. Germination frequency of encapsulated somatic embryos, dehydrated with 2 different protocols. Encapsulation in 1% alginate plus 0.6% gellan gum, beads made with 100 mM CaCl₂. Bars = standard deviation

higher a_w than the first one, at low R.H. (results not shown). The germination frequency of encapsulated embryos, dehydrated down to 15% R.H., stabilized around 45%, while only 25% of them germinated when dehydrated down to 35% R.H. by the first protocol. Senaratna *et al.* (1989) and Tetteroo *et al.* (1995) showed the importance of a slow dehydration for the survival of naked somatic embryos. Our first protocol corresponded to an a_w decrease of 1.3×10^{-2} h^{-1}, while the second one allowed only a decrease of 0.95×10^{-2} h^{-1}, comparable to the humidity decrease used by Tetteroo *et al.* (1995).

Osmotic pressure, ABA, proline, chilling and heat shock were found to enhance desiccation tolerance of somatic embryos, and sometimes their reserve accumulation (results not shown). Combinations of several treatments may be of great interest to enhance desiccation tolerance of somatic embryos (Kim and Janick, 1990; Anandarajah and McKersie, 1990; Saranga *et al.*, 1992), as well as slow rehydration (Tetterroo *et al.*, 1995).

Conclusion

Survival and germination of encapsulated embryos may depend on many factors. We characterized beads hardness and diffusivity before any dehydration. Beads hardness was increased by alginate and calcium chloride concentrations, and by addition of minerals to the matrix. In most cases, the consequence is a decrease of germination frequency, but the easiest manipulation of beads. A compromise must be found between these two objectives. Kersulec *et al.* (1993) showed the importance of O_2-diffusivity for respiration of encapsulated somatic embryos. Sucrose is the main carbohydrate source for heterotrophic somatic embryos. Its diffusivity in hydrogels, loaded or not with minerals or polysaccharides, may be an element of choice for *in vivo* conversion of encapsulated somatic embryos.

Encapsulation by itself can slow the dehydration of somatic embryos, and preserve part of their viability. Desiccation tolerance of encapsulated somatic embryos can be enhanced using a slower dehydration method, particularly between 70 and 45% R.H. Combination of maturation treatments, as well as a slow rehydration process (Tetteroo *et al.*, 1995), may provide perfect synthetic seeds. They would be preserved for long times, planted *in vivo*, and regenerate normal and healthy plants.

Acknowledgements

Gellanate was kindly provided by Dr L. Doner (USDA, Philadelphie, USA). Thanks to the Division des Technologies Agro-Industrielles of UTC for lending their rheometer, and to the Biopôle Végétal de Picardie for their financial support.

References

Anandarajah, K. and McKersie, B.D. 1990. *Plant Cell Reports* 9: 451–455.

Crank, J., 1975. *The Mathematics of Diffusion,* Second Edition, pp. 93. Oxford: University Press.

Cheetham, P.S.J., Blunt, K.W. and Bucke, C. 1979. *Biotechnology and Bioengineering* 21: 2155–2168.

Etienne, H., Montoro, P., Michaux-Ferrire, N. and Carron, M.P. 1993. *Journal of Experimental Botany* 44: 1613–1619.

Janick, J., Kitto, S. and Kim, Y.H. 1989. *In Vitro Cell Development and Biology* 25: 1167–1172.

Kersulec, A., Bazinet, C., Corbineau, F., Cme, D., Barbotin, J.-N., Hervagault, J.-F. and Thomas, D. 1993. *Biomaterials, Artificial Cells and Immobilization Biotechnology* 21: 375–381.

Kim, Y.H. and Janick, J. 1990. *Acta Horticulurae* 280: 23–28.

Kim, Y.H. and Janick, J. 1991. *Plant Cell Tissue and Organ Culture* 24: 83–89.

Kitto, S. and Janick, J. 1985a. *Journal of the American Society for Horticultural Science* 110: 277–282.

Kitto, S. and Janick, J. 1985b. *Journal of the American Society for Horticultural Science* 110: 283–286.

Lecouteux, C., Tessereau, H., Florin, B., Courtois, D. and Pétiard, V. 1992. *Comptes-Rendus de l'Académie des Sciences de Paris, Série III* 314: 423–428.

Murashige, T. and Skoog, F. 1962. *Physiologia Plantarum* 15: 473–497.

Murashige, T., Barz, W., Reinhard, E. and Zenk, M.H. 1977. In : *Applied and Fundamental Aspects of Plant Cell, Tissue and Organ Culture*, pp. 392–403 (eds J. Reinert and Y.P.S. Bajaj). Berlin : Springer-Verlag.

Pu, H.T. and Yang, R.Y.K. 1988. *Biotechnology and Bioengineering* 32: 891–896.

Redenbaugh, K., Paash, B.D., Nichol, J.W., Kossler, M.E., Viss, P.R. and Walker, K.A. 1986. *Bio/Technology* 4: 797–801.

Saranga, Y., Kim, Y.H. and Janick, J. 1992. *Journal of the American Society for Horticultural Science* 117: 342–345.

Senaratna, T., McKersie, B.D. and Bowley, S.R. 1989. *Plant Science* 65:253–259.

Shigeta, J., Mori, T., Toda, K. and Ohtake, H. 1990. *Biotechnology Techniques* 4: 21–24.

Tanaka, H., Matsumura, M. and Veliky, I.A. 1984. *Biotechnology and Bioengineering* 26: 53–58.

Tetteroo, F.A.A., Hoekstra, F.A. and Karssen, C.M. 1995. *Journal of Plant Physiology* 145: 349–356.

Timbert, R., Barbotin, J.N., Kersulec, A., Bazinet, C. and Thomas, D. 1995. *Biotechnology and Bioengineering* 46: 573–578.

8. Cryopreservation of Somatic Embryoids of *Phoenix dactylifera*

D.J. MYCOCK[1], P. BERJAK[2], N.W. PAMMENTER[2] and
C.W. VERTUCCI[3]

[1]*Department of Botany, University of the Witwatersrand, Johannesburg;* [2]*Department of Biology, University of Natal, Durban, South Africa;* [3]*USDA, ARS, National Seed Storage Laboratory, Fort Collins, CO, USA*

Abstract

Late globular/early torpedo stage date palm embryoids can continue normal growth and development after cryopreservation provided they are pre-treated with a cryoprotectant mixture of glycerol and sucrose and then dried to water contents in the range of 0.4–$0.7\,g.g^{-1}$. The embryoids were frozen by direct immersion in liquid nitrogen. Although further drying allows for 100% recovery, growth is in the form of unorganized callus. Tissues frozen at extremely rapid rates (by immersion in liquid freon) retain cytoskeletal structures, whereas material frozen at slower rates (in liquid nitrogen) appear to lose this subcellular system. The slow recovery rate of material frozen in liquid nitrogen could, in part, be due to reconstitution of this subcellular matrix.

Introduction

Phoenix dactylifera L. (date palm) is an extremely important crop in the middle east and northern Africa (Janick *et al.*, 1981). The species is dioecious and over the last two thousand years female strains which are high yielding, drought tolerant and fast growing, have been selected (Janick *et al.*, 1981). As the fruit of the date is highly priced these selected characteristics make the species an ideal cash crop for subsistence farmers. For this reason the crop has also been introduced into the southern African region and it is now grown in the Limpopo and Zambezi valleys. However, the seeds of date palm are short lived and as a consequence the species cannot be effectively long-term stored (Janick *et al.*, 1981).

A number of the methods associated with plant *in vitro* technology offer feasible storage alternatives for the difficult-to-store categories of germplasm e.g recalcitrant seeds and vegetatively propagated crops. *In vitro* technology has the distinct advantage that a variety of cells and organized tissues may be stored (Bajaj, 1983). In addition, this technology is responsible for the generation of novel and new genetic resources, through, for example, somaclonal variation, and these too must be maintained (Mycock *et al.*, 1991). At a very broad level plant tissue culture can be seen to revolve around two developmental processes, organogenesis and embryogenesis (George, 1993). In terms of plant conserva-

R.H. Ellis, M. Black, A.J. Murdoch, T.D. Hong (eds.), Basic and Applied Aspects of Seed Biology, pp. 75–82.

tion, callus and the various developmental stages of somatic embryogenesis are ideal candidates for cryostorage (Grout, 1986; Mycock *et al.*, 1995).

Effective cryopreservation requires the empirical optimisation of numerous variables, including (amongst others) sample size, sample water content, the use of cryoprotectants and the rates of freezing and thawing. Previous work has shown that somatic embryoids of date can be retrieved from cryopreservation (with a 60% survival) provided the material is treated with cryoprotectants, partially dried and frozen by direct immersion in liquid nitrogen (Mycock *et al.*, 1995). The present contribution is an extension of those studies and details the optimisation of the cryopreservation procedure.

Materials and Methods

Generation of Somatic Embryoids

The meristematic tissue of 3-year old plants was utilized to generate callus and all the developmental stages of somatic embryogenesis (Tisserat, 1984). Late globular–early torpedo stage embryoids were isolated and subjected to the various cryopreservation procedures.

Effect of Various Cryoprotectants

The following cryoprotectants were tested: dimethylsulphoxide (DMSO), sucrose and sorbitol and each was combined with glycerol. The material was first soaked for 5 to 30 min in a 5% (v/v) glycerol solution containing one of: 5% (v/v) DMSO/0.5 M sucrose/0.5 M sorbitol. The embryoids were then transferred for equivalent times to more concentrated 10% glycerol solutions each of which contained one of the following: 10% DMSO, 1.0 M sucrose, 1.0 M sorbitol, as appropriate.

Effect of Drying

The effects of drying on the embryoids that had been treated with the cryoprotectants were tested by desiccating the tissues in the air stream of a laminar flow bench. Flow rate was kept constant at 10.5 msig.

Water Content Determinations

Prior to, and after all treatments, embryoid water content was determined gravimetrically after drying at 80°C. As cryoprotectants have a high affinity for water, dry mass was determined at regular intervals until constant. Water content was expressed as g H_2O g^{-1} (dry mass basis for 20 embryoids).

Cryopreservation using cryoprotectant, drying and plunge-freezing in mesh baskets.

Fifty embryoids were pre-treated by immersion for 15 min in a 5% (v/v) glycerol solution containing 0.5 M sucrose, followed by a more concentrated cryoprotectant solution (10% glycerol containing 1.0 M sucrose) for a further 15 min. After treatment the embryoids were dried in a laminar flow to various water contents. The samples were then placed in stainless steel mesh baskets and immersed directly in liquid nitrogen. The samples were retrieved from liquid nitrogen after 1 h and immersed directly into the liquid Murashige and Skoog (1962) medium containing 4% sucrose at 37°C, for 1 min. After this, the embryoids were aseptically plated onto the solid version of the same medium and allowed to develop at 25°C in a 16 h light 8 h dark regime.

Differential Scanning Calorimetry

Pre-treated individual embryoids (see above) were sealed in DSC pans and frozen at a rate of 200°C min^{-1}. This freezing rate was achieved by precooling the sample holder of a Perkin Elmer DSC 7 to –160°C then loading the sample (which came to equilibrium with the holder in less than a minute). Warming thermograms were recorded at a scanning rate of 10°C min^{-1}.

Electron Microscopy

Freeze fracture replicas of material frozen at 200°C min^{-1} were produced in a JEOL 9010C. Samples were fractured at –100°C, recooled to –170°C, shadow-coated with platinum and rotary coated with carbon. In the sub-zero state the samples were attached to gold grids using 1% Lexan. After warming to ambient, the samples were cleaned with a serial dilution of chromsulphuric acid (5–100%), then delexanized with 1,2-dichloroethane. By way of comparison replicas were also produced of material frozen in liquid freon 12 (freezing rate of 5000°K sec^{-1} [Elder *et al.*, 1982]).

Results and Discussion

Callus was derived from meristematic tissue of a three-year-old plant and all the stages of somatic embryoid formation were identified and isolated. Nascent mature embryoids could be germinated and grown into plantlets which have subsequently been hardened. Sample size plays an important rôle in successful cryopreservation, generally the smaller the better (Kartha, 1985; Mycock *et al.*, 1991). It was because of their size that late globular–early torpedo stage embryoids were used in the present investigation. These were small structures weighing approximately 1 ± 0.3 mg (wet mass) and were less than 1 mm in length, consequently they could be isolated with ease and handled without damage.

Apart from size, a number of other parameters must also be taken into consideration when preparing material for cryopreservation, e.g. the use of cryoprotectants, sample water content and the rate of freezing and thawing (Kartha, 1985; Mycock *et al.*, 1991).

Much emphasis has been placed on the use of cryoprotectants in the preparation of biological material for cryopreservation. Cryoprotectants are thought to depress both the freezing and supercooling points of pure water, thereby reducing the amount of water in the tissue available to freeze (Kartha, 1985). It has been found that cryoprotectants are generally more effective when used in combination rather than singly (Finkle and Ulrich, 1979; Mycock *et al.*, 1991; 1995). However, some of these substances can be poisonous to plant tissues, the extent of toxicity varying with the type and concentration of cryoprotectant and the plant species (Withers, 1985). It is therefore necessary to determine empirically the correct combination and concentration for each species and if necessary remove the cryoprotectant after retrieval of the plant tissue from liquid nitrogen.

Growth of the embryoids was adversely affected by the cryoprotectant combinations, DMSO and glycerol and sorbitol and glycerol (Table 1). However, when glycerol was combined with sucrose this cryoprotectant mixture appeared to have no effect on deleterious embryoid development except when samples were treated for 1 h. Therefore in order to obtain maximum effect, without apparently influencing growth, the 30 min exposure time was utilized in all subsequent cryopreservation experiments.

Cryoprotectants alone are seldom sufficient to prepare plant tissues for cryopreservation and it is often also necessary to dehydrate the tissues. Drying of the somatic embryoids to water contents above 0.1 $g.g^{-1}$ had no effect on embryoid viability, but at water contents below this, viability was completely lost (Table 2). This indicated the limits of desiccation tolerance of the tissues at this developmental stage.

Once the apparently correct cryoprotectant treatment and the limits of sample dehydration had been determined, material that had been exposed to

Table 1. The effect of various exposure times to cryoprotectant mixtures containing glycerol, and DMSO, sucrose and sorbitol on growth of date palm somatic embryoids. Results are expressed as a percentage of 30 embryoids treated and assessed 30 days after treatment

	Percent continued growth		
Total time in cryoprotectant (min)	DMSO & glycerol	Sucrose & glycerol	Sorbitol & glycerol
10	57	100	67
20	0	100	47
30	0	100	10
60	0	100[a]	0

[a]embryoids swollen

Table 2. Effect of drying on embryoid continued growth. Samples were dried in a laminar flow bench

Drying time (min)	Water content range (g.H$_2$O g^{-1})	Percent survival
0	1.2–2.0[a]	100
5	0.7–1.2	100
10	0.4–0.7	100
15	0.1–0.4	100
30	>0.1	0

[a]control

30 min in cryoprotectant mixture and then dried to various water contents above 0.1 g.g^{-1} was cryopreserved. The samples were frozen by plunging them, in wire mesh baskets, directly into liquid nitrogen. Previous work has shown that this procedure freezes the material at a rate of the order of 200°C.min^{-1} (Berjak *et al.*, 1995). If the material was not dried at all after treatment with the cryoprotectant mixture all the embryos were killed by the freezing and/or thawing processes (Table 3). Considering the high water content this was not unexpected: during freezing the water within the tissues would presumably have formed ice crystals which would have lethally damaged the tissue. This was confirmed by the lack of ultrastructural detail in the frozen state and the presence of a large melting peak in the thermogram (Figs. 1 and 2). Drying samples down to the range of 0.7–1.2 g.g^{-1} allowed for a 32% survival rate (Table 3). Although ultrastructural detail was discernible in this material, the ground substance was clumped, with what are interpreted to be small, discrete ice crystals (Fig. 3). Presumably it was this ice that caused the melting peak of samples in this water content range (Fig. 1). Further reduction of the water

Table 3. Percentage survival of date palm somatic embryoids after cryopreservation. The material was pre-treated with cryoprotectants, dried to various water contents and then immersed, in wire mesh baskets, directly into liquid nitrogen

Water content range (g. H$_2$O g^{-1})	Percent survival
<1.2[a]	0
0.7–1.2	32
0.4–0.7	90[b]
0.1–0.4	100[c]

[a]control; [b]organized growth and development; [c]unorganized callus growth

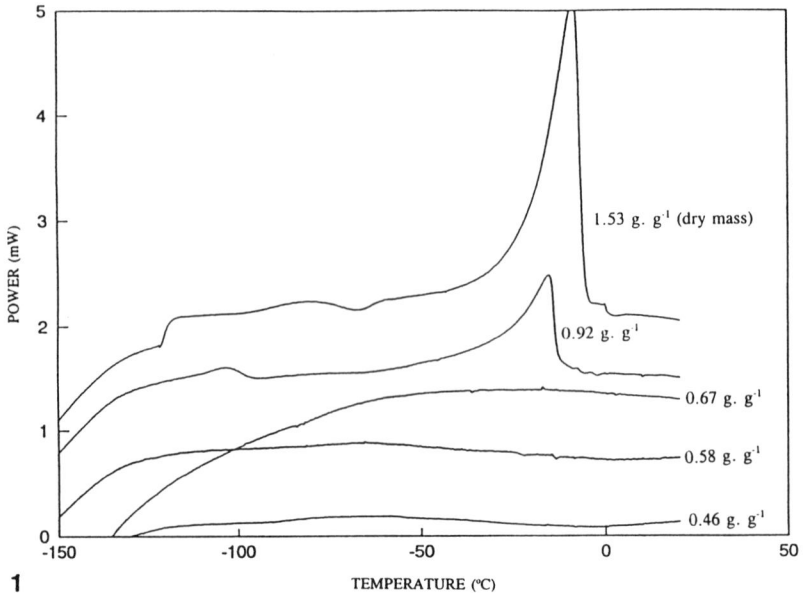

Figure 1. Warming thermograms of date palm somatic embryoids with various water contents and frozen at 200°C min^{-1}

content to the range of 0.4–0.7 g.g^{-1} allowed for 90% recovery (Table 3). The cells of these frozen embryos demonstrated typical ultrastructural detail and the ground substance was smooth (Fig. 4). There was also no discernible ice melting curve in the thermogram (Fig. 1).

More importantly, recovery in these cases was in the form of normal growth and development into a plantlet. Complete recovery (100%) was achieved when the tissues were dehydrated below 0.4 g.g^{-1} (Table 3). However, recovery from cryopreservation of this dry material was in the form of unorganized callus.

Normal recovery is not a usual event when cryopreserving somatic embryos, as the material often forms callus which may regenerate via secondary embryogenesis. Thus the presently observed normal development was significant and indicated that for date palm somatic embryoids there is an optimum water content range for successful cryopreservation. Further, since the unfrozen control material could be dried to water contents as low as 0.1 g.g^{-1} without apparent damage (Table 2), it appears that desiccation effects are initiated at higher water contents (0.1–0.4 g.g^{-1}) and are exacerbated by the freezing and/or thawing processes.

Although the cryopreserved material survived (water content range 0.4–0.7 g.g^{-1}), the rate at which it developed was much slower compared with the unfrozen controls (retarded by 4–6 weeks). This was presumably due to repair

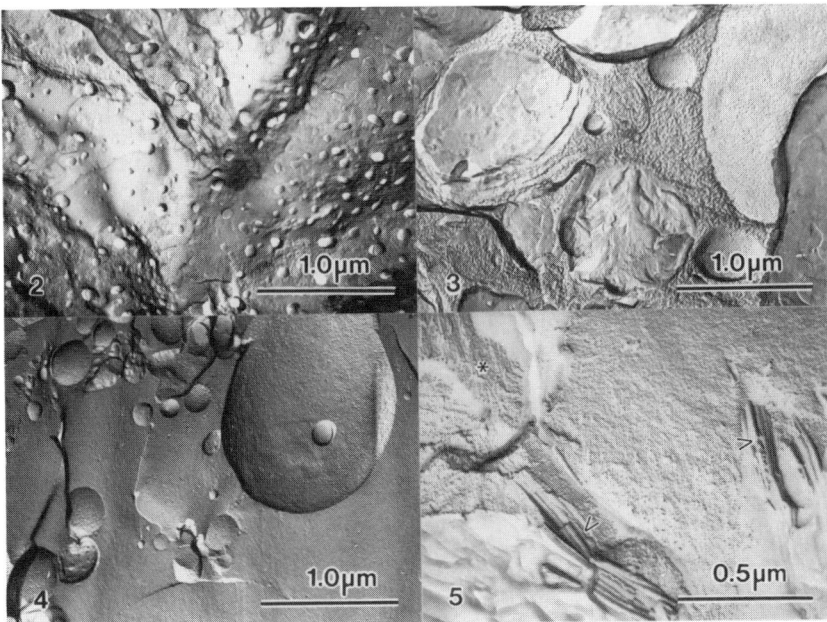

Figure 2. Ice crystal formation in date palm embryoid with a water content above 1.2 g.g^{-1}. Note that ultrastructural detail was not discernible

Figure 3. Intracellular ice crystal formation in date palm embryoid with a water content in the range 0.7–1.2 g.g^{-1}

Figure 4. Ultrastructural detail was discernible in material with a water content in the range 0.4–0.7 g.g^{-1}, there was no evidence of ice crystal formation

Figure 5. Microtubules (<) and microfilaments (*) were observed in date palm embryoids, having a water content between 0.4 and 0.7 g.g^{-1}, and frozen in liquid freon

processes that were necessary before continued growth and development.

By way of comparison samples were also frozen in liquid freon 12 (freezing rate of 5000°K.s^{-1} [Elder *et al.*, 1982]) and subcellular detail of this material was dramatically enhanced. Preservation was, in fact, so good that cytoskeletal elements were clearly discernible (Fig. 5), a feature not observed in the material frozen at the slower rate. The eukaryotic cytoskeleton, which is generally composed of the three elements, *viz.* microtubules, microfilaments and intermediate filaments, can react extremely rapidly to external stimuli (Wolfe, 1993). Microtubules and microfilaments of the cytoskeleton are also extremely sensitive to lowered temperature and increased intracellular ion concentrations (Wolfe, 1993). In the present situation, the date palm somatic embryoids were frozen over a minute, which, when compared with the freezing rate achieved from liquid freon, was relatively slow. It can therefore be proposed that during this minute the cytoskeleton could have been adversely affected by both the lowered temperatures and increased ion concentrations caused by freezing of the water. As a result one of the repair processes necessary in subsequent development might have been the re-establishment of a functional cytoskeleton. This system too, is likely to have been adversely affected by over-dehydration below 0.4 $g.g^{-1}$. Dismantling of an internal supporting system in the absence of substantially strengthened cell walls, could facilitate the assumption of an essentially minimum energy, spherical cell conformation, which might be the trigger for undifferentiated callus formation.

In order to facilitate continued normal development after cryopreservation, date palm somatic embryoids should be treated with a glycerol and sucrose mixture and then dried to water contents between 0.4 and 0.7 $g.g^{-1}$. Excessive drying (between 0.1 and 0.4 $g.g^{-1}$), whilst allowing 100% recovery, appears to induce desiccation damage, as when the date palm somatic embryoids are too dry they recover from cryopreservation as callus. Collectively the present results strengthen the previously reported proposal that sample water content and freezing rate are the principle factors controlling successful cryopreservation (Wesley-Smith *et al.*, 1992; Mycock *et al.*, 1995).

References

Bajaj, Y.P.S. 1983. In: *Plant cell culture in crop improvement*, pp. 19–41 (eds S.K. Sen and K.L. Giles). New York: Plenum Press.

Berjak, P., Mycock, D.J., Watt, P., Wesley-Smith, J. and Hope, B. 1995. In: *Biotechnology in agriculture and forestry*, Vol 32. Cryopreservation of Plant Germplasm, pp. 292–307 (ed. Y.P.S. Bajaj). Berlin, Germany: Springer-Verlag.

Elder, H.Y., Gray, C.C., Jardine, A.G., Chapman, J.N. and Biddlecombe, W.H. 1982. *Journal of Microscopy* 126: 45–61.

Finkle, B.J. and Ulrich, J. 1979. *Plant Physiology* 63: 598–604.

George, E. 1993. *Plant propagation by tissue culture: the technology*, Second Edition. Edington, England: Exegetics Ltd.

Grout, B.M.W. 1986. In: *Plant tissue culture and its agricultural applications*, pp. 303–309, (ed L.A. Withers). London: Butterworths.

Janick, J., Schery, R.W., Woods, F.W. and Ruttan, V.W. 1981. *Plant science: an introduction to world crops*, San Francisco: W.H. Freeman and company.

Kartha, K.K. 1985. *Cryopreservation of plant cells and organs*. Boca Raton, USA: CRC Press.

Murashige, T. and Skoog, E. 1962. *Physiologia Plantarum* 15: 473–497.

Mycock, D.J., Watt, M.P. and Berjak, P. 1991. *Journal of Plant Physiology* 138: 728–733.

Mycock, D.J., Wesley-Smith, J. and Berjak, P. 1995. *Annals of Botany* 75: 331–336.

Tisserat, B. 1984. In: *Cell culture and somatic cell genetics of plants*. Vol 1 *Laboratory procedures and their applications*, pp. 74–81 (ed. I. Vasil). Orlando, USA: Academic Press.

Wesley-Smith, J., Vertucci, C.W., Berjak, P., Pammenter, N.W. and Crane, J. 1992. *Journal of Plant Physiology* 140: 596–604.

Withers, L. 1985. In: *Cryopreservation of plant cells and organs*, pp. 243–267 (ed. K.K. Kartha). Boca Raton, USA: CRC Press.

Wolfe, S.L. 1993. *Molecular and cellular biology*, pp. 496–523. Belmont, USA: Wadsworth Publishing Company.

9. Patenting in the Seed World

L. SCOTT

Cruikshank & Fairweather, 19 Royal Exchange Square, Glasgow, G1 3AE, Scotland, UK

Abstract

What a patent is, what it can do. When patenting is indicated and why it can be important for industry. Advantages and disadvantages of patenting versus secrecy.

Illustration of the relevance of patenting advances in seed technology with examples of claims relating to seed treatment processes, seed product forms and somatic embryogenesis.

General overview of numbers of published patent applications/granted patents generated by some of the largest seed companies in Europe, USA and Japan.

Introduction

Intellectual property and rights derived therefrom take various forms; however, the only form with which this paper is concerned is patents and patent "rights".

Protection of ideas in industry may take two forms: that of patenting and secrecy. The general aim of these two types of protection is to provide freedom for operation in a commercial context.

Patents and Secrecy

A patent is a bargain between the state and the patentee. The bargain is essentially thus: the patentee is granted a monopoly for a limited period of time defined by the state in exchange for as full a disclosure of the invention to the public in the application as is possible at the date of filing. It is important to note that the monopoly is defined by the scope of the claims of any granted patents. Thus the bargain gives the patentee an exclusivity on the commercial use of the invention for a limited time (generally about 20 years).

Patents give the patentee a patent right. This is an exclusive negative right which may exclude others from making or prevents others from commercially using the invention. A granted patent does not exclude everyone from the particular field in which the patent is filed, it merely serves as a warning that a particular aspect of the field is covered and, as such, if the patentee exercises the patent right it may help keep others out of the area defined by the scope of the claims. As such, the patent may be used to obtained royalties from competitors, may be used to cross-licence technology, or may be used simply to exclude others from the market place etc.

Patents may be indicated if coverage is available for a broad, generic, non-

R.H. Ellis, M. Black, A.J. Murdoch, T.D. Hong (eds.), Basic and Applied Aspects of Seed Biology, pp. 83–90.
© *1997 Kluwer Academic Publishers, Dordrecht. Printed in Great Britain.*

trivial invention. "Broad" and "generic" refer to the scope of the monopoly as defined by the claims. It is almost always desirable to attempt to obtain broad, generic protection since such protection serves to make it difficult for competitors to design around the claims. The word "non-trivial" indicates that the invention must be non-trivial in a commercial sense. Thus a very simple invention having large commercial importance (e.g. satisfies a demand and/or is easy to make, easy to sell etc.) may not be a non-trivial invention. On the other hand, a highly complex invention may give rise to a complex piece of equipment or a spectacularly well-performing seed type. However, if such advances are not commercially viable (e.g. they are too expensive to make and/or implement or the market will simply not support sales) it may be that patents may not be indicated. It is very much a question of balance in a commercial context as to whether or not patents should be applied for.

Patents may also be indicated if there is a high risk of independent discovery or development of the same invention by competitors. For example, if breaking seed dormancy became fashionable in a commercial sense for any particular seed type or types and a special process of breaking seed dormancy was developed, then it may be worthwhile protecting that process and if at all possible the seeds having reduced seed dormancy characteristics.

Patenting may be indicated if secrecy is difficult to maintain or not possible. For example, should a seed product be readily definable from an analysis of the seed product or the process which gave rise to the product be definable from the seed product, it may be advisable to patent the process and the product, if possible.

Generally the commercial prospects should outweigh the costs of seeking patent protection. For example, where the annual sales of a product are anticipated to be in the order of hundreds of thousands of pounds per annum or tens of thousands of pounds over many years, then it may be that seeking patent protection is or may be appropriate. However, where the development of a product is fast and end sales thereof are going to be for a limited period of time, then the potential requirement for patenting should be weighed up against the commercial wisdom of seeking patent protection. In both instances, the decision to patent is generally a commercial one based on inputs from at least three separate sources: research, marketing and management (strategists). The types of questions which may be asked include: Does the commercial life and development time scale fit that of the patenting process? Is there any strategic and/or tactical advantage in filing for patent protection? Are patents costs justified?

Patenting may also help prevent copying since third parties once put on notice that there is a patent application in a particular area may be more reluctant to simply copy the product and possibly more inclined to negotiate.

Patents may also be used to secure return on investment. Where it takes a number of years to develop products for the market, for example, the breeding of transgenes into plants, patents may help secure return on investment in the sense that royalties may be obtained and/or indeed competitors may be simply blocked from entering the market place. In the latter situation, it could be envisaged or anticipated that a valid, strong patent could act to keep major competitors at bay, while a strong marketing position is acquired and/or strengthened.

Patents may also strengthen the case for funding of particular projects. For example, Government bodies may be more inclined to provide money for projects on which some work has already been done and which has resulted in the filing of a patent application(s). Such applications give the governmental body a clearer idea that the requestee for grant money is in fact seriously contemplating commercial avenues of endeavour, and may well have lined up a collaborating body.

Secrecy, as mentioned above may be indicated if it is at all possible. For example, where the invention is not analysable from the product secrecy may be available. Coupled with that, there may be a low chance of independent invention, or, patent protection as is available is considered to be either weak or perhaps not available in a relevant sense. Again secrecy may be indicated if the product has a short commercial life relative to the time needed to obtain patent protection. In such instances, it may be better to rely on other intellectual property rights such as trademarks and/or designs, if appropriate. Another consideration which may make a potential patentee lean towards secrecy is in the instance where patenting would mean publishing trade secrets which would otherwise have remained secure. The most famous example of this is probably the recipe for Coca-Cola which has been a trade secret for decades. It is clear that even when secrecy is indicated it may be advisable to file for defensive reasons, for example, when there is a perceived risk of being blocked by competitors. And finally on secrecy, secrecy may only be available if the cost of maintaining secrecy is not too high. For example, secrecy may entail the provision of guards, security systems, dogs, wire fences, and the use of legal instruments e.g. employee/employer contracts, which may include clauses designed to safeguard secrecy.

The importance of patenting to industry and by extension perhaps to academic institutions is that it can provide and has been clearly shown to provide a competitive advantage. The competitive advantage is that the competitive position in terms of selling and buying in the market place are maintained or improved and the considerable amount of money invested in research and development may be safeguarded. Thus market share can be improved if there is corresponding adequate patent protection in place. In such cases royalty payments may be sought from competitors in exchange for exclusive or non-exclusive licences. Finally, patent protection may assist in obtaining access to inventions belonging to others. For example, if one had a patented seed product form of a particular type and the seed product required a particular seed coating technology developed and patented by somebody else and that person wanted to license this technology then there may be solid ground for discussion.

So in conclusion, protection means, in the context of this paper applying for patents for non-trivial inventions which may provide the patentee with an exclusive patent right which right is a negative right i.e. a right which may be exercised to keep others off of your patented ground. Alternatively, weighing up the pros and cons of secrecy versus patentability may mean maintaining secrecy rather than putting in place patent protection which may give rise to undesirable consequences, for example, the publication of trade secrets.

Overview of Published Patents/Applications in the Plant and Seed World

Reference is made to Table 1 which relates to published patent applications or patents which have been filed by a number of seed companies over the period 1985 to 1991. The figures are not limited to seed product forms *per se*. Clearly, some of the major players, for example, Plant Genetic Systems, Monsanto, Calgene and Ciba Geigy are active in the patents arena in the plant and seed world. It is clear that Calgene has a number of granted U.S. patents in the area as has Monsanto and Ciba Geigy and not far behind is Plant Genetic Systems. It is noted that all of these companies have about the same number of published European patent applications over the given period. There is no breakdown shown here of patents versus published applications with respect to the European situation. Going down the list it is interesting to note that Mitsubishi, a Japanese company big in the seed world filed mostly in Japan, whereas Sakata (not included on this list) has filed very little anywhere in the world and has certainly not filed very much in Japan: the number was about two or three applications for the period under review. It is interesting to note that Pioneer has at least 70 granted American patents in the plant and seed world.

Examples of Claims for Seed Products and Somatic Embryogenesis Applications

Finally, some examples of both granted claims and claims in suit are provided to illustrate the relevance of patenting to the seed world.

In EP 309551 B1, a granted European patent, there is provided a method of osmoconditioning seeds characterised in that the mixing is with aeration and particulate solid matrix material, said particulate solid matrix material being nonpathogenic to the seeds and friable enough to permit its mechanical separation from treated seed.

The novelty conferring characteristics of this method are *inter alia* that a particulate solid matrix material is admixed with the seeds such that the matrix material is nonpathogenic to the seeds and is friable enough to permit its mechanical separation from treated seed. Clearly, if a method of seed priming did not involve a particulate solid matrix material such a method would lie outside the scope of this claim.

Granted European patent 202879 B1 relates to a high viability seed lot of a plant species other than one having a seminal root system. It would appear that some of the novelty conferring elements of the claim are seeds having emerged

Table 1. Published patent applications/patents 1985–1991

Company	State	Applications published 1985–1991
Plant genetic systems	EP	49
	JP	35
	US	13*
	WO	36
Calgene	EP	40
	JP	25
	US	63*
	WO	28
Monsanto	EP	42
	JP	28
	US	43*
	WO	11
Ciba Geigy	EP	65
	JP	50
	US	31*
	WO	8
Nickerson	EP	5
	JP	1
	US	2*
	WO	6
Agrigenetics/Lubrizol	EP	17
	JP	7
	US	20*
	WO	16
Upjohn	EP	13
	JP	11
	US	17*
	WO	13
Dekalb	EP	3
	JP	3
	US	16*
	WO	4
Mitsubishi	EP	11
	JP	83
	US	2*
	WO	–
Pioneer	EP	31
	JP	11
	US	70*
	WO	13

*Figures for US reflect published, granted patents

Claims: Seed treatments and somatic embryogenesis examples

EP 309551

1. Method of osmoconditioning seeds ...... characterised in that the mixing is with aeration and particulate solid matrix material/ ...... said particulate solid matrix material being nonpathogenic to the seeds/ ...... friable enough to permit its mechanical separation from treated seeds.

EP 202879

1. A highly viability seed lot of a plant species other than one having a seminal root system/ ...... seeds having emerged radicles/ ...... having a moisture content at which radicle development is suspended without loss of seed viability.

WO 94/05145

1. Pregerminated seeds comprising desiccation tolerant emerged radicles.

EP 254569

1. Method of priming seed ...... maintaining the seed in a state of stirring motion ...... releasing water into the chamber at a controlled rate ...... distributing water evenly amongst the seed ...... surface of seeds dry enough such that seeds remain free flowing ...... water content of seed is progressively and uniformly raised to a desired max. value ...... for a period sufficient to produce seed priming.

WO 90/01058

1. Method of plant cell somatic embryogenesis comprising
a) inducing somatic embryo formation by contacting plant callus tissue with an induction medium ...... with at least one synthetic auxin analogue.

b) regenerating the somatic embryos by contacting said induced tissue with a regeneration medium ......

EP 608716

1. A method of promoting pro-embryogenic mass (PEM) formation ...... characterised in that non-callus explant material is placed in contact with a liquid plant tissue culture medium including an effective amount of an auxin or mixture of auxins sufficient to induce PEM formation.

radicals having a moisture content at which radical development is suspended without loss of seed viability. It is clear that in order to interpret this claim, as in other claims of other granted patents, one would have to look deeply into the description to find out what is meant by such terms as "high viability seed lot", "plant species not having a seminal root system", "radical development is suspended", "without loss of seed viability". Clearly, this claim requires careful analysis.

PCT application WO 94/05145 relates to quite simply pregerminated seeds comprising desiccation tolerant emerged radicals. There are two elements to this claim: the first is naturally enough "pregerminated seeds" and the other is that they have "desiccation tolerant emerged radicals". So, as for other patent claims the specification needs careful analysis to find out what is meant by "desiccation tolerant emerged radicals".

Granted European patent 254569 B1 is to a method of seed priming wherein there are a number of elements which would need to be considered when interpreting the claim. For example, "in a state of stirring motion", "releasing water into the chamber at a controlled rate", "distributing water evenly amongst the seed" such that the "seeds remain free-flowing" and "the water content of seed is progressively and uniformly raised for a period sufficient to produce seed priming". Again the description has to be analysed to find out what is meant by those elements indicated above.

There now follow two examples relating to somatic embryogenesis. Both relate to patent applications only. I believe WO 90/01058 is now withdrawn. Briefly, this patent application was to a method for plant cell somatic embryogenesis comprising inducing somatic embryo formation by contacting plant callus tissue with an induction medium "containing at least one synthetic auxin analogue". It appears that perhaps the novelty conferring element is the addition of this synthetic auxin analogue.

European patent application 608716 relates to a method of promoting pro-embryogenic mass formation characterised in that, and here the novelty conferring features; "non-callus explant material" is placed in contact with a "liquid plant tissue culture medium" including an "effective amount of an auxin or mixture of auxins" "sufficient to induce PEM formation". Again, what is meant by a "non-callus explant material" and the other elements of the claim as indicated above, would have to be investigated. Please note, however, that both these somatic embryogenesis examples relate to non-granted claims and as such do not necessarily reflect the scope of the claims which may result at grant.

Concluding Remarks

This chapter on patenting in the seed world has been limited to addressing issues which seed companies commonly have to address and is also limited in the sense that the subject matter principally relates to seeds technology *per se*. No effort is made to talk about transgenic plants or issues relating to the obtaining of patent protection for plants *per se* or indeed any mention of the patenting of genes or transgenic material. Naturally, the issues discussed herein relating to patenting versus secrecy in the seed technology world apply to such aspects of the seed and plant world.

10. From Encapsulated Somatic Embryo to Plantlet Regeneration

C. BAZINET[1,2], C. DÜRR[1], G. RICHARD[1] and J.-N. BARBOTIN[2]

[1]Unité d'Agronomie de Laon, INRA, rue Fernand-Christ, 02007 Laon; [2]Laboratoire de Technologie Enzymatique, URA 1442 CNRS, Université de Technologie de Compiègne, B.P. 649, 60206 Compiègne, France

Abstract

Growth conditions (growth substratum, maturation and sucrose supply) for the regeneration of encapsulated somatic carrot embryos into plantlets were tested. The germination rate and frequency of morphologically normal and abnormal plantlets were recorded. Germination rates were high (80–95%) and identical in all growth substrata. There was no conversion into plantlets on the liquid medium. The total conversion rates into plantlets on agar and sand were the same (35–40%), but there were significant differences in the rate of conversion into morphologically normal plantlets (20% on agar and 40% on sand), due to secondary embryogenesis on the agar. Rates of conversion into normal plantlets were still lower when a maturation step was included prior to sowing. The time required for seedling development was much longer than for seedlings from natural seeds. Plantlet production from these embryos also requires a carbohydrate supply in the growth medium. Thus, regeneration should not be evaluated by the germination rates alone, but by the rate of conversion into normal plantlets. A sterilized sand growth substratum supplied with carbon and other nutrients appears to be most appropriate for plant regeneration.

Introduction

Synthetic seeds are currently tested by measuring germination and/or regeneration on solid or liquid media. Protusion of the radicle from the coating (germination rate) and root or shoot elongation are the criteria most often used to estimate this growth (Buchheim et al., 1989; Uozumi et al., 1992; Shigeta et al., 1993; Shigeta and Sato, 1994). The plantlet stage is not often precisely defined when plantlet regeneration is observed. The number of morphologically normal plantlets is often low (Redenbaugh et al., 1987; Saranga and Janick, 1991; Ghosh and Sen, 1994; Merkle and Watson-Pauley, 1994).

This paper examined growth conditions and plantlet features that can be used to estimate conversion into normal plantlets. All tests were made on encapsulated somatic carrot embryos (Daucus carota L.) prepared by a standardized synthetic seed production process, with or without a maturation step. Agitated and air-bubbled liquid medium, semi-solid (agar) and solid (sand) media were tested under sterile conditions and with a supply of nutritive solution and sucrose. The sand medium without sucrose was also tested. Germination rate and morphologically normal and abnormal conversion rates into plantlets were recorded.

R.H. Ellis, M. Black, A.J. Murdoch, T.D. Hong (eds.), Basic and Applied Aspects of Seed Biology, pp. 91–94.
© 1997 Kluwer Academic Publishers, Dordrecht. Printed in Great Britain.

Materials and Methods

Production of Encapsulated Embryos

Carrot seeds (cv Nanco, hybrid F1) were surface sterilized and germinated at 25°C on Petri dishes. Radicle segments (2–4 mm) were cut and laid on solid Murashige and Skoog (1962) (MS) medium containing 2,4-dichlorophenoxy-acetic acid (2,4-D, 4.5 μM), kinetin (2 μM) and sucrose (40 g.l^{-1}), and solidified with agar (7.5 g.l^{-1}, Difco). Calluses were removed after 4 weeks and subcultured every month under 16 h light/8 h dark conditions at 25°C. Cell suspensions were prepared from callus in a liquid MS basal medium containing 2,4-D, kinetin and sucrose (20 g.l^{-1}). Cells were inoculated at 50 g.l^{-1} density and cell suspensions were subcultured on fresh medium every 14 days. After two weeks of culture on a gyratory shaker (110 rpm), the cell suspension was passed through stainless steel mesh screens. Embryos were initiated with cell aggregates of 50–85 μm diameter, at a density of 5 to 10 g.l^{-1}, in a hormone free liquid basal MS medium with sucrose (20 g.l^{-1}). Torpedo-stage embryos (1000–500 μm) were mixed in sterile sodium alginate solution (1%, high viscosity; Sigma) and placed in sterilized 50 mM calcium chloride for 20 min at room temperature. Encapsulated somatic embryos were collected on a nylon screen, and kept in MS liquid for 12 h. Encapsulated embryos were matured by placing them in an Erlenmeyer flask in 100 g.l^{-1} sucrose, 1 mM glutamine and 1 μM ABA (Iida et al., 1992) for one week without agitation at 25°C. They were immediately used in experiments.

Experiments

Encapsulated somatic embryos were grown on sterilized diluted (1/2) MS medium containing 20 g.l^{-1} sucrose in flasks, in plant culture tubes (7.5 g.l^{-1}, agar-agar) or in pots filled with sand. Some embryos were also tested on the sand without sucrose. Experiments were run in a growth chamber with at 20°C, a 12-hour photoperiod (330 μmoles m^{-2}.s^{-1}) and an atmospheric relative humidity of 80%. Flasks used for experiments in liquid medium were agitated (110 rpm) and bubbled with air (flow rate 4.5 cm^3.min^{-1}). Embryos (30) were placed in each flask with 50 ml nutrient solution and two flasks were sampled on each observation date (4, 7, 14 and 21 days after sowing). One encapsulated embryo was layed on each plant culture for agar medium. No destructive observations were performed on 60 untreated and 30 treated (maturation) embryos (4, 25, 34, 45 and 52 days after sowing). Five encapsulated embryos were sown in each sterilized Magenta plant pot (20 mm sowing depth), filled with sterilized sand (Fontainebleau sand, 150-210 μm, volumic mass 1.43 g.cm^{-3}). Destructive observations were performed on six pots 7, 14, 28 and 52 days after sowing. The sand water content was kept at 0.2 g.g^{-1} (water potential –2 kPa). These conditions provided sufficient water and did not limit oxygen diffusion (Tamet et al., 1994). Conversion into plantlet was defined by the presence of root and leaf, even if growth was abnormal. Normal conversion was considered to have occurred when the plantlet morphology was comparable to that of a natural seedling. Conversion frequencies were calculated as a percentage of the germinated embryos.

Results and Discussion

Germination rates were maximal at 4–7 days, and were greater than 80% in all treatments (Table 1). These high germination rates are in agreement with previous reports (Shigeta *et al.*, 1990; Kersulec *et al.*, 1993; Ghosh and Sen, 1994). They were similar for all the growth media containing 20 g.l^{-1} sucrose, for matured or unmatured embryos. The germination rate was slighttly lower in sand without sucrose.

Plantlet regeneration rates were much lower than germination rates and varied between treatments (4–50%). Leaves appeared on the first seedlings 14–25 days after sowing, and the most advanced plantlets at 52 days had 3–6 leaves. Seedling growth was much slower than that of seedlings from natural seeds (Tamet *et al.*, 1994). No conversion occurred in liquid growth medium (Table 1). The embryos remained without leaves and with one long radicle without secondary roots. Plantlet regeneration increased until 52 days on agar and sand. The total regeneration rates on both substrata with maturation were similar.

Table 1. Effect of growth substratum, a maturation step and sucrose concentration on the germination and total and normal regeneration frequencies of encapsulated somatic embryos of *Daucus carota* L. 52 days after sowing

Support	Sucrose (g.l^{-1})	Maturation step	Germination (%) [z]	Plant regeneration (%) [y] Total	Normal	Abnormal
Liquid	20	No	7 a		–	–
	20	Yes	9 a		–	–
Agar	20	No	9 a	3 a	22	13
	20	Yes	9 a	5 b	7	43
Sand	20	No	9 a	4 c	40	0
	20	Yes	8 a	d	4	0
	0	No	7 b		–	–
	0	Yes	8 b		–	–

[z]Values followed by the same letter are not significantly different ($p = 0.05$) according to LSD

[y]Values followed by the same letter are not significantly different after χ^2 tests ($p = 0.05$) performed for each date considering three seedlings classes (no leaf, normal and abnormal plantlets)

The normal regeneration rate on agar was lower than on sand, due to secondary embryogenesis. These secondary embryos appeared on agar three weeks after sowing. The maturation treatment increased the frequency of abnormal seedlings on agar. It decreased the total and normal regeneration rates on sand. Secondary embryogenesis was assumed to be induced by ABA (Iida et al., 1992) or by high sucrose concentration (Smith and Krikorian, 1988) during the maturation step. These results suggest that the environmental conditions also have an effect on secondary embryogenesis, since no secondary embryos were produced on sand with the maturation step. No regeneration into plantlets was observed on sand without added sucrose. A carbon supply appears to be necessary prior to the autotrophic growth.

The germination rate or embryo early growth are not sufficient to estimate the effects of a synthetic seed production process. Conversion into morphologically normal plantlets with leaves should be monitored and assessed, using an appropriate growth substratum such as sand with a nutrient supply. Such standardized procedures for evaluating conversion would greatly help the assessement of the effects of the many modifications proposed to improve synthetic seed production.

Acknowledgements

The work was carried out with financial support of the "Biopôle Végétal de Picardie" (France). The authors thank C. Rytter and V. Dufrênes-Devillers for their technical assistance and Owen Parkes for checking the English text.

References

Buchheim, J.A., Colburn, S.M. and Ranch, J.P. 1989. *Plant Physiology* 89: 768–775.
Ghosh, B. and Sen, S. 1994. *Plant Cell Reports* 13: 381–385.
Iida, Y., Watabe, K.-I., Kamada, H. and Harada, H. 1992. *Journal of Plant Physiology* 140: 356–360.
Kersulec, A., Bazinet, C., Corbineau, F., Côme, D., Barbotin, J.-N., Hervagault, J.-F. and Thomas, D. 1993. *Biomaterials, Artificials Cells and Immobilization Biotechnology* 21: 375–381.
Merkle, S.A. and Watson-Pauley, B.A. 1994. *HortScience* 29: 1186–1188.
Murashige, T. and Skoog, G. 1962. *Physiologia Plantarum* 15: 473–497.
Redenbaugh, K., Viss, P., Slade, D. and Fujii, J.A. 1987. In: *Plant Biology: Plant Tissue and Cell Culture* pp. 473–493 (eds. C.E. Green, D.A. Somers, W.P. Hackett and D.D. Biesboer). New York: Alan R. Liss Press, Inc..
Saranga, Y. and Janick, J. 1991. *HortScience* 26: 1335.
Shigeta, J., Mori, T., Toda, K. and Ohtake, H. 1990. *Biotechnology Techniques* 4: 21–24.
Shigeta, J., Mori, T. and Sato, K. 1993. *Biotechnology Techniques* 7: 165–168.
Shigeta, J.-I. and Sato, K. 1994. *Plant Science* 102: 109–115.
Smith, D.L. and Krikorian, A.D. 1988. *Plant Science* 58: 103–110.
Tamet, V., Dürr, C. and Boiffin, J. 1994. *Acta Horticulturae* 354: 17–25.
Uozumi, N., Nakashimada, Y., Kato, Y. and Kobayashi, T. 1992. *Journal of Fermentation and Bioengineering* 74: 21–26.

11. Raffinose Series Oligosaccharides and Desiccation Tolerance of Developing Viviparous Maize Embryos

P. BRENAC[1], M.E. SMITH[2] and R.L. OBENDORF[1]

[1]Seed Biology, Department of Soil, Crop and Atmospheric Sciences, 619 Bradfield Hall; [2]Department of Plant Breeding and Biometry, 524 Bradfield Hall, Cornell University, Ithaca, New York 14853-1901, USA

Abstract

The maize *Viviparous-1* (*Vp1*) gene is associated with biosynthesis of carotenoids, abscisic acid (ABA), embryo storage proteins, and other factors related to quiescence during embryo maturation. Acquisition of desiccation tolerance in developing maize embryos requires a threshold sucrose to raffinose mass ratio of 10 to 1 or lower. Three mutant *vp1* alleles were used to determine if the *Vp1* gene regulates raffinose accumulation and the onset of desiccation tolerance in isolated embryos. Embryos of wild type *Vp1-R* kernels (purple, non-viviparous) acquired tolerance to slow drying at 18 to 22 days after pollination (DAP) and to fast drying at 22 to 30 DAP, coincident with a sucrose to raffinose mass ratio of 10 to 1 or less. Embryos of mutant *vp1-R* kernels (yellow, viviparous) accumulated only small amounts of raffinose, the sucrose to raffinose ratio never dropped to 10 to 1, and embryos never acquired desiccation tolerance. Developing embryos of the modified mutant alleles *vp1-McWhirter* (ivory, non-viviparous) and *vp1-1695* (yellow, non-viviparous) and wild type kernels (*Vp1-McWhirter*, yellow; *Vp1-1695*, purple) became desiccation tolerant coincident with the sucrose to raffinose mass ratio reaching a value of 10 to 1 or lower. Our results suggest that the acquisition of desiccation tolerance in developing maize embryos is related to a sucrose to raffinose mass ratio of 10 to 1 or lower and that raffinose biosynthesis occurs in the absence of a fully functional *Vp1* gene product.

Introduction

Vivipary in maize (*Zea mays* L.) is under the control of nine known loci. *Viviparous-1* (*vp1*) is a seed-specific mutation that affects the development and maturation of maize kernels. Mutant embryos do not become quiescent but undergo germination and seedling development while the kernel is still attached to the ear (Eyster, 1924; Mangelsdorf, 1926; Robertson, 1955). The *Vp1* gene is one of the eight genes whose expression is required for the biosynthesis of anthocyanins (Coe and Neuffer, 1977). Unlike the other viviparous mutants, *vp1* mutant embryos do not respond to ABA, but ABA biosynthesis (Neill *et al.*, 1987) and ABA metabolism (Robichaud and Sussex, 1987) are unaffected. While most of the 18 known mutant alleles of *vp1* are viviparous and anthocyaninless, certain alleles (*vp1-McWhirter*, *vp1-1695*, *vp1-C821708*, and *vp1-A1*) prevent anthocyanin synthesis but produce normal, non-viviparous

R.H. Ellis, M. Black, A.J. Murdoch, T.D. Hong (eds.), Basic and Applied Aspects of Seed Biology, pp. 95–101.

seeds (McCarty *et al.,* 1989a,b; McCarty and Carson, 1991; Hattori *et al.,* 1992).

The acquisition of desiccation tolerance in seeds is typically related to an accumulation of soluble sugars and maturation proteins (Blackman *et al.,* 1992, 1995; Horbowicz and Obendorf, 1994). Maize kernels accumulate soluble carbohydrates mostly as sucrose and raffinose (Kuo *et al.,* 1988). Raffinose is associated with desiccation tolerance and storability of maize seed (Koster and Leopold, 1988; Bernal-Lugo *et al.,* 1993; Horbowicz and Obendorf, 1994). Recently, we have found that desiccation tolerance of whole kernels and isolated embryos of Cornell 175 hybrid maize occurs when the mass ratio of sucrose to raffinose reaches a threshold level of less than 10 to 1 (Brenac *et al.,* submitted).

The objective of this study was to determine if the *Vp1* gene regulates raffinose accumulation and acquisition of desiccation tolerance during the development and maturation of isolated embryos of *vp1* mutants. Three alleles of *vp1* mutants were used, including *vp1-R* (the standard *vp1* mutant, viviparous), *vp1-McWhirter*, and *vp1-1695* (both non-viviparous). Mutants were compared to their respective wild type.

Materials and Methods

Three alleles of maize (*Zea mays* L.) *vp1* mutants were used in this study: *vp1-R*, *vp1-McWhirter* (D.R. McCarty, University of Florida, Gainesville, FL, USA) and *vp1-1695*, also known as *vp1-Coe* (E. Coe, University of Missouri, Columbia, MO, USA). All mutant stocks had predominantly W22 inbred background. Kernels heterozygous for *vp1* were generated in nurseries at Homestead, FL, USA during winter 1993/1994 and at Aurora, NY, USA during summer 1994.

Embryos were collected in 1994/1995 from greenhouse-grown self pollinated plants at 12 to 40 days after pollination (DAP). Wild type and mutant kernels from heterozygous plants were harvested from the same ears and classified based on aleurone color (purple/yellow for *vp1-R* and *vp1-1695*; yellow/ivory for *vp1-McWhirter*). Samples collected before the difference of pigmentation was visible (18 to 22 DAP) were randomly selected.

Harvested kernels were surface sterilized with 2.5% (v/v) bleach in water and rinsed three times in sterile water. Embryos isolated under sterile conditions were germinated fresh or tested for desiccation tolerance after rapid drying or after slow drying. For each treatment, five embryos for each of three replicate ears were used for germination tests. Two other embryos for each of three replicate ears were sampled fresh and after rapid or slow drying treatments, and analysed for soluble carbohydrates as described below.

Germination tests were conducted in 30- or 60-mm-diameter Petri dishes at 25°C in the dark. Five isolated embryos were placed scutellum down on filter paper saturated with a minimal-nutrients medium supplemented with 88 mM sucrose, 55 µM *myo*-inositol and 0.24 µM thiamine-HCL (Rivin and Grudt, 1991; Thomann *et al.,* 1992). Germination was defined as an increase in length of the radicle and coleoptile of more than 5 mm and was recorded daily. Percent germination at 7 d is reported. For the fast drying (FD) treatment (the desiccation tolerance test), isolated embryos were placed scutellum down on dry filter paper disks in Petri dishes and dried rapidly at 12% RH over a saturated solution of LiCl at 22°C. After 6 d, two embryos were analysed for soluble carbohydrates. Five embryos were humidified overnight at 98% RH and tested for germination as described above. For the slow drying (SD) treatment, immature embryos were placed scutellum down on dry filter paper disks and dried slowly during a 7d period by transferring embryos daily to sequentially decreasing relative humidity conditions (92.5, 87, 75, 51, 45, 32.5, and 12% RH) maintained in sealed desiccators over saturated salt solutions of KNO_3, Na_2CO_3, NaCl, $Mg(NO_3)_2$, K_2CO_3, $MgCl_2$, and LiCl, respectively. Two embryos were analysed for soluble carbohydrates. Five embryos were humidified overnight and tested for germination. Soluble carbohydrates in isolated embryos, sampled both before and after drying treatments, were assayed by high resolution gas chromatography as previously described (Horbowicz and Obendorf, 1994) for three replicate ears harvested from 12 to 40 DAP.

Results

Fructose, glucose, sorbitol, *myo*-inositol, sucrose, maltose, raffinose, and four unknowns were detected in the samples. However, this paper presents only the results of the sucrose to raffinose mass ratio, which we have shown is closely related to the acquisition of desiccation tolerance (Brenac *et al.*, submitted).

Embryos from wild type *Vp1-R* kernels (purple, non-viviparous) germinated fresh at 12 to 15 DAP (Fig. 1A). Acquisition of desiccation tolerance occurred at 18 to 22 DAP after slow drying and at 22 to 30 DAP after fast drying (the desiccation tolerance test *sensu stricto*). For both of these drying treatments, germination was coincident with a sucrose to raffinose mass ratio of 10 to 1 or lower (Fig. 1B). Embryos from mutant *vp1-R* kernels (yellow, viviparous), the reference allele of *vp1*, germinated fresh at 12 to 18 DAP but never acquired tolerance to slow drying or fast drying (Fig. 1C). The sucrose to raffinose mass ratio never dropped below the value of 10 to 1 (Fig. 1D). Vivipary was obvious at 26 to 30 DAP.

Embryos from wild type *Vp1-McWhirter* kernels (yellow, non-viviparous) germinated fresh at 12 to 15 DAP (Fig. 2A). Desiccation tolerance to slow drying occurred at the same time as for the wild type *Vp1-R* (18 to 22 DAP) but slightly later after a fast drying treatment (30 to 40 DAP). For both drying treatments, acquisition of desiccation tolerance occurred when the sucrose to raffinose mass ratio reached the value of 10 to 1 or lower (Fig. 2B). The patterns for the mutant embryos from *vp1-McWhirter* kernels (ivory, non-viviparous) were not significantly different from those of the wild type (Fig. 2C, D).

Embryos from wild type *Vp11695* kernels (purple, non-viviparous) germinated fresh at 15 to 18 DAP (Fig. 3A). Germination occurred at 18 to 26 DAP

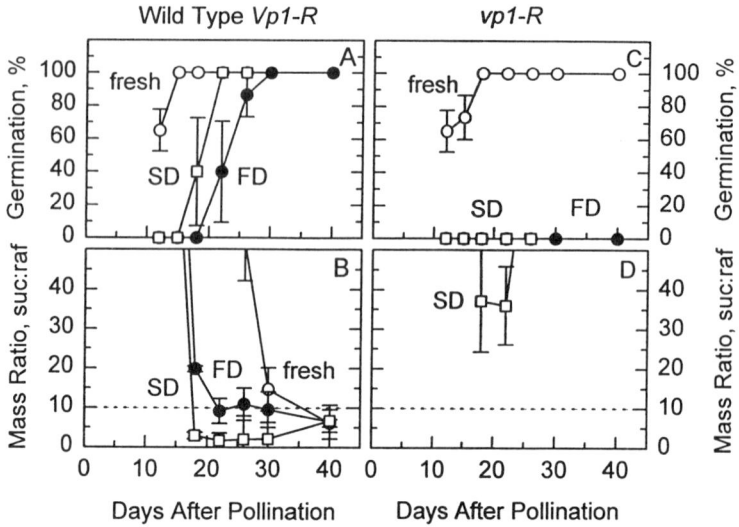

Figure 1. (A,B) Wild type *Vp1-R*. (C,D) Mutant *vp1-R*. (A,C) Embryo desiccation tolerance as a function of DAP. (B,D) Mass ratio of sucrose to raffinose as a function of DAP. Isolated embryos were germinated or analysed fresh (○), after a fast drying treatment (●), or after a slow drying treatment (□). Values represent the mean±s.e. of the mean of three replicates at each harvest

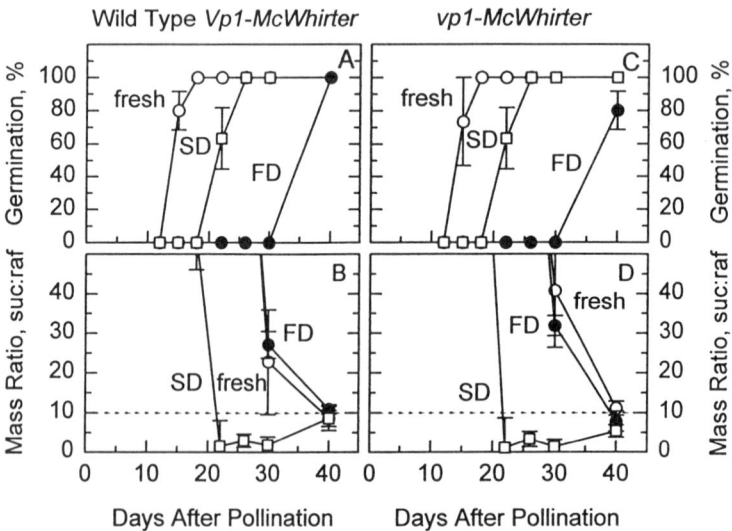

Figure 2. (A,B) Wild type *Vp1-McWhirter*. (C,D) Mutant *vp1-McWhirter*. (A,C) Embryo desiccation tolerance as a function of DAP. (B,D) Mass ratio of sucrose to raffinose as a function of DAP. Isolated embryos were germinated or analysed fresh (○), after a fast drying treatment (●), or after a slow drying treatment (□). Values represent the mean±s.e. of the mean of three replicates at each harvest

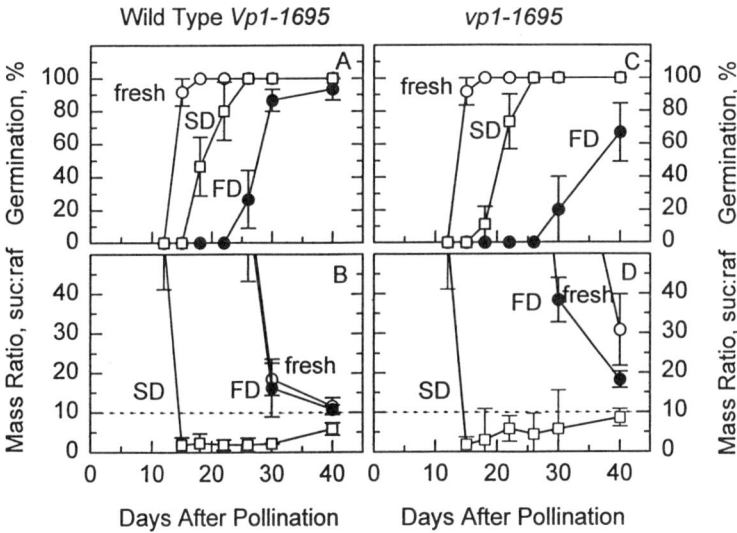

Figure 3. (A,B) Wild type *Vp1-1695*. (C,D) Mutant *vp1-1695*. (A,C) Embryo desiccation tolerance as a function of DAP. (B,D) Mass ratio of sucrose to raffinose as a function of DAP. Isolated embryos were germinated or analysed fresh (○), after a fast drying treatment (●), or after a slow drying treatment (□). Values represent the mean±s.e. of the mean of three replicates at each harvest

after slow drying and at 26 to 40 DAP after fast drying. Under each of these drying regimes, the sucrose to raffinose mass ratio dropped to 10 to 1 or lower when germination occurred (Fig. 3B). Germination of the mutant embryos from *vp1-1695* kernels (yellow, non-viviparous) occurred at the same time as for the wild type embryos for the freshly harvested and slow dried samples (Fig. 3C). Acquisition of desiccation tolerance after the fast drying treatment was slightly delayed and was coincident with a decrease in the sucrose to raffinose mass ratio to 10 to 1 or lower.

Discussion

Our results clearly show that the acquisition of desiccation tolerance during development and maturation of viviparous maize embryos is associated with an accumulation of raffinose and, more specifically, with a sucrose to raffinose mass ratio of 10 to 1 or lower, the threshold value required for germination after drying treatments. In addition to the three alleles of *vp1* investigated, an excellent relationship was observed for whole kernels and isolated embryos of field-grown maize hybrid Cornell 175 (Brenac *et al.*, submitted). Bochicchio *et al.* (1994) reported maize embryo desiccation tolerance in one treatment with a sucrose to raffinose ratio of 40 to 1, but survival was reduced. The delay between germination of fresh samples and decrease of their sucrose to raffinose mass ratio indicates that accumulation of raffinose is not required for germination of freshly harvested embryos. The ability of freshly isolated embryos to germinate (Neill *et al.,* 1987; Bochicchio *et al.,* 1991) is associated with biosynthesis of developmental proteins during germination (Oishi and Bewley, 1992) and could be due to altered ABA metabolism (Neill *et al.,* 1987).

The initial hypothesis of our study was that the differences observed for the acquisition of desiccation tolerance are based on differences in expression among the alleles of *Vp1*. *Vp1* is a seed-specific gene, with no homology to any other known gene, that encodes a 2500 nucleotide mRNA (McCarty *et al.,* 1989a, b). The resulting VP1 protein, a 691 amino acid protein with no similarities to known protein sequences, acts as a transcriptional factor and is required for expression of the maturation program in developing maize seeds (McCarty *et al.,* 1991; McCarty, 1995). The *vp1* mutant allele is characterized by DNA rearrangements including 5 insertions and 1 deletion, and is not transcribed into mRNA (McCarty *et al.,* 1989a,b; McCarty and Carson, 1991). Our results for *vp1-R* are in agreement with the hypothesis that *Vp1* gene may regulate raffinose biosynthesis and the acquisition of desiccation tolerance. Despite the coincidence between lack of desiccation tolerance and a sucrose to raffinose mass ratio that never drops to 10 to 1, small quantities of raffinose were found in these mutant embryos, but much less than in the wild type. Presence of raffinose in the *vp1-R* mutant embryos may, in this case, indicate that raffinose biosynthesis is not totally controlled by the VP1 regulatory protein.

The *vp1-McWhirter* and *vp1-1695* mutants are known to be non-viviparous (McCarty *et al.,* 1989a,b; McCarty and Carson, 1991; Hattori *et al.,* 1992). Molecular analysis showed that DNA of the mutant *vp1-McWhirter* contains a 4000 base pair insertion and is transcribed into mRNA that is 200–300 nucleotides shorter than the wild type mRNA. The resulting VP1 protein is truncated by loss of ~150 amino acids at the C-terminal end (McCarty *et al.,* 1989a; McCarty and Carson, 1991). The modification present in the *vp1-1695* mutant is still unknown. Our data show that embryos carrying these alleles acquire desiccation tolerance coincident with a decrease of the sucrose to raffinose mass ratio to 10 to 1 or lower. Thus, a modification of the *vp1* allele or of its expression allows raffinose accumulation to a sucrose to raffinose mass ratio of 10 to 1 or lower and allows the embryos to become desiccation tolerant.

In conclusion, our results suggest that the acquisition of desiccation tolerance in developing viviparous maize embryos is related to a sucrose to raffinose mass ratio of 10 to 1 or lower and may be regulated by the *Vp1* gene. The expression of a full-length VP1 regulatory protein is not required. Raffinose biosynthesis occurs in the absence of a fully functional *Vp1* gene product.

Acknowledgements

This research was conducted as part of Western Regional Research Project W-168 and funded by Pioneer Hi-Bred International, Inc. We thank D.R. McCarty and E. Coe for mutant maize seedstocks, S. Norman and D. Lathwell for assistance with plant pollination, and I. Castel and M. Withey for assistance with the embryo cultures and analysis of soluble carbohydrates.

References

Bernal-Lugo, I., Diaz de Leon, F., Castillo, A. and Leopold, A.C. 1993. In: *Proceedings of the Fourth International Workshop on Seeds: Basic and Applied Aspects of Seed Biology, Angers, France, 20–24 July 1992*, pp. 789–792 (ed. D. Côme and F. Corbineau). Paris: ASFIS.

Blackman, S.A., Obendorf, R.L. and Leopold, A.C. 1992. *Plant Physiology* 100: 225–230.

Blackman, S.A., Obendorf, R.L. and Leopold, A.C. 1995. *Physiologia Plantarum* 93: 630–638.

Bochicchio, A., Vazzana, C., Velasco, R., Singh, M. and Bartels, D. 1991. *Maydica* 36: 11–16.

Bochicchio, A., Rizzi, E., Balconi, C., Vernieri, P. and Vazzana, C. 1994. *Seed Science Research* 4: 123–126.

Coe, E.H. and Neuffer, M.G. 1977. In: *Corn and Corn Improvement*, pp. 111–223 (ed. G.F. Sprague). Madison, WI: American Society of Agronomy.

Eyster, W.E. 1924. *The American Naturalist* 58: 436–439.

Hattori, T., Vasil, V., Rosenkrans, L., Hannah, L.C., McCarty, D.R. and Vasil, I.K. 1992. *Genes and Development* 6: 609–618.

Horbowicz, M. and Obendorf, R.L. 1994. *Seed Science Research* 4: 385–405.

Koster, K.L. and Leopold, A.C. 1988. *Plant Physiology* 88: 829–832.

Kuo, T.M., VanMiddlesworth, J.F. and Wolf, W.J. 1988. *Journal of Agricultural and Food Chemistry* 36: 32–36.

Mangelsdorf, P.C. 1926. *Connecticut Experiment Station Bulletin* 279: 513–614.

McCarty, D.R., Carson, C.B., Lazar, M. and Simonds, S.C. 1989a. *Developmental Genetics* 10: 473–481.

McCarty, D.R., Carson, C.B., Stinard, P.S. and Robertson, D.S. 1989b. *Plant Cell* 1: 523–532.

McCarty, D.R., Hattori, T., Carson, C.B., Vasil, V., Lazar, M. and Vasil, I.K. 1991. *Cell* 66: 895–905.

McCarty, D.R. 1995. *Annual Review of Plant Physiology and Plant Molecular Biology* 46: 71–93.

McCarty, D.R. and Carson, C.B. 1991. *Physiologia Plantarum* 81: 267–272.

Neill, S.J., Horgan, R. and Rees, A.F. 1987. *Planta* 171: 358–364.

Oishi, M.Y. and Bewley, J.D. 1992. *Journal of Experimental Botany* 43: 759–767.

Rivin, C.J. and Grudt, T. 1991. *Plant Physiology* 95: 358–365.

Robertson, D.S. 1955. *Genetics* 40: 705–760.

Robichaud, C.S. and Sussex, I.M. 1987. *Journal of Plant Physiology* 130: 181–188.

Thomann, E.B., Sollinger, J., White, C. and Rivin, C.J. 1992. *Plant Physiology* 99: 607–614.

12. Morphological and Physiological Changes Associated with Desiccation in Maize Embryos

J.S. BURRIS, J.M. PETERSON, A.J. PERDOMO and D.S. FENG

Seed Science Center, Iowa State University, Ames, Iowa 50011, USA

Abstract

This study proposes to further clarify the effect of drying temperature and rate on vigour, membrane integrity, mitochondrial competence and synthetic processes in susceptible genotypes. Hybrid maize of the seed parent B73 by H99xH95 as pollen parent was harvested periodically, husked and dried under four temperature regimes to 12% mc providing drying times from 4 to 300 h. Standard germination results were generally high and cold test values were similar regardless of temperature or rate. Dry weights from seedlings grown at 25°C were responsive to drying temperature and rate while cold test dry weights showed little difference except for the rapid drying treatment in which root development was consistently depressed more than shoot. Membrane integrity was not compromised in seed dried in normal thinlayer drying but the rapidly dried material exhibited a linear increase in leakage with time. Oxygen consumption by imbibing embryos was reduced by high drying temperature but not rate. Incomplete alignment of lipid bodies along the cell wall following rapid drying treatments as compared to complete alignment in all others suggests roles in both drying and imbibition. Percentage of a 66 kDa protein (associated with maturation) in proportion to total non heat-denatured protein increased in response to drying but not at the rapid rate.

Introduction

Considerable research and a national and international symposia have been devoted to the topic of seed desiccation (Black, 1994). General concepts of reserve deposition, acquisition of desiccation tolerance, transition from developmental to germinative modes and membrane stabilization are well accepted. The details of the regulation of these events and their genetic components are poorly understood. Modern hybrid maize seed production attempts to manage the latter stages of maturation and to control the desiccation process by artificial drying. The impact of genotype, harvest maturity, drying temperature and a preconditioning process (exposure to <40°C and <60% RH) have been described (Navratil and Burris, 1984; Herter and Burris, 1989; Chen and Burris, 1990). This study proposed to 1) further clarify the relationship of drying environment and morphological and metabolic changes in susceptible genotypes; 2) investigate the impact of drying rate and temperature on subsequent seed quality and vigour and to suggest a histological explanation for these responses; and 3) to relate the impact of these physical parameters on the accumulation of a heat stable desiccation induced protein in embryonic axes.

R.H. Ellis, M. Black, A.J. Murdoch, T.D. Hong (eds.), Basic and Applied Aspects of Seed Biology, pp. 103–111.
© *1997 Kluwer Academic Publishers, Dordrecht. Printed in Great Britain.*

Materials and Methods

Hybrid maize seed was produced at Iowa State University, Ames, IA, using the inbred line B73 as the seed parent, and single cross H99xH95 as the pollen parent. Random ear samples were harvested to obtain ears with seed moisture contents ranging from 30–36% (w/w). Except where noted, ear samples were brought into the laboratory, husked, and immediately placed in thinlayer dryers (35 or 45°C), vertical rack (23°C), or carefully hand shelled and placed in a fluidized bed dryer (35°C) and dried to 12% moisture content (mc). Subsequently, shelled seed was placed in paper bags and stored in conditioned storage at 10°C and 50% relative humidity (RH).

Seed quality was measured by standard germination and a soil free cold test (Loeffler *et al.*, 1985) carried out on three replicates of 50 seeds per sample. Shoots and roots were removed from the seedlings, weighed and dried (oven method, 105°C for 24 h) to obtain seedling dry weights. Conductivity of leachate was measured on three replicates of 50 seeds per sample and measured with the Genesis 2000 (Wavefront Inc. Ann Arbor, MI) at 1, 5, 8, 12 and 24 h.

Oxygen uptake by excised embryonic axes was measured polarographically (Hansatech, Cambridge, England) by using air-saturated water at 25°C (250 mM O_2). Samples were maintained on water-saturated sterile blotters between measurements.

Transmission electron microscopy (TEM) was done on embryonic axes fixed in 3% glutaraldehyde/3% paraformaldehyde in 0.1 M sodium cacodylate at pH 7.2 for 18 h at 4°C. Samples were stained in 4% aqueous uranyl acetate, dehydrated, and embedded in Spurr's embedding medium. Sections were examined and photographed with a JEOL 1200EX STEM (Peabody, MA).

To obtain SDS-PAGE protein profiles, ten embryonic axes per sample were excised, weighed and homogenized in grinding buffer (50 mM Tris, 5 mM β-mercaptoethanol, and 1 mM phenylmethanesulphonyl fluoride at pH 7.5) using mortar and pestle on ice. Following centrifugation at 20 000 × g for 20 min at 4°C, the supernatants were removed and heated to 100°C for at least 10 min and centrifuged as before. SDS-PAGE of non-heat denatured protein extracts was performed according to the method of Laemmli (1970). Gels were stained with Coomassie blue R-250 and analyzed by image analysis (Oberlin Scientific, Oberlin, OH).

Results

Seed Quality

Though drying treatments represented radical contrasts in temperature and rate, conventional measures of seed quality showed little response to drying treatment, except for the rapidly dried material harvested at 36.5% mc (Table 1). Samples were retested three times with consistent results. Dry weights of

Table 1. Influence of drying rate and harvest moisture in 1994 of hybrid B73x(H99xH95) on warm and cold germination values and the seedling dry weights produced in each test

Drying treatment	Harvest moisture (%)	Warm germ. (%)	Seed dry wt. (mg/sd)	S/R ratio	Cold germ. (%)	Seed dry wt.	S/R ratio
Room 23°C	36.1	100	60	1.33	100	83	0.96
Thinlayer 35°C	35.1	100	56	1.48	100	80	0.87
Thinlayer 45°C	36.3	97	43	1.79	99	70	1.09
Rapid I 35°C	36.5	61	22	3.11	87	35	1.77
Rapid II 35°C	34.5	92	39	2.65	94	56	1.36
SD	0.05	7.24	15	1.03	3.19	9	0.38

seedlings from seed dried on the ear at 35°C (thinlayer) and at 23°C (room conditions) produced by the warm germination test were similar. The 45°C (thinlayer) treatment resulted in a reduction in vigour while the rapidly dried seed was most damaged, particularly that material harvested at 36.5% mc. The two percentage points difference in mc between Rapid I and II represent four days between harvest dates and demonstrate the sensitivity of drying rate and injury level to the harvest mc. The distribution of seedling dry matter between shoot and root tissue confirms the previously reported sensitivity of root development to dryer injury with substantially higher ratios in the rapidly dried samples as compared to the 45°C treatment which was only slightly higher than the Room or 35°C thinlayer treatments. Trends similar to the warm germination were exhibited by the cold germination test values. The higher germination under cold conditions for the rapidly dried may have resulted from the additional time in test afforded the cold test.

Seedling dry weight results from the cold tests were greater than those from the warm test for all treatments and showed similar ranking. With the additional time in testing, root tissue may recover resulting in lower shoot/root ratios. However, the rapidly dried seed remained unable to recover in spite of additional time.

Membrane Damage

Membrane damage as reflected in electrolyte leakage (conductivity of steep water) was used to estimate the relationship of rate and temperature of the desiccation process and the histological differences observed (Fig. 1). Although previous studies had shown an increase in leakage with increased drying temperature there was little difference between the three temperatures when

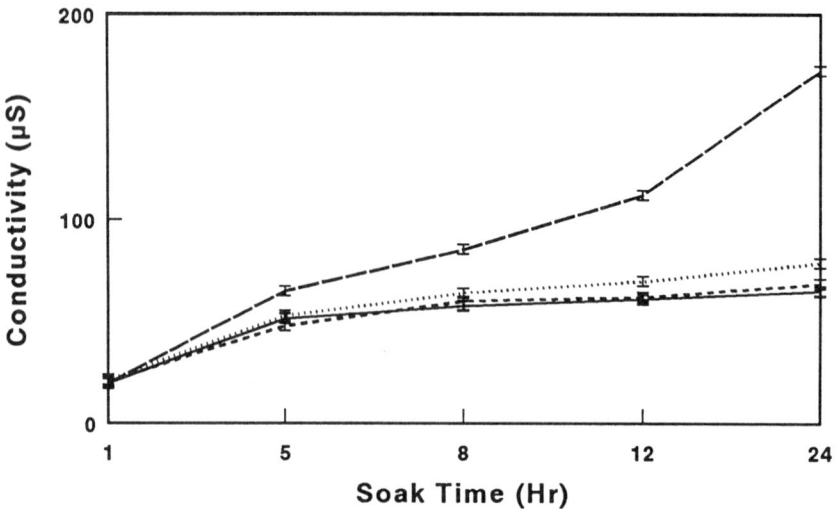

Figure 1. Effect of drying treatment on membrane leakage at different soaking times.
—— room; - - - - 35°C; ······· 45°C; – – – rapid

drying required more than 48 h. The dramatic increases associated with the rapidly dried material suggests the importance of the lipid body alignment adjacent to the cell wall as contributing to control of membrane leakage. At harvest the lipid bodies were randomly distributed throughout the cytoplasm and lacked definition (Fig. 4). As the seed matured or dried, some degree of organization became apparent unless the drying process was very fast. Drying at 35, 45°C or room conditions resulted in similar rates of leakage irrespective of the fact that room temperature drying required 4–6 times longer than the 35 or 45°C drying. Although 45°C drying resulted in some increase in leakage, this injury was very minor and may not be associated with the same sources of electrolytes as in the rapidly dried material. The nearly linear increase in leakage from the rapid dried seed is indicative of a severely damaged membrane system. It may be assumed that the capacity for osmotic regulation would be adversely affected. Severe leakage may in this way contribute to the depressed seedling vigour expressed in both the warm and cold germination tests for the rapidly dried seed as well as contribute to the promotion of seedbed microorganisms.

Respiratory Competence

Results (Fig. 2) are consistent with previous reports that seed dried at 45°C exhibited low rates of oxygen consumption. This was apparent within one hour of imbibition and continued through the six hour measurements. The one hour rates for both the fastest and slowest drying rates were similar and higher than

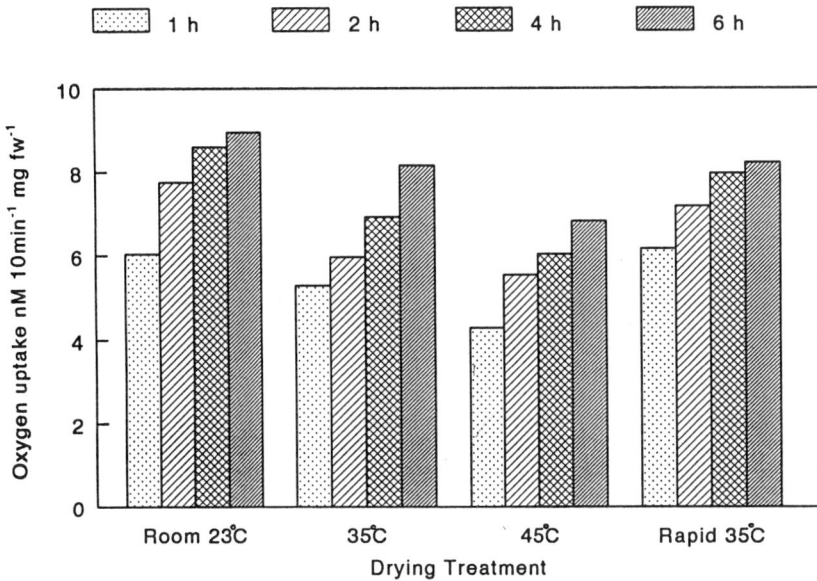

Figure 2. Effect of drying treatment on oxygen uptake by excised embryos at different imbibition times

either 35 or 45°C dried seed. Following six hours of imbibition, the 23°C treatment exhibited the highest rate of oxygen consumption while the two 35°C treatments regardless of drying rate exhibited similar oxygen uptake. The temperature effect is consistent with earlier results from this laboratory indicating that mitochondrial damage results from high temperature drying (Madden and Burris, 1995). The mitochondria from the 45°C drying treatment exhibited large areas of electron transparent ground substance and few visible cristae. In contrast, the 35°C material had well defined cristae and both the inner and outer membranes appeared to be intact. As imbibition time increased, the seed from the 35°C dryer exhibited normal mitochondrial development while the high temperature treated mitochondria appeared to deteriorate.

Synthetic Capacity

Analysis of SDS-PAGE profiles of water soluble, non heat-denatured proteins from seed dried at 23°C or rapidly at 35°C indicate that the 66 kDa protein is induced by drying and accumulated at all temperature combinations but less so in the rapid drying treatment (Table 2 and Fig. 3). Thus, in addition to the requirement for drying there is also a time element to accumulation which was maximal under the room conditions (300 h).

An overall increase in the non heat-denatured fraction of axis protein as well

Table 2. Influence of drying rate on the percentage of a 66 kDa protein present in the water soluble and heat stable protein fraction of the maize embryo

Drying treatment	Percent of heat soluble protein
Room 23°C	7.27%
Thinlayer 35°C	7.73%
Thinlayer 45°C	6.34%
Rapid 35°C	5.53%

as an increase in a 66 kDa protein in proportion to the total non heat-denatured protein have been observed during both natural maturation and in response to artificial drying in high moisture seed (Peterson, 1993). Relationship of the 66 kDa protein to the LEA family of proteins has been proposed, a possible role in desiccation protection is under continuing investigation. Providing 35°C conditions but at a relative humidity greater than 90% resulted in very little protein accumulation and no preconditioning response. This confirms the need for a moisture gradient for the elicitation of preconditioning (Herter and Burris, 1989) and for the associated protein accumulation.

Histology

Early work in this laboratory reported the impact of high temperature drying on mitochondrial competence but was unable to address the effect of drying rate alone (Madden, 1995). Under relatively low mc conditions, the mitochondria did not appear to be damaged by drying rate. Mitochondria were abundant in the rapidly dried seed and their outer membranes appeared to be intact. The most striking histological difference evident in the rapidly dried seed is the lack of lipid body migration to the cell wall (Fig. 4). Marked differences in lipid body migration were apparent in seed dried at moderate rates, i.e., 23°C, 35°C thinlayer or 45°C thinlayer and those dried rapidly at 35°C. The lipid bodies appeared to move out of the cytoplasm and became tightly aligned to the plasmalemma in seed which has been dried at a moderate rate. The rapidly dried seed continued to exhibit a random distribution of the lipid bodies throughout the cytoplasm with only moderate alignment (Perdomo, 1995). Migration and alignment of the lipid bodies may contribute to the physical regulation of moisture loss from desiccating seed or upon imbibition moderate the uptake of water and its associated consequences. Such responses are clearly shown in the quality tests and the conductivity results presented earlier.

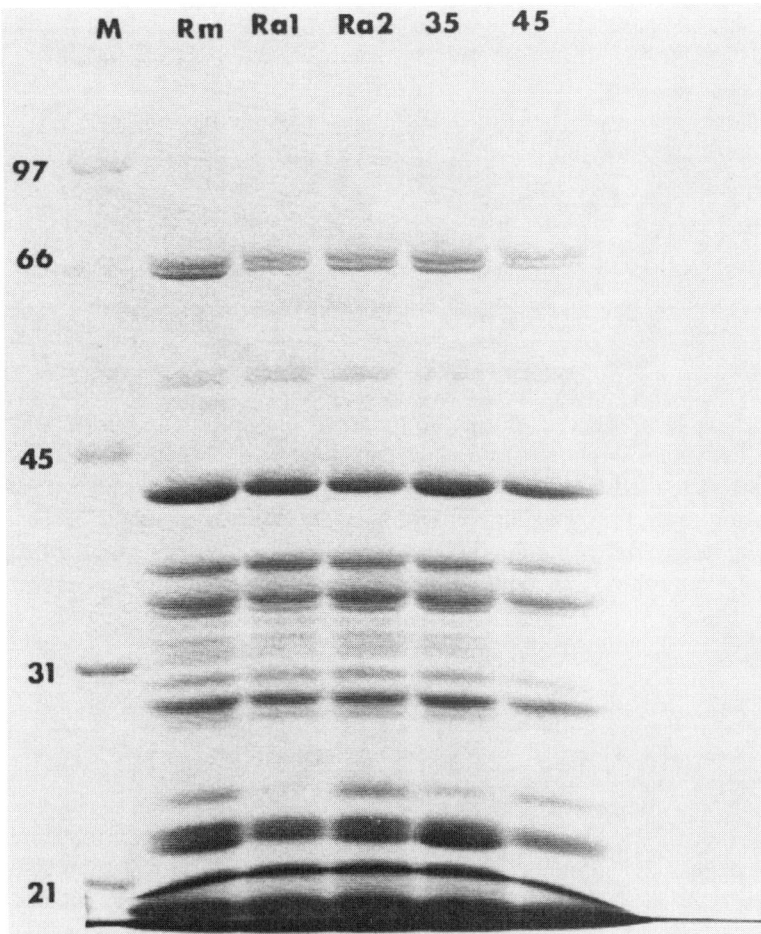

Figure 3. SDS-PAGE gel of water soluble, non heat-denatured proteins accumulated in embryonic axes in response to various dying treatments

Summary

Work in this laboratory has focused on understanding maturation and the drying of maize seed. Maize acquires desiccation tolerance very soon after fertilization; however, attempts to describe the days after pollination (DAP) required to reach maximum vigour is only legitimate in the context of the conditions the seed is exposed to after being removed from the parent plant. In addition to the classical desiccation response, we have reported the gradual

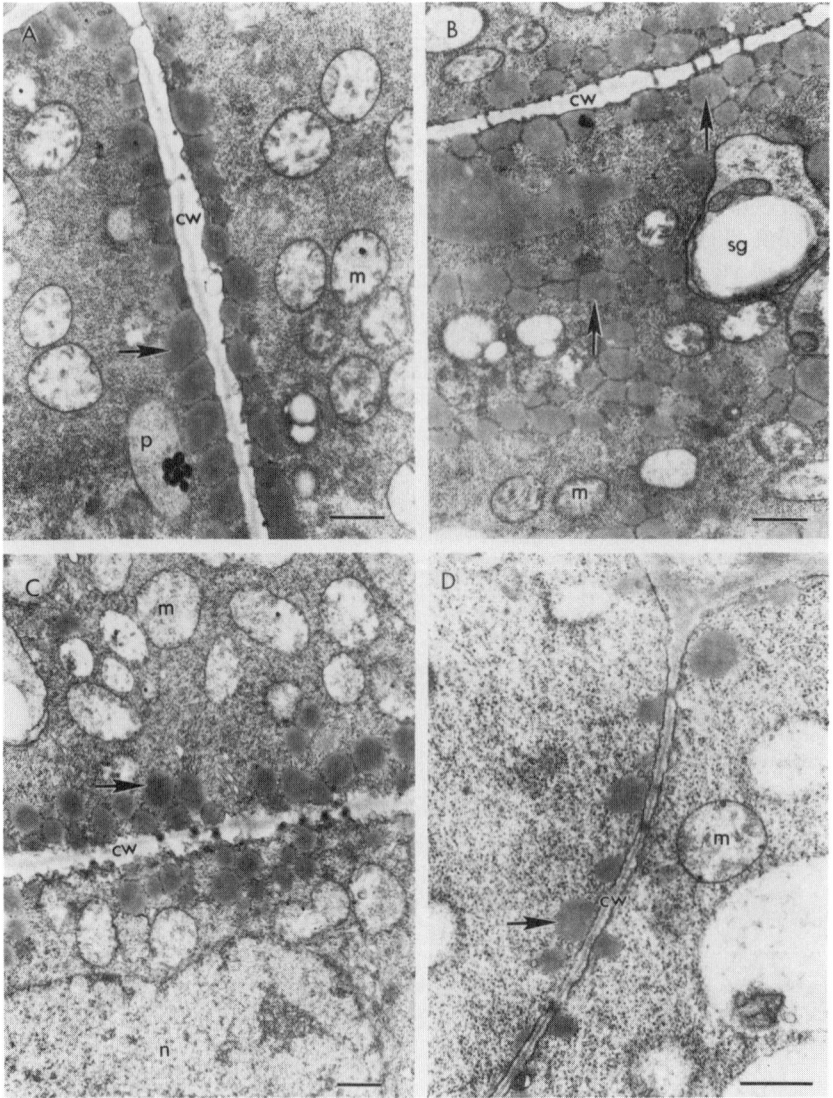

Figure 4. TEM micrographs of maize embryonic radicle tissue after different drying treatments: (A) dry seed following slow drying at room temperature; (B) dry seed following rapid drying at 35°C, in a fluidized bed; (C) following 25 h of imbibition of seed dried at room temperature; (D) following 24 h of imbibition of seed dried rapidly at 35°C in a fluidized bed. Arrows indicate the position of the lipid bodies, cw = cell wall, p = plastid, m = mitochondria, n = nucleus, and bars = 0.5 mm

acquisition of tolerance to high temperature drying with natural maturation or in response to artificial drying conditions which include temperatures of less than 40°C and humidities of less than 75%. Until recently it was not possible to separate the effect of high temperature drying from rate of drying. High temperature would appear to impact the competency of the conserved mitochondria and to a lesser extent impact membrane integrity. Rapid drying precludes migration of the lipid bodies to the plasmalemma thus limiting the beneficial role these lipid bodies may play in regulating moisture loss. Cells without aligned lipid bodies may encounter unregulated imbibition and thus reduced vigour because of membrane failure associated with damage. Rapid drying if it occurs at a safe temperature does not damage the mitochondria and the rapid rate of water uptake associated with the rapidly dried seed may actually accelerate the initial activation of the mitochondria. It should be noted that the rates of drying used in the rapid treatment are not possible if the seed remains on the cob during the drying process. The protein changes identified with natural maturation and the preconditioning process were present in rapidly dried material but failed to reach the levels found at 23°C or 35°C and are only slightly lower than those found at 45°C.

References

Black, M. 1994. *Seed Science Research* 4(2): 109–256.
Chen, Y. and Burris, J.S. 1990. *Crop Science* 30: 971–975.
Herter, U. and Burris, J.S. 1989. *Canadian Journal of Plant Science* 69: 775–789.
Laemmli, U.K. 1970. *Nature* 227: 680–685.
Loeffler, N.L., Meier, J.L. and Burris, J.S. 1985. *Seed Science and Technology* 13: 653–658.
Madden, R.F. and Burris, J.S. 1995. *Crop Science* 35: 1661–1667.
Navratil, R.J. and Burris, J.S. 1984. *Canadian Journal of Plant Science* 64: 487–496.
Perdomo, A.J. 1995. *Histochemical, physiological and ultrastructural changes in the maize embryo during drying.* PhD. diss. Iowa State Univ., Ames, Iowa. 82 pg.
Peterson, J.M. 1993. *Desiccation induced maturation proteins in the embryonic axes of maize.* MS thesis. Iowa State Univ., Ames, Iowa. 52 pg.

13. Fertilization and Seed Formation of *Cyclamen persicum* Mill.

A. EWALD and H.-G. SCHWENKEL

Institute for Vegetable and Ornamental Crops, Department of Floriculture, Mittelhaeuser Strasse, D-99189 Kuehnhausen, Germany

Abstract

Up to now there is little knowledge about the fertilization processes in *Cyclamen* and even about the reasons for low or missing seed set. In our studies, we found that after pollination the number of pollen tubes in the style varied from 0–100. Eight to nine hours after pollination the pollen starts to germinate. After 5 days only a few pollen tubes reached the ovules of the free central placenta. Typical mechanisms of incompatibility on the stigma or in the style could be not observed, so there are other reasons for low seed set. Repeated pollination resulted in a higher yield, but this is mainly due to a lower rate of aborted inflorescences. Additionally there were big differences between single plants.

Introduction

Cyclamen persicum belonging to the *Primulaceae* is an obligatory outcrossing crop and exclusively propagated by seeds (Maatsch, 1971). Breeding and commercial seed production of this economic important pot flower is often hampered by a low capacity for seed formation. Factors for that are still unknown (Gruendler, 1987). To overcome this problem we first studied the processes of fertilization and seed formation.

Materials and Methods

Flowering plants from diploid cultivars ('Lachsscharlach', 'Dunkellachs', 'Kleine Dresdnerin Rosemarie') and from a seed parent (RS1), which were grown in a greenhouse during the flowering season in the winter time or in a climate chamber during summer (12 h 35 klx, 20/16°C), were used. Before anthesis, flowers were emasculated and pollinated, and then covered with tin foil. In another experiment after 3 days flowers were pollinated again. The pollen mixture from the plants used was collected from closed anthers. The viability of 300 pollen grains per sample was estimated using a fluorescence staining technique (Heslop-Harrison and Heslop-Harrison, 1970) modified by Kison (1979). Pollen tube growth was studied from 10 pollinated flowers each after 8, 12, 16, 24, 48, 72, 96, 120, 144, 168, 192 and 216 h after pollinating and subsequently staining (Kho and Baer, 1968). For the later study of seed

R.H. Ellis, M. Black, A.J. Murdoch, T.D. Hong (eds.), Basic and Applied Aspects of Seed Biology, pp. 113–118.
© *1997 Kluwer Academic Publishers, Dordrecht. Printed in Great Britain.*

development inflorescences were cut off and fixed in a Carnoy-solution (60% ethanol, 30% di-chlor-methan, 10% acetic acid) 0, 3, 5, 8, 10, 12, 15, 30 and 60 days after pollination and then embedded in glycol-methyl-acrylate (Dohm *et al.*, 1991). Slices of 4 of μm thickness were cut with a rotocut-microtom and stained with periodic-acid-Schiff procedure (Feder and O'Brien, 1968) followed by counterstaining with aniline blue (O'Brien and McCully, 1981).

To determine seed set, the seed pods were harvested 90 days after pollination. Correlation analysis between number of pollen tubes in the style and in the ovules was carried out using χ^2-test.

Results

The gynaecium of *Cyclamen* consists of 5 carpels grown together surrounding the free central placentation and forming the pistil. There is no connection between the ovary and the pistil. The style, an open pipe, ends with the ring-shaped stigma. The pollen tubes grow along this pipe at the inner wall. In the first experiments we found up to 300 pollen grains sticking at the surface of the distal end of the pistil after pollination once by hand (unpublished data). The viability of pollen fertility ranged from 15–40% between the experiments.

The course of pollen tube growth is given in Table 1. The germination of the pollen grains starts about 8 hours after pollination. Typical incompatibility reactions on the stigma or in the style could not be observed. After 5 days the first fast growing tubes reached the ovules in the distal region of the placenta. Only a few tubes had penetrated the micropyle (Fig. 1).

Histological studies showed that in the early stages of flower development ovules arose from domes at the surface of the placenta starting from the base to the distal end. Nevertheless at the beginning of anthesis there were no differences in the developmental stages between the ovules of the same flower. Because the moment of penetrating the micropyle varied in every case and the frequency of detectable events of fertilization in the proved plant material was

Table 1. Course of pollen tube growth in *Cyclamen* 'Lachsscharlach' and 'Dunkellachs' grown under controlled conditions (12 h 35 klx, 20°/16°C)

Time span after pollinations (h)	Observations
8–9	germination of the pollen grains
11	pollen tube growth (length 300–400 μm)
48	pollen tubes reached the end of the style (3000–4000 μm)
72	first pollen tubes at the surface of the ovary
120	a few pollen tubes have penetrated some ovules

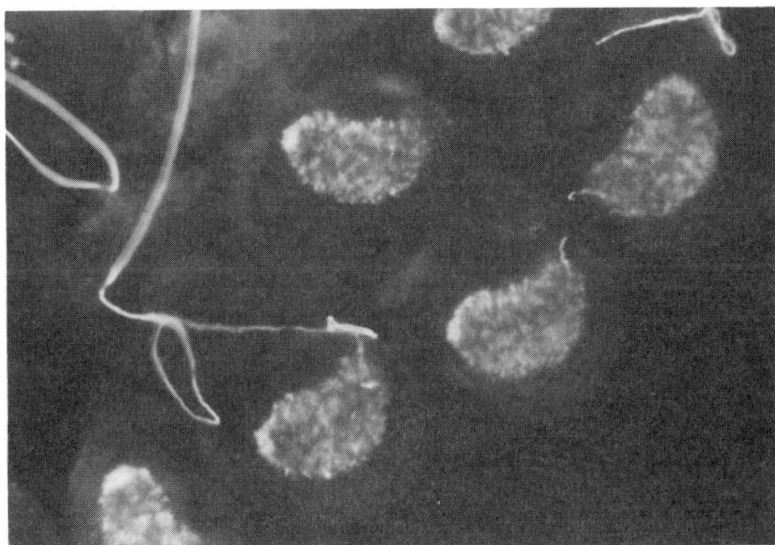

Figure 1. Pollen tubes and ovules in the outer region of the placenta

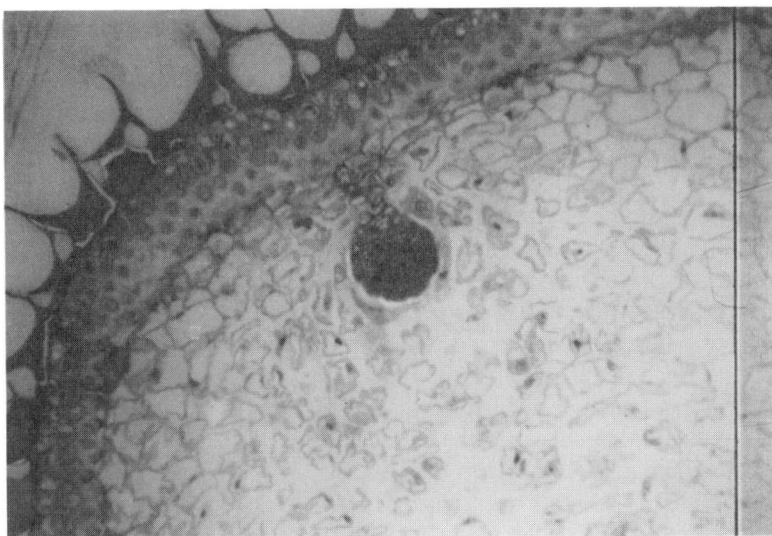

Figure 2. Zygotic embryo in an ovule 20 days after pollination

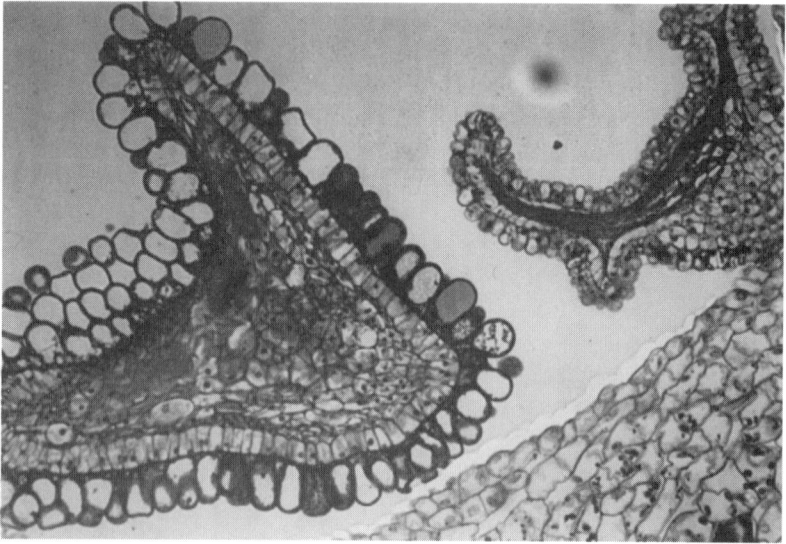

Figure 3. Aborted ovules 20 days after pollination

very low, early steps of the fertilization processes were not found. Twenty days after pollination some ovules with globular embryos were detectable (Fig. 2), as well as irregular shaped ovules (Fig. 3). In the latter case it is a matter of aborted ovules, because the tissue, enriched with tannins, has collapsed. It could not be found out whether those ovules were penetrated before by pollen tubes. Sixty days after pollination, seed formation and embryo development has nearly finished.

Quantitative studies between the results of single and repeated pollination resulted in nearly a doubled number of pollen tubes in the style (Table 2). Also the number of pistils with only a few or even no pollen tubes decreased and the number of penetrated ovules increased.

Repeated pollination also led to a higher seed set (Table 3). Parthenocarpic seed pods were found after single and repeated pollination.

The number of seeds varied from 0–47 in flowers pollinated once and 0–50 in flowers treated twice. Because the number of seeds per pod after repeated pollination increased negligibly, the higher yield was mainly due to the decreased number of aborted inflorescences. Additionally there were high differences in seed set between single plants (Table 4). The mean number of seeds differed from 0.0 to at maximum 30.6. Differences between the treatments were due to varying quality and quantity of pollen on the pistil.

Table 2. Mean number of pollen tubes of RS1 in the style 10 days after pollination

Number of pollen tubes	Flowers pollinated once	Flowers pollinated twice
In the style	15.7	30.2
Penetrating the micropyle	2.9	8.5

Table 3. Seed set from flowers of RS1 pollinated once or twice ($n = 270$)

Number of	Single pollination	Repeated pollination
Seed pods	58	99
Grains (in total)	612	1197
Mean number of seeds per capsule	10.4	12.2

Table 4. Mean number of seeds per pod from plants RS1 ($n = 11$–25 flowers per treatment)

Plant ID	Single pollination	Repeated pollination
1	13.0	18.0
2	0.0	0.0
3	12.6	16.7
4	17.7	13.0
5	1.9	2.8
6	12.4	12.3
7	1.5	0.7
8	22.1	30.6
9	10.5	11.0
10	12.2	9.0
11	8.7	10.0
12	21.6	21.0
13	7.5	12.0
14	8.5	7.0
15	10.6	13.8
16	10.7	16.1
17	12.0	11.2

Discussion

Since up to 300 viable pollen grains stick on the surface of the stigma and there are 160–240 ovules per inflorescence seed set in *Cyclamen persicum* should be higher. One of the reasons for lower yield than 27–71 on average as reported by Gruendler (1987) in our experiments might be due to an inbreeding effect, because we used homozygous plants of an inbred line. But also other factors might be involved like the existence of unripe ovules found in *Spinacia* (Wilms, 1981) or the failure of a chemotropic response of the ovules in *Brassica napus* (Pechan, 1988). Typical incompatibility reactions in the pistil also exist in the family Primulaceae like *Primula elatior* (Schou, 1984), *P. obconica* or *P. veris* (Richards, 1986) resp., but were not observed in *Cyclamen*. A late acting mechanism in the ovary as was found in *Anchusa officinalis* (Schou and Philipp, 1983) can also not be excluded.

References

Dohm, A., Schwenkel, H.-G. and Grunewaldt, J. 1991. *Gartenbau-wissenschaft* 56: 58–65.

Feder, N. and O'Brien, T.P. 1968. *American Journal of Botany* 55: 123–142.

Gruendler, I.-G. 1987. *Das Alpenveilchen.* Berlin: VEB DeutscherLandwirtschaftsverlag.

Heslop-Harrison, J. and Heslop-Harrison, Y. 1970. *Stain Technology* 45 (3): 115–120.

Kho, Y. and Baer, J. 1968. *Euphytica* 17: 298–302

Kison, H.U. 1979. *Akad. Landw. Wiss. DDR* Thesis, Quedlinburg.

Maatsch, R. 1971. *Cyclamen.* Berlin and Hamburg: Paul Parey Verlag.

O'Brien, T.P. and McCully, M.E. 1981. *The Study of Plant Structure.* Oxford: Blackwell Science Publications.

Pechan, P.M. 1988. *Annals of Botany* 61: 201–207.

Schou, O. 1984. *Protoplasma* 121: 99–113.

Richards, A.J. 1986. *Plant breeding systems.* London: Allen and Unwin.

Schou, O. and Philipp, M. 1983. In: *Pollen* (eds D.L Mulcahy and E. Ottaviano) Elsevier.

Wilms, H.J. 1981. *Acta Botanica Neerlandica* 30: 101–102.

14. Galactosyl Cyclitols and Raffinose Family Oligosaccharides in Relation to Desiccation Tolerance of Pea and Soyabean Seedlings

R.J. GÓRECKI[1] and R.L. OBENDORF[2]

[1]Department of Plant Physiology and Biochemistry, Olsztyn University of Agriculture and Technology, Kortowo, 10-718 Olsztyn, Poland; [2]Seed Biology, Department of Soil, Crop and Atmospheric Sciences, 619 Bradfield Hall, Cornell University, Ithaca, New York 14853-1901, USA

Abstract

The loss of desiccation tolerance of pea and soyabean seedlings and associated changes in their soluble carbohydrates were studied. For the desiccation tolerance test, intact seedlings were dried rapidly at 12% relative humidity, or slowly in a series of decreasing relative humidities, followed by rehydration. Soluble carbohydrates were assayed before and after drying. Survival and regrowth of shoot, hypocotyl and root tissues were recorded. Soyabean hypocotyl and root tissues lost desiccation tolerance within 36 h during germination in association with the loss of raffinose, stachyose, galactopinitol A, galactopinitol B, ciceritol, and fagopyritol B_1. In contrast, fructose, glucose and sorbitol accumulated. Sucrose levels increased during desiccation, but there was no effect of drying on changes of other soluble carbohydrates. Pea root tissues lost desiccation tolerance during the first 36 h of germination. Unlike roots, 80% of pea epicotyls survived the slow drying treatment and 40% survived the fast drying treatment of seedlings at up to 96 h after imbibition. Sucrose levels increased 5 to 10 fold in root and hypocotyl tissues and even more in epicotyls during desiccation. Glucose and fructose increased during germination and remained elevated after drying. The data indicate that monosaccharides may induce desiccation damage whereas oligosaccharides are not a prerequisite for desiccation tolerance of pea seedling epicotyls.

Introduction

A remarkable feature of orthodox seeds is their ability to withstand extreme desiccation and yet retain capability for germination after storage. Acquisition of desiccation tolerance in maturing seeds involves several phenomena including accumulation of oligosaccharides (Blackman *et al.*, 1992; Horbowicz and Obendorf, 1994a), maturation proteins (Blackman *et al.*, 1991; 1995; Bewley and Oliver, 1992), free radical-scavenging systems (Hendry, 1993; Smirnoff, 1993; Finch-Savage *et al.*, 1994) and changes in membrane phospholipids (Vertucci and Farrant, 1995).

Accumulation of di- and oligosaccharides has been reported to be the major factor in the acquisition of desiccation tolerance. These sugars may confer seed desiccation tolerance through several ways (Vertucci and Farrant, 1995): 1) hydroxyl groups of sugars may substitute for water in membranes and proteins

R.H. Ellis, M. Black, A.J. Murdoch, T.D. Hong (eds.), Basic and Applied Aspects of Seed Biology, pp. 119–128.

(Clegg, 1986; Crowe *et al.*, 1992), 2) raffinose and stachyose can prevent the crystallisation of sucrose and upon drying, this may permit formation of a stable glassy state preventing membrane fusion (Caffrey *et al.*, 1988; Koster, 1991; Koster *et al.*, 1994), 3) sucrose and other polyols may act as scavengers of free radicals (Smirnoff and Cumbes, 1989; Smirnoff, 1993; Orthen *et al.*, 1994).

Soyabean and yellow lupin seeds acquire desiccation tolerance during maturation in association with the yellowing of axis tissues and accumulation of oligosaccharides, mainly stachyose (Blackman *et al.*, 1991; 1992; Górecki *et al.*, unpublished). Desiccation tolerance is lost during germination and correlates with loss of oligosaccharides and increased monosaccharides (Koster and Leopold, 1988).

In addition to sucrose and oligosaccharides, mature legume seeds also contain cyclitols and galactosyl cyclitols (Horbowicz and Obendorf, 1994a; Górecki *et al.*, 1995). Except for galactinol, the galactosyl cyclitols are often ignored as unknown during analysis of soluble carbohydrates in relation to desiccation tolerance. Recently, it was determined that galactosyl cyclitols substitute for the oligosaccharides in promoting desiccation tolerance in buckwheat seeds (Horbowicz and Obendorf, 1994a,b). In the current work, we studied the loss of desiccation tolerance of pea and soyabean seedlings and associated changes in their sugars, cyclitols and galactosyl cyclitols.

Materials and Methods

Pea (*Pisum sativum* L. cultivar 'Frolic'; Rogers Brothers Seed Company, Boise, ID, U.S.A.), and soyabean (*Glycine max* (L.) Merrill cv. 'Chippewa 64'; greenhouse-grown seeds dried in pods at 22°C and ambient relative humidity (RH)) seeds were humidified over water for 18 h. Four to six replications of 10 to 20 seeds each were germinated on wet germination paper towels (Anchor Paper Company, St. Paul, MN, U.S.A.) for 12 to 96 h at 25°C in the dark. Germination and length of axis tissues were recorded at each harvest.

For fast drying treatments, seedlings were dried for 8 days at 12% RH and 25°C over a saturated solution of LiCl. For slow drying treatments, seedlings were dried over 7 days by transferring daily to progressively lower RHs (92.5, 87, 75, 51, 45, 32.5 and 12%) maintained over saturated salt solutions of KNO_3, Na_2CO_3, NaCl, $Mg(NO_3)_2$, K_2CO_3, $MgCl_2$, and LiCl (for 8 days), respectively. After 8 days at 12% RH, both fast and slow dried seedlings were humidified over water for 16 h at 25°C and rehydrated in wet paper towel rolls as described above. Survival and normal growth of epicotyl, hypocotyl and radicle tissues were recorded at 5 (soyabean) or 6 (pea) days after rehydration. Ability to resume growth was a test for desiccation tolerance.

At harvest times, before desiccation and after rapid or slow desiccation, seeds and seedlings were separated into their respective parts: cotyledons, epicotyl, hypocotyl, and radicle, and stored at −80°C until analysis. Tissues were homogenized in ethanol:water (1:1, v/v) containing phenyl α-D-glucoside as internal standard. After heating to denature hydrolases, the supernatant was passed through a 10 000 Mr cut-off filter, and dried. TMS-carbohydrates were analysed by high resolution chromatography as described previously (Horbowicz and Obendorf, 1994a) and compared with authentic standards as available.

Results and Discussion

Germination, as measured by radicle emergence, began at 24 h in both pea and soyabean seeds (Figs. 1A and 2A). At 36 h nearly all seeds had completed germination. Both drying treatments resulted in a decrease in water concentra-

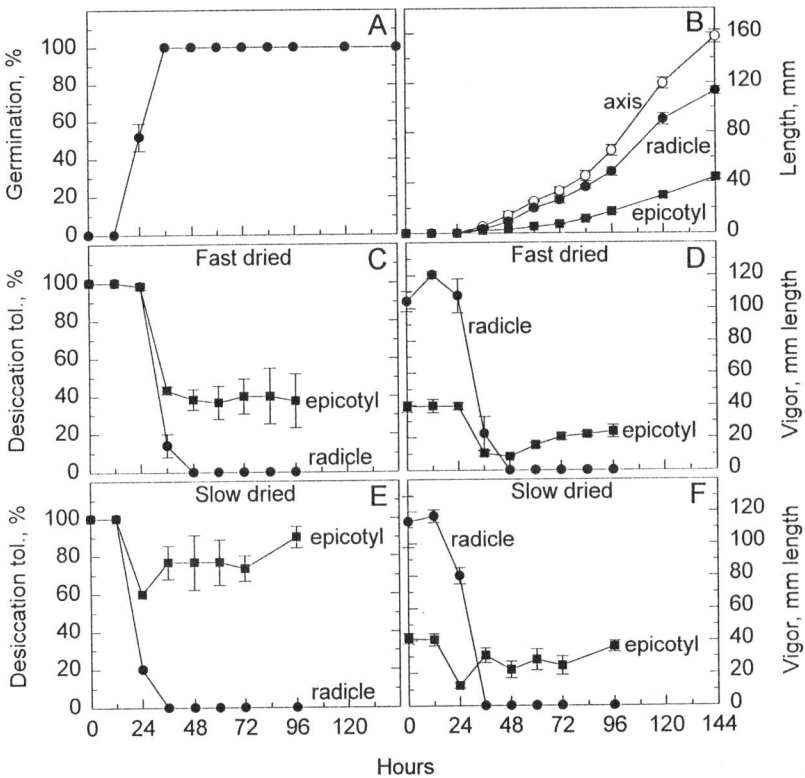

Figure 1. Pea seedling (A) germination and (B) length of axis, radicle and epicotyl as a function of hours after imbibition. (C,E) Desiccation tolerance and (D,F) length of radicle and epicotyl at 6 days after rehydration of (C,E) fast dried or (D,F) slow dried seedlings as a function of seedling age when dried. Vertical bars represent s.e. of the mean

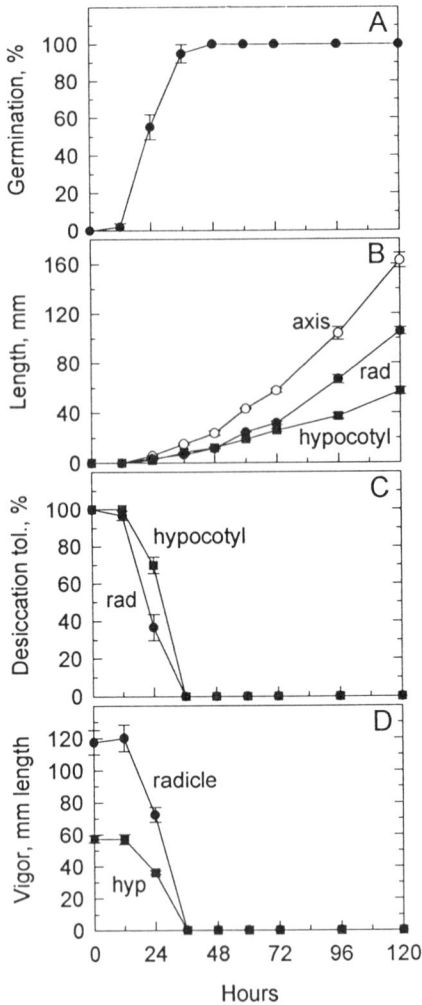

Figure 2. Soyabean seedling (A) germination and (B) length of axis, radicle and hypocotyl as a function of hours after imbibition. (C) Desiccation tolerance and (D) length of radicle and hypocotyl at 5 days after rehydration of fast dried seedlings as a function of seedling age when dried. Vertical bars represent s.e. of the mean

tion in seedlings to about 6%. Pea seeds which were imbibed for 12 and 24 h and subsequently rapidly dehydrated did not suffer impaired radicle and epicotyl regrowth (Figs. 1C,D). As the duration of imbibition time increased to 36 h, radicle tissues lost desiccation tolerance. Unlike roots, 40% of pea epicotyls survived the fast drying treatments and 80% epicotyl survived the slow drying treatment (Figs. 1C,E). The vigour of epicotyls was also higher after the slow drying treatment than after fast drying (Figs. 1D,F). When the radicle was killed, most of the surviving epicotyls formed adventitious roots. Similarly, soyabean radicles were more sensitive to desiccation than hypocotyls (Fig. 2). However, within 36 h of imbibition, the entire seedling lost desiccation tolerance.

Differences in the desiccation tolerance between tissues of pea and soyabean seedlings were also reported by Senaratna and McKersie (1983) and Koster and Leopold (1988). These authors found an inverse relationship between dehydration sensitivity and electrolyte leakage, an indicator of membrane damage.

Reducing sugars were not present in dry pea seeds (Fig. 3). Sucrose plus the raffinose family of oligosaccharides accounted for 94% of the total soluble carbohydrates in axis tissues and 92% in cotyledons. The galactosyl sucrose oligosaccharides were readily metabolised following seed imbibition (Fig. 4). The disappearance of raffinose, stachyose and verbascose from axis tissues was associated with the onset of germination and the loss of desiccation tolerance. Drying treatments stimulated the accumulation of small amounts of raffinose in

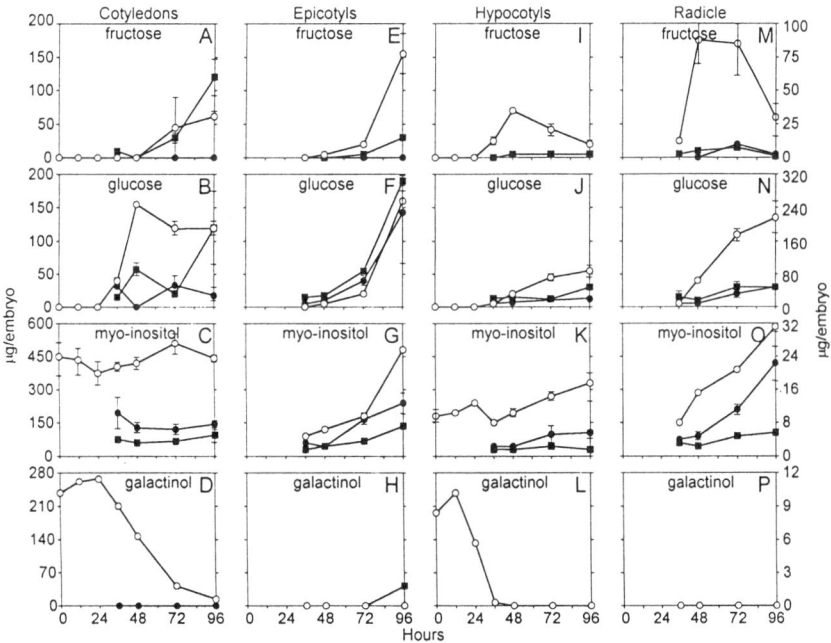

Figure 3. Pea seedling fructose, glucose, *myo*-inositol, and galactinol in (A–D) cotyledons, (E–H) epicotyls, (I–L) hypocotyls, and (M–P) radicles as a function of seedling age before drying (open circles), after fast drying (closed circles), and after slow drying (closed squares). Whole axis tissues were assayed at 0, 12, and 24 h before drying. Vertical bars represent s.e. of the mean

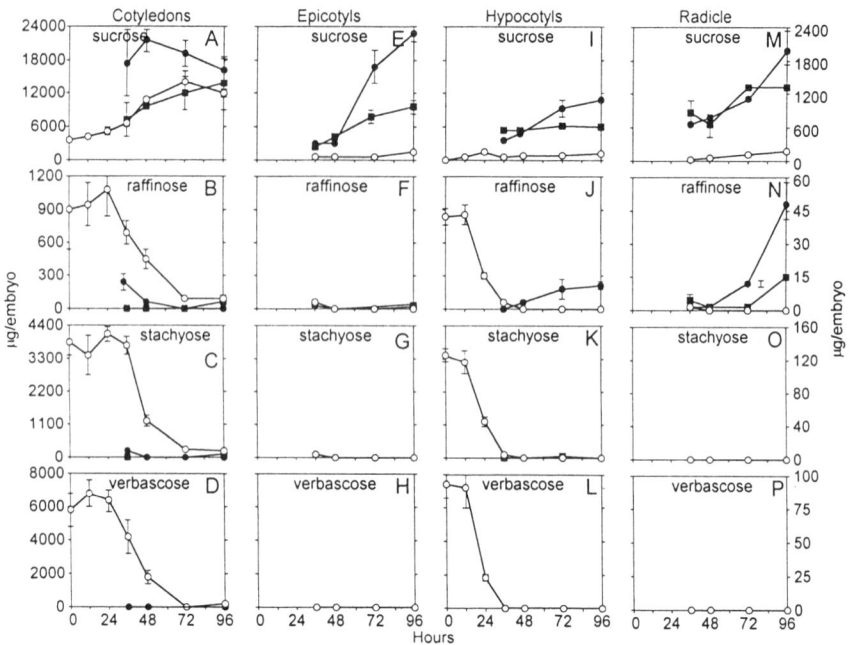

Figure 4. Pea seedling sucrose, raffinose, stachyose and verbascose in (A–D) cotyledons, (E–H) epicotyls, (I–L) hypocotyls, and (M–P) radicles as a function of seedling age before drying (open circles), after fast drying (closed circles), and after slow drying (closed squares). Whole axis tissues were assayed at 0, 12, and 24 h before drying. Vertical bars represent s.e. of the mean

epicotyls, hypocotyls and radicles, but this was not enough to protect radicle tissues from desiccation. Glucose and fructose increased during germination in whole seedlings and remained elevated after drying (Fig. 3). Sucrose levels increased 5 to 10 fold in root and hypocotyl tissues and even more in epicotyls during desiccation (Fig. 4). This reflected the mobilization of oligosaccharides. Finally, the content of *myo*-inositol remained high during germination probably due to hydrolysis of galactinol (Fig. 3).

In soyabean seedlings, sucrose was the predominant soluble carbohydrate, and increased during desiccation (Fig. 5). Loss of desiccation tolerance of soyabean hypocotyl and radicle tissues was associated with the loss of raffinose, stachyose, galactopinitol A, galactopinitol B, ciceritol, and fagopyritol B_1 (Figs. 5–7). Fagopyritol B_1 has been shown to be associated with acquisition of desiccation tolerance in buckwheat (Horbowicz and Obendorf, 1994b). Pinitol is known as a prominent drought stress protectant in leaves of legumes (Keller and Ludlow, 1993) and free radical scavenger *in vitro* (Orthen *et al.,* 1994). Pinitol accumulated in hypocotyls and radicles during germination, reflecting

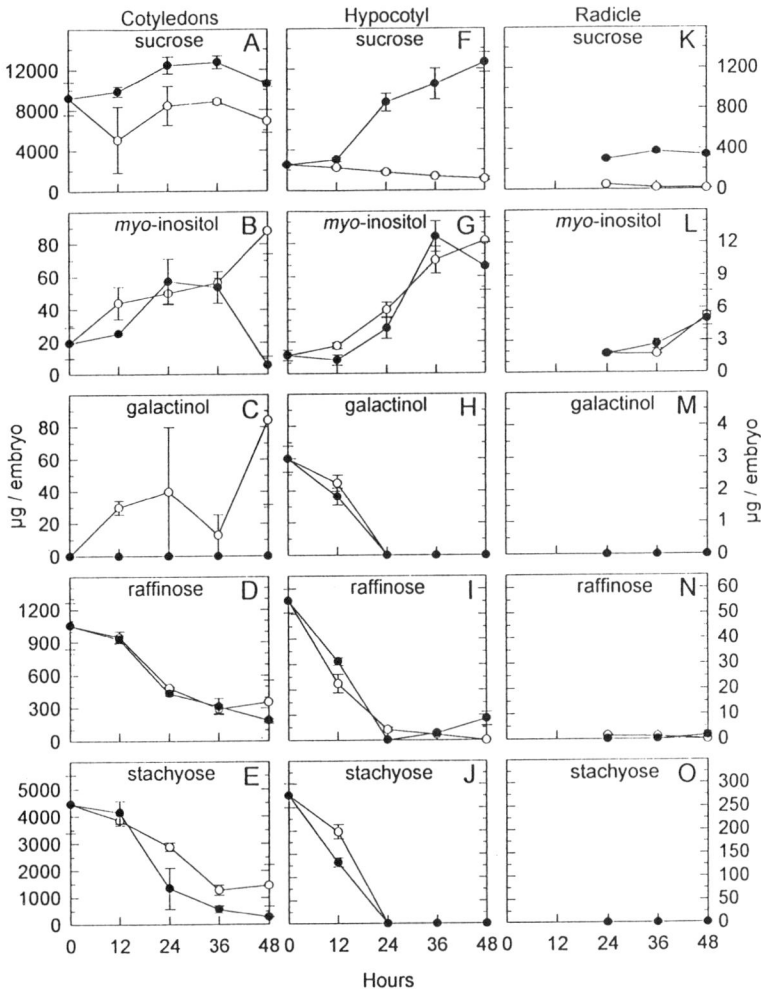

Figure 5. Soyabean seedling sucrose, *myo*-inositol, galactinol, raffinose, and stachyose in (A–E) cotyledons, (F–J) hypocotyls, and (K–O) radicles as a function of seedling age before drying (open circles) and after fast drying (closed circles). Whole axis tissues were assayed at 0 and 12 h before drying. Vertical bars represent s.e. of the mean

the degradation of galactopinitols, but changed little during desiccation (Fig. 6). The reducing sugars, fructose and glucose, and sorbitol (an acylic polyol), increased upon germination (Fig. 7). Sorbitol is an intermediate in the interconversion of glucose to fructose (Kuo *et al.*, 1990). Sorbitol, fructose and glucose decreased during drying.

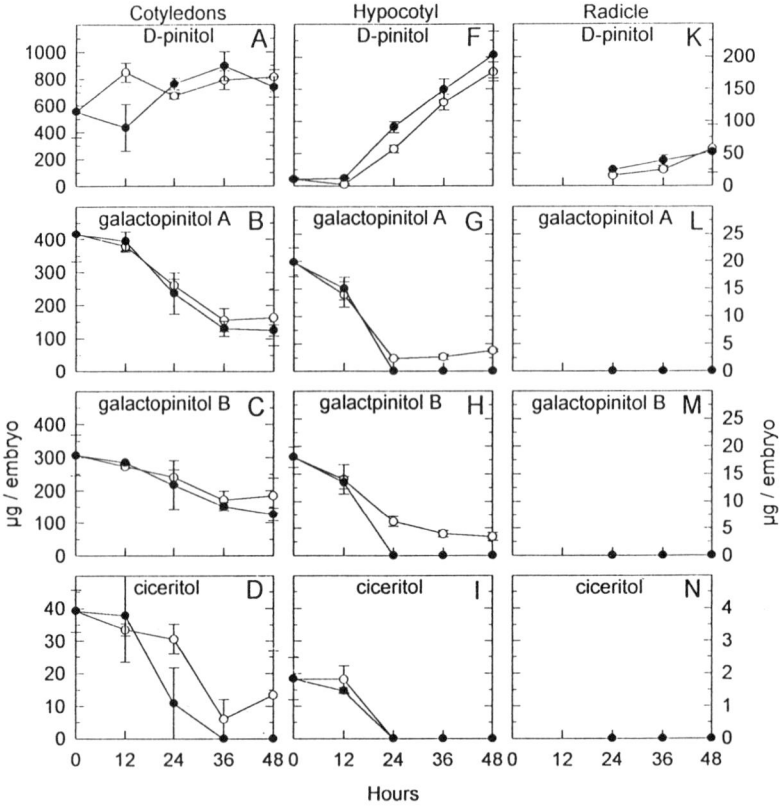

Figure 6. Soyabean seedling D-pinitol, galactopinitol A, galactopinitol B, and ciceritol in (A–D) cotyledons, (F–I) hypocotyls, and (K–N) radicles as a function of seedling age before drying (open circles) and after fast drying (closed circles). Whole axis tissues were assayed at 0 and 12 h before drying. Vertical bars represent s.e. of the mean

We conclude that the loss of desiccation tolerance of soyabean seedlings coincides with the loss of raffinose, stachyose and the galactosyl cyclitols in axis tissues. A high level of sucrose alone is not sufficient for acquisition of desiccation tolerance because soyabean roots and hypocotyls do not acquire desiccation tolerance. The data also indicate that oligosaccharides are not a prerequisite for desiccation tolerance of pea seedling epicotyls. On the other hand, monosaccharides may induce desiccation damage. The involvement of fructose and glucose in desiccation damage was also suggested recently by Hoekstra *et al.* (1994).

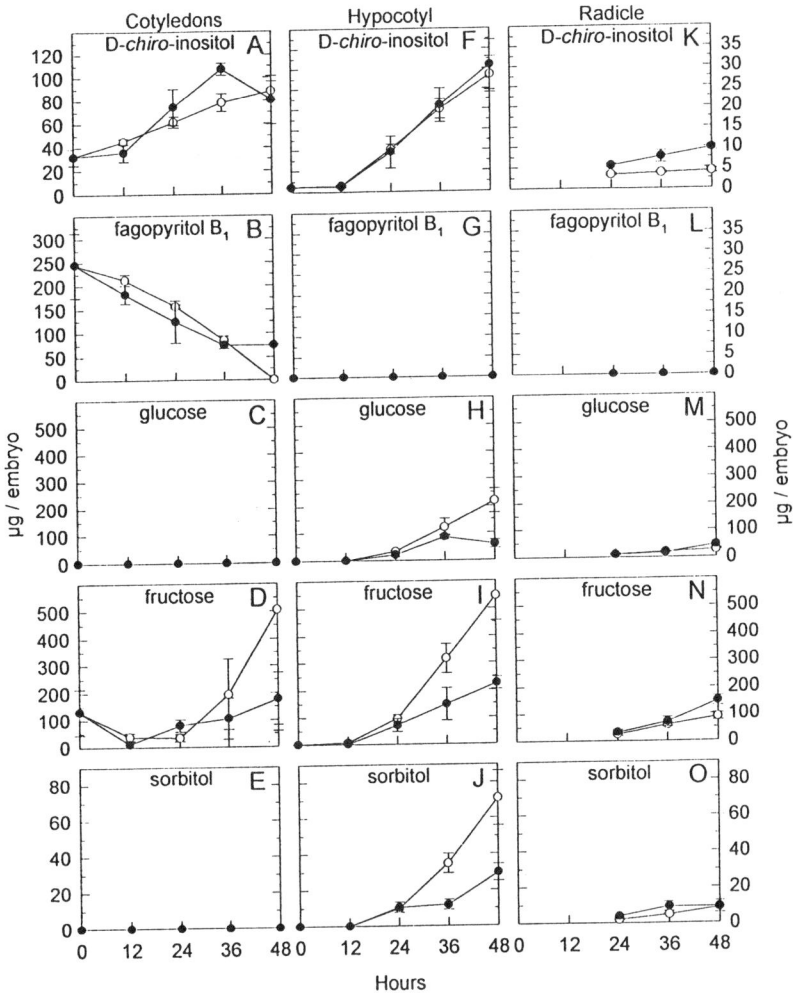

Figure 7. Soyabean seedling D-*chiro*-inositol, fagopyritol B₁, glucose, fructose, and sorbitol in (A–E) cotyledons, (F–J) hypocotyls, and (K–O) radicles as a function of seedling age before drying (open circles) and after fast drying (closed circles). Whole axis tissues were assayed at 0 and 12 h before drying. Vertical bars represent s.e. of the mean

Ackowledgements

This research was conducted as part of Western Regional Research Project W-168 and funded in part by grants from the Polish Committee for Scientific Research (KBN) project No. 5 S302 033, The Kosciuszko Foundation, and Pioneer Hi-Bred International, Inc.

References

Bewley, J.D. and Oliver, M.J. 1992. In: *Water and Life: Comparative Analysis of Water Relationships at the Organismic, Cellular and Molecular Levels,* pp. 141–160 (eds. G.N. Somero, C.B. Osmond and C.L. Bolis). Berlin; New York: Springer-Verlag.

Blackman, S.A., Wettlaufer, S.H., Obendorf, R.L. and Leopold, A.C. 1991. *Plant Physiology* 96: 868–874.

Blackman, S.A., Obendorf, R.L. and Leopold, A.C. 1992. *Plant Physiology* 100: 225–230.

Blackman, S.A., Obendorf, R.L. and Leopold, A.C. 1995. *Physiologia Plantarum* 93: 630–638.

Caffrey, M., Fonseca, V. and Leopold, A.C. 1988. *Plant Physiology* 86: 754–758.

Clegg, J.S. 1986. In: *Membranes, Metabolism, and Dry Organisms,* pp. 169–187 (ed. A.C. Leopold). Ithaca: Comstock Publishing Associates, Cornell University.

Crowe, J.H., Hoekstra, F.A. and Crowe, L.M. 1992. *Annual Review of Physiology* 579–599.

Finch-Savage, W.E., Hendry, G.A.F. and Atherton, N.M. 1994. *Proceedings of the Royal Society of Edinburgh Section B (Biological Sciences)* 102B: 257–260.

Górecki, R.J., Brenac, P., Clapham, W.M., Willcott, J.B. and Obendorf, R.L. 1995. *Crop Science* (accepted).

Hendry, G.A.F. 1993. *Seed Science Research* 3: 141–153.

Hoekstra, F.A., Haigh, A.M., Tetteroo, F.A.A. and van Roekel, T. 1994. *Seed Science Research* 4: 143–147.

Horbowicz, M. and Obendorf, R.L. 1994a. *Seed Science Research* 4: 385–405.

Horbowicz, M. and Obendorf, R.L. 1994b. *Plant Physiology* 105: S-164.

Keller, F. and Ludlow, M.M. 1993. *Journal of Experimental Botany* 44: 1351–1359.

Koster, K.L. 1991. *Plant Physiology* 96: 302–304.

Koster, K.L., Webb, M.S., Bryant, G. and Lynch, D.V. 1994. *Biochimica et Biophysica Acta* 1193: 143–150.

Koster, K.L. and Leopold, A.C. 1988. *Plant Physiology* 88: 829–832.

Kuo, T.M., Doehlert, D.C. and Crawford, C.G. 1990. *Plant Physiology* 93: 1514–1520.

Orthen, B., Popp, M. and Smirnoff, N. 1994. *Proceedings of the Royal Society of Edinburgh Section B (Biological Sciences)* 102: 269–272.

Senaratna, T. and McKersie, B.D. 1983. *Plant Physiology* 72: 620–624.

Smirnoff, N. 1993. *New Phytologist* 125: 27–58.

Smirnoff, N. and Cumbes, Q.J. 1989. *Phytochemistry* 28: 1057–1060.

Vertucci, C.W. and Farrant, J.M. 1995. In: *Seed Development and Germination,* pp. 237–271 (eds. J. Kigel and G. Galili). New York: Marcel Dekker, Inc.

15. Maternal Effects on *Callistephus chinensis* Seed Yield and Quality

M. GRZESIK, K. GÓRNIK and M.G. CHOJNOWSKI

Research Institute of Pomology and Floriculture, Pomologiczna 18, 96–100 Skierniewice, Poland

Abstract

Germination, ethylene release, *in vivo* ACC oxidase activity and conductivity were assessed immediately after harvest and during six months of storage of *Callistephus chinensis* seeds, taken at different dates from primary, secondary and tertiary capitulums, were examined.

The mature seeds harvested in early autumn and from the primary or secondary capitulums were of highest volume and germinated better than those obtained in late autumn and from the tertiary capitulums. The best temperature for germination of seeds immediately after harvest was 5–15°C. The range of temperatures favourable for germination broadened with passage of storage time up to six months. The results of electrolyte leakage, ethylene release and *in vivo* ACC oxidase activity measurements confirmed these findings.

Introduction

Quantity and quality of crops can be affected by the environmental conditions during seed development and maturation. They are also determined by varieties, plant growth, density and architecture of plants (Vis, 1980; Matous, 1986; Kobza, 1986). Estimation of yield component contribution conducted by sequential yield component analysis (SYCA) shows that in sweet pea the highest contribution to final yield arises from the number of pods per branch, followed by the ratio of total lateral outgrowth to height, the number of seeds per pod and seed weight (Chojnowski and Grzesik, 1994). In the some other species the highest quality seeds in terms of germination are assumed to be produced on the main shoots (Duczmal, 1989). In research of Kobza (1986) with *Callistephus chinensis* or in investigations conducted by Carette and Laurent (1989) with *Cichorium endivia* favourable plant architecture for producing high quality crops was obtained by increasing plant density. Field harvests showed that the highest population gave the highest yield and a satisfactory germination rate, although the seed weight in chicory was decreased.

The aim of this study was to determine the quantity and quality of seeds, harvested on different dates from the different parts of China aster plants, in order to identify the most economical method of harvesting.

R.H. Ellis, M. Black, A.J. Murdoch, T.D. Hong (eds.), Basic and Applied Aspects of Seed Biology, pp. 129–135.
© *1997 Kluwer Academic Publishers, Dordrecht. Printed in Great Britain.*

Materials and Methods

Material

Plants of *Callistephus chinensis* Nees. cv. Aleksandra were grown in Szymanów, Central Poland. Mature seeds were harvested separately from the primary, secondary and tertiary capitulums on 19 and 29 October 1993. Seeds were dried at room conditions, weighed and then stored for six months at a temperature of 3°C and 30% relative humidity (RH).

Germination Tests

Germinability was tested in 5 cm Petri dishes (100 achenes per dish in three replicates), on a layer of filter paper moistened with distilled water. Germination assays were carried out at 5, 10, 15, 20, 25, 30 and 35°C in darkness. A seed was regarded as germinated when the radicle had pierced the pericarp.

*Measurements of Ethylene, EFE (*in vivo *ACC Oxidase) Activity Determination and Conductivity Test*

Ethylene release, *in vivo* ACC oxidase activity and electrolyte leakage were evaluated after 0, 1, 2, 3, 4, 5 and 6 months of storage.

After preimbibition in the distilled water in the darkness for 24, 48 or 72 h at 25°C, 10 seeds were placed in 10 ml flasks in 0.5 ml of distilled water or 10^{-3}M ACC. The flasks were closed tightly and after two hours of incubation at the same conditions, production of C_2H_4 was determined in 1 ml samples of flask air. The Hewlett Packard 5890 gas chromatograph was equipped with a flame ionization detector and an activated alumina column. Nitrogen was used as a carrier gas and oven temperature was 100°C. All results are means of five measurements.

Electrolyte leakage was measured after soaking seeds for 1, 2, 3 and 4 h in 3 ml of distilled water at 20°C. Four replications of 50 seeds were used for this conductivity test.

Results and Discussion

The results obtained show that mature seeds of China aster 'Aleksandra' harvested in early autumn yielded more and better quality seeds than those which were taken later (Fig. 1). In practice, the late harvest would probably be unprofitable, if the cost of processing is considered. The highest yield in Aleksandra variety was produced by secondary capitulums (Fig. 1).

The seeds harvested from the primary capitulum germinated better than secondary and tertiary seeds at temperatures of 10–30°C (Fig. 2). Freshly harvested seeds germinated best at temperatures of 5–15°C in contrast to the

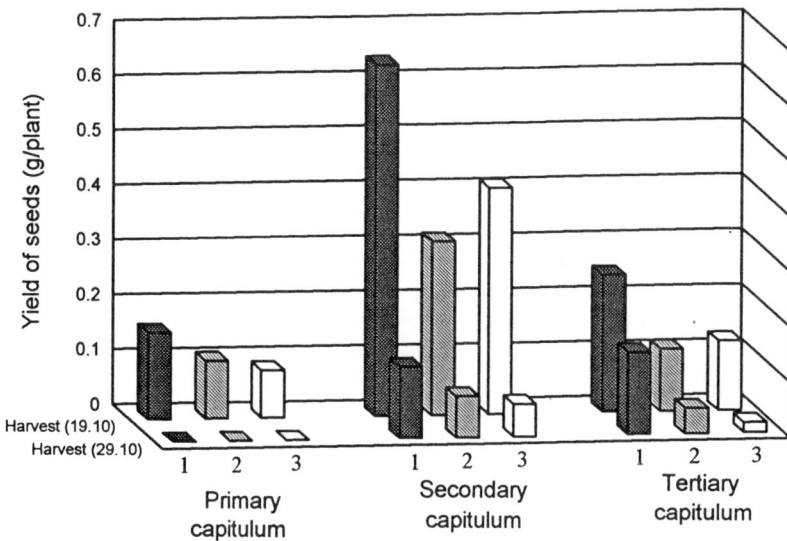

Figure 1. Yield of *Callistephus chinensis* 'Aleksandra' seed harvested on 19 and 29 October 1993 from the primary, secondary and tertiary capitulums. 1, total yield; 2, germinable non-stored seeds; 3, germinable six month stored seeds

recommendation of 20°C in the ISTA rules (1985). At 20°C, the germination level often failed to reach that required by trade rules. Mature seeds harvested on 19 October 1993 germinated better than those obtained from the plants 10 days later especially if those from the tertiary capitulum are considered (Fig. 2).

Germination, particularly at 20° and 35°C increased with storage period (Fig. 3), probably due to dormancy release.

Measurements of electrolyte leakage, ethylene release and *in vivo* ACC oxidase activity tend to correspond with seed germinability (Figs. 4, 5; cf. Fig. 1) confirming the results of other experiments with *Impatiens balsamina* and *Matthiola bicornis*. For example, deterioration of seeds stored in unfavourable conditions reduced the range of temperature for germination, lowered EFE activity and increased conductivity (Grzesik and Chojnowski, 1994). Similar dependencies were obtained by Khan (1994) in seeds of some vegetable species. Such findings suggest it may be possible to use these tests for the evaluation of seed vigour (Figs. 4, 5).

Mature seeds of *Callistephus chinensis* 'Aleksandra' can, therefore, be harvested at a single time from the primary and secondary capitulum, which means it could be possible to harvest them mechanically.

(A)

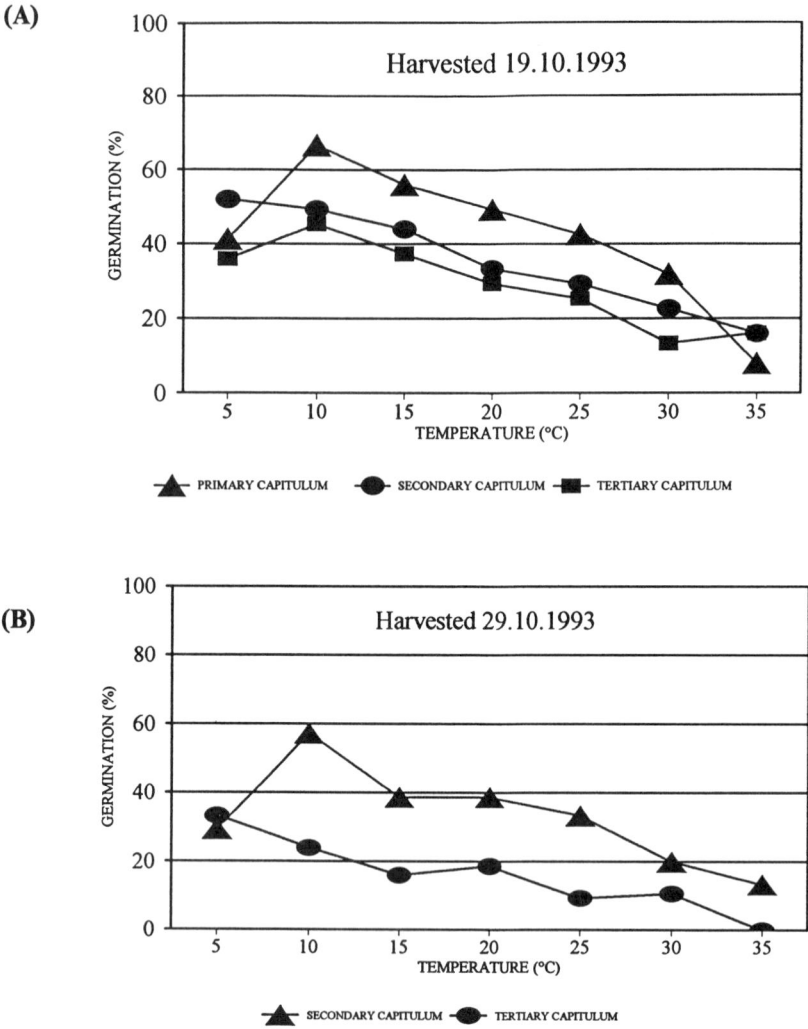

(B)

Figure 2. The effect of temperature on the germination of seeds harvested from different capitula of *Callistephus chinensis* 'Aleksandra' plants. Mature seeds were harvested on (A) 19 October 1993 and (B) 29 October 1993. Note symbols are used differently in A and B

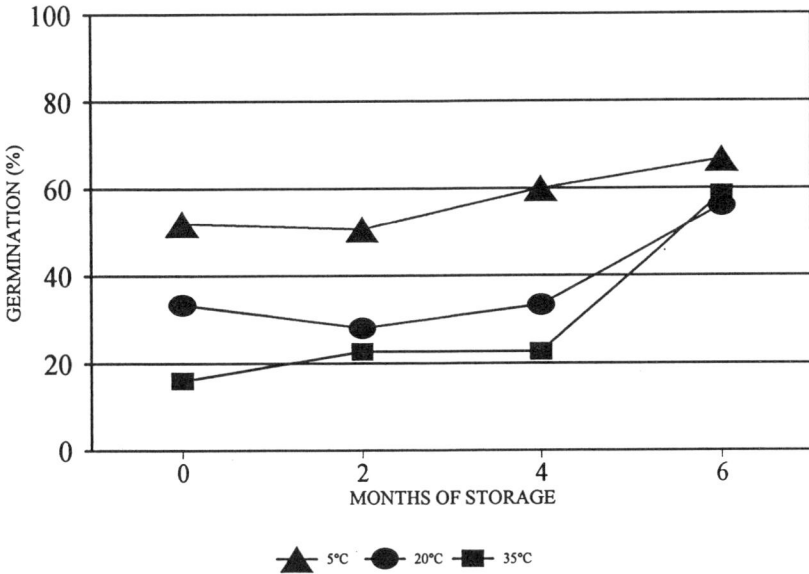

Figure 3. Germination of *Callistephus chinensis* 'Aleksandra' seeds at 5, 20 and 35°C harvested from the secondary capitulums and stored for six months

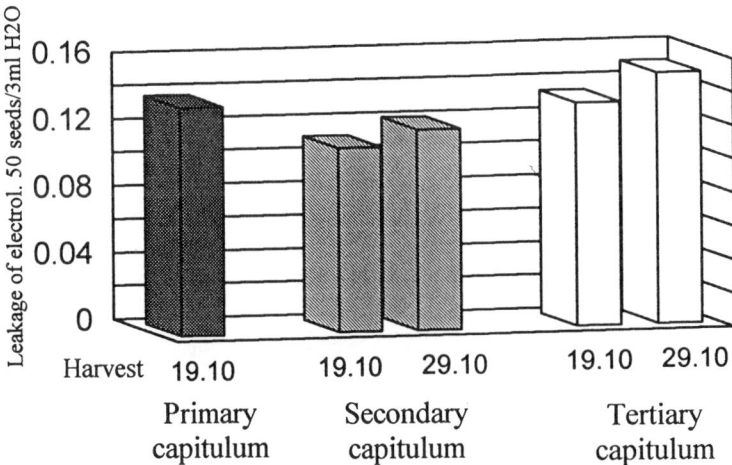

Figure 4. Leakage of electrolytes from 50 seeds soaked in 3 ml H_2O, harvested on 19 and 29 October 1993 from different parts of plants

(A)

In vivo ACC oxidase activity

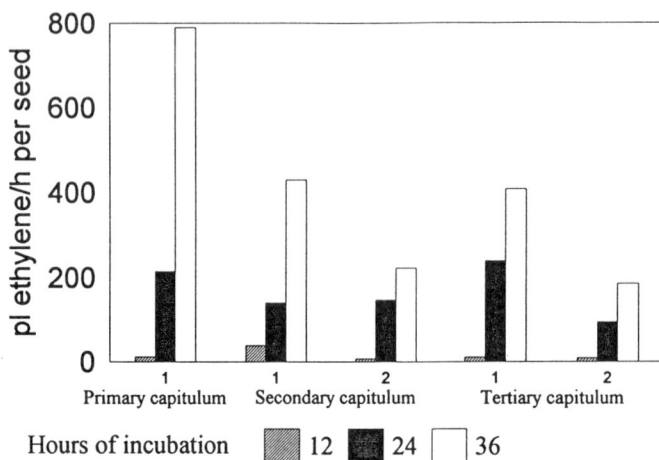

(B)

Ethylene release

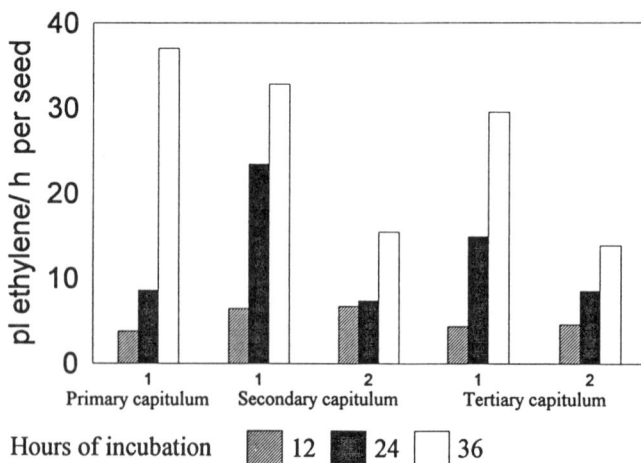

Figure 5. (A) *In vivo* ACC oxidase activity and (B) Ethylene release in aster seeds harvested in (1) 19 and (2) 29 October1993 from primary, secondary and tertiary capitulums

Acknowledgment

The investigations were sponsored by KBN, Grant Nr PBO 132/S3/93/05.

References

Carette, B. and Laurent, E. 1989. *Acta Horticulturae* 253: 31–44.
Chojnowski, M. and Grzesik, M. 1994. *Breeding and propagation of flower and their production and trade,* pp. 110–114 (ed. F. Kobza). University of Brno.
Duczmal, K.W. 1989. *Biuletyn Instytutu Hodowli i Aklimatyzacji Roslin* 169: 49–58.
Grzesik, M. and Chojnowski, M. 1994. *ISTA/ISHS Symposium. Technological Advances in Variety and Seed Research. Book of Abstracts* p.33 (eds W.J. van der Burg and R.J. Bino). ISTA, ISHS, Wageningen.
ISTA. 1985. *International Seed Testing Association. International Rules,* ISTA Zurich, Switzerland.
Kobza, F. 1986. *Eucarpia. Breeding and propagation of ornamental plants.* Lednice: 69–74.
Khan, 1994. *Journal of the American Society for Horticultural Science* 119: 5.
Matous, M. 1986. *Eucarpia. Breeding and propagation of ornamental plant.* Lednice: 64–68.
Vis, C. 1980. *Seed Science and Technology* 8: 495–503.

16. The Influence of JA-Me and ABA on Induction of Somatic Embryogenesis in *Medicago sativa* L. Tissue Cultures

J. KEPCZYNSKI and I. FLOREK

Department of Plant Physiology, University of Szczecin, Felczaka 3a, 71–412 Szczecin, Poland

Abstract

The influence of methyl jasmonate (JA-Me), a new plant growth regulator, and abscisic acid (ABA) on growth of callus and somatic embryogenesis in petiole-derived tissue cultures of *Medicago sativa* has been investigated.

A two-step protocol based on SH medium for somatic embryogenesis was used in the experiments. JA-Me and ABA (5, 50, 450 μM) were added to induction medium before autoclaving. The weight of plant tissue (petioles with callus) after 10 days of incubation on induction medium, prior to transfer to differentiation medium, was determined. After 7 weeks on differentiation medium, cultures were weighed and somatic embryos were counted.

Both JA-Me and ABA are considered to be inhibitors of the induction of callus growth and somatic embryogenesis in *Medicago sativa* cultures. An exposure to those inhibitors during induction stage reduced the number of somatic embryos obtained. JA-Me showed to be less active in the inhibition of callus growth and somatic embryogenesis than ABA.

Introduction

Methyl jasmonate (JA-Me) is known as an odoriferous compoud in the essential oil of *Jasminum grandiflorum* L. and *Rosmarinus officinalis* L. (Demole *et al.*, 1962; Crabalona, 1967). The jasmonates have been proposed as naturally occuring plant growth regulators due to their wide distribution (Meyer *et al.*, 1984) and effects on many physiological processes (Parthier, 1990; Parthier *et al.*, 1991). It has been suggested that the action of the jasmonates in the regulation of some processes such as seed germination, leaf senescence, chlorophyll degradation and tuberisation may be similar to that of abscisic acid (ABA) (Corbineau *et al.*, 1988; Kepczynski and Bialecka, 1994; Staswick, 1992; Ueda *et al.*, 1981). There are few reports demonstrating the influence of jasmonates or ABA on growth and development in plant tissue cultures. According to data reported by Ueda and Kato (1982) both JA-Me and ABA effectively inhibit cytokinin-induced soyabean callus growth. Methyl jasmonate exhibited an ABA-like effect by inhibiting callus formation from rice anthers (Yeh *et al.*, 1995). On the other hand it was observed that JA enhanced cell division and microcallus development in protoplast cultures of *Solanum tuberosum* (Ravnicar *et al.*, 1992) and stimulated the development of shoots on

R.H. Ellis, M. Black, A.J. Murdoch, T.D. Hong (eds.), Basic and Applied Aspects of Seed Biology, pp. 137–140.
© *1997 Kluwer Academic Publishers, Dordrecht. Printed in Great Britain.*

isolated meristems (Ravnicar and Gogala, 1990).

The aim of the study was to determine the effects of JA-Me in comparison to ABA on induction of somatic embryogenesis in *Medicago sativa* L. petiole-derived tissue cultures.

Materials and Methods

Plants of *Medicago sativa* were grown in a growth chamber under a 16 h photoperiod. A two stage protocol for somatic embryogenesis was used in the experiments. Petiole-derived explants of about 15 mm in length (5 per Petri dish) were incubated for 10 d on an induction medium containing 5 mg/l 2,4-dichlorophenoxyacetic acid (2,4-D) and 1 mg/l kinetin. Then they were transferred to a differentiation medium without these regulators. JA-Me or ABA (5, 50, 450 µM) were added to the induction medium before autoclaving.

The weight of plant tissue (petioles with callus) after 10 days of incubation on the induction medium, prior to transfer to the differentiation medium, was determined. After 7 weeks on the differentiation medium, cultures were weighed and somatic embryos were counted.

The results (presented as mean values) were analyzed by the Duncan test.

Results and Conclusions

The weight of tissue cultures, measured after 10 d incubation on the induction medium, was lower in the presence of JA-Me or ABA than in control (Fig. 1). There was observed a negative correlation between the rate of growth inhibition and the level of JA-Me or ABA applied.

The addition of JA-Me to the induction medium had nearly no effect on subsequent growth of callus on the differentiation medium without growth regulators (Fig. 2). ABA caused a reduction in callus growth on the differentiation medium when applied during the induction stage at the concentrations of 50 or 450 µM. The exposure to various concentrations of JA-Me or ABA during the induction stage reduced the number of somatic embryos obtained (Fig. 3). Both compounds were already inhibitory at 5 µM. The application of ABA at 450 µM during the induction stage caused virtually no somatic embryos to be obtained.

Both methyl jasmonate and abscisic acid can be considered as inhibitors of induction of callus growth and somatic embryogenesis in *Medicago sativa* L. tissue cultures. JA-Me, however, seems to be less active than ABA.

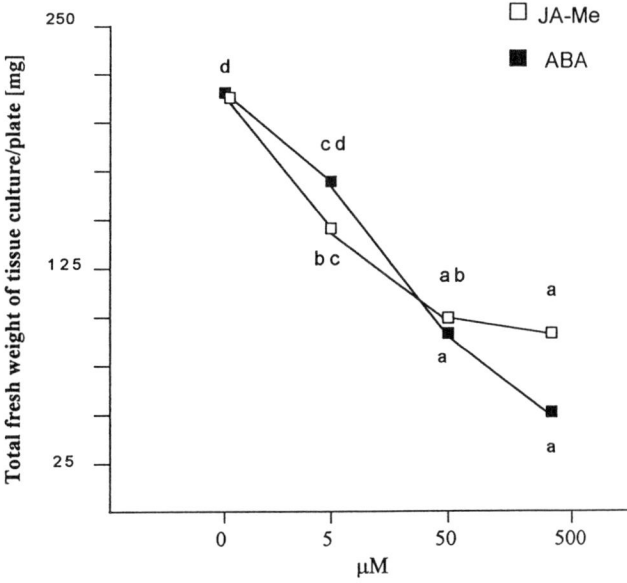

Figure 1. The effect of JA-Me or ABA on growth in *Medicago sativa* L. tissue cultures. The weight of tissue cultures was determined after 10 d incubation on induction medium prior to transfer to differentiation medium

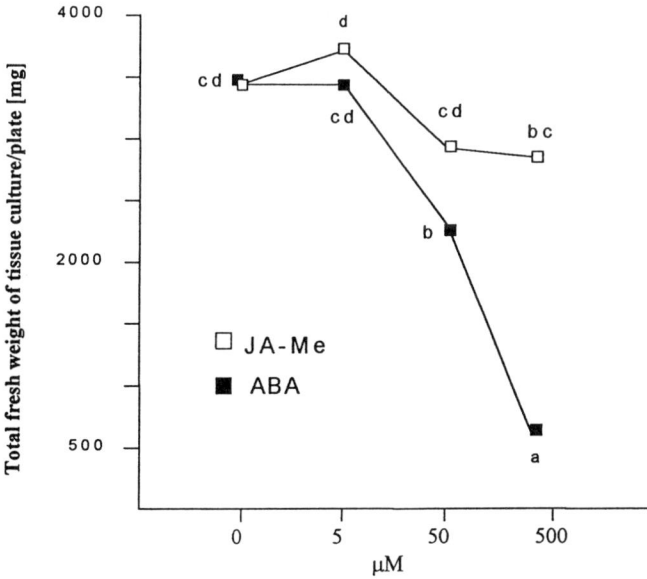

Figure 2. The effect of JA-Me or ABA applied during induction stage of somatic embryogenesis on subsequent growth of callus on differentiation medium in *Medicago sativa* L. tissue cultures

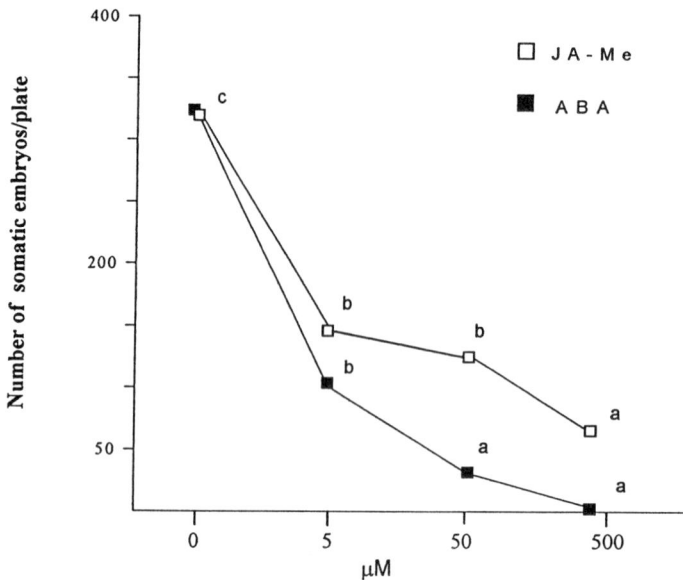

Figure 3. The effect of JA-Me or ABA on induction of somatic embryogenesis in *Medicago sativa* L. tissue cultures

References

Corbineau, F., Rudnicki, R.M. and Come, D. 1988. *Journal of Plant Growth Regulation* 7: 157–169.

Crabalona, L. 1967. *Comptes Rendus de l'Academie des Sciences (Paris) Serie C* 264: 2074–2076.

Demole, E., Lederer, E. and Mercier, D. 1962. *Helvetica Chimica Acta* 45: 675–685.

Kepczynski, J. and Bialecka, B. 1994. *Journal of Plant Growth Regulation* 14: 211–216.

Meyer, A., Miersch, O., Buttner, C., Dathe, W. and Sembdner, G. 1984. *Journal of Plant Growth Regulation* 3: 1–8.

Parthier, B. 1990. *Journal of Plant Growth Regulation* 9: 57–63.

Parthier, B., Brukner, C., Dathe, W., Hause, B., Herrmann, G., Knofel, H.D., Kramell, H.M., Kramell, R., Lehmann, J., Miersch, O., Reinbothe, S., Sembdner, G., Wasternack, C. and Zur Nieden, U. 1991 In: *Progress in Plant Growth Regulation. Proceedings of the 14th International Conference on Plant Growth Substances* pp. 276–286 (eds C.M. Karssen , L.C. Van Loon and D. Vreugdenhil). Dordrecht: Kluwer Academic Publishers.

Ravnicar, M. and Gogala, N. 1990. *Journal of Plant Growth Regulation* 9: 233–236.

Ravnicar, M., Vilhar, B. and Gogala, N. 1992. *Journal of Plant Growth Regulation* 12: 29–33.

Staswick, P.E. 1992. *Plant Physiology* 99: 804–807.

Ueda, J., Kato, J., Yamane, H. and Takahashi, N. 1981. *Physiologia Plantarum* 52: 305–309.

Ueda, J. and Kato, J. 1982. *Physiologia Plantarum* 54: 249–252.

Yeh, C.-C., Tsay, H.-S., Yeh, J.-H., Tsai, F.-Y., Shih, C.-Y. and Kao, C.-H. 1995. *Journal of Plant Growth Regulation* 14: 23–28.

17. Ultrastructural and Chemical Characterization of *Bulnesia* Seeds

S. MALDONADO[1], C. LIMA[2], M. ETCHART[2], V. LAINEZ[1] and R.M. DE LEDERKREMER[2]

[1]*Institute de Recursos Biologicos, INTA 1712, Castelar;* [2]*Departamento de Química Orgánica, Facultad de Ciencias Exactas y Naturales, Universidad de Buenos Aires, Pabellón 2, Ciudad Universitaria, 1428, Buenos Aires, Argentina*

Abstract

Bulnesia, a genus of the Zygophyllaceae comprises two subgenera, subgenus *Bulnesia* and subgenus *Gonopterodendron*. One representative species of each subgenus was studied: *B. schickendantzii* and *B. bonariensis* respectively. *B. schickendantzii* endosperm consists of four to six cell layers. Cell walls are composed of a middle lamella, a very thick outer wall and a thin inner wall. *B. bonariensis* endosperm is reduced to one layer of cells. The cell walls are formed by a middle lamella and a thinner wall. Protein bodies are observed in the embryo of both species and in the endosperm of *B. schickendantzii*. Lipid vesicles are present in embryo and endosperm of both species. Phytin crystals, present in the protein bodies of *B. schickendantzii* are rare in the protein bodies of *B. bonariensis*. *B. schickendantzii* endosperm contains the majority of sugars and proteins. In *B. bonariensis* the higher percentage of carbohydrate and protein is found in the embryo. Determination of neutral sugars was performed after acid hydrolysis. Sugars were analyzed as alditol acetates by gas–liquid chromatography. Arabinose, xylose, galactose and glucose are the main monosaccharides but the ratio differs in tissues of both species. Soluble proteins were analyzed by polyacrylamide gel electrophoresis showing a different profile in tissues of both species. Very few bands are glycosylated. The differences observed between both species encourage further comparative studies on *Bulnesia* germination.

Introduction

Bulnesia, a South American genus of Zygophyllaceae, includes eight species, five of them native of Argentina. The genus had been divided in two subgenera: the subgenus *Bulnesia* consists of four species, *B. retama, B. chilensis, B. foliosa* and *B. schickendantzii*, mainly characterized by endospermic seeds and the subgenus *Gonopterodendron* composed of *B. arborea, B. carrapo, B. sarmientoi* and *B. bonariensis*, characterized by non-endospermic seeds (Palacios and Hunziker, 1984).

Up to date there are no studies of *Bulnesia* seeds or references indicating the existence of different types of seeds in species of this genus.

We now report the first studies on seeds of a representative species of each subgenus, i.e. *B. schickendantzii* (subgenus *Bulnesia*) and *B. bonariensis*

R.H. Ellis, M. Black, A.J. Murdoch, T.D. Hong (eds.), Basic and Applied Aspects of Seed Biology, pp. 141–152.
© *1997 Kluwer Academic Publishers, Dordrecht. Printed in Great Britain.*

(subgenus *Gonopterodendron*). We have analyzed the ultrastructure of the embryo and the endosperm of the seeds and we have also determined the content of total sugar, soluble proteins and cell-wall neutral sugars of both tissues. A correlation between ultrastructural and chemical features with the subgeneric division is proposed.

Materials and Methods

Material

Bulnesia bonariensis. Argentina, Prov. La Rioja, Dep. Independencia. JHH 12491. Seeds from 30 individuals.

Bulnesia schickendantzii. Argentina, Prov. La Rioja, Dep. San Blas de los Sauces. JHH 11989. Seeds from 12 individuals. Prov. Catamarca, Dep. Tinogasta. JHH 11991. Seeds from 21 individuals.

Specimen Preparation

Small blocks of tissue cut from different regions of seeds were fixed for 4–8 h in 2.5% glutaraldehyde in 0.1 M phosphate buffer pH 7.5 at 4°C. For light microscopy the fixed tissue was dehydrated through an ethanol series and embedded in 2-butoxyethanol methacrylate (JB4 Polyscience, Inc., Warrington PA). Both fresh unfixed tissue and tissue that had been embedded for TEM were also used for light microscopy. The staining procedure used included acid fuchsin and toluidine blue O (Feder and O'Brien, 1968); Coomassie brilliant blue (Pearse, 1985); fast green FCF (Fulcher *et al.*, 1972); iodine–potassium iodide and sulphuric acid; periodic acid–Schiff (PAS) O'Brien and McCully, 1981) and Sudan black B (Bronner, 1975).

For TEMs tissue fixed in aldehyde as described above was post-fixed in 1% OsO_4 in the same buffer for 4 h, dehydrated in Spurr's resin. Sections were mounted on grids coated with Formvar and then carbon, stained in uranyl acetate followed by lead citrate, and examined in a JEOL 1200EX II transmission electron microscope.

For SEM, whole seeds or pieces of seeds, cut longitudinally and transversely, were mounted onto aluminium stubs, gold coated and viewed with a JEOL 35 scanning electron microscope.

Protein and Carbohydrate Extraction

The endosperm and embryo were mechanically separated from 46 seeds of *B. schickendantzii* and 36 seeds of *B. bonariensis* previously soaked in alcohol. The tissues were air-dried until they reached a constant weight. The following values were obtained: *B. schickendantzii* endosperm 115.6 mg, *B. schickendantzii* embryo 32.27 mg, *B. bonariensis* endosperm 54.84 mg, *B. bonariensis* embryo

901.3 mg. The tissues were ground into a fine powder in a mortar and the sugar content was determined. The tissues were further defatted three times with chloroform: methanol (2:1), sonicated for 10 min and centrifuged for 15 min at 10 000 RPM. The pellet was air-dried, suspended in 0.1 M citrate buffer (pH 6) 2N NaCl, sonicated for 10 min with stirring for 1 h at room temperature. The extract was separated after centrifugation at 10000 RPM for 20 min. The treatment was repeated twice. The supernatants were dialyzed against water for 36 h and concentrated under reduced pressure. The protein content was determined and further analyzed by sodium dodecyl sulphate polyacrylamide gel electrophoresis (SDS, PAGE). The precipitates were washed with water and air-dried to the following constant dry weights: *B. schickendantzii* endosperm 30.44 mg, *B. schickendantzii* embryo 12.81 mg, *B. bonariensis* endosperm 23.25 mg and *B. bonariensis* embryo 242.58 mg. A sample (5 mg) of each tissue was hydrolyzed under different conditions (see below). The monosaccharides obtained were analyzed as alditol acetates by gas–liquid chromatography.

Analytical Method

Total sugar was determined by the phenol–sulphuric acid method with glucose as standard (Dubois *et al.,* 1956).

Protein content was determined using bovine serum albumin as standard (Bradford, 1976). SDS-PAGE was performed on 12% polyacrylamide slab gels in the presence of 0.1% SDS (Studier, 1973). A sample of 40 µg of protein was loaded in each well. Gels were stained with Coomassie blue R 250 (Korn and Wright, 1973). BIO-RAD Prestained SDS-PAGE Standard (Low Range) were used as marker proteins. For glycoprotein detection gels were stained with the periodic acid/Schiff reagent (Korn and Wright, 1973).

Sugar Composition

Component sugars were determined after acid hydrolysis of the samples with 2 M trifluoroacetic acid (TFA 5 mg/ml) at 105°C for 3 h. Inositol was used as inner standard. The insoluble precipitate was separated by centrifugation at 6000 RPM for 10 min. The solution was evaporated under reduced pressure with repeated additions of water (condition a). The residue was air-dried and further hydrolyzed with 1.3 ml TFA (99%+ purity) with stirring at 37°C for 18 h. An aliquot (0.5 ml) was transferred to a glass vial. Water (1 ml) and inositol were added. This sample, for hemicellulose hydrolysis, was heated at 100°C for 30 min and then evaporated to dryness (condition b). A second aliquot (0.5 ml) was transferred to a similar vial, water (0.1 ml) was added and heated for 30 min. The addition of water (0.1 ml) was repeated at 30 min intervals until a total volume of 0.5 ml was added, inositol standard was used as above and heating continued for a further 2 h before evaporation to dryness (condition c) (Morrison, 1988).

Capillary gas–liquid chromatography (glc) was performed with a Hewlett-

Packard 5890 gas chromatograph with nitrogen as the carrier gas. The following conditions were used: SP 2330 fused silica column (0.25 mm × 15 m), flow rate 4 ml/min, t_i 250°C, t_c 220°C, t_d 250°C. Sugars were analyzed as alditol acetates (Sloneker, 1972).

Results

Histology

The two areas of food reserves in *Bulnesia* seeds are the endosperm and the embryo (Figs. 1 and 3).

Endosperm

In *B. schickendantzii* the endosperm consists of four to six layers of living cells (Figs. 1 and 2). Endosperm cells have very thick cell walls except in the areas of primary pit fields. At the electron microscope level, the walls consist of three distinct layers: middle lamella, thickened outer wall and thin granular inner wall (Figs. 2 and 5). Cell walls are the site of polysaccharide storage. The cells themselves store abundant lipids and proteins in the form of lipid vesicles and protein bodies (Figs. 2 and 5). Protein bodies consist of a proteinaceous matrix containing one or more globoid crystals (Figs. 2 and 5). The number and size of globoid crystals vary between different protein bodies in the same endosperm cell. Globoid crystals, which are hard and do not infiltrate well with resin, were often removed during sectioning, leaving a hole in the section (Figs. 2 and 5). No ribosomes or endomembranes are discernible in the mature endosperm. Vacuoles are present occasionally.

In *B. bonariensis* the endosperm consists of one to two layers of living cells (Fig. 3). Cell walls differentiate two layers: middle lamella and comparatively thinner cell walls. The cells store lipids in the form of lipid vesicles (Fig. 7). Cytoplasm stains positively with the protein stains but protein bodies are absent. Mitochondria and ribosomes are discernible (Fig. 7).

In both species, cells of the outermost layer of endosperm next to the seed coat have a layer of cuticle on the surface of the outer cell walls.

Cell walls are stained with the PAS, fast green, and toluidine blue reagents. The proteinaceous regions stained positively with the Coomassie blue, fast green FCF, and acid fuchsin. Globoid crystals stain metachromatically with toluidine blue. The cuticle and the storage lipid vesicles stain with Sudan black.

Embryo

In the seeds of both species of *Bulnesia*, the embryo consists of a hypocotyl–radicle axis and two cotyledons. The tissues of the cotyledons and axis are still meristematic, but their position and cytological characteristics indicate their

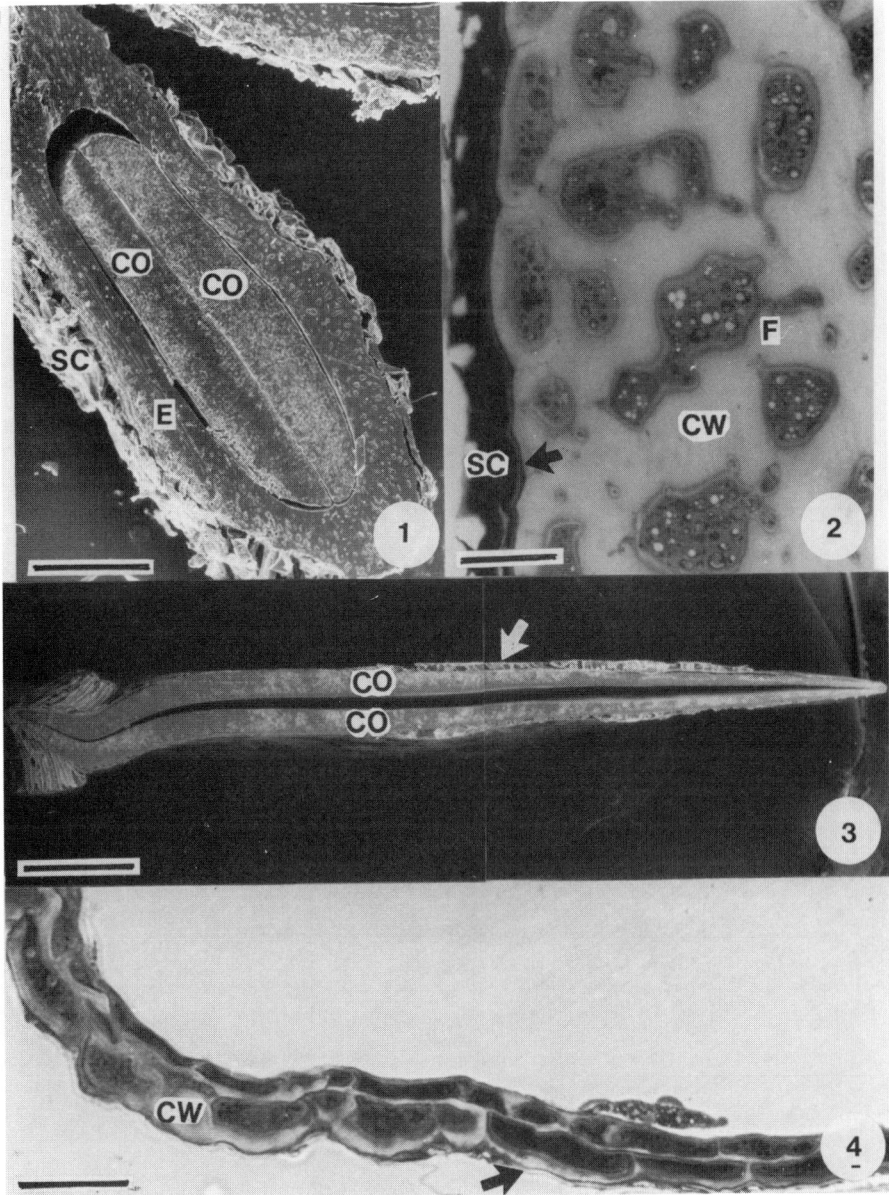

Figures 1–4. Fig. 1. Scanning electron micrograph of a transverse section of a *B.schickendantzii* seed. CO, cotyledon; E, endosperm, SC, seed coat. Scale bar = 500 µm. *Fig. 2.* Light microscopy of a transverse section of the endosperm of *B. schickendantzii*. CW, cell wall; F, primary pit field; SC, seed coat. Endosperm cells show protein bodies and nucleus. On the outer cell layer both the cuticle layer (black arrow) and the cell walls are present. The hard globoid crystals may have chipped out during sectioning, thereby forming a hole inside protein bodies. Scale bar = 100 µm. *Fig. 3.* Scanning electron micrograph of a transverse section of a *B. bonariensis* seed. CO, cotyledon; white arrow indicates seed coat plus endosperm. Scale bar = 1 mm. *Fig. 4.* Light microscopy of a section of the endosperm of *B. bonariensis*. CW, cell wall; black arrow indicates cuticle layer. Scale bar = 70 µm

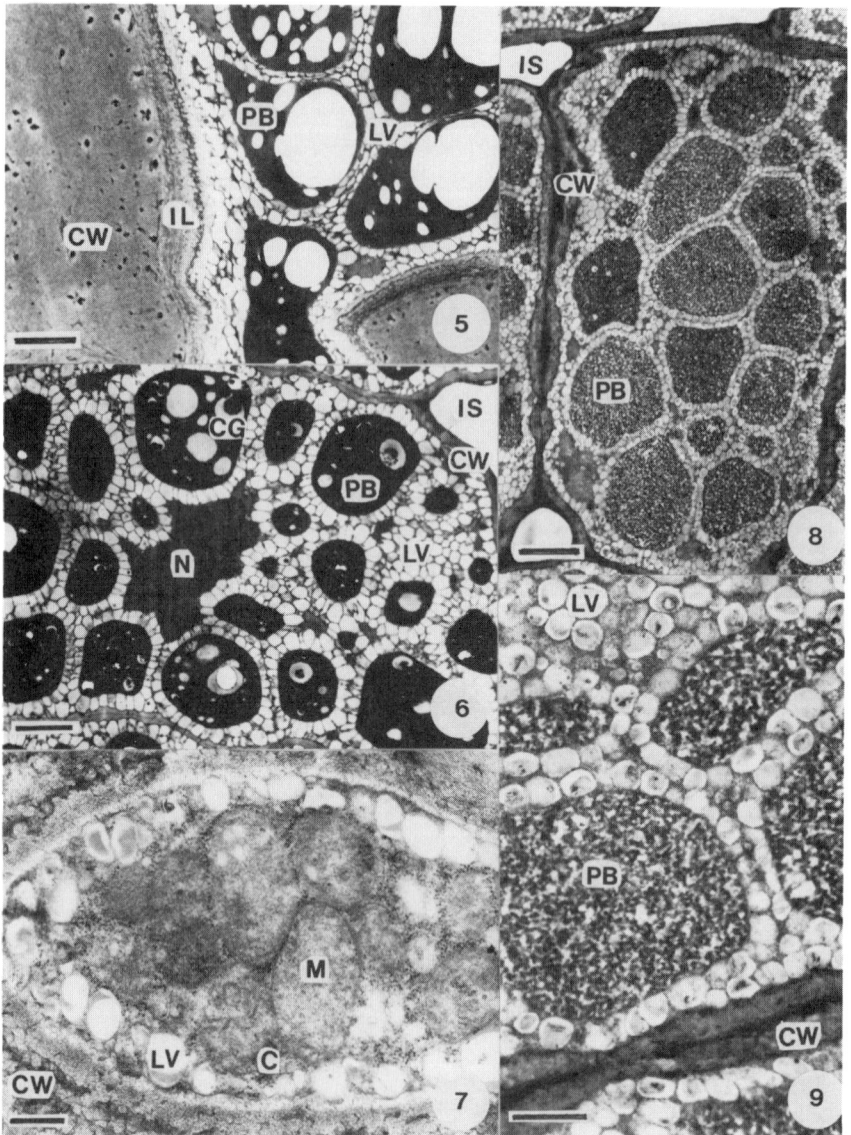

Figures 5–9. Transmission electron micrographs. *Fig. 5.* Cell of the endosperm of *B. schick-endantzii.* CW, cell wall; IL, inner layer of the cell wall; LV, lipid vesicle; PB, protein body . Globoid crystals have chipped out during sectioning, forming holes in the sections. Scale bar = 2 μm. *Fig. 6.* Cell of the palisade tissue of a cotyledon of *B. schickendantzii.* CG, crystal globoid; CW, cell wall; LV, lipid vesicles; N, nucleus; PB protein body. Scale bar = 2 μm. *Fig. 7.* Cell of the endosperm of *B. bonariensis.* C, cytoplasm; CW, cell wall: LV, lipid vesicle; M, mitochondrion. Ribosomes are seen in the cytoplasm. Scale bar = 2 μm. *Fig. 8.* Cell of the embryo of *B. bonariensis.* CW, cell wall; IS, intercellular space; PB, protein body. Scale bar = 1 μm. *Fig. 9.* Detail of Fig. 8. CW, cell wall; LV. lipid vesicle; PB protein body. Scale bar = 1 μm

fate in the developing seedling. The mesophyll (palisade and spongy tissues), protoderm and procambium are distinguished in the cotyledons and the protoderm, procambium and ground meritstem are distinguished in the axis.

All embryo cells have thin primary cell walls (Figs. 6, 8 and 9) that stain with the PAS reaction, fast green and toluidine blue. No starch was present in the embryo, as determined by a lack of staining with iodine–potassium iodide. Protein bodies and lipid vesicles occupy most of the cytoplasm of all embryo tissues (Figs. 6, 8 and 9). The presence of lipid was determined by staining with Sudan black B, whereas the presence of protein in the protein bodies was determined by staining with Coomassie blue, acid fuchsin and fast green FCF. In *B. schickendantzii*, protein bodies contain two or more phytin crystals in a dense proteinaceous matrix. In *B. bonariensis* the proteinaceous matrix is loose, frequently lacking phytin crystals. Protein crystalloids are absent in the protein bodies of both species.

Carbohydrate Analysis

The total content of carbohydrate was different in the tissues of both species. *B. schickendantzii* shows more sugar in the endosperm (730.18 µg/seed) than in the embryo (96.6 µg/seed). *B. bonariensis* presents a higher content of sugars in the embryo (5466 µg/seed) than in the endosperm (1741 µg/seed) (Fig. 10).

Sugar analysis by glc was performed after acid hydrolysis under different

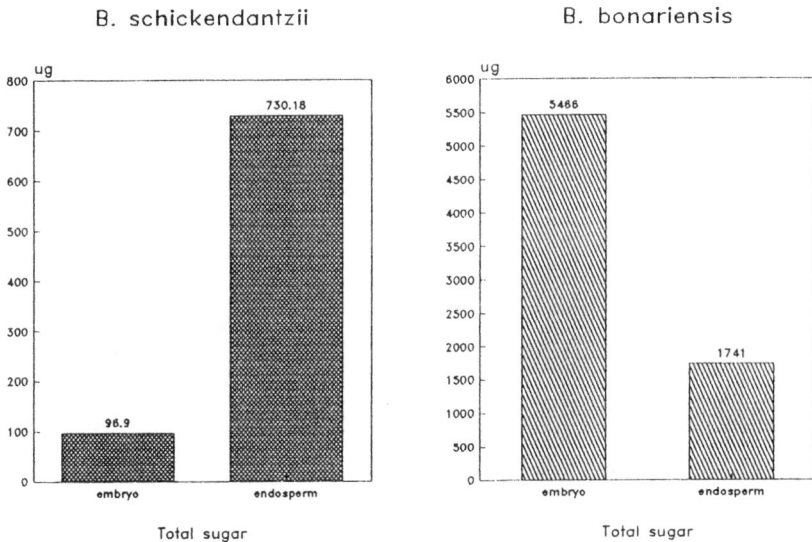

Figure 10. Total sugar (µg/seed) was determined by the phenol–sulphuric acid method with glucose as standard

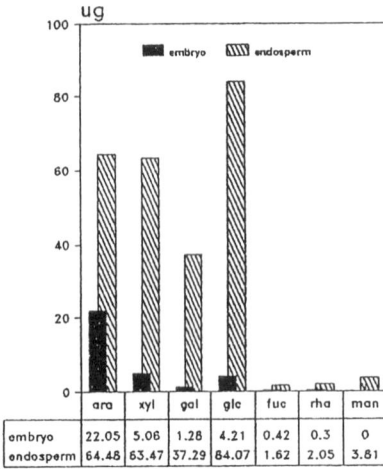

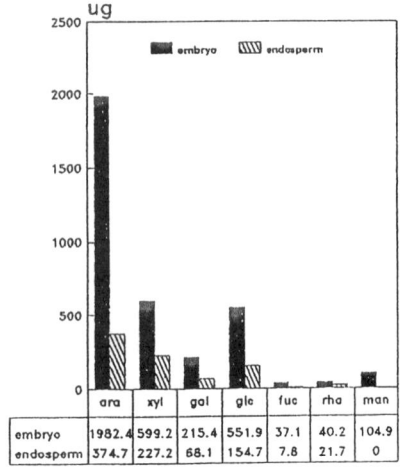

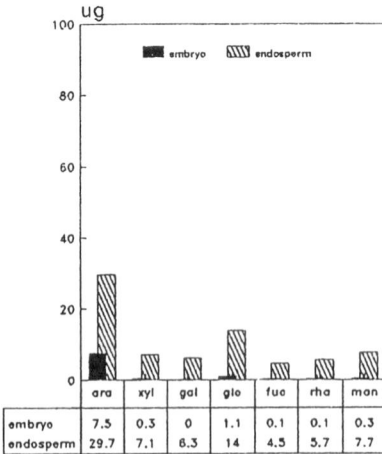

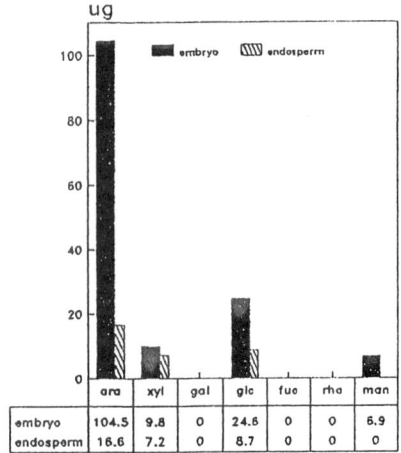

Figure 11. Determination of cell wall neutral sugars (μg/seed) after acid hydrolysis under three conditions: **(a)**, 2 M trifluoroacetic acid (TFA) at 105°C, 3 h; **(b)**, TFA 99%, 37°C, 18 h. Aliquot (0.5 ml), water (1 ml) added, at 100°C, 30 min; **(c)**, aliquot (0.5 ml), water (0.5 ml) added, at 100°C, 2 h 30 min. Sugars were analyzed as alditol acetates by gas–liquid chromatography. Inositol was used as inner standard

(c)

B. schickendantzii   B. bonariensis

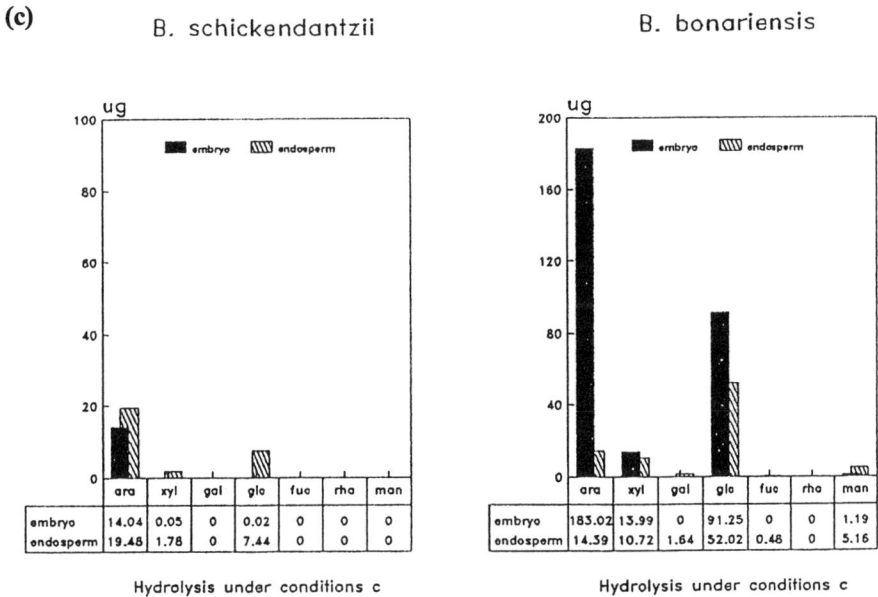

embryo	ara	xyl	gal	glo	fuo	rha	man
embryo	14.04	0.05	0	0.02	0	0	0
endosperm	19.48	1.78	0	7.44	0	0	0

Hydrolysis under conditions c

	ara	xyl	gal	glo	fuo	rha	man
embryo	183.02	13.99	0	91.25	0	0	1.19
endosperm	14.59	10.72	1.64	52.02	0.48	0	5.16

Hydrolysis under conditions c

Figure 11 (cont.)

conditions. Under condition **a**, the endosperm and embryo of both species presented arabinose, xylose, galactose and glucose as main monosaccharides. Mannose, fucose and rhamnose appeared in lower amounts (Fig. 11). Further hydrolysis under condition **b** showed the same carbohydrates as those under condition **a** but in different proportions (Fig. 11). Finally, the hydrolysis under condition **c** presented only arabinose, xylose and glucose (Fig. 11).

Protein Analysis

With regard to the protein content, *B. schickendantzii* showed 0.035 mg/seed in the embryo and 0.049 mg/seed in the endosperm (Fig. 12). *B. bonariensis* embryo contained 1905 mg/seed whereas the endosperm showed 124 mg/seed (Fig. 12).

The electrophoretic analysis showed that the protein profile of the tissues differed in both species. In *B. schickendantzii* the embryo proteins separated into eight bands ranging between 22.3 kDa and 109.5 kDa (Fig. 13, lane A). For convenience the bands were designated: a (109.5 kDa), b (73.2 kDa), c (66.0 kDa), d (53.0 kDa), e (18.0 kDa), h (35.4 kDa), i (30.9 kDa), k (22.3 kDa). The endosperm proteins resolved into nine bands ranging from 19.7 kDa to 106.0 kDa (Fig. 13, lane **B**). They were designated as follows: a (106.0 kDa), b (73.2

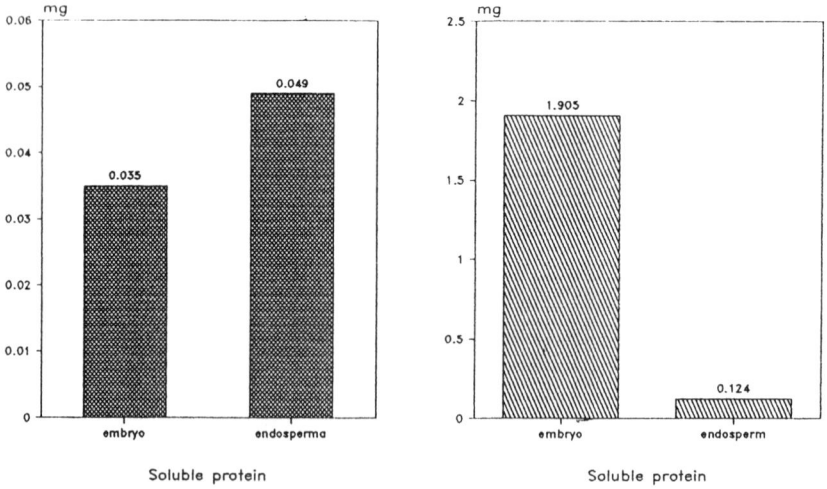

Figure 12. Soluble protein content (mg/seed) isolated with buffer citrate 0.1 M, pH 6, 2 M NaCl. Proteins were determined by the Bradford assay using bovine serum albumin as standard

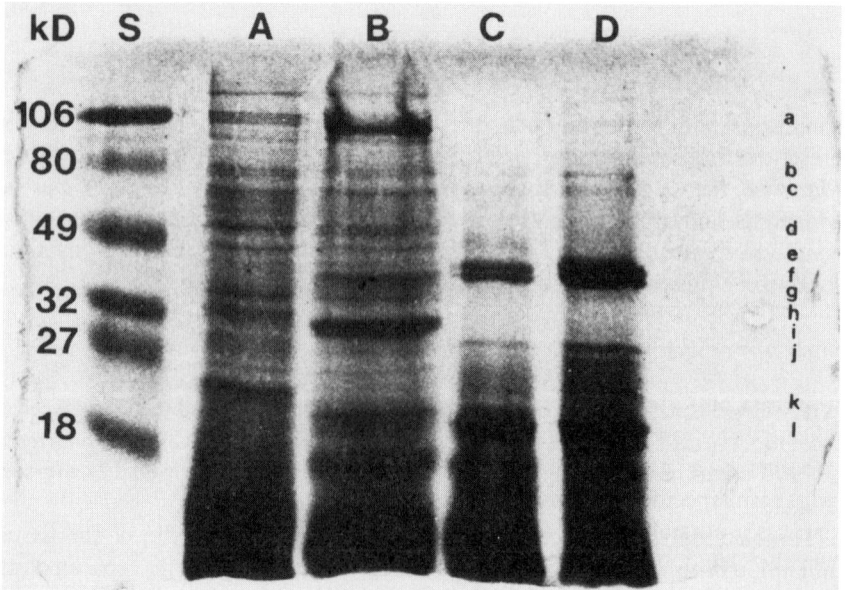

Figure 13. Polyacrylamide gel electrophoresis was performed on 12% polyacrylamide with 0.1% SDS. Gel was stained with Coomassie blue. Lane S: mol wt standards; lane A: *B. schickendantzii* embryo proteins; lane B: *B. schickendantzii* endosperm proteins; lane C: *B. bonariensis* embryo proteins; and lane D: *B. bonariensis* endosperm proteins. Small letters indicate protein bands referred to in the text.

kDa), c (66.0 kDa), d (53.0 kDa), e (48.0 kDa), g (40.2 kDa), h (35.4 kDa), i (30.9 kDa) and l (19.7 kDa). Both tissues had in common the bands a, b, c, d, e, h and i. The endosperm presented a band at around 40 kDa (protein g) and at around 20 kDa (protein l) which were not found in the embryo. The embryo, as well, showed a low mol. wt. band at 22.3 kDa (protein k). Only two bands were glycosylated, protein d in the embryo and protein g in the endosperm. In *B. bonariensis* the embryo protein separated into four bands ranging between 19.4 kDa and 42.6 kDa (Fig. 13, lane **C**). There were observed darkly stained bands at 42.6 kDa (protein f) and at 40.2 kDa (protein g), a lightly stained band at 28.1 kDa (protein j) and a low molecular weight band at 19.4 kDa (protein l). The endosperm proteins resolved into six bands ranging from 19.9 kDa to 77.6 kDa (Fig. 13, lane **D**). The profile showed darkly stained bands at 42.6 kDa (protein f), 40.2 kDa (protein g), and at 19.9 kDa (protein l). Lightly stained bands appeared at 77.6 kDa (protein b), at 67.1 kDa (protein c) and at 28.1 kDa (protein j). Proteins f and g appeared in both tissues as prominent bands. Protein j was the single glycosylated protein, seen in both tissues but in the endosperm was stronger stained than in the embryo. A low molecular weight protein (l) occurred in both tissues. Proteins b and c of high molecular weight were found only in the endosperm.

Discussion

This is the first report about seeds differing in the structure and the type of reserve accumulation in species of the same genus of Angiosperms. Palacios and Hunziker (1984) have described a taxonomic revision of the genus *Bulnesia* proposing the division in two subgenera on the account of the presence or absence of endosperm in the seed: subgenus *Bulnesia* having albuminous seeds and subgenus *Gonopterodendron* characterized by exalbuminous seeds.

We have studied one representative species of each subgenus, i.e. *B. schickendantzii* (subgenus *Bulnesia*) and *B. bonariensis* (subgenus *Gonopterodendron*), and have found that both species have endosperm in the mature seeds although in a very different proportion. This, together with the histological and chemical properties of both species seems to justify the division in two subgenera.

Four common endosperm-specific bands were observed in both species (proteins b, c, g). According to De Mason *et al.* (1985), the presence of endosperm-specific proteins is consistent with the hypothesis that the endosperm is a possible source of hydrolytic enzymes involved in autocatalysis. On the other hand, *B. bonariensis* embryo and *B. schickendantzii* endosperm, the most important tissues in reserve storage, presented two common bands (protein g and l).

Some protein bands were specific for each species: *B. schickendantzii* showed protein a, d, e, h, i, and l which did not appear in *B. bonariensis* and protein f and j were present only in *B. bonariensis*. These observations suggest that these proteins could be possible taxonomic-specific markers. No common bands were found in the embryos of the two species.

In the embryo and endosperm of *B. schickendantzii* and in the embryo of *B. bonariensis* the major extractable proteins were derived from the protein bodies; in those tissues, cytoplasm is very reduced. In *B. bonariensis* protein bodies are absent in the endosperm cells and extractable proteins are from organelles and cytoplasm (Chandra Sekhar *et al.*, 1988). At the ultrastructural level this tissue showed dense cytoplasm with ribosomes, mitochondria and ER.

Sugar analysis was performed after acid hydrolysis under different conditions. As expected, the proportion of sugars was higher in the endosperm than in the embryo for *B. schickendantzii* and the reverse was true for *B. bonariensis*. The endosperm and embryo of both species presented arabinose, xylose, galactose, and glucose as the main monosaccharides. Mannose, fucose and rhamnose appeared in a lower amount. Monosaccharide composition, expressed as µg of monosaccharide obtained per mg of dried tissue, showed that storage tissues contained arabinose and glucose as the most abundant. However, the ratio differed in each tissue. In *B. schickendantzii* arabinose and glucose appeared in the same proportion, whereas in *B. bonariensis*, the amount of arabinose was three times higher than glucose. This suggests that the cell walls of the embryo are constituted mainly of hemicellulosic polysaccharides and accords to Cleland (1971) and Roland *et al.* (1977), who consider that the presence of hemicellulose in primary cell walls plays an active role in elongation or extension processes during growth.

References

Bradford, M. 1976. *Analytical Biochemistry* 72: 248–254.

Bronner, R. 1975. *Stain Technology* 50: 1–4.

Chandra Sekhar, K.N. and De Mason, D.A. 1988. *American Journal of Botany* 75: 323–329.

Cleland, R. 1971. *Annual Review Plant Physiology* 21: 235.

Dubois, M., Gilles, K.A., Hamilton, J.K. Rebers, P.A. and Smith, F. 1956. *Analytical Chemistry* 28: 350–356.

De Mason, D.A., Sexton, R., Gorman, M. and Reid, J.S.G. 1985. *Protoplasma* 126:159–167.

Feder, R.G., and O'Brien, T.P. 1968. *American Journal of Botany* 55: 123–142.

Fulcher, R.G.N., O'Brien, T.P. and Simmonds, D.H. 1972. *Australian Journal of Biological Science* 25: 487–497.

Korn, E.D. and Wright, P.L. 1973. *Journal of Biological Chemistry* 248: 439.

Morrison, I.M. 1988. *Phytochemistry* 27: 1097–1100.

O'Brien, T.P. and McCully, M.E. 1981. *The Study of Plant Structure, Principles and Selected Methods*. Melbourne: Thermacarphy Pty. Ltd.

Palacios, R. and Hunziker, J.H. 1984. *Darwiniana* 21: 299–320.

Pearse, A.G.E. 1985. *Histochemistry. Theoretical and Applied*. 4th ed. New York: Churchill Livingstone.

Roland, J.C., Vain, B. and Reis, D. 1977. *Protoplasma* 91: 125.

Sloneker, J.H. 1972. *Methods in Carbohydrate Chemistry* 6: 20–24.

Studier, F.W. 1973. *Journal of Molecular Biology* 79: 237–248.

18. Role of Acyltransferases in the Effect of ABA and Osmoticum on Oil Deposition during Seed Development

R. RODRÍGUEZ-SOTRES[1], F. PACHECO-MOISÉS[1] and M. BLACK[2]

[1]Department of Biochemistry, Chemistry Faculty, Universidad Nacional Autónoma de México, Cd. Universitaria, Mexico City 04510, Mexico; [2]Division of Life Sciences, King's College London, University of London, Campden Hill Road, Kensington, London W8 7AH, UK

Abstract

Abscisic acid and water potential have been shown to regulate triacylglycerol (TAG) synthesis in developing wheat embryos (R. Rodríguez-Sotres and M. Black. 1994). To extend these observations, isolated maize embryos of 25 days were incubated in dilute basal medium or in basal medium supplemented with ABA, high mannitol, or both. At the end of the incubation period embryos were collected and ground to obtain either the total lipid fraction or a microsomal fraction. In some cases the embryos were fed with 2-[^{14}C]-acetate for 5 h prior to extraction. Results show that maize embryos stop synthesizing TAG after 24 h or more of incubation in dilute medium: changes after 12 h were not detected. TAG synthesis capacity is maintained in those embryos incubated in the presence of ABA, mannitol or both. The specific activity of the glycerol-3-phosphate acyltransferase, the lysophosphatidate acyltransferase and the diacylglycerol acyltransferase (DAG-AT) present in microsomal preparations from these embryos changed parallel to the observed TAG synthesis capacity, indicating that ABA and water potential are regulating their activity. The dose–response curves of the TAG accumulation and DAG-AT response to ABA and low water potential fell within values considered to be of physiological significance. We discuss these results in terms of the possible role of ABA and the osmotic environment in seed development.

Introduction

Abscisic acid (ABA) and the osmotic potential of the environment have been shown to regulate a number of responses related to the development of seeds and, in particular, to the reserve deposition processes (Bewley and Black, 1994). In a recent paper (Rodríguez-Sotres and Black, 1994), we reported that ABA and a low osmotic potential (LOP) are able to induce and mantain the synthesis of triacylglycerol (TAG) in isolated immature wheat embryos. Observations reported by other authors seem to indicate that this response is not exclusive to wheat (Finkelstein and Sommerville, 1989; Holbrook et al., 1992). Therefore, to extend these observations to other cereals, we decided to investigate if TAG synthesis is regulated by the levels of ABA and the osmotic environment in maize embryos as well. In addition, we decided to investigate if the activity of

R.H. Ellis, M. Black, A.J. Murdoch, T.D. Hong (eds.), Basic and Applied Aspects of Seed Biology, pp. 153–158.
© 1997 Kluwer Academic Publishers, Dordrecht. Printed in Great Britain.

the acyltransferases involved in the synthesis of TAG (acyl-CoA:glycerol-3-*sn*-phosphate-1-*sn*-acyltransferase, E.C. 2.3.1.15, Acyl-CoA:1-*sn*-lysophosphatidic acid-2-*sn*-acyltransferase, E.C. 2.3.1.51, and acyl-CoA:1,2-*sn*-diacylglycerol-3-*sn*-acyltransferase, E.C. 2.3.1.20) was changing in response to these two factors in a way that accounts, at least partly, for the observed changes in the TAG synthesis capacity of the embryos.

Material and Methods

Plants of maize (*Zea mays* crystal-yellow c5) were grown under field conditions in an experimental field of the National Institute for Agricultural and Forest Research (INIFAP) at Zacatepec, Morelos, Mexico. Seeds were a kind gift from Ing. Hugo Cordoba of the International Center for Development of Maize and Wheat at Texcoco, Mex., Mexico (CIMMYT). Young cobs in the maize plant were covered with paper bags (Lawson pollination bags, Delaware, USA) and manually pollinated after female flowering. The age of the seeds is reported as days post-pollination (dpp). At 23–25 dpp the cobs were removed from the plant and transported to the laboratory. During transportation (about 2 h) the cobs were kept on cotton pads wetted in a solution containing 20% polyethyleneglycol 6000, 30 mM sucrose and 10 µM ABA.

Seeds were removed from the cobs and surface sterilized by inmersion in a solution containing 20% polyethyleneglycol 6000 (PEG) and 1% commercial chlorine bleach for 15 minutes. After thorough washing with PEG solution the embryos were isolated from the seeds in a laminar flow bench.

A treatment consisted of 15 isolated maize embryos incubated in sterile 50 ml Erlenmeyer flask with 2.5 ml of a basal medium containing Murashige Skoog macro and micro nutrients and 4 mM morpholinoethanesulphonic acid (MES) at pH 5.5. Where indicated incubation medium was supplemented with 30 mM sucrose, 10 µM ABA, 600 mM mannitol or a combination of them. At the end of the indicated times the embryos were collected and ground in a mortar either in chloroform:methanol 1:1 or in Tris buffer 0.1 M pH 5.5. In some cases the embryos were fed with ^{14}C-acetate (NEM Dupont, 2 µCi/flask) for 5 h at the end of the indicated time and prior to extraction.

Total lipid fractions from incubated embryos were extracted as described by Rodríguez-Sotres and Black (1994) and microsomal fractions were prepared as described in Cao and Huang (1986).

The activity of the glycerol-3-phosphate acyltransferase (G3P-AT) was determined by the incorporation of ^{14}C]-glycerol-3-phosphate (NEM, Dupont) into chloroform soluble products (Bertrams and Heinz, 1981). Reaction mixture contained 0.1 M Tris buffer pH 7.5, 5 mM ^{14}C]-glycerol-3-phosphate (0.5 µCi), 5 mM $MgCl_2$, 40 µM oleoyl-CoA, and 5–10 µg of microsomal protein in a final volume of 0.15 ml. The reaction was stopped with 1 ml of chloroform:methanol 1:1 and the chloroformic phase was recovered, dried and counted by liquid scintillation. Lysophosphatidate acyltransferase (LPA-AT) was determined by the incorporation of 1-[^{14}C]-oleoyl from labelled oleoyl CoA (NEM, Dupont) into phosphatidic acid (Oo and Huang, 1989). The typical reaction mixture contained 0.1 M Tris buffer pH 7.5, 2 mM MgCl$_2$, 0.3 mM lysophosphatidic acid, 0.03 mM 1-[^{14}C]-oleoyl-CoA (0.2 µCi) and 5–20 µg of microsomal protein in a total volume of 0.15 ml. Reaction was stopped by the addition of 1 ml chloroform:methanol 1:1 and the chloroform phase was recovered, dried and applied to a silica gel thin layer chromatography (TLC) plate. The plate was developed with chloroform:methanol:acetic acid:water 25:15:4:2 and the phosphatidic acid spot was stained with iodine vapours, scraped from the plates and counted by liquid scintillation. Diacylglycerol acyltransferase (DAG-AT) was measured by the incorporation of labelled oleoyl-CoA into triacylglycerol (Cao and Huang, 1986). A standard reaction mixture was made of 0.1 M Tris buffer pH 7.4, 30 µM 1-[^{14}C]-oleoyl-CoA (0.2 µCi), 2 mM MgCl$_2$, 0.33 mM dipalmitoyl-diacylglycerol previously emulsified in Tween 20 (final Tween concentration in the assay was 0.02%), and 20–30 µg of microsomal protein in a final volume of 0.15 ml. The reaction was stopped as before and the products were separated by TLC with hexane:diethylether:acetic acid 80:20:1; lipid spots were detected with iodine and diacylglycerol spots were scraped from the plates and counted by liquid scintillation.

Determination of TAG content was performed as described in Rodríguez-Sotres and Black (1993). Determination of radioactivity incorporation into TAG was done as described in Rodríguez-Sotres and Black (1994).

Results and Discussion

Figure 1 shows the TAG content (panel III) and TAG synthesis capacity (panel II) of isolated maize embryos after a 96 h incubation in either basal dilute medium (B) or basal medium suplemented with 30 mM sucrose (BS), sucrose plus 10 µM ABA (BSA), sucrose plus 500 mM mannitol (BSM) or sucrose plus ABA plus mannitol (BSMA). From this result it is clear that embryos isolated from the seeds rapidly lose their ability to synthesize and accumulate triacylglycerols unless ABA or a low osmotic potential is provided. In fact synthetic capacity is much reduced after the first 24 h (panel I), but not after 12 h (not shown). This response is very similar to the one found for wheat embryos (Rodríguez-Sotres and Black, 1994), and since maize is a more oily embryo, the amount and the incorporation of labelled acetate into triacylglycerol is higher than that in wheat. We also found that contrary to what was reported for wheat embryos, there is a significant reduction of TAG content in those embryos incubated in dilute buffer. The concentration of ABA required for these effects was about 4 µM (Fig. 2-I). This concentration is 8 times larger than the one required by wheat embryos, but it has been reported that cultures of maize cells metabolize up to 90% of the exogenously added ABA in 24 h (Balsevich *et al.*, 1994), which could account for the difference. The concentration of mannitol required was about 450 mM (Fig. 2-II); this concentration gives an osmotic

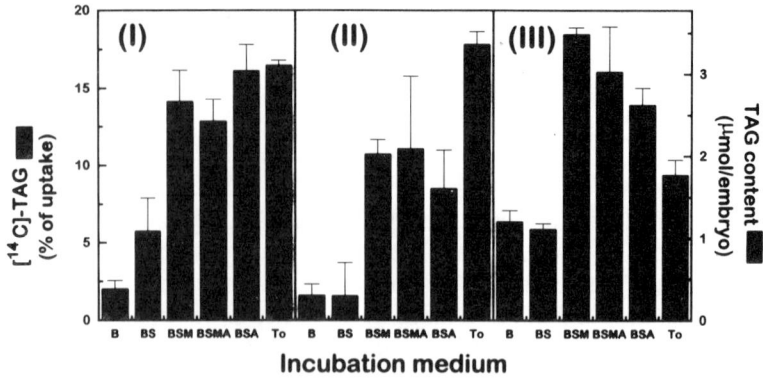

Figure 1. (I and II) TAG synthesis capacity of isolated and pre-treated maize embryos of 23 dpp determined as the amount of [^{14}C]acetate incorporated into TAG in a 5 h feeding period given inmediately after the end of the indicated treatment, I) incubation time 24 h; II) incubation time 96 h; III) total TAG content of the embryos after the treatment. B, diluted basal medium; BS, basal medium plus 30 mM sucrose; BSA, basal medium plus sucrose plus 10 μM ABA; BSM, basal medium plus sucrose plus 600 mM mannitol; BSMA basal medium with sucrose, ABA and mannitol. T$_0$, embryos were either fed with labelled acetate and ground, or simply ground for TAG extraction, immediately after isolation

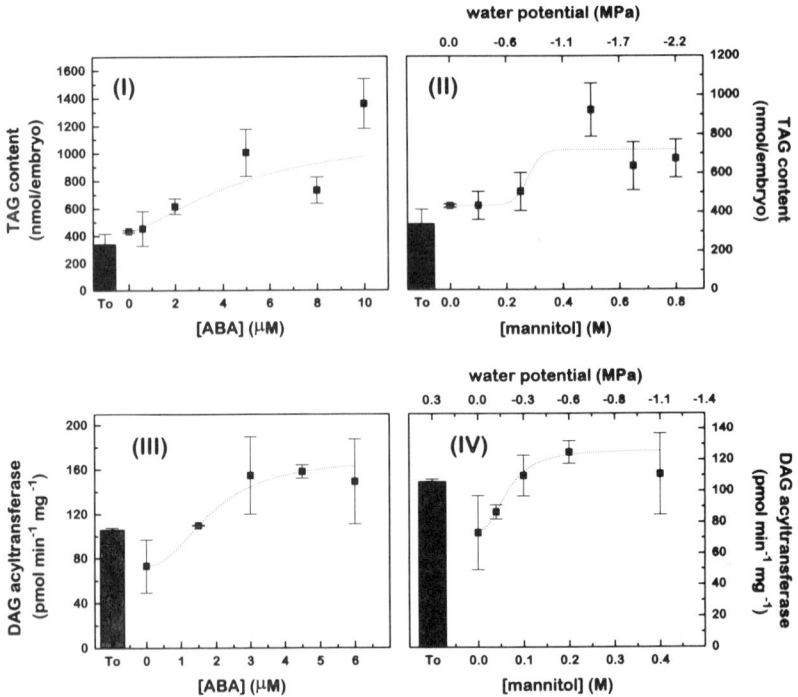

Figure 2. Dose–response curves of TAG accumulation (I and II) or diacylglycerol transferase activity (III and IV) of isolated and pre-treated maize embryos of 23 dpp to ABA and mannitol concentration. The embryos were extracted for TAG after 96 h (I and II) or 24 h (III and IV) incubation in basal medium plus 30 mM sucrose and the indicated concentrations of ABA (I and III) or mannitol (II and IV). T$_0$ values correspond to the freshly isolated embryos. Other conditions as in Figure 1

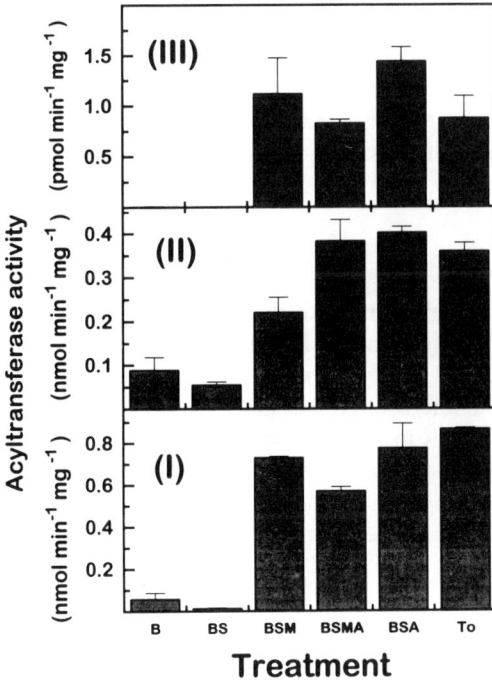

Figure 3. Acyltransferase specific activities measured in the microsomal fractions obtained from the isolated 23 dpp maize embryos that were incubated for 24 h as described in Figure 1. I) glycerol-3-phosphate acyltransferase activities, II) lysophosphatidic acid acyltransferase activities, III) diacylglycerol acyltransferase activities. Incubation conditions as in Figure 1

potential of about −1.2 MPa, a value very similar to the one determined in the case of wheat (Rodríguez-Sotres and Black, 1994).

Based on the above results we propose that ABA and osmotic potential are playing an important role in the regulation of TAG deposition during seed development; these factors are probably fulfilling this role on various seeds (Bewley and Black, 1994). Therefore we decided to explore the possible changes in appropriate enzyme activities.

TAG synthesis occurs in two parts, one is fatty acid synthesis which takes place mainly in the plastids, the other is the incorporation of acyl moieties into the glycerol phosphate molecule, one at a time, to build the triacylglycerol. We examined possible changes in the activity of the enzymes participating in this second stage of TAG synthesis. We therefore measured the acyltransferase activities in the microsomal fractions from treated and untreated embryos after a 24 h incubation period. The results are shown in Figure 3. It is evident that the activity of these three enzymes parallels the changes in TAG synthesis capacity that are shown in Fig. 1-I. Therefore, the presence of ABA and/or a low osmotic potencial is a requirement of the embryos in order to maintain an elevated rate of TAG synthesis. As before, panels III and IV in Figure 2 show that the amount of ABA or the osmotic potential required to elicit the observed response is in agreement with the data presented in Figure 2 panels I and II and the values found are likely to be of physiological significance (Bewley and Black, 1994).

Conclusions

Here we present evidence indicative of the role that ABA and the osmotic environment of the seed play in the regulation of the accumulation of triacylglycerol in cereal embryos. In addition, our data support the involvement of the activity of the microsomal (presumably from the endoplasmic reticulum) acyltransferase-enzymes in the observed regulation of TAG synthesis. We can not yet answer the question as to whether these factors are acting at transcription, translation or at protein activity level. Current research in our laboratories is trying to address this matter.

Acknowledgements

This work has been partly financed by the EEC joint research project CICT93/ 0335. We also thank Ing. Hugo Cordoba, from the CIMMYT, Mexico for the seeds of maize and Ing. Alberto Trujillo from INIFAP, Zacatepec, Mor, Mexico, for the use of their plant growth facilities and his help.

References

Balsevich, J.J., Cutler, A.J., Lamb, N., Friesen, L.J., Kurz, E.U., Perras, M.R. and Abrams, S.R. 1994. *Plant Physiology* 106: 135–142.
Bertrams, M and Heinz, E. 1981. *Plant Physiology* 68: 653–657.
Bewley, J.D. and Black, M. 1994. *Seeds: Physiology of Development and Germination*, Second Edition, pp. 445. New York, London: Plenum Press.
Cao, Y.-Z. and Huang, A.H.C. 1986. *Plant Physiology* 82: 813–820.
Finkelstein, R. and Somerville, C. 1989. *Plant Science* 61: 213–217.
Holbrook, L.A., Magus, J.R. and Taylor, D.C. 1992. *Plant Science* 84: 99–115.
Oo, K.-Ch. and Huang, A.C.H. 1989. *Plant Physiology* 91: 1288–1295.
Rodríguez-Sotres, R. and Black, M. 1993. *Phytochemical Analysis* 4: 68–71.
Rodríguez-Sotres, R. and Black, M. 1994. *Planta* 192: 9–15.

19. The Phytochrome Family and their Roles in the Regulation of Seed Germination

P.F. DEVLIN, K. HALLIDAY and G.C. WHITELAM

Department of Botany, University of Leicester, University Road, Leicester LE1 7RH, UK

Abstract

Light is an important factor influencing the germination of seeds of many plant species. Phytochrome is well known to be involved in the perception of the light signals that influence seed germination. Higher plants possess multiple, discrete molecular species of phytochrome, the apoproteins of which are encoded by a small family of divergent genes. In *Arabidopsis thaliana*, five distinct apophytochrome-encoding genes have been identified (*PHYA-PHYE*). There is increasing evidence that different phytochromes play different roles in photomorphogenesis. The analysis of photomorphogenic mutants, particularly those that are deficient in individual phytochromes or which over-express specific *PHY* cDNAs, has been especially useful in defining the roles of the members of the phytochrome gene family. Analysis of such mutants has revealed that phytochrome B, a low abundance, light-stable phytochrome, plays the major role in the photocontrol of *Arabidopsis* seed germination. However, phytochrome A, the light-labile phytochrome species that predominates in etiolated plant tissues, and another, as yet unidentified, phytochrome, also participate in the photocontrol of *Arabidopsis* seed germination.

Introduction

It has long been established that light can be an important factor that influences the germination of seeds of many plant species. Whereas the seeds of some species germinate equally well in the presence or absence of light, the germination of some can be promoted by light whilst in others germination is inhibited by light. Phytochrome is well known to be involved in the perception of light signals that influence seed germination. Indeed, the photoreversible effects of brief pulses of red and far-red light on lettuce seed germination played a pivotal role in the early characterisation of phytochrome. The promotive effect of a red light pulse, the reversal of this effect by a subsequent far-red light pulse and the repeatable red/far-red photoreversibility led to the initial proposal that the active photoreceptor was reversibly photochromic and could exist as an inactive red-absorbing form, Pr, and an active far-red light-absorbing form, Pfr.

It is now known that a brief pulse of red light can promote the germination of seeds of a large number of plant species and that far-red light has an inhibitory effect (see Frankland and Taylorson, 1983). Far-red light can not only negate the promotive effect of red light pulses in seeds that require light for germination

R.H. Ellis, M. Black, A.J. Murdoch, T.D. Hong (eds.), Basic and Applied Aspects of Seed Biology, pp. 159–171.
© *1997 Kluwer Academic Publishers, Dordrecht. Printed in Great Britain.*

but it can also be inhibitory to some seeds that germinate in the dark. The finding that a pulse of far-red light leads to the inhibition of, for example, tomato seed germination, and that the effect of far-red light is reversed by subsequent red light irradiation indicates that the presence of Pfr in the seed is a requirement for dark germination of these seeds. Red/far-red reversibility is the classical hallmark of phytochrome action and responses of this type are considered to reflect the so-called low fluence response (LFR) mode of phytochrome action (Mancinelli, 1994).

Two additional actions of far-red light on seed germination have been frequently observed. For the seeds of many plant species, germination can be inhibited at a late stage by the action of prolonged far-red light. This action, which can often be observed in seeds that are no longer responsive to a saturating pulse of far-red, is characterised by an absolute requirement for continuous, prolonged irradiation and by a dependency on far-red fluence rate (see Cone and Kendrick, 1986). Furthermore, action spectra for the inhibitory effect of prolonged far-red light display peaks at around 710-720 nm. These characteristics indicate the operation of a far-red high irradiance response (FR-HIR); a phytochrome response mode that also operates in etiolated plant tissues (Mancinelli, 1994). The third action of far-red light is observed in highly sensitive seeds in which a far-red light pulse cannot fully reverse the promotive effects of a red light pulse because far-red light alone significantly promotes germination relative to germination in the dark (see Cone and Kendrick, 1986). It is proposed that in such seeds only small amounts of Pfr, such as those that remain after far-red light, are needed to promote germination. Phytochrome responses of this type have also been observed in etiolated plant tissues and they probably reflect the operation of the so-called very low fluence response (VLFR) mode of phytochrome action (Mancinelli, 1994).

The Phytochrome Family and Phytochrome Mutants

Phytochrome Genes and Proteins

The presence, amount, direction, duration and quality of incident light radiation all regulate the growth and development of plants. These regulatory light cues are perceived by a series of signal-transducing photoreceptors, the best characterised of which are the phytochrome family. The phytochromes are reversibly photochromic, red/far-red light absorbing biliproteins. Higher plants possess multiple, discrete molecular species of phytochrome, the apoproteins of which are encoded by a small family of divergent genes (see Quail, 1994). In *Arabidopsis thaliana*, five distinct apophytochrome-encoding genes have been identified, and full-length cDNAs for all of these have been cloned and sequenced (Sharrock and Quail, 1989; Clack *et al.*, 1994). From comparisons of peptide sequences and deduced amino acid sequences it is known that the apoprotein moiety of the well-characterised, light-labile phytochrome species that predominates in etiolated plant tissues, is encoded by the *PHYA* gene. This light-labile phytochrome, formerly called Type 1 phytochrome, is now referred to as phytochrome A. The other *PHY (B–E)* genes encode the apoproteins of phytochromes B–E, which are lower abundance and more light-stable (or Type 2) species of the photoreceptor. Counterparts of *PHYA, PHYB* and other *PHY* genes have been isolated from several other plant species (Quail, 1994). In tomato there may be as many as 9 *PHY* genes (L.H. Pratt, personal communication).

The phytochrome A protein has been isolated from a number of etiolated plant sources and its molecular properties have been determined. Phytochrome A is a dimer of two identical subunits of molecular weight ~124 kDa. Each monomer possesses a single tetrapyrrole chromophore prosthetic group (phytochromobilin) attached via a thio-ether bond to a conserved cysteine residue in the amino-terminal portion of the protein. Absorption of a photon of light by the chromophore induces a reversible isomerisation of the tetrapyrrole and a concomitant reversible alteration in the conformation of the phytochrome protein moiety. It is assumed that the photoreversible conformational changes in the phytochrome protein that occur upon transformation between Pr and Pfr are the basis of the differential activities of the two forms of phytochrome. From analysis of deduced amino acid sequences, the apoproteins of all five of the *Arabidopsis* phytochromes appear to have similar overall structures. However, that different members of the phytochrome family display marked diversity among their amino acid sequences is consistent with the suggestion that each phytochrome may play a discrete role in photomorphogenesis (Smith and Whitelam, 1990; Whitelam and Harberd, 1994).

Photomorphogenic Mutants

The analysis of photomorphogenic mutants, particularly those that are deficient in individual phytochromes or which over-express specific *PHY* cDNAs, has been especially useful in defining the roles of the members of the phytochrome gene family (see Whitelam and Harberd, 1994).

Phytochrome Chromophore Mutants

A number of mutants that display a global deficiency in all functional phytochromes, due to lesions in phytochrome chromophore biosynthesis, have been described. The *Arabidopsis hy1* and *hy2* mutants which, compared with wild type plants, have elongated hypocotyls following growth under red, far-red, blue and white light conditions belong to this class (Whitelam and Harberd, 1994). Whilst the analysis of these mutants is of limited value in defining the roles of individual phytochrome family members, these are useful mutants for establishing the actions of phytochromes in general. For example, the insensitivity of the hypocotyls of the *Arabidopsis hy1* and *hy2* mutants to both prolonged red and prolonged far-red light indicates that one or more phytochromes, and not other photoreceptors, mediate the plant's responses to these different wavelengths. Phenotypically similar mutants to these have been isolated in tomato and tobacco (Koornneef and Kendrick, 1994).

Phytochrome B-Deficient Mutants

Mutants selectively deficient in phytochrome B activity have been isolated in several plant species. *Arabidopsis hy3* mutants display an elongated hypocotyl when grown under prolonged red and white light conditions but are indistinguishable from wild type seedlings when grown under prolonged far-red light. This suggests that phytochrome B plays a major role in mediating seedling responses to red light and that some other phytochrome mediates the effects of prolonged far-red light. Immunochemical analysis of the phytochromes in *hy3* mutant seedlings has shown that these mutants are deficient in phytochrome B, but accumulate wild-type levels of phytochrome A. Further analysis of a number of *hy3* alleles showed that they contain mutations within the *PHYB* structural gene (Reed *et al.*, 1993). These observations suggest that the phenotype conferred by the *hy3* mutations is due to a deficiency of functional phytochrome B, and these alleles are now designated *phyB-1*, *phyB-2* etc. (Reed *et al.*, 1993; Quail *et al.*, 1994). Mutants deficient in phytochrome B, phytochrome B-like, or light stable phytochromes possibly equivalent to the *Arabidopsis* phytochrome B, have been identified in other plant species and include the long hypocotyl, *lh* mutant of cucumber (Lopez-Juez *et al.*, 1992), the elongated internode, *ein* mutant of *Brassica rapa* (Devlin *et al.*, 1992), the ma_3^R mutant of sorghum (Childs *et al.*, 1991), the *lv* mutant of pea (Weller *et al.*, 1995) and the *tri* mutant of tomato (van Tuinen *et al.*, 1995a). Etiolated seedlings of all of these mutants display a reduced sensitivity to red light, with respect to the inhibition of axis extension, but show normal responsiveness to prolonged far-red light.

As well as displaying a marked reduction in elongation growth in response to red or white light, the cotyledons of etiolated *phyB*, *lh* and *ein* mutants display reduced expansion and, in the case of *phyB* and *ein*, reduced pigmentation under these same light conditions. These observations indicate that phytochrome B plays a role in at least some aspects of de-etiolation.

White light-grown seedlings of phytochrome B-deficient mutants display increased stem and petiole elongation and increased apical dominance. Also, in contrast to their respective wild-types, the *phyB*, *lh* and *ein* mutants do not show significant growth promotions by end-of-day far-red light treatments. Furthermore, all three of these mutants respond only poorly to reductions in red:far-red ratio (Whitelam and Smith, 1991; Devlin *et al.*, 1992). In fact, the elongated growth habit of phytochrome B-deficient mutants resembles the growth responses of wild type seedlings to low red:far-red ratio light conditions; the shade avoidance syndrome (Whitelam and Harberd, 1994). For these reasons, phytochrome B is thought to play a role, along with other phytochromes, in mediating the shade avoidance responses of wild type seedlings.

The flowering responses of *phyB* or *phyB*-like mutants of several species have been studied. The *ein* mutant of the long day plant *Brassica rapa* and the *lh* mutant of the day neutral plant cucumber show early flowering compared with their respective wild types, but the effects are rather small. In the long day plant *Arabidopsis*, the *phyB* mutant displays very marked early flowering following growth under both long day and short day conditions (e.g. Goto *et al.*, 1991). However, even though the reduction in flowering time appears slightly more marked under short days, the *phyB* mutant is still responsive to daylength.

It is possible that there is a relationship between phytochrome B and the gibberellins (GAs) in the regulation of plant growth. This relationship is suggested by the fact that two mutants now known to be deficient for phytochrome B-like or light-stable phytochrome species (*ein* in *Brassica*, ma_3^R in sorghum) were originally identified as gibberellin overproducing mutants. It has also been suggested that phytochrome B deficiency may be correlated with a change in sensitivity to GAs rather than with increased production of GAs. Thus, it has recently been shown that the cucumber *lh* mutant, which does not have elevated endogenous GAs, displays an increased responsiveness to applied GAs compared with wild type seedlings (Lopez-Juez *et al.*, 1995).

Phytochrome B Over-expressers

Transgenic *Arabidopsis* and tobacco seedlings that over-express *PHYB* cDNAs have also been created (Wagner *et al.*, 1991; McCormac *et al.*, 1993a; Halliday and Whitelam, unpublished). In both cases the transgenic seedlings display increased sensitivity to red light with respect to the inhibition of hypocotyl elongation and the promotion of cotyledon expansion. This increased sensitivity to red light of *PHYB* over-expressers contrasts with the reduced red-light sensitivity of phytochrome B-deficient mutants. Significantly, etiolated seedlings of transgenic *PHYB* over-expressing lines show normal sensitivity to far-red light (McCormac *et al.*, 1993a; Halliday and Whitelam, unpublished), consistent with the view that the effects of prolonged far-red light on etiolated seedlings are not mediated by phytochrome B. Light-grown *PHYB* over-expressing lines are darker green than their corresponding wild types and show reduced stem and petiole elongation (McCormac *et al.*, 1993a; Halliday and Whitelam, unpublished). This phenotype contrasts with the spindly and pale green phenotype that is common among phytochrome B-deficient mutants.

Not all of the altered photoresponses of *PHYB* over-expressers are the opposite of the behaviour of phytochrome B-deficient mutants. For example, the *Arabidopsis phyB* mutant is early flowering, under both long- and short-day conditions (Goto *et al.*, 1991; Whitelam and Smith, 1991). Significantly, *Arabidopsis* seedlings that have been transformed with *PHYB* cDNAs are also early flowering, compared with wild type seedlings, under these same conditions (Bagnall *et al.*, 1995). It seems possible that effects of this type are the result of ectopic expression of the introduced *PHY* gene, since in most cases transgene expression is driven by the cauliflower mosaic virus 35S promoter and not by the native *PHY* gene promoter. These examples serve to indicate that over-expression phenotypes are not always a reliable indicator of the roles of endogenous phytochrome species.

Phytochrome A-Deficient Mutants

Mutants selectively deficient in phytochrome A activity have been described for *Arabidopsis* and tomato (e.g. Whitelam *et al.*, 1993; van Tuinen *et al.*, 1995b). *Arabidopsis phyA* (formerly *fhy2* = *hy8*, = fre1) mutants represent mutations at the *PHYA* locus (Whitelam *et al.*, 1993; Whitelam and Harberd, 1994). Etiolated seedlings of null alleles of this mutant display a complete insensitivity to prolonged far-red light (i.e. the FR-HIR) with respect to the inhibition of hypocotyl elongation (e.g. Whitelam *et al.*, 1993). This indicates that phyto-chrome A is the photoreceptor mediating the FR-HIR, a finding consistent with physiological interpretations of the this response mode (see Smith and White-lam, 1990). The hypocotyls of etiolated *phyA* seedlings display a normal sensitivity to prolonged red light in accordance with the proposal that phytochrome B plays the major role in seedling responses to red light (see Quail *et al.*, 1995). When grown in white light, *phyA* mutants display a more or less wild-type phenotype (Whitelam *et al.*, 1993), but show increased elongation growth responses to reductions in red:far-red ratio (Johnson *et al.*, 1994). Etiolated seedlings of the phytochrome A-deficient tomato mutant (*fri*) is similarly insensitive to far-red light and light-grown *fri* seedlings display a more or less wild type phenotype (van Tuinen *et al.*, 1995b).

The phytochrome A-deficient mutant of *Arabidopsis*, a quantitative long day plant, has been studied in relation to photoperiodic floral induction (e.g. Johnson *et al.*, 1994). Compared with wild type plants, *phyA* seedlings are relatively insensitive to low-fluence day extensions (Johnson *et al.*, 1994). Furthermore, under 8 h short day conditions, *phyA* mutants flower later than wild-type plants (Johnson *et al.*, 1994). These observations implicate phyto-chrome A in daylength perception in *Arabidopsis*. From physiological studies, a light-labile phytochrome, such as phytochrome A, had previously been impli-cated in photoperiodic time measurement in long day plants. Since *phyA* mutants are not completely daylength insensitive, other phytochromes are also presumed to play a role in daylength measurement.

Phytochrome A Over-expressers

Transgenic *PHYA* over-expression lines have been created for *Arabidopsis*, tobacco, tomato and several other plant species. These lines are characterised by a very marked increased sensitivity to far-red light of etiolated seedlings. Furthermore, in contrast with wild type seedlings, *PHYA* over-expression lines maintain the FR-HIR following de-etiolation (e.g. Whitelam *et al.*, 1992). These findings are fully consistent with the proposal that phytochrome A is the photoreceptor mediating seedling responses to prolonged far-red light.

Whereas light-grown phytochrome A-deficient mutants display a more-or-less wild type phenotype, light-grown *PHYA* over-expressers are darker green, dwarfed and show reduced apical dominance compared with their wild types.

Some of the photoresponses of *PHYA* over-expressers appear to be the opposite of the corresponding responses of phytochrome A-deficient mutants. Thus, whereas phytochrome A-deficient *phyA* mutants of *Arabidopsis* show increased elongation growth responses, compared with wild type plants, to reductions in red:far-red ratio (Johnson *et al.*, 1994), *PHYA* over-expressers typically display reduced elongation growth responses compared with wild types (Whitelam *et al.*, 1992). Also, whereas *phyA* mutant seedlings show late flowering compared with wild type plants, *Arabidopsis* seedlings that over-express a *PHYA* cDNA flower earlier than wild type seedlings (Bagnall *et al.*, 1995). This latter observation provides additional, circumstantial evidence that phytochrome A plays a role in the regulation of flowering time.

Transgenic *Arabidopsis* seedlings expressing cereal *PHYA* cDNAs display a marked increased responsivity to red light, compared with wild-type seedlings (Whitelam *et al.*, 1992). This appears to be at odds with the finding that *Arabidopsis phyA* mutant seedlings are indistinguishable from wild-type seedlings with respect to responsivity to red light (Whitelam *et al.*, 1993). However, there are situations (such as in the *phyAphyB* double mutant – see below) where the *phyA* mutation does lead to reduced responsivity to red light (Reed *et al.*, 1994).

Phytochrome A/Phytochrome B Interactions and Double Mutants

From the study of monogenic mutants deficient in either phytochrome A or phytochrome B it appears that these two photoreceptors play fairly discrete roles in seedling photomorphogenesis. Thus, it seems that phytochrome A is involved in the perception of prolonged far-red light whilst phytochrome B mediates perception of prolonged red light. This apparent "division of labour" has interesting consequences for seedlings exposed to polychromatic light containing both red and far-red wavelengths. For example, under conditions of white light supplemented with far-red light, the actions of phytochromes A and B on elongation growth are antagonistic (see Johnson *et al.*, 1994). Under these light conditions phytochrome A action, which is determined by the amount of far-red light, is to cause **increased** inhibition of elongation growth (via a high-irradiance type response). In contrast, phytochrome B action (which seems to be determined by Pfr concentration) will be reduced, as the Pfr to Pr photoreaction is favoured, and this will lead to **reduced** inhibition of elongation growth (Johnson *et al.*, 1994).

The actions of phytochromes A and B are not always antagonistic and the analysis of mutants deficient in both of these phytochromes has revealed that their roles can overlap. For example, Reed *et al.* (1994) have shown that for *Arabidopsis* seedlings exposed to red or white light the *phyAphyB* double mutants have significantly longer hypocotyls than monogenic *phyB* mutants.

The response of the monogenic *phyA* mutant to red- or white light is indistinguishable from that of the wild type. Therefore, in the *phyB* background, phytochrome A affects elongation growth under red and white light conditions and the action of phytochrome A in this response is not only seen in prolonged far-red light. Other examples where phytochromes A and B appear to mediate the same response to the same light signal in *Arabidopsis* include the expansion of cotyledons under prolonged red light and the induction of *CAB* gene (encoding light-harvesting chlorophyll a/b-binding protein) expression by a pulse of red light (Reed *et al.*, 1994). In both cases, the monogenic *phyA* or *phyB* mutant responds well to red light whereas the *phyAphyB* double mutant shows a significantly impaired response. This indicates that in the absence of either phytochrome A or B, the remaining phytochrome is capable of mediating these responses to red light.

Phytochromes and Germination in *Arabidopsis* Seeds

Arabidopsis is a good model species in which to study the roles of the different phytochromes in the control of seed germination. Many phytochrome-related mutants and transgenic lines are available in *Arabidopsis* and the photocontrol of seed germination in this species is well established. Action spectra for the promotion and inhibition of *Arabidopsis* seed germination have been determined and these indicate the operation of the phytochrome system. Furthermore, it has been known for almost twenty five years that the spectral quality under which the parent plant was grown could influence the subsequent dark germination of seeds (e.g. McCullough and Shropshire, 1970). McCullough and Shropshire (1970) found that the seeds (ecotype Estland) taken from plants grown under white fluorescent light (which would have a high red:far-red ratio) displayed high germination frequencies in the dark whereas seed taken from plants grown under incandescent-rich light (which would have a low red:far-red ratio) had a light-requirement for germination. The authors suggested that for seed developing under fluorescent light a high proportion of Pfr would be established in the seeds thereby precluding a light requirement. The availability of mutant genotypes of *Arabidopsis* has allowed the identification of the phytochromes that mediate the various effects of light on germination.

Phytochrome B Plays a Major Role in the Regulation of Germination

Several independent studies from this and other laboratories have established that phytochrome B plays a major role controlling the germination of *Arabidopsis* seeds in response to brief red light treatment and in the dark. The data in Table 1 show that whereas the germination of wild type seeds is significantly promoted by a 5 min pulse of red light, seeds of the *phyB* mutant are almost unresponsive to this light treatment. In this instance, where light treatments are given to seeds immediately after a 19 h imbibition period,

Table 1. Germination frequencies of phytochrome-related mutants

Light	Germination (%)[a]				
	Wild type	*phyA*	*phyB*	*phyAphyB*	*hy1*
Dark	4.0±0.5	4.0±0.8	0.8±0.4	0.6±0.4	2.4±0.6
5 min Red	30.6±4.9	34.0±2.6	2.6±0.4	2.2±0.6	9.4±1.0
cRed	99.6±0.3	80.8±0.9	97.4±0.8	69.6±1.4	85.0±2.1
5 min Far-red[b]	0.0±0	0.6±0.3	0.6±0.3	0.8±0.4	1.2±0.2
cFar-red[c]	3.0±1.0	1.2±0.6	20.8±1.0	0.6±0.3	6.0±0.5
cWhite	99.8±0.2	85.8±2.0	96.2±1.4	90.0±1.0	91.4±1.1

[a]Seeds were sown on 0.6% agarose in mineral salts medium and the plates incubated for 19 h at 4°C then 1 h at 20°C in the dark before light treatment. Germination was assessed, as radicle emergence, 4 days after the onset of light treatments. Data are the means of at least four replicates each of 100 seeds

[b]Narrow band far-red light provided by a 740 nm interference filter

[c]Broad band far-red light provided by a bank of light emitting diodes

phytochrome A appears to play no part in the seeds' response to red light. Thus, seeds of the *phyA* mutant show a promotion of germination in response to a red light pulse that is the same as that of the wild type. As expected, seeds of the *phyAphyB* double mutant are also unresponsive to a light pulse and seeds of the *hy1* mutant, which is defective in the synthesis of the phytochrome chromophore, also respond poorly to a red light pulse (Table 1). Reduced germination responses of *phyB* seeds to red light pulses have also been reported by Shinomura *et al.* (1994) and by Botto *et al.* (1995).

Although the seed batches of the various genotypes used in the experiments reported in Table 1 show low dark germination it is nevertheless apparent that germination in the dark is lower in the *phyB* and *phyAphyB* mutants than in wild type or *phyA* seeds. Since the germination of wild type and *phyA* seeds is reduced (compared with seeds maintained in the dark) by a brief pulse of far-red light, whereas the germination of *phyB* and *phyAphyB* mutant seeds is not, this suggests that Pfr$_B$ is present in wild type (and *phyA*) seeds in the dark and can promote germination. Further evidence that seed Pfr$_B$ can promote *Arabidopsis* seed germination in the dark has come from the analysis of transgenic plants that over-express *PHY* cDNAs (McCormac *et al.*, 1993b). The seeds of transgenic plants expressing a *PHYB* cDNA, under control of the cauliflower mosaic virus, show very high dark germination irrespective of the red:far-red ratio under which the parent plants were grown. In contrast, the germination of wild type seeds was significantly influenced by the light conditions under which the seed developed, whilst *phyB* seeds always showed very low dark germination (McCormac *et al.*, 1993b). Seeds derived from transgenic plants expressing a *PHYA* cDNA showed germination characteristics that were similar to wild type seeds.

Phytochrome A also Plays a Role in the Regulation of Germination

Wild type *Arabidopsis* seeds exposed to continuous red light for 4 days show essentially 100% germination, as do seeds of the *phyB* mutant (Table 1). However, germination of *phyA* seeds, and *phyAphyB* seeds, under continuous red light is significantly lower, suggesting that in wild type (and *phyB*) seeds phytochrome A is playing a role in promoting germination.

Previously, phytochrome A had been implicated in the photocontrol of germination of *Arabidopsis* seeds that had been allowed to imbibe in darkness for a few days prior to exposure to light. For example, Johnson *et al.* (1994) showed that the germination of wild type seeds, that had been allowed to imbibe in the dark for two days, was promoted by brief or prolonged exposure to broad band far-red light. These germination-promoting effects were not seen in seeds of the *phyA* mutant. The promotive effect of a far-red pulse is suggestive of the operation of a VLFR whilst the action of prolonged far-red is taken to indicate the operation of a FR-HIR. The promotion of germination by continuous far-red, via a high irradiance reaction, contrasts with the frequently reported inhibitory effect of this light regime on seed germination (see Cone and Kendrick, 1986). Thus, for *Arabidopsis* seeds the FR-HIR acts in the same way to an inductive red light pulse.

The data in Table 2 show that a period of dark imbibition prior to light treatment significantly increases the effect of continuous far-red light on the germination of wild type and *phyB* seeds. In these seed batches germination of the wild type is not promoted by a brief pulse of far-red light only by continuous irradiation. Seeds of the *phyA* mutant are, as expected, unresponsive to continuous far-red light. The requirement for a dark period for wild type seeds to display a response to continuous far-red light may reflect the need for phytochrome A synthesis to occur. It has been shown that phytochrome A is not immunochemically-detectable in dry *Arabidopsis* seeds (Shinomura *et al.* 1994).

The Pr form of Phytochrome B Inhibits Germination?

The promotory effect of continuous far-red light on germination is significantly more pronounced in seeds of the *phyB* mutant compared with wild type seeds (Tables 1 and 2). Thus, *phyB* seeds show a marked FR-HIR promotion of germination even without a period of dark imbibition (Table 1). A similar difference in the responses of wild type and *phyB* seeds to continuous far-red light has been observed by Shinomura *et al.* (1994) and Reed *et al.* (1994). In order to account for the difference, Shinomura *et al.* (1994) proposed that, whereas in both wild type and *phyB* seeds phytochrome A action (induced by continuous far-red) promotes germination, in wild-type seeds phytochrome B is converted to Pr_B by the far-red light and this suppresses germination. In other words it was proposed that some biological activity resides in the Pr form of phytochrome B; the supposed inactive form of phytochrome. Whilst the data do

Table 2. Incubation of imbibed seeds in the dark increases germination frequency in response to continuous far-red light

	Germination (%)[a]		
Treatment	Wild type	*phyB*	*phyA*
Dark	4.0±0.5	0.8±0.4	4.0±0.8
4 d cFar-red	6.0±1.2	33.8±2.9	1.2±0.6
1 d Dark + 4 d cFar-red[b]	12.2±1.4	43.4±2.0	–
2 d Dark + 4 d cFar-red	19.4±1.4	40.8±2.8	5.0±0.8
2 d Dark + 5 min Far-red[b] + 4 d Dark	4.6±0.7	–	4.6±1.0
6 d cFar-red	11.4±1.6	30.4±1.9	–

[a]Experimental protocol as for Table 1
[b]Broad band far-red light provided by a bank of light emitting diodes

appear to support this proposal there may be alternative explanations for the observed differences in responses of wild type and *phyB* seeds to continuous far-red light. It is apparent from Table 2 that the effectiveness of continuous far-red light in promoting germination increases when seeds are allowed to imbibe in the dark prior to irradiation. This is true for both wild type and *phyB* seeds and may be related to *de novo* synthesis of phytochrome A. It is possible that in *phyB* seeds the phytochrome A-mediated high irradiance response is displayed even in seeds that have not been incubated in the dark. In other words, the phytochrome B status of the seeds could influence the extent of the phytochrome A action rather than Pr_B acting antagonistically to phytochrome A action. This could be because phytochrome A levels may be higher in the seeds of the *phyB* mutant. It has previously been shown that in germinating oat caryopses phytochrome A synthesis was inhibited by the action of a light-stable (phytochrome B-like) phytochrome (Thomas *et al.*, 1989). Furthermore, the proposal that Pr_B suppresses *Arabidopsis* seed germination is not consistent with the germination promoting effects of a pulse of far-red light seen in wild type seeds that have been allowed to imbibe in the dark for a few days (Johnson *et al.*, 1994). Such a pulse would convert any phytochrome B to the Pr form, yet germination is promoted relative to seeds maintained in the dark.

Novel Phytochromes are Involved in the Photoregulation of Germination

In addition to the actions of phytochromes A and B, at least one other phytochrome also appears to be involved in the photocontrol of *Arabidopsis* seed germination. From Table 1 it can be seen that whereas *phyAphyB* seeds are unresponsive to a brief pulse of red light, they display a marked promotion of germination in response to continuous red light. This indicates the operation of

Table 3. A novel phytochrome promotes germination in response to prolonged red light irradiation

Treatment	Germination (%)[a]	
	Wild type	*phyAphyB*
Dark	4.0 ± 0.5	0.6 ± 0.4
5 min Red	30.8 ± 2.7	2.4 ± 0.6
1 h Red	40.6 ± 2.9	1.8 ± 0.6
8 h Red	84.0 ± 2.8	20.0 ± 1.7
8 × 5 min Red[b]	80.4 ± 2.0	12.8 ± 2.0
24 h Red	99.0 ± 0.5	58.8 ± 3.1
48 h Red	99.8 ± 0.2	86.4 ± 2.3
96 h Red	97.6 ± 0.3	88.0 ± 1.9

[a]Experimental protocol as for Table 1

[b]Light pulses were separated by 55 min periods of darkness

another red light photoreceptor; most probably a phytochrome. The data in Table 3 show that action of this phytochrome requires prolonged red light and that germination increases with increased duration of red light treatment. There seems to be a time-dependent component to the action of this other phytochrome. So, whilst 60 min of continuous red light is ineffective in promoting the germination of *phyAphyB* seeds, eight 5 min duration pulses of red light delivered at one hourly intervals does lead to a significant germination response. This requirement for prolonged irradiation with a partial substitution by light pulses is typical of red high-irradiance responses of etiolated seedlings. The current lack of loss-of-function mutations for the other phytochromes of *Arabidopsis* (i.e. phytochromes C, D and E) means that the identity of this other phytochrome cannot yet be determined.

Conclusions

The analysis of the germination responses of several *Arabidopsis* genotypes with altered expression of phytochromes has revealed the action of at least three members of the family. Phytochrome B controls germination in the dark and in response to red light pulses. Phytochrome A regulates germination in response to brief or prolonged exposure to far-red light and a third phytochrome is able to promote germination in response to prolonged red light treatments.

References

Bagnall, D., King, R. W., Whitelam, G. C., Boylan, M. T., Wagner, D. and Quail, P. H. 1995. *Plant Physiology* 108: 1495–1503.

Botto, J. F., Sanchez, R. A. and Casal, J. J. 1995. *Journal of Plant Physiology* 146: 307–312.

Childs, K. L., Pratt, L. H. and Morgan, P. W. 1991. *Plant Physiology* 97: 714–719.

Clack, T., Matthews, S. and Sharrock, R. A. 1994. *Plant Molecular Biology* 25: 413–427.

Cone, J. W. and Kendrick, R. E. 1986. In: *Photomorphogenesis in Plants*, pp 443–465 (eds. R. E. Kendrick and G. H. M. Kronenberg). Dordrecht: Martinus Nijhoff.

Devlin, P. F., Rood, S. B., Somers, D. E., Quail, P. H. and Whitelam, G. C. 1992. *Plant Physiology* 100: 1442–1447.

Frankland, B. and Taylorson, R. 1983. In: *Photomorphogenesis, Encylopedia of Plant Physiology, vol 16*, pp 428–456 (eds. W. Shropshire, Jr and H. Mohr). Berlin: Springer–Verlag.

Goto, N., Kumagai, T. and Koornneef, M. 1991. *Physiologia Plantarum* 83: 209–215.

Johnson, E., Bradley, J. M., Harberd, N. H. and Whitelam, G. C. 1994. *Plant Physiology* 105: 141–149.

Koornneef, M. and Kendrick, R. E. 1994. In: *Photomorphogenesis in Plants, 2nd edition*, pp 601–628 (eds. R. E. Kendrick and G. H. M. Kronenberg). Dordrecht: Kluwer.

Lopez-Juez, E., Nagatani, A., Tomizawa, K-I., Deak, M., Kern, R., Kendrick, R. E. and Furuya, M. 1992. *Plant Cell* 4: 241–251.

Lopez-Juez, E., Kobayashi, M., Sakurai, A., Kamiya, Y. and Kendrick, R. E. 1995. *Plant Physiology* 107: 131–140.

Mancinelli, A. L. 1994. In: *Photomorphogenesis in Plants, 2nd edition*, pp 211–269 (eds. R. E. Kendrick and G. H. M. Kronenberg). Dordrecht: Kluwer.

McCormac, A. C., Wagner, D., Boylan, M. T., Quail, P. H., Smith, H. and Whitelam, G. C. 1993a. *Plant Journal* 4: 19–27.

McCormac, A. C., Smith, H. and Whitelam, G. C. 1993b. *Planta* 191: 386-393.

McCullough, J. M. and Shropshire, W., Jr. 1970. *Plant and Cell Physiology* 11: 139–148.

Quail, P. H. 1994. In: *Photomorphogenesis in Plants, 2nd edition*, pp 71–103 (eds. R. E. Kendrick and G. H. M. Kronenberg). Dordrecht: Kluwer.

Quail, P. H., Boylan, M. T., Parks, B. M., Short, T. W., Xu, Y. and Wagner, D. 1995. *Science* 268: 675–680.

Quail, P. H., Briggs, W. R., Chory, J., Hangarter, R. P., Harberd, N. P., Kendrick, R. E., Koornneef, M., Parks, B., Sharrock, R. A., Schäfer, E., Thompson, W. F. and Whitelam, G. C. 1994. *Plant Cell* 6: 468–471.

Reed, J. W., Nagpal, P., Poole, D. S., Furuya, M. and Chory, J. 1993. *Plant Cell* 5: 147–157.

Reed, J.W., Nagatani, A., Elich, T. D., Fagan, M. and Chory, J. 1994. *Plant Physiology* 104: 1139–1149.

Sharrock, R. A. and Quail, P. H. 1989. *Genes and Development* 3: 1745–1757.

Shinomura, T., Nagatani, A., Chory, J. and Furuya, M. 1994. *Plant Physiology* 104: 363–371.

Smith, H. and Whitelam, G. C. 1990. *Plant Cell and Environment* 13: 695–707.

Thomas, B., Penn, S. E. and Jordan, B. R. 1989. *Journal of Experimental Botany* 40: 1299–1304.

van Tuinen, A., Kerckhoffs, L. H. J., Nagatani, A., Kendrick, R. E. and Koornneef, M. 1995a. *Plant Physiology* 108: 939–947.

van Tuinen, A., Kerckhoffs, L. H. J., Nagatani, A., Kendrick, R. E. and Koornneef, M. 1995b. *Molecular and General Genetics* 246: 133–141.

Wagner, D., Tepperman, J. M. and Quail, P. H. 1991. *Plant Cell* 3: 1275–1288.

Weller, J. L., Nagatani, A., Kendrick, R. E., Murfet, I. and Reid, J. B. 1995. *Plant Physiology* 108: 525–532.

Whitelam, G. C. and Harberd, N. P. 1994. *Plant Cell and Environment* 17: 615–625.

Whitelam, G. C., Johnson, E., Peng, J., Carol, P., Anderson, M. L., Cowl, J. and Harberd, N. H. 1993. *Plant Cell* 5: 757–768.

Whitelam, G.C., McCormac, A. C., Boylan, M. T. and Quail, P. H. 1992. *Photochemistry and Photobiology* 56: 617–622.

Whitelam, G.C. and Smith, H. 1991. *Journal of Plant Physiology* 139:119–125.

20. The Effect of Molybdenum on Seed Dormancy

A.L.P. CAIRNS[1], A.T. MODI[1], A.K. COWAN[2] and J.H. KRITZINGER[1]

[1]Department of Agronomy, [2]Department of Horticultural Science, Faculty of Agriculture, University of Natal, P.O. Box X01, Scottsville, 3209, Pietermaritzburg, South Africa

Abstract

Foliar applications of molybdenum (Mo) on wheat grown in sand culture and irrigated with Mo-free nutrient solution increased ABA content of the grain which in turn led to an increased dormancy. Alpha-amylase development in field-grown grain subject to simulated rain was reduced by Mo application. Similarly, the production of late maturity amylase (LMA) was prevented by foliar Mo applications. Embryos excised from seed of Mo-treated plants showed enhanced sensitivity to exogenously applied ABA. Molybdenum also played a key role in nitrogen metabolism and led to a significant increase in protein content of field-grown grain. The use of Mo to reinforce dormancy and to boost protein content is discussed.

Introduction

The first report on the effect of molybdenum (Mo) on dormancy was published by Tanner (1978) working on maize in Zimbabwe. He found that maize growing on acid soils where Mo was deficient tended to germinate precociously on the cob. This problem could be alleviated by foliar applications of Mo or by soil liming. He ascribed this lack of dormancy in the Mo-deficient seeds as being mediated by an increase in seed nitrate levels caused by a lack of activity of nitrate reductase which is a molybdo-enzyme. Similar results were reported by Farwell et al. (1991). However, neither of these two groups could show a relationship between seed nitrate content and dormancy. Cairns and Kritzinger (1992), working in sand culture, showed that a Mo-deficiency in wheat caused a lack of seed dormancy associated with a reduced level of ABA. Modi and Cairns (1994) substantiated these findings and also showed an increased sensitivity to exogenous ABA in embryos derived from Mo-treated plants.

The involvement of Mo in the synthesis of ABA has been postulated by Walker-Simmons et al. (1989). Cowan and Richardson (1995) postulated that 1′,4′-trans-ABAdiol was the immediate precursor of ABA in citrus fruit and that Mo is required for this conversion to take place. They reported a linear increase in the conversion of ABAdiol to ABA over the Mo concentration range of 1–100 μM. The present paper summarizes the results of work carried out on foliar Mo application on dormancy, ABA content, α-amylase development and protein content of both sand culture grown and field-grown wheat. The role of Mo in lessening the damage caused by late maturity α-amylase development and pre-harvest sprouting is also discussed.

R.H. Ellis, M. Black, A.J. Murdoch, T.D. Hong (eds.), Basic and Applied Aspects of Seed Biology, pp. 173–180.

Materials and Methods

Growing Conditions

Wheat was grown in free-draining, 110 cm diameter pots filled with quartz sand which had been acid washed according to the method of Modi and Cairns (1994). All treatments were replicated at least four times. Plants were irrigated with a modified Mo-free Hoagland solution as described by Modi (1993) and grown in a growth room at 20/15°C day/night temperature under a 12 h photoperiod. In 1992, replicated field trials were carried out at four sites in South Africa viz. Bethlehem (28° 14'S 28° 18'E), Pietermaritzburg (29° 35'S 30° 25'E), Stellenbosch, (33° 58'S 18° 50'E) and Winterton (28° 45'S 29° 32'E). In 1993 trials were conducted only at the Ukulinga and Winterton sites.

Molybdenum Application

Plants were treated at the flag leaf stage with various concentrations of Mo supplied as sodium molybdate ($Na_2MoO_4.2H_2O$). The Mo was applied to field trials using a CP3 knapsack sprayer delivering a spray volume of approximately 200 1 ha^{-1}. In pot trials the plants were sprayed to a point just before run-off using a hand-held domestic sprayer. Tween 20 at 0.5% was used as a wetting agent for all foliar applications.

Harvesting

Plants grown under controlled environment conditions were tagged at anthesis and harvested 60 days afterwards when grain moisture was approximately 12%. Field trials were hand threshed at a grain moisture level of approximately 15% and stored at –30°C to maintain dormancy.

Germination Tests and Abscisic Acid Determination

Seed germination was tested as described by Modi and Cairns (1994) and germination was expressed as a germination index calculated according to the method of Walker-Simmons (1988). Abscisic acid analysis and sensitivity of excised embryos to abscisic acid were determined as described by Modi and Cairns (1994).

α-Amylase Determination

Ears harvested from the field trials were subjected to simulated rain as described by Modi and Cairns (1995). α-Amylase was extracted by a method described by Nicholls (1979) and enzyme activity was assayed using the "Phadabas" tablet method (Barnes and Blakeney, 1974).

Protein Analysis

Protein analysis was carried out by the South African Wheat Board on an Inframatic 8111 NAR protein analyser in accordance with the manufacturers instructions.

Results and Discussion

Germination and ABA Content

Germination and ABA content of grain harvested from plants which had been sprayed at the flag leaf stage with various concentrations of Mo show clearly that dormancy increased with increasing Mo concentration (Fig. 1). The increase in dormancy was linear over the concentration range of Mo applied whereas the ABA level was little affected by Mo concentrations up to 200 mg l^{-1}. At Mo concentrations of 400 and 600 mg l^{-1}, there was a sharp increase in grain ABA levels.

These findings suggest that foliar applications of Mo may increase dormancy by stimulating the synthesis of ABA and corroborate similar findings by Cairns and Kritzinger (1992). Although it is assumed that the synthesis of ABA is to a large extent regulated by a molybdo-enzyme (Walker-Simmons *et al.*, 1989; Cowan and Richardson, 1995), this is the first time that seed ABA levels have been shown to increase with high levels of Mo application. Indeed the levels

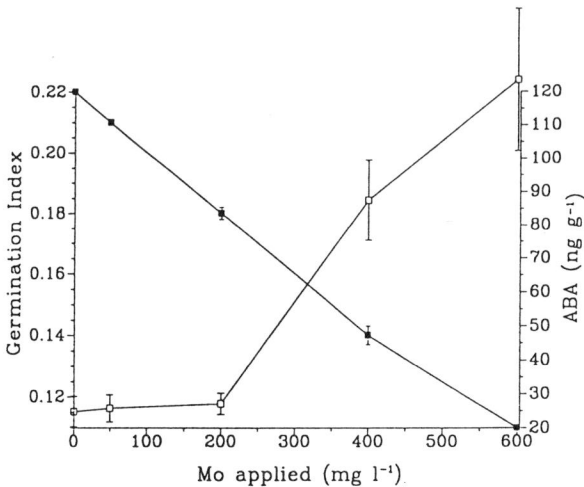

Figure 1. Effect of foliar applications of molybdenum on germination index (Walker-Simmons, 1988) and ABA content of wheat. (Open symbols: ABA; closed symbols: germination index)

Table 1. Effect of foliar applications of Mo to wild oat (*Avena fatua*, Montana 73) on germination and seed ABA content

Treatment	Germination index	ABA ($ng\ g^{-1}$)
−Mo	0.387	580
+Mo (100 mg l^{-1}	0.002	950

used here are far in excess of the 100 mg l^{-1} which is generally the rate used to correct molybdenum deficiency under acid soil conditions. This would seem to indicate that foliar applications of Mo can reinforce grain dormancy even in areas where Mo is not considered to be deficient in the soil.

Further corroborative evidence on the involvement of Mo in the induction of seed dormancy was obtained by applying a foliar application of Mo to the very dormant wild oat (*Avena fatua*) line, Montana 73. Both Mo-treated and control batches of seed were totally dormant at harvest but differences in dormancy level were expressed by germinating both batches of seed in 10^{-4} M GA$_3$ (Table 1).

Sensitivity of Excised Embryos to Exogenous ABA

Excised embryos of field grown wheat (cv. SST 66) from control and Mo-treated plants were tested for sensitivity to exogenously applied ABA. Germination of these embryos in the presence of increasing concentrations of ABA is shown in Figure 2. It is clear that embryos from Mo-sprayed plants showed enhanced sensitivity to inhibition of germination imposed by ABA. However, the ABA content of the embryos was not determined and it is possible that the sensitivity to ABA-induced inhibition of growth was enhanced in the embryos from Mo-treated plants due to higher endogenous ABA levels rather than increased physiological sensitivity of the ABA target tissues in the embryo.

α-Amylase Activity

Molybdenum was applied to six wheat cultivars grown in field trials at four different sites in South Africa during 1992. After harvest the wheat was subjected to 12 h and 24 h of simulated rain and the α-amylase content of the seed determined. Although there was no difference in the α-amylase content of the sound wheat which had not been subjected to simulated rain, there was a highly significant effect of the Mo application on the α-amylase activity which developed after 12 h and 24 h of simulated rain (Table 2). Over all four sites, and with both rainfall treatments, the Mo application resulted in an approximately 30% decrease in α-amylase formed.

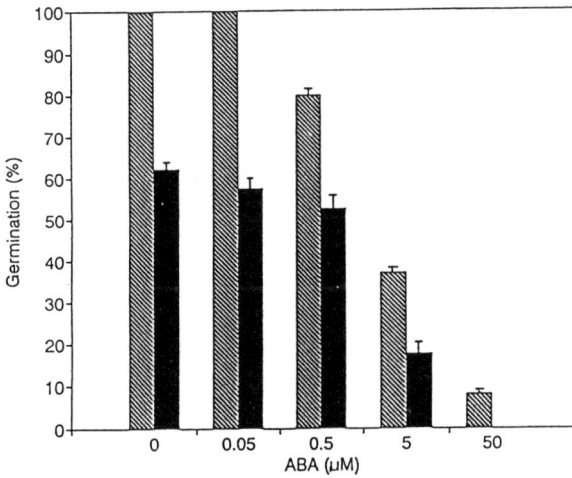

Figure 2. Effect of foliar applications of molybdenum on the germination of excised wheat embryos in the presence of ABA (solid: with molybdenum; hatched: no molybdenum)

Table 2. Effect of foliar applications of molybdenum on α-amylase development of wheat subjected to simulated rain. (Pooled results of six cultivars planted at four sites in 1992)

	α-Amylase activity (EU g^{-1}) Simulated rain (h)		
Treatment	0	24	48
−Mo	11.2[a]*	32.8[c]	62.5[e]
+Mo (100 mg L^{-1})	10.9[a]	23.3[b]	43.5[d]

*Means not sharing the same letter are significantly different

Late Maturity Amylase

In the 1992 trials it was noticed that some cultivars had abnormally high α-amylase levels in the absence of simulated rain. This phenomenon was referred to as late-maturity α-amylase (LMA) by Mares and Mrva (1992). Several cultivars that differed in the amount of LMA formed were planted at two sites in 1993 and sprayed with Mo. The α-amylase levels, at harvest, of four of the cultivars are shown in Table 3.

Mrva and Mares (1995) found that this genetic defect was controlled by a single recessive gene expressed in the triploid aleuron. They suggest that the high

Table 3. The effect of foliar applications of Mo on the development of late maturity α-amylase development in grain of four cultivars grown at two sites in 1993

| | α-Amylase activity (EU g^{-1}) | | | |
| | Cultivars | | | |
Treatment	Adam Tas	SST 86	Palmiet	SST 66
–Mo	28.6^{b*}	6.15^{a}	12.2^{c}	9.8^{c}
+Mo$^{#}$	12.6^{c}	9.0^{c}	12.3^{c}	8.0^{c}

*Means not sharing the same letter are significantly different
$^{#}$Mo applied at 100 mg l^{-1}

α-amylase typical of this sort of genetic defect, is a consequence of a stimulation of the aleuron similar to that which occurs during germination. This phenomenon only occurs in some seasons apparently in response to a specific set of environmental conditions (Gale and Lenton, 1987). However in some cultivars such as Spica and Lerma 52, the production of LMA occurs under most, if not all, growing conditions, although the production of LMA is exacerbated under humid conditions (Mares and Gale, 1990). These authors also found that the synthesis of LMA involves α-amylase isozymes controlled by genes on the group 6 chromosome and is controlled by a single recessive gene which is modulated by environmental and genetic background. It is interesting to note that in the present study SST 86 showed similar high LMA levels at both environments tested. However Adam Tas only showed high LMA levels at Winterton (results not shown). It would appear from the results presented in Table 3 that Mo application can completely prevent the production of LMA in cultivars that are subject to this genetic defect. However, these results represent only one season's work and need to be corroborated by further work using a wider range of cultivars known to be subject to this genetic defect. However, if Mo can indeed turn off the genes responsible for LMA it will represent a way of salvaging many promising high-yielding new lines which otherwise would have had to be rejected due to this defect.

Protein Content

The effect of a foliar application of 600 mg L^{-1} Mo applied at different stages on the grain protein content and yield of SST 66 is shown in Table 4. All the Mo treatments increased the protein content of the wheat but Mo application at the three leaf stage gave the highest increase in protein content. This trial was carried out on soil which had zero acid saturation and a calcium content in

Table 4. The effect of application stage of foliar applications of Mo on grain yield and protein content of wheat (SST 66)

Molybdenum application stage[#]	Yield (kg ha^{-1})	Protein (%)
Control	3154	13.33[a*]
Seed treatment	3228	14.17[b]
3 leaf stage	3309	14.59[b]
6 leaf stage	3439	13.89[ab]
Flag leaf stage	3000	14.25[b]
Seed+3+6+flag stages	2891	13.76[ab]

*Means not sharing the same letter are significantly different
[#]Mo applied at 600 mg l^{-1}

excess of 3000 mg l^{-1}. Despite the fact that Mo deficiency would not have been expected to appear on these soils, foliar application of the microelement increased protein content. Cairns and Kritzinger (1992) observed improved levels of grain protein content when a total of 200 mg l^{-1} Mo was applied to cultivar RL4137 in a split application at the sixth and flag leaf stages. The increased protein content was assumed to be via increased activity of the molybdo-enzyme, nitrate reductase which, according to Layzell (1990) is the rate-determining step in the nitrogen assimilation pathway of the plant. However, Modi (1993) could find no increase in grain protein content when he applied 100 ppm Mo at the flag leaf stage on six cultivars growing at four sites. He however found a significant decrease in the nitrate content in grain harvested from Mo-sprayed plants (results not shown). The results shown in Table 4 would seem to indicate that while 100 mg l^{-1} Mo might be sufficient to boost the dormancy of wheat, higher levels of Mo will be needed to increase protein content. The optimum stage of application of Mo for increased grain protein content would also seem to be different from that needed to induce maximum dormancy. Modi (1993) showed that the flag leaf stage was the optimum application stage for dormancy induction, while the present study indicates that the three leaf stage would be optimum for the increase of protein.

Overall the effect of Mo on grain yield was just not significant but a distinct trend could be observed with yield peaking with the application of Mo at the 6-leaf stage. The slightly lower protein content of the wheat at this stage can be ascribed to be at least partly due to a dilution effect of protein at the higher yield. Further work will have to be carried out to determine the optimum Mo application stage for maximum economic benefit for the producer.

Conclusions

In a wide ranging set of experiments under both controlled environment and field conditions, it has been conclusively shown that Mo can be used to increase dormancy in wheat and by implication to reduce to the risk of pre-harvest sprouting damage in wheat. This, together with the probability that Mo can increase wheat protein content has resulted in the spraying of Mo on several commercial wheat crops in Natal. Wheat is grown in the winter under irrigation in Natal and sprouting damage is a sporadic but serious problem when summer rains fall just before harvest. Protein content of wheat grown in Natal is low ($< 11.5\%$) and if Mo can consistently increase the grain protein content Mo application would become even more commercially viable. However, much more work needs to be done on both prevention of sprouting damage and boosting of protein before any firm recommendations can be made. The dosage rate currently recommended is $25\,g$ $MoO_4.2H_2O$ per hectare and it is usually applied together with other routine crop spraying operations.

References

Barnes, W.C. and Blakeney, A.B. 1974. *Die Staerke* 26: 193–197.

Cairns, A.L.P. and Kritzinger, J.H. 1992. *Plant and Soil* 145: 295–297.

Cowan, A.K. and Richardson, G. 1995. *15th International Symposium on Plant Growth Substances.* Abstract 124.

Farwell, A.J., Farina, M.P.W. and Channon, P. 1991. In: *Plant–Soil Interactions at Low pH.* pp.355–361 (eds. R.J. Wright, V.C. Baligar and R.P. Murrman). Dordrecht: Kluwer Academic.

Gale, M.D. and Lenton, J.R. 1987. *Aspects of Applied Biology* 15:115–124.

Layzell, D.B. 1990. In: *Plant Physiology, Biochemistry and Molecular Biology*, pp.389–406 (eds Dennis, D.T. and Turpin, D.H.). New York: John Wiley and Sons.

Mares, D.J. and Gale, M.D. 1990. In: *Proceedings of the Fifth International Symposium on Pre-Harvest Sprouting in Cereals*, pp. 183–184 (eds Ringlund, K., Mosleth, E. and Mares, D.J.). Boulder, CO, USA: Westview Press.

Mares, D.J. and Mrva, K. 1992. In: *Proceedings of the Sixth International Symposium on Pre-Harvest Sprouting in Cereals*, pp. 178–184 (eds Walker-Simmons, M.K. and Ried, J.L.). Boulder, CO, USA: Westview Press.

Modi, A.T. 1993. *M.Sc. Agric Thesis, University of Natal, Pietermaritzburg.*

Modi, A.T. and Cairns, A.L.P. 1994. *Seed Science Research* 4: 329–333.

Modi, A.T. and Cairns, A.L.P. 1995. *South African Journal of Plant and Soil* 12: 108–111.

Mrva, K. and Mares, D.J. 1995. *Proceedings of the Seventh International Symposium on Pre-Harvest Sprouting in Cereals.* In press. Boulder, CO: Westview Press.

Nicholls, P.B. 1979. *Australian Journal of Plant Physiology* 6: 229–240.

Tanner, P.D. 1978. *Plant and Soil* 49: 427–432.

Walker-Simmons, M.K. 1988. *Plant Physiology* 84: 61–66.

Walker-Simmons, M.K.,Kudrna, D.A. and Warner, R.L. 1989. *Plant Physiology* 90: 728–733.

21. Moisture Content Controls the Effectiveness of Dormancy Breakage in *Abies nordmanniana* (Steven) Spach Seeds

M. JENSEN

Danish Institute of Plant and Soil Science, Department of Ornamentals, Research Centre Aarslev, Kirstinebjergvej 10, 5792 Aarslev, Denmark

Abstract

In an attempt to overcome the lack of control of germination during traditional prechilling of *Abies nordmanniana* seeds, an investigation on the effect of reduced seed moisture content (m.c.) on the rate of dormancy breakage and radicle protrusion was carried out. Reduced m.c. delayed radicle protrusion and reduced the fraction of germinated seeds. At 33% m.c. fresh weight basis (f.w.) no radicle protrusion was seen even after a prolonged period at +4°C. Germination capacity (GC), and mean time to germination (MGT), were calculated from seeds prechilled at eight m.c. levels for different durations and then germinated at 25°C. Reduced m.c. resulted in a decrease in the rate of dormancy breaking, quantified as a lower increase in GC or lower reduction in MGT during the first weeks of prechilling. The critical lower m.c. threshold for dormancy breaking was around 23%. Combining the results it was concluded that prechilling of *Abies nordmanniana* seeds at a controlled m.c. of 33% allowed the release of dormancy in all seeds in the seedlot without any germination during the prechilling treatment. By introducing a higher degree of control the new method means improved germination capacity and an easier and more reliable dormancy breaking method.

Introduction

The term seed dormancy in the species *Abies nordmanniana* is traditionally used to describe the mechanisms that delay germination at favourable temperatures and moisture conditions. Information on dormancy breaking and germination of *Abies nordmanniana* seeds is limited but treatment of intact cones at +4°C enhances germination capacity (Muller, 1971) and prechilling seeds for a few weeks generally improves performance (Franklin, 1974). Comparing the characteristics of accumulated germination curves for a dormant and a non-dormant seedlot, the dormancy is displayed partly as a slower germination and partly as a lower final germination percentage. Most commercial seedlots only contain a small fraction of seeds having a level of dormancy that inhibits germination (Hoyer, 1995; Knudsen, 1995); most seeds can germinate at temperatures between 20–25°C, but will do so at a slow rate.

Dormancy breaking of *Abies nordmanniana* seeds in Denmark is traditionally done by prechilling fully-hydrated seeds at 4°C for 3 to 6 weeks (Nyholm, 1986).

R.H. Ellis, M. Black, A.J. Murdoch, T.D. Hong (eds.), Basic and Applied Aspects of Seed Biology, pp. 181–190.
© *1997 Kluwer Academic Publishers, Dordrecht. Printed in Great Britain.*

Considering the facts that non-dormant, fully-imbibed seeds germinate well at +4°C and that there is a significant difference in the level of dormancy between single seeds in a seedlot, the traditional dormancy-breaking method does not control germination. The percentage of non-dormant, rapidly germinating seeds is increased when prechilling is prolonged. However the number of seeds that germinate during the treatment also increases and since these will not survive normal sowing procedures, this makes a prolongation of the prechilling treatment inappropriate.

One possibility of acquiring control of germination while allowing loss in dormancy to proceed is the application of reduced seed moisture content during prechilling. This method has been investigated in a number of conifer species (Jones and Gosling, 1994; Gosling and Rigg, 1990; De Matos Malavasi *et al.*, 1985). Within the genus *Abies* research on prechilling at reduced moisture contents has been carried out by Tanaka and Edwards (1986), Edwards (1986), and Leadem (1986; 1989).

Most of those studies only investigated a few m.c. levels, which only provide a rough description of how the rate of dormancy breaking is affected by moisture content. Precise cardinal moisture contents such as critical minimum m.c. and optimum m.c. for dormancy breaking have not been determined. This information is essential for developing an optimal prechilling method and improving our ability to predict prechilling efficiency.

This study investigated in a systematic way the effect of seed moisture content on radicle protrusion ability and dormancy breaking effectiveness in *Abies nordmanniana* seeds.

Materials and Methods

The *Abies nordmanniana* seedlot (No. H29 harvested 1993/94), provenance Ambrolauri, Tlugi from Caucasian Georgia, where this species is native, was supplied by the Danish Land Development Society. Seeds were given a 24 h soak in a container continuously flooded with cold aerated tapwater, and redried to the required m.c. The m.c. at full hydration (MC_h), the freshweight of the seed sample at full hydration (FW_h) and the target moisture content (TMC) investigated were entered into the following formula, and the target fresh weight (TFW) of the sample at the target m.c. determined.

$$TFW = \frac{FW_h - (MC_h \times FW_h / 100)}{1 - (TMC/100)}$$

Before drying the seeds were soaked in a 0.3% (w/w) aqueous solution of benomyl fungicide (Benlate) for 5 min. The seeds were dried in open trays in ambient laboratory temperature and humidity.

When a sample reached its target fresh weight, determined by frequent weighing, it was placed in a metal container closed with a lid that allowed oxygen exchange but reduced loss of moisture from the seeds. The volume of seeds to the volume of air in the container was initially about 1:3. The containers were stored in a cold store room at +4°C.

In the first experiment, designed to investigate the critical m.c. for radicle protrusion at +4°C, seeds were equilibrated to eight different m.c. values, 28, 32, 33, 34, 35, 36, 39, and 41% m.c. and prechilled for 11 weeks at +4°C in Petri dishes (10 × 1.5 cm) in tightly closed plastic bags. Four replicates of 100 seeds were tested at each m.c. Every week the number of seeds with a protruding radicle was recorded.

In the second experiment, designed to show the effect of m.c. on the rate of dormancy breaking, seeds were equilibrated to eight different m.c. levels, 8, 14, 18, 21, 25, 32, 34 and 44% m.c. (f.w.) and prechilled for 20 weeks at +4°C in the metal containers described above. Samples of 500 seeds were drawn each week or every second week. From this sample four replicates of 25 seeds were immediately taken for m.c. determination by oven drying for 24 h at 103 ± 2°C. The rest of the sample was divided into 4 replicates of 100 seeds and placed on top of moist blotting paper in a Petawawa (PNFI) germination box (Wang and Ackerman, 1983) containing 300 ml of distilled water supplied to the paper by filter paper wicks. The time for the seeds to reach full hydration from the different equilibrated moisture contents was included in the recorded time to germination. The number of seeds that had germinated during the prechill treatment was recorded before starting the germination test and then removed.

ISTA (1985) prescribes a double germination test with no prechill or 21 days prechill at 3–5°C followed by germinating for 28 days at 20/30°C. The germination test in this study was carried out at +25°C (constant) in the dark except when recording germination. Germinated seeds, defined as seeds with a radicle longer than 3 mm, were counted every second or third day during a 30-day period, after which germination was negligible. Germinated seeds were removed from the germination box immediately after recording.

Germination capacity, including seeds that germinated during the prechill treatment, was calculated on a filled seed basis according to a cutting-test of non-germinated seeds performed at the end of each germination test. Mean germination time for seeds germinating during the germination test only was calculated using the formula of Jones and Gosling (1994) taking into account that recording was done on non-consecutive days.

GC and MGT data from different prechill durations for each m.c. level were described by linear or polynomial regression lines. The slopes of these curves at any duration interval describe the relative changes in GC and MGT and can be used as markers for dormancy breaking activity. The changes are greatest during the first weeks of prechilling and decline when the maximum prechilling effect is almost reached. Consequently estimates for GC and MGT for 0 and 4 weeks of prechilling were calculated from the regression curves and presented as evidence of the efficacy of dormancy breaking at different moisture contents in a non-prechill-saturated seedlot.

Results

Non-prechilled seeds germinated 5–7 days later and reached a final germination capacity about 20% lower than seeds prechilled for 5 weeks at +4°C and 44% m.c. (Fig. 1). The dormancy release curve of a seed sample hydrated to 44% m.c.

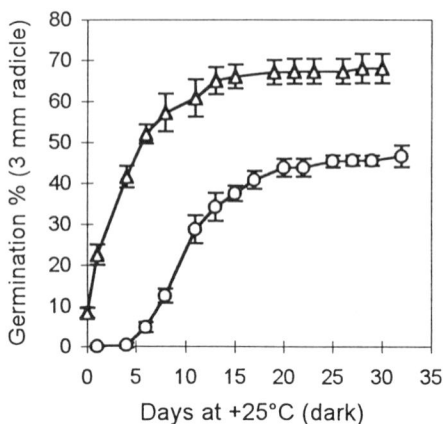

Figure 1. Accumulated germination curves for seeds either non-prechilled (circles) or 5 weeks prechilled (triangles) at + 4°C and 44% m.c., germinated at constant 25°C. Each data point is the mean of four replicates of 100 seeds, ± standard deviation (s.d.)

as related to the germination capacity of this sample after 30 days at 25°C showed a steep increase in GC in the first weeks of prechilling at 4°C followed by a gradual plateauing as the potential prechilling effect was reached (Fig. 2 upper curve). The non-dormant seeds that had the lowest thermal time requirements began germinating during prechilling at 4°C after about 4 weeks, and GC reached a plateau after approximately 14 weeks of prechilling (Fig. 2 lower curve). The germination of seeds during prechilling ended up 10–15% lower than prechilled seeds transferred to a germination test at 25°C for 30 days.

The ability to germinate (radicle protrusion) during prechilling at 4°C depended on the m.c. of the seeds (Fig. 3). The lower the m.c. the smaller the fraction of seeds germinated and the slower the germination. Germination percentage decreased almost linearly with decreasing m.c. ending up in a critical lower m.c. for radicle protrusion around 33–34%. At moisture contents at or below 33% no germination was observed even after prolonged prechilling periods (data not shown).

Germination capacity of non-prechilled seeds was strongly affected by the equilibration m.c. they had when placed on the moist blot paper (Fig. 4). A m.c.

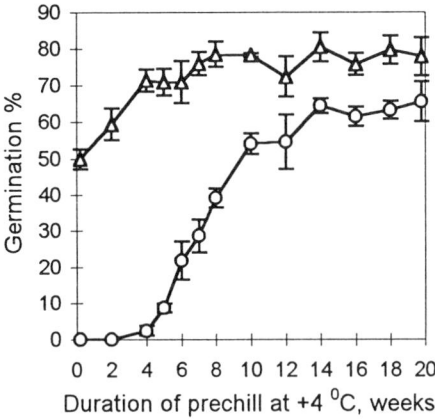

Figure 2. Germination capacity of seeds germinating during prechilling (circles) at +4°C (44% m.c.) after different prechill durations and total percentage of seeds (triangles) germinated during prechilling at +4°C and during a subsequent germination test at 25°C for 30 days. Each data point is the mean of four replicates of 100 seeds, $\pm$ s.d.

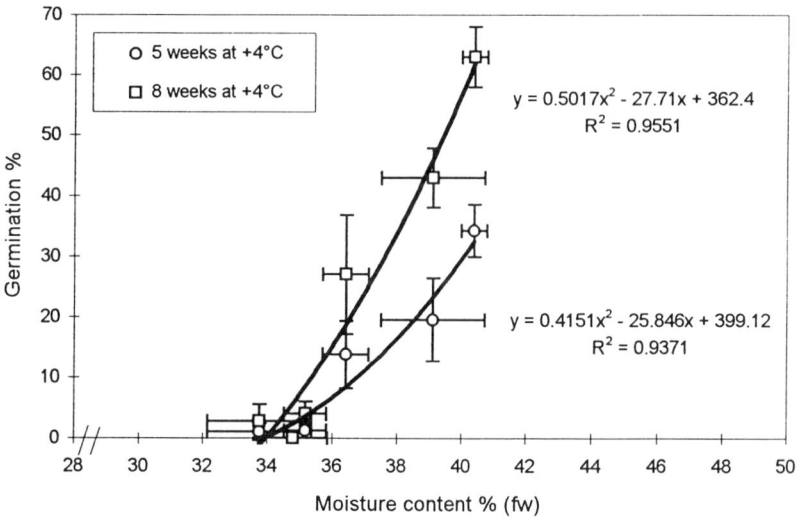

Figure 3. Accumulated germination capacity of seeds held at different constant moisture contents after 5 (circles) and 8 weeks (squares) at +4°C. Each GC data point is the mean of four replicates of 100 seeds $\pm$ s.d. and moisture contents are the mean of four replicates of 25 seeds $\pm$ s.d. taken after one week at +4°C

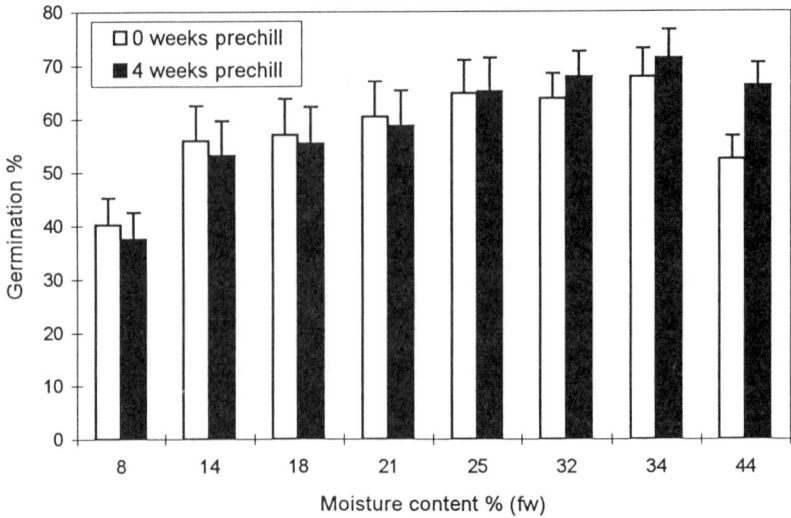

Figure 4. Estimated germination capacity of seeds prechilled at different constant moisture contents at +4°C for 0 or 4 weeks, and then placed on top of moist blot paper and germinated at 25°C for 30 days. GC estimates were generated from regressions of original GC data at each m.c. level. Error bars are the standard errors of GC estimates from the regression. Moisture contents are the rounded mean m.c. within the regression interval

of 8% resulted in the lowest GC. Increased m.c. gave higher GC up to an optimal m.c. around 34%. A moisture content above 40% resulted in a reduction in GC. Comparing non-prechilled and 4 weeks prechilled seeds, it was shown that seeds prechilled at 8–21% m.c. and entered into the germination test at this m.c., produced decreasing GC, whereas m.c. from 25–44% gave increasing GC. The largest increase was seen at 44% m.c. and the critical lower m.c. for increase in GC was around 23%. A m.c. of 34% gave the highest GC during the 4 weeks prechill period. Statistical analysis of the regression curves showed that all slopes except for 25% m.c. were significantly different from zero ($p < 0.05$) indicating significant changes in GC during prechilling.

Only small differences were found between MGT for seeds placed on the blot paper at different m.c. levels at 0 weeks of prechilling (Fig. 5). Seeds equilibrated to 8% m.c. germinated 2–3 days slower than seeds at higher m.c. The fastest germination for non-prechilled seeds was at 32% m.c. Comparing non-prechilled seeds with seeds prechilled for 4 weeks, seeds held at 8 to 21% m.c. during prechilling produced slightly increasing MGT (regression analysis for 14 and 21% m.c. data showed significant increases in MGT ($p < 0.05$)) whereas seeds at a m.c. from 25 to 44% produced decreasing MGT (regressions showed highly significant reductions in MGT ($p < 0.0001$)).

Prechilling of seeds at 44% m.c. gave the largest reduction in MGT and

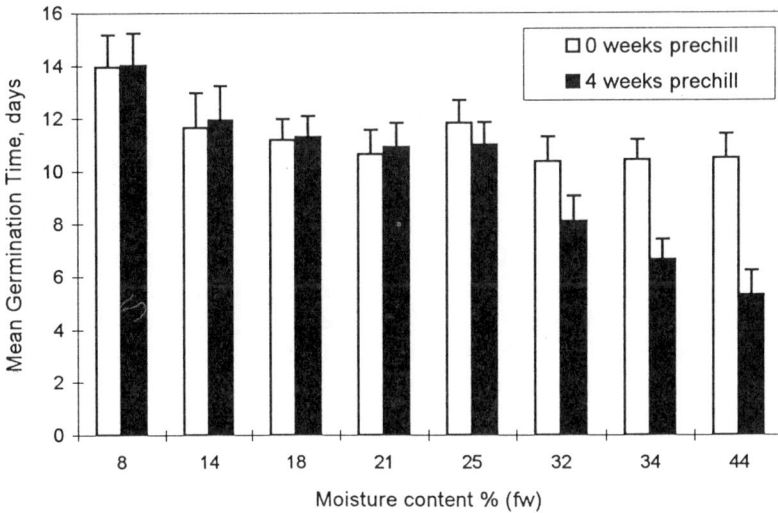

Figure 5. Estimated mean germination time of seeds prechilled at different constant moisture contents at +4°C for 0 or 4 weeks and then placed on top of moist blot paper and germinated at 25°C for 30 days. MGT estimates were generated from regressions of original MGT data at each m.c. level. Error bars are the standard errors of MGT estimates from the regression. Moisture contents are the rounded mean m.c. within the regression interval

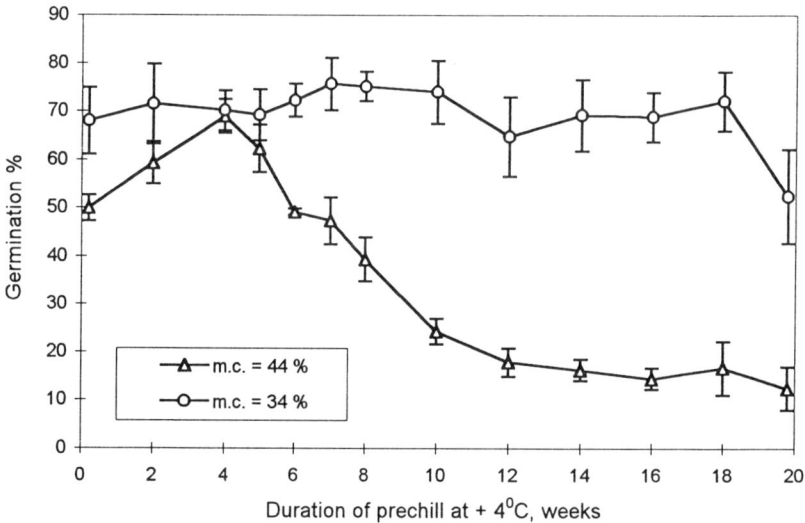

Figure 6. Percentage of seeds germinated only during a germination test at 25°C for 30 days following a prechilling at +4°C for different durations at 34% (circles) or 44% (triangles) m.c. Seeds which germinated during the prechill treatment are not included in these data. Data are means of four replicates of 100 seeds ±s.d.

decreasing the seed m.c. reduced the reduction in MGT. Critical m.c. for reduction of MGT was found to be around 23%. Seeds prechilled at 44% m.c. had the lowest MGT after the 4 weeks of prechilling.

As pointed out earlier germinated seeds with elongated radicles will not produce seedlings following standard nursery sowing. A comparison of prechilling seeds at 34% or at 44% m.c. for different durations was consequently made only on those seeds that germinated at 25°C after the end of the prechilling treatment, i.e. total GC – GC during prechilling (Fig. 6). Prechilling of seeds at 34% m.c. gave higher GC at all durations except after 4 weeks of treatment. GC after prechilling at 44% m.c. was strongly dependent on the duration of the treatment and was seriously reduced after a prolonged prechilling. Prechilling at 34% m.c. was much less sensitive to the duration of the prechilling and gave uniform high GC for an extended period of prechilling.

Discussion

The delay of germination and the lower final germination capacity of the dormant seeds are typical expressions of seed dormancy in *Abies nordmanniana*. Changes in both characteristics can thus be used as markers for the rate of dormancy release.

The problem of seeds germinating during the prechilling treatment is encountered in many conifers and broadleaved trees and is therefore an important problem in production. In a few species, *Pseudotsuga menziesii*, germination at 2–4°C is delayed for several months (Edwards, 1986), which allows a prolonged prechilling period with little risk of germination. The ability to germinate at very low temperatures seems to be a general feature of north temperate species of *Abies* (Franklin and Kruger, 1968). Seeds of *Abies nordmanniana* hydrated to 44% m.c. begin to germinate during prechilling at 4°C after 4 weeks and *A. grandis, A. lasiocarpa* and *A. amabilis* kept at 45% m.c. began germinating after 3 months of prechilling at 2°C (Edwards, 1986). Controlling germination during prechilling by lowering the prechill temperature is therefore not practical.

The use of a reduced m.c. to restrict germination is a universal method that can be used on most species without problems. The critical m.c. for radicle protrusion depends on the biochemical composition of the seeds in a species and is therefore a unique character. The critical m.c. around 33–34% m.c. found in this study is very close to those indicated for the three *Abies* species mentioned above (Edwards, 1982; Tanaka and Edwards, 1986). This suggests a similarity in seed composition or moisture control of radicle protrusion within this genus. Species from other genera are known to have different critical moisture contents for radicle protrusion (Jones and Gosling, 1994).

The general reaction of prechilling seeds at moisture contents below the critical m.c. for radicle protrusion found in this study is similar to what has been found in other *Abies* species as well as in other conifer species. Very low moisture contents were not effective in breaking dormancy, whereas a m.c. just below the critical m.c. for radicle protrusion seems to give the highest rate of dormancy breaking (Edwards, 1982; 1986; Gosling and Rigg, 1990). The critical lower m.c. threshold for dormancy breaking in *Abies nordmanniana* seeds was indicated to be around 23%. Results from Edwards (1982) indicates that the critical lower m.c. for dormancy breaking in other *Abies* species is in close agreement with the results found for *Abies nordmanniana* in this study. This suggests a high degree of similarity in the regulation of dormancy breaking activity by moisture content. In comparison it has been shown that dormancy breakage in *Picea sitchensis* can occur at a m.c. as low as 15–20% f.w. (Gosling and Rigg, 1990) but at a very slow rate. This indicates that differences in cardinal m.c. for dormancy breakage is to be expected, whereas the general reaction pattern to reduced m.c. could be the same for a number of conifer species. A high rate of dormancy breakage is interesting for growers, but of equal or higher importance is that a prechilling method is easy to manage and provides predictable seed performance following prechilling of seeds. Prechilling of seeds at reduced moisture contents seems to offer a number of these advantages.

Acknowledgements

The author wishes to thank Professor J.D. Bewley for valuable advice in this study, Lisa Sargeant for help in germination testing, and Dr. Bruce Downie for inspiring discussions. Thanks also to PNFI for making the germination boxes available. This study was carried out while the author was a visiting Ph.D. student at The Department of Botany, University of Guelph, Canada.

References

De Matos Malavasi, M., Stafford, S.G. and Lavender, D.P. 1985. *Annales Sciences Forestières* 42: 371–384.
Edwards, D.G.W. 1986. *Journal of Seed Technology* 10: 151–171.
Edwards, D.G.W. 1982. In: *Proceedings of IUFRO International Symposium on Forest Tree Seed Storage*, September 23–27, 1980, Petawawa National Forestry Institute, Chalk River, Ontario, pp. 195–203 (eds B.S.P. Wang and J.A. Pitel). Ottava, Ontario: Canadian Forestry Service.
Franklin, J.F. and Kruger, K.W. 1968. *Journal of Forestry* 66: 416–417.
Franklin, J.F. 1974. In: *Seeds of Woody Plants In The United States, Agricultural Handbook No. 450*, pp. 168–183 (ed. C.S. Schopmeyer). Washington DC: USDA.
Gosling, P.G. and Rigg, P. 1990. *Seed Science and Technology* 18: 337–343.
Hoyer, H. 1995. Personal communication. The Danish Land Development Society, Seed Laboratory, Krogaardsvej 6, Tvilum, DK-Faarvang, Denmark.
ISTA (International Seed Testing Association). 1985. International rules for seed testing. Annexes 1985. *Seed Science and Technology* 13: 356–513.
Jones, S.K. and Gosling, P.G. 1994. *New Forests* 8: 309–321.
Knudsen, F. 1995. Personal communication. Tree Improvement Station, Krogerupvej 21, DK-3050 Humlebaek.
Leadem, C.L. 1986. *Canadian Journal of Forest Research* 16, 755–760.

Leadem,C.L. 1989. *Stratification and Quality Assesment of Abies lasiocarpa Seeds.* The Canada-British Columbia Forest Resource Development Agreement, FRDA Report 095. Ministry of Forests and Forestry Canada. and Canada/BC Economic & Regional Development Agreement.

Muller, C. 1971. *Biologie et Forêt,* F23-4: 436–439.

Nyholm, I. 1986. *Haandbog i Frøbehandling.* Published and distributed by Dansk Planteskoleejer Forening, Hvidkjærvej 29, 5250 Odense V, Denmark.

Tanaka, Y. and Edwards, D.G.W. 1986. *Seed Science and Technology* 14: 457–464.

Wang, B.S.P. and Ackerman, F. 1983. *A New Germination Box for Tree Seed Testing.* Information Report PI-X-27, Petawawa National Forestry Institute, Chalk River, Ontario, Canadian Forestry Service, Environment Canada.

22. Primary Dormancy in Tomato. Further Studies with the *sitiens* Mutant

H.W.M. HILHORST

Department of Plant Physiology, Wageningen Agricultural University, Arboretumlaan 4, NL-6703 BD Wageningen, The Netherlands

Abstract

Over the past ten years the non-dormant ABA-deficient *sitiens* mutant of tomato (*Lycopersicon esculentum* cv. Moneymaker) has been used to study primary dormancy. Much of the attention has been focussed on the role of endosperm softening, a prerequisite for radicle protrusion, and control of this process by plant hormones. Other studies were directed at water relations of the embryo with regard to embryo expansion and radicle elongation. Only recently studies of a possible role of the seed coat were initiated. A short summary will be given of these results and what they have taught us about the mechanism and control of tomato seed dormancy.

Introduction

General

Many higher plant species produce seeds that are dormant upon shedding and dispersal. This prevents the newly formed seeds from germinating under end-of-season conditions that are adverse to successful growth. This type of dormancy is often called primary dormancy, solely to indicate that it is imposed during seed development. Thus, it is by no means a qualification referring to mechanism or location, but one of timing of occurrence.

Primary dormancy is one of the many survival strategies of higher plants. It serves a spread of risk, both in time and place and is therefore common among wild species. Many of our modern crop species, which are the result of intensive breeding programs, lack dormancy since it is obviously an unwanted characteristic. However, dormancy in crop species may recur when germination and seedling growth take place under stress conditions, such as drought or/and extreme temperatures. Low, slow and asynchronous germination can be the result. In this sense it may be argued that dormancy is on a similar scale as vigour. It is therefore no wonder that most of the research of dormancy is presently carried out on crop species.

It is now generally accepted, although not conclusively proven, that abscisic acid (ABA) plays a pivotal role in the establishment of primary dormancy. It has been shown for a large number of species that the characteristic transient rise in ABA content at approximately mid-development coincides with the imposition

R.H. Ellis, M. Black, A.J. Murdoch, T.D. Hong (eds.), Basic and Applied Aspects of Seed Biology, pp. 191–201.

of dormancy. However, it also coincides with several other major events of seed development, such as dehydration, solidification of endosperm, testa formation and seed abscission. This makes establishment of a causal relationship between ABA and dormancy complicated. A major step forward in seed dormancy research was made with the introduction of ABA-deficient mutants. These mutants are now available for *Arabidopsis*, maize, pea, and tomato. Many of these mutants have in common that they are wilty and that they lack dormancy. This provided direct proof that the absence of ABA during seed development results in a lack of dormancy in the mature seed. Consequently, studies have aimed at elucidating the mechanism of ABA action in the development of dormancy.

Some Characteristics of Normal and ABA-deficient Tomato Seeds

The mature tomato seed consists of an embryo which is completely invested by several layers of thick walled endosperm cells. The endosperm is surrounded by a testa of 3–5 cell layers thick, derived from the integuments, and presumably consisting of dead tissue. The outer layer of the testa contains numerous hairs which are formed from the outer epidermis of the integument (Smith, 1935).

Tomato seeds develop and mature in the moist environment of the fruit. The final water content of the seeds is high enough to allow germination. Thus, precocious growth of the embryo must be suppressed from as early as 20 days after anthesis when embryogenesis *sensu stricto* has been completed and seeds are fully germinable (Berry and Bewley, 1991).

Seed and fruit development are not clearly affected in the *sit*^w mutant (Groot *et al.*, 1991). The most conspicuous differences compared with the normal seeds occur during late maturation. Mutant seeds do not develop dormancy whereas normal seeds acquire variable levels of dormancy. In addition, mutant seeds often show vivipary in overripe fruits from aproximately 70 days after anthesis onwards. These observations have led to the hypothesis that ABA plays a role in the prevention of precocious germination and is required to induce primary dormancy (Groot and Karssen, 1992). However, also ABA is thought to be involved in several other events during seed development, such as storage protein synthesis and induction of desiccation tolerance (Black, 1991), all occurring during approximately the same period of seed maturation. Thus, the question may be raised whether ABA affects precocious germination and dormancy in a direct manner or through the other events.

In this contribution a summary will be given of the information that has so far been obtained from studies with the ABA-deficient *sitiens* mutant of tomato, with respect to viviparous germination, dormancy and its location within the seed.

The Characterization of Tomato Seed Dormancy

Freshly harvested tomato seeds show a considerable variation in their level of dormancy. There are indications that elevated temperatures during development are conducive to dormancy in tomato (Groot and Karssen, 1992; Y. Liu, personal communication). Dormancy can be partly relieved by pre-chilling at 2°C and irradiation with red light. However, dry storage for several weeks is most efficient in removing dormancy (Groot and Karssen, 1992). Seeds from the sit^w mutant are never dormant and germinate readily in water without light. As mentioned above, seeds that appear non-dormant under optimal conditions may show signs of dormancy under stress conditions. When germination of normal seeds is compared with seeds from the mutant under osmotic stress in a range of polyethylene glycol (PEG) solutions, a substantial difference in resistance to the applied stress becomes apparent (Fig. 1). Mutant seeds are able to germinate at lower (more negative) solute potential than normal seeds. This difference may be regarded as an expression of a deeper dormancy in the normal seeds. Dry storage does not remove this dormancy, not even after several years (Hilhorst, 1995). It is this feature that makes the tomato seed an attractive system for dormancy studies which may also be relevant to seed practice.

The Location of Tomato Seed Dormancy

The Mechanism of Germination

In simple physical terms, germination of the tomato seed depends on the net result of two opposing forces: the thrusting force of the embryo and the resistance of the surrounding endosperm and testa. In comparison to the whole seed, the embryo has a negative water potential, of the order of –2.5 MPa (Haigh and Barlow, 1987). The mechanical restraint of the surrounding tissues prevents water uptake by the embryo. As the embryo does not accumulate solutes, nor increase its pressure potential prior to germination, the control of germination appears to lie in the events which lead to weakening of the mechanical restraint of endosperm or/and testa. Indeed, endosperm softening occurs prior to radicle protrusion. By means of measuring the puncture force required to break through the endosperm and testa layers, Groot *et al.* (1988) have convincingly shown that the puncture force decreases from approximately 0.60 Newtons (N) to 0.20 N before radicle protrusion can occur. The remaining restraint is probably caused by the testa. Presumably, endosperm weakening occurs through hydrolysis of the cell walls. The enzyme endo-β-mannanase has been implicated in the endosperm degradation of a number of species, including lettuce (Leung *et al.*, 1979), fenugreek (Reid *et al.*, 1977), and tomato (Groot *et al.*, 1988). Interestingly, activity of this enzyme is controlled by gibberellins which induce its activity as well as endosperm weakening, and promote germination, and ABA which acts as an antagonist on all three processes (Groot *et al.*, 1988).

In summary, a block to tomato seed germination may theoretically be located in all of the seed's tissues: in the embryo by controlling water potential, in the endosperm by controlling cell wall hydrolysis, and in the testa as a passive, restraining force, since its cells are dead. In the following paragraphs evidence for either of these possible sites of dormancy will be discussed.

The Embryo as Site of Dormancy

A first indication that in tomato dormancy is not located in the embryo was obtained from a simple experiment in which the micropylar end of the endosperm and testa were removed, thus relieving the embryo from its restraint (Groot and Karssen, 1992; Fig. 1). The germination response to the osmotic stress of the detipped normal seeds became similar to that of the intact *sit*^{w} seeds. Interestingly, detipping of the *sit*^{w} seeds did not affect the response (Fig. 1). This suggests that in the *sit*^{w} seeds the mechanical restraint was not limiting to germination under these conditions, but another process, possibly the

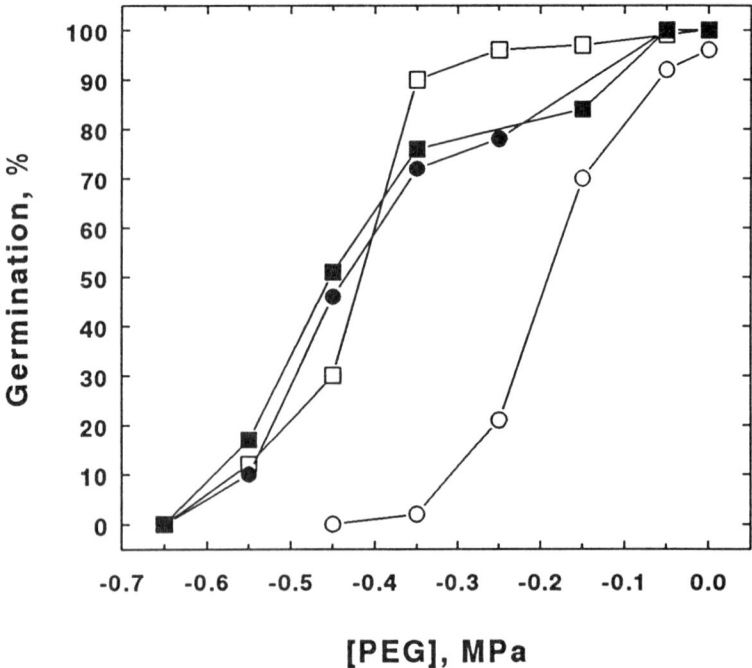

Figure 1. Germination of intact (○) and detipped (●) normal seeds and of intact (□) and detipped (■) *sit*^{w} seeds over a range of external osmotic potentials. From Hilhorst and Downie, 1996

Table 1. Water relations and soluble carbohydrates in embryos of normal and *sit^w* tomato embryos. Water and osmotic potentials were measured 4 h before germination of the first seed. Carbohydrates determined by PAD-HPLC in extracts from embryos from dry seeds. From Hilhorst and Downie (1996)

	Normal	*sit^w*
Water potential, H_2O imbibed (MPa)	−1.44	−1.51
Water potential, PEG imbibed (MPa)	−1.59	−1.61
Osmotic potential, H_2O imbibed (MPa)	−1.72	−1.75
Osmotic potential, PEG imbibed (MPa)	−1.91	−2.07
Soluble sugar content (µg/mg dry weight)	23.9	29.0

embryo expansion. Measurements of embryonic water and osmotic potentials in water and −0.3 MPa PEG supported the suggestion that the block to germination was not situated in the embryo. No major differences were found between both genotypes, either in water or in −0.3 MPa PEG (Table 1). The osmotic potential of *sit^w* embryos in PEG was slightly more negative than that of the normal embryos, but could not account for the dissimilar germination. An analysis of the soluble sugars present in the embryo under normal and stress conditions revealed that the small difference in embryonic osmotic potentials was likely to be caused by the higher level of soluble sugars in the *sit^w* embryo (Table 1). These findings are in agreement with reports of altered carbohydrate metabolism in ABA-deficient mutants of *Arabidopsis thaliana* (DeBruijn *et al.*, 1993). In these mutants the synthesis of long-chain fatty acids from carbohydrate sources is impeded, resulting in substantially higher sugar levels.

In seeds of *Brassica napus* exogenous ABA inhibited cell elongation without affecting the osmotic potential of the seeds (Schopfer and Plachy, 1985). It was concluded that ABA could act at the site of wall loosening, presumably by controlling wall extensibility. To explain the persistent difference in resistance to osmotic stress, Groot and Karssen (1992) adopted this suggestion for tomato. They argued that mature seeds of both genotypes contained too little ABA to account for the difference in germination performance. Hence, the difference must originate from the stage of development during which *sit^w* seeds contain less than 3% of the ABA content of the normal seeds. However, so far, evidence for a greater cell wall extensibility in mutant embryos is lacking. The fact that normal and mutant water potentials are similar (Table 1) strongly contradicts this possibility.

The Endosperm as Site of Dormancy

There is some consensus now that one of the major control points in tomato seed germination is the weakening of the endosperm. Haigh and Barlow (1987) suggested this from their observation that the embryo water potential was sufficiently low for expansion. They could not detect a lowering of the embryo osmotic potential or a build up of turgor prior to germination. It was concluded that embryo water content was restricted by the constraint on embryo expansion caused by the enclosing endosperm. Ni and Bradford (1993) reached a similar conclusion, deduced from the effects of ABA and GA on tomato seed germination under osmotic stress. GA increased and ABA decreased the germinating fraction of the seed population by influencing the relative frequency distribution of the so-called base water potentials. The base water potential of a given seed is that potential that just inhibits germination. These observations were congruent with the aforementioned effect of these hormones on endosperm weakening.

Endosperm weakening prior to germination has been reported for *Capsicum annuum* seeds (Watkins and Cantliffe, 1983), *Datura ferox* (Sanchez *et al.*, 1990) and tomato (Groot and Karssen, 1987). The process has been studied in some detail in tomato. The cell walls of tomato seed endosperms contain galacto-mannans or/and galactoglucomannans (Groot *et al.*, 1988). The mannan back-bone can be cleaved by the enzyme endo-β-mannanase (EC 3.2.1.78) which might weaken the mechanical restraint of the endosperm and allow radicle protrusion. Activity of this enzyme increases prior to germination and is controlled by GA. GA-deficient tomato seeds of the *gib1* mutant do not germinate on water nor display mannanase activity (Groot *et al.*, 1988). Application of GA both induces enzyme activity and germination. This is suggestive of a key role for this enzyme in germination. Further studies revealed that mannanase activity increased only in the micropylar endosperm before radicle protrusion whereas activity in the rest of the endosperm tissues did not increase until after radicle protrusion (Nonogaki *et al.*, 1995). This corroborates strongly with a role for mannanase in endosperm softening.

Mannanase activities were measured in endosperm extracts from normal and *sit*^{w} seeds in water and PEG to determine if mannanase activity could be the limiting factor in the germination of normal seeds on –0.3 MPa PEG. Activities were indeed higher in the *sit*^{w} endosperms under all conditions (Fig. 2). Highest activities were measured in dry seeds. This raises the possibility that the enzyme is also active during seed development. In –0.3 MPa PEG mannanase activities were significantly lower than in water. This suggests that galacto-mannan breakdown is inhibited by osmotic stress. It confirms the results of Spyropoulos and Reid (1988) who found that water stress suppressed mannanase activity in fenugreek seeds. However, a lower galacto-mannan degrading activity does not account for the inability of the normal tomato seeds to germinate on –0.3 MPa PEG. It may be anticipated that the germination rate is slower due to slower endosperm softening. Even under the considerable osmotic stress of a –1.1 MPa

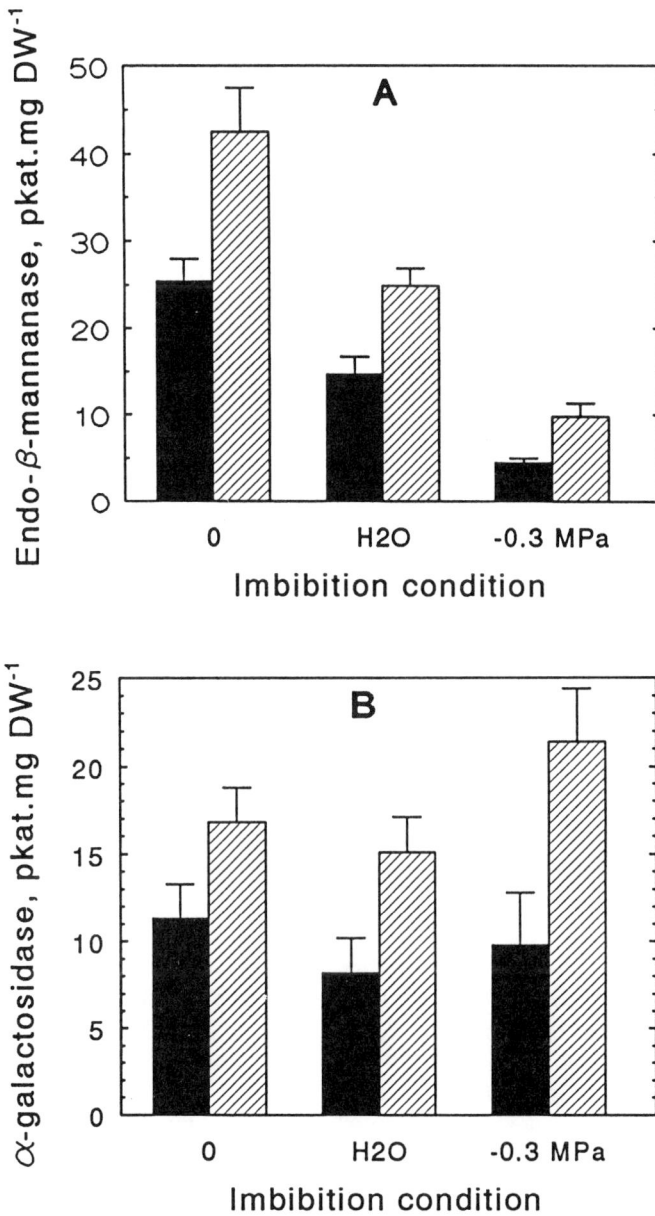

Figure 2. A) Endo-β-mannanase and B) α-galactosidase activities in endosperm extracts from normal (solid bars) and sit^w (hatched bars) seeds. Conditions: "0" dry seeds; "H₂O": water imbibed; "−0.3 MPa": imbibed in −0.3 MPa PEG. All imbibitions were stopped 4 h before radicle protrusion of the first seed. From Hilhorst and Downie, 1996

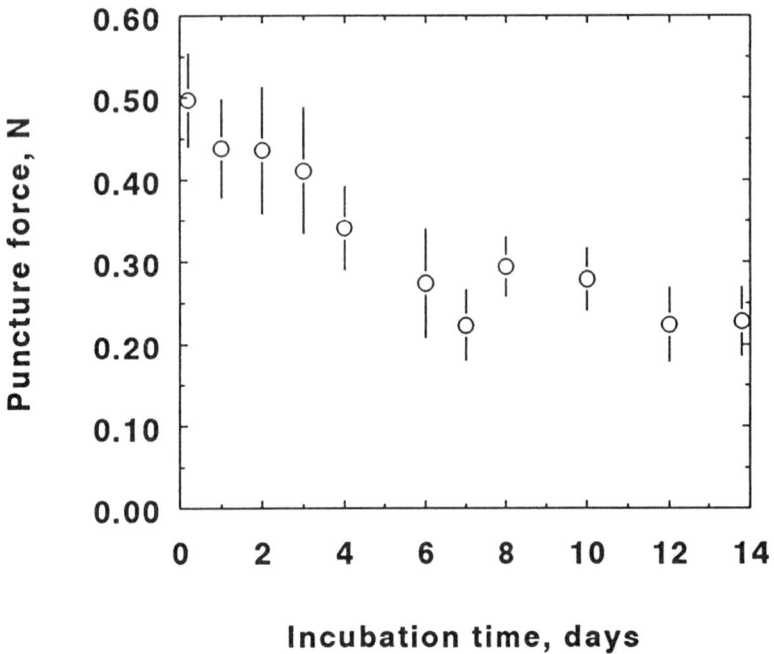

Figure 3. Puncture force required to break through the endosperm + testa opposing the radicle tip of normal seeds during priming in –1.1 MPa PEG. Karssen and Heimgarten, original data

PEG solution endosperm weakening occurs in tomato (Fig. 3). Yet radicle protrusion does not follow.

In summary, endosperm softening is an obvious prerequisite for tomato seed germination. Yet, endosperm breakdown seems not decisive for radicle protrusion. It occurs both in normal and *sit^w* endosperms but only *sit^w* seeds germinate on –0.3 MPa PEG (Fig. 1).

The Testa as Site of Dormancy

The failure to find possible sites of tomato seed dormancy in either embryo or endosperm prompted an exploration of some characteristics of the testa that could contribute to dormancy. A first simple approach was to repeat the 'detipping' experiment described above, but now with only the testa removed, leaving the endosperm fully intact. Surprisingly, stripping off the micropylar end of the testa resulted in a germination performance of the normal seeds on –0.3 MPa which was similar to that of the *sit^w* seeds (Fig. 4). Scanning electron microscopy of dry seeds corroborated these findings. The testa of normal seeds

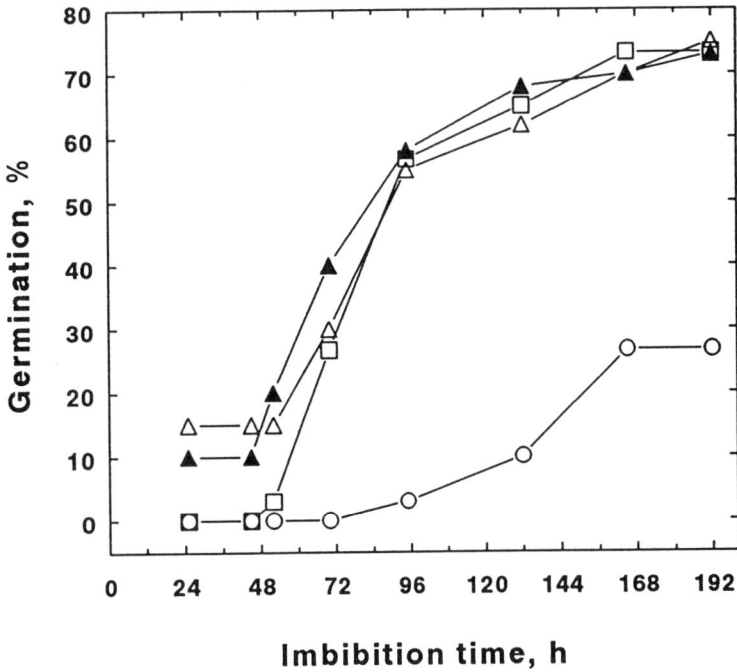

Figure 4. Germination progress curves in –0.3 MPa PEG solution of intact normal (○) and *sit^w* (□) tomato seeds, and of normal (△) and *sit^w* (▲) seeds with the micropylar end of the testa removed. From Hilhorst and Downie, 1996

consists of 5 cell layers whereas *sit^w* seeds have testa of only one cell layer thickness (Hilhorst and Downie, 1996). If we assume that the thinner testa has a lower mechanical strength it may explain the response to osmotic stress of the normal and *sit^w* seeds. In both genotypes endosperm weakening occurs at –0.3 MPa. After the completion of endosperm degradation both the normal and the *sit^w* embryo experience a residual resistance from the testa, in addition to the osmotic restraint imposed by the PEG. Apparently, the higher restraint of the testa together with the solute potential of the medium prevented embryo expansion in the normal seed whereas in the *sit^w* seeds the (lower) restraint could still be overcome by the embryo water potential, allowing radicle protrusion. Removing the testa removes the difference in restraint and results in a similar response since embryo water potentials do not differ between both genotypes. The seed shape mutants *ats* and the transparent *ttg* from *Arabidopsis thaliana*, possessing thinner testas, also germinated faster than normal seeds (Léon-Kloosterziel *et al.*, 1994). This suggests that, in general, testa structure of seeds with thin and dead seed coats may be more important in the germination process than is currently believed. It remains to be seen whether the thinner testa of the sit^w seeds is a direct result of the absence of ABA in the tomato mutant. If so, ABA may play a role in integument or/and testa formation.

Can a Thinner Testa Account for Vivipary in ABA-deficient Seeds?

Apart from a thinner testa, mature dried seeds from the sit^w mutant displayed other characteristics which were not found in normal seeds. Mutant seeds consistently showed free space around the embryo and often the radicle tip was swollen (Hilhorst and Downie, 1996). These features have also been described for normal tomato seeds primed in PEG (Liu *et al.*, 1994). X-ray analysis of tomato seeds showed that the induction of free space during priming only occurred in seeds that had been dehydrated before. It is quite possible that during priming endosperm hydrolysis takes place, since endosperm resistance has been shown to decrease in PEG (Fig. 3). This would result in the induction of free space. Consequently, the mechanical restraint is locally removed and may lead to water uptake and, hence, expansion of the embryo. It is tempting to speculate that sit^w seeds 'prime' in the fruit during late maturation. However, it is not clear why normal seeds do not do this. At late maturation tomato seed germination is virtually insensitive to ABA (Berry and Bewley, 1992). Moreover, the osmotic potential of the locular tissues surrounding the seeds is similar in normal and mutant seeds (Liu *et al.*, 1994) and endosperms of both genotypes contain mannanase activity (Fig. 2). It suggests that the absence of ABA during seed development, as in the sit^w mutant, may eliminate the normally required drying for free space formation. Thus, the occurrence of vivipary in sit^w seeds and its absence in normal seeds cannot be fully explained by the thinner sit^w testa.

Conclusions

An important conclusion that can be drawn from this short overview is that the testa may play a decisive role in the germination of the tomato seed and possibly also plays a role in viviparous germination. The tomato seed does not possess a hard or thick seed coat and its dormancy will thus not be categorized as a typical coat imposed dormancy. However, there are more examples of (thin) seed coat modifications that influence germination.

The results described above make clear that in the tomato seed all seed parts, including the dead testa, contribute to successful germination. Testa resistance may be regarded as a minimum restraint that the embryo must overcome in order to protrude. This resistance determines if a seed will complete germination. Embryos not able to generate sufficient force to penetrate the testa will not protrude, as shown above for the normal seeds on −0.3 MPa PEG (Fig. 1). In series with testa resistance is the resistance of the endosperm. This resistance controls the rate of germination since other physiological processes, such as induction of hydrolytic activity in the endosperm and water uptake by the embryo, must first take place (Leviatov *et al.* 1994; Hilhorst and Downie, 1996). These processes require time to proceed to a point sufficient for radicle protrusion.

References

Berry T. and Bewley, J.D. 1991. *Planta* 186: 27–34.

Berry, T. and Bewley, J.D. 1992. *Plant Physiology* 100: 951–957.

Black, M. 1991. In: *Abscisic Acid. Physiology and Biochemistry.* pp. 99–124 (eds. W.J. Davies and H.G. Jones). Oxford: BIOS Scientific Publishers.

DeBruijn, S.M., Ooms, J.J.J., Basra, A.S., Van Lammeren, A.A.M. and Vreugdenhil, D. 1993. In: *Basic and Applied Aspects of Seed Biology. Proceedings of the Fourth International Workshop on Seeds*, pp.103–108 (eds. D. Côme and F. Corbineau). Paris: ASFIS.

Groot, S.P.C. and Karssen, C.M. 1987. *Planta* 171: 525–531.

Groot, S.P.C. and Karssen, C.M. 1992. *Plant Physiology* 99: 952–958.

Groot, S.P.C., Kieliszewska-Rokicka, B., Vermeer, E. and Karssen, C.M. 1988. *Planta* 174: 500–504.

Groot, S.P.C., VanYperen, I.I. and Karssen, C.M. 1991. *Physiologia Plantarum* 81: 73–78.

Haigh, A.M. and Barlow, E.W.R. 1987. *Australian Journal of Plant Physiology* 14: 485–492.

Hilhorst, H.W.M. 1995. *Seed Science Research* 5: 61–73.

Hilhorst, H.W.M. and Downie, B. 1996. *Journal of Experimental Botany* 47:89–97

Léon-Kloosterziel, K.M., Keijzer, C. and Koornneef, M. 1994. *Plant Cell* 6: 385–392.

Leung, D.W.M., Reid, J.S.J. and Bewley, J.D. 1979. *Planta* 146:335–341.

Leviatov, S., Shoseyov, O. and Wolf, S. 1994. *Scientia Horticulturae* 56: 197–206.

Liu, Y. Bergervoet, J.H.W., RicDeVos, C.H., Hilhorst, H.W.M., Kraak, H.L., Karssen, C.M. and Bino, R.J. 1994. *Planta* 194: 368–373.

Ni, B.-R. and Bradford, K.J. 1993. *Plant Physiology* 101: 607–617.

Nonogaki, H., Nomaguchi, M. and Morohashi, Y. 1995. *Physiologia Plantarum* 94: 328–334.

Reid, J.S.G., Davies, C. and Meier, H. 1977. *Planta* 133:219–222.

Sánchez, R.A., Sunell, S., Labavitch, J.M. and Bonner, B.A. 1990. *Plant Physiology* 93: 89–97.

Schopfer, P. and Plachy, C. 1985. *Plant Physiology* 77: 676–686.

Smith, O. 1935. *Cornell University Agricultural Station Memo* 184: 3–16.

Spyropoulos, C.G. and Reid, J.S.G. 1988. *Planta* 174: 473–478.

Watkins, J.T. and Cantliffe, D.J. 1983. *Plant Physiology* 72: 146–150.

23. Seed Dormancy and Responses of Seeds to Phytohormones in *Nicotiana plumbaginifolia*

M. JULLIEN and D. BOUINOT

Unité de Recherche Biologie des Semences INRA/INA P-G, Institut National Agronomique Paris-Grignon, 16 rue Claude Bernard, 75231 Paris Cedex 05, France

Abstract

We describe some characteristics of the dormancy in *N. plumbaginifolia* wild type seeds and ABA deficient *I217* mutant seeds. The effects of the main phytohormones on germination were considered in relation to dormancy. It appeared that the depth of dormancy was correlated with variation in the sensitivity to every phytohormone, except ethylene. In particular, some cytokinins exerted a strong inhibitory effect on the germination of the slow germinating wild type dormant seeds which totally disappeared when seeds afterripened. Germination of the ABA deficient mutant was resistant to high levels of auxin and cytokinin but not ABA. Lastly, the very efficient effect on dormancy breaking of the ABA biosynthesis inhibitors norflurazon and fluridone implied an important role of ABA synthesis on the maintenance of dormancy in imbibed seeds.

Introduction

Nicotiana plumbaginifolia is a wild species of Nicotianae which is used for genetic studies in plant development. Numerous hormone mutants have been isolated in this species, auxin autotrophic (Fracheboud and King, 1991), auxin resistant (Bitoun *et al.*,1990; Blonstein *et al.*, 1991b), cytokinin resistant (Jullien *et al.*, 1990; Blonstein *et al.*, 1991a). Some of these mutants were later found to be abscisic acid deficient (Parry *et al.*, 1991; Rousselin *et al.*, 1992). Gibberellin deficient mutants are also available (Traas *et al.*, 1995; Rousselin, unpublished). Elsewhere, phytochrome deficient mutants have been obtained (Kraepiel *et al.*, 1994a). The isolation of these various mutants could give new tools for the study of seed biology. For instance, the ABA deficient *I217* mutant is early germinating (Rousselin *et al.*, 1992) and the phytochrome deficient *pew1* mutant which germinates slowly contains a higher ABA level than the wild type (Kraepiel *et al.*, 1994b), which suggests a role of abscisic acid in dormancy and photosensitivity in this species.

The role of phytohormones in seed biology is well documented (Bewley and Black, 1994). ABA induces dormancy during seed maturation in some species, e.g. *Arabidopsis thaliana* (Hilhorst and Karssen, 1992). In others, such as *Helianthus annuus*, a continuous synthesis of ABA appears to induce but also to maintain dormancy (Bianco *et al.*, 1994), which is in agreement with the well documented inhibitory effect of ABA on seed or embryo germination. Various

R.H. Ellis, M. Black, A.J. Murdoch, T.D. Hong (eds.), Basic and Applied Aspects of Seed Biology, pp. 203–214.
© *1997 Kluwer Academic Publishers, Dordrecht. Printed in Great Britain.*

hormones, especially gibberellins but also in some species cytokinins and ethylene can break dormancy (Thomas, 1992). If the variations in the endogenous level of hormones are probably of great importance in inducing or breaking dormancy, many observations also indicate that variations of sensitivity to the hormones should be considered. For instance, the effect of ABA on the inhibition of the germination of seeds varied with the depth of dormancy (Walker-Simmons, 1987; Morris *et al.*, 1989; Corbineau *et al.*, 1992). Variations of sensitivity of the aleurone tissue to GA have also been correlated to dormancy of caryopses (Schuurink *et al.*, 1992).

Here, we will describe the characteristics of the dormancy and the effects of the main phytohormones and of the ABA biosynthesis inhibitors norflurazon and fluridone on the germination of *N. plumbaginifolia* wild type dormant, non-dormant seeds and ABA deficient mutant seeds.

Material and Methods

The *Nicotiana plumbaginifolia* seeds used were derived from the homozygous line PbH1D (Bourgin *et al.*, 1979) here called the wild type (WT) and from the mutant *1217* selected by Bitoun *et al.* (1990) after mutagenesis of this line. This mutant is impaired in the last step of ABA synthesis (Parry *et al.*, 1991; Rousselin *et al.*, 1992). The corresponding plants were grown in greenhouse in the same conditions of light, temperature and nutrition. Seeds were collected just before the mature capsules became dehiscent and stored dry in a culture room with dim light, 24°C constant temperature and about 50–60% relative humidity.

All the germination experiments were performed with seeds surface sterilized in an aqueous solution of the commercial disinfectant Domestos (Lever) 20% v/v for 1 h, carefully washed, dried again and then stored in the same conditions as above. We verified that this operation was without effect on the level of seed dormancy. For germination experiments, 25 sterile seeds were sown in 5 cm diameter Petri dishes containing 7 ml of agar solidified B medium (Bourgin *et al.*, 1979). Dishes were sealed with parafilm and then incubated for 20–30 days in a culture chamber with short day photoperiod 8 h/16 h (85 $\mu Em^{-2}s^{-1}$) associated with a 24/20°C thermoperiod. For all experiments, 4 dishes per condition were used, that is 100 seeds. Germination measurements were made with the help of a lens.

Hormones when not water soluble were furnished as DMSO solutions. This solvent exerted no effects on *N. plumbaginifolia* seed germination at concentrations lower than 1% v/v. N^6-Benzyladenine (BA), 1-naphthalenacetic acid (NAA), gibberellic acid (GA3), abscisic acid (ABA) and ethephon came from Sigma. Zeatin, N-(chloro-2 pyridyl-4)-N'-phenylurea (CPPU) and fluridone (Dowelanco France) were gifts of respectively M. Laloue (Versailles) and M.T. Lepage (Nice). Norflurazon was a gift of Sandoz France.

Results

Dormancy Characters in N. plumbaginifolia

Seeds of *Nicotiana plumbaginifolia* are of small size (about 0.4×0.3 mm), and they contain a well developed endosperm, which at germination is generally pierced by the radicle after the disruption of the testa.

The dormancy characteristics of *N. plumbaginifolia* seeds have never been depicted. Dry freshly harvested seeds present a typical dormancy which is expressed by the long and very heterogeneous time necessary for seeds to germinate (Fig. 1). The *I217* mutants seeds of the same age are not dormant, they germinate quickly and homogeneously (Fig. 1). As can be seen in Figure 2, dormancy is suppressed by afterripening. Freshly harvested seeds (14 days old) need 16 days to give 50% germination (gt 50%) whereas the same sample stored dry for 6 or 18 months reach this value after 4.5 days. Longer storage becomes then detrimental, such that seeds start to lose viability and surviving seeds germinate more slowly. Accordingly, we consider that *N. plumbaginifolia* seeds stored dry for no more than 2–3 months are dormant (D seeds) and non-dormant (ND seeds) if stored for 6–20 months.

There is no dormancy in the ABA deficient mutant *I217* whose freshly harvested seeds germinate a little earlier than the ND wild type seeds (compare

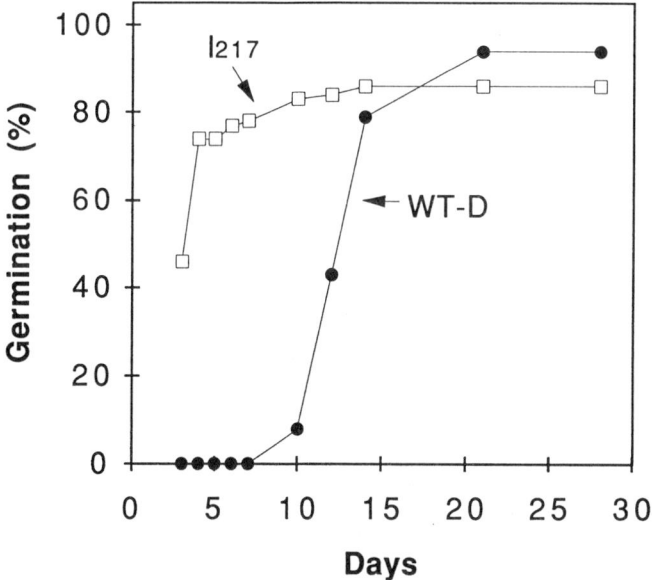

Figure 1. Kinetics of germination of two *N. plumbaginifolia* genotypes. Seeds were stored 1.5 months after harvest before sowing, –●– wild type dormant seeds (WT-D), –□– mutant *I217*

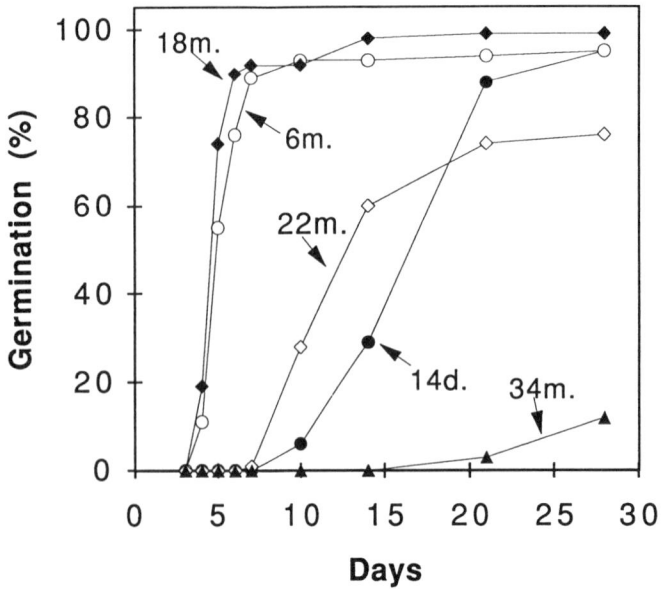

Figure 2. Effect of the duration of dry storage on the kinetics of germination of *N. plumbaginifolia* wild type seeds. The indicated numbers were the age of seeds in days (d) or months (m), that is the duration of dry storage before sowing

Figures 1 and 2). Germination of D and ND wild type seeds is strictly light-dependent (not shown).

Influence of Phytohormones on the Germination of Dormant and Non-dormant Seeds.

The germination of D seeds is greatly stimulated by gibberellic acid (Fig. 3a) whereas ND (Fig. 3 b) and *I217*seeds (not shown) are much less sensitive to this hormone. 100 µM GA_3 appears to be the optimal concentration which gives to D seeds a germination progress curve very close that of the ND seeds.

Ethylene (furnished as ethephon) exerts no effect on the germination of *N. plumbaginifolia* D and ND seeds (not shown).

Low to medium concentrations of auxin (NAA) slightly advance the germination of D seeds (Fig. 4a) but concentrations of NAA higher than 10 µM are inhibitory (Figs. 4a,c). Germination of ND seeds and *I217* seeds is on the contrary less sensitive to auxin (Fig 4b,c): low to medium NAA concentrations exert no stimulatory effect and the toxicity of the highest concentrations (100 µM) is lower.

Cytokinins BA and CPPU (but not zeatin) are very potent inhibitors of the germination of D seeds at concentrations as low as 3–5 µM but are almost

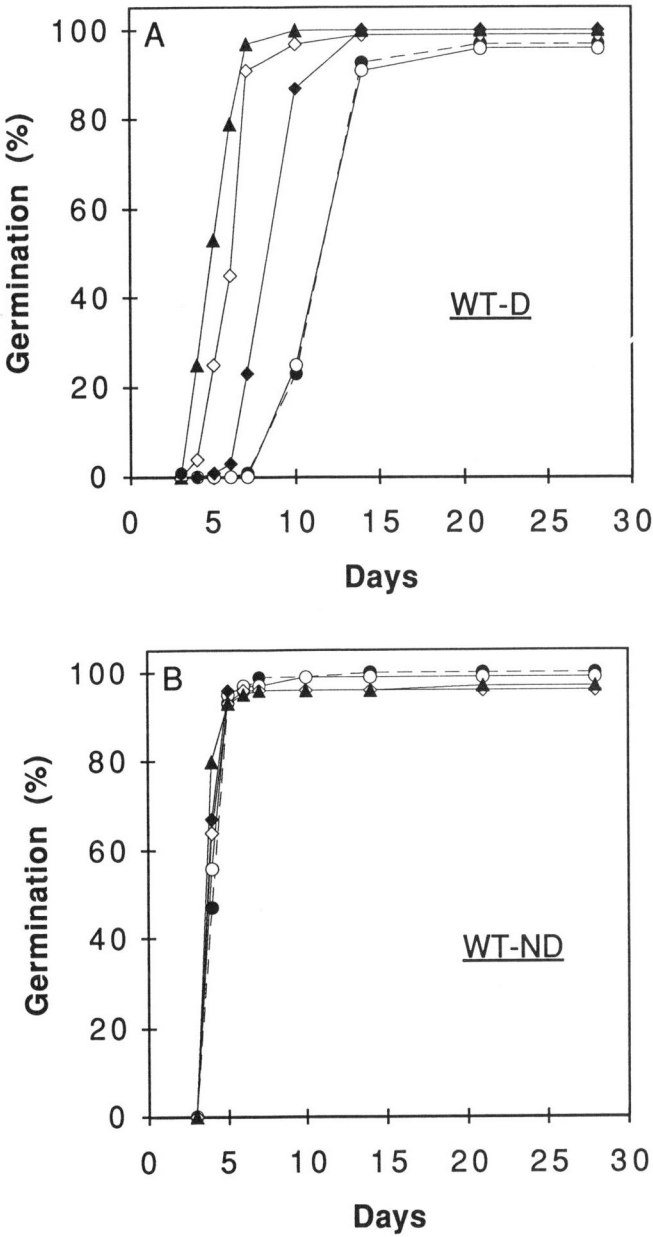

Figure 3. Effect of gibberellic acid on the germination of dormant and non-dormant wild type (WT) seeds.
(A): WT-D (dormant seeds), B: WT-ND (non-dormant seeds). GA_3 (μM): $-\bullet-$ 0, $-\bigcirc-$ 3, $-\blacklozenge-$ 10, $-\diamondsuit-$ 30, $-\blacktriangle-$ 100. D and ND seeds were respectively stored 1 and 16 months after harvest

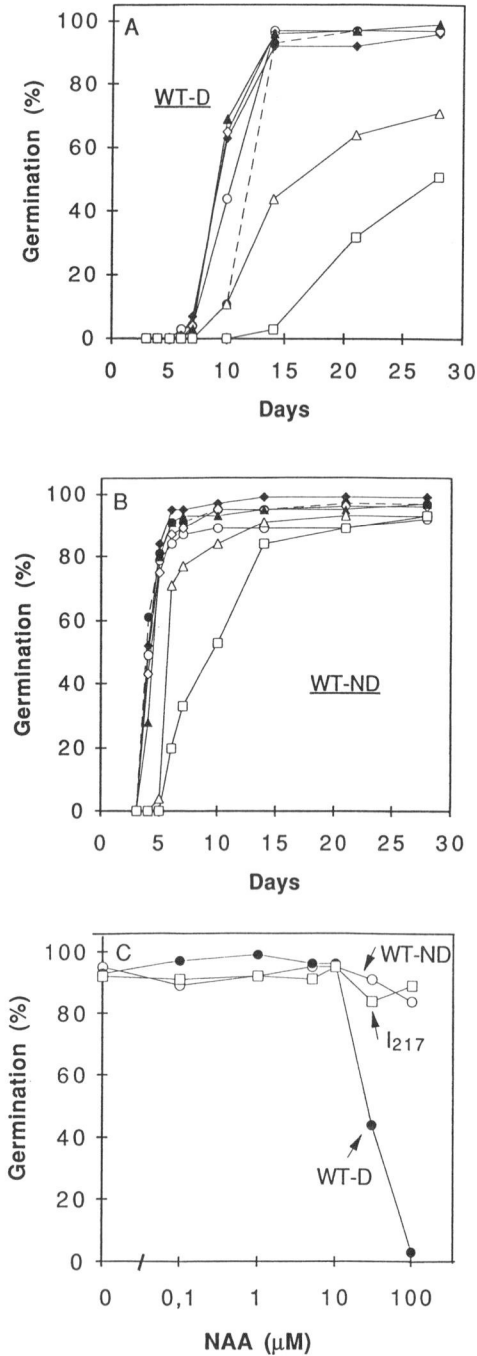

Figure 4. Effect of the auxin NAA on the germination of dormant, non-dormant wild type and mutant *I217* seeds. A: WT-D seeds, B: WT-ND seeds. NAA (μM) –●– 0, –○– 0.1, –◆– 1, –◇– 3, –▲– 10, –△– 30, –□– 100. C: germination values 14 days after sowing. –●– WT-D, –○– WT-ND, –□– mutant *I217*. WT-D and *I217* seeds were stored 3 months and WT-ND seeds 12 months after harvest

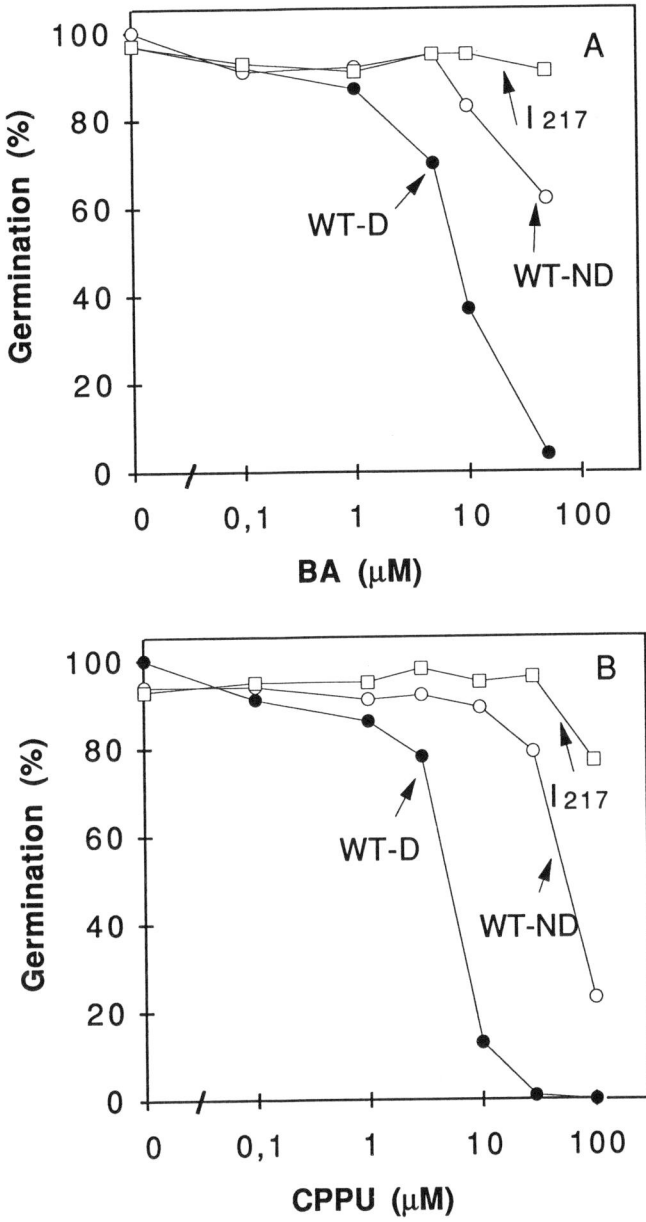

Figure 5. Effect of some cytokinins on the germination of dormant, non-dormant, wild type seeds and mutant *I217* seeds. A: BA, B: CPPU. Germination values 14 days after sowing. –●– WT-D, –○– WT-ND, –□– mutant *I217*. Length of storage as Fig. 4

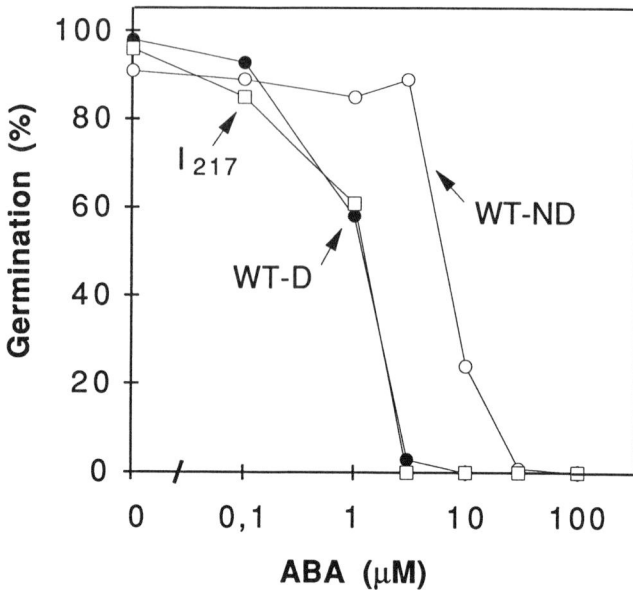

Figure 6. Effect of abscisic acid on the germination of dormant, non-dormant, wild type seeds and mutant *I217* seeds. Germination values 14 days after sowing: –●– WT-D, –○– WT-ND, –□– mutant *I217*. Length of storage as Fig. 4

ineffective on ND seeds and *I217* seeds, except at the very high concentration of 50–100 μM (Figs. 5a,b). D seeds inhibited by cytokinins are neither dead nor in secondary dormancy; when transferred to basal medium they germinate as the untreated control (not shown). Last, benzyladenine (10 μM) given along with GA3 (100 μM) is no more inhibitory for D seed germination (not shown).

Abscisic acid very efficiently inhibits the germination of D seeds at concentrations around 3 μM when ND seeds need 30 μM to be inhibited (Fig. 6). Surprisingly the *I217* seeds retain a very high sensitivity to ABA (Fig. 6). As for cytokinin, ABA-inhibited D seeds are neither dead nor in secondary dormancy; when transferred to basal medium they germinate as the untreated control (not shown). Howewer, contrary to what we observed for cytokinins, ABA (10 μM) given along with GA$_3$ (100 μM) stays inhibitory for D seed germination (not shown).

Influence of Fluridone and Norflurazon on the Germination of Dormant and Non-dormant Seeds

The inhibitors of ABA synthesis fluridone and norflurazon stimulate very efficiently the germination of D seeds. Fluridone is clearly more efficient than norflurazon as 10 μM of the former is as effective as 100 μM of the latter in

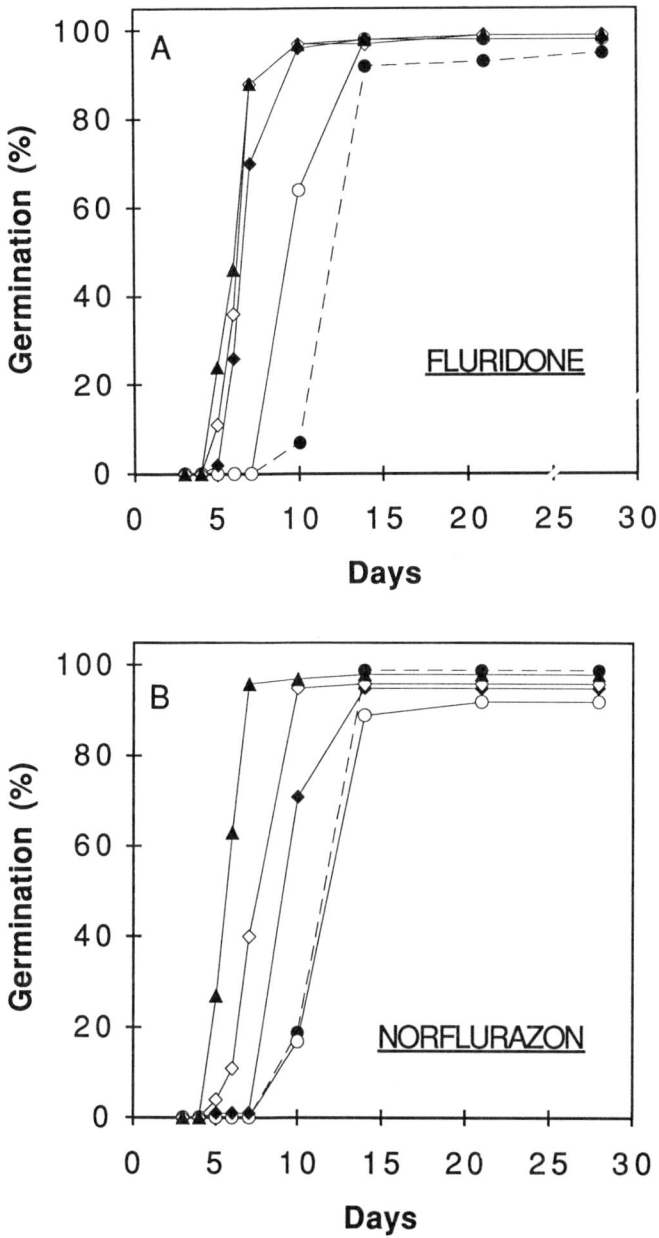

Figure 7. Effect of fluridone and norflurazon on the germination of dormant wild type seeds. A: fluridone. B: norflurazon (μM): $-\bullet-$ 0, $-\bigcirc-$ 0.1, $-\blacklozenge-$ 1, $-\lozenge-$ 10, $-\blacktriangle-$ 100. Seeds stored 3 months after harvest

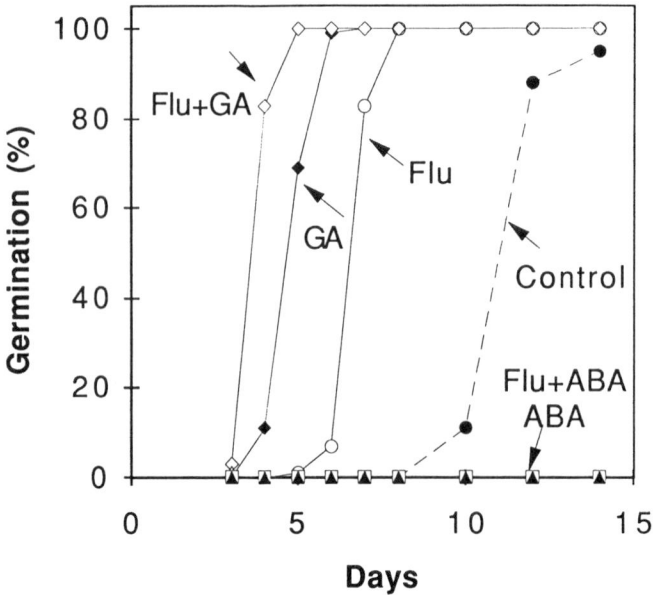

Figure 8. Combined effects of fluridone, gibberellic acid and abscisic acid on the germination of dormant wild type seeds. Concentrations (µM): fluridone (flu) 100, GA$_3$ 100, ABA, 30. $-\bullet-$ control, $-\bigcirc-$ flu, $-\blacklozenge-$ GA, $-\Diamond-$ flu.GA, $-\blacktriangle-$ ABA, $-\square-$ flu+ABA. Seeds stored one month after harvest

breaking dormancy (Figs. 7a,b). These products exert almost no effect on the germination of ND seeds and *I217* seeds (not shown). Plantlets developed at every concentration of these inhibitors are completely white with a hypocotyl of normal size and very small cotyledons.

Fluridone (100 µM) is less efficient than GA$_3$ (100 µM) in breaking dormancy of D seeds, the mean germination times being respectively 6.5 d and 4.5 d (Fig. 8). Moreover, fluridone and GA$_3$ can act together in breaking dormancy as seeds sown on the mixture have a mean germination time of 3.5 d. Lastly, the effect of fluridone (100 µM) on germination is completely reversed in presence of ABA (30 µM).

Discussion

Dormant seeds of *N. plumbaginifolia* are sensitive to gibberellin, but they need a very high concentration of GA$_3$ to hasten germination. The stimulatory effect of gibberellin is almost totally supressed for afterripened non-dormant seeds (Fig. 3) and for the mutant *I217* seeds which present no dormancy. As *I217* is ABA deficient, we can postulate that the loss in sensitivity to GA$_3$ of afterripened wild type seeds could be related to a decrease of the level of ABA in dry seeds or during the imbibition of seeds.

Dormant seeds of *N. plumbaginifolia* are very sensitive to auxin, low concentrations have a small hastening effect on germination and high levels are inhibitory (Figs. 4a,b). Such an inhibitory effect has been known for a long time with indole-3-acetic acid (review by Khan, 1971). These effects are scarcely detected with ND or *1217* seeds which could be considered as 'resistant' to auxin in terms of germination (Fig. 4c). This result probably explains why selective screenings for mutants resistant to the germinative inhibitory effect of auxin has led to ABA deficient mutants (Bitoun *et al.*, 1990).

Cytokinins strongly inhibit the germination of D seeds and are less inhibitory for ND and especially *1217* seeds (Fig. 5). Cytokinins and particularly the N-substituted phenylureas stimulate the germination in the dark for light requiring seeds (Thomas, 1992). More generally, cytokinins are considered effective alone or in connection with gibberellin in breaking dormancy in various species (Khan, 1971). So our results which show that BA and CPPU reinforce the dormancy of *N. plumbaginifolia* seeds are rather surprising. As CPPU is a cytokinin which is metabolized to a lesser extent than BA, the observed effect could probably not be explained by the production of toxic metabolites of cytokinin. Moreover the cytokinin-inhibited seeds germinate quickly when transferred to a basal medium. It is well known that the efficiency of different cytokinins in various physiological responses is highly variable. The inefficiency of zeatin in our context has probably to be linked to the intense metabolism of this cytokinin in plants (McGaw and Burch, 1995). ND wild type seeds and *1217* seeds are very less sensitive to cytokinins than D seeds (Fig. 5). This could mean that some cytokinins exert an important effect in the control of germination in *N. plumbaginifolia* seeds in connection with ABA.

ABA strongly inhibits the germination of D seeds and *1217* seeds and is really less inhibitory for ND seeds (Fig. 6). The result for the wild type is in agreement with various data cited in the introduction which have shown that the sensitivity of seeds to ABA decreases when dormancy is alleviated. That the *1217* non-dormant mutants seeds are more sensitive to ABA than the ND wild type seeds is then surprising. The same situation has been described in the GA deficient *ga1* mutants of *Arabidopsis thaliana* whose seeds are more responsive to the stimulatory effects of GA than the wild type (Derkx *et al.*, 1994) and could be explained by some up-regulation of a receptor.

Fluridone and norflurazon are efficient inhibitors of the biosynthesis of carotenoids and ABA (Moore and Smith, 1984; Henson, 1984). Fluridone treatments of developing seeds of sunflower decrease their endogeneous ABA level and suppress the induction of dormancy (Le Page-Degivry and Garello, 1992); moreover, a continuous synthesis of ABA is necessary for the expression of dormancy of imbibed seeds (Bianco *et al.*, 1994). A very similar result has been obtained in dormant microtubers of potato where fluridone treatment reduced amounts of free ABA and induced precocious loss of dormancy (Suttle and Hultstrand, 1994). It is clear from our results that norflurazon and fluridone treatments break the dormancy of *N. plumbaginifolia* seeds (Figs. 7a,b), but are without effect on ND and *I217* seeds. This confirms that synthesis of ABA in imbibed seeds is necessary to maintain dormancy. Fluridone is less efficient than GA_3, the two compounds have additive effects and ABA counteracts the dormancy breaking effect of fluridone (Fig. 8).

So it appears that afterripening of *N. plumbaginifolia* dormant seeds leads to a general decrease of sensitivity of imbibed seeds to the stimulatory compounds GA_3 and fluridone and to the inhibitory compounds ABA, cytokinins and to a lesser extent auxins. We can reasonably hypothesize that these variations are related to a decrease by dry storage of the synthesis of inhibitory compounds, such as ABA and perhaps cytokinins during imbibition. Direct measurement of the concentrations in various phytohormones in imbibed dormant or afterripened seeds would test this hypothesis.

References

Bewley, J.D. and Black M. 1994. *Seeds: Physiology of Development and Germination* 2nd Edition, pp. 445p. New York, London: Plenum Press.

Bianco, J. Garello, G. and Le Page-Degivry, M.T. 1994. *Seed Science Research* 4: 57–62.

Bitoun, R., Rousselin, P. and Caboche, M. 1990. *Molecular and General Genetics* 220: 234–249.

Blonstein, A.D., Parry, A.D., Horgan, H. and King, P.J. 1991a. *Planta* 183: 244–250.

Blonstein, A.D., Stirnberg, P. and King, P.J. 1991b. *Molecular and General Genetics* 228:361–371.

Bourgin, J.P., Chupeau, Y. and Missonier, C. 1979. *Physiologia Plantarum* 45: 288–292.

Corbineau, F. and Côme, D. 1992. In: *Quatrième rencontre internationale sur les semences,* Angers, 20–24 juillet 1992, pp 581–589, eds. D. Côme and F. Corbineau.

Derkx, M.P.M., Vermeer, E. and Karssen, C. M. 1994. *Plant Growth Regulation* 15: 223–234.

Fracheboud, Y. and King, P.J. 1991. *Molecular and General Genetics* 227: 397–400.

Hilhorst, H.W.M. and Karssen,C.M. 1992. *Plant Growth Regulation* 11: 225–238.

Henson, I.E. 1984. *Zeitschrift für Pflanzenphysiologie* 114: 5–43.

Jullien, M., Lesueur, D., Laloue, M. and Caboche, M. 1990. In: *Physiology and Biochemistry of Cytokinin in Plants,* pp.157–162 (eds D.N.S. Mok and K.M. Kamine). Pays-Bas: Academic Publishing.

Karssen, C.M., Zagorsky, S., Kepczynski, J. and Groot, S.P.C. 1989. *Annals of Botany* 63: 1–80.

Khan, A.A. 1971. *Science* 171: 853–859.

Kraepiel, Y., Jullien, M., Cordonnier-Pratt, M.M. and Pratt, L. 1994a. *Molecular and General Genetics* 242: 559–565.

Kraepiel, Y., Rousselin, P., Sotta, B., Kerhoas, L., Einhorn, J., Caboche, M. and Miginiac, E. 1994b. *The Plant Journal* 6: 665–672.

Le Page-Degivry, M.T. and Garello, G. 1992. *Plant Physiology* 98: 1386–1390.

McGaw, B.A. and Burch, L.R. 1995. In: *Plant Hormones. Physiology, Biochemistry and Molecular Biology,* pp98–117 (ed. P.J. Davies). Dordrecht: Kluwer Academic Publishers.

Moore, R. and Smith, J.D. 1984. *Planta* 162: 342–344.

Morris, C.F., Moffat, J.M., Sears, R.G. and Paulsen, G.M. 1989. *Plant Physiology* 90: 643–647.

Parry, A.D., Blonstein, A.D., Babiano, M.J., King, P.J. and Horgan, R. 1991. *Planta* 183: 237–243

Rousselin, P., Kraepiel, Y., Maldiney, R., Miginiac, E. and Caboche, M. 1992. *Theoretical and Applied Genetics* 85: 213–221.

Schuurink, R.C., Sedee, N.J.A. and Wang M. 1992. *Plant Physiology* 100: 1834–1839.

Suttle, J.C. and Hultstrand, J.F. 1994. *Plant Physiology* 105: 891–896.

Thomas, T.H. 1992. *Plant Growth Regulation* 11: 239–248.

Traas, I., Laufs, P., Jullien, M. and Caboche, M. 1995. *The Plant Journal* 7: 785–796.

Walker-Simmons, M. 1987. *Plant Physiology* 84: 61–66.

24. ABA Involvement in the Psychrolabile Dormancy of *Fagus* Embryo

M.T. LE PAGE-DEGIVRY, P. BARTHE, J. BIANCO and G. GARELLO

Laboratoire de Physiologie végétale, Faculté des Sciences, Université de Nice – Sophia Antipolis, 06108 Nice Cedex 2, France

Abstract

Embryos of *Fagus sylvatica* isolated from fruits at harvest were dormant. After a cold-treatment in restricted water conditions, embryos were able to germinate at a percentage which increased with chilling duration. Embryo ABA content decreased during the dormancy-releasing treatment at 4°C; it also decreased in the same proportions during embryo culture at 23°C, a temperature allowing dormancy to be strongly expressed. It thus appears that embryo ABA content was not correlated with the physiological potentialities of the embryo. The significant decrease in ABA level observed during culture at 23°C could be explained by a very rapid ABA metabolism, revealed by the results of [³H]ABA feeding experiments. Also, when fluridone, an inhibitor of carotenoid synthesis, was applied directly to axes, dormant embryos were able to germinate after a one-week culture. The comparison between axis ABA content in the absence or in the presence of fluridone after a 6-day culture gave an estimation of the axis capacity for a *de novo* ABA synthesis. Consequently, it appears that the psychrolabile dormancy of *Fagus* is associated with the ability of the axis to synthesize its own ABA.

Introduction

When hydrated seeds fail to respond to favourable environmental conditions, the origin of the seed dormancy could reside either within the embryo itself or in the structures surrounding the embryo. In this latter case, called coat-imposed dormancy, the stimulation of growth can only be successful when the growth potential of the embryos is stronger than the mechanical restraint of the layers surrounding the root. Over the past decade, work with hormone-deficient and hormone-insensitive mutants of *Arabidopsis thaliana* and *Lycopersicum esculentum* (reviewed in Hilhorst and Karssen, 1992) provided a considerable body of evidence for the involvement of hormones in germination and coat-imposed dormancy. ABA played a pivotal role during the development of primary dormancy which is induced by ABA during maturation (Karssen *et al.*, 1983); gibberellins were involved in the induction of germination (Karssen and Laka, 1985). Moreover changes in sensitivity to these hormones occurred during induction and breaking of dormancy (Derkx and Karssen, 1993).

A first study of an embryo dormancy was done in *Helianthus annuus* (Le Page-Degivry *et al.*, 1990). At harvest, this species exhibited an embryo

R.H. Ellis, M. Black, A.J. Murdoch, T.D. Hong (eds.), Basic and Applied Aspects of Seed Biology, pp. 215–224.
© *1997 Kluwer Academic Publishers, Dordrecht. Printed in Great Britain.*

dormancy which could be eliminated by storage in dry conditions. Using fluridone, a pyridinone inhibitor of carotenoid biosynthesis, it was possible to manipulate ABA levels in the maternal head tissue (Le Page-Degivry *et al.*, 1993) or in the embryo itself (Le Page-Degivry and Garello, 1992). It could therefore be demonstrated that this embryo dormancy was also induced by ABA during seed development. Moreover, the maintenance of embryo dormancy was associated with the ability of the embryonic axis to synthesize ABA; the release from dormancy was correlated with a decrease in ABA biosynthesis capacity during dry storage (Bianco *et al.*, 1994).

However, to be released, this type of embryo dormancy does not require any discontinuity in the environmental conditions. It can be questioned whether similar mechanisms are involved in the case of seeds which do not germinate unless they are chilled for several weeks at temperatures around 4°C, i.e. in the case of psychrolabile embryo dormancy. The results reported hereafter deal with the typical psychrolabile seed dormancy of *Fagus sylvatica*. In this species, Muller and Bonnet-Masimbert (1989) proposed a method of prechilling at 4°C without medium, at 30% moisture content; this precise moisture content is enough for the removal of dormancy but does not allow seeds to germinate. After the study of the *in vitro* behaviour of embryos isolated from dormant or prechilled seeds, we aimed to determine how ABA levels, ABA metabolism and ABA biosynthesis capacity were correlated with the physiological potentialities of the embryo.

Material and Methods

Plant Material

Two seed lots were used in 1993; they originated from the Tree Improvement Station – Humleback, Denmark. In 1994, a third lot was provided by the Office National des Forêts – Sécherie de la Joux, France.

For all these lots, dormancy-breakage treatment was performed in the Centre National de la Recherche Forestière, Champenoux, France, as described by Muller and Bonnet-Masimbert (1989). The duration required for a classical cold stratification with medium, described as X (Suszka and Zieta, 1977) differed from one seed lot to another.

Instead of this classical cold stratification with a medium, the nuts were prechilled without medium, at 30% (fresh weight basis) moisture content; a duration of two weeks longer (X + two weeks) than the classic cold stratification method, allows seed germination to be observed with alternating conditions of 14 h of darkness at 5°C and 10 h of artificial light at 15°C (C. Muller, personal communication).

Embryo Culture in vitro

Twenty-four embryos aseptically isolated, were cultured on water agar (6 g/l). The cultures were maintained at 23°C under white fluorescent light (45 W/cm², 16 h/d). An embryo was considered to have germinated when elongation of the radicle reached 2 mm.

Fluridone (1-methyl-3 phenyl 5-[3-(trifluoromethyl) phenyl]-4-(1H)-pyridinone), generously provided by Dow Elanco, was applied to the axis in 10% (v/v) acetone–water, in quantities of 20 µl per embryo, at a concentration of 100 µg/ml.

ABA Extraction and Quantification by Radioimmunoassay

Fifty axes or fifty pairs of cotyledons were ground in a mortar with chilled 80% methanol containing 2,6-di-t-butyl-4-methyl phenol as anti-oxidant (100 mg/l). The homogenate was stirred for 2 h at 4°C and centrifuged for 10 min at 2000g. The pellet was reextracted twice with the same volume of cold 80% methanol. The supernatants were collected and evaporated under reduced pressure at 40°C. The remaining aqueous extract was adjusted to pH 3.0 and extracted four times with diethyl ether (v/v). ABA was quantified by RIA performed exactly as described previously (Le Page-Degivry and Garello, 1992).

Feeding of [³H]-ABA to Embryos

The natural (+)-[³H]-ABA enantiomer was prepared as described by Barthe *et al.* (1993). The mixture of the two enantiomers (±)-[³H]-ABA (1.85–3.7 TBq/mmol, Amersham, UK) was converted to its methyl esters with ethereal diazomethane. HPLC was carried out on a column packed with chiralcel OD. The mobile phase employed was isopropanol-hexane (1:9) at a flow rate of 0.5 ml.min^{-1}. Determination of the radioactivity in the eluate was monitored with an on-line scintillation counter (model Flo-one/Beta series A-200 Radiomatic). The (+) ABA methyl ester eluted at 15.3 min. It was converted to the corresponding acid by saponification with 2 M KOH-Et OH (1:2), followed by acidification and extraction.

(+)-[³H]-ABA was fed to *Fagus* isolated embryos through their cotyledons (10 embryos for each experiment). Each embryo received 12.2 picomoles of (+)-[³H]-ABA dissolved in 10 µl water which were absorbed in 1 h. After this pulse, water or ABA (10 µM) was added in a quantity sufficient to be available during the whole incubation duration (6, 24 or 48 h).

ABA was extracted from embryos as mentioned above. Free metabolites of ABA, extracted from the aqueous phase (pH 3.0) with diethyl ether, were isolated by thin layer chromatography. After elution of the plates with toluene–ethyl acetate–acetic acid (50–30–4), peaks were identified by comparison of their Rf with those of authentic standards. The proportion of non-metabolised ABA can therefore be calculated in percentage of total radioactivity absorbed.

Results

In vitro *Behaviour of Isolated Embryos*

The behaviour of isolated embryos was first studied at 23°C, when water uptake was performed through both cotyledons immersed in the culture medium. At harvest or after dry storage, isolated embryos germinated very slowly, reaching a low percentage (about 25%) after a one-month culture (Figs. 1A,a). After the cold-treatment, embryos isolated from wet seeds were able to germinate as soon as the first week of culture; the germination percentage increased with prechilling duration (Fig. 1A,b,c). Developmental behaviour was also modified

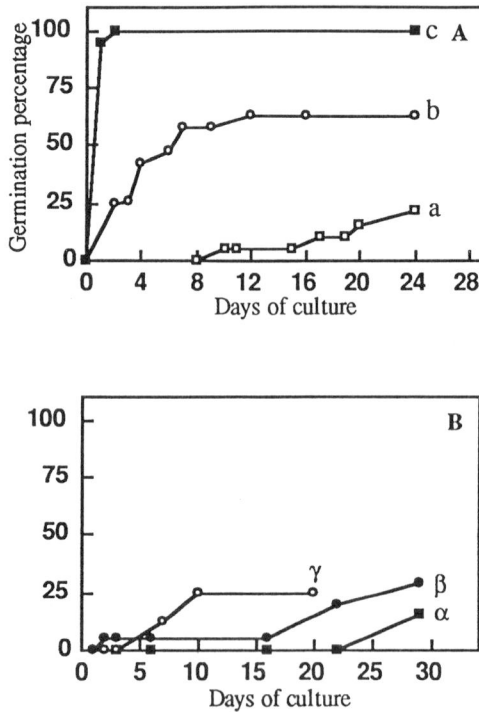

Figure 1. Changes, during culture on water agar, at 23°C, in the germination percentages of embryos isolated from beechnuts
A – according to the prechilling treatment duration: untreated (a), X + 2 weeks (b), X + 7 weeks (c). Imbibition was performed through both cotyledons immersed in the medium
B – according to imbibition modality: through either radicle (α), or one cotyledon laid flat on the medium (β) or both cotyledons immersed in the medium (γ). Embryos were isolated from untreated beechnuts

by prechilling treatment. At harvest, after one month of culture, a great heterogeneity was observed in embryo development, from the total absence of growth to normal plantlets. After the cold-treatment, the development of plantlets was normal and homogeneous.

Such a behaviour changed with embryo culture conditions. Germination percentages could be increased by lowering the temperature of culture to 14°C (data not shown) or decreased when imbibition was performed through only one cotyledon laid flat on the medium or through the radicle (Fig. 1B).

ABA Levels and Physiological Behaviour of Embryos

ABA content of dormant embryos and its distribution between the different parts of the embryo are shown in Table 1.

Table 1. ABA content in axis and cotyledons of dormant *Fagus* embryos. (Data are means of 8 individual estimations ± SE)

| | ABA ng/embryo | | | ABA ng/g F.W. | |
	Total	Axis	Cotyledons	Axis	Cotyledons
Lot 1993a	13.6±1.4	1.9±0.2	11.7±1.2	595±60	90±9
Lot 1993b	12.4±1.3	2.2±0.2	10.2±1.0	651±65	85±8

ABA content was about 13 ng per embryo for the two seedlots: the greater part was localised in the cotyledons whereas the axis only contained about 2 ng. But when results were reported in fresh weight, ABA level was much higher in the small axis (600 ng per gram of fresh weight) than in the large cotyledons (85–90 ng per gram of fresh weight).

During cold treatment, ABA content (Table 2) decreased in the embryo, both in axes and in cotyledons. This decrease intervened as early as X+3 weeks of treatment, but even after a long duration of prechilling (X+7) a non-negligible ABA level was still present.

When dormant embryos were cultured on water agar at 23°C, their ABA content (Table 3) also decreased, both in axes and in cotyledons. After a 10-day culture, embryo ABA level was about half of the original level.

Table 2. Changes in axis and cotyledon ABA content with the duration of the cold treatment: (a = lot 1993a; b = lot 1993b). (Data are means of 8 individual estimations ± SE)

Seed cold-treatment (4°C) duration	ABA content					
	ng/embryo		ng/organ			
			Axis		Cotyledons	
	a	b	a	b	a	b
0	13.6±1.4	12.4±1.3	1.9 ±0.2	2.2 ±0.2	11.7±1.2	10.2±1.0
X+3	9.2±0.9		0.74±0.07		8.5±0.8	
X+7	8.8±0.9	6.3±0.6	0.55±0.06	0.80±0.08	8.2±0.8	5.5±0.6

Table 3. Changes in axis and cotyledon ABA content of dormant embryos during *in vitro* culture on water agar at 23°C   (a = lot 1993a; b = lot 1993b). (Data are means of 8 individual estimations ± SE)

Embryo culture (23°C) duration (days)	ABA content					
	ng/embryo		ng/organ			
			Axis		Cotyledons	
	a	b	a	b	a	b
0	13.6±1.4	12.4± 1.3	1.9 ±0.2	2.2 ±0.2	11.7±1.2	10.2±1.0
10	7.9±0.8	6.5±0.08	0.77±0.08	1.27±0.1	7.1±0.7	5.2±0.5

Metabolism of (+)-[^{3}H]-ABA

In order to estimate the importance of ABA catabolism, (+)-[^{3}H]-ABA was fed to embryos through their cotyledons for a pulse of one hour; incubation was then performed in the presence of either water or cold ABA 10 µM. [^{3}H]-ABA was catabolised by both oxidation and conjugation processes. Metabolism was very active as soon as dormant embryos were imbibed at 23°C since a large part of ABA was already catabolized after a 6-h incubation and a very slight amount of non-metabolized ABA (Table 4) was present after 48 h of culture.

Table 4. Changes with time in non-metabolised [³H]-ABA in embryos incubated either in water or in the presence of 10 μM cold (+) ABA after a 1-h pulse of [³H]-ABA (lot 1994). Results are expressed in percentage of the total radioactivity absorbed

	1.2 picomoles (+) [³H]-ABA (1-h pulse)	
	Then water	Then 10 μM (+) ABA
6 h	46.2	50.1
24 h	9.3	10.8
48 h	4.8	4.1

Manipulation of ABA Biosynthesis by Fluridone Treatment

It is possible to manipulate ABA biosynthesis by applying fluridone, an inhibitor of carotenoid biosynthesis. Germination was strongly stimulated by this treatment (Fig. 2) whatever the modality of water uptake. However, the germination percentage was much higher when both cotyledons were immersed in the medium.

When ABA was quantified in cotyledons of both control and fluridone-treated embryos after a 6-d culture, i.e. just before germination occurred, the large decrease already described in Table 3 was again observed and was similar

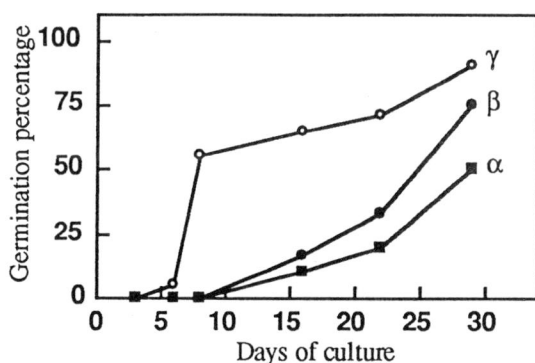

Figure 2. Changes, during culture on water agar at 23°C, in the germination percentages of embryos isolated from dormant beechnuts after a treatment by fluridone. Imbibition was performed through either radicle (α), or one cotyledon laid flat on the medium (β) or both cotyledons immersed in the culture medium (γ)

Table 5. Changes in axis ABA content of dormant embryos after 6 days of culture in the presence or absence of fluridone (lot 1993a). (Data are means of 8 individual estimations ± SE)

	ABA (pg/axis)		
	Fluridone-treated	Untreated	ABA synthesis
Exp. 1	150 ± 15	720 ± 75	570 ± 55
Exp. 2	170 ± 20	430 ± 45	260 ± 30

whether fluridone was present or not. However, in axes, a significant difference appeared between both conditions (Table 5); in fluridone-treated axes, the ABA level was low while in the absence of treatment, the higher amount of ABA was due to *de novo* biosynthesis.

Discussion

When embryos isolated from non-treated beechnuts were cultivated *in vitro*, they showed the characteristic features of a psychrolabile embryo dormancy. They exhibited the typical heterogeneity in development described, first by Flemion (1931) for dormant embryos of *Sorbus aucuparia*, then for several species belonging to different families of woody dicots. Moreover, germination percentages and modalities of development were dependent on culture conditions: dormancy is more or less expressed according to temperature in culture or to water uptake modalities as described by Barthe and Bulard (1982) for *Pyrus malus* embryos. Prechilling treatment at 30% water content caused the transition from heterogeneous to homogeneous development and a widening of the temperature range over which germination could proceed; a lengthening of the treatment allowed germination to be obtained even at 23°C continuously maintained, a condition especially unfavourable for the germination of this species.

During this prechilling, a decrease in ABA level occurred but this decrease took place early and did not become more accentuated when treatment was prolonged. Such a shift between ABA decrease (which was total at X+3 weeks) and the acquisition of aptitude for germination (which was total only at X+7 weeks) was already reported for *Pyrus malus* seeds (Rudnicki,1969). Moreover a similar decrease in ABA level was also observed when dormant embryos were cultured at 23°C. Such a lack of correlation between changes in ABA content and physiological behaviour was emphasized by Balboa-Zavala and Dennis (1977) in apple seeds and led authors (Walton, 1980–81) to question the involvement of ABA in dormancy control; this conclusion does not take into account that ABA level is the result of a rapid turn-over where the amount of preexistent ABA in dry seed is only one component, catabolism and *de novo* biosynthesis playing an important role. Feeding [³H]-ABA to embryos proved that ABA was rapidly metabolized by dormant embryos at 23°C: it can therefore be understood that ABA, preexistent in dry embryos, strongly decreased during the first days of culture. Also, it is possible to modify the endogenous ABA level using fluridone which decreased ABA level in the axis. It therefore appears that, as in *Zea mays* embryos (Gage *et al.*, 1989), in sunflower axis (Garello and Le Page-Degivry, 1992) or dormant barley grains (Wang *et al.*, 1995), ABA was synthesized via the indirect pathway requiring a xanthophyll precursor. Such fluridone-treated dormant embryos became able to germinate, demonstrating that the suppression of ABA synthesis in the axis allowed germination to occur. Such a correlation between *in situ* ABA synthesis and maintenance of embryo dormancy was already demonstrated for several species where dormancy can be released by dry storage. In sunflower, the induction and the maintenance of dormancy was associated with ABA synthesis in the axis (Le Page-Degivry and Garello, 1992); moreover, drying treatment led to the suppression of the capacity for ABA synthesis (Bianco *et al.*, 1994). Wang *et al.* (1995) also demonstrated that a *de novo* synthesis of ABA took place in embryos isolated from dormant barley grains during incubation but not in embryos isolated from non-dormant grains. Our work gives the first evidence that the continued presence of ABA due to its continued *de novo* biosynthesis is necessary for the maintenance of a psychrolabile dormancy.

Acknowledgements

This work was supported by an ECC grant (PL92-1667). The authors thank C. Muller for supplying prechilled beechnuts and H. Le Bris for excellent technical assistance.

References

Balboa-Zavala, O. and Dennis, F.G. 1977. *Journal of American Horticultural Science* 102: 633–637.
Barthe, P. and Bulard, C. 1982. *New Phytologist* 91: 517–529.
Barthe, P., Hogge, L.R., Abrams, R. and Le Page-Degivry, M.T. 1993. *Phytochemistry* 34: 645–648.
Bianco, J., Garello, G. and Le Page-Degivry, M.T. 1994. *Seed Science Research* 4: 57–62.
Derkx, M.P.M. and Karssen, C.M. 1993. *Physiologia Plantarum* 89: 360–368.
Flemion, F. 1931. *Contributions from the Boyce Thompson Institute* 3: 413–439.
Hilhorst, H.W.M. and Karssen, C.M. 1992. *Plant Growth Regulation* 11: 225–238.
Karssen, C.M. and Laka, E. 1985. In: *Plant Growth Substances* pp. 315–323 (ed. M. Bopp). Berlin: Springer-Verlag.
Karssen, C.M., Brinkhorst-van der Swan, D.L.C., Breekland, A.E. and Koornneef, M. 1983. *Planta* 157: 158–165.
Le Page-Degivry, M.T. and Garello, G. 1992. *Plant Physiology* 98: 1386–1390.

Le Page-Degivry, M.T., Barthe, P. and Garello, G. 1990. *Plant Physiology* 92: 1164–1168.
Le Page-Degivry, M.T., Barthe, P., Bianco, J. and Garello, G. 1993. In: *Fourth International Workshop on Seed. Basic and Applied Aspects of Seed Biology, Angers, France, 1992* pp. 615–625 (eds D. Come and F. Corbineau). Paris : ASFIS.
Muller, C. and Bonnet-Masimbert, M. 1989. *Seed Science and Technology* 17: 15–26.
Rudnicki, R. 1969. *Planta* 86: 63–68.
Suszka, B. and Zieta, L. 1977. *Arboretum Kornickie* 22: 237–255.
Walton, D.C. 1980–81. *Israel Journal of Botany* 29: 168–180.
Wang, M., Heimovaara-Dijkstra, S. and Van Duijn, B. 1995. *Planta* 195: 586–592.

25. Abscisic Acid-Mediated Responses in Seeds Involving Protein Kinases and Phosphatases

S.D. VERHEY and M.K. WALKER-SIMMONS

US Department of Agriculture, Agriculture Research Service, 209 Johnson Hall, Washington State University, Pullman, WA 99164-6420, USA

Abstract

In this report we discuss recent developments in our work on PKABA1, an abscisic acid (ABA) upregulated protein kinase expressed in wheat seed embryos. We also describe other protein kinases, including ATCDPK1 and ATCDPK2, and Aspk9 and Aspk10; and a protein phosphatase, *abi1*, that have been linked to abscisic acid-mediated effects in seeds.

Introduction

Our understanding of ABA-mediated processes involving protein phosphorylation-dephosphorylation events during seed development and maturation is just emerging. In developing seeds ABA levels peak at the onset of desiccation resulting in the induction of many ABA-responsive genes. This increase in ABA has been associated with acquisition of desiccation tolerance and dormancy in seeds. Involvement of protein phosphorylation–dephosphorylation processes in these ABA-mediated events are indicated by the identification of ABA-regulated protein kinases and by genetic evidence suggesting that protein phosphatases are involved in ABA responses. During embryo maturation, cell division cycle control is crucial, and protein kinases such as $p34^{cdc2}$ are known to be involved (Hemerly *et al.*, 1993). Key roles for protein phosphorylation in seeds are anticipated because of the ease of reversibility, the demonstrated ability of protein phosphorylation to respond to internal and external stimuli and because, as we will discuss here, elements of protein phosphorylation–dephosphorylation signal transduction systems have been linked with ABA responses. We will spend a minimum of space describing signal transduction processes in plants, as several reviews are available (Verhey and Lomax, 1993; Giraudat *et al.*, 1994; Chrispeels *et al.*, 1995).

Two main technical approaches have led to our current comprehension of potential roles for protein kinases and phosphatases in seeds. The first approach is based on the recognition that all protein kinases have in common a large protein primary sequence domain, consisting of a series of twelve subdomains. Each subdomain contains one or more amino acids that are invariant in all of the hundreds of protein kinases identified to date, and numerous residues that are highly conserved (Hanks and Quinn, 1991). With knowledge of these sequence hallmarks it has been possible to design various types of cloning

R.H. Ellis, M. Black, A.J. Murdoch, T.D. Hong (eds.), Basic and Applied Aspects of Seed Biology, pp. 225–233.
© *1997 Kluwer Academic Publishers, Dordrecht. Printed in Great Britain.*

strategies for use in obtaining clones of kinases associated with specific physiological conditions mediated by ABA. The second main approach is the use of hormone mutants, a strategy that has been particularly fruitful for ABA (Hilhorst and Karssen, 1992; Reid, 1993).

Our laboratory is characterizing ABA-responsive protein kinase genes expressed in seeds. We are determining the role of these protein kinases in regulation of seed dormancy and in environmental stress responses during seed germination. Our initial interest in ABA-mediated responses came from observations that dormant seed embryos are far more sensitive than after-ripened embryos to ABA as a germination inhibitor (Walker-Simmons, 1987). In wheat seeds we found that as seed dormancy dissipates with afterripening or postharvest storage time, embryonic sensitivity to ABA decreases. No differences in ABA levels of dormant and nondormant seeds were found (Walker-Simmons, 1987). Genetic studies have provided even stronger evidence that ABA is involved in the development and perhaps the maintenance of seed dormancy. ABA-deficient or nonresponsive mutants of arabidopsis and other types of seeds exhibit reduced levels of seed dormancy (for recent review see Hilhorst, 1995). All this evidence has encouraged us to seek to identify ABA-responsive genes, especially protein kinase genes, with potential regulatory functions in dormant seeds. In this report we review our recent progress. We also note other recent examples indicating that protein phosphorylation steps are involved in ABA-mediated events in seeds.

ABA-Responsive Protein Kinase – PKABA Subfamily

PKABA1, an ABA-responsive Protein Kinase mRNA

We have identified PKABA1, an ABA-responsive protein kinase mRNA, by using degenerate oligonucleotides corresponding to the conserved catalytic regions of eukaryotic protein kinases (Anderberg and Walker-Simmons, 1992). A probe was prepared by polymerase chain reaction amplification (PCR) to conserved kinase subdomains VIb to VIII. A wheat embryo cDNA library was screened with the PCR probe and two potential protein kinase clones were identified. One clone, PKABA1, was found to correspond to an ABA-responsive protein kinase mRNA. Application of ABA to isolated wheat embryos caused an increase in PKABA1 levels. Expression of the message is highly sensitive to ABA with concentrations as low as 0.1 mM causing a significant increase in PKABA1 mRNA levels. The message likewise responds to increases in ABA resulting from dehydration or cold treatment of germinating seedlings (Holappa and Walker-Simmons, 1995).

The PKABA1 cDNA sequence contains all the features of serine/threonine protein kinases, including homology with all 12 conserved regions of the kinase catalytic domain (Fig. 2). A novel feature of the PKABA1 sequence is a stretch of 13 acidic amino acids outside of the catalytic domain near the C-terminal

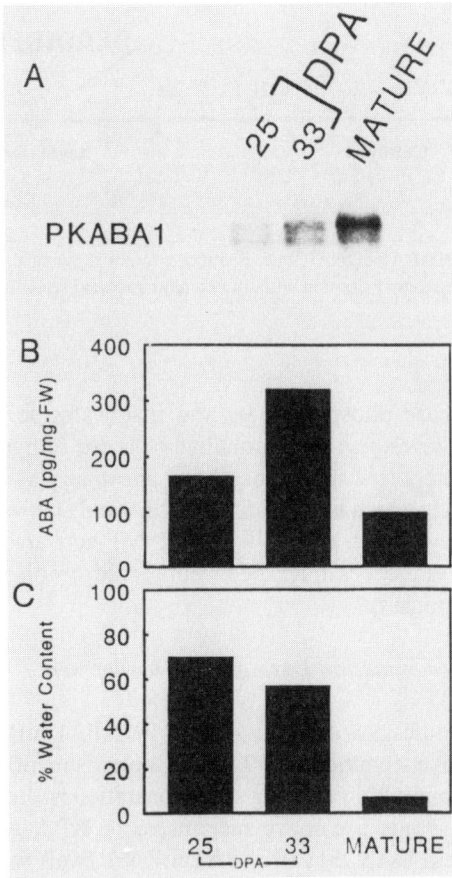

Figure 1. (A) PKABA1 mRNA levels in wheat embryos from developing (25 and 33 days post anthesis) and mature seeds. Each lane contained 1 mg of poly(A)⁺RNA. (B) Embryonic ABA and (C) Relative water content of embryos. From Anderberg and Walker-Simmons (1992)

end. Since the PKABA1 cDNA sequence was originally reported, protein kinases with a similar sequence have been reported in soyabean, rapeseed, arabidopsis and ice plant (Holappa and Walker-Simmons, 1995).

Interestingly, the PKABA1 kinase subfamily has the closest sequence similarity to other types of protein kinases involved in physiological responses to stress including the yeast SNF1· and the mammalian AMP-activated protein kinases (reviewed by Hardie *et al.*, 1994). Yeast SNF1 protein kinases are involved in responses to nutrient stress when cells become starved for glucose. Plant homologues of the yeast SNF1 protein kinases have been identified (Hardie *et al.*, 1994), but their function is not yet clear. Mammalian AMP-

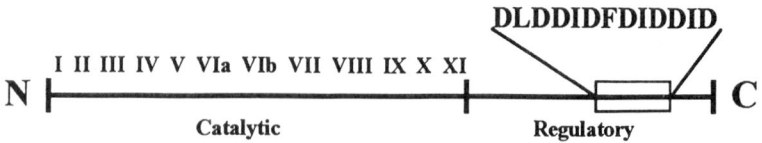

Figure 2. Model of the PKABA1 cDNA clone: deduced protein showing the conserved catalytic subdomains, I–XI, and regulatory domain with an acidic amino acid stretch

activated protein kinase phosphorylates and inactivates several biosynthetic enzymes when AMP levels rise in mammalian cells due to hypoxia, heat shock and other types of stress that inhibit oxidative phosphorylation. The sequence similarity of PKABA1, which is regulated by ABA and environmental stress at the transcriptional level, with the SNF1 and AMP-activated protein kinases suggests that these sequences have been conserved among protein kinases involved in stress responses.

PKABA1 Message Accumulation During Seed Development

As ABA levels rise in desiccating wheat seeds, PKABA1 mRNA accumulates (Anderberg and Walker-Simmons, 1992). A comparison of ABA levels and PKABA1 mRNA accumulation during seed maturation is shown in Figure 1.

Because dormant seeds are more responsive to ABA as a germination inhibitor it might seem likely that PKABA1 mRNA levels would be higher in dormant compared to nondormant seeds at maturity. However, we have not detected significant differences in PKABA1 mRNA levels in dormant and nondormant seeds of wheat. Differences in PKABA1 mRNA levels do occur when dormant and nondormant seeds are imbibed. PKABA1 mRNA levels are maintained at elevated levels in imbibed dormant seeds that remain growth-arrested. In contrast PKABA1 mRNA levels decline in germinating seeds.

Modes of Regulation of PKABA1 and Other Kinases

Transcriptional regulation by ABA of the PKABA1 kinase is indicated by ABA upregulation of the kinase mRNA in isolated embryos and in dehydrated seedlings. ABA regulation at the transcriptional level may also be occurring in hydrated dormant seeds.

Transcriptional regulation is unusual among protein kinases, but not unique. Taking an approach similar to ours, using PCR with degenerate primers to produce kinase-specific probes, Urao *et al.* (1994) have reported two calcium-dependent protein kinases, ATCDPK1 and ATCDPK2, that are induced by drought and salt stress – but not ABA – in arabidopsis. Calcium-dependent protein kinases are a class of protein kinase unique to plants and fungi that consist of a kinase catalytic domain fused to a calmodulin-like calcium-binding domain. A comparable approach was used to identify protein kinases in germinating seeds of *Avena sativa* that are responsive to gibberellic acid (GA), an ABA antagonist (Huttly and Phillips, 1995). These messages, Aspk9 and Aspk10, are affected oppositely by the application of GA: Aspk9 expression is higher in tissue incubated in the absence of exogenous GA, while Aspk10 expression is increased in the presence of GA. Interestingly, Aspk9 bears strong homology to the mitogen activated protein (MAP) kinase family (also known as extracellularly regulated kinases, or ERK). Aspk10 most resembles a ribosomal protein kinase first identified in arabidopsis (Huttly and Phillips, 1995). Another message, Aspk1, has sequence homology to PKABA1, but is not affected by GA.

The observation that PKABA1 is regulated by transcription also opens the possibility for regulation at the translational level. There are several points at which translation can be regulated, including initiation, elongation, and termination. For example, extensive secondary structure in the promoter region of the mRNA can negatively affect ribosome assembly and translation initiation (Mathews, 1986). It is interesting to note that sequence analysis of the promoter region of a genomic clone corresponding to PKABA1 indicates the potential for significant secondary structure formation (Holappa and Walker-Simmons, unpublished data). Analysis of possible translational regulation requires antibodies to PKABA1. We have recently prepared antibodies that recognize recombinant PKABA1 protein and experiments are underway to analyse translational regulation of the kinase.

While transcriptional regulation has been described for a few plant kinases, and translational regulation is possible as well, the most common types of regulation are biochemical (e.g. phosphorylation–dephosphorylation, second messengers) and substrate-level (Verhey and Lomax, 1993). Regulation at one or more of these levels may be even more important than transcriptional or translational regulation. Experiments aimed at measuring PKABA1 at the biochemical level are in progress, including expression and biochemical characterization of the recombinant protein kinase. It is worth noting, in discussing the possibility of substrate level regulation, that the maize ABA-responsive protein Rab17 is phosphorylated, and that Rab17 phosphorylation is greatest in mature maize embryos (Goday *et al.*, 1994). Phosphorylation affects the ability of Rab17 to bind nuclear localization signals (Goday *et al.*, 1994). Nuclear changes in protein phosphorylation in response to applied ABA have been reported in chromatin-associated proteins of *Lemna* nuclei (Chapman *et al.*, 1975).

Other Genes involved in ABA-Mediated Processes in Seeds

One of the more straight-forward approaches to finding mutants with altered ABA responses in seeds is to screen seeds from mutagenized arabidopsis plants for the ability to germinate in the presence of (normally) inhibitory levels of ABA. Several ABA insensitive mutants have been isolated in this way, and one, *abi1* (Koornneef *et al.*, 1984), is particularly noteworthy in terms of phosphorylation-dephosphorylation events.

As expected from the ability to germinate despite high exogenous ABA, seeds from *abi1* plants exhibit reduced dormancy. There are also vegetative effects in mutant plants, including wiltiness stemming from increased stomatal aperture, and the ability to form roots in the presence of high concentrations of exogenous ABA (Leung *et al.*, 1994). This range of phenotypes spanning seeds and vegetative growth suggests that the defect in *abi1* plants is fairly early in the ABA signalling pathway (Leung *et al.*, 1994). In addition, the mutation is dominant: *abi1/+* heterozygotes display the same kind of phenotypes as homozygous mutant plants (Koornneef *et al.*, 1984).

The *abi1* locus was recently independently cloned by two groups (Leung *et al.*, 1994; Meyer *et al.*, 1994). Sequence analysis by both groups shows that the ABI1 protein contains extensive sequence similarity with protein phosphatase 2C (PP2C) from both rat and yeast. In addition to the PP2C region at the C-terminal end of the deduced polypeptide, the N-terminal region contains sequences similar to EF-hand, calcium-binding sites suggesting the protein might be able to bind calcium. This arrangement is strongly reminiscent of that found in calcium-dependent protein kinases (Harper *et al.*, 1993). The apparent source of the *abi1* lesion is a point mutation that changes the wild-type glycine[180] to an aspartic acid residue (Leung *et al.*, 1994; Meyer *et al.*, 1994). Further biochemical analysis of both forms of the ABI1 protein is needed to identify the basis for the dominant effect of the mutant protein, although the ABI1 protein is evidently able to act as a phosphatase (Meyer *et al.*, 1994). Both groups speculate that ABA could inactivate the phosphatase in wild-type plants, while in mutants the phosphatase might be irreversibly active (Leung *et al.*, 1994; Meyer *et al.*, 1994). This suggests that, in cells in which Ca^{2+} levels are elevated by ABA, calcium might inactivate the wild-type phosphatase. Alternatively, the mutation could cause a change in substrate specificity (Leung *et al.*, 1994; Meyer *et al.*, 1994).

Summary and Discussion

In this report we have discussed examples of the two components of phosphorylation-dephosphorylation systems, protein kinases and protein phosphatases, that are each linked to the effect of ABA on seeds. PKABA1 is a protein kinase that is upregulated by ABA, while ATCDPK1 and ATCDPK2 are induced by stress apparently without the agency of ABA. Levels of two others, Aspk9 and

```
┌─────────────────────────────────────────────────┐
│                    DORMANT                        │
│                                                   │
│  • Embryos are ABA sensitive                      │
│  • Protein phosphorylation promoted by:           │
│    - high PKABA1 mRNA                              │
│    - reduced ABI1 protein phosphatase activity     │
└─────────────────────────────────────────────────┘

                    afterripening
                         ↓

┌─────────────────────────────────────────────────┐
│                  NONDORMANT                       │
│                                                   │
│  • Embryos are ABA insensitive                    │
│  • Protein dephosphorylation promoted by:         │
│    - reduced PKABA1 mRNA levels                    │
│    - increased ABI1 protein phosphatase activity   │
└─────────────────────────────────────────────────┘
```

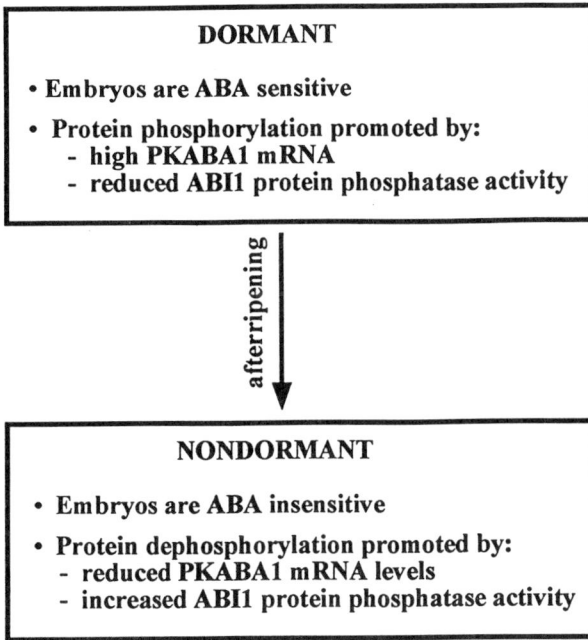

Figure 3. Scheme showing differences in ABA sensitivity and protein phosphorylation–dephosphorylation activities in dormant and nondormant seeds

Aspk10 are affected by gibberellic acid. The level of PKABA1 message remains high in growth-arrested cells such as imbibed dormant embryos, and declines in germinating embryos. On the dephosphorylation side of the equation, *abil* encodes a protein phosphatase, and there is structural evidence that Ca^{2+} could affect the activity of the enzyme, possibly in a negative way.

A basis, if any, for a relationship between the products of either PKABA1 or *abil* and physiological events such as cell cycle control or states such as dormancy remains to be demonstrated. Still, with representatives of each of the directions of the phosphorylation-dephosphorylation cycle juxtaposed as they are here, it is tempting to make a few observations.

It is conceivable, for example, that any of the kinases and/or the phosphatase discussed here could be involved in regulation of the cell cycle. As an illustration, *snf1-D* mutants of *Saccharomyces cerevisiae* fail to arrest in mid-G1 phase, instead continuing to grow and bud, becoming twice as large as wild-type cells (Hardie *et al.*, 1994). Plant homologues of SNF1 have been identified in several species of plants, including rye, barley, and arabidopsis (Hardie *et al.*, 1994). SNF1 homologues or other kinases such as PKABA1 could be involved with either the regulation of expression or of the biochemical activity of cell cycle control components such as p34^{cdc2}, a protein kinase which is required for the G1/S transition (Hemerly *et al.*, 1993). The same sort of reasoning holds for phosphatases such as the product of *abi1* (Giraudat *et al.*, 1994). It is interesting to note that promoter activity of the arabidopsis homologue of p34^{cdc2} is completely inhibited by ABA (Hemerly *et al.*, 1993).

An increase in phosphorylation may be one aspect of establishment and maintenance of dormancy (Trewavas, 1987) and/or desiccation tolerance in seeds. A reduction in ABI1 enzymatic activity in cells where increased ABA levels lead to an increase in Ca^{2+} levels fits with this model (Fig. 3). So, too, does an increase in PKABA1 levels in ABA-treated cells. Changes in the phosphorylation states of putative ABA receptors or other upstream regulatory proteins could change sensitivity to ABA. Obviously, numerous other protein kinases could be involved as well, as could other phosphatases. According to this model, kinase inhibitors or phosphatase activators are predicted to reduce dormancy and stress tolerance.

Characterization of protein kinases and protein phosphatases in seeds is just beginning. The functional roles of these enzymes are not yet known, but it does appear that major breakthroughs regarding the importance of protein phosphorylation/dephosphorylation in control of cell division, acquisition of desiccation tolerance and regulation of seed dormancy are just around the corner.

Acknowledgement

Portions of this work were supported by the U.S. Department of Agriculture-National Research Initiative Competitive Grant Program, Grant No. 94-37100-0313.

References

Anderberg, R.J. and Walker-Simmons, M.K. 1992. *Proceedings of the National Academy of Sciences USA* 89: 10183–10187.

Chapman, K.S.R., Trewavas, A. and van Loon, L.C. 1975. *Plant Physiology* 55:293–296.

Chrispeels, M.J., Green, P.J. and Nasrallah, J.B. 1995. *Plant Cell* 7: 237–248.

Giraudat, J., Parcy, F., Bertauche, N., Gosti, F., Leung, J., Morris, P.-C., Bouvier-Durand, M. and Vartanian, N. 1994. *Plant Molecular Biology* 26: 1557–1577.

Goday, A., Jensen, A.B., Culianez-Macia, F.A., Alba, M.M., Figueras, M., Serratosa, J., Torrent, M. and Pages, M. 1994a. *Plant Cell* 6: 351–360.

Goday, A., Sanchez-Martinez, D., Gomez, J., Puigdomenech, P. and Pages, M. 1994b. *Plant Physiology* 88: 564–569.

Hanks, S.K. and Quinn, A.M. 1991. *Methods in Enzymology* 200: 38–61.

Hardie, D.G., Carling, D. and Halford, N. 1994. *Seminars in Cell Biology* 5: 409–416.

Harper, J.F., Binder, B.M. and Sussman, M.R. 1993. *Biochemistry* 32: 3282–3290.

Hemerly, A.S., Ferreira, P., de Almeida Engler, J., Van Montagu, M., Engler, G. and Inze, D. 1993. *Plant Cell* 5: 1711–1723.

Hilhorst, H.M.W. 1995. *Seed Science Research* 5: 61–73.

Hilhorst, H.W.M. and Karssen, C.M. 1992. *Journal of Plant Growth Regulation* 11: 225–238.

Holappa, L.D. and Walker-Simmons, M.K. 1995. *Plant Physiology* 108: 1203–1210.

Huttly, A.K. and Phillips, A.L. 1995. *Plant Molecular Biology* 27: 1043–1052.

Koornneef, M., Reuling, G. and Karssen, C.M. 1984. *Physiologia Plantarum* 61: 377–383.

Leung, J., Bouvier-Durand, M., Morris, P.-C., Guerrier, D., Chefdor, F. and Giraudat, J. 1994. *Science* 264: 1448–1452.

Mathews, M.B. 1986. *Translational Control.* 192 pp. Cold Spring Harbor, N.Y., Cold Spring Harbor Laboratory.

Meyer, K., Leube, M.P. and Grill, E. 1994. *Science* 264: 1452–1455.

Reid, J.B. 1993. *Journal of Plant Growth Regulation* 12: 207–226.

Trewavas, A.J. 1987. *Bioessays* 6: 87–92.

Urao, T., Katagiri, T., Mizoguchi, T., Yamaguchi-Shinozaki, K., Hayashida, N. and Shinozaki, K. 1994. *Molecular General Genetics* 244: 331–340.

Verhey, S.D. and Lomax, T.L. 1993. *Journal of Plant Growth Regulation* 12: 179–195.

Walker-Simmons, M. 1987. *Plant Physiology* 84: 61–66.

26. Dormancy in Sitka Spruce Seeds

S.K. JONES[1], P.G. GOSLING[1] and R.H. ELLIS[2]

[1]*Forestry Commission, Forest Research Station, Alice Holt Lodge, Wrecclesham, Farnham, Surrey, GU10 4LH;* [2]*Department of Agriculture, The University of Reading, P.O. Box 236, Reading, Berkshire, RG6 6AT, UK*

Abstract

Sitka spruce (*Picea sitchensis* [Bong.] Carr.) seeds exhibit conditional (relative) dormancy, whereby conditionally-dormant seeds will not germinate in tests at 10°C but will germinate at 20°C. Such dormancy can be removed in Sitka spruce seeds by using a moist cold treatment (prechill) at about 3°C. The conditional dormancy is removed progressively by extending the prechill period. Following a suitable prechill period, Sitka spruce seeds acquire the ability to germinate in tests at 10°C and also germinate faster at the optimal test temperature of 20°C. Variation in these two variables among seed samples are correlated and can be used to determine the level of conditional dormancy in Sitka spruce seedlots. The effects of seed development on both the ability of Sitka spruce seeds to germinate and their level of dormancy was also studied, as was the effect of a commercial seed extraction procedure on dormancy. Implications for the commercial collection of, processing of, and presowing treatments to Sitka spruce seeds are discussed.

Introduction

Seeds of commercial conifer species like Sitka spruce (*Picea sitchensis* [Bong.] Carr.) are extracted from tightly shut cones harvested before the seeds are naturally shed, which makes for easier collection and helps to maximise seed yields (Gordon, 1992). The tightly shut cones are usually allowed to dry and partially open under ambient conditions (10–15°C) for two to four months. To ensure the cones are fully open, they are then exposed to forced air at 30–40°C for between 4 and 8 h in a tunnel kiln. The result is cones with flared back scales which are then tumbled to release the winged seeds. The seeds are subsequently de-winged, cleaned, dried to a storage moisture content of between 6 and 8% (fresh weight basis), packaged into air-tight containers and stored at between –18 and +5°C until needed (Gordon, 1992; Jones, 1995). The resulting commercial seedlot is ready for nursery sowing but like most conifer seeds, Sitka spruce seeds exhibit conditional (relative) dormancy. That is, seeds only germinate well at optimal temperatures. Such dormancy can be removed by a moist cold treatment (prechill) which allows germination over a much wider range of temperatures.

In the case of conditionally-dormant Sitka spruce seeds, the optimum temperature for germination is 20°C but UK grown seedlings are usually

R.H. Ellis, M. Black, A.J. Murdoch, T.D. Hong (eds.), Basic and Applied Aspects of Seed Biology, pp. 235–244.
© *1997 Kluwer Academic Publishers, Dordrecht. Printed in Great Britain.*

produced by sowing seeds on the open nursery in the spring when soil temperatures are *circa* 10 to 12°C. Therefore the performance at sub-optimal temperatures, such as 10°C, is very important. By using a Target Moisture Content (TMC) prechill the cold period can be safely extended to 12 or 18 weeks without the possibility of seeds germinating during the prechill (Jones and Gosling, 1994). This allows the seeds to achieve their maximum germination even at a 10°C germination temperature. The prechill also speeds up the rate of germination and uniformity of germination at the optimal germination temperature of 20°C, and so both germination capacity at 10°C and rate of germination at 20°C can be used to measure the level of dormancy. In the following experiments the effect of cone collection date, and hence seed maturity, on the ability of Sitka spruce seeds to germinate was studied. The level of dormancy was also estimated at the different collection dates. The effect on dormancy of a 35°C kilning treatment was subsequently studied. Implications for the commercial collection of, processing of, and presowing treatments to Sitka spruce seeds are discussed.

Materials and Methods

Experiment 1. The Effect of Collection Date on Seed Germination and Dormancy

Sitka spruce cones were collected from the Forestry Commission's clone bank at Ledmore, Perthshire, Scotland (long. 3° 33′ W, lat. 56° 29′ N, UK seed region 20) on seven dates during 1992 (9 July, 7 August, 2 September, 18 September, 30 September, 27 October, and 25 November). The cones were open-pollinated, pollination occurring from mid- to end-May. A pollination date of 24 May was assumed, providing values of 46, 75, 101, 117, 129, 156 and 185 days after pollination (DAP) for the respective sample dates. Between five and six cones were harvested per date from each of two clones (Forestry Commission codes 1809 and 1996).

Normal extraction methods where cones are air dried at ambient, then kiln dried at 30–40°C prior to seed extraction were not used as the moisture content of the seeds and their ability to germinate prior to desiccation needed to be determined. Cones were therefore dissected and seeds removed from individual cone scales by hand.

X-ray analysis of seed-like structures from the 46 DAP sample showed that almost 50% of all seed-like structures were empty. To avoid bias in samples only full seeds were put through moisture content and germination tests. Full seeds per cone were detected by X-ray analysis (see ISTA, 1991). Full seeds were then selected by comparison between the actual seed-like structures and their images on the processed X-ray. Results from the X-ray analysis were also used to determine the anatomical maturity of seeds.

Fresh weight, dry weight and moisture content (fresh weight basis) were determined for the individual cones and samples of 10 full seeds. Seeds were germinated at a constant 20°C, in the dark, on top of moist filter paper in plastic germination boxes (as described in Gosling, 1988), with or without a 3-week top-of-paper prechill in the dark at 4°C. Due to the limited number of seeds available only one replicate of 50 seeds per sample time per clone were used.

Standard errors are shown on the figures, where they are larger than the symbols, for cone and seed, dry weights and moisture contents. For germination data the standard errors based on the theoretical binomial distribution are shown. Mean germination time (MGT) was calculated as described in Jones and Gosling (1994).

Experiment 2. The Effect of a 35°C Kilning Treatment on Dormancy

Open pollinated cones from three clones (Forestry Commission codes 1505, 2019 and 2122) were collected from the Forestry Commissions Seed Orchard at Ledmore, Perthshire, Scotland (the clones used in Experiment 1 were not available). The cones were collected on one date only, 20 September 1993 (corresponding to a normal collection time and 119 DAP).

Cones were randomly split into two groups. One group was oven dried at 35°C for 16 h which caused the cone scales to flare back (i.e. open the cones) and allow seed release. Cones were exposed to 35°C soon after collection in a deliberate attempt to induce a deeper conditional dormancy in the seeds. The seeds were not extracted at this stage but cones and seeds were stored in polythene bags (62.5 μm thick) in a controlled temperature room at 20°C ±2°C and 50±5% RH, for five months until extraction of both groups. The other group of cones was left at ambient temperatures within the Forestry Commission's seed extractory (*circa* 10–15°C) in open trays, to allow natural opening of the cones and then extracted along with the first group in early February 1994, a total of five months after collection. The two treatments are referred to as (i) 35°C and (ii) ambient.

Germination tests on full seeds were also carried out as outlined in Experiment 1, except that seeds from each clone were germinated at both 10°C and 20°C in the dark, with and without a 3-week prechill at 4°C in the dark. Germination capacity (%) was based on seeds taken from five replicate cones. Due to the variable number of full seeds produced per cone the means produced were weighted using the square root of the number of seeds per cone before analysis of transformed data. The results at 20°C were then analysed using analysis of variance with the angular transformation for germination capacity (%) and $\log_{10}$ for mean germination time. For clarity only untransformed data are presented. The 10°C results were not analysed due to the unbalanced nature of the data (i.e. zero values for the untreated control germinations at 10°C) but non-weighted, untransformed means are presented for comparison with the results at 20°C.

Results and Discussion

The X-ray analyses in Experiment 1 allowed the stage of anatomical maturity to be estimated for Sitka spruce seeds (Fig. 1). Anatomical maturity is the stage when the gap between the megagametophyte plus embryo tissues and the seed coat has just disappeared. Similar definitions were used for maturity using X-ray results for radiata pine seeds (Rimbawanto *et al.*, 1989) and slash pine seeds (Barnett, 1976). At 46 DAP both clones had distinct gaps (*circa* 0.1 mm) between the seed coat and the embryo and megagametophyte tissues. By 75 DAP the gaps were 0.05 mm or less, appearing as a halo around embryo and megagametophyte tissues. Seeds were completely filled and considered anatomically mature by 101 DAP. This developmental stage coincided with the

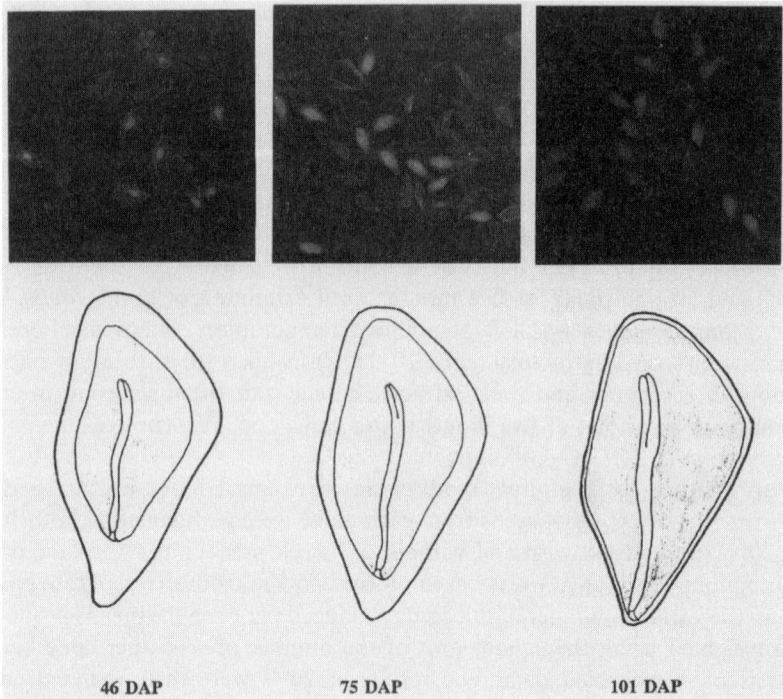

46 DAP   75 DAP   101 DAP

Figure 1 Experiment 1. Photographs of the X-ray results of seeds from clone 1809 at 46, 75 and 101 DAP (days after pollination) with diagrammatic representations of the stages of anatomical maturity underneath. At 46 DAP there was a *circa* 0.1 mm gap between the seed coat and megagametophyte tissues plus embryo. By 75 DAP the gap was much smaller appearing as a halo less than 0.05 mm around the embryonic tissues. At 101 DAP the gap had disappeared. The results were similar for clone 1996

attainment of maximum dry weight in clone 1996 (Fig. 2d). Clone 1996 produced the heavier seeds at about 3.3 mg each, while clone 1809 seeds weighed about 2.5 mg each (dry weight) (Figs. 2c,d). These are within the normal range of weights quoted for Sitka spruce seeds (Phillipson, 1987; Chaisurisri *et al.*, 1992; Gordon, 1992;). Seed moisture content (Figs. 2c,d) was always lower than cone moisture content (Figs. 2a,b) and dropped rapidly between 46 and 76 DAP, from 36% to 14% for 1809 and 39% to 19% for 1996 corresponding to the later stages of seed filling. Cone moisture content decreased more slowly and did not approach equivalent seed moisture contents until 156 DAP, when the cone scales had naturally flexed back, so most of the seeds were shed before harvesting.

Combining the observations on anatomical development with seed weight changes, seed filling must have been completed (mass maturity) at between 75 and 101 DAP in both cases. By deriving the intercept of a line between the dry weights for 46 and 75 DAP, with a line drawn through the dry weights for 101, 117 and 129 DAP, mass maturity can be crudely estimated to be about 80 DAP for clone 1809 and about 85 DAP for clone 1996, indicated by an arrow in Figures 2c,d.

No seed from either clone was capable of germinating at 20°C until 75 DAP (Figs. 2e,f). For clone 1809 germination of non-prechilled, non-dried seeds increased consistently from 64% at 75 DAP to 94% at 156 DAP. For clone 1996 germination of non-prechilled, non-dried seeds increased from 66% at 75 DAP to 84% at 117 DAP, but thereafter decreased to 74% at 129 DAP (last sample for clone 1996). Prechilling had little effect on either germination capacity (Figs. 2e,f) or mean germination time (Figs. 2g,h) determined at 20°C, except to reduce the latter in clone 1809 in the first and last harvests (75 and 185 DAP).

Hand-extracted Sitka spruce seeds from Experiment 1 were relatively non-dormant, germinating readily at 20°C, with quite low mean germination time values of between 8 and 16 d. UK-extracted commercial seedlots would typically show more dormancy than these hand-extracted seeds. For example, non-prechilled commercial seedlots would typically have mean germination times of 14 to 16 d when tested at 20°C (Jones, 1995). Differences in dormancy could be due to the extraction methods, as cones in Experiment 1 were deliberately not exposed to extraction kiln temperatures. Consequently in Experiment 2 the hypothesis that exposure of cones to a kilning temperature of 35°C induces dormancy in Sitka spruce seeds was tested.

After extraction by either the 35°C or ambient method, seed moisture contents were similar for all clones. Seed moisture contents ranged from 6–7%, typical moisture contents for commercial seed storage. Dry weights of extracted seeds showed clonal differences. The heaviest seeds were from clone 2122 and the lightest from 2019; i.e. following the pattern for cone weights (data not shown). X-ray analysis of seeds in Experiment 2 showed that all three clones had anatomically mature seeds, i.e. no gap between seed contents and seed coat was detected. The dry weights for seeds from clones 2122 and 1505 covered the

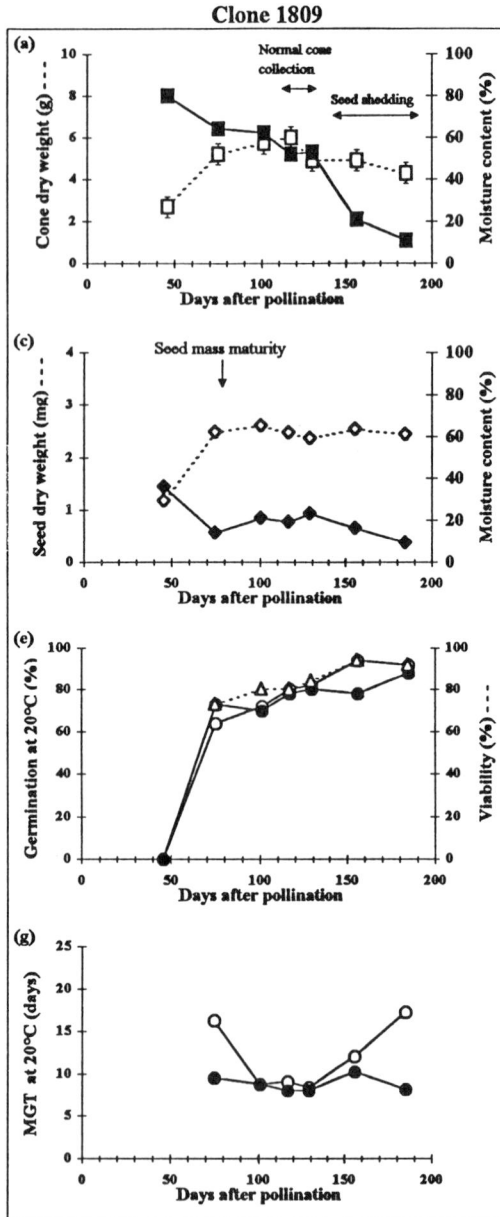

Figure 2 Experiment 1. Changes in seed and cone characteristics (data from 5–6 cones) between 46 and 185 DAP (days after pollination) for clone 1809 (a, c, e, g) and clone 1996 (b, d, f, h). (a & b) Cone dry weight (g) (□) and moisture content (%) (■). Typical times for cone collection and seed shedding are indicated and apply to all figures. (c & d) Seed dry weight (mg) (◇) and moisture content (%) (◆), based on 10 seeds per collection date for 5–6 cones. (e & f) Germination (%) after 42d at 20°C in the dark for non-prechilled (O) and 3-week prechilled seeds (●).

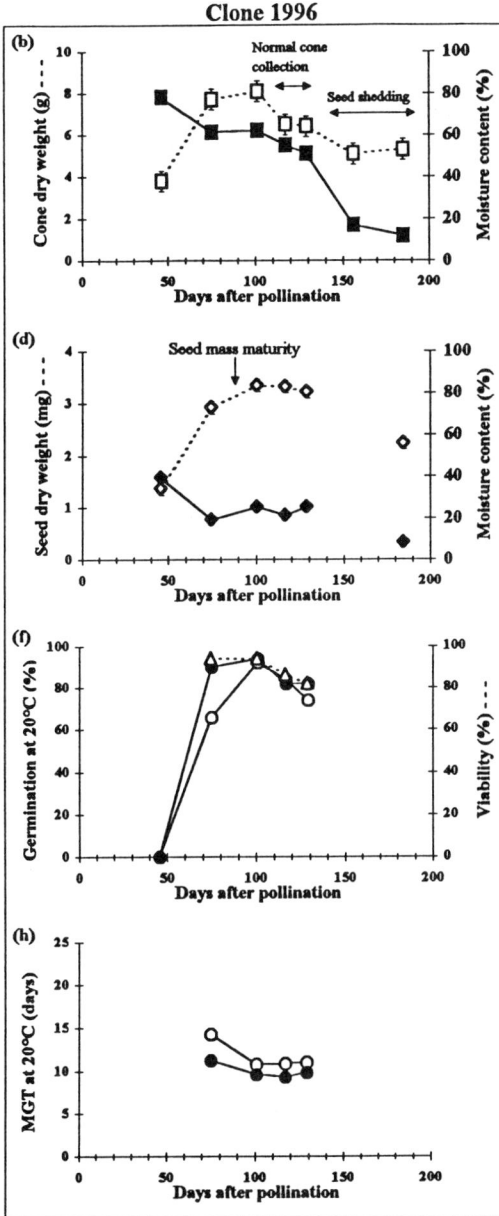

Clone 1996

Figure 2 Experiment 1 (cont). All seeds were tested immediately after collection and hand extraction without further desiccation, i.e. at the seed moisture contents shown in c&d. Seed viability for each sample is also shown (△). Each data point is based on 50 full seeds, except at 46 DAP for clones 1809 (31 seeds) and 1996 (43 seeds). (g & h) Mean germination time (MGT, days) values for tests shown in (e & f), for non-prechilled (O) and 3-week prechilled (●) seeds. MGTs for collection dates with zero germination cannot be calculated.

Table 1. Experiment 2. Germination (%) and mean germination time (MGT, days) for seeds of clones 1505, 2019 and 2122 germinated at either 10°C or 20°C in the dark, without (NPC) or with (PC) a 3-week top-of-paper prechill at 3°C. Viability (%) is a mean value across temperature, NPC and PC treatments. Seeds were extracted from cones exposed to '35°C' for 16 h soon after being collected then stored at ambient for 5 months until extracted, or stored at 'ambient' for 5 months until extracted. Data is based on variable numbers of seeds from 5 cones per treatment

| | | 10°C germination temperature | | | | 20°C germination temperature | | | |
| | Viability (%) | Germination (%) | | MGT (d) | | Germination (%) | | MGT (d) | |
		NPC	PC	NPC	PC	NPC	PC	NPC	PC
Clone 1505									
35°C extraction	80	0	9	–	28.9	54	79	20.3	9.2
Ambient extraction	85	0	50	–	29.1	86	76	14.8	7.7
Clone 2019									
35°C extraction	87	0	6	–	34.7	84	83	13.1	8.4
Ambient extraction	90	0	27	–	30.3	91	87	11.9	7.4
Clone 2122									
35°C extraction	94	0	1	–	37.5	80	84	15.6	9.0
Ambient extraction	96	0	15	–	28.0	94	95	12.6	7.5

range 1.95 to 3.41 mg per seed, similar to values found for mature seeds in Experiment 1 and the range found by Chaisurisri *et al.* (1992) for seeds of 18 Sitka spruce clones of 2.0 to 3.1 mg per seed.

The germination of untreated and prechilled seeds from 35°C extracted and ambient extracted cones were tested at both 10 and 20°C. Seeds from 35°C extracted, open pollinated cones had significantly lower and slower germination than ambient extracted seeds, especially at 10°C (Table 1). The analyses of variance for angular transformed germination capacity and $\log_{10}$ mean germination time were only done for the 20°C germination temperature, with separate analyses for each clone. There was a significant interaction between the effects of prechilling and extraction treatment for clone 1505. But the most significant result was that, for all three clones, the 35°C extracted seeds had lower germination than ambient extracted seeds ($p < 0.01$ for clone 1505, $p < 0.10$ for clone 2019 and $p < 0.001$ for clone 2122). Germination rate as measured by mean germination time was significantly slower ($p < 0.05$ or 0.001) in 35°C extracted seeds (meaned across prechill treatments) compared with hand-extracted seeds, and for untreated seeds (meaned across extraction treatments) compared with prechilled seeds, for all except clone 2122. Since seed viability was similar for both ambient and 35°C extraction seeds (Table 1), the differences in germination capacity at 10°C and mean germination time at 20°C reflect differences in conditional dormancy but this could be removed by a prechill treatment.

Seeds were not killed by the kilning treatment, however. This observation supports the idea that the best quality Sitka spruce seeds come from cones allowed to dry under cover with ambient conditions prior to kilning (Gordon and Waddell, unpublished, 1974/75). Gordon (1979) suggested that commercial seedlots imported from the USA and Canada had greater dormancy compared to UK-seedlots. This could be due to variation in extraction procedures. A similar observation was made by Løken (1959) comparing USA and Norwegian sources of Sitka spruce seeds. Equally the differences could be due to genetic variation. The clones used in Experiment 2 showed some variation in level of dormancy, while those tested from Canadian seed orchards by Chaisurisri *et al.* (1992) showed much wider variation in dormancy status.

These results raise a question-mark over current UK-extraction methods since exposure to 35°C kilning treatments seems to deepen the dormancy of Sitka spruce seeds. Kilning is, however, necessary for the complete extraction of seeds from cones and any dormancy imposed can still be released by applying a suitable prechill treatment. Further experiments investigating the length of exposure to both kilning and other temperatures would be worthwhile.

Conclusions

Anatomical maturity and maximum germination capacity are reached at similar times in Sitka spruce seeds. Exposure of cones to extraction kiln temperatures of *circa* 35°C soon after collection may aid seed extraction but causes a deeper seed dormancy. Any deepening of seed dormancy caused by kilning at 35°C can be overcome by a prechill but prechilling may harm anatomically immature seeds (Jones, 1995). The results support the current practice of collecting Sitka spruce cones in mid to late September and storing intact cones at ambient temperatures under cover to finish any development before exposing them to kilning temperatures. The range of seed dormancy found in some Sitka spruce seedlots is likely to be the result of a combination of factors including genetic make-up and extraction method.

Acknowledgements

We thank all the staff at the Forestry Commission's Research Stations for their help in this study especially Bill Brown, Yvonne Samuel, Dr Elizabeth Major and Richard Napper.

References

Barnett, J.P. 1976. *USDA Forest Service Research Paper SO-122*, Southern Forest Experiment Station, USA, pp.11.

Chaisurisri, N.C., Edwards, D.G.W. and El-Kassaby, Y.A. 1992. *Silvae Genetica* 41(6): 348–355.

Gordon, A.G. 1979. In: *Proceedings of Flowering and Seed Development in Trees: a Symposium*, p. 362 (ed. F. T. Bonner). Southern Forest Experiment Station, Starkville, Mississippi, USA.

Gordon, A.G. 1992. *Seed Manual for Forest Trees*. Forestry Commission Bulletin 83. HMSO, London.

Gosling, P.G. 1988. *Journal of Seed Technology* 12(1): 90–98.

ISTA (International Seed Testing Association) 1991. *Tree and Shrub Seed Handbook*. (eds. A.G. Gordon, P.G. Gosling and B.S.P. Wang) ISTA, Zurich, Switzerland.

Jones, S.K. 1995. Ph.D. Thesis. The University of Reading, U.K.

Jones, S.K. and Gosling, P.G. 1994. *New Forests* 8: 541–547.

Løken, A. 1959. *Årsskrift 1958-1959 Norske Skogplanteskoler* 47–50.

Phillipson, J.J. 1987. *Forest Ecology and Management* 19: 147–157.

Rimbawanto, A., Coolbear, P. and Firth, A. 1989. *Seed Science and Technology* 17(2): 399–411.

27. The Influence of Embryo Restraint During Dormancy Loss and Germination of *Fraxinus excelsior* Seeds

W.E. FINCH-SAVAGE and H.A. CLAY

Horticulture Research International, Wellesbourne, Warwick, CV35 9EF, UK

Abstract

Evidence is presented which suggests that the embryo is constrained by its enclosing tissues, the endosperm and seed coat layers, and this constraint plays an important role in the control of germination in the deeply-dormant fruit of *Fraxinus excelsior* L. The puncture force required to penetrate the endosperm/seed coat layer opposing the radicle declined during the stratification treatment used to break dormancy. This change was independent of embryo growth and occurred in advance of dormancy loss during the cold (3°C) period of stratification, but it did not occur when seeds were kept at constant 15°C. The results suggest that a localized weakening of the seed coat/endosperm layers in front of the radicle is an essential part of the process of dormancy loss during cold stratification and subsequent germination of *F. excelsior* seeds.

Introduction

The fruit of ash is a single winged indehiscent samara with a thin pericarp containing the oval flattened seed. Beneath the pericarp a thin seed coat closely adheres to the endosperm which completely surrounds the embryo. The fruit is deeply dormant having a physiologically dormant embryo and a further apparently independent dormancy imposed by the surrounding tissues (Nikolaeva, 1969). In addition the pericarp can limit oxygen availability to further delay germination (Villiers and Wareing, 1964).

Stratification treatment is necessary to break dormancy although the specific requirements can vary with seed origin (Nikolaeva and Vorobeva, 1978). In most cases, moist seeds need to be exposed first to warm temperature treatment and then to a period at lower temperature, the combined treatment time can be 32 weeks or more (Suszka *et al.*, 1994). The seed is shed with a fully differentiated embryo and growth occurs during stratification both by cell division and limited expansion (Villiers and Wareing, 1964). When embryo growth is sufficiently stimulated in the initial warm phase it will continue during the subsequent cold phase so that by the end of stratification the embryo is equal to the length of the endosperm (Nikolaeva, 1969). As stratification proceeds, the embryo gains the ability to germinate if isolated. However, the intact seed remains dormant because of some influence of the seed coat/endosperm layers. This influence can be overcome only by prolonged cold treatment or the partial

R.H. Ellis, M. Black, A.J. Murdoch, T.D. Hong (eds.), Basic and Applied Aspects of Seed Biology, pp. 245–253.
© *1997 Kluwer Academic Publishers, Dordrecht. Printed in Great Britain.*

removal of these layers (Nikolaeva, 1969).

Dormancy in *Fraxinus excelsior* has been the subject of a number of studies (reviewed in Nikolaeva, 1969; 1977; Bewley and Black, 1982) but the control of embryo dormancy and, in particular, the nature of the influence of the enclosing tissues remains obscure. The enclosing tissues may affect dormancy by limiting oxygen availability, by containing inhibitors and/or by preventing inhibitors leaching from the embryo (Nikolaeva, 1969). However, more recently the importance of physical constraint of the embryo by enclosing tissues in the control of germination in non-dormant endospermic annuals such as celery (van der Toorn and Karssen, 1992) and tomato (Groot and Karssen, 1987) has been demonstrated. The purpose of the present work was to see if embryo constraint by the pericarp/seed coat/endosperm layers influences dormancy loss and germination of deeply-dormant *F. excelsior* seeds.

Materials and Methods

Fruits of *Fraxinus excelsior* L. harvested from a single tree on 17 October 1994 were dried to 10% moisture content (fresh weight basis) and then imbibed in water for 48 h. Fruits were then added to moist peat and sand in the ratio 1:1:1 in trays within polyethylene sacks loosely folded to minimize moisture loss, but to allow aeration. The trays were placed for 16 weeks at $15 \pm 1°C$ followed by 16 weeks at $3 \pm 1°C$ and the contents were mixed regularly during this stratification treatment (Suszka *et al.*, 1994). The trays were then placed in an alternating 3 and 25°C (16 h:8 h) regime in the dark to germinate. Throughout this period seed samples (4 replicates of 5 seeds) were removed at intervals for weight and embryo length determinations.

Water Relations

Moisture content was determined on 4 replicates of 5 seeds by the low constant temperature method at 103°C for 17 h (Anon, 1993). Water potentials (ψ_w) and osmotic potentials (ψ_π) were measured using thermocouple psychrometers (Wescor, Logan UT, USA) in chambers of different sizes appropriate to the tissue. At least 3 replicates of 25, 30 and 5, fruits, seeds and embryos respectively were used per measurement. Before measurement, fruits were imbibed until there was no further increase in fresh weight. Precautions were taken during dissection of the samples to minimize evaporation of water and readings were taken when vapour-pressure equilibrium was reached as shown by repeated measurements. ψ_π was measured in the same sample as ψ_w following freezing and thawing of the sample in microcentrifuge tubes. Pressure potentials (ψ_p, ψ_t) were calculated according to the following formulae:

1   ψ_p embryo $= \psi_w$ embryo $- \psi_\pi$ embryo

2   ψ_t endosperm $+$ seed coat $= \psi_w$ seed $- \psi_w$ embryo

Puncture Force Determination

The radicle of the *F. excelsior* embryo is situated close to one end of the seed and remains in this position as the embryo grows during stratification treatment. The terminal 4 mm of a fully imbibed seed, which contained the radicle and hypocotyl, was cut from the rest of the seed. The radicle/hypocotyl was removed from the seed end and replaced with a 0.7 mm diameter steel probe which had its tip rounded to the shape of the radicle tip. The probe was attached to the crosshead of an Instron 4301 mechanical testing instrument (Instron Ltd, High Wycombe, UK) and the force required to drive the probe through the endosperm and seed coat opposing the radicle tip was recorded. The probe moved at a constant speed of 1 mm min^{-1}. The seed end was held in a fixed position that provided no restriction on the movement of the probe through the seed tissues.

Results and Discussion

Throughout stratification the embryo within the seed increased in length and in both fresh and dry weight (Fig. 1). The rate of growth was similar in both the warm and cold phases of stratification showing that a factor more dominant than temperature was limiting embryo growth.

Water Relations

When placed on moist filter paper the intact fruit became fully imbibed within 80 h (Fig. 2). Seeds removed from the imbibed fruit had a mean moisture content of 53.3% and imbibed no further water when placed on moist filter paper suggesting that there was no barrier to imbibition. However, embryos removed from fully imbibed seeds at both the beginning and end of stratification imbibed water rapidly when placed on moist filter paper (Fig. 3). This imbibition lasted only minutes and then embryo moisture content remained constant. Embryo growth began within 24 h in stratified seeds and moisture content began to increase again (data not shown). As the imbibition of water does not appear to be limited by the seed tissues, the data suggest that the embryos may be physically constrained by the surrounding seed coat/endosperm. It is suggested that removal of this restraint allows the embryos to imbibe more water and expand.

Further evidence for embryo restraint comes from water relations data determined in thermocouple psychrometers. Fully imbibed untreated fruits and the seed removed from those fruits were at water potential equilibrium (*c.* -0.02 MPa)

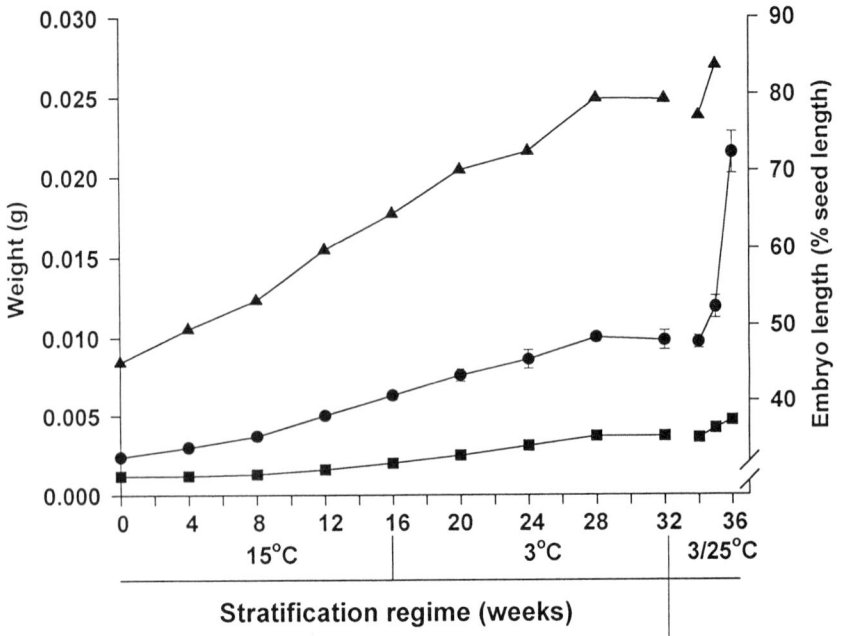

Figure 1. Embryo growth during stratification and subsequent germination. The final values for fresh and dry weight are from germinated seeds only, whereas, all previous values are from ungerminated seeds. (▲) embryo length (●) fresh and (■) dry weight. S.E. are shown where they are larger than the symbols

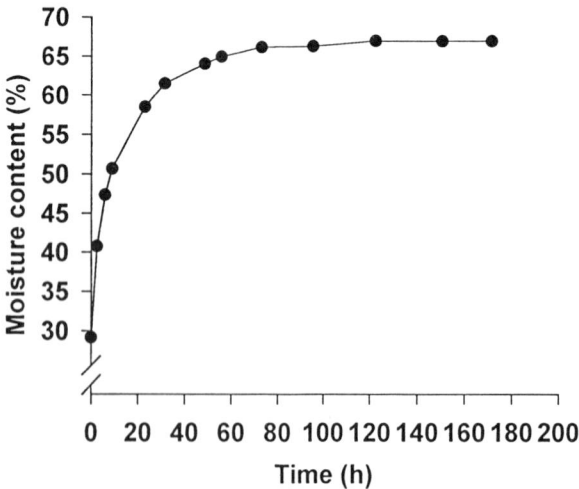

Figure 2. The imbibition of *F. excelsior* fruits on moist filter paper following humidification. S.E. smaller than the symbols

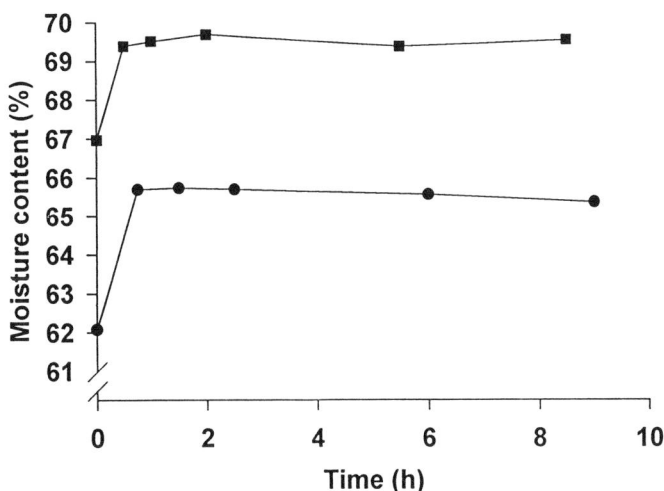

Figure 3. The imbibition of excised embryos from untreated (●) and stratified (■) seeds on moist filter paper. S.E. smaller than the symbols

Table 1. Water relations of ash fruit. Values are provided in MPa with s.e. in parenthesis

	Untreated	Stratified
ψ_w Fruit	−0.02	−*
ψ_w Seed	−0.03	−0.02
ψ_w Embryo	−0.93 (0.057)	−0.79 (0.050)
ψ_w Embryo (imbibed)	−0.01	−0.02
ψ_π Embryo	−1.45 (0.049)	−1.34 (0.066)
ψ_p Embryo	0.52	0.55
ψ_t Endosperm + seed coat	0.90	0.77

*the pericarp is broken during treatment to expose the seed

with water (Table 1). However, the water potential of embryos excised from these fruits was considerably lower (−0.93 MPa). Since the whole fruit appeared to be in water potential equilibrium (Table 1), the lower water potential in the embryo probably results from a drop in turgor pressure on release from the constraining seed tissues (see equation 1). The pericarp appears to exert no pressure upon the seed as there was no difference in water potential between the fruit and the seed within it (Table 1). This suggests that the restraining tissue is the combined seed coat/endosperm layer. The extent of the pressure potential exerted by this layer, that is to say the resistance to embryo expansion, is estimated as 0.9 MPa from equation 2 (Table 1). Such a restraint is likely to severely limit embryo growth rate as shown for instance in celery by van de Toorn and Karssen (1992) and may account for the slow growth rate observed in Figure 1.

Welbaum and Bradford (1990) suggest that for radicle/hypocotyl growth and therefore seed germination to occur one or more of three things must take place: solute accumulation in the embryo which leads to increased turgor; weakening of the tissues surrounding the embryo; cell-wall relaxation and development of a water potential gradient for water uptake by the embryo. Although essential for growth, when the embryo is under restraint, the latter is likely to have little effect on its ability to escape restraint and germinate. Following stratification the osmotic potential of the embryo was not significantly different to that in embryos from untreated fruits suggesting no significant solute accumulation (Table 1). Pressure potential of the seed coat/endosperm layer was lower in stratified seeds, probably as a result of transfer of endosperm reserves to the growing embryo, nevertheless it remained above 80% of the value in untreated seeds (Table 1). These results suggest that a more subtle or localized change was taking place to bring about germination.

Puncture Force Required for Germination

In endospermic seeds of several annual species evidence has been provided that localized endosperm weakening at the radicle tip precedes germination (e.g. Watkins and Cantliffe, 1983; Groot and Karssen, 1987). The force required to penetrate the seed coat/endosperm layers in front of the radicle of *F. excelsior* was determined with an Instron mechanical testing instrument. Figure 4 shows that in untreated seeds a range of forces was encountered (c. 0.5–1.0 N). At the end of stratification the mean force required to puncture the covers had declined significantly (untreated 0.681 ± 0.0166 N, stratified 0.637 ± 0.0184 N). By the time some seeds had begun to germinate (21%) following transfer to a regime of alternating temperatures (3 and 25°C) there was a further decline in mean puncture force (0.587 ± 0.0178 N) of the seeds remaining ungerminated. It is important to note that in *F. excelsior* dormancy loss is far from synchronous within the seed population. At the stage of treatment where some seeds have germinated few of the remaining seeds would germinate if they were placed at constant 15°C. These seeds would therefore be considered to be still dormant. An alternating temperature regime was used to stimulate germination as it allows for continued cold treatment (3°C) of the seeds which remain dormant. The inset to Figure 4 shows the same data redrawn to illustrate more clearly the distribution of puncture forces. It can be seen that the whole distribution moves to lower puncture forces during treatment suggesting that these changes begin to occur while the seeds are still dormant. Thus the localized weakening of the enclosing tissues should be considered part of dormancy loss rather than a part of germination following dormancy loss.

In *F. excelsior* seeds when the puncture force recorded was below 0.44 N the

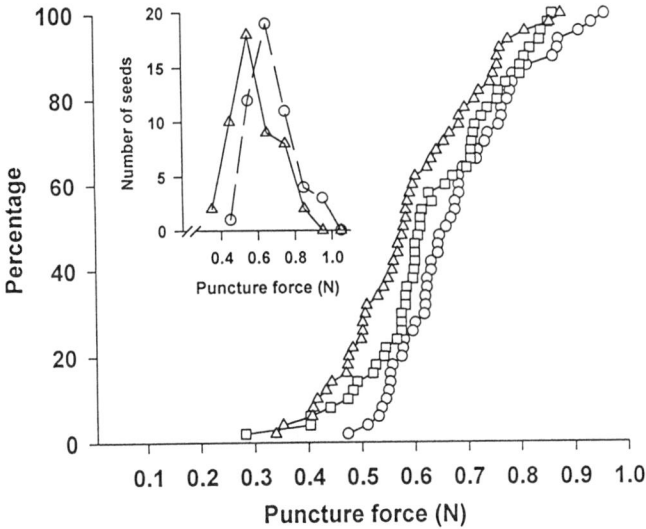

Figure 4. Accumulative distributions of the forces required to puncture the seed coat/endosperm layers opposing the radicle. (○) untreated seeds, (□) stratified seeds, (△) ungerminated seeds from a population where germination had begun (21%) after stratification. The inset shows the same data redrawn

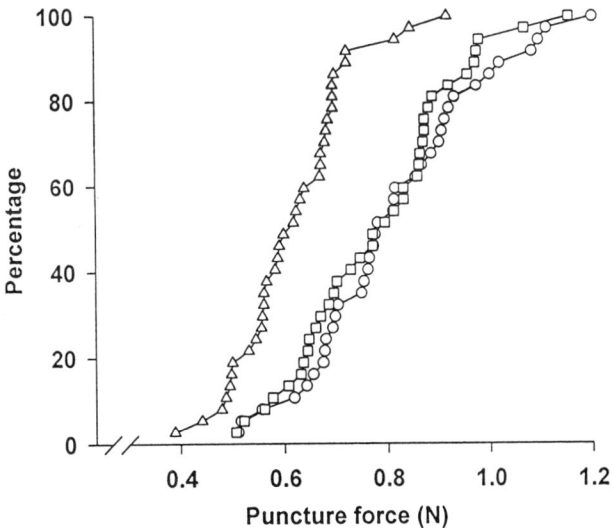

Figure 5. Accumulative distributions of the forces required to puncture the seed coat/endosperm layers opposing the radicle. (○) untreated seeds, (□) moist seeds kept at 15°C for 52 weeks, (△) moist seeds kept at 3°C for 52 weeks.

radicle and endosperm were visible through the seed coat suggesting seeds were close to germination. This is similar to values reported in pepper seeds where puncture force declined to 0.3–0.4 N before germination took place (Watkins and Cantliffe, 1983).

Further evidence for the involvement of seed coat/endosperm weakening in dormancy loss is presented in Figure 5. In this experiment three seed lots were compared: untreated seeds at harvest, moist seeds kept for 12 months at 15°C and moist seeds kept for 12 months at 3°C. In most *F. excelsior* seed lots embryo growth within the seed has first to be stimulated by warmer temperatures before it will occur at lower temperatures during stratification (Nikolaeva, 1977). As a consequence of this the embryos within seeds kept at 15°C had grown to fill the space within the endosperm (embryo 95% of seed length), whereas, the embryos in seeds kept at 3°C had not grown and were the same size as those in untreated control seeds (embryo 50% of seed length). In contrast, the puncture force required by the radicle to emerge from the seed coat/endosperm layers remained unchanged in seeds kept at 15°C compared to the untreated control (untreated, 0.8134 ± 0.0276 N; 15°C, 0.7845 ± 0.0254 N, Fig. 5), whereas, it had declined significantly in seeds kept at 3°C (0.6188 ± 0.0183 N). Although an excised embryo test showed that seeds kept at both 15 and 3°C were viable (data not shown) neither was able to germinate and remained dormant.

These data agree with the hypothesis outlined by Nikolaeva (1969) that there are two sets of dormancy breaking mechanisms (largely unknown), one in the embryo which occurs mainly under warm conditions and results in dormancy loss and growth of the embryo and a second set of mechanisms, occurring exclusively in the cold, which brings about changes in the surrounding tissues to allow germination of the whole seed/fruit. The results here suggest that a localized weakening of the seed coat/endosperm layers in front of the radicle is an important part of the changes resulting in dormancy loss and germination during cold stratification of *F. excelsior* seeds.

We thank the Ministry of Agriculture Fisheries and Food for funding this work.

References

Anon, 1993. *Seed Science and Technology* 13:338–341.
Bewley, J.D. and Black, M. 1982. *Physiology and Biochemistry of Seeds in Relation to germination.* Vol. 2, *Viability, Dormancy and Environmental Control.* pp. 339. Berlin, Heidelberg and New York: Springer-Verlag.
Groot, S.P.C. and Karssen, C.M. 1987. *Planta* 171:525–531.
Nikolaeva, M.G. 1969. *Physiology of deep dormancy in seeds*, pp. 219. Jerusalem: Israel Program for Scientific Translations.
Nikolaeva, M.G. 1977. *The Physiology and Biochemistry of Seed Dormancy and Germination.* pp. 51–74 (ed. A.A. Khan). Amsterdam, New York and Oxford: North-Holland Publishing Company.
Nikolaeva, M.G. and Vorobeva, N.S. 1978. *Botanicheskii Zhurnal* 63:1155–1167.

Suszka, B., Muller, C. and Bonnet-Masimbert, M. 1994. *Graines des Feuillus Forestiers: De la Récolte au Semis.* pp. 195–212. Paris: INRA.

van der Toorn, P. and Karssen, C.M. 1992. *Physiologia Plantarum* 84: 593–599.

Villiers, T.A. and Wareing, P.F. 1964. *Journal of Experimental Botany* 15: 359–361.

Watkins, J.T. and Cantliffe, D.J. 1983. *Plant Physiology* 72:146–150.

Welbaum, G.E. and Bradford, K.J. 1990. *Plant Physiology* 92:1046–1052.

28. Molecular Changes Associated with Dormancy-breakage in Douglas Fir Tree Seeds

S.B. JARVIS, M.A. TAYLOR and H.V. DAVIES

Department of Cellular and Environmental Physiology, Scottish Crop Research Institute, Invergowrie, Dundee DD5 5DA, UK

Abstract

Douglas fir (*Pseudotsuga menziesii*) seeds require a period of cold (4°C) moist incubation (stratification) to break dormancy. During stratification there are changes in gene expression which have been identified using two dimensional gel electrophoresis of labelled *in vitro* translation products. Differential screening and subtractive hybridisation methodology have been used in order to isolate cold-induced genes from imbibed dormant Douglas fir seeds. This paper describes the isolation and characterisation of cDNA clones encoding three such genes. Sequence homology has identified these genes as Late Embryogenesis Abundant protein (LEA) encoding genes, each clone encoding a LEA of a different class. Northern analysis has shown that these display differential expression during dormancy-breakage. DF65 showed strong sequence similarity to a dehydrin (class 2 LEA gene), and is expressed in seeds treated at 4°C but not in those seeds incubated at 20°C. DF6 and DF77, which were highly similar to class 1 and class 3 LEA genes respectively, were expressed in seeds incubated at 4°C over the four weeks of treatment, whereas in seeds incubated at 20°C, the expression level decreased after one week of treatment. The possible role these LEA proteins might play in the dormancy-breakage of this species is discussed.

Introduction

Douglas fir (*Pseudotsuga menziesii*) seeds like many seeds of temperate tree species, require a period of cold (4°C) moist incubation (stratification) to break dormancy (Bewley and Black, 1994). After six weeks at 4°C approximately 73% of seeds will germinate given suitable conditions. Seeds exhibit an imposed-dormancy and embryos excised from dormant seeds are capable of germination. In the intact seed it is the tissues surrounding the embryo which prevent germination. Two-dimensional electrophoresis of *in vitro* translation products has demonstrated that during the period of stratification there are changes in gene expression (Taylor *et al.*, 1993). These changes were only detected in samples from stratified seeds and not in those which had been incubated at 20°C. These experiments also demonstrated that differences could be detected after one week at 4°C.

Changes in gene expression have been linked to dormancy-breakage in other seeds. In dormant wheat caryopses, Morris *et al.* (1989) demonstrated the expression of LEA protein genes. Like Douglas fir seeds, these seeds also exhibit

R.H. Ellis, M. Black, A.J. Murdoch, T.D. Hong (eds.), Basic and Applied Aspects of Seed Biology, pp. 255–260.
© *1997 Kluwer Academic Publishers, Dordrecht. Printed in Great Britain.*

an imposed-dormancy. LEA proteins are typically synthesised during seed development just prior to the dessication stage (Galau *et al.*, 1991). However they have also been detected in mature plants which have been subjected to stresses such as cold, dessication, salt or the application of ABA (reviewed in Skriver and Mundy, 1990). Their role is not clear but Morris *et al.* (1991) have proposed that in the dormant wheat caryopses they might play a role in determining internal hydration states.

The techniques of differential screening and subtractive hybridisation, used to investigate changes in gene expression in Douglas fir, were designed to identify cold-induced genes which could be involved in the dormancy-breaking mechanism. Genes involved in the maintenance of the seed, and/or in the preparation of the seed for subsequent germination would not be isolated. cDNAs of three genes have been isolated which exhibit enhanced expression after one week at 4°C. From sequence similarities it appears that all three represent different classes of LEA protein genes. The changes in level of expression of these clones during dormancy-breakage have been investigated and their possible role in dormancy-breakage is discussed.

Materials and Methods

Seeds of Douglas fir (Lot number 89 (797)030) were obtained from the Forestry Commission Farnham, Surrey, UK. Seeds were air-dried and stored in air-tight containers at 4°C. Seeds were incubated and total RNA was extracted as described in Taylor *et al.* (1993). Northern analysis was performed as described in Jarvis *et al.* (1995). The protocol for subtractive hybridisation was modified from that of D. Baulcombe (John Innes Institute, Norwich). RNA from seeds incubated for one week at 20°C was subtracted from RNA from seeds incubated for one week at 4°C. This method and the protocols used for the subsequent isolation of cDNA clones are described fully in Jarvis *et al.* (1995). DNA sequences for both strands were obtained using cycle sequencing (DyeDeoxy Terminator kit, Perkin Elmer) and a 373 automated DNA sequencer (Applied Biosystems). DNA sequence analysis was carried out using software available on the SEQNET Computational Molecular Biology Facility at the SERC Daresbury Laboratory UK. Database searching was carried out using the BLAST server developed by the NCBI at the National Library of Medicine.

Results

Analysis of Steady-state Expression Levels.

Three clones were isolated by differential screening and subtractive hybridisation which were expressed at higher levels in one week chilled seeds compared with one week warm-incubated seeds. Northern blots were used to study the steady-state expression level of the three genes during dormancy-breakage at 4°C and during a timecourse of seeds maintained at 20°C (Fig. 1). All three clones were present in the dry seed; however within 1 h of imbibition the transcripts could not be detected. After 4 h of chilling DF6 and DF65 transcripts could be detected. DF77 could only be detected after 8 h of chilling. The steady-state expression level of all three genes increased with prolonged chilling. DF65 expression was maximal after one week reaching 3 times that detected in the dry seed. DF6 and DF77 expression increased and was maintained between one and four weeks at 3 and 4 fold (respectively) greater than that seen in the dry seed. At 20°C the transcript for DF65 could not be detected. DF6 and DF77 expression levels increased after one week by 3 and 4 fold respectively but subsequently levels fell by 90% and remained low for the following five weeks of treatment.

Sequence Analysis

Using the subtracted PCR fragments as probes a 2 week chilled cDNA library was screened in order to isolate full length clones. Three clones were isolated for DF65, one for DF6 and four for DF77. Table 1 shows the properties of these clones and their predicted proteins. The two DF65 clones were found to be 82% identical. The predicted proteins contain high proportions of hydrophilic amino acids; glutamine, lysine, and glycine accounting for 32% of the amino acid composition. Cysteine and tryptophan are absent. A search of the databases

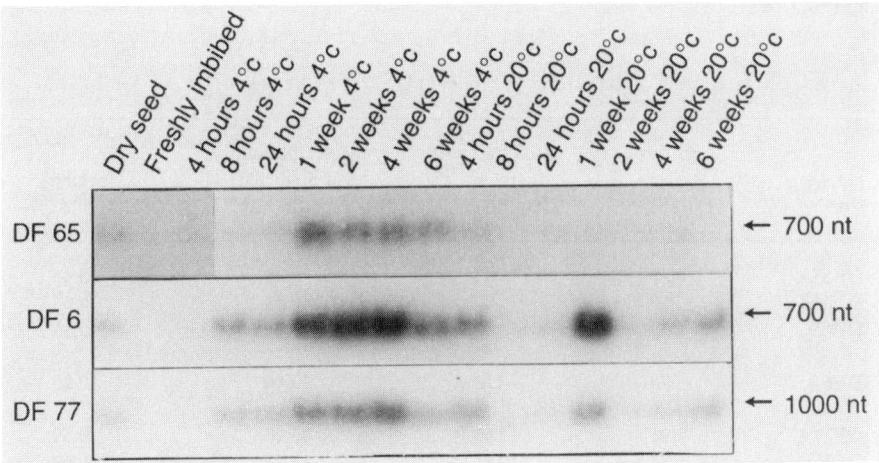

Figure 1. RNA blot analysis of DF clones during chilling induced dormancy-breakage and warm-incubation. Total RNA was isolated from intact Douglas fir tree seeds before and during incubation, for up to 6 weeks at 4°C and 20°C. RNA was analysed on 1.2% denaturing agarose gel. RNA was capillary blotted onto Hybond N (Amersham), hybridised with labelled pDF clones and autoradiographed. Three clones used were DF65, DF6 and DF77

revealed significant similarities to a number of dehydrin proteins (Table 2). The most similar was a cold-induced stress protein from *Poncirus trifoliata* (unpublished), a lipase from *Zea mays* (unpublished) and a dehydrin from pea (*Pisum sativum*) (Robertson and Chandler, 1994). Characteristic of dehyrin proteins, DF65 contains two 17 amino acid repeats containing a lysine-rich motif of 8 amino acids. The generally accepted lysine rich consensus sequence KIKEKLPG was not conserved absolutely (Galau and Close, 1992) (Table 3). More gymnosperm dehydrin sequences will have to be obtained to determine if these differences are significant.

The protein encoded by DF6 is highly hydrophilic, containing a high proportion of lysine, glutamic acid and threonine. Cysteine and tryptophan are absent. A search of the database revealed that DF6 was most similar to class 1 LEA proteins (Table 2) such as an embryonic abundant protein from sunflower seeds (*Helianthus annuus*) (Almoguera and Jordano, 1992) and a LEA protein homologue from *Arabidopsis thaliana* (unpublished). The conserved 20 amino acid motif present in many class 1 LEA proteins was absent.

The proteins encoded by DF77 are also highly hydrophilic, containing high proportions of lysine, glutamic acid and threonine. Cysteine and tryptophan are absent. The greatest similarities identified with the database search were to class 3 LEA proteins (Table 2), including a seed protein from cotton (*Gossypium hirsutum*) (Baker et al., 1988) and an embryonic protein from carrot (*Daucus carota*) (Seffens et al., 1990). DF77A and DF77B contain three 11'mer repeat motifs whereas DF77C and DF77D contained four. These repeated motifs were similar in properties to those already reported for group 3 LEA proteins (Dure et al., 1989).

Table 1. Characteristics of the pDF65, pDF6 and pDF77 cDNA clones

cDNA clone	Insert size (bp)[a]	Coding region (amino acids)	Molecular mass of predicted protein (kD)[b]	Isoelectric point[b]
DF65A	687	83	8.99	10.45
DF65B	710	82	8.88	10.64
DF65C	732	82	8.84	10.35
DF6	690	113	12.40	10.25
DF77A	896	153	15.91	9.59
DF77B	884	153	15.92	9.59
DF77C	823	164	17.01	8.97
DF77D	783	164	17.01	9.84

[a]Insert sizes including poly (A) residues have been determined from the sequence data
[b]The molecular mass and pI are calculated for the putative protein

Table 2. Database similarities to DF65, DF6 and DF77

DF clone	Clone	Similarity (identity) (%)[a]	Plant	Reference
DF65	L39004	64 (39)	*Poncirus trifoliata*	Unpublished
	L35913	56 (36)	*Zea mays*	Unpublished
	S25121	35 (57)	*Pisum sativum*	Robertson and Chandler (1994)
DF6	S23528	54 (43)	*Helianthus annuus*	Almoguera and Jordano (1992)
	S38452	54 (34)	*Arabidopsis thaliana*	Unpublished
DF77	P13939	56 (35)	*Gossypium hirsutum*	Baker *et al.* (1988)
	L35913	49 (29)	*Daucus carota*	Seffens *et al.* (1990)

[a]The GCG GAP programme was used to determine percentage similarities and identities to the Douglas fir cloned ORFs

Table 3. Comparison of the two DF65 dehydrin-like motifs with the class 2 LEA consensus sequence for monocots and dicots (as reported by Galau and Close (1992))

	Amino acid sequence[a]
Consensus monocots and dicots	**EKKGIMDKIKEKLPGQH**
Consensus DF65 I[b]	**EKAGLLDKIKQMLPGGQ**
Consensus DF65 II[c]	**QKPGMLDKIKAKIPGLH**

[a]Letters in bold face show amino acids that are identical to the consensus sequence
[b]Consensus amino acid sequence for amino acids 30–47 (I) is derived from the three DF65 ORFs
[c]Consensus amino acid sequence for amino acids 67–83 (II) is derived from the three DF65 ORFs

Discussion

Analysis of mRNA extracted from stratifying Douglas fir seeds showed that there were differences in gene expression between seeds which had been chilled as compared to seeds maintained at 20°C. In this study, we have cloned three of the genes which are differentially expressed, as these are good candidates for having roles in cold-induced dormancy-breakage. The three genes cloned were found to encode proteins with sequence similarities to LEA proteins. Based on the classification of Dure *et al.* (1989), each cDNA clone encoded a LEA protein belonging to a different class.

LEA proteins are not usually present in hydrated mature seeds and the proteins are absent from germinating seeds (Baker *et al.*, 1988). However in dormant wheat, Morris *et al.* (1991) have shown that LEA protein transcripts were maintained in the embryos of hydrated but not the germinating seeds. In Douglas fir all three LEA transcripts were found in seeds after 4 to 8 h of imbibition. The steady state levels then increased and remained high during the first 4 weeks of the chilling period. Towards the end of the chilling period, the steady state transcript levels of the three LEA transcripts decreased correlating with a decrease in the depth of dormancy of the seeds. DF65 was only detected in chilled seeds. DF77 and DF6 were detected in seeds held at 20°C but the level of transcript fell greatly after one week.

No definitive function has been ascribed to LEA proteins but because of their highly conserved sequence motifs and high composition of hydrophilic amino acids, it is thought they might have a common function. The most likely role for these proteins involves control of internal hydration states. They could provide dessication tolerance to the seed during dispersal (Baker *et al.*, 1988) and regulate the uptake of water during imbibition (Raynal *et al.*, 1989). In the dormant seed they could provide protection against repeated wetting and drying cycles over winter (Curry and Walker-Simmons, 1993). However the induction of DF65 specifically in dormant seeds subjected to chilling implies a more direct role in dormancy-breakage. In this instance, the presence of LEA proteins may cause parts of the seed to be subjected to water stress by binding available water, as was proposed for dormant wheat seeds (Morris *et al.*, 1991).

In summary, we have isolated the first LEA protein genes from gymnosperms and shown that they are expressed in hydrated seeds. We have also demonstrated that their expression is enhanced on stratification and decreases as the depth of dormancy decreases, suggesting that these LEA genes may have a role to play in cold-induced dormancy-breakage.

References

Almoguera, C. and Jordano, J. 1992. *Plant Molecular Biology* 19: 781–792.

Baker, J., Steele, C. and Dure III, L. 1988. *Plant Molecular Biology* 11: 277–291.

Bewley, J.D. and Black, M. 1994. *Seeds. Physiology of Development and Germination*, Second Edition, pp. 445. New York, London: Plenum Press.

Curry, J. and Walker-Simmons, M.K. 1993. *Plant Molecular Biology* 21: 907–912.

Dure III, L., Crouch, M., Harada, J., Ho, T.D., Mundy, J., Quatrano, R., Thomas, T. and Sung, Z.R. 1989. *Plant Molecular Biology* 12: 475–486.

Galau, G.A. and Close, T.J. 1992. *Plant Physiology* 98: 1523–1525.

Galau, G.A., Hughes, D.W. and Dure III, L. 1991. *Physiologia Plantarum* 81: 280–288.

Jarvis, S.B., Taylor, M.A., MacLeod, M.R. and Davies, H.V. 1995. *Journal of Plant Physiology* In Press.

Morris, C.F., Moffatt, J.M., Sears, R.G. and Paulsen, G.M. 1989. *Plant Physiology* 95: 643–647.

Raynal, M., Depigny, D., Cooke, R. and Delseny, M. 1989. *Plant Physiology* 91: 829–836.

Robertson, M. and Chandleer, P.M. 1994. *Plant Molecular Biology* 26: 805–816.

Seffens, W.S., Almoguera, C., Wilde, H.D., Vonder Haar, R.A. and Thomas, T.L. 1990. *Developmental Genetics* 11:65–76.

Skriver, K. and Mundy, J. 1990. *Plant Cell* 2: 503–512.

Taylor, M.A., Davies, H.V., Smith, S.B., Abruzzese, A. and Gosling, P.G. 1993. *Journal of Plant Physiology* 142: 120–123.

29. The Involvement of Microbes and Enzymes in the Pretreatment of Woody Seeds to Overcome Dormancy

D.R. MORPETH[1], A.M. HALL[1] and F.J. CULLUM[2]

[1]Environmental Sciences, University of Hertfordshire, College Lane, Hatfield, Hertfordshire AL10 9AB; [2]Writtle College, Chelmsford, Essex CM1 3RR, UK

Abstract

The addition of a compost activator, 'Garotta', to an otherwise 'normal' commercial stratification of *Rosa corymbifera* 'Laxa' has been found to greatly enhance the percentage germination at the end of this period. Germination in this study was seen to rise from 21 to 81% in the field and more markedly from 10 to 88% in the laboratory. Commercial stratification occurs over 24 weeks for this species, consisting of 12 weeks at 25°C followed by 12 weeks at 4°C. Studies have found that whilst the cold period is critical, i.e. it cannot be shortened, the warm period can be reduced by at least 6 weeks. During this warm period microbes are encouraged by providing near ideal conditions for their growth, warmth, moisture and a food source ('Garotta'). It is suspected that the combination of these conditions produces the enhanced germination.

Introduction

Woody tree seed often exhibits dormancy which usually prevents synchronous germination at the wrong time of year. This obviously has advantages for survival, but from a commercial point of view is undesirable. When seed is dispersed naturally, the hard seed coat is exposed to microbial decay in the soil, causing the seed coat to be weakened and any inhibitors can then be degraded or leached from the seed (Trumble, 1937; Campbell, 1985; Bewley and Black, 1994; Mayer and Poljakoff-Mayber, 1975; Bradbeer, 1988). Seeds usually shed in the autumn may not start decaying until the following spring and summer when temperatures rise to stimulate microbial growth (Jackson and Blundell, 1963) and will then germinate in the second spring following a cold period over the winter (Crocker, 1948).

Traditional stratification attempts to mimic the conditions experienced in nature; however this can often lead to poor germination both in quantity and quality. Such commercial stratification is based on guidelines for individual species (Lines, 1987), and it is therefore not surprising to find great variation in germination from year to year when such parameters as prevailing weather conditions influence the state of dormancy within an individual tree (Rolston, 1978).

This paper reports a novel pretreatment for woody seeds using a proprietary compost activator, 'Garotta', and discusses the probable mechanisms involved.

R.H. Ellis, M. Black, A.J. Murdoch, T.D. Hong (eds.), Basic and Applied Aspects of Seed Biology, pp. 261–277.
© 1997 Kluwer Academic Publishers, Dordrecht. Printed in Great Britain.

The aim of this research is to identify the process leading to enhanced germination with *Rosa corymbifera* 'Laxa' and translate this into a significant contribution to the industry.

Materials and Methods

Seed Supply

Rosa corymbifera 'Laxa' hips were harvested from the stock bushes at Writtle College, Chelmsford, Essex and from stock plants at Wheatcroft Nurseries, Nottingham. Hips were picked when red and firm and manually crushed prior to soaking in tap water. After three days the achenes were recovered from the softened fruit through a series of sieves. Achenes were washed in running tap water before being allowed to air dry on the bench.

Stratification Procedure

A standard ratio of seed, 'Garotta' and moist vermiculite (water added to give a water holding capacity of 70%) was adopted based on initial work by Cullum *et al.* (1990). 10 g moist seed (soaked for 24 h in sterile tap water) was mixed with 25 g moist vermiculite. 1g of 'Garotta' was added for each 10g moist seed to give the 'Garotta' treatment whilst the control had no such addition (i.e. the current 'commercial' procedure).

Material used for laboratory experiments was set up in rigid 2-litre plastic containers, whilst that destined for field trials was stratified in plastic bags of similar volume (like the commercial growers). All treatments were given 12 consecutive weeks at 25°C followed by a further 12 weeks at 4°C unless otherwise stated. During the warm period the contents of the bags/containers were shaken or stirred to allow aeration.

After the stratification period was completed four replicates of 100 seeds were taken for germination tests. Laboratory testing was carried out in Petri dishes whilst the field trials were conducted in pre-prepared seed beds at Oakover Nurseries, Kent.

Cotton Strip Assay

Burial Test Fabric (BS 2576) was purchased from Shirley Dyeing and Finishing Limited, Cheshire in 10 metre bolts. Strips were cut and frayed to give a final length of 20 cm and width of 25 mm. Strips were inserted into the stratification media (2-litre containers) using a modified method of Latter and Howson (1977).

Five strips were placed in each stratification treatment and incubated for seven days. These strips were then removed and loose media shaken off. If appropriate 5 fresh strips were inserted into the containers and stratification continued. Removed strips were frozen in aluminium foil until sufficient numbers were available to test.

Tensile strength of the cotton strips was measured on a tensometer (Hounslow, 600 N beam). Strips were soaked with water and wrung to remove excess prior to testing.

Fluorescein Diacetate (FDA) Hydrolysis

Samples (1 g) from the stratification mix of each treatment were processed according to the following protocol. 1 g of well mixed sample (referred to in the text as '1 g sample') was transferred to a 50 ml conical flask and 10 ml of sodium phosphate buffer (pH 7.6) was added to each. FDA hydrolysis was initiated by the addition of 200 μl of a 1 mg/ml FDA solution to give a final FDA concentration of 20 μg/ml. The flask was capped with parafilm and incubated at 37°C for 30 min in a shaking water bath (approx. 180 strokes/min).

The sample was removed and 10 ml of acetone added to terminate the reaction. 1.5 ml of fluid sample was centrifuged at 13 000 rpm for 30 s before the absorbance was read at 490 nm using a spectrophotometer.

After seven weeks small pieces of cotton (2 cm^2) were also inserted into the stratification mix to allow FDA analysis. This was to avoid the measurement of esterases from the seed which the 1 g samples would contain (FDA hydrolysis has been used for viability testing (Pritchard, 1985)).

Microbial Counts

A serial dilution series was used with an initial 1 g sample taken from the stratification mix. Plates were made using nutrient agar (NA) and Potato Dextrose Agar (PDA) to select for bacteria and fungi respectively. Once set the plates were incubated at 22°C for 48 h before colonies were counted.

Results

Stratification Procedure

In all stratification protocols with *Rosa corymbifera* 'Laxa' the addition of the activator greatly enhanced the germination of the seed batch from less than 21% to over 80% in both the field and the laboratory (Table 1).

Other observations during the stratification found characteristic changes in the appearance of the seed. After 6 weeks of the warm period of stratification the seed treated with the activator darkened considerably compared to the control which remained the light brown of fresh seed. It was also observed that control seed remained firmly intact around the suture, whilst the majority of seed from the activator treatment had split. This splitting has become an indicator for successful germination.

Table 1. Germination after 14 days of control and 'Garotta' treated *Rosa corymbifera* 'Laxa' achenes following stratification

		% Germination	Standard deviation
Laboratory[a]	Control	10.17	3.49
	Garotta	87.67	9.61
Field[b]	Control	20.81	5.77
	Garotta	80.56	10.34

[a]Average of six replicates of 100 seeds
[b]Average of four replicates of 400 seeds

Cotton Strip Assay

Tensile strength is rapidly lost in the treatment containing the compost activator, reaching a peak of nearly 50% loss after only three weeks (Fig.1). The strips in the control treatments show little in the way of loss in strength, a maximum loss of only 15% was recorded after 4 weeks.

Once the stratification was continued in the cold (4°C), the loss in tensile strength in both activator and control treatments was negligible (see Fig. 1).

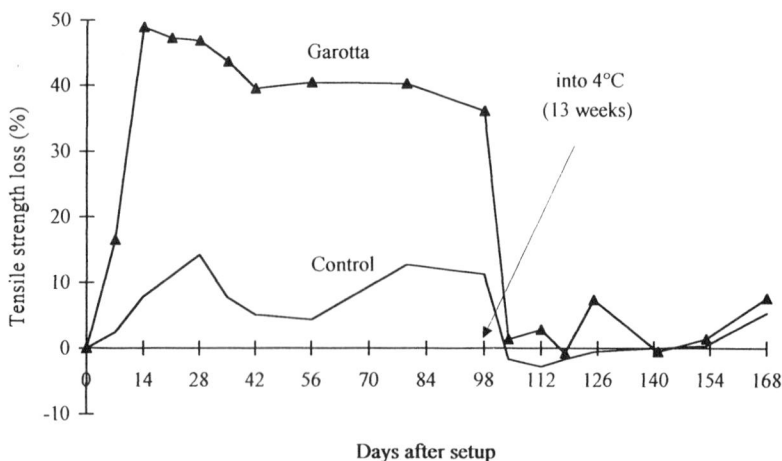

Figure 1. Loss in tensile strength of cotton strips during stratification *of Rosa corymbifera* 'Laxa'. Each point is the combination of three replicates (i.e. mean of 15 strips), exposed for 7 days to the medium

Fluorescein Diacetate (FDA) Hydrolysis

The hydrolysis of FDA is brought about by the cleaving of the diacetate by esterases. Thus esterases produced within the system by microbes will result in such a reaction, and this can be seen in Figure 2.

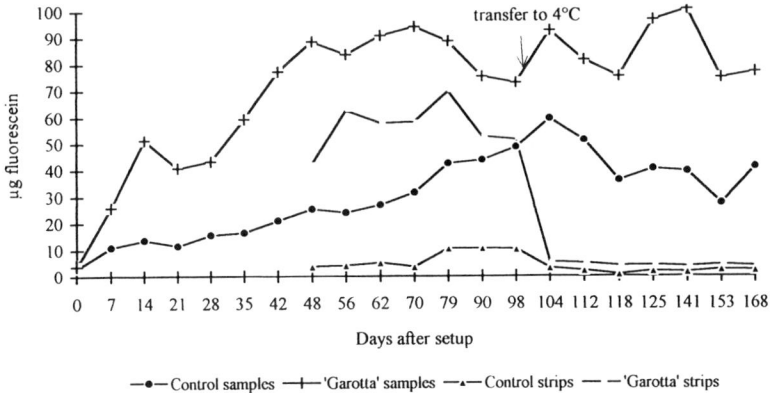

Figure 2. FDA results for the 1g samples and cotton strips during the stratification of *Rosa corymbifera* 'Laxa'. Results are shown for 1g samples of stratification medium and after 7 weeks for small cotton strips in the media

During the course of this experiment it was also found that esterases produced by the seeds could influence the result; in fact the procedure has been used as an orchid seed viability test (Pritchard, 1985). This led to the addition of small pieces of cotton being introduced to the stratification media for FDA testing. These results are also shown in Figure 2.

The graph shows that the treatment with the activator consistently has a much higher metabolic activity than the control, both for the 1 g samples and cotton strips. However a difference is seen once the treatment enters the cold period of the stratification. The 1 g samples continue to maintain their FDA activity, whilst that for the cotton squares dropped to a level where no activity was measured.

Microbial Counts

Microbial counts during the stratification of *Rosa corymbifera* 'Laxa' show an increase in both fungi and bacteria during the first four weeks of the warm period (Figure 3). This is enhanced in the presence of the activator.

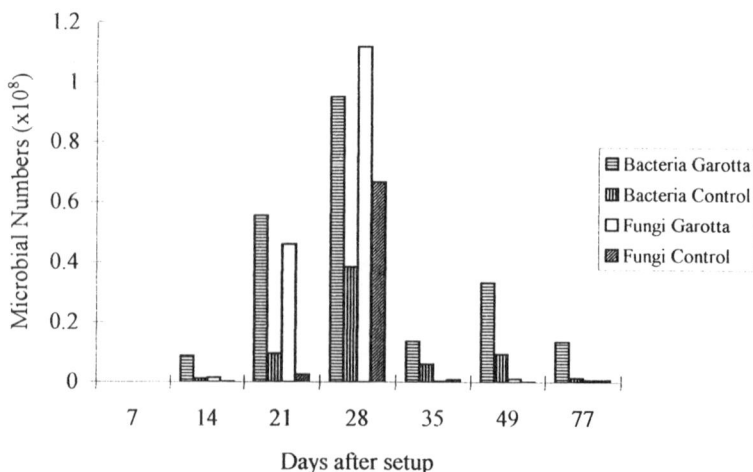

Figure 3. Fungal and bacterial counts during the warm period of stratification of *Rosa corymbifera* 'Laxa'

Discussion

Germination of *Rosa corymbifera* 'Laxa' is greatly enhanced when the commercial stratification protocol has an addition of a compost activator. It was thought that perhaps this was due to the involvement of microorganisms and their associated extracellular enzymes. This was suspected due to the darkening of the seed coat during the procedure, coupled with the splitting of the seed coat and enhanced microbial numbers in the presence of the activator.

During commercial stratification a grower may introduce fungicides to combat potentially harmful microbes. Interestingly, there has been no observed detrimental effects from encouraging the microbes in this system; on the contrary very recent results have shown a decrease in germination if the seed is surface sterilised prior to stratification. However work so far has not conclusively shown it to be purely the microbial action causing the enhanced germination.

Cotton strips lose half of their strength when placed in the activator treatment, but less than 15% in the control. This is either due to enzymatic action on the cellulose fibres, or chemical action from the activator. Recent work has found that the activator alone will cause weakening of the strips. Enzymatic and chemical studies are currently being carried out to investigate the nature of this loss in tensile strength.

Measuring metabolic activity in the stratification mix using the FDA assay showed a much higher activity in the presence of the activator over control, both when measured on samples of the mix and indirectly using cotton. Whilst it is likely that such activity is partially due to the activator and possibly the seed, it has also been found that FDA can also be cleaved by a suspension of the activator. Further work will elucidate the exact contribution made by each of these components.

Microbial numbers increase with time during the stratification period. These microorganisms are introduced to the system on the seed; seed is sterile in the hip, but gains a microbial loading when extracted. Stratification with surface sterilised seed in the presence of the activator still shows a marked increase in germination over control, although such seed does not attain the same percent germination as the non sterile seed. It has proved very difficult to keep such a system totally sterile for the entire stratification, partly due to the irregular surface of the seed coat. Prolonged exposure to sterilant could cause damage to the embryo, as *Rosa corymbifera* 'Laxa' is permeable to water.

It is evident that a novel stratification procedure has been successfully introduced for certain woody tree seed, but what remains uncertain is the exact mechanism involved. Current research is looking at the mechanism and once sufficient understanding is obtained then the protocol will be adapted to cover many more species.

Acknowledgements

This work is funded by the Horticulture Development Council (HDC).

References

Bewley, J.D. and Black, M. 1994. *Seeds. Physiology of Development and Germination*, Second Edition, pp. 445. New York, London: Plenum Press.
Bradbeer, J.W. 1988. *Seed Dormancy and Germination*, First Edition, pp. 146. Glasgow: Blackie.
Campbell, R. 1985. *Plant Microbiology*, First Edition, pp. 191. London: Edward Arnold Ltd.
Crocker, W. 1948. *Growth of Plants. Twenty Years' Research at Boyce Thompson Institute*, pp. 459. New York: Reinhold Publishing Corp.
Cullum, F.J., Bradley, S.J. and Williams, M.E. 1990. *Combined Proceedings of the IPPS* 40: 244–250.
Jackson, G.A.D. and Blundell, J.B. 1963. *Journal of Horticultural Science* 38: 310–320.
Latter, P.M. and Howson, G. 1977. *Pedobiologia* 17: 145–155.
Lines, R. 1987. *Forestry Commission Bulletin 66*, pp. 61. London: HMSO.
Mayer, A.M. and Poljakoff-Mayber, A. 1975. *The Germination of Seeds*, Second Edition, pp. 192. Oxford: Pergamon Press.
Pritchard, H.W. 1985. *Plant, Cell and Environment* 8: 727–730.
Rolston, M.P. 1978. *Botanical Review* 44 (3): 365–396.
Trumble, H.C. 1937. *Journal of the Department of Agriculture, South Australia* 40: 779–786.

30. Dormancy Breaking and Short-term Storage of Pretreated *Fagus sylvatica* Seeds

M.P.M. DERKX and M.K. JOUSTRA

Applied Research for Nursery Stock, Rijneveld 153, P.O. Box 118, 2770 AC Boskoop, The Netherlands

Abstract

Seeds of *Fagus sylvatica* with a seed moisture content (m.c.) of 30% showed a widening in the temperature range of germination during the first 16–20 weeks of pretreatment at 3°C. After 24 weeks this range became narrower again. Effects of dehydration and dry storage on germination capacity depended on the provenance, the duration of pretreatment preceding dehydration, the seed m.c. after dehydration, the storage temperature and the duration of storage. When pretreated seeds with a m.c. of 30% were stored at –2°C, no decrease in germination capacity was observed during at least four months. It is postulated that seeds withstand dehydration during the phase of dormancy breakage, whereas dehydration after completion of that phase results in irreversible damage in the metabolically active seeds, that otherwise may start early germinative events, although radicle protrusion is prevented.

Introduction

The European beech (*Fagus sylvatica* L.) is an important ornamental and timber species and is often used in reforestation plantings. Freshly harvested seeds only germinate at low temperatures (2–5 °C). This germination is slow. A moist cold-temperature treatment is required to overcome dormancy. In traditional nursery practice seeds are stratified in moist sand outside during the winter and seeds are sown in spring. Field emergence is often low and irregular, since prematurely germinated seeds may get killed during sowing, whereas dormancy of other seeds has not been completely released. Premature germination during a dormancy-releasing treatment can be prevented by controlling the moisture content (m.c.) of the seeds (Suszka and Zieta, 1976; 1977). This is easiest when seeds are pretreated without medium. Previous studies showed that a seed m.c. of 30% on a fresh weight basis is optimal when dormancy breakage is aimed at, while premature germination has to be prevented (Suszka, 1979; Suszka and Kluczynska, 1980). The optimal duration of pretreatment may vary between seed lots and between years (Suszka, 1979; Muller and Bonnet-Masimbert, 1989; Gosling, 1991). Durations between 4 and 20 weeks have been reported.

When conditions for sowing are not appropriate after completion of the dormancy-releasing treatment, seeds have to be stored for some time. There is some disagreement about storage of beechnuts. It is sometimes suggested that pretreated seeds should be dried to a m.c. of 10% before short-term storage at

R.H. Ellis, M. Black, A.J. Murdoch, T.D. Hong (eds.), Basic and Applied Aspects of Seed Biology, pp. 269–278.
© *1997 Kluwer Academic Publishers, Dordrecht. Printed in Great Britain.*

sub-zero temperatures (Suszka and Zieta, 1976). However, a loss in germination capacity of 30% at a test temperature of 20°C was not uncommon. In the past, it was thought that beechnuts could not withstand drying without injury (Holmes and Buszewicz, 1958). Later studies revealed that factors like drying temperature (LaCroix, 1986) and rate of imbibition of dry seeds (Poulsen, 1993a) are very critical. Since beechnuts are not as easy to store as the seeds of many other species, Gosling (1991) suggested storage characteristics between those of recalcitrant and orthodox seeds (see Roberts, 1973 for definitions). Recently, Poulsen (1993c) calculated viability equations for beechnuts and demonstrated an orthodox seed behaviour. However, she studied storage characteristics of non-pretreated seeds.

It was the aim of the present study to determine whether the optimal duration of pretreatment is very critical. For this purpose, seeds of different provenances were pretreated over various durations, up to one year. At regular intervals germination capacity was tested over a range of temperatures. As long as this range becomes wider, dormancy breakage is not complete (Vegis, 1964; Karssen, 1982). Furthermore, best conditions for short-term storage after pretreatment were studied. It was also investigated whether effects of dehydration and storage under several conditions depended on the duration of pretreatment preceding storage.

Materials and Methods

Seeds

Three provenances of *Fagus sylvatica* L. were used: NL2.1-Gooi (The Netherlands), WDN bukowy (Poland) and Gråsten, F413 (Denmark). The seeds (the structure referred to as 'seed' is actually a fruit and includes the pericarp) were collected from the ground in autumn 1993. Dutch and Polish seeds were obtained from Boevé Seeds, Boskoop, The Netherlands and Danish seeds from the Tree Improvement Station, Humlebaek, Denmark. After arrival viability of the seeds (tetrazolium test) was 94, 83 and 97% for Dutch, Polish and Danish seeds, respectively.

Moisture Content Determination

After arrival the moisture content (m.c.) of the seeds was determined by drying samples of 5 g at 105°C for 17 h. Dutch, Polish and Danish seeds had a m.c. of 31.1, 26.4 and 20.6%, respectively. Moisture contents are expressed on a fresh weight basis. The m.c. of all seed lots was adjusted to 30%.

Pretreatment and Storage Conditions

Portions of 4 kg seeds with a m.c. of 30% were pretreated at 3°C without medium in perforated plastic bags. At weekly intervals the seeds were mixed and the bags were weighed to check the m.c. If required, water was added.

After pretreatment at 3°C for 8, 12 or 16 weeks portions of seeds were transferred to storage conditions. Seeds were dried to m.c. of 9 or 16% or not dried and stored at –2 or +3°C in perforated plastic bags. Drying of the seeds occurred in a cabinet with controlled temperature and relative humidity (RH) (Van den Berg K890L, Montfoort, The Netherlands). A m.c. of 9% was obtained by drying the seeds during 7 days at 17°C, 45% RH, a m.c. of 16% by drying at 17°C, 75% RH.

Germination Conditions

After several intervals of pretreatment and storage, germination was tested. Four replicates of 50 seeds were sown in plastic boxes on one layer of thick paper (ZH/224, Schut B.V., Heelsum, The Netherlands), moistened with distilled water. Germination was tested at 3, 10, 15, 17, 20, 25 or 30°C in temperature-controlled cabinets (Navep, Super-720, Groningen, The Netherlands). Except for the germination counts, that were done in daylight twice a week, the boxes were kept in darkness. A radicle length of at least 2 mm was the criterion for germination. Germination tests were finished when germination was expected to be complete.

Final percentages of germination were calculated, as well as mean germination times (MGT), using the formula quoted by Bewley and Black (1985) and modified by Jones and Gosling (1994). This modification was done since germination was recorded on non-consecutive days. The actual number ($N_{t(x)}$) of seeds that was recorded on day $t_{(x)}$ was multiplied by the day number half way between $t_{(x)}$ and the previous day number on which germination was recorded ($t_{(x-1)}$). The sum of these values was divided by the total number of seeds germinated at the end of the germination test ($S(N_{t(x)})$).

$$MGT = \Sigma(N_{t(x)} \cdot (t_{(x)} - t_{(x-1)})/2) \, / \Sigma N_{t(x)}$$

Statistical Analysis

Germination percentages were transformed using the angular transformation. For separate factors, as well as for combinations of factors, data were analysed by analysis of variance.

Results

Effects of Pretreatment

During pretreatment at 3°C, seeds with a m.c. of 30% did not germinate. However, dormancy was released, as became visible after transfer of the seeds to germination conditions, in which the presence of water was not limiting anymore. Figure 1 shows results of the Dutch provenance. Freshly harvested seeds hardly germinated at temperatures of 10°C or above. Germination at 3°C was slow and could be attributed to dormancy breakage, followed by germination. During the first 16–20 weeks of pretreatment the temperature range of germination became wider (Fig. 1A), resulting in germination percentages of 80–90% at temperatures between 3 and 20°C, and about 60% germination at 25°C. After 24–28 weeks of pretreatment germination percentages started to decline (Fig. 1B). Since the provenances from Poland and Denmark showed similar tendencies, these results will not be presented. Freshly harvested Danish seeds were somewhat less dormant than seeds of the other provenances.

The mean germination time decreased during the first 24 weeks of pretreatment and increased thereafter (Fig. 2).

Effects of Storage of Pretreated Seeds

After pretreatments of 8, 12 and 16 weeks seeds were dried to different m.c. and stored at –2 or +3°C during 0–16 weeks. Effects of dehydration and storage will be demonstrated on Dutch (Figs. 3A,C) and Polish seeds (Figs. 3B,D). Danish seeds reacted similarly to Polish seeds. Dutch seeds that were pretreated at 3°C for 12 weeks, dehydrated to m.c. of 9 and 16% and stored at –2°C during 8 weeks showed a small reduction in germination at high test temperatures (Fig. 3A). Storage at +3°C (Fig. 3C) reduced germination more. A seed m.c. of 9% during storage was more detrimental to germination than a m.c. of 16%.

In Polish seeds dehydration and dry storage had more dramatic effects than in

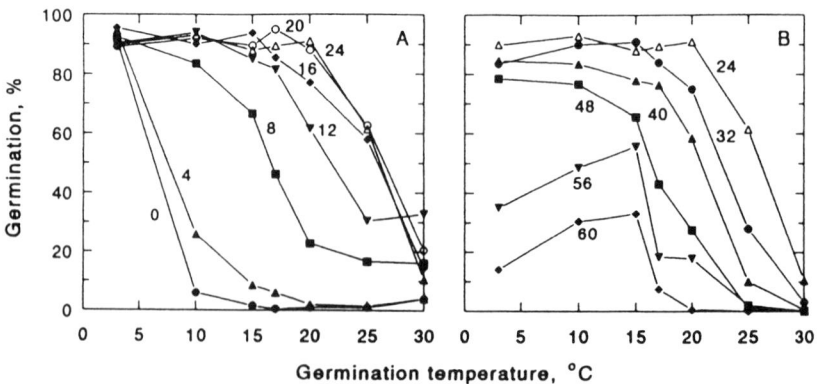

Figure 1. The effect of pretreatment on the temperature range of germination of Dutch *F. sylvatica* seeds. Seeds with a m.c. of 30% were pretreated without medium at 3°C. After 0 (●), 4 (▲), 8 (■), 12 (▼), 16 (◆), 20 (○), 24 (△), 32 (●), 40 (▲), 48 (■), 56 (▼) and 60 (◆) weeks of pretreatment, germination was tested in darkness at constant temperatures. (A) shows 0–24 weeks pretreatment and (B) 24–60 weeks

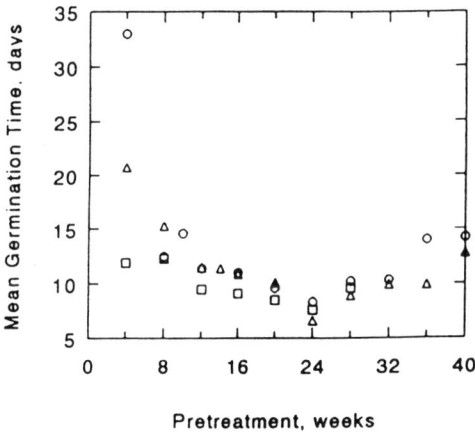

Figure 2. The effect of pretreatment on mean germination times of Dutch (O), Polish (△) and Danish (□) seeds of *F. sylvatica*. Pretreatments as described in Figure 1. Germination was tested at 3°C

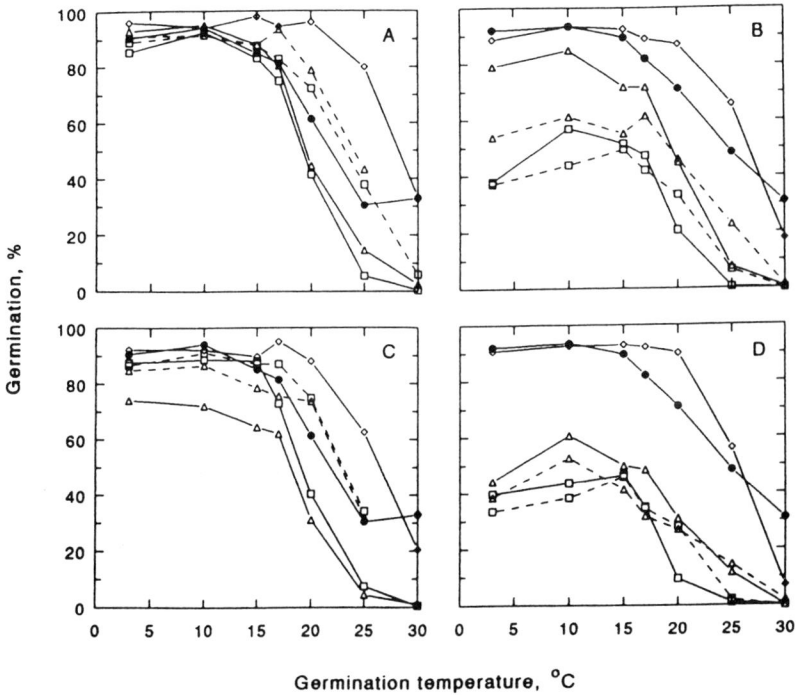

Figure 3. Effects of dehydration and dry storage on germination capacity of pretreated *F. sylvatica* seeds. Dutch (A,C) and Polish (B,D) seeds with a m.c. of 30% were pretreated without medium at 3°C for 12 weeks (●). Thereafter the seeds were dried to m.c. of 9 or 16%, or not dried. The seeds were then stored at –2 (A,B) or +3°C (C,D) for 8 weeks (△, 9%; □, 16%; ◇, 30%). After storage, dried seeds were rehydrated to a m.c. of 30% and pretreated at 3°C for another 4 weeks (broken lines)

Dutch seeds. A large reduction in germination was seen at both storage temperatures and at both m.c. (Figs. 3B,D). Like the Dutch seeds, germination capacity declined most during storage at +3°C. In contrast with the Dutch seeds, storage with a m.c. of 16% was most detrimental.

When pretreated seeds were not dried, but stored with a m.c. of 30% at –2°C during 8 weeks a remarkable increase in germination capacity was observed (Figs. 3A,B). Storage at +3°C, which means prolongation of the pretreatment, unsurprisingly also had a positive effect on germination capacity (Figs. 3C,D).

In order to see whether a reduction in germination capacity caused by dry storage could be reversed, seed m.c. was re-adjusted to 30% and the seeds got a second pretreatment at +3°C during 4 weeks. The small reduction in germination capacity in Dutch seeds could indeed be reversed (Figs. 3A,C), whereas the larger reduction in Polish seeds could not be reversed (Figs. 3B,D).

Effects of dehydration and dry storage also depended on the duration of the pretreatment preceding storage. This is demonstrated in Figure 4 for Dutch seeds. When seeds pretreated during 8 weeks were dried to a m.c. of 9%, no decrease in germination capacity was observed. Storage of these seeds at –2°C during maximally 16 weeks also did not reduce germination (Fig. 4A). However, when less dormant seeds were dried, germination capacity of the seeds declined (Figs. 4B,C). A further decline occurred during 16 weeks storage.

Discussion

Changes in Germination Capacity

The present results clearly show that a cold pretreatment of 16–20 weeks is required to get optimal dormancy breakage in freshly harvested *F. sylvatica* seeds (Figs. 1 and 2). This was true for all tested provenances. The enormous widening of the temperature range of germination has the important implication that temperature conditions at the time of sowing in spring are not very critical. Since the optimal duration of pretreatment may vary between seed lots, Suszka (1979) and Muller and Bonnet-Masimbert (1989) tried to develop a general method to get insight into the time required for dormancy breakage. In this method, they determined the time that is required to achieve 10% germination at 3°C in a stratification medium, and called this 'x'. Extending this period 'x' with 2 extra weeks ('x+2') gave satisfactory dormancy breakage in seeds pretreated without medium. For the provenances used in our study, 'x' values were calculated for germination at 3°C, either on moist paper or on vermiculite. For Dutch, Polish and Danish seeds 'x' values were 9, 5 and 5 weeks, respectively. Extending this period with 2 extra weeks gave far from optimal dormancy breakage, as can best be seen at sub-optimal germination temperatures (Fig. 1).

In contrast with the study by Muller and Bonnet-Masimbert (1989), the optimal duration of pretreatment was not very critical, since the temperature

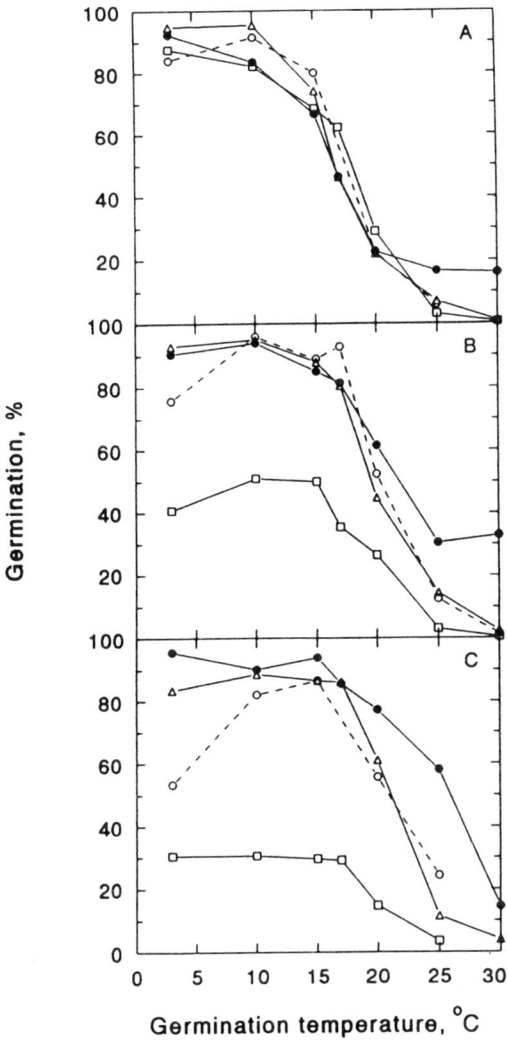

Figure 4. Effects of dehydration and dry storage on germination capacity of Dutch *F. sylvatica* seeds, pretreated at 3°C for 8 (A), 12 (B) or 16 (C) weeks. Pretreatment and germination tests as in Figure 1. Pretreated seeds were dehydrated to a m.c. of 9% and stored at –2°C. Germination was tested after pretreatment (●), after dehydration (○) and after dry storage for 8 (△) and 16 (□) weeks

range of germination only started to narrow after about 24 weeks of pretreatment (Fig. 1B). It is most likely that the decline in germination can be attributed to a decrease in viability, although some induction of secondary dormancy cannot be fully excluded after these long pretreatments. After about 40 weeks of cold pretreatment, germination capacity also started to decline at low temperatures. Root tips looked damaged after these long pretreatments.

Dehydration and Dry Storage of Pretreated Seeds

Effects of dehydration and dry storage depended on the duration of pretreatment preceding storage (Fig. 4). Increasing the duration of pretreatment reduced desiccation tolerance and storability of the seeds. This phenomenon has been described before (Tao and Fu, 1993). An explanation has not been given, so far. It may be hypothesized that dormancy breakage is complete after some time of pretreatment, and thereafter early germinative events preparing the seed for radicle protrusion may start, since the m.c. of the seeds (30%) allows metabolic activity (Poulsen, 1993b). It may be proposed that seeds that are dehydrated during the phase of dormancy breakage, withstand dehydration and dry storage, whereas dehydration during the phase of early germinative events causes damage, that cannot be repaired.

Regarding the m.c., pretreated seeds of beech did not show consistent storage behaviour. In Dutch seeds storage with a m.c. of 9% reduced germination significantly more than storage with a m.c. of 16% (Figs. 3A,C), whereas the opposite was seen in Polish (Figs. 3B,D) and Danish seeds (data not shown). Although only results after one combination of pretreatment (12 weeks) and storage (8 weeks) are presented, these conclusions are valid for other combinations as well. In general, storage at +3°C reduced germination more than storage at –2°C. This is consistent with storage physiology of many species.

Storage of seeds with a high m.c. (30%) at sub-zero temperatures (–2°C) did not reduce germination. Surprisingly, even an increase in germination capacity was found in seeds that got a non-optimal dormancy-breaking pretreatment before storage (Figs. 3A,B).

Although seed longevity of highly-dormant, non-pretreated seeds may be described by orthodox seed storage behaviour (Poulsen, 1993c), our study shows that longevity of low-dormant seeds does not conform with orthodox seed storage behaviour. This difference in behaviour depending on the degree of dormancy may explain discrepancies in literature. We agree with Gosling (1991) that beechnuts exhibit storage characteristics between those of orthodox and recalcitrant seeds. An intermediate category of seed storage behaviour was also found in other species, like *Coffea arabica* (Ellis *et al.*, 1990). Finch-Savage (1992) suggested that there is a continuous scale of desiccation tolerance across species.

Nursery Practice

To get good dormancy breakage of freshly harvested seeds, a cold pretreatment of 16 weeks can be advised. Seed m.c. during this treatment should be maintained at 30%. When conditions for sowing are not appropriate after the dormancy-releasing treatment is complete, seeds can best be stored with a m.c. of 30% at a temperature just below 0°C. Alternatively, pretreatment at 3°C may be prolonged with another 8–12 weeks. Dehydration of low-dormant seeds is not recommended.

When long-term storage over more than one winter is wanted, dehydration of the seeds cannot be avoided. The seeds may be dried directly after harvest (Suszka and Zieta, 1977) or after dormancy breaking (Muller, 1993). In the latter case a sub-optimal dormancy-breaking treatment is preferred rather than an optimal one due to dehydration damage. In this case, the determination of 'x' values (Suszka, 1979) may give satisfactory results.

Acknowledgements

The authors thank Mr. G.A. Houtman for technical assistance and Mr. J.H.M. Sieverink for statistical analysis of the results.

References

Bewley, J.D. and Black, M. 1985. *Seeds: Physiology of Development and Germination*, pp. 4–5. New York: Plenum Press.

Ellis, R.H., Hong, T.D. and Roberts, E.H. 1990. *Journal of Experimental Botany* 1: 1167–1174.

Finch-Savage, W.E. 1992. *Seed Science Research* 2: 17–22.

Gosling, P.G. 1991. *Forestry* 64: 51–59.

Holmes, G.D. and Buszewicz, G. 1958. *Forestry Abstracts* 19: 1–31.

Jones, S.K. and Gosling, P.G. 1994. *New Forests* 8: 309–321.

LaCroix, Ph. 1986. *Revue Forestière Française* 3: 205–212.

Karssen, C.M. 1982. In: *The Physiology and Biochemistry of Seed Development, Dormancy and Germination*, pp. 243–270 (ed. A.A. Khan). Amsterdam: Elsevier Biomedical Press.

Muller, C. 1993. In: *Dormancy and barriers to germination. Proceedings of an international symposium of IUFRO Project Group P2-04-00*, pp. 79–85 (ed. D.G.W. Edwards).

Muller, C. and Bonnet-Masimbert, M. 1989. *Seed Science and Technology* 17: 15–26.

Poulsen, K.M. 1993a. In: *Internationales Symposium über Forstsaatgut. Munster-Uelzen, Niedersachsen*, pp. 91–95.

Poulsen, K.M. 1993b. In: *Internationales Symposium über Forstsaatgut. Munster-Uelzen, Niedersachsen*, pp. 96–102.

Poulsen, K.M. 1993c. *Seed Science and Technology* 21: 327–337.

Roberts, E.H. 1973. *Seed Science and Technology* 1: 499–514.

Suszka, B. 1979. *Arboretum Kórnickie* 24: 111–135.

Suszka, B. and Zieta, L. 1976. *Arboretum Kórnickie* 21: 279–296.

Suszka, B. and Zieta, L. 1977. *Arboretum Kórnickie* 22: 237–255.

Suszka, B. and Kluczynska, A. 1980. *Arboretum Kórnickie* 25: 231–255.

Tao, K.L. and Fu, S. 1993. In *Basic and Applied Aspects of Seed Biology. Proceedings of the Fourth International Workshop on Seeds*, pp. 781–788 (eds. D. Côme and F. Corbineau). Paris: ASFIS.

Vegis, A. 1964. *Annual Review Plant Physiology* 15: 185–224.

31. Germination Requirements and Dormancy in *Festuca gigantea* (L.) Vill. Populations

K.H. CHORLTON, A.H. MARSHALL and I.D. THOMAS

Institute of Grassland and Environmental Research, Plas Gogerddan, Aberystwyth, Dyfed SY23 3EB, UK

Abstract

Festuca gigantea (L.) Vill. (Giant fescue) is a wild species found in woodland margins and on riverbanks throughout Europe. It has large leaves of high nutritive value and is used in breeding programmes which aim to broaden the adaptability of ryegrasses (*Lolium perenne* L. and *L. multiflorum* L.), introducing desirable characteristics from fescue species through intergeneric hybridisation. As well as its agronomic attributes, *F. gigantea* also has some specific germination characteristics; a high level of seed dormancy, a high temperature threshold for germination (28°C) and a relatively slow rate of germination. Such characteristics may limit its use. The germination requirements of populations collected from throughout Europe have been studied. Variation in optimal germination temperature and rate of germination have been identified. A pre-chill of 4 weeks at 4°C prior to germination at 28°C has reduced dormancy and increased the rate of germination of all populations examined. Storage at 4°C for up to 12 months reduced dormancy but subsequent storage for 30 months had little further effect on reducing dormancy. The germination requirements of this species are considered and implications for breeding discussed.

Introduction

One of the aims of grass breeding programmes at the Institute of Grassland and Environmental Research (IGER) is to broaden the adaptability of ryegrasses by introducing desirable characteristics from fescue species through intergeneric hybridisation. *Festuca pratensis* Huds. and *F. arundinacea* Schreb. have both been used as the fescue parent in this breeding programme. Interest in the relatively wild species *F. gigantea* (L.) Vill. (Giant fescue) has also arisen due to its recent use as another of the fescue parents in the *Lolium/Festuca* hybridisation breeding programme (Thomas and Humphreys, 1991). *F. gigantea* is found growing in shaded woodland margins and clearings throughout Europe and temperate Asia. It is nearly always found in damp or wet habitats by ditches, riverbanks and drainage channels. Each genotype within a population produces large amounts of seed which sheds readily when ripe. Populations of *F. gigantea* occur as discrete groups of closely growing plants quite often separated from other *F. gigantea* populations by several kilometres. It has a widespread distribution in the British Isles but is less frequent in Eire and Scotland and absent from the Western Isles and Northern Scotland (Perring and Walters,

R.H. Ellis, M. Black, A.J. Murdoch, T.D. Hong (eds.), Basic and Applied Aspects of Seed Biology, pp. 279–287.
© *1997 Kluwer Academic Publishers, Dordrecht. Printed in Great Britain.*

1962). It occurs throughout most of Europe except the extreme north, but is rare in the Iberian peninsula and the Mediterranean region (Tutin *et al.*, 1980).

F. gigantea has some valuable agronomic characteristics. It produces a mass of very large soft leaves of high nutritive value during mid-summer when ryegrasses tend to be less productive and more susceptible to drought. However, it also has some disadvantages, particularly associated with seed dormancy and slow sward establishment (Humphreys *et al.*, 1989) which is a potential constraint to its use in a hybridisation programme. Preliminary studies by Tyler (1990) showed that populations of *F. gigantea* obtained through seed exchange had a temperature optimum of approximately 28°C and were slow to germinate, although some form of pre-chill prior to germination could increase the rate of germination. Unfortunately the background of the seed used in this study was not always complete. The Plant Genetic Resources Unit at IGER, Aberystwyth carries out plant collecting expeditions throughout Europe and North Africa. These expeditions aim to collect a range of forage grass and legume populations for use in present and future research and breeding programmes. A programme of work was therefore originated in 1990 by the Genetic Resources Unit to add *F. gigantea* to the range of species collected. The aim was to identify populations with a rapid rate of germination and a low temperature threshold for germination. These would then be available for incorporation into the breeding programme. This paper describes the germination characteristics of material collected during a plant collecting expedition to Poland in 1990 and some of the factors influencing the germination behaviour of this material.

Materials and Methods

Variation in Germination Characteristics

In 1990, the Plant Genetic Resources Unit of IGER carried out a plant collecting expedition in south-east Poland (Chorlton *et al.*, 1996). Ten populations of *F. gigantea* (populations A to J) were collected along a 170 km east–west transect in the Beskidy region of the Carpathian mountains (Fig. 1 and Table 1). Seed samples were collected because every genotype in the populations found had set seed and a seed collection gave a representative sample of each population. The heads were threshed on site and the seeds placed in manila envelopes in a desiccator containing silica gel. The seed drying process was started at the point of collection. On return to IGER the seed samples were cleaned (by sieving and column blowing) to remove chaff, light seed and broken stalks. When the seed had dried down to 5% moisture content in the laboratory at an ambient temperature of 20°C, a random sample of 100 seed per population was used to produce plants for subsequent regeneration (on 28 January 1991), and the remainder of the original seed was stored in the genebank. The seeds were germinated at 25°C constant for 21 days, followed by 14 days at 0°C and finally 14 days at 25°C with two replicates of 50 seeds per replicate on damp

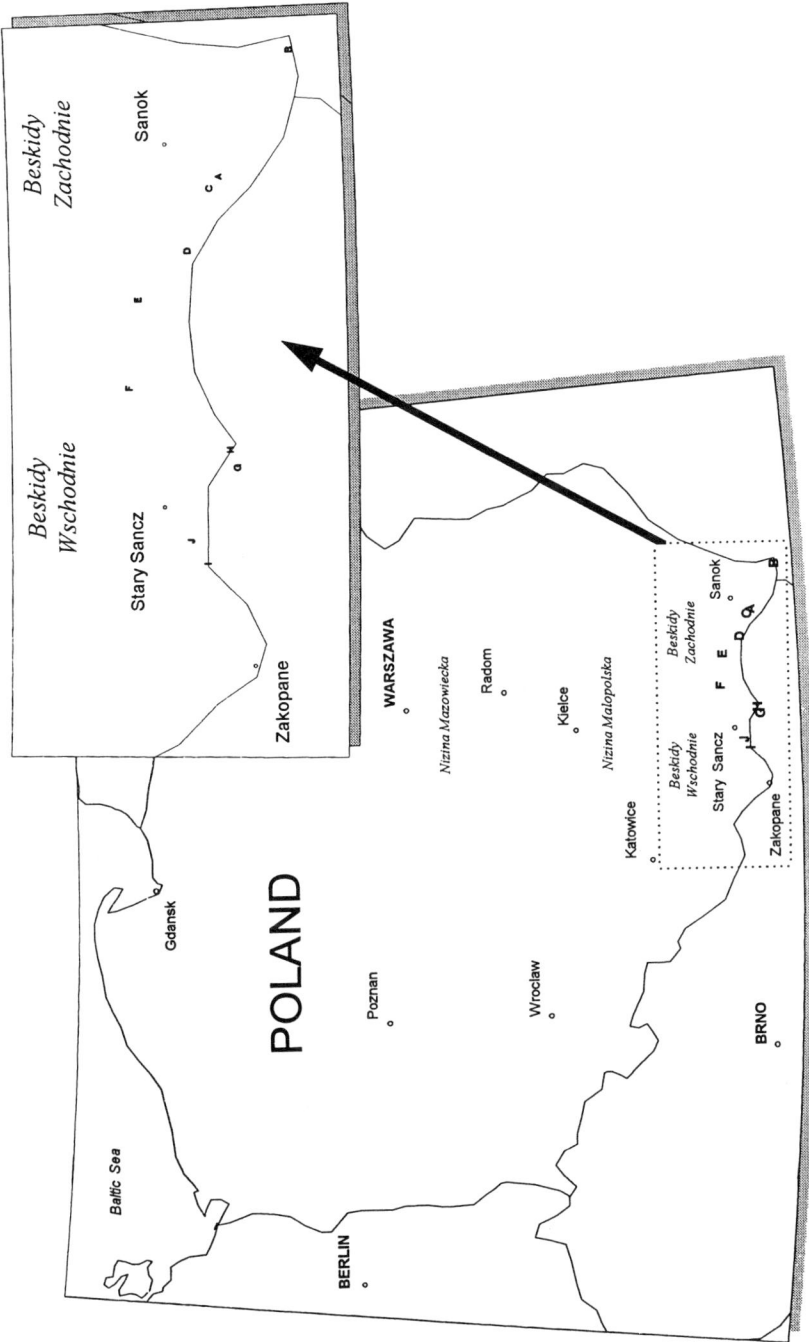

Figure 1. Distribution of *F. gigantea* populations (A to K) collected in Poland, 1990

Table 1. Festuca gigantea populations: Collection data

Accession number	Pop.	Country	Atlas location	Location	Alt. m.a.s.l.
BS 4265	A	Poland	Sanok	(Beskid Niski) Komancza, near petrol station rest area	475
BS 4268	B	Poland	Cisna	(Bieszczady Gorgany) Brzegi Gorne, 26 km E of Cisna	725
BS 4269	C	Poland	Sanok	(Beskid Niski) 3 km S of Wislok Wielki, SW of Sanok	500
BS 4270	D	Poland	Dukla	(Beskid Niski) Stasiana, 2 km E of Tylawa, S of Dukla	500
BS 4271	E	Poland	Dukla	(Beskid Niski) Nowy Zmigord, 15 km W of Dukla	300
BS 4272	F	Poland	Gorlice	(Beskid Niski) on road to Symbark, 5 km SW of Gorlice	350
BS 4273	G	Poland	Stary Sacz	(Beskid Sadeckie) Czaczow, near Barnowiec, SE of Nowy Sacz	500
BS 4274	H	Poland	Muszyna	(Beskid Sadeckie) Milik, 4 km W of Muszyna	600
BS 4275	I	Poland	Stary Sacz	(Pieniny) Majerz Meadow, 3 km E of Czorsztyn, on edge of Rezerwat Lasek, SE of Stary Sacz	660
BS 4277	J	Poland	Stary Sacz	(Pieniny) Kroscienko, N. Dunajcem, SW of Stary Sacz	418
BS 4286	K	Poland	Suwalki	Suwaalki, Czerwony Swor	–
BS 4287	L	Poland	Kielce	Kielce, Huta Stara	–
BS 4288	M	Soviet	–	Chechen-Ingush	–
BS 4289	N	Ukraine	–	Doneck Region	–

filter paper in 'repli-dishes' in an incubator. For each population, thirty seedlings were taken at random to grow on for bulk regeneration of seed in 1992. The regeneration of each population followed a standard procedure (Tyler, 1981). The seed obtained was stored at 0°C and 5% moisture content in hermetically sealed foil-laminate packets in the genebank on 30 November 1993.

Four additional populations (populations K, L, M and N, Table 1) obtained through seed exchange were also regenerated in the same way.

The optimal germination temperature of the fourteen populations was evaluated using a temperature gradient plate. The plate was set up with a single continuous gradient, the 'cold end' maintained at 20°C and the 'warm end' at 32°C. Single sheets (46 × 52 cm) of Whatman 3mm chromatography paper were placed on the bar and moistened. Twenty five seeds from each of the 14 populations were placed in each of the 14 compartments along the gradient. Germination (radicle emergence > 2 mm) was monitored on a daily basis, with seed removed as they germinated, beginning on 2 March 1993 until 8 April 1993 when no further germination was observed. The experiment was repeated from 13 April 1993 until 19 May 1993. Data was analysed using a probit analysis procedure to calculate the time to 50% germination from which the optimal germination temperature was derived.

Effect of Pre-chill Treatment on Germination

Three populations of *F. gigantea* of diverse origin (Table 2), which had been regenerated at IGER, were used to examine the effect of pre-chill on germination. This included two populations obtained during a plant collection expedition and one population obtained through seed exchange. In 1991, 100 seeds of each of the three populations were germinated and approximately 20 plants of each population grown on and seed obtained in 1992 using the same procedure as described above. The seed was dried using the technique described earlier and the seed from the twenty plants bulked and placed in the cold store at 0°C. A germination test was carried out at harvest (before the seed was placed in the cold store) and again after twelve and thirty months storage. Seed was removed from the cold store and set to germinate in multi-compartment trays. Twenty five seeds were set to germinate per tray, so that each of the twenty five compartments contained one seed to which was added 10 ml distilled water. The trays were then enclosed in polythene bags. The seeds of all populations were subject to four pre-chill treatments; placed in cold at 4°C for 28 days, 14

Table 2. Origin of three populations of *F. gigantea* used to examine the effect of pre-chill treatment on germination

Accession number	Collection number	Location
BS 3992	DEUAUT 86:50:1	Bavarian alps, Unterholz
BS 3993	DEUAUT 86:56:1	Upper Rhine Valley, Rust
BS 4062	Seed exchange list 1555	Botanic garden sample (Stuttgart University), no information on origin

days, 3 days and 0 days (control) and then placed in an incubator maintained at 28°C with no light. Germination was assessed weekly, when seeds with a radicle of >2mm were removed, until twelve weeks after the beginning of germination when no further germinated seed was recorded. There were four replicate trays of each prechill treatment for each of the three populations.

Results

Variation In Germination Characteristics

The optimal germination temperature and number of days to 50% germination of the ten *F. gigantea* populations collected in Poland, four populations obtained through seed exchange and seed of *F. pratensis* and *L. perenne* are shown in Figure 2. All populations of *F. gigantea* were slower to germinate than *F. pratensis* and *L. perenne*, most requiring at least 15 days at their optimal temperature to attain 50% germination. Only population N germinated at a faster rate, requiring 8 days to achieve 50% germination, but even then germinated at a slower rate than the two control species. Population N also had one of the highest optimal germination temperatures. There was considerable variation between populations in the optimal germination temperature. Three populations (populations B, E and F) had an optimal temperature of c. 20°C whilst the other populations had optimal germination temperatures of up to 26°C. Population H had the highest optimal germination temperature of c. 30°C.

Effect of Pre-chill Treatment on Germination

In general, the germination of the three populations examined showed a similar response to the pre-chill treatments. The overall effect of these treatments and storage on the germination of *F. gigantea* (mean of the three populations) is shown in Figure 3. Storage at 4°C for 12 months significantly increased the rate of germination and the final percentage germination compared to freshly harvested seed; however even after storage *F. gigantea* still required 7 weeks to achieve 50% germination. Storage for up to 30 months had no further significant effect on the rate of germination or final percentage germination (data not shown). A pre-chill period at 4°C prior to germination at 28°C was observed to influence the rate of germination but the effectiveness of the pre-chill was dependent upon its duration. A pre-chill period of 3 days had no significant effect on the rate of germination compared to the control (no pre-chill); however a 14 day pre-chill significantly increased the rate of germination. A 28 day pre-chill resulted in the fastest rate of germination, with 90% germination being achieved within 2 weeks of transfer to 28°C.

Few significant differences were observed between the three populations in their response to the pre-chill treatments. Differences between the populations

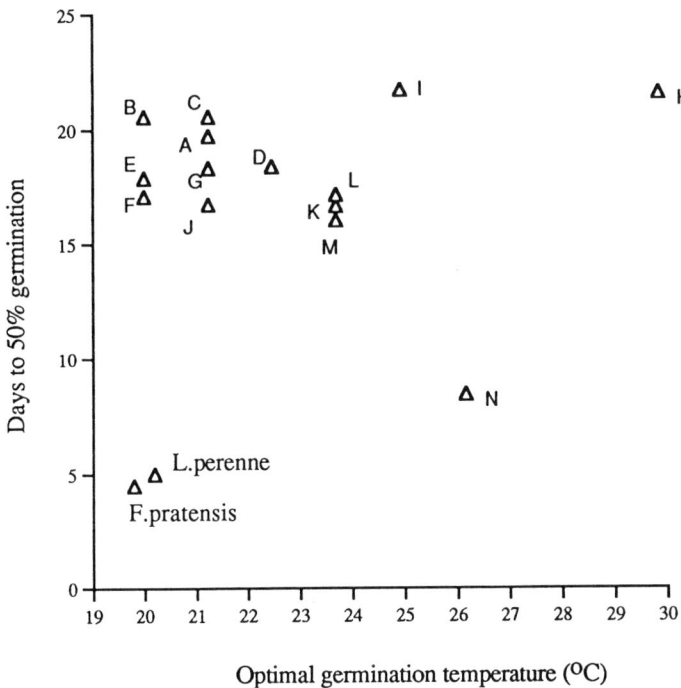

Figure 2. Optimal germination temperature (°C) and days to 50% germination of ten populations (populations A to J) of *F. gigantea* collected in Poland and four populations (K, L, M and N) obtained through seed exchange. *L. perenne* and *F. pratensis* are shown for comparison

were only observed in their rate of germination (days to 50% germination) in response to the no pre-chill (control) and the 3 day pre-chill treatments where BS 4062 (control, 42 days; 3 day pre-chill, 35 days) had a faster rate of germination (days to 50% germination) than both BS 3992 (control, 49 days; 3 day pre-chill, 42 days) and BS 3993 (control, 54 days; 3 day pre-chill, 52 days). Pre-chill treatments produced no difference between populations in the final percentage germination.

Discussion

The agronomic characteristics of *F. gigantea* make it a valuable species in the intergeneric hybridisation breeding programme being undertaken at IGER. However, in common with many agriculturally undeveloped species *F. gigantea* possesses a degree of seed dormancy. Hybrids between *F. gigantea* and ryegrass have been shown to retain some of the germination characteristics of *F. gigantea*

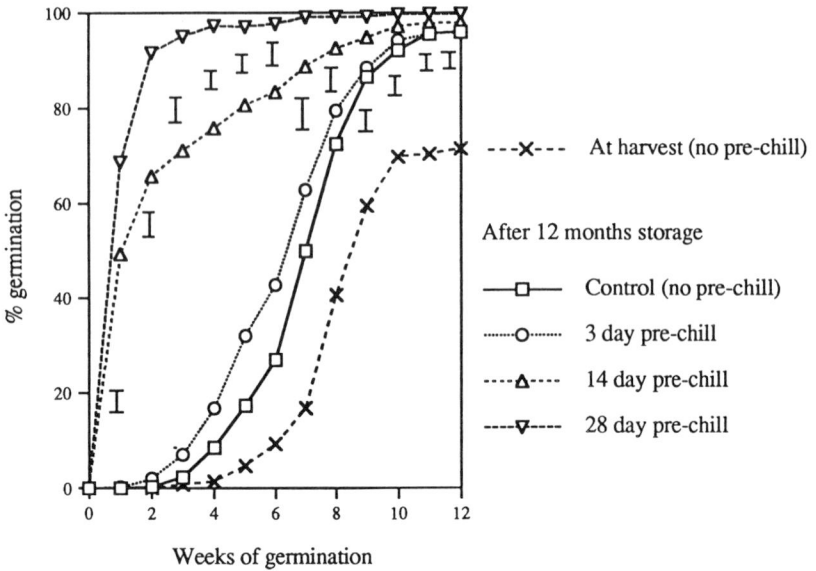

Figure 3. Germination of *F. gigantea* (mean of three populations) at 28 °C at harvest, and after 12 months storage at 4°C and four pre-chill treatments (no pre-chill, 3 day pre-chill, 14 day pre-chill and 28 day pre-chill at 4°C). Vertical bars represent LSD at $p = 0.05$

exhibiting a slower rate of germination and a higher optimal germination temperature than ryegrass (Humphreys *et al.*, 1989). These researchers suggested that screening of *F. gigantea* populations from diverse areas for their germination characteristics might identify populations more suitable for incorporation into future breeding programmes. Evidence from this current work suggests that there is considerable variation among populations of *F. gigantea* collected in Poland and within other material examined, for both the rate of germination and the optimal germination temperature. Populations with low optimal germination temperatures (c. 20°C) have been identified. Although the germination rate of these populations was higher than *L. perenne*, there was some variation among these populations for this characteristic. This will provide valuable information to the breeder and ensure that the most appropriate populations, which have their germination characteristics quantified are used in further crossing programmes. Further studies will examine populations from other parts of Europe to seek those populations which combine a faster rate of germination and a low optimal germination temperature.

An additional objective of this work was to identify some of the factors influencing the dormancy and germination of *F. gigantea*. Both a period of dry storage and a pre-chill at 4°C prior to germination at 28°C increased the germination of all populations of *F. gigantea* examined. Storage of dry seed for up to 12 months increased the rate of and final percentage germination compared to freshly harvested seed; however the increase though significant was small. No further increase in rate or final percentage germination was observed by further storage for up to 30 months. Grime *et al.* (1981) reported that freshly collected seed of *F. gigantea* has a low germination rate and that storage for 3, 6 and 12 months appeared to inhibit germination equally but increased the germination rate. The present results indicate that whilst storage can improve germination, any increase in germination after prolonged storage will be relatively small. Tyler (1990) suggested that a pre-chill prior to germination is essential to reduce seed dormancy and increase germination. From this current experiment, a 14 day pre-chill at 4°C is necessary to increase the rate of germination to a reasonable level. However a further increase in the rate of germination can be obtained with a 28 day pre-chill. Germination after this pre-chill period was rapid and there was a high final percentage germination. The extent to which the duration of the pre-chill may influence the optimal germination temperature has not as yet been studied, but will form the basis of future experiments.

F. gigantea is among the latest flowering temperate grasses (Hubbard, 1976). Tyler (1990) observed that the date of inflorescence emergence was highly positively correlated with the depth of dormancy and suggested that high seed dormancy was associated with late flowering and is an adaptive mechanism preventing autumn germination and ensuring winter survival. A requirement for a long period of cold prior to germination can also be interpreted as a further adaptive mechanism ensuring that *F. gigantea* will survive until spring.

References

Chorlton, K.H., Thomas, I.D. Bowen, D.W., Bulinska-Radomska, Z. and Gorski, M. 1996. *Genetic Resources and Crop Evolution* 43: 69–77.

Grime, J.P., Mason, G., Curtis, A.V., Rodman, J., Band, S.R., Mowforth, M.A.G., Neal, A.M. and Shaw, S. 1981. *Journal of Ecology* 69: 1017–1059.

Hubbard, C.E. 1976. *Grasses*. London: Penguin.

Humphreys, M.O., Thomas, H. and Tyler, B.F. 1989. *Proceedings of the XVI International Grassland Congress*, Nice, France, 1989, 253–254.

Perring, F.H. and Walters, S.M. 1962. *Atlas of the British Flora*. London: Nelson.

Thomas, H. and Humphreys, M.O. 1991. *Journal of Agricultural Science* (Cambridge) 117: 1–8.

Tutin, T.G., Heywood, V.H., Burges, N.A., Moore, D.M., Valentine, D.H., Walters, S.M. and Webb, D.A. 1980. *Flora Europaea. Vol.5*, pp. 132. Cambridge University Press.

Tyler, B.F. 1981. In. *Proceedings of the C.E.C./Eucarpia seminar*, Nyborg, Denmark, 15–17 July, 1981. pp 69–78 (eds E. Porceddu and G. Jenkins).

Tyler, B.F. 1990. *Report of the Welsh Plant Breeding Station for 1989*, pp. 11.

32. QSAR Modelling of Dormancy-Breaking Chemicals

M.A. COHN

Department of Plant Pathology & Crop Physiology, Life Sciences Building, Louisiana State University Agricultural Center, Baton Rouge LA 70803, USA

Abstract

Previously, the activity of dormancy-breaking chemicals of red rice (*Oryza sativa* L.) has been correlated with octanol/water partition coefficients (Log Ko/w). However, this model was chosen empirically. In this study, quantitative structure vs. biological activity relationships (QSAR) were analyzed more rigorously using Molecular Analysis Pro, a PC-based computational chemistry program. The entire set of alcohols, carboxylic acids, esters, aldehydes and ketones ($n = 42$) could still be modeled to Log Ko/w ($r^2 = 0.61$, $p < 0.0001$) similar to previously published results ($n = 25$). A stable, two-variable model adding density as a bulk steric factor better described the data ($r^2 = 0.85$). However, free energy changes per methylene group for homologous series of linear alcohols or monocarboxylic acids were 639 cal/mole and 236 cal/mole, respectively, for dormancy-breaking activity. Therefore, alcohols and carboxylic acids were also evaluated separately. While either monocarboxylic acids or primary alcohols could each be described by Log Ko/w, addition/substitution of descriptors for size, shape, density, or dipole moment provided improved QSAR models. These new QSARs provide further guidance for designing dormancy-breaking chemicals of enhanced efficacy.

Introduction

Weed control for crop production employs herbicides, cultivation, and crop rotation. However, even with intensive effort, weed problems are chronic because of weed seed reserves that remain dormant but viable in the soil. Each year producers can expect to spend significant time and money for weed control. Therefore, weed seed dormancy limits the success of agricultural weed control practices, adding to the yearly fixed costs of production. No existing weed control treatment is generally effective in directly eradicating dormant weed seeds from the soil.

Field treatment with dormancy-breaking chemicals would deplete the buried weed seed reserves. Coupling this practice with the judicious use of herbicides would reduce the population of emerging weed seedlings and prevent new seed formation. Consequently, the frequency of herbicide applications currently required for weed control will be significantly reduced. This will not only lower the fixed costs and enhance the agricultural profitability, but also will favorably impact the environment by reducing chemical inputs long-term. However,

R.H. Ellis, M. Black, A.J. Murdoch, T.D. Hong (eds.), Basic and Applied Aspects of Seed Biology, pp. 289–295.
© *1997 Kluwer Academic Publishers, Dordrecht. Printed in Great Britain.*

existing dormancy-breaking treatments are unusable in the field environment due to logistical, economic, or toxicological problems.

Until recently the most serious barrier to establishing safe and effective dormancy-breaking technology for field application has been the total lack of logical, structural design principles for biological activity. Results acquired during the past decade indicate that efficacy of the dormancy-breaking chemicals is a function of their lipid solubility, degree of dissociation, as well as the nature and position of functional groups (reviewed in Cohn, 1989; Cohn, 1993). In general, potencies were acids >aldehydes >esters >hydroxy-acids >primary alcohols. Empirically-derived computational QSAR comparisons indicated that the functional group contribution was independent of lipophilicity (Cohn *et al.*, 1989).

With the advent of more powerful desktop computer hardware and software, it is now possible to rapidly (relatively) build molecular structures, calculate a variety of physical descriptors, and model the potential role of these parameters to bioassay data. The objective of the research summarized in this report was to examine a large data set ($n = 42$) and a variety of data subsets of dormancy-breaking chemicals by regression analysis protocols to confirm the role of lipophilicity as a key feature of activity and to identify other chemical descriptors that might explain the functional group effect.

Materials and Methods

Dehulled dormant red rice (*Oryza sativa* L.) caryopses were challenged with a 24-h pulse of dormancy-breaking chemical at a series of concentrations followed by a 7-d incubation on water-saturated filter paper as described by Cohn (1989 and references therein). The concentration of each chemical eliciting 50% germination (G50) was used as the bioassay index of activity. Chemical structures were drawn and minimized to their lowest energies, physical descriptors calculated, and various databases constructed using Molecular Modeling Pro (WindowChem Software, Fairfield, CA, USA), a PC-based computational chemistry program. Regression analysis and other mathematical operations were performed using the companion program, Molecular Analysis Pro. More than three dozen descriptors of lipophilicity, steric and electronic parameters were evaluated. Experimentally derived partition coefficients were used and obtained from Sangster (1989), Hansch and Leo (1979), or by determinations in this laboratory. Free energy change per methylene group was calculated as outlined by Herrmann (1962).

Results and Discussion

Cohn *et al.* (1989) showed a correlation between lipophilicity and dormancy-breaking activity for a variety of carboxylic acids, alcohols, aldehydes and esters:

(1) Log G50 = –0.73 (Log Ko/w) + 1.86
     $n = 25; r^2 = 0.62, p < 0.001$

Even though Log Ko/w was chosen empirically for this early analysis, more rigorous modeling with the currently available tools confirmed and extended the validity of this model:

(2) Log G50 = –0.69 (Log Ko/w) (± 0.08) + 1.98 (± 0.11)
     $n = 42; r^2 = 0.67; s = 0.56; F = 81, p < 0.0001$

Chemicals with increasing lipophilicity broke dormancy at lower concentrations (Fig. 1). Expanding the model in this form to two or more variables yielded little improvement. However, transformation of the response parameter simplified the relationship:

(3) Log (MW/G50) = 0.79 Log Ko/w (± 0.08)
     $n = 42; r^2 = 0.71; s = 0.59; F = 92, p < 0.0001$

where MW = molecular weight. Expanding the transformed model with a bulk steric factor accounted for more of the variance:

(4) Log (MW/G50) = 0.89 Log Ko/w (± 0.06) – 1.69/density (± 0.28) + 1.76 (± 0.32)
     $n = 42; r^2 = 0.85; s = 0.43; F = 108, p < 0.0001$

This model is very stable to subtractive analysis when randomly selected groups of 4 compounds are removed from the sample, and the regression is recalculated.

Linear Primary Alcohols and Monocarboxylic Acids

Descriptors of lipophilicity (Log Ko/w), shape (Kappa 2) (Kier, 1985) or size (valence 3) (Kier and Hall, 1986) accurately model the activity of linear primary alcohols or monocarboxylic acids as separate groups with $r^2 > 0.89$. Addition of a density term to these two equations is not appropriate at this time for two reasons: (a) the sample size for each group is too small ($n = 8$ for each); (b) when density is included anyway, the model coefficients for the term are not statistically different from zero. But, each chemical class has a very different response curve *vs.* Log Ko/w (Fig. 2) as confirmed by a test of homogeneity of regression coefficients (Cohn *et al.*, 1989). Yet, Log Ko/w values for each class are very similar (Fig. 2, inset).

The slopes of curves in Figure 2 can be used to calculate the change in free energy per methylene group ($\Delta\Delta F$) for each class of compound. The $\Delta\Delta F$ for dormancy-breaking by linear, primary alcohols was 639 cal/mole. This high value suggests the need for a highly hydrophobic environment for activity

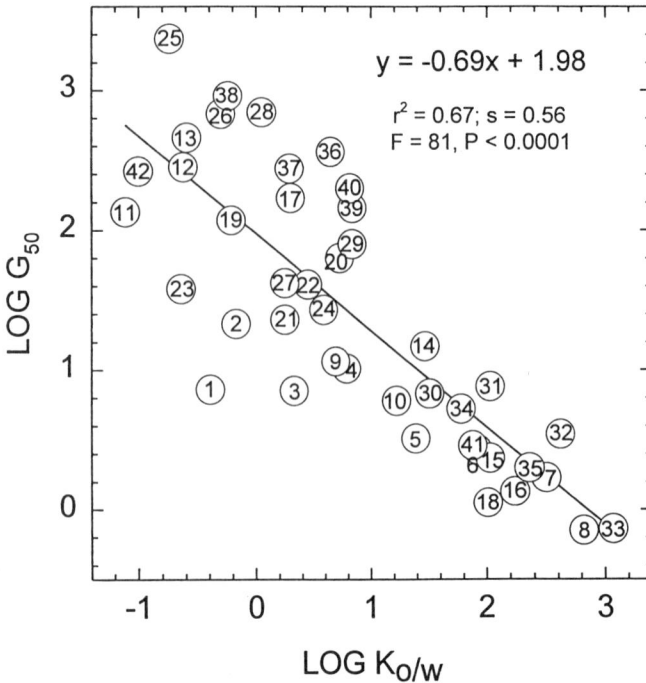

Figure 1. Correlation between chemical lipophilicity (Log Ko/w) and concentrations required for 50% germination. (1) formic acid; (2) acetic acid; (3) propionic acid; (4) butyric acid; (5) valeric acid; (6) caproic acid; (7) heptanoic acid; (8) octanoic acid; (9) isobutyric acid; (10) isovaleric acid; (11) glycolic acid; (12) lactic acid; (13) succinic acid; (14) trimethylacetic acid; (15) benzoic acid; (16) salicylic acid; (17) dimethadione; (18) cyclohexanemonocarboxylic acid; (19) methyl formate; (20) ethyl acetate; (21) methyl propionate; (22) acetaldehyde; (23) butyrolactone; (24) propional-dehyde; (25) methanol; (26) ethanol; (27) propanol; (28) isopropanol; (29) butanol; (30) pentanol; (31) hexanol; (32) heptanol; (33) octanol; (34) cyclohexyl-methanol; (35) 2-cyclohexylethanol; (36) 2-butanol; (37) 2-butanone; (38) acetone; (39) 2-pentanone; (40) 3-pentanone; (41) cyclohexane-carboxaldehyde; (42) 1,5-pentanediol

(Mattson *et al.*, 1970) and is in line with $\Delta\Delta F$ values of as large as 590 cal/mole for the interaction of alcohol dehydrogenase with its linear substrates (calcu-lated from data in Hansch *et al.*, 1972) and the $\Delta\Delta F$ values for partitioning of alcohols between octanol and water itself (777 cal/mole). In contrast, the $\Delta\Delta F$ of dormancy-breaking for monocarboxylic acids is only 236 cal/mole, which is dramatically different from the partitioning of these acids between octanol and water (680 cal/mole). Generally, then, the energy barrier for dormancy-breaking by –COOH is much lower than for the corresponding alcohols.

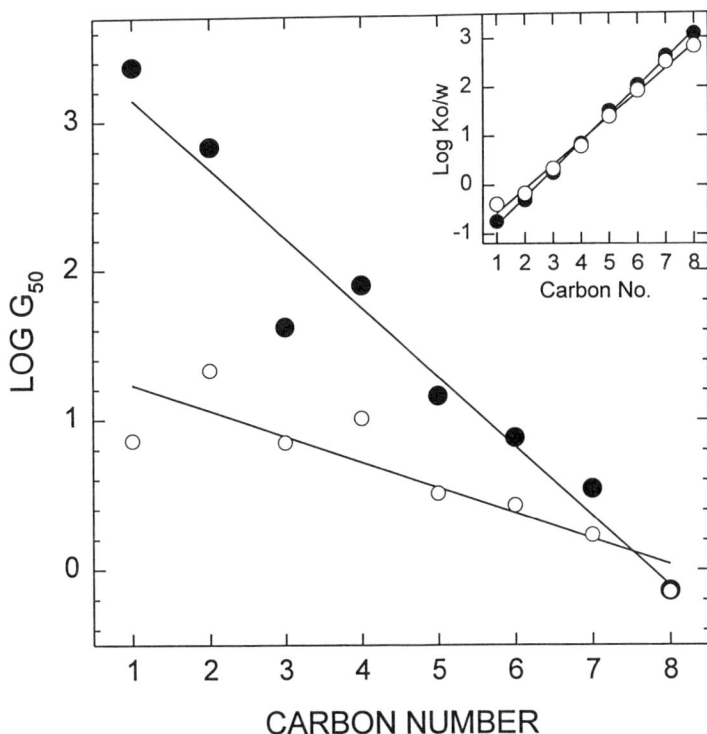

Figure 2. Correlation between carbon chain length of linear, primary alcohols (●; y = –0.46x + 3.61, $r^2 = 0.95$) or linear monocarboxylic acids (○; y = –0.17x + 1.40, $r^2 = 0.78$) and the chemical concentration required to elicit 50% germination for each chemical. Inset: Correlation between carbon chain length of linear, primary alcohols or linear monocarboxylic acids and Log Ko/w for each chemical

Weak Acids and Alcohols with More Complex Characteristics

Increasing the complexity of the acid data set by adding isovaleric, isobutyric, benzoic cyclohexylcarboxylic, and trimethylacetic acids reduces the tightness of the one parameter correlations. While all 3 descriptors yield significant one parameter models (r^2: Log Ko/w = 0.81, V3 = 0.85, K2 = 0.71), a two parameter model combining aspects of size and shape gives the nicest picture:

(5) Log (MW/G50) = 0.64(V3)(± 0.10) + 0.13(K2)(± 0.03) + 0.38 (± 0.10)
    $n = 13$; $r^2 = 0.94$; s = 0.15; F = 77, $p < 0.0001$

Increasing the complexity of the alcohol data set by adding isopropanol, 2-butanol, cyclohexylmethanol, and 2-cyclohexylethanol does not reduce the significance of Log Ko/w ($r^2 = 0.93$) as an important one parameter model. A two parameter model of size and shape is also appropriate for this alcohol dataset:

(6) $\text{Log (MW/G50)} = 0.94(\text{V3})(\pm 0.15) + 0.32(\text{K2})(\pm 0.05) - 1.62\ (\pm 0.18)$
$n = 12;\ r^2 = 0.95;\ s = 0.32;\ F = 91,\ p < 0.0001$

Variation in the carbon chain length clearly dominates with each model for both the –COOH and the alcohols resulting in the inclusion of the same descriptive parameters. Perhaps the role of –OH groups can be ascertained by examining a mixed set of simple –COOH and hydroxyacids? Evidence still suggests that Log Ko/w is the most reasonable single descriptor ($r^2 = 0.86$; $s = 0.34$). However, the hydroxy acids tend to be stronger dipoles than their corresponding monocarboxylic acid. Therefore, dipole moment (dp) in the model improves the fit of the data. Transformation was necessary from examination of the residuals of the original equation:

(7) $\text{Log(MW/G50)} = 0.80(\text{Log Ko/w})(\pm 0.06) - 1.09\ \text{Log(MW/dp)}(\pm 0.28) + 2.23(\pm 0.48)$
$n = 17;\ r^2 = 0.93;\ s = 0.24;\ F = 100,\ p < 0.0001$

This model did not accurately describe the data set consisting only of simple monocarboxylic acids or the alcohol data set.

Combining the data sets represented by Equations 6 and 7 yields a group consisting of all –COOH containing molecules and all –OH containing molecules excepting the diol. For this sample ($n = 29$), the best one parameter is again Log Ko/w ($r^2 = 0.73$). The two parameter model (representing 70% of the total available n) is very similar to Equation 4 for all active chemicals:

(8) $\text{Log(MW/G50)} = 0.92(\text{Log Ko/w})(\pm 0.06) - 1.83/\text{density}(\pm 0.29) + 1.90\ (\pm 0.32)$
$n = 29;\ r^2 = 0.89;\ s = 0.38;\ F = 107,\ p < 0.0001$

Comments and Conclusions

The relationship between dormancy-breaking activity of organic chemicals and Log Ko/w has been confirmed using more rigorous methods and a wider variety of bioassay response sets than was available previously. The contributions of steric factors (size, shape, density) were in evidence in a variety of cases. However, due to the dominance of carbon chain size and shape, the effects of dormancy-breaking agents with differing functional groups could not be modelled with the current approach. It will be necessary to look at a variety of functional groups on a small, constant-sized carbon skeleton to progress further.

However, the only difference in dormancy-breaking action of acids vs. alcohols, suggested by the slopes of the response curves (Fig. 2), may be the requirement for the alcohols to be metabolized to the corresponding weak acid in red rice seeds as suggested by Cohn *et al.* (1989), and confirmed by the tissue pH measurements of Footitt and Cohn (1992), 13-C NMR studies of propanol metabolism by Footitt *et al.* (1995), and structure vs. activity bioassays (Cohn *et al.*, 1991). In addition, dormant oat seeds do not respond to dormancy-breaking alcohols in the presence of the alcohol dehydrogenase inhibitor, 4-methylpyrazole (Corbineau *et al.*, 1991). Application of 4-methylpyrazole to red rice also inhibits the alcohol-induced germination response (Lin and Cohn, in preparation).

While QSAR studies do not currently provide an understanding of the mechanism by which organic chemicals break dormancy, the importance of lipophilicity should enhance the design and testing of new chemicals with enhanced efficacy. QSARs provide a working, rational framework to guide the evaluation of dormancy-breaking chemicals.

References

Cohn, M.A. 1989. In: *Recent Advances in the Development and Germination of Seeds. Proceedings of the Third International Workshop on Seeds*, pp. 261–267 (ed. R.B. Taylorson). New York, London: Plenum Press.

Cohn, M.A. 1993. *Search* 28: 1–6.

Cohn, M.A., Church, D.F., Ranken, J. and Sanchez, V. 1991. *Plant Physiology* 96: S-63.

Cohn, M.A., Jones, K.L., Chiles, L.A. and Church, D.F. 1989. *Plant Physiology* 89: 879–882.

Corbineau, F., Gouble, B., Lecat, S. and Come, D. 1991. *Seed Science Research* 1: 21–28.

Footitt, S. and Cohn, M.A. 1992. *Plant Physiology* 100: 1196–1202.

Footitt, S., Vargas, D. and Cohn, M.A. 1995. *Physiologia Plantarum* 94: 667–671.

Hansch, C. and Leo, A. 1979. *Substituent Constants for Correlation Analysis in Chemistry and Biology*. New York, Chichester: John Wiley & Sons.

Hansch, C., Schaeffer J. and Kerley, R. 1972. *Journal of Biological Chemistry* 247: 4703–4710.

Herrmann, K.W. 1962. *Journal of Physical Chemistry* 66: 295–300.

Kier, L.B. 1985. *Quantitative Structure–Activity Relationships* 4: 109–116.

Kier, L.B. and Hall, L.H. 1986. *Molecular Connectivity in Structure–Activity Analysis*. New York, Chichester: John Wiley & Sons.

Mattson, F.H., Volpenhein R.A. and Benjamin, L. 1970. *Journal of Biological Chemistry* 245: 5335–5340.

Sangster, J. 1989. *Journal of Physical and Chemical Reference Data* 18:1111–1200.

33. Natural Far Red Irradiation and Weed Seed Persistence in the Soil

A. DOROSZEWSKI

Institute of Soil Science and Plant Cultivation, Osada Palacowa, 24–100 Pulawy, Poland

Abstract

It may be assumed that seed reactions to light are an indication of adaptation to environmental conditions. Thus, it is necessary to accept the fact that these reactions can be fully observed only in natural conditions. The results presented concerning the germination of seeds of different photoblastic types were obtained in experiments carried out in natural soil conditions. The aim of the experiments was to describe the role of the secondary seed dormancy enforced by far red irradiation on seed persistence in the soil. Seed dormancy was enforced by placing Petri dishes with seeds under a canopy of *Parthenocissus quinquefolia* leaves. After inducing dormancy by natural far red, the seeds were transferred under a cover of rhubarb leaves and placed in prepared rows in the soil, where they were buried. These operations were performed on cloudy days in the afternoon. The following plants were sown in such a way: *Amaranthus retroflexus, Apera spica-venti, Crepis capillaris, Lactuca serriola, Verbascum lychnitis, Oenothera biennis*. After periods of 1 to 24 months the upper layer of the soil was removed. The seedlings that emerged from far red inhibited and control seeds were counted after several months. Many more seedlings were obtained from inhibited seeds; the differences for 5 species were statistically significant. Similar results were obtained with seeds buried in nylon bags.

Abbreviations: PP, positively photoblastic; NP, negatively photoblastic; I, indifferent; P_{FR}, far red absorbing form of phytochrome; P_R, red absorbing form of phytochrome; P_{FR+R}, total amount of phytochrome; R, red; FR, far red

Introduction

The external conditions may be favourable and yet viable seeds are not always able to start germination because of dormancy. Light can either promote or inhibit seed germination. It is the phytochrome that determines whether the germination will start or will be suppressed under the prevailing radiation. However, the details of the manner in which the phytochrome mechanism operates and the possibility of the involvement of other light receptors are still the matter of controversy (Gaba and Black, 1979; Senger, 1980; Volotovsky, 1987).

The change of the state of phytochrome can be brought about by photoconversion following exposure of seeds to red ($P_R \rightarrow P_{FR}$) or far red ($P_{FR} \rightarrow P_R$) light. In nature a large portion of red radiation that passes through green leaves

R.H. Ellis, M. Black, A.J. Murdoch, T.D. Hong (eds.), Basic and Applied Aspects of Seed Biology, pp. 297–302.
© *1997 Kluwer Academic Publishers, Dordrecht. Printed in Great Britain.*

is absorbed whereas far red is allowed to go through. As a result, the ratio of FR to R under the leaves is high and, consequently, the value of P_{FR}/P_{FR+R} is small. Thus, under those conditions seed germination either does not occur at all or is very low. Based on a survey of 310 species (Górski *et al.,* 1977; Górski, 1983) it was found that all positively photoblastic (PP) species that require light to germinate are inhibited by radiation transmitted through the leaves. Conversely, only some seeds designated as group I (indifferent to light and germinating both in light and dark) are inhibited when germinated under a leaf canopy. Similar results were obtained by Silvertown (1980) for 27 species and by Fenner (1980) for 18 East African weeds.

The inhibition by far red which in nature occurs under leaf canopy keeps the seeds from germinating under conditions where a seedling would not have a chance of survival because of competition from developed plants.

Results are reported from an experiment which was aimed at answering the following question. What is the significance of the secondary seed dormancy brought on by exposure to far red for the survival of seeds in the soil?

Methods

Seeds of the following PP (positively photoblastic) species were used in the experiment: *Apera spica-venti* (L.) P. B., *Lactuca serriola* Torner, *Oenothera biennis* L., *Verbascum lychnitis* L.

Seeds of I (indifferent) species were those of *Crepis capillaris* (L.) Wallr., seeds of *Amaranthus retroflexus* L. including two NP (negatively photoblastic) batches and one I batch.

Seed dormancy was forced by placing the seeds in Petri dishes under the leaf canopy of *Parthenocissus quinquefolia.* The seeds placed under the leaves received ca. 2–5% of the photosynthetically active radiation reaching the upper surface of the canopy. The far red to red ratio under the canopy of *Parthenocissus* is ca. 6 to 1. The seeds were exposed for 17 days to these conditions and scarcely germinated (0–10%).

After being inhibited by natural far red under *Parthenocissus* canopy, on a cloudy day or in the evening the seeds were transferred under the cover of rhubarb leaves to a site at which they were to be buried in previously prepared ditches 30 cm deep and 1 metre wide. Unexposed air-dry seeds from the same batches as were used as controls. The seeds were sown in rows: *Amaranthus retroflexus* – 10 rows of control seeds and 10 rows of inhibited seeds, 25 seeds being sown in a row, *Apera spica-venti* – 12 rows of control and 12 rows of inhibited seeds, 30 seeds in a row, *Crepis capillaris, Lactuca serriola, Verbascum lychnitis, Oenothera biennis* – 6 rows of control and 6 of inhibited, 25 seeds in a row. After being placed at the bottom of the ditch the seeds were covered with a plastic net of 7×7 mm mesh and with the soil on top of the net. While burying the seeds care was taken to preserve the natural order of the layers of the soil. The soil was a weak loamy sand with a thick humus layer. The surface of the soil was covered with a dense sward of a lawn. The layer of soil and the net were removed after periods of up to 24 months, to expose the seeds. Only those seedlings which emerged precisely within a row were taken into calculation.

Some inhibited seeds of *Amaranthus retroflexus* were placed in six 1×0.5 mm-mesh nylon bags (100 seeds in each bag). The bags were then buried at a depth of 20 cm in soil in pots 45.0 cm high and 22.5 cm in diameter.

Further similar bags of seeds of *Amaranthus retroflexus* and *Crepis capillaris* were buried in the soil at a depth of 30 cm in a bower all the walls of which were made up of *Parthenocissus* leaves.

After being exhumed, the seeds were scored for germination by applying slight pressure to their seed cover. Ungerminated seeds were transferred to Petri dishes and placed in darkness or in diffuse light.

Results

Of the *Amaranthus retroflexus* seeds that showed behaviour typical of that species i.e. germinated in the dark and failed to germinate in light (Table 1, Table 2 items 2a, 2b) germination in the soil was mostly restricted to those which had not been induced with far red even when seeds were buried in the soil for 6 months. Only with seeds buried for 1 month (Table 2 item 2b) did the controls show germination similar to that of far red-induced seeds. After the seeds were exhumed (Table 2) or following the removal of the top layer of the soil (Table 1) inhibited seeds germinated to a much greater extent than the controls.

Table 1. Germination of seeds (%) following exhumation from a ditch (one-row average). Seeds were exposed to a natural source of far red (FR) light or not exposed (C) prior to burial

Species	Photo-blastism	Length of burial period in the soil (months)									
		1		8		9		20		24	
		C	FR	C	FR	C	FR	C	FR	C	FR
Amaranthus retroflexus	NP							8.8[a]	25.6[b]		
Apera spica-venti	PP									9.0[a]	13.0[b]
Crepis capillaris	I	12.8[a]	43.2[b]	20.8[a]	40.8[b]						
Lactuca serriola	PP					5.2[a]	16.0[b]				
Verbascum lychnitis	PP					12.8[a]	23.6[b]				
Oenothera biennis	PP			36.8[a]	32.8[a]						

Significance levels: values with different suffix letters significantly differ at 0.05 according to *t*- (Student) test and non-parametric tests (of signs, ranks)

C, control; FR, induced

Table 2. Germination of seeds buried in the soil in nylon bags. Seeds were exposed to a natural source of far red (FR) light or not exposed (C) prior to burial

No.	Species	Photo-blastism	Burial site	Germination treatment	1		3		9		16	
					C	FR	C	FR	C	FR	C	FR
1	*Amaranthus*	I	Pot	In soil					89.7ᵃ	16.0ᵇ		
				In D after exhumation					0.5ᵃ	6.8ᵃ		
				In diff. after exhumation					0.5ᵃ	5.5ᵃ		
2a	*Amaranthus*	NP	Pot	In soil	22.3ᵃ	2.8ᵇ						
				In D after exhumation	26.0ᵃ	48.0ᵃ						
				In diff. after exhumation	7.2ᵃ	10.5ᵃ						
2b	*Amaranthus*	NP	Bower	In soil	5.3ᵃ	5.8ᵃ	48.7ᵃ	4.8ᵇ			88.8ᵃ	10.7ᵇ
				In D after exhumation	9.5ᵃ	75.5ᵇ	5.5ᵃ	82.0ᵇ			6.5ᵃ	77.7ᵇ
				In diff. after exhumation	7.5ᵃ	58.3ᵇ	8.0ᵃ	26.3ᵃ			11.8ᵃ	72.7ᵇ
3	*Crepis*	I	Bower	In soil	5.1ᵃ	2.3ᵃ	13.7ᵃ	3.7ᵇ			22.2ᵃ	3.3ᵇ
				In D after exhumation	67.7ᵃ	18.0ᵇ	43.0ᵃ	71.0ᵇ			71.5ᵃ	94.0ᵃ
				In diff. after exhumation	93.0ᵃ	95.2ᵃ	80.2ᵃ	94.2ᵃ			80.0ᵃ	97.0ᵇ

Significance levels: values with different suffix letters significantly differ at 0.05 according to *t*- (Student) test and non-parametric tests (of signs, ranks)

-C, control; FR, induced; D, dark; diff., diffuse irradiation

The batch of 12-year old *Amaranthus* seeds (Table 2 item 1) which showed full germination both in darkness and in light was found to behave in the soil and after being exhumed in exactly the same way as other seed batches which showed a typical negative photoblastic response i.e. more controls than inhibited seeds germinated while buried in the soil and, conversely, more inhibited seeds than controls germinated following exhumation.

The seeds buried in the soil over the longest period were those of *Apera spicaventi* (PP). After being buried for two years and following the removal of the top layer of the soil more far-red induced seeds germinated compared to the controls (Table 1). Similar differences were observed for those *Lactuca serriola* and *Verbascum lychnitis* seeds (PP) which remained buried in the soil for 9 months (Table 1). After being buried for 8 months far red-inhibited *Oenothera* seeds (PP) showed germination similar to that of the controls.

The seeds of *Crepis capillaris* represented those with an indifferent type of response to light. In the majority of cases, while buried in the soil more control than inhibited seeds were found to germinate (Table 2). Conversely, after exhumation or after the removal of the top layer of the soil more inhibited seeds than the controls germinated as a rule (Tables 1, 2).

Discussion

Far red radiation reaching the seeds after being transmitted through green leaves carries a message for seeds at the bottom of the vegetation stand that there are other competitor plants close by. The seeds which possess the phytochrome mechanism do not germinate if they are at the bottom of the stand because they 'know' that the seedlings will have no chance of survival.

Induction of the seeds with far red before burying them prevents them from germinating in the soil. There were statistically significant differences in germination between induced and control seeds for 5 species. It is only for *Oenothera* that such differences were not found. As could be expected, the most pronounced differences occurred for the photoblastically indifferent species *Crepis capillaris* and for the negatively photoblastic *Amaranthus retroflexus*. With the photoblastically positive seeds, the cessation of light alone causes inhibition which makes the effect of FR less clear.

The inhibition of seeds in the soil can be forced by many factors (Grzesiuk, 1967; Wesson and Wareing, 1969; Roberts, 1972). The environmental significance of FR seems, in this case, to consist in the prevention of seed germination before the dormancy induced by soil factors sets in (Górski and Rybicki, 1985). The dormancy can be overcome only after the seeds have been uncovered and exposed to light. It is true even of those seeds which did not show any response to radiation before being buried (Wesson and Wareing, 1969). Seeds in the soil are known to be able to preserve their viability for very long periods (Barton, 1965). The experiments by Beal (Kivilaan and Bandurski, 1981) revealed that seeds of some species remained viable for as many as 100 years after burial.

Unlike those kept in dry storage, moistened buried seeds are likely to undergo repair processes (Villiers, 1975). Since the dormancy of seeds in the soil can be broken by exposure to radiation i.e. through the alteration of the phytochrome it is reasonable to assume that the metabolism of buried seeds, even if they do not show photoblastism before burial, is to some extent controlled by the phytochrome.

References

Barton, L.V. 1965. In: *Encyclopedia of Plant Physiology*, pp. 1058–1085, XV (2), (ed. W. Ruhland). Berlin-Heidelberg-New York: Springer Verlag.

Fenner, M. 1980. *Weed Research* 20: 135–138.

Gaba, V. and Black, M. 1979. *Nature* (London) 278:51–54.

Górski, T. 1983. *Zeszyty Problemowe Postępów Nauk Rolniczych* 258: 171–183.

Górski, T., Górska, K. and Nowicki, J. 1977. *Flora* 166: 249–259.

Górski, T. and Rybicki, J. 1985. *Pamiêtnik Pulawski* 85: 29–40.

Grzesiuk, S. 1967. *Fizjologia Nasion,* pp. 523. Warszawa: Państwowe WydawnictwoRolnicze i Lene.

Kivilaan, A. and Bandurski, R.S. 1981. *American Journal of Botany* 68: 1290–1292.

Roberts, E.H., 1972. In: *Viability of seeds*, pp. 321–359 (ed. E.H. Roberts). London: Chapman and Hall.

Senger, H. 1980. *The Blue Light Syndrome*, pp. 511 (ed. H. Senger). Berlin-Heidelberg-New York: Springer Verlag.

Silvertown, J.W. 1980. *The New Phytologist* 85: 109–118.

Villiers, T.A. 1975 In: *Crop Genetic Resources for Today and Tomorrow. Genetic Maintenance of Seeds in Imbibed Storage,* pp. 492 (eds. Frankel O.H. and Hawkes J.G.). Cambridge: Cambridge University Press.

Volotovsky, I.D. 1987. *Plant Physiology* (Moscow) 34: 644–655.

Wesson, G. and Wareing, P.F. 1969. *Journal of Experimental Botany* 20: 402–413.

34. Effect of Ethephon on Dormancy Breaking in Beechnuts

E. FALLERI, C. MULLER and E. LAROPPE

INRA Centre de Nancy, Unité de Recherche sur les Semences Forestières, 54280 Champenoux, France

Abstract

The ability of ethephon to reduce the cold requirement of dormant beechnuts was investigated. Several pretreatment methods were tested applying ethephon solutions at different concentrations and different cold durations to 3 seedlots. Ethephon strongly stimulated both germination percentage and rate whatever the cold duration but ethephon application without chilling did not completely overcome dormancy. In particular 0.7 mM ethephon reduced the cold requirement by approximately 4 weeks in comparison with the classical pretreatment. Ethephon's stimulatory effect was still present after drying seeds to 8% moisture content (m.c.) and 4 months storage.

Introduction

Beechnuts (*Fagus sylvatica* L. seeds) as many other forest seeds show a very deep and heterogeneous degree of dormancy varying from year to year, among and within seedlots (Edwards, 1980). Long cold pretreatments are the most common techniques to overcome dormancy and they are usually performed with medium (classical stratification) or without medium (Suszka, 1974; Edwards, 1986; Leadem, 1986; 1989). Both methods involve estimating the number of chilling weeks to obtain 10% germination, typically 3–8 weeks, thus allowing comparisons among seedlots with a different degree of dormancy. Pretreatment without medium shows several advantages compared to classical stratification; it provides fast and uniform germination, it prevents germination during chilling and it can be applied either before or after seed storage (Muller and Bonnet-Masimbert, 1989; Muller *et al.*, 1990). Chemical compounds can be applied easily during the imbibition phase. Nevertheless this is a time-consuming method, its duration for beechnuts ranging from 1 to 3 months.

In order to reduce the duration of this dormancy breaking period, the ability of ethephon to decrease the cold requirement of dormant beechnuts was investigated. Ethephon (2-chloroethylphosphonic acid) is an ethylene releasing compound and ethylene effects on seed germination are well documented; it promotes germination in many species as *Lactuca sativa* (Abeles and Lonski, 1969; Abeles, 1986), *Arachis hypogaea* (Ketring, 1977), *Amaranthus caudatus* (Kepczynsky and Karssen, 1985), *Amaranthus retroflexus* (Schonbeck and Egley, 1981), *Xanthium pennsylvanicum* (Esashi and Katoh, 1975; Katoh and Esashi, 1975) and *Helianthus annuus* (Corbineau *et al.*, 1990). In contrast several

R.H. Ellis, M. Black, A.J. Murdoch, T.D. Hong (eds.), Basic and Applied Aspects of Seed Biology, pp. 303–309.
© *1997 Kluwer Academic Publishers, Dordrecht. Printed in Great Britain.*

species do not respond to ethylene; Olatoye and Hall (1973) reported that only 5 of 12 weed species were stimulated by this gas.

The objectives of this study were twofold. First to identify the most effective ethephon concentration and chilling period in relation to the pretreatment duration, secondly to investigate possible adverse effects of drying and storage on seeds imbibed with ethephon.

Materials and Methods

Testing Different Concentrations of Ethephon Solutions

Seeds were soaked in ethephon solutions at different concentrations (0.7, 0.07 and 0.007 mM) for 4–5 h at 4°C. After draining of excess solution but no surface drying on absorbent paper, they were placed in plastic boxes for 24–48 h to complete the uptake of the solution still present on the seeds' surface and to reach 30% moisture content (m.c.). Moisture content was determined on fresh weight basis by oven-drying two samples of 10 seeds each at 103°C for a period of 17 h (ISTA, 1985). When the m.c. of some seeds was lower than 30% it was adjusted by adding small volumes of ethephon solutions and waiting for further uptake. Seeds imbibed on water to 30% m.c. and chilled without medium for 0 and 7 weeks were used as controls (controls 1 and 2 respectively). For each treatment four 50-seed replicates were germinated in plastic boxes on top of moist filter paper at 15°C in the dark. Germination tests were performed in an incubator with a control accuracy of ±1°C. Germination was assessed daily over a 30 day period and seeds were considered to have germinated when the radicle had protruded by at least one third of seed length.

Comparison between Cold and Cold + Ethephon Pretreatments

Seeds from 3 seedlots were hydrated to 30% m.c. with 0.7 mM ethephon solution and then incubated without medium at 4°C for 3, 5, 7 and 9 weeks. Seed imbibed with water to 30% m.c. and pretreated for the same cold periods were used as controls. For each treatment four 50-seed replicates were germinated at 15°C in the dark.

Comparison between Long Cold Treatment and Ethephon plus Short Cold Treatment, Before and After Storage

Beechnuts from 3 seedlots were pretreated by two different methods:

- hydration to 30% m.c. with 0.7 mM ethephon solution plus chilling for 3 weeks;

- imbibition to 30% m.c. with water plus chilling for 7 weeks.

All seeds were then dried to 8% m.c. and stored at –7°C for a period of 4 months. For both experimental conditions and for each seedlot four 50-seed replicates were germinated at 15°C in the dark. Germination percentage and rate were also tested after storage to check for possible deterioration caused by drying or by storage.

Germination percentage (GP, %) was calculated each day and as the final value after 30 days. Mean germination time (MGT, days) was calculated as follows: MGT $= \Sigma n_i d_i/n$, where n is the total number of germinated seeds during the 30-days germination test, n_i is the number of germinated seeds on day d_i and i is the days during germination period (between 0 and 30) (Younsheng and Sziklai, 1985). All percentage data were analysed by factorial analysis after angular transformation i.e. arcsin $\sqrt{\%}$. Mean values were compared with tukey test calculated with q 0.01. All differences cited in the text are significant at the 1% level.

Results

Testing Different Concentrations of Ethephon Solutions

Ethephon strongly stimulated germination of dormant beechnuts but it did not completely break dormancy (Table 1). Application of increasing ethephon concentrations progressively enhanced germination capacity. Comparisons with controls 1 and 2 showed that germination of seeds pretreated with ethephon at 0.7 mM was significantly higher than for untreated seeds (control 1) but lower than 7-week cold pretreated seeds (control 2). Increasing concentrations of ethephon also produced a gradual but significant reduction in mean germination time compared to untreated seeds (control 1, Table 1). As with germination percentage, control 2 again proved to be the most effective, its germination rate being the highest.

Table 1. Effect of different ethephon concentrations on germination percentage (GP) and mean germination time (MGT)

Treatments	GP (%)	MGT (days)
0 weeks cold (control 1)	30.5 a[a]	15.3 a
0.007 mM ethephon	31.0 a	15.1 ab
0.07 mM ethephon	34.5 ab	13.4 abc
0.7 mM ethephon	52.0 bc	11.0 bc
7 weeks cold (control 2)	63.5 c	10.0 c

[a]Values in the same column followed by the same letter(s) are not significantly different ($p = 0.01$)

Comparison Between Cold and Cold+Ethephon Pretreatment

Both 0.7 mM ethephon and water imbibed seeds showed higher germination percentages after cold pretreatments up to 7 weeks (Table 2); 9-week chilling did not produce any additional significant increase in germination. Significantly higher germination was recorded for cold+ethephon pretreated seeds whatever the chilling period. Differences between the two kinds of pretreatment were more marked for short cold periods. Germination of seeds imbibed in ethephon and chilled for only three weeks was similar to seeds chilled with water for 7 weeks (73.2 and 74% respectively).

Table 2. Effect of cold and cold+ethephon pretreatments on beechnut germination percentage after different cold periods

Treatments	Cold pretreatments (weeks)				Average
	3	5	7	9	
Cold	38.4 a	59.6 a	74.0 a	72.6 a	61.2 a[a]
Cold+ethephon	73.2 b	73.2 b	82.6 b	80.0 b	77.2 b
Average	55.8 A[b]	66.4 B	78.4 C	76.4 C	

[a]Values in the same column followed by the same letter are not significantly different ($p = 0.01$)
[b]Values in the same row followed by the same capital letter are not significantly different ($p = 0.01$)

Comparison Between Long Cold Treatment and Ethephon + Short Cold Treatment, Before and After Storage

On average for both pretreatments, highly significant differences in GP were observed among the 3 seedlots while MGT values did not differ statistically (Table 3).

No seedlot exhibited a significant difference in germination when the two pretreatments – ethephon+3-week cold and 7-week cold treatments – were applied (Fig. 1A) even though in the ethephon method the cold period was reduced by half. No significant variation in germination response was observed after 4 months storage (Fig. 1B). Application of ethephon+3-week cold caused a significant decrease in mean germination time (i.e. an increase in germination rate) in comparison with 7-week cold treatment (Fig. 2A) This stimulatory effect on germination rate was still present after storage (Fig. 2B).

Table 3. Germination percentage (GP) and mean germination time (MGT) in 3 seedlots of beechnuts

Seedlots	GP (%)	MGT (days)
1	82.5 a[a]	9.4 a
2	78.1 a	9.3 a
3	66.6 b	8.9 a

[a]Values in the same column followed by the same letter are not significantly different ($p = 0.01$)

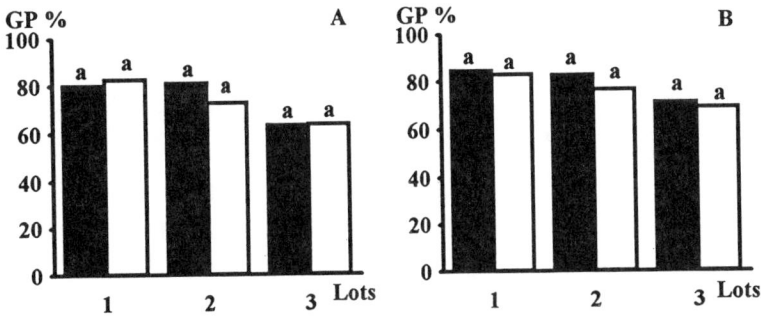

Figure 1. Germination percentage (GP) in beechnut seedlots 1, 2 and 3 pretreated with ethephon+3-week cold (■) and with 7-week cold (□), before (A) and after storage (B). Values for a given seedlot and treatment with the same letter (a) show no significant differences ($p = 0.01$)

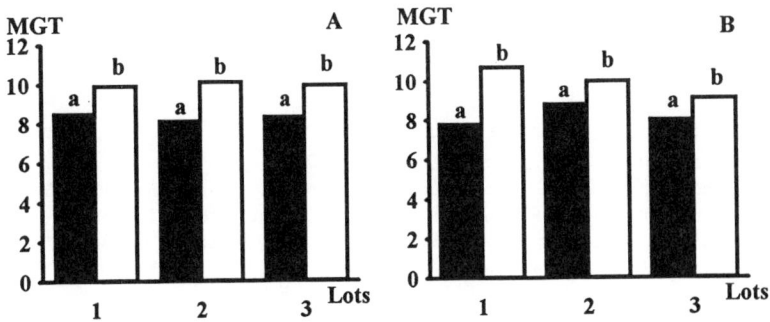

Figure 2. Mean germination time (MGT) in beechnut seedlots 1, 2 and 3 pretreated with ethephon+3-week cold (■) and with 7-week cold (□), before (A) and after storage (B). Values for a given seedlot and treatment with different letters (a–b) are significantly different ($p = 0.01$)

Discussion

Germination of dormant beechnuts was strongly stimulated by ethephon pretreatments in particular when a 0.7 mM solution was used. Similar results were reported for apple embryos where the application of 100 mg/l (0.7 mM) ethephon in the stratification medium increased germination (Sinska and Gladon, 1984). To optimize dormancy breaking both cold and ethephon treatments were needed. The joint action of these two factors was so effective that a dramatic decrease in cold requirement occurred. An interaction of ethylene and other factors has been observed elsewhere; germination was stimulated in combination with increased oxygen tension (Katoh and Esashi, 1975; Esashi et al., 1987). In dormant sunflower embryos ethylene allowed germination in hypoxia (Côme and Corbineau, 1992) and it enabled cocklebur seeds to germinate under water stress (Esashi et al., 1989). Likewise in *Amaranthus caudatus* ethylene overcame germination inhibition imposed by osmotic agents (Kepczynski and Karssen, 1985; Kepzynsky, 1986). In contrast, Hamilton (1972) reported that in seeds of woody species (*Eleagnus angustifolia*, *Taxodium disticum* and *Sheperdia argentea*) soaking in ethephon solutions increased germination capacity only when it was not associated with cold treatments.

In the present study ethephon application could replace 4 weeks of cold treatment thus reducing the chilling time by more than half. Ethephon + short cold treatment also led to faster germination rates. Seed drying to 8% m.c. and storage for a period of 4 months did not affect the germination response of ethephon imbibed seeds. This method to obtain dormancy release could be advantageous especially for deeply dormant seedlots also avoiding possible viability loss of a certain number of seeds during long cold moist pretreatments. Considering that ethephon formulations provide a convenient way to apply ethylene without needing gas-tight containers (De Wilde, 1971) and that short cold treatments allow more flexibility, an application to nursery procedures could be expected.

References

Abeles, F.B. 1986. *Plant Physiology* 81: 780–787.
Abeles, F.B. and Lonski, J. 1969. *Plant Physiology* 44: 277–280.
Côme, D. and Corbineau, F. 1992. In: *Advances in the Science and Technologies of Seeds*, pp. 288–298 (eds F. Jiariu and A.A. Khan). Beijing, New York: Science Press.
Corbineau, F., Bagniol, S. and Côme, D. 1990. *Israel Journal of Botany* 39: 313–325.
De Wilde, R.C. 1971. *HortScience* 4: 12–18.
Edwards, D.G.V. 1980. *Seed Science and Technology* 8: 625–657.
Edwards, D.G.V. 1986. *Journal of Seed Technology* 10(2): 151–171.
Esashi, Y. and Katoh, H. 1975. *Plant and Cell Physiology* 16: 707–718.
Esashi, Y., Hase, S. and Kojima, K. 1987. *Journal of Experimental Botany* 38: 702–710.
Esashi, Y., Abe, Y., Ashino, H., Ishizawa, K. and Saitoh, K. 1989. *Plant, Cell and Environment* 12:183–190.

Hamilton, D.F. 1972. *International Plant Propagation Society Proceedings* 22: 368–373.

ISTA (International Seed Testing Association). 1985. *Seed Science and Technology* 13: 356–513.

Katoh, H. and Esashi, Y. 1975. *Plant and Cell Physiology* 16: 687–696.

Kepczynski, J. 1986. *Physiologia Plantarum* 67: 588–591.

Kepczynski, J. and Karssen, C.M. 1985. *Physiologia Plantarum* 63: 49–52.

Ketring, D.L. 1977. In: *The Physiology and Biochemistry of Seed Dormancy and Germination*, pp. 157–178 (ed. A.A. Khan). Amsterdam: North Holland.

Leadem, C. 1986. *Canadian Journal of Forest Research* 16(4): 755–760.

Leadem, C. 1989. *Canada-British Columbia Forest Resource Development Agreement*, pp.18. FRDA Report 095.

Muller, C. and Bonnet-Masimbert, M. 1989. *Seed Science and Technology* 17: 15–26.

Muller, C., Bonnet-Masimbert, M. and Laroppe, E. 1990. *Revue Forestière Francaise* 42(3): 329–345.

Olatoye, S.T. and Hall, M.A. 1973. In: *Seed Ecology*, pp. 233–240 (ed. W. Heydecker). London: Butterworths.

Schonbeck, M.W. and Egley, G.H. 1981. *Plant, Cell and Environment* 4: 229–235.

Sinska, I. and Gladon, F. 1984. *HortScience* 19: 73–75.

Suszka, B. 1974. *Arboretum Kornickie* 19: 105–128.

Younsheng, C. and Sziklai, O. 1985. *Forest Ecology and Management* 10: 269–281.

35. Endogenous Gibberellins and Dormancy in Beechnuts

H. FERNANDEZ[1], P. DOUMAS[1], E. FALLERI[2], C. MULLER[2] and
M. BONNET-MASIMBERT[1]

[1] *I.N.R.A., Station d'Amélioration des Arbres Forestièrs, Centre de Recherches d'Orléans, 45160 Ardon;*
[2] *I.N.R.A., Laboratoire de Semences Forestières, Centre de Recherches de Nancy, 54280 Champenoux, France*

Abstract

Attempts to break dormancy of beechnuts (*Fagus sylvatica* L.) by soaking them in GA3 and GA4/7 solutions were successfully performed. The ability of these exogenous gibberellins (GAs) to break dormancy suggests a possible role for endogenous GAs in this process. Endogenous GAs were extracted from dormant and non dormant seeds and identified by gas chromatography–mass spectrometry (GC–MS) with full scans. Fourteen GAs were detected: GA1, GA2, GA3, GA4, GA7, GA9, GA15, GA19, GA20, GA29, GA36, GA37, GA51 and GA54. Levels of some endogenous GAs (GA1, GA3, GA8, GA9, GA19 and GA20) from seeds treated with different dormancy breaking treatments were analysed separately in cotyledons and embryonic axis. After purification of the extracts, GAs were quantified by GC-selected ion monitoring (GC-SIM) with deuterated GAs as internal standards. The results showed that GAs corresponding to the 13-OH pathway seemed to be involved in dormancy breaking. Along the process, accumulation of GA3 in the embryonic axis took place at the same time as an increase of GA20 and a decrease of GA19 in the cotyledons. The quantitative differences between dormant and non dormant seeds in some of the analysed GAs and the positive effect of GA applications may suggest that GAs are directly involved in the dormancy breaking process.

Introduction

Seeds of several species show dormancy, and are unable to germinate even in conditions which are perfectly adequate for germination. According to Côme (1989) seed dormancy could be considered as a mechanism to regulate germination but from a practical point of view, it results in an enormous waste of seeds and in poor yields in the nurseries (Muller and Bonnet-Masimbert, 1989). In recent years a lot of research in our group has focused on dormancy breaking in order to allow a very fast and grouped seedling emergence in the nursery (Muller and Bonnet-Masimbert, 1989; Muller *et al.* 1990; Suszka *et al.* 1994).

Breaking of dormancy in beechnuts (*Fagus sylvatica*) can be released by the classical cold stratification (i.e. prechilling with a moist medium) (Mayer and Poljakoff-Mayber, 1963). It is well established that the action of low temperature

R.H. Ellis, M. Black, A.J. Murdoch, T.D. Hong (eds.), Basic and Applied Aspects of Seed Biology, pp. 311–321.
© *1997 Kluwer Academic Publishers, Dordrecht. Printed in Great Britain.*

can be partially or totally substituted by a brief gibberellic acid (GA3) treatment, including in beechnuts (Bonnet-Masimbert and Muller, 1976) but the possible role of gibberellins (GAs) in dormancy is far from clear. Different approaches have been applied to this question. In *Corylus avellana* seeds Williams and Bradbeer (1974) observed that chilling slightly increases GA1 and GA9 levels. Experiments using inhibitors of GA biosynthesis on celery (*Apium graveolens*) seeds suggested that *de novo* synthesis of GAs is a prerequisite for the release of dormancy (Thomas, 1989) but the possibility that inhibition of germination may be due to actions of growth retardants cannot be excluded. Using both ABA and GA *Arabidopsis* mutants, Hilhorst and Karssen (1992) concluded that synthesis of GAs is not necessary for the release of dormancy. Identifying endogenous GAs and elucidating the pathways for their synthesis is a first step in understanding the roles that GAs play in dormancy. In this respect, the new methods described here offer the ability to detect and quantify minute amounts of plant hormones.

The positive effect of GAs, such as GA3 and GA4/7 mixture, on dormancy breaking in other species (reviewed by Arora *et al.*, 1990) has inspired us to explore in depth the involvement of these compounds in the dormancy of beechnuts. Besides exogenous application, identification and quantification of the endogenous GAs were performed on dormant and non dormant seeds.

Materials and Methods

Exogenous Gibberellin Applications

Beechnuts of a Danish seedlot (number 22) collected in 1993 were pretreated by rehydration up to 30% moisture content (m.c.) by soaking them during three hours either in water (control) or in GA3 or GA4/7 solutions at different concentrations (10, 30, 100, 300 and 1000 mg/l for GA3; and 10, 30, 100 and 300 mg/l for GA4/7). After the pretreatments, the germination test was done at constant 15°C with a photoperiod of 8 h light–16 h dark.

Identification of Endogenous Gibberellins

Plant Material

Beechnuts were rehydrated up to 30% m.c. by soaking them in water and pretreated for seven weeks at 3°C. The teguments of dormant and non dormant seeds were discarded. The embryos (axis + cotyledons) were dropped in liquid nitrogen before storing at –80°C.

Extraction and Purification

After grinding at 4°C, samples of 80 g fresh matter were homogenized for 1 min with a blender (Ultraturax) in 400 ml methanol 80% and 100 ml 0.05 M phosphate buffer, pH 8 with 0.2 g of BHT as antioxidant. The homogenates were extracted by stirring for 12 h at 4°C in darkness followed by a 10 min sonication. After filtering, the residue was re-extracted for 2 h with 200 ml methanol in the same conditions as above. Both filtrates were combined and evaporated under reduced pressure at 35°C. The residual aqueous phase was adjusted to a volume of 100 ml with 0.05 M phosphate buffer, pH 8 and was acidified to pH 2.7 with 6 M HCl. The acidified phase was then extracted 5 times with 100 ml ethyl acetate and 2 times with 100 ml n-butanol. Organic solvent extracts were combined and concentrated to 10 ml under reduced pressure at 35°C. Extracts were mixed with 1 g of celite, then dried and poured on an activated SiO_2 column which was eluted with 120 ml ethyl acetate:hexane (95:5 v/v) saturated with 0.5 M formic acid and an equal volume fraction was collected. This fraction was dried on a flask evaporator. Samples were purified on reverse phase HPLC using a column, 250×4.6 mm i.d., packed with 5 µm Lichrospher 100 RP-18. The mobile phase was a linear gradient of 10% to 73% methanol in an aqueous solution with 1% acetic acid at a flow rate of 1.5 ml min^{-1} (Koshioka *et al.*, 1983). Eighty 1-min fractions were collected and evaporated in a concentrator (Savant Instrument Co, USA). Aliquots (1:20) of each fraction were thereafter tested for GA-like activity with the Tan-ginbozu dwarf rice micro-drop bioassay (Murakami, 1968) allowing the identification of biologically active fractions. In addition, 1:20 aliquots of each fraction were methylated with ethereal diazomethane and tested for binding to polyclonal antibodies raised against GA3 (anti-GA3) (Bounaix and Doumas, 1995). Fractions exhibiting GA-like activity were combined and then purified on normal phase HPLC using a column, 150×3.9 mm i.d., packed with 5 µm Nucleosil NO_2. The HPLC procedure has already been described by Odén *et al.* (1987). The mobile phase was a linear gradient from n-heptane saturated with 0.5 M formic acid in water to 98.5% ethyl acetate, 1% acetic acid and 0.5% formic acid over 60 min and then isocratic elution for 20 min, at a flow rate of 2 ml min^{-1}. Eighty 1-min fractions were collected, evaporated and assayed at 1:20 aliquots on the micro-drop bioassay. In addition, 1:20 aliquots of each fraction were methylated with ethereal diazomethane and tested with anti-GA3 (Bounaix and Doumas, 1995).

Gas Chromatography–Mass Spectrometry of Gibberellins

Fractions which were both biologically active and immunoreactive were dried and analyzed by gas chromatography–mass spectrometry (GC–MS). Samples were methylated using ethereal diazomethane and trimethylsilylated using 50 µl N,O bis (trimethylsilyl) trifluoroacetamide with 1% (v:v) trimethylchlorosilane for 30 min at 65°C. Samples were taken to dryness and dissolved in chloroform before injecting splitless onto a fused silica glass capillary column, AT-1 chemical bonded phase 0.25 μm, 0.25 mm (id) × 25 m long (Heliflex) installed in a MFC 500 GC (Fisons Instruments, UK) connected to QMD 1000 MS (Fisons Instruments, UK). Injection and interface temperatures were respectively 270°C and 280°C. The column temperature was maintained at 60°C for 1 min, then increased by 20°C min^{-1} to 200°C, followed by 4°C min^{-1} to 250°C. Electron energy was 70 eV. Samples were run in the full scan mode, scanning between m/z 40 to 650.

Quantification of Endogenous Gibberellins

Plant Material

Beechnuts were pretreated by the application of the hormones in association with cold. Seeds were rehydrated to 30% m.c. in water or in solutions GA3 (100 and 300 mg/l) or ethephon (100 mg/l) and then the seeds were maintained at 3°C for three weeks for hormone treated seeds or three and seven weeks for water treated seeds. After the pretreatments, germination was tested at constant 15°C and alternating 3/20°C (16 h/8 h), with light for 8 h per day. Results are expressed as germination percentage (GP) and mean germination time (MGT)[1]. The whole seeds were dropped in liquid nitrogen to remove the teguments and to separate the embryonic axis and cotyledons. The plant material (cotyledons and embryonic axis) was thereafter separately stored at −80°C until analyzed.

Extraction and Purification

Analysis of GAs was done on embryonic axis and cotyledons. Samples (300 mg) were homogenized in 15 ml of refrigerated methanol (80%) combined with tritiated GAs (GA1, GA4, GA8 and GA20) and 100 ng of deuterated GAs (GA1, GA3, GA8, GA9, GA19 and GA20) (100 ng). They were extracted overnight in darkness at 4°C. Each extract was filtered off and the tissue debris were washed with 5 ml 80% methanol. Then, it was purified through a C18 column Sep-pack to remove the pigments. The eluate was absorbed on to 0.5 g of celite, purified by step-elution silicic acid (SiO$_2$) partition chromatography, and then fractionated by reverse-phase C18 (250 × 4.6 mm, lichrospher 100 RP-18, Merck, 5 μm) HPLC. The mobile phase was a linear gradient of 10 to 73% methanol in an aqueous solution with 1% acetic acid for 85 min at 1 ml min^{-1} flow rate. Sixty 1 ml fractions were collected and the radioactivity levels measured for each fraction with a scintillation counter (Beckman LS 1100, USA).

---

[1] MGT = $(\Sigma n_i t_i)$/N with n_i = number of germinated seeds at t_i days and N = total germinated seeds.

Quantitative Analysis by GC-SIM

The fractions containing radioactivity were collected, pooled, evaporated to dryness, and methylated as above. GC-MS procedures were similar to those used for GA identification. The identity of eluted GAs, according to appropriate retention times, was verified by monitoring two or three diagnostic ions of both endogenous and deuterated GAs. Levels of endogenous GAs were determined by measuring the abundance of ion pairs: 506/508 amu for GA1, 504/506 amu for GA3, 594/596 amu for GA8, both 298/300 and 270/272 amu for GA9, 434/436, 374/376 and 375/377 amu for GA19, and 418/420 amu for GA20. The ratio of integrated peak areas between endogenous (^{1}H)GAs and deuterated (^{2}H)GAs was used to calculate the amount of endogenous GAs.

Results

Exogenous Gibberellin Applications and Breaking of Dormancy

Germination progress curves for beechnuts pretreated with several concentrations of GA3 and GA4/7 are presented in Figure 1. The two kinds of GAs enhance the germination percentage. With GA4/7, the optimal concentration for breaking of dormancy is 100 mg/l, as opposed to 300 and 1000 mg/l for GA3. However, with GA3, the results are slightly lower than those obtained with GA4/7. For comparison with the second experiment (Table 1) note that mean germination times are 20.2 and 19.5 for GA3 100 mg/l and GA3 300

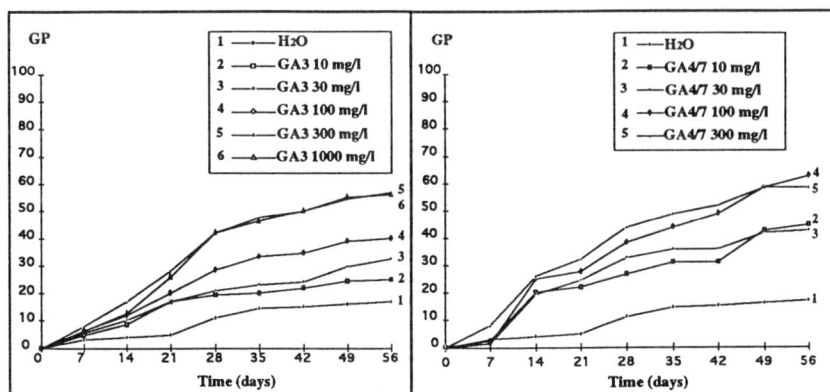

Figure 1. Effect of exogenous applications of GA3 or GA4/7 mixture applications at different concentrations on germination of *Fagus sylvatica* L. seeds. Germination test at 15°C. (GP: germination percent)

mg/l, respectively.

Identification of Endogenous Gibberellins

Some of the HPLC fractions from dormant and non dormant seeds recognised as biologically active were ascertained by the anti-GA3 immunoanalysis. From these fractions, GA1-methyl ester trimethylsilyl (GA1-MeTMSi), GA3-MeTM-Si, GA4-MeTMSi, GA7-MeTMSi, GA9-Me, GA15-Me, GA19-MeTMSi and GA20-MeTMSi were identified by comparison with retention characteristics and full scan mass spectra of authentic standards. The identity of GA2-MeTMSi, GA29-MeTMSi, GA36-MeTMSi, GA37-MeTMSi, GA51-MeTMSi and GA54-MeTMSi was confirmed by comparison with standards from the literature (Gaskin and MacMillan, 1991). The amounts of other biologically active and immunoreactive GA-like compounds were too small or too contaminated to give clean full scan mass spectra.

Quantification of Endogenous Gibberellins

Differences of dormancy release, expressed by germination percentage were found among the different pretreatments (Table 1). These differences are especially clear at 15°C, a temperature which better reveals the dormancy status of the seeds than 3/20°C for which the cold phase can complete the dormancy breaking treatment. These data show that three weeks of cold treatment alone (H_2O+3°C) is clearly insufficient to break the dormancy totally. It is also

Table 1. Effect of different hormonal treatments in association with the cold on dormancy breaking in *Fagus sylvatica* seeds

| | Germination temperature | | | | |
| | 15°C | | | 3/20°C | |
Treatments	GP	MGT		GP	MGT
Dormant seed	11	19.4		21	24.5
100GA3+3C	81	11.8		86	12.3
300GA3+3C	78	12.4		81.5	13.8
100ETH+3C	78	9.1		81	11.3
H_2O+3C	54.5	11.4		64	16.6
H_2O+7C	68	8.0		63.5	7.6

100GA3+3C = Imbibition in 100 mg/l GA3 plus three weeks at 3°C; 300GA3+3C Imbibition in 300 mg/l GA3 plus three weeks at 3°C; 100ETH+3C = Imbibition in 100 mg/l ethephon plus three weeks at 3°C; H_2O+3C = Imbibition in water plus three weeks at 3°C; and H_2O+7C = Imbibition in water plus seven weeks at 3°C. Data obtained after 35 days of germination at constant 15°C or 3/20°C (16 h/8 h)

possible that seven weeks are not enough. Imbibition of seeds in GA3 or ethephon improved germination significantly in comparison to water alone. No differences were found between GA3 and ethephon applications.

The data from quantification analyses of the above mentioned GAs show differences in GA1, GA3, GA19 and GA20 levels between dormant and non dormant beechnuts (Figs. 2a–d) but no differences were found for GA8 and GA9 levels (Figs. 2e,f). After the application of the dormancy breaking treatment, the major effect is related to large variation of GA3 concentrations between cotyledons and embryonic axis. In non dormant seeds, GA3 was reduced in the cotyledons and increased in the embryonic axis where the amount reached the microgram order, especially for those seeds treated with ethephon (Fig. 2a); the levels of the other GAs were always in the order of nanograms. The levels of GA20, always higher in cotyledons than in the embryonic axis, were significantly higher after the application of the dormancy breaking treatments. No differences were found among those treatments (Fig. 2b). With regard to GA19, the amount of this compound decreased in the cotyledons and increased in the embryonic axis of non dormant seeds, especially for seeds treated with ethephon (Fig. 2c). The GA1 levels increased in both the embryonic axis and cotyledons of non dormant seeds, and they were the highest in the embryonic axis from seeds treated with 300 mg/l GA3 (Fig. 2d). The levels of GA8 were always higher in cotyledons than in the embryonic axis. In the embryonic axis from seeds treated with ethephon this value was low (Fig. 2e). Finally, the levels of GA9 were very low (0–10 ng) with regard to the other GAs analysed, and no strong differences were found either between the embryonic axis and cotyledons or between treatments (Fig. 2f)

Discussion

The positive effect of GAs on dormancy breaking in beechnuts (Bonnet-Masimbert and Muller, 1976) is confirmed by the present results. The comparative quantification of endogenous GAs in dormant and non dormant seeds supports arguments for the implication of GAs on this process and reveals possible biosynthesis of these growth regulators, whatever the pretreatment used to break dormancy. If there is an abundant literature on GAs and seeds, most of the papers deal with maturation and development of seeds (Pharis and King, 1985) and very few deal with dormancy.

First of all, although exogenous applications of GA3 or GA7 enhance the germination percentage (Fig. 1) this effect is much higher when GAs are associated with cold treatments (Table 1). This stimulation is accompanied by an increased rate of germination, i.e. much lower mean germination times (cf. Table 1). Thus GAs can only partially substitute for cold requirements of dormant seeds.

The fact that the 14 identified GAs in beechnuts are involved in the three GA biosynthetic pathways described in plants (MacMillan, 1984), allows us to

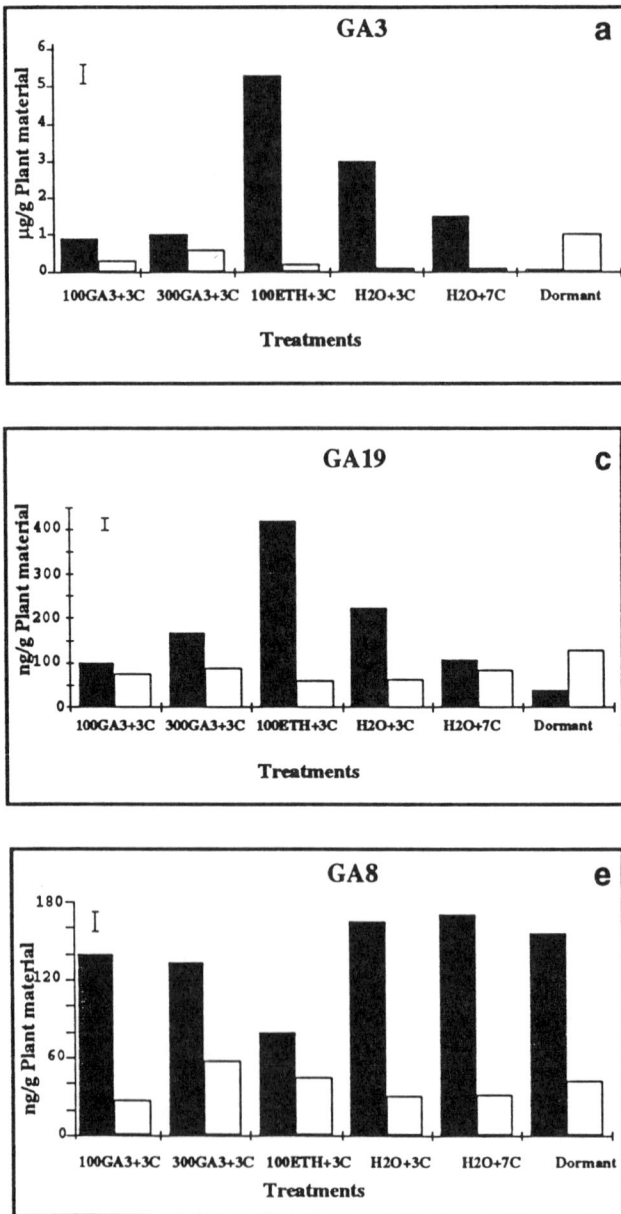

Figure 2. Levels of endogenous GAs in embryonic axis (solid bars) and cotyledons (open bars) of dormant and non dormant beechnuts. Five treatments were applied to break dormancy, using hormones in association with the cold. (For the treatments see Table 1). The bars indicate SE

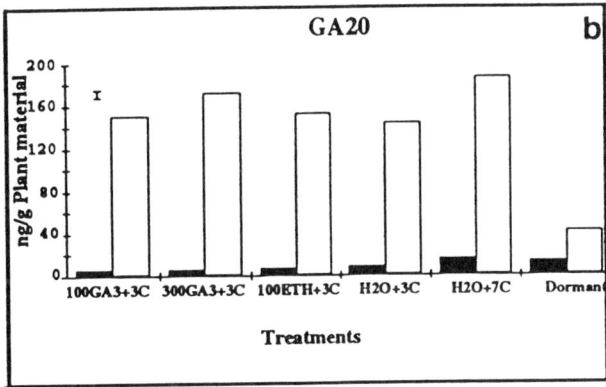

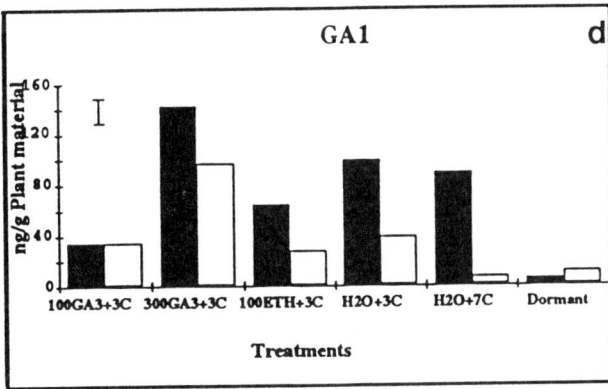

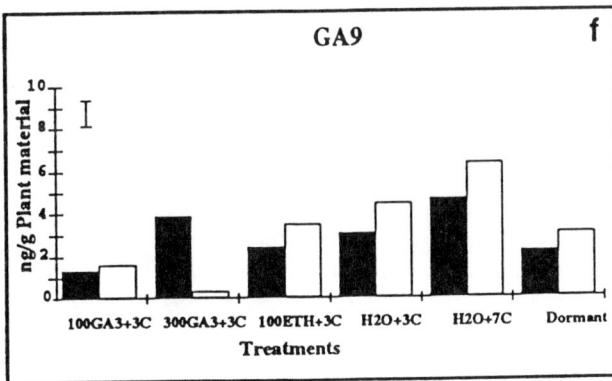

Figure 2. (cont.)

suppose that all those pathways can operate in *F. sylvatica* seeds. On the other hand, the increase in the levels of GA19, GA20, GA3 and GA1 in non dormant seeds reveals that the metabolism of GAs is activated as a consequence of dormancy release. The results reported here focus on the possible implication of the 13-OH GAs in this process.

We observed dramatic GA variations at low temperature (3°C) on moderately moist seeds (30% m.c.) as opposed to Williams *et al.* (1974) who detected strong variations in GA1 and GA9 only after the fully imbibed non dormant hazel seeds were transfered to favorable germination conditions (20°C).

In beechnuts, differences in GA19, GA20, GA3 and GA1 levels between cotyledons and the embryonic axis suggest that these two parts are involved in the 13-OH GA metabolic steps (GA19→GA20→GA1 and GA3, according to MacMillan, 1984). The increase of GA20 and the decrease of GA19 observed in the cotyledons of non dormant seeds, suggest that the conversion GA19→GA20 could occur in this organ. The final metabolic steps to the active forms GA1 and GA3 (the latter accumulate in concentrations of micrograms in the embryonic axis (Fig. 2a) could be carried out either in the cotyledon with translocation to the embryonic axis, or in the embryonic axis itself after translocation of the precursors. We must also consider the possiblity that all these metabolic steps could occur in the embryonic axis as we observed here an increase of GA19 (a known precursor of GA3 and GA1).

Ethylene (released from the ethephon treatment) effectively broke dormancy, yielding germination percentages similar to those obtained from the GA3 applications and faster germination (Table 1). However, the increase of GA19 and GA3 within embryonic axis as well as the decrease of the inactive GA8 by ethephon application were larger than by GA3 application (Fig. 2) . Ethylene releases dormancy in seeds of several species (Bewley and Black, 1982). Corbineau *et al.* (1989) established that the ability of seeds to convert ACC to ethylene decreases during the development and the dehydration of seeds and that non dormant seeds are practically unable to synthesize ethylene until germination and growth occurs. The imbibition of seeds with ethylene gives the highest levels of some GAs, so ethylene may play a role, possibly through an interaction with the levels of endogenous GAs, in the regulation of both dormancy and, consequently, germination.

In the case of GA3 treatments, the exogenous application of this growth regulator increases the amount of this compound in the embryonic axis from non dormant seeds but we cannot distinguish exogenous GA3 from biosynthesized GA3. Curiously, the GA3 content in the embryonic axis of GA3-treated seeds was lower than in the embryonic axis of seeds having received the other treatments. However, if GA3 is involved, this level was enough to break dormancy and to yield similar germination percentages to these obtained using ethephon. With regard to GA1, its highest amount in the embryonic axis was observed in the seeds treated with the higher GA3 concentration. Can the exogenous GA3 application induce the GA3→GA1 conversion in beechnuts? This step of the 13-OH pathway is not established up to now (Fujioka *et al.*,

1990).

This work shows for the first time strong variations in endogenous GAs (and possibly in biosynthesis of GAs) during dormancy release in beechnuts. However, several questions remain unanswered. Where could GA3 be synthesized? What are the relations between the embryonic axis and cotyledons with respect to GA metabolism during dormancy breaking? What is the relationship between the cold, ethylene and GA metabolism? Why are the higher levels of GA19 and GA3 in the embryonic axis of seeds treated with water and cold than those treated with exogenous GA associated with a lower germination percentage? To answer all these questions would require a large study of the metabolism of GAs. It would result in a better understanding of the control of both dormancy and germination processes.

References

Arora, S.K., Banerjee, M.K., Partap, P.S. and Srivastava, V.K. 1990. *Hormonal Regulation of Plant Growth and Development*, pp.125–148 (ed. S.S. Purohit). Agro Botanical Publishers (India), Bikaner 334 001.

Bewley, J.D. and Black, M. 1982. *Physiology and Biochemistry of Seeds in Relation to Germination* pp. 339. Berlin, Heidelberg, New York: Springer-Verlag.

Bonnet-Masimbert, M. and Muller, C. 1976. *Canadian Journal of Forest Research* 6: 281–285.

Bounaix, C. and Doumas, P. 1995. *Plant Growth Regulation* 17: 7–13.

Côme, D. 1989. Quelques Aspects de la Régulation Métabolique des Dormances. *Communication au Colloque Biologie des Semences.* 13–16 March 1989. Angers (France).

Corbineau, F., Rudnicki, R. M. and Côme, D. 1989. *Plant Growth Regulation* 8: 105–115.

Fujioka, S., Yamane, H., Spray, S.R., Phinney, B.O., Gaskin, P., MacMillan, J. and Takahashi, N. 1990. *Plant Physiology* 94: 127–131.

Gaskin, P. and MacMillan, J. 1991. *GC-MS of the Gibberellins and Related Compounds: Methodology and a Library of Spectra*, Bristol, UK: Cantock's Enterprises.

Hilhorst, H.W.M. and Karssen, C.M. 1992. *Plant Growth Regulation* 11: 225–238.

Koshioka, M., Harada, J., Takeno, K., Noma, M., Sassa, T., Ogiyama, K., Taylor, J.S., Rood, S. B., Legge, R.L. and Pharis, R.P. 1983. *Journal of Chromatography* 256: 101–115.

MacMillan, J. 1984. In: *The Biosynthesis and Metabolism of Plant Hormones*, pp.1–16 (ed. A. Crozier and J.R. Hillman). Cambridge: Cambridge University Press.

Mayer, A.M. and Poljakoff-Mayber, A. 1963. *The Germination of Seeds*, pp. 192. New York: Pergamon Press.

Muller, C. and Bonnet-Masimbert, M. 1989a. *Annales des Sciences Forestières* 46 suppl.: 92s–94s.

Muller, C. and Bonnet-Masimbert, M. 1989b. *Seed Science and Technology* 17: 15–26.

Murakami, Y. 1968. *Botanical Magazine* 81: 33–43.

Odén, P.C., Schwenen, L. and Graebe, J. E. 1987. *Journal of Chromatography* 464: 195–200.

Pharis, R.P. and King, R.W. 1985. *Annual Review of Plant Physiology.* 36: 517–568.

Suszka, B., Muller, C. and Bonnet-Masimbert, M. 1994. *Graines des Feuillus Forestières: de la Récolte au Semis.* pp. 292. I.N.R.A. Editions Paris.

Thomas, T.H. 1992. *Plant Growth Regulation* 11: 239–248.

Williams, P.M. and Bradbeer, J.W. 1974. *Planta* 117: 101–108.

36. Molecular Approach to the Role of ABA and GA$_3$ in the Dormancy of *Fagus sylvatica* Seeds

G. NICOLÁS, C. NICOLÁS and D. RODRÍGUEZ

Dpto. Biología Vegetal, Facultad de Biología, Univ. Salamanca, 37007 Salamanca, Spain

Abstract

The dormancy of *Fagus sylvatica* seeds is eliminated by cold treatment at 4°C over a period longer than 8 weeks. Application of GA$_3$ is able to substitute for the cold treatment, while the application of ABA reverses the effect of low temperature and GA$_3$ maintaining the seeds in the dormant state. *In vitro* translation of the mRNAs isolated from these seeds shows the presence of some polypeptides in dormant and in ABA-treated seeds, which tend to disappear during cold or GA$_3$ treatments. By differential screening of a cDNA library constructed from RNA of ABA-treated seeds, some positively ABA-induced clones have been isolated. A search of the sequence data bases showed that one of these clones is a Glycine Rich Protein (GRP) with a possible consensus sequence type RNA-binding protein.The role of this gene product, which seems to be up-regulated by ABA and repressed by GA$_3$, in the dormancy of beechnuts, is discussed.

Introduction

Seed dormancy is an adaptive mechanism to ensure plant survival, and in many seeds can be overcome by chilling, light, plant hormones, temperature and osmotic shock (Schneider and Gifford, 1994). Although several hypotheses have been advanced to explain the action of these factors (Bewley and Black, 1994) the molecular and the biochemical processes underlining seed dormancy remain virtually unknown (Goldmark *et al.*, 1992).

The original hormonal theory according to which dormancy would be controlled both by inhibitors, such as ABA, and by activators, such as gibberellins, cytokinins and ethylene, has been questioned in the light of the results of experiments on ABA and/or GA$_3$ deficient mutants of *Arabidopsis*, in which gibberellin synthesis has been found to be unnecessary for the breaking of dormancy (Karssen and Lacka, 1985). Despite this, exogenous GA$_3$ has proved to be effective in breaking dormancy and in substituting for the requirement of cold stratification in many seeds (Powell, 1987). Moreover, GA$_3$ stimulates germination and reverses the effects of ABA in these and other processes antagonistically affecting the gene expresion (Jacobsen and Beach, 1985; Rodríguez *et al.*, 1987; Cuming *et al.*, 1994).

In the present work we studied the type of dormancy exhibited by *Fagus sylvatica*, the regulatory functions of ABA and GA$_3$ in the maintenance and breaking of dormancy and the role of both hormones on the expression of a

R.H. Ellis, M. Black, A.J. Murdoch, T.D. Hong (eds.), Basic and Applied Aspects of Seed Biology, pp. 323–333.
© *1997 Kluwer Academic Publishers, Dordrecht. Printed in Great Britain.*

Glycine-Rich Protein (GRP) which seems to be involved in the dormancy of beechnuts.

Material and Methods

Plant Material

Seeds of *Fagus sylvatica* were obtained from the Danish State Forestry Tree Improvement Station. Seeds were dried to a moisture content of 21% and stored at –4°C in sealed jars.

Prechill and Germination Conditions

Seeds were briefly surface-sterilized in 1% sodium hypochlorite, thoroughly rinsed in sterile water, and then placed in plastic trays on moistened filter paper at either 4 or 15°C. For some of the experiments the pericarp was removed. The media used for the imbibition of the seeds were sterile water, 50 μM ABA, 100 μM ABA, 100 μM GA₃ and 100 μM GA₃ + 100 μM ABA. After 1 to 6 weeks the percentages of germination were calculated and the seeds collected, frozen in liquid nitrogen and stored at –70°C until later use.

RNA Extraction and in vitro Translation

After the removal of the pericarp and testa, the seeds (3 g) from the different times and treatments were ground in liquid nitrogen and the RNA extracted using Qiagen-pack 500 cartridge, following the manufacturer's protocol. RNA was quantified by measuring the absorbance at 260 nm. The 260:280 absorbance ratio was in the range 1.8 to 2.1, indicative of pure RNA preparation. Intact ribosomal RNA bands were observed on ethidium bromide stained agarose denaturing gels. Total cellular RNA was translated in vitro using a rabbit reticulocyte lysate system (Amersham), as described by Colorado *et al.* (1991).

Preparation and Screening of the cDNA Library

Double stranded cDNA was synthesized from poly A⁺ RNA extracted from embryos of *Fagus sylvatica* seeds incubated for 2 weeks in the presence of 100 μM ABA using a cDNA synthesis kit (Stratagene) and following the manufacturer's instructions. The recombinant phages were packaged *in vitro* using Gigapack (Stratagene). Approximately 7×10^5 primary recombinants were obtained. The library was later amplified after Maniatis *et al.* (1982). Differential screening of the cDNA library was carried out by preparing plaque lifts on nylon membranes (Hybond-N, Amersham) and hybridizing the lifts with ³²P labelled ss-cDNA probes prepared against poly A+ RNA obtained from embryos incubated either in water at 4°C for 6 weeks or in ABA for 2 weeks at 4°C. Plaques that showed hybridization with the probes from embryos incubated in ABA and not in water were purified and the recombinant cDNA was excised from the phage in pBluescript SK (–) using the biological rescue recommended by the suppliers (Stratagene).

Northern Analysis

Total RNA (10 μg) was fractionated on denaturing formaldehyde agarose gels and Northern analysis was carried out after Colorado *et al.* (1994).

cDNA Sequence Analysis

Both strands of the GRP cDNA were sequenced by the dideoxy chain termination method (Sanger *et al.,* 1977) using T7 Sequencing kit (Pharmacia). Labelling and termination reactions were as described in the supplier's instructions. The sequence of the cDNA was read visually and the data analysed using the DNA Strider program.

Results

Table 1 shows the percentages of germination after the seeds had been subjected to different treatments producing the breaking or maintenance of dormancy. Cold stratification at 4°C led to an increase in germination with respect to seeds sown at 15°C which did not germinate even after 6 weeks, reflecting their dormant state.

Removal of the pericarp in both treatments (4 and 15°C) increases the percentages of germination, indicating that the seed coat is partly responsible for dormancy in these seeds. The addition of ABA reversed the effects of cold while the addition of GA₃ was able to substitute for the cold treatment in breaking dormancy. The effects of GA₃ and ABA were seen to counteract each other when they were added jointly.

Electrophoretic separation of the polypeptides resulting from *in vitro* translation (Fig. 1A) shows the appearance of two polypeptides designated A and B with molecular weights of ca. 24 and 22 kDa respectively in dormant seeds, and these polypeptides disappeared after 2 weeks at 4°C which abolished dormancy and permitted seed germination (Table 1). The levels of these polypeptides were very abundant under conditions which maintain the seeds dormant (ABA, 15°C). In contrast, these polypeptides disappeared after only 2 weeks of treatment with GA₃, whereas following treatment with GA₃ + ABA they persisted after 3 weeks, although their levels tended to decline (Fig. 1B).

By differential screening of a cDNA library constructed from mRNA of ABA-treated seeds, some positively ABA-induced clones have been isolated. One of them of about 0.7 kb (data not shown) seems to be involved in the

Table 1. Effects of temperature, ABA and GA$_3$ on the germination of dormant seeds of *Fagus sylvatica*. Data are means of 3 to 4 experiments ±SD

Treatments	% of germination during the time (weeks)					
	1	2	3	4	5	6
Whole seed (4°C)	0	0	7±2.1	22±4.3	32±3.8	40±5.5
Without pericarp (4°C)	0	0	16±3.9	30±4.0	53±6.3	68±8.4
Without pericarp + 50 µM ABA (4°C)	0	0	0	6±1.0	8±1.0	9±1.0
Without pericarp + 100 µM ABA (4°C)	0	0	0	3±0.6	4±0.5	4±0.8
Without pericarp + 100 µM GA$_3$ (4°C)	0	9±1.9	50±6.1	83±8.8	97±1.7	ND*
Without pericarp + 100 µM GA$_3$ + 100 µM ABA (4°C)	0	3±0.6	6±1.8	25±3.3	33±4.1	47±3.2
Whole seeds (15°C)	0	0	0	0	0	0
Without pericarp (15°C)	13±2.2	6±1.1	12±1.5	19±3.7	29±3.2	42±6.4
Without pericarp + 100 µM GA$_3$ (15°C)	2±0.4	41±4.1	68±5.9	96±1.5	ND	ND
Without pericarp + 100 µM GA$_3$ + 100 µM ABA (15°C)		18±2.5	25±3.9	38±4.6	50±5.0	66±3.6

*Not determined

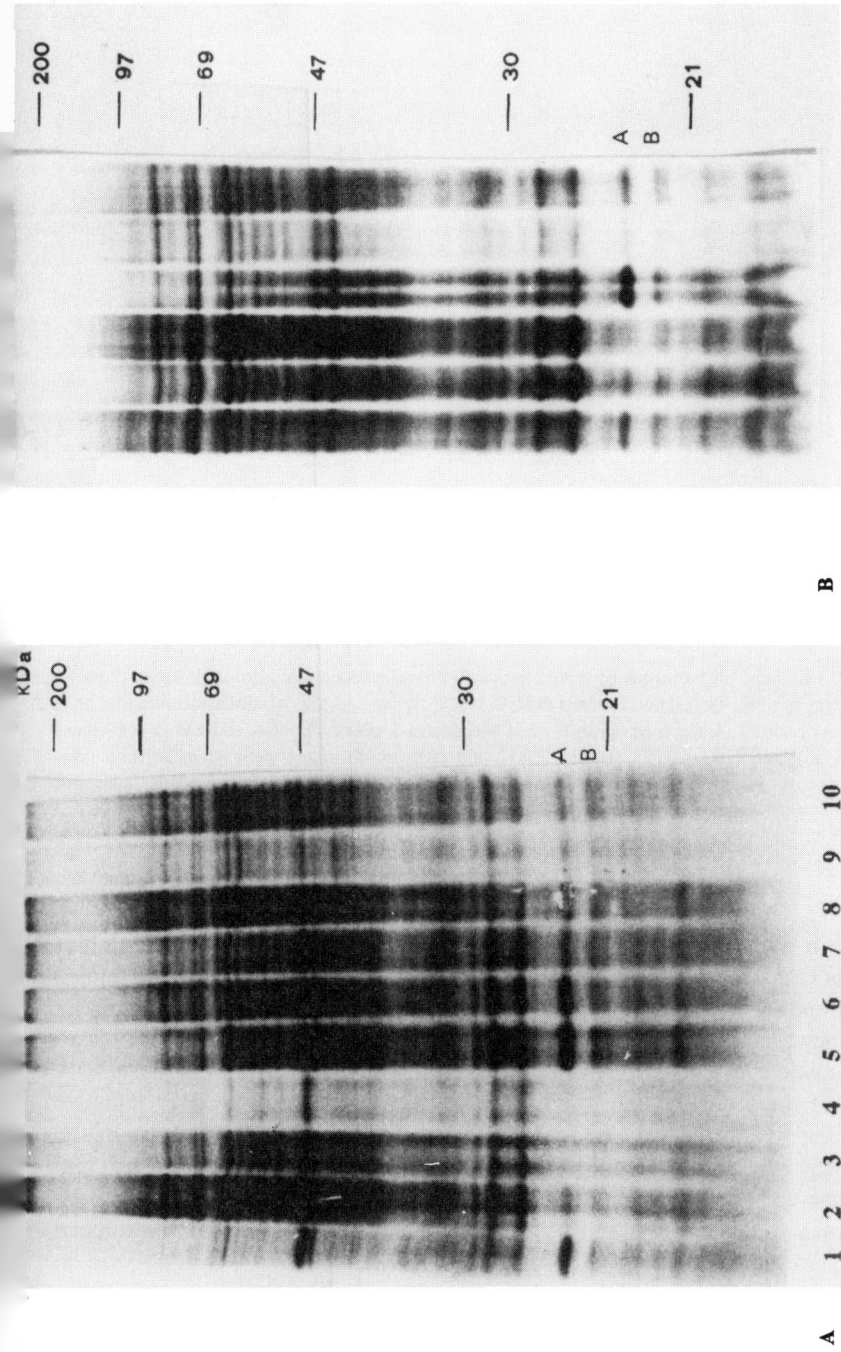

Figure 1. (A) Autoradiography of *in vitro* translation products of RNAs isolated from *Fagus sylvatica* dormant seeds (lane 1), and dormant seeds sown for 2, 4 and 6 weeks in H₂O at 4°C (lanes 2, 3 and 4), in 100 μM ABA at 4°C (lanes 5, 6 and 7) and in H₂0 at 15°C (lanes 8, 9 and 10). (B) Autoradiography of *in vitro* translation products of RNAs isolated from *Fagus sylvatica* dormant seeds sown for 1, 2 and 3 weeks in 100 μM GA₃ at 4°C (lanes 1, 2 and 3) and in 100 μM GA₃ + 100 μM ABA at 4°C (lanes 4, 5 and 6)

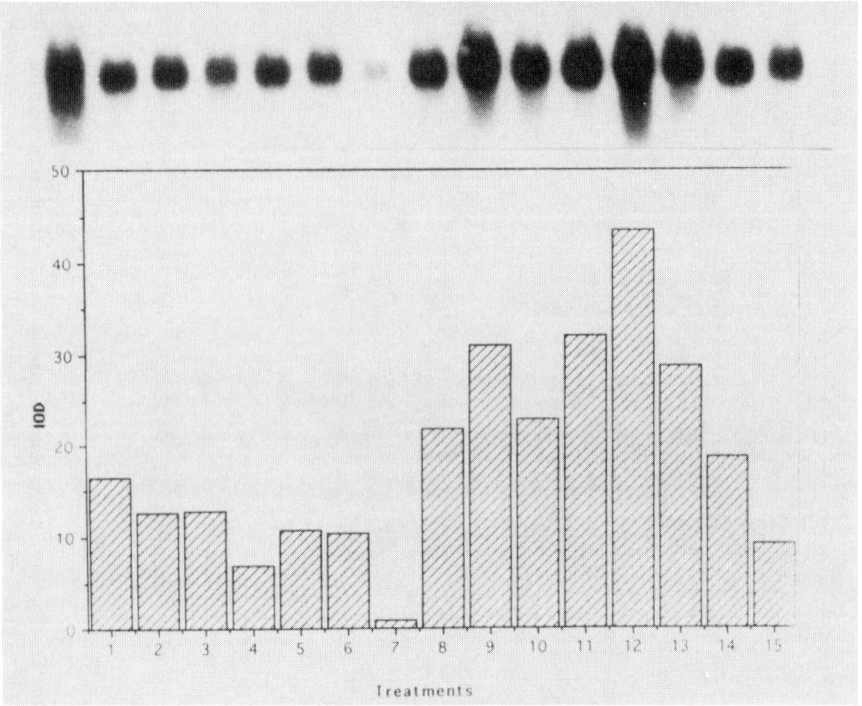

Figure 2. Northern blot analysis of RNAs isolated from *Fagus sylvatica* dormant seeds (lane 1), dormant seeds sown for 2, 4 and 6 weeks in H_2O at 4°C (lanes 2, 3 and 4), dormant seeds without pericarp sown for 2, 4 and 6 weeks in H_2O at 4°C (lanes 5, 6 and 7), ABA 100 μM at 4°C (lanes 8, 9 and 10), and H_2O at 15°C (lanes 11, 12 and 13); non-dormant seeds (lane 14) and non-dormant seeds sown for 5 days in H_2O at 15°C (lane 15). 10 μg RNA was used per lane and hybridized with a probe from the GRP clone

induction and maintenance of dormancy in beechnuts. The Northern blot analysis (Figs. 2 and 3) shows that the mRNA corresponding to this clone is very abundant under dormant conditions, but in those conditions which break dormancy this mRNA tends to disappear. It should be stressed that there is an antagonistic effect of ABA and GA_3 in respect of the levels of this mRNA. A search of the sequence data bases showed that this clone is a GRP. Figure 4 shows the nucleotide sequence and the predicted amino-acid sequence of the clone. The beechnut protein has the carboxy-terminal glycine-rich domain and the conserved RNA-binding domain, and shows a high degree of similarity with other plant GRPs (63% identity with a GRP-RNA binding and ABA-induced protein from maize; data not shown).

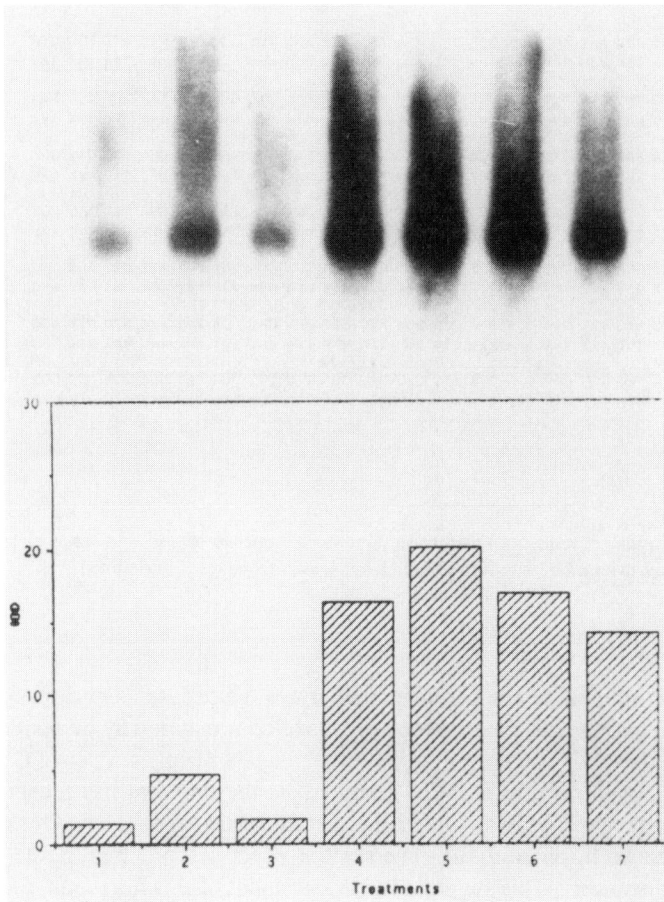

Figure 3. Northern blot analysis of RNAs isolated from *Fagus sylvatica* dormant seeds sown for 1, 2 and 3 weeks in 100 μM GA₃ at 4°C (lanes 1, 2 and 3), in 100 μM GA₃ + 100 μM ABA at 4°C (lanes 4, 5 and 6) and 2 weeks in ABA 100 μM (7) as a control. 10 μg RNA was used per lane and hybridized with a probe from the GRP clone

Discussion

Dormancy constitutes an intrinsic block to germination and the seed coats seem to play an important role in this process. Additionally, to be released from dormancy, the seeds must experience certain environmental factors or must undergo certain metabolic changes (Bewley and Black, 1994). The results of the present work show that *Fagus sylvatica* seeds display an endogenous dormancy that can be released by cold treatment over a certain length of time (Table 1).

```
1/1                                    31/11
CCG CGA TTT CTG AAT AGC ATT TTA TCA AGA AAA ATG AAT TCT AGG GCT TTT ATA TTC CTA
pro arg phe leu asn ser ile leu ser arg lys met asn ser arg ala phe ile phe leu
61/21                                  91/31
GCT CTT CTG TTT GCA TCT GTT CTA CTC ATC TCC TCA GCT GTG GCG ACT AAG ACA TCC AAA
ala leu leu phe ala ser val leu leu ile ser ser ala val ala thr lys thr ser lys
121/41                                 151/51
GAT GAG GAA AAA CCA GAA GAA TCA AAC CCG GTA GAT GAT ACA AAG TAT GGT GGG TAC GGA
asp glu glu lys pro glu glu ser asn pro val asp asp thr lys tyr gly gly tyr gly
181/61                                 211/71
GGC CAT TAT GGT GGT GGA CAT GGA GGA GGT TAT GGA GGA GGG CAT GGA GGA GGG TAT GGT
gly his tyr gly gly gly his gly gly gly tyr gly gly gly his gly gly gly tyr gly
241/81                                 271/91
GGT GGA CAT GGA GGG CGT GGT GGT GGT GGT GGT GGA CGT GGA GGA GGT TTT GTC ACC AAG
gly gly his gly gly arg gly gly gly gly gly gly arg gly gly gly phe val thr lys
301/101                                331/111
CGT GAA ACC ATG CAC GTA GCA GAA GAA TCA ATC GGT AGA GAT ACA AGT ATG GTG GTT ACG
arg glu thr met his val ala glu glu ser ile gly arg asp thr ser met val val thr
361/121                                391/131
GTG GAG GCA TGG TGG CAT GGA GGC ATG GAG GAG GCA TGG TGG TGG CAT GGA GGA GGC GGT
val glu ala trp trp his gly gly met glu glu ala trp trp trp his gly gly gly gly
421/141
GGT GGT
gly gly
```

Figure 4. Nucleotide sequence and deduced amino acid sequence of a cDNA for a glycine-rich protein from *Fagus sylvatica* . The RNA binding consensus sequence is underlined

Furthermore, removal of the pericarp accelerates the release from dormancy in the seeds, indicating that the presence of a hard coat may partly be responsible for dormancy.

Among endogenous inhibitors of germination the one that has received the most attention regarding its involvement in seed dormancy is ABA. Our results show that the addition of ABA reverses the effect of cold treatment on the breaking of dormancy. This suggests the involvement of this hormonal factor in the maintenance of dormancy in these seeds. ABA has been shown to play an important role in many of the processes related to the formation, germination and dormancy of seeds, eliciting changes in RNA and protein synthesis (Finkelstein *et al.*, 1985), tolerance to desiccation (Dure *et al.,* 1989; Colorado *et al.,* 1995) and the inhibition of precocious germination (Quatrano, 1987). These and other findings point to the involvement of ABA in the regulation of gene expression in seeds and in the blocking of germination.

The results of the *in vitro* translation of mRNAs isolated from beechnuts (Fig. 1) show that ABA stimulates the presence of at least two polypeptides that disappear during the cold-induced dormancy release and that they are very abundant in dormant seeds. These findings are consistent with those reported by other authors in the sense that the role of ABA in dormancy would not be to suppress the expression of certain genes but rather to induce the expression of certain genes related to the blocking of germination (Ried and Walker-Simmons, 1990; Morris *et al.*, 1991; Goldmark *et al.*, 1992). It should be stressed that under conditions that maintain dormancy, the levels of the polypeptides

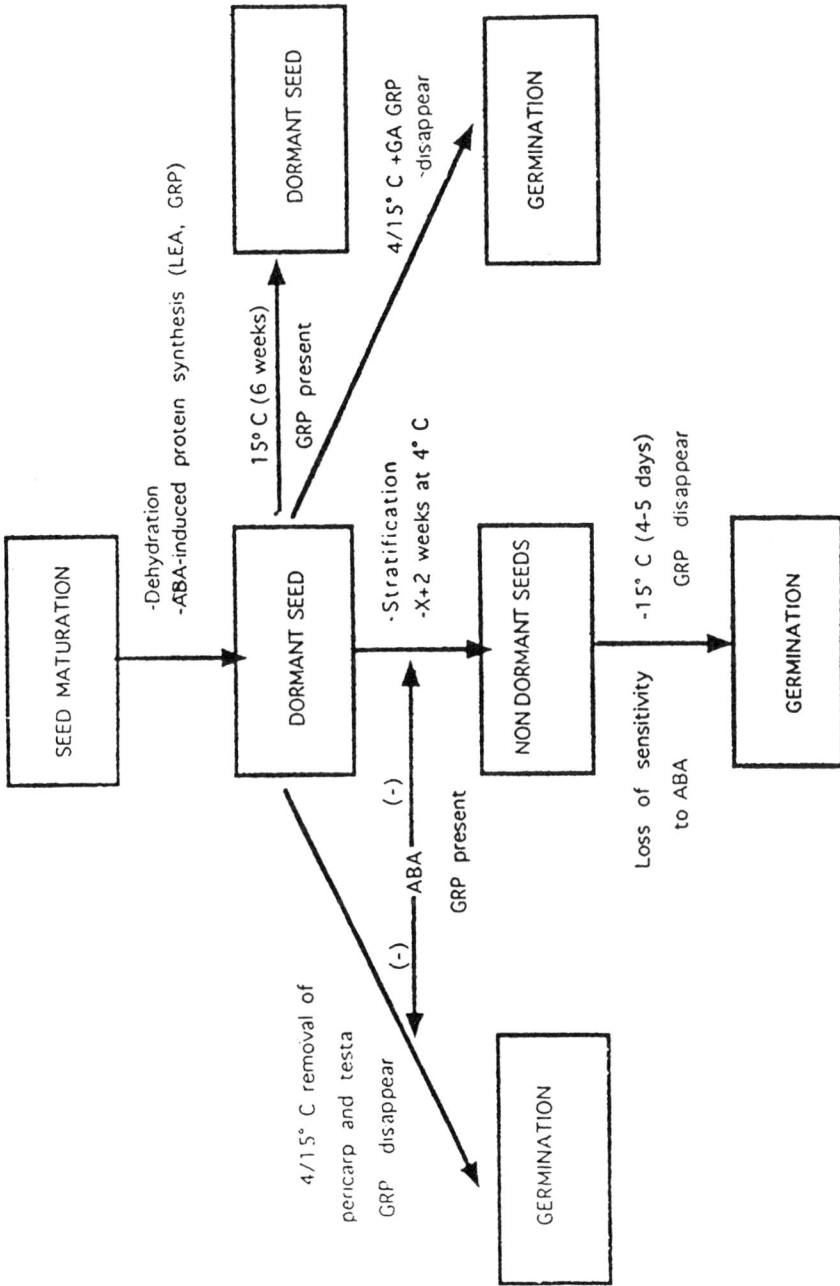

Figure 5. Model based on current results about the role of ABA and GA₃ in the dormancy of seeds of *Fagus sylvatica*. Explanation in the text

supposedly induced by ABA persist or even increase and the addition of GA₃ accelerates the disappearance of the polypeptides regulated by ABA.

The isolation and characterization of a GRP by differential screening of a cDNA library constructed with poly A^+ RNA from 2 weeks ABA-treated seeds seem to be very interesting. Northern analysis shows how this protein, which is very abundant in dormant seeds and under conditions that maintain the seeds dormant but decreases under dormant-breaking treatments, seems to be up-regulated by ABA and repressed by GA₃. Similarly Cuming et al. (1994) have recently described that the expression of dehydrin genes in barley are induced by ABA and inhibited by GA₃.

In the past few years, a novel class of eukaryotic proteins (GRP) has been identified whose members have been implicated in a number of diverse cellular processes requiring RNA recognition (Nocker and Viestra, 1993). To date several cDNA encoding putative glycine-rich RNA binding proteins have been isolated from plants: two from Sorghum vulgaris (Cretin and Puigdomenech, 1990), an ABA-induced one from Zea mays (Gomez et al., 1988) a wound-inducible one from Daucus carota (Sturm, 1992) and two from Arabidopsis thaliana (Nocker and Viestra, 1993). The role of these proteins in plants is not yet known, although the hormonal regulation of the maize gene (Gomez et al., 1988) and the response of the carrot gene to wounding (Sturm, 1992) suggest a role in stress response.

In Figure 5, we present a working model in which we summarize our preliminary proposals about dormancy breakage in Fagus sylvatica. Striking differences in GRP transcript levels appears upon hydration of dormant and non-dormant seeds (Figs. 2 and 3). Upon imbibition, the transcript levels are maintained or even increase under the conditions which kept the seeds dormant (ABA, 15°C) but rapidly decline and disappear in non-dormant seeds and under the conditions which break dormancy (stratification, removal of pericarp and testa, GA₃) when seeds subsequently germinate. It should be stressed that non-dormant seeds have lost their sensitivity to ABA and germinate even in the presence of the hormone (data not shown).

Our findings provide additional evidence supporting the hypothesis that both GA₃ and ABA would be involved in the breaking of seed dormancy, and further contribute to the scanty knowledge on the hormonal regulation of forest species, in particular Fagus sylvatica in which for the first time a molecular approach to the causes of dormancy has been conducted.

References

Bewley, J.D. and Black, M. 1994. Seeds. Physiology of Development and Germination, Second Edition, pp 199. NY: Plenum Press.

Colorado, P., Nicolás, G. and Rodríguez, D. 1991. Physiologia Plantarum 83: 457–462.

Colorado, P., Rodríguez, A., Nicolás, G. and Rodríguez, D. 1994. Physiologia Plantarum 91: 461–467.

Colorado, P., Nicolás, C., Nicolás, G. and Rodríguez, D. 1995. Physiologia Plantarum 94: 1–6.

Cretin, C. and Puigdomenech, P. 1990. *Plant Molecular Biology* 21: 695–699.

Cuming, A.C., Robertson, T., Close, T. and Chandler, C. 1994. *Journal of Experimental Botany* 45: supplement.

Dure, L.III., Crouch, M.L., Harada, J., Mundy, J., Quatrano, R., Thomas, T. and Sung, Z.R. 1989. *Plant Molecular Biology* 12: 475–486.

Finkelstein, R., Tenbarge, K., Shumway, J.E. and Crouch, M.L. 1985. *Plant Physiology* 78: 630–636.

Goldmark, P.J., Curry, J., Morris, C.F. and Walker-Simmons, M.K. 1992. *Plant Molecular Biology* 19: 433–441.

Gómez, J., Sánchez-Martín, D., Stiefel, V., Rigau, J., Puigdomenech, P. and Pages, M. 1988. *Nature* 334: 262–264.

Jacobsen, J.V. and Beach, L.R. 1985. *Nature* 316: 275.

Karssen, C.M. and Lacka, E. 1985. In: *Plant Growth Substances* 1985. pp. 315. (ed. M. Bopp). Berlin, Heidelberg: Springer-Verlag.

Maniatis, T., Fritsch, E.F. and Sambrook, J. 1982. *Molecular Cloning: A Laboratory* Manual. NY: Cold Spring Harbor.

Morris, C.F., Andenberg, R.J., Goldmark, P.J. and Walker-Simmons, M.K. 1991. *Plant Physiology* 95: 814–821.

Nocker, S. and Viestra, D. 1993. *Plant Molecular Biology* 21: 695–699.

Powell, L. 1987. *Plant Hormones and their Role in Plant Growth and Development.* pp. 539. (ed P.J. Davies). Dordrecht: Martinus Nijhof Publishers.

Quatrano R.S. 1987. *Plant Hormones and their Role in Plant Growth and Development.* pp. 494. (ed P.J. Davies). Dordrecht: Martinus Nijhof Publishers.

Ried, J.L. and Walker-Simmons, M.K. 1990. *Plant Physiology* 93: 662–667.

Rodríguez, D., Dommes, J. and Northcote, D.H. 1987. *Plant Molecular Biology* 9: 227–235.

Sanger, F., Nicklen, S. and Coulson, A.R. 1977. *Proceedings of the National Academy of Sciences USA* 74: 5463–5467.

Schneider, W.L. and Gifford, D.J. 1994. *Physiologia Plantarum* 90: 246–252.

Sturm, A. 1992. *Plant Physiology* 99: 1689–1692.

37. Controlled Stratification of *Prunus avium* L. Seeds

A. NOWAG, H. PINNOW and W. SPETHMANN

University of Hannover, Institute for Fruit and Nursery, Am Steinberg 3, 31157 Sarstedt and Forest Seed Centre Oerrel, Forstweg 5, 29633 Munster, Germany

Abstract

In a Ph.D. thesis at the University of Hannover and the Forest Seed Centre Oerrel, Germany, three stratification methods were compared concerning plant yield/kg seeds and content and consumption of sugars, starch and lipids in seeds of wild cherry at harvest and during stratification. Seeds from different provenances, harvest years, stages of ripeness and single trees were investigated. Two stratification methods gave a yield of about 2500 plants/kg seeds for seed lots with a viability of 80% and more. Sugar reserves (glucose, fructose, saccharose) built up during the last 2–3 weeks of fruit-ripeness. There was no difference between single trees from one provenance concerning sugar content at harvest and consumption during stratification. No difference could be shown in sugar consumption during stratification in seeds from different provenances in the same harvest year, but differences were detected within one provenance harvested in different years.

Introduction

Wild cherry seeds (*Prunus avium* L.) are dormant at harvest. To break dormancy so that the seeds will germinate when sown in spring, stratification is necessary. For this, the seeds are mixed with a medium, kept moist and placed outside (= uncontrolled stratification, a common nursery method) or subjected to a certain temperature for a certain time period (= controlled stratification). Stratification of cherry seeds takes about 5–6 months.

With uncontrolled stratification nurseries normally achieve a yield of 800–1200 plants/kg seeds from a seed lot with a viability of 80% and more. In such a seed lot there are 5000–6000 viable seeds/kg. This gives a poor plants:viable seeds ratio.

Normally the stratification period of 5–6 month is not a problem, since the fruits are harvested in July and there is enough time for stratification until March or April. But control of the seeds during stratification takes personnel time, and a shorter stratification period would lower the costs, just as a higher plant yield would.

Until now, one did not achieve the same result with the same stratification method for cherry seeds from one provenance but different harvest years. This suggests that the climatic conditions during flowering, fruit set and fruit-ripening and the input into the seeds of reserve compounds such as sugars, starch and lipids may affect the success of stratification.

The aim of this Ph.D. thesis at the University of Hannover and the Forest Seed Centre Oerrel, Germany, supervised by Prof. Dr Spethmann, is to clarify if

R.H. Ellis, M. Black, A.J. Murdoch, T.D. Hong (eds.), Basic and Applied Aspects of Seed Biology, pp. 335–338.
© *1997 Kluwer Academic Publishers, Dordrecht. Printed in Great Britain.*

and how the reserve substances (sugars, starch and lipids) differ between provenances, harvest years, single trees and different harvest times and how they change during different stratification methods and during storage at −5°C. Until now only the sugars glucose, fructose and saccharose have been analysed. Another aim is to shorten the stratification period.

Harvest and After-harvest Treatment

Cherry fruits are harvested in July by pounding. Nets are placed under the trees, the tree-climbers ascend the trees and pound the branches until the fruits fall down. Leaves and small branches are sorted out by ground personnel (Nowag and Gille, 1993).

After harvest, the fruits are stored in water until maceration. Maceration has to follow fast after harvest, because inhibitory substances in the fruit flesh increases dormancy depth in the seeds. Fruit flesh and stones are seperated with a machine at the central storing house in Oerrel and thereafter the fruit flesh is washed away with large amounts of water. The stones are kept in water over night after maceration and are dried the next day.

Drying and Storage

If the cherry seeds are to be sold immediately after harvest, they are only dried superficially to a moisture content (m.c.) of about 20% (fresh weight, fw). For storage the seeds are dried to a m. c. of 8–10%. Drying takes place in a climatic chamber at +18°C and 20% R.H. The seeds are stored in cans with 6 kg in each at −5°C.

Before the start of stratification in October the stones have to be acclimatised for 1–2 days and imbibed for at least 24 h. After this they have a m.c. of about 20% and during the first week of stratification this increases to about 30%.

Materials and Methods

Three stratification methods were used:

1. Two weeks +20°C, 2 weeks + 4°C, 2 weeks +20°C and 12–16 weeks +4°C (until germination)

2. Two weeks +20°C, 6 weeks +4°C, 2 weeks +20°C, 2 weeks +4°C, 2 weeks +20°C, 4–6 weeks +4°C (until germination)

3. Method 1 plus compost activator in a concentration of 1:20 (1 part activator, 20 parts seeds).

Methods 1 and 2 are used according to the results of Suszka (1976). Stratification with compost activator shortened the time needed for *Rosa corymbifera* seeds from about 12–18 to 6 months (Cullum *et al.*, 1991). Samples for moisture content, sugar, starch and lipid analyses were taken before and after stratification and after each warm period. The sugar contents were determined in mg/g dry weight; the moisture content was based on fresh weight. Viability and weight of 1000 seeds was determined before and after stratification. Per cent germinated seeds was determined just before sowing. In the laboratory, cracking of the stones and maximum germination capacity was determined. In 1992 three provenances (Bovenden, Palsterkamp, Schöningen) in Lower Saxony were harvested. One part was stratified in 1992/93, a second part in 1993/94, a third in 1994/95 and the rest will be treated in 1995/96. No harvest was possible in 1993. In 1994 only one provenance from 1992, Bovenden, gave a good harvest. Therefore two new provenances were chosen (Grohnde and Kattenbühl). In 1995 harvest was possible in Grohnde and Palsterkamp and a small amount of seeds were collected in Bovenden.

Results

Addition of compost-activator to the stratification medium did not shorten the stratification period. The stones cracked at the same time as stones treated with method I without activator, and the seeds germinated simultanEously. However, plant yield from variants treated with activator was much lower than for other variants.

Unripe (green) harvested fruits had only very small amounts of sugar in the seeds. Half-ripe (yellow-pink) fruits had almost the same amount of sugar in their seeds as full-ripe (dark red-black) harvested fruits.

No differences in the amount of sugar at harvest and consumption during stratification in seeds of three single trees were detected. Germination capacity in the laboratory was similar for all trees, but seeds from one tree cracked and germinated about 2 weeks earlier than seeds from other trees.

The amount at harvest of the sugars glucose, fructose and saccharose in cherry seeds from different provenances but one harvest year was almost equal. Also the picture of sugar consumption and re-building during different stratification methods was the same. In seeds from one provenance but different harvest years, the amount of sugar at harvest differed, and so did consumption and re-building during stratification. The amount of sugars changed very little during storage for 1 and 2 years. Plant yield in variants treated with method I and II was about 2500 plants per kg seeds for seed lots with a viability of 85% and more. No difference could be found between harvest years. Storage for 1 or 2 years did not affect plant yield. Maximum germination capacity in the laboratory was always much higher than germination or plant development in the seed bed. Between 50–70% of germination capable seeds developed into a plant.

Conclusions

Compost-activator does not enhance dormancy breaking in seeds of *Prunus avium* L. Two methods give satisfactory plant yield with about 2500 plants/kg seeds for seed lots with a viability of 80% and more. Maximum germination in the laboratory is not reached under field conditions, probably due to changing environment outside. Reserves of glucose, fructose and saccharose are built up during the last 2–3 weeks of fruit-ripeness.

The amount of glucose, fructose and saccharose at harvest and the consumption and re-building of these sugars during stratification does not explain dormancy breaking or germination capacity. HeterogenEous germination and plant emergence is due to varying dormancy depth in *Prunus avium* L. seeds from different trees.

References

Cullum, F.J., Bradley, S.J. and Williams, M. 1991. Improved germination of *Rosa corymbifera* 'Laxa' seed using a compost activator. *Combined Proceedings of the International Plant Propagators' Society* 40: 244–250.

Nowag, A. and Gille, K. 1993. Saatguternte und Aufbereitung in der Praxis. *Forest tree seed harvest and treatment in practise*. Proceedings of the International Symposium about Forest Tree Seeds. Forest Seed Centre Oerrel, Germany.

Suszka, B. 1976. Increase of germination capacity of mazzard cherry (*Prunus avium* L.) seeds through the induction of secondary dormancy. *Arboretum Kornickie* 21: 257–270.

38. Studies on the Persistence of Rape Seeds (*Brassica napus* L.), Emphasizing their Response to Light

C. PEKRUN, F. LÓPEZ-GRANADOS and P.J.W. LUTMAN

IACR Rothamsted, Harpenden, Herts, AL5 2JQ, UK

Abstract

Rape seed showed a range of responses to light. When tested under standard germination conditions, they were unresponsive to light but under unfavourable germination conditions they tended to be inhibited by light. After prolonged exposure to sub-optimal germination conditions in darkness rape seed exhibited light sensitivity. Their germinability in darkness was considerably reduced, whilst their germinability in light was high. This light sensitivity was subsequently lost. Seeds being transferred from darkness to light during the germination test were not able to react to light any more. So, during germination tests in darkness seeds developed skotodormancy. Generally, there was no change of germinability when rape seed were imbibed under sub-optimal germination conditions in light. In one cultivar light inhibition was imposed but seeds of this cultivar remained highly germinable in darkness as well. So, after imbibition in light rape seed were never dormant in darkness. A pot experiment carried out to test the conclusions of the previous laboratory experiments under a more natural environment confirmed the strong impact of the light environment on the ability of rape seed to persist in darkness. It also confirmed genotypic differences identified in the laboratory studies.

Introduction

Rape seed can be induced into secondary dormancy and thus persist in the soil for several years (Schlink, 1994). Emergence of rape seedlings can cause considerable weed problems (Lutman, 1993), not only in other crops but also in rape itself. In areas with cold winters, dense populations of volunteer plants can result in a low overwinter survival of the sown crop due to strong competition and resulting weak plants (Blanck, 1989). Rape from previous growing seasons can cause infestation problems when varieties with differing qualities are grown in one rotation, e.g. rape for industrial uses on the set-aside land and rape for human consumption in the normal rotation. Infestations of this kind are likely to become more frequent in the near future as a large range of rape types are being bred at the moment (Carruthers, 1995). Apart from the direct agronomic impact, long-term persistence of rape seed (up to 10 years: Sauermann, 1993) has to be taken into account in risk assessment of genetically modified rape.

The origin of volunteer seedlings IS seed shed before and during harvest. Seed losses of up to 500 kg/ha or about 10% of the yield seem to be quite common

R.H. Ellis, M. Black, A.J. Murdoch, T.D. Hong (eds.), Basic and Applied Aspects of Seed Biology, pp. 339–347.

(Lutman, 1993; Brown *et al.*, 1995; Price *et al.*, 1996). This equals about 10 000 seeds/m^2. If only a small proportion of these seeds becomes dormant a large number of seeds will be added to the soil seed bank by a single harvest. So, every attempt has to be made to reduce the number of seeds induced into secondary dormancy. Soil cultivation may have an impact on the induction of secondary dormancy. It influences the position of the seed in the soil and so the environmental conditions to which they are exposed. An important factor that is influenced by soil cultivation is the light environment. Studies were undertaken to test the effect of light environment on the induction of secondary dormancy in rape seed.

Materials and Methods

Laboratory Experiments

The general procedure of these experiments consisted of an incubation period under sub-optimal germination conditions and a subsequent germination test. The effect of the incubation period could be measured by comparing the germinability of untreated, dry stored seed with the germinability of treated seed. During the incubation period the seeds were exposed to either white light, far red light or darkness. The germination test was carried out in both white light and darkness. All work on seed potentially being light sensitive was carried out in a dark room under a green safety light.

Incubation Period

Either 50 or 100 seed, according to the experiment, were imbibed in 9 cm Petri dishes on 2 layers of Whatman No. 1 filter paper in either 8 ml of an aqueous solution of polyethylene glycol 6000 (PEG 6000), which generated a water potential of –15 bars (Michel and Kaufmann, 1973) or in demineralized water. To avoid evaporation losses the Petri dishes were wrapped in transparent polythene bags. They were then placed in a light and temperature controlled incubator providing 20°C and white light for 4 weeks or in a cold room at 6°C for 5 days (far red light experiment).

Germination Test

Treated or non-treated seed were placed in fresh 9 cm Petri dishes with 2 layers of filter paper and 7 ml of demineralized water. The Petri dishes were then placed in light and temperature controlled incubators providing white light at 12°C and 20°C. Seeds not germinating during the germination test were exposed to a 0.2% solution of gibberellic acid or they were stratified at 2 ± 2°C for 3 days and retested at 20°C in light.

Light Sources

White light was provided by 35 W incandescent lamps. At the level of the Petri dishes the photon fluence rate emitted by white light (330–1100 nm) was measured to range between 5 and 10 μmol m^{-2} s^{-1}, 25–28% of the radiation consisting of red light (600–700 nm) and 2–3% of far red light (720–780 nm). For the **far red light** a Black 901 Cryllex filter was used. The far red light transmitted through this filter had a photon fluence rate of 21 μmol m^{-2} s^{-1}. Seed were kept in **darkness** by placing Petri dishes in black polythene bags or by wrapping them into aluminium foil. The **green safety light** was created by filtering white light through a Green 139 Lee filter to produce light within wavelengths between 500 and 600 nm.

Greenhouse Experiment

Air dried and sieved (2 mm) loam soil (40% sand, 37% silt, 22% clay, 1% gravel) was mixed with 10% (by volume) of water ($\psi < -15$ bars) and left to equilibrate for 2 weeks. 16-cm pots were then filled with 1.3 kg of soil and 1000 seeds spread either at a depth of 5 cm or on the soil surface. To avoid moisture losses, the pots were wrapped in transparent polythene and placed in a greenhouse for 4 weeks at 20/16°C with a 16 h day. After that, soil cultivation and subsequent rain was simulated by thoroughly mixing the seed into the soil and heavily watering the pots. All seedlings that subsequently emerged were removed. When emergence was complete the pots were sieved and the number of persistent seeds counted.

Results

Laboratory Experiments

Germination of Untreated Seeds

Regular germination tests on stored seed of Bienvenu, Jet Neuf, Liglandor and Rubin showed that during storage from one harvest to the next, the germinability of the seeds did not change (data not shown). Between 98 and 100% of the stored seeds germinated, the germinability being equally high in light and in darkness, at 12°C and at 20°C. A slight inhibition by light was observed in one cultivar (Liglandor) when seeds were tested at 12°C (average germination percentage at 12°C in light: 95%).

Germination after Imbibition under Water Stress in Darkness or White Light

Rape seed imbibed in a PEG 6000 solution in darkness for 4 weeks were clearly light sensitive (compare initial 12D with 12L and 20D with 20L treatments in Fig. 1). Their germinability in light was still very high but their germinability in darkness was reduced. By transferring seeds from darkness to light little or no additional germination was recorded. Transferring them from 12°C to 20°C caused further seed to germinate. Full germination was achieved after a 3-day stratification treatment. Testing seeds in a 0.2% gibberellic acid-solution also resulted in full germination.

The effect of light environment on the germinability in darkness is shown in Figure 2. Seeds were imbibed in a PEG 6000 solution in either darkness or light and then tested for their germinability in darkness and light. Both treatments resulted in a high germinability in light and so results of this germination test are not presented. The germination test in darkness revealed a large difference between the two treatments: there was a high percentage of dormant seeds after imbibition in darkness and none, or almost none, after imbibition in light. These results were confirmed with a further 8 genotypes. Only one genotype developed secondary dormancy in the light. Seed of Liglandor reacted with a further imposition of light inhibition. After the treatment up to 35% of the seed were inhibited by white light (data not shown). Their germination in darkness remained high.

Seed transferred from light to darkness during imbibition in a PEG 6000

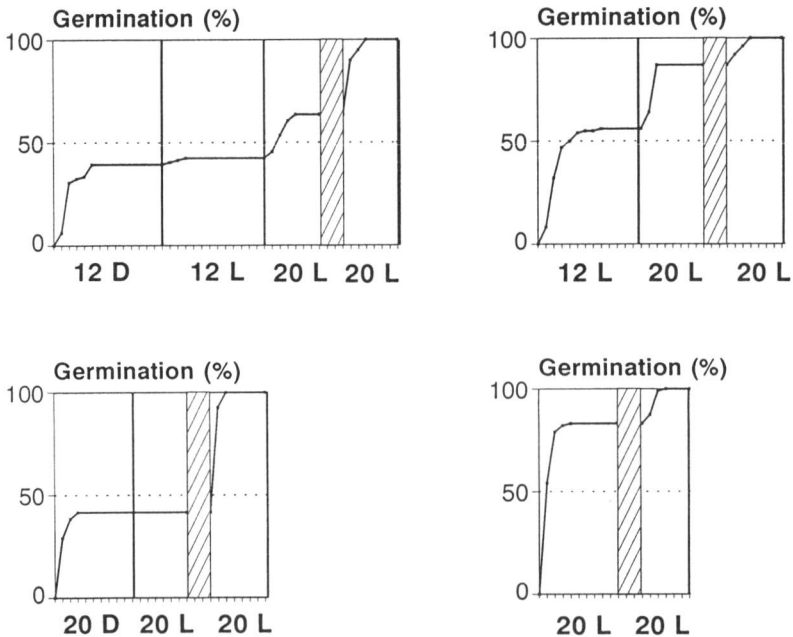

Figure 1. Germination after 4 weeks imbibition in a PEG 6000 solution ($\psi = -15$ bars) at 20°C in darkness. Seed was tested at 12°C darkness (12 D), 12°C light (12 L), 20°C darkness (20 D) or 20°C light (20 L). After the germination test the seed was stratified at 2 ± 2°C for 3 days (hatched bar) and retested at 20°C in light. Cv. Liglandor

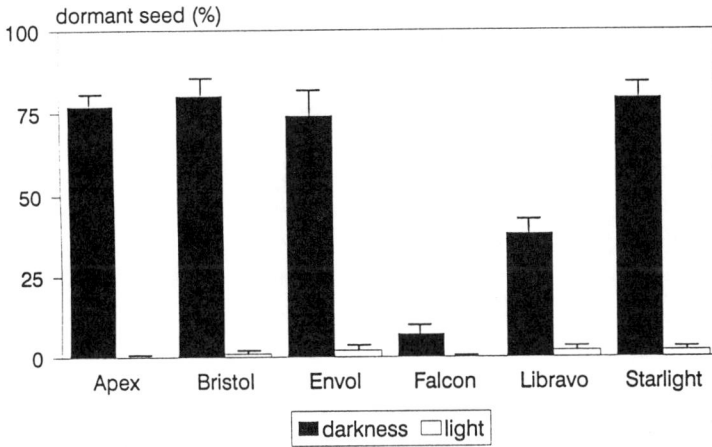

Figure 2. Percentage of dormant seed after 4 weeks imbibition in a PEG 6000 solution ($\psi = -15$ bars) at 20°C in either white light or darkness. Subsequent germination test: 20°C darkness. Error bars represent s.e.m., 5 replicates

solution showed decreasing percentages of dormant seed with increasing imbibition time in light prior to darkness (Fig. 3).

Germination after Imbibition in Far Red Light

Far red light was more effective in imposing dormancy than darkness (Table 1). Imbibition in water, instead of imbibition in a PEG 6000 solution, further increased the response. The same response was seen with both cultivars tested.

Greenhouse Experiment: Persistence in Relation to Position of the Seed in the Soil

Seed that had been lying on the soil surface for 4 weeks subsequently showed a low potential to persist. In contrast, seed that had been buried 5 cm deep showed some persistence (Fig. 4). So, during burial the seed had built up secondary dormancy. The varietal differences found in the parallel Petri dish experiment (see Fig. 2) were also reflected in this pot experiment.

Moisture contents differed between seed lying on and in the soil. During the simulated drought period, seed lying on the soil surface acquired a moisture content of 12%, seed buried in the soil reached a moisture content of 21%.

dormant seed (%)

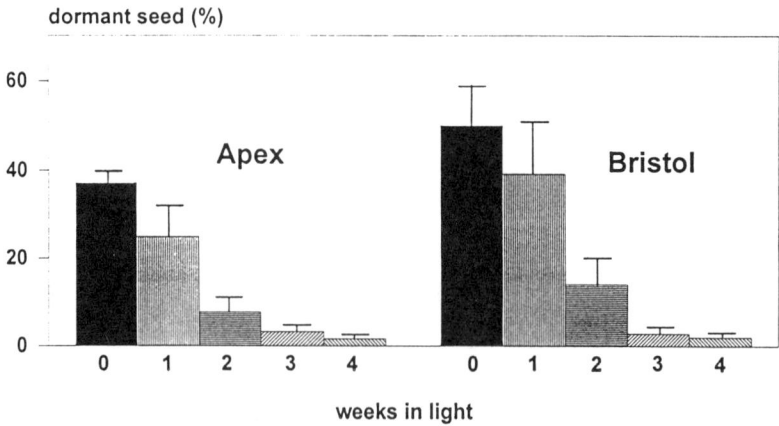

weeks in light

Figure 3. Percentage of dormant seed after 4 weeks of imbibition in a PEG 6000 solution ($\psi = -15$ bars) at 20°C in light and subsequently in darkness. Varying exposure to light and darkness: 0 weeks light/4 weeks darkness, 1 week light/3 weeks darkness, 2 weeks light/2 weeks darkness, etc. Subsequent germination test: 20°C darkness. Error bars represent s.e.m., 5 replicates

Table 1. Percentage of dormant seed after 5 days exposure to darkness or far red light at 6°C. Seed were either imbibed in a PEG 6000 solution ($\psi = -15$ bars) or in demineralized water. The following germination test was done at 12°C and 20°C in darkness (average shown). ( ) logit transformed data, s.e.d. = standard error of the difference

	PEG darkness	PEG far red light	H_2O far red light	s.e.d.
Falcon	1.8 (–2.11)	3.1 (–1.83)	16.4 (–1.13)	(0.078)
Libravo	7.6 (–1.49)	10.3 (–1.34)	17.8 (–0.92)	(0.112)

Discussion

Although rape seed, in general, are considered to be unaffected by light (ISTA, 1993), they can show quite a range of responses. Fresh or dry stored seed are highly germinable in light and in darkness, when they are tested under conditions of unlimited water and oxygen supply and at suitable temperatures. Under sub-optimal germination conditions, such as water stress (see Pekrun, 1994) or low temperatures, they tend to be inhibited by white light. Light inhibition under sub-optimal germination conditions has also been reported in rape seed by Bazanska and Lewak (1986). It has been observed in many species when the embryo is suffering some constraint (Bewley and Black, 1994).

After rape seed have been exposed to water stress or low temperatures in

persistent seed (%)

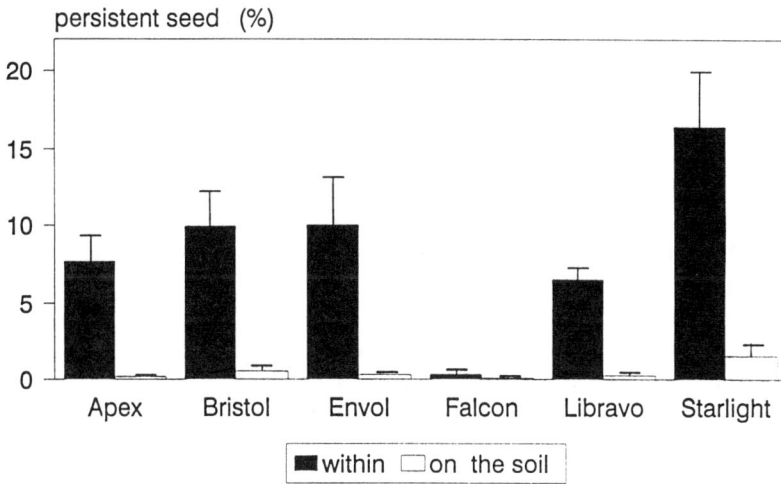

Figure 4. Percentage of persistent seed after 4 weeks exposure to a dry loam soil ($\psi < -15$ bars). Subsequently seed were incorporated into the soil by thorough mixing. Persistence was tested by watering the pots and subsequently assessing the number of non-germinated seed. Error bars represent s.e.m., 4 replicates

darkness for a prolonged amount of time they exhibit light sensitivity. Induction of light sensitivity is a very common feature in weed seed buried in the soil (Wesson and Wareing, 1969; Taylorson, 1972) and has also been assessed in rape seed recovered from the soil (Schlink, 1994). However, although most weed seed have primary dormancy at the time of shedding, rape seed are highly germinable. So, they need to experience some environmental constraint to build up light sensitivity. In several studies, it has been shown that prolonged exposure to water stress (Kahn, 1960; Khan and Karssen, 1980), unfavourable temperatures (Borthwick *et al.*, 1954) or an oxygen-deficient atmosphere (Vidaver and Hsiao, 1975), can lead to the induction of light sensitivity. Exposure to any of these factors is effective in rape seed (for imbibition under oxygen deficiency see Pekrun and Baeumer, 1992; Pekrun, 1994).

Sub-optimal germination conditions and darkness seem not to be the only requirements for the induction of light sensitivity. Another requirement seems to be an adequate water content of the seeds. Induction of light sensitivity is due to the dark thermal reversion of Pfr (Borthwick *et al.*, 1954). Maximal activity of the phytochrome system is linked to an optimum water content of the seed (Berrie *et al.*, 1974; Frankland, 1976; Samimy and Khan, 1983). This optimum water content might not have been reached during the five days treatment at 6°C in the far red light experiment, especially when the water uptake was further restricted by a PEG 6000 solution of –15 bars. Imbibition in water generated much higher percentages of dormant seed and might have resulted in a clearer difference between far red light and darkness, which would be expected if the phytochrome system is involved.

Although seed were light sensitive directly after the imbibition treatment in darkness, they quickly lost their ability to react to light during the germination test in darkness (Fig. 1). While exposed to non-restricting germination conditions in the dark, seed developed skotodormancy (see Bewley and Black, 1994). This phenomenon was found in all experiments and has also been identified in rape seed recovered from the soil after a longer period of persistence (Schlink, 1994; Pekrun, 1994).

Imbibition in light did not reduce the seeds' germinability in darkness. After four weeks of imbibition in a PEG 6000-solution and white light the seed of fourteen genotypes out of the fifteen tested did not change their germinability at all. Seed of the cultivar Liglandor, like those of *Rumex crispus* (Samimy and Khan, 1983), developed further light inhibition. Their germinability in darkness, however, remained high as well. So, rape seed never developed the ability to persist in the dark after prolonged exposure to sub-optimal germination conditions in the light.

Imbibition under water stress conditions for four weeks with varying exposure times to light and darkness revealed that the percentage of dormant seed decreased with increasing exposure to white light prior to darkness. The percentage of dormant seed was related to the time the seed were exposed to darkness. This is consistent with earlier studies where seed were exposed to sub-optimal germination conditions in darkness for various amounts of time and then were directly tested in a germination test (Pekrun and Baeumer, 1992; Pekrun, 1994). It can be explained by the time dependency of the dark thermal reversion of Pfr (Borthwick *et al.*, 1954).

Results generated in laboratory experiments seem to be quite useful for understanding the behaviour of rape seed in soil. The greenhouse experiment presented here clearly paralleled the results of the laboratory experiment. The only difference was the level of percentages of dormant seed. The percentages of dormant seed were much lower in the greenhouse experiment. This was possibly due to the very low water content of the seed during the simulated drought period. Seed in the soil acquired a water content of only 21%. In contrast, rape seed imbibing in a PEG 6000 solution with $\psi = -15$ bars at 20°C acquired a water content of 33% (Pekrun, 1994).

The ability to build up secondary dormancy clearly varied between genotypes. This shows that farmers can influence future volunteer problems by growing cultivars with a low dormancy potential. It also shows that breeding towards low dormancy genotypes would be possible.

Conclusions

As rape seed do not acquire the ability to persist in the soil after being exposed to light, shed seed should be left on the soil surface rather than buried into the soil. As the percentage of dormant seed increases with prolonged exposure to darkness, post-harvest cultivation should be delayed as long as possible. Research is needed to test whether seed lying under a thick layer of straw can acquire light sensitivity, as after rape harvest a large part of the field is covered with straw.

Acknowledgements

The authors gratefully acknowledge financial support by the Home Grown Cereals Authority and the Ministry of Agriculture, Fisheries and Food of the UK and the Spanish Ministry of Agriculture. IACR Rothamsted receives grant-aided support from the Biotechnology and Biological Science Research Council of the United Kingdom.

References

Bazanska, J. and Lewak, S. 1986. *Acta Physiologiae Plantarum* 8: 145–149.
Bewley, J.D. and Black, M. 1994. Seeds: *Physiology of Development and Germination*, Second Edition, pp. 445. New York, London: Plenum Press.
Blanck, K.-D. 1989. *Top-agrar Extraheft. Rapsanbau für Könner.* 94–98.
Borthwick, H.A., Hendricks, S.B., Toole, E.H. and Toole, V.K. 1954. *Botanical Gazette* 115: 205–225.
Brown, J., Erickson, D.A., Davis, J.B. and Brown, A.P. 1995. *Proceedings of the 9th International Rape Seed Congress.* 339–341.
Carruthers, S.P. 1995. *Proceedings of the 9th International Rapeseed Congress.* 1327–1331.
Frankland, B. 1976. In: *Light and Plant Development.* pp. 477–491 (ed. H. Smith). London: Butterworths.
ISTA (International Seed Testing Association). 1993. *Seed Science and Technology* 21: Supplement.
Kahn, A. 1960. *Plant Physiology* 35: 1–7.
Khan, A.A. and Karssen, C.M. 1980. *Plant Physiology* 66: 175–181.
Lutman, P.J.W. 1993. *Aspects of Applied Biology* 35: 29–36.
Michel, B.E. and Kaufmann, M.R. 1973. *Plant Physiology* 51: 914–916.
Pekrun, C. 1994. Ph.D. Thesis. University of Göttingen, Germany.
Pekrun, C. and Baeumer, K. 1992. *Mitteilungen der Gesellschaft für Pflanzenbauwissenschaften* 5: 83–86.
Price, J.S., Hobson, R.N., Neale, M.A. and Bruce, D.M. 1996. *Journal of Agricultural Engineering Research* (in press).
Samimy, C. and Khan, A.A. 1983. *Weed Science* 31: 153–158.
Sauermann, W. 1993. *Raps* 11: 82–86.
Schlink, S. 1994. Ph.D. Thesis. University of Göttingen. Germany.
Taylorson, R.B. 1972. *Weed Science* 20: 417–422.
Vidaver, W. and Hsiao, A.I. 1975. *Canadian Journal of Botany* 53: 2557–2560.
Wesson, G. and Wareing, P.F. 1969. *Journal of Experimental Botany* 20: 414–425.

39. The Hydrotime Concept in Seed Germination and Dormancy

K.J. BRADFORD

Department of Vegetable Crops, University of California, Davis, CA 95616–8631, USA

Abstract

The hydrotime concept has been developed to describe the relationship between water potential (ψ) and seed germination rates. In analogy with thermal time, hydrotime is the accumulated ψ units (MPa) above a base or threshold value (ψ_b) multiplied by the time of imbibition at that ψ. The germination rate (inverse of time to radicle emergence) increases linearly as the seed ψ increases above ψ_b. As ψ_b values vary among individual seeds, the amount by which the ψ exceeds ψ_b also varies among seeds, resulting in the typical right-skewed sigmoid germination time courses. Factors which influence seed germination kinetics alter the mean ψ_b of the seed lot, with higher (more positive) values resulting in an inhibition of germination or imposition of dormancy and lower (more negative) values being associated with rapid germination and high vigour. The induction or alleviation of dormancy involves physiological shifts in the ψ_b distribution, allowing a seed population to track both long-term and short-term environmental conditions to enhance seedling survival.

Introduction

A key question in germination physiology is how seeds integrate the signals from their environment to determine when to initiate radicle growth and commit to seedling development. This is a critical 'decision' for seeds, as the likelihood of seedling survival is dependent upon the subsequent availability of adequate water, temperature, light and nutrients to support plant growth. Given the large numbers of seeds produced by most species, the selection pressures of a seasonally predictable but locally chaotic environment, and both genetic and physiological variation in germination and dormancy characteristics, it is not surprising that a myriad of germination behaviours have evolved that are closely in tune with their native environments. It may not be possible, therefore, to identify a single physiological mechanism that can account for all of the exuberant individuality that is expressed in the germination ecology of diverse species. Nonetheless, it seems likely that water is so essential to successful seedling establishment that virtually all mesophytic and xerophytic species will have evolved mechanisms to gauge the water potential or content of their environment and attune their germination physiology to ensure that seedlings venture forth only if there is a high probability of adequate water being available for their initial establishment.

R.H. Ellis, M. Black, A.J. Murdoch, T.D. Hong (eds.), Basic and Applied Aspects of Seed Biology, pp. 349–360.
© *1997 Kluwer Academic Publishers, Dordrecht. Printed in Great Britain.*

In this paper, the hydrotime concept will be introduced as a unifying model to understand the patterns of germination that occur as seed populations enter and leave environmentally induced dormant or inhibited states. As has been pointed out by Gordon (1973), there is a characteristic pattern of germination timing as dormancy is lost in a seed population. There is almost always a relationship between the final germination percentage (i.e. the percentage of nondormant seeds) and the time required for radicle emergence to occur in the nondormant fraction. That is, as a dormancy-breaking treatment (afterripening, stratification, etc.) is extended, not only do additional seeds become capable of completing germination, but seeds which already have this capacity are able to complete germination more rapidly. Gordon (1973) proposed that there was a 'resistance to germination' that was still present even in 'nondormant' seeds (i.e. seeds that could eventually complete germination) which could be detected by variation in their germination rates. Thus, one could say that a dormancy continuum exists on both sides of the germinability divide. Among the nongerminable seeds, this is detectable by the different extents of dormancy-breaking treatments required to allow germination; among the germinable seeds, it is detectable by variation in the time required to complete germination. This paper will focus on describing how the hydrotime model can account for this general pattern of seed behaviour.

The Hydrotime Concept

The hydrotime concept was first proposed by Gummerson (1986) to account for the effects of reduced water potential (ψ) on germination. In analogy with thermal time or degree-days, Gummerson (1986) proposed that the time to germination is related to the magnitude of the difference between the seed or environment ψ and the water potential threshold for radicle emergence (ψ_b). In thermal time, the degrees in excess of a base or threshold temperature (T_b), multiplied by the time to a developmental event (e.g. radicle emergence) is a constant. Thus, many biological events can be normalized on a common degree-days time scale once the T_b is determined for the particular process. In the case of seeds, radicle emergence occurs at different times for different seeds, so the thermal time to germination must be defined for a particular germination percentage or fraction in the population (see Bradford, 1995). Interestingly, Gummerson (1986) showed that in the case of germination responses to reduced ψ, the total hydrotime (MPa-days) to radicle emergence was the same for all seeds in the population, but that individual seeds varied in their threshold ψ at which radicle emergence would be prevented. The following symbolism describes the basis of the hydrotime model (Bradford, 1990):

$$\theta_H = (\psi - \psi_b\,(g))\,t_g \qquad (1)$$

where θ_H is the hydrotime constant (MPa-days), ψ is the actual seed water

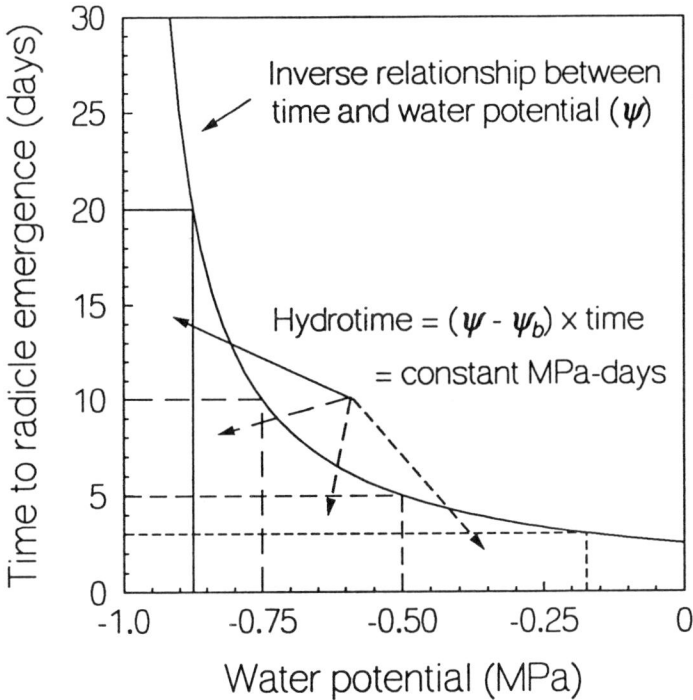

Figure 1. Relationship between time to germination and water potential (ψ). Assuming a base or threshold ψ (ψ_b) of –1.0 MPa, the curve shows the inverse relationship between the time to radicle emergence (t_g) and $\psi - \psi_b$. The rectangles illustrate that the same total hydrotime (θ_H, MPa-days) is required for germination in all cases

potential, ψ_b *(g)* is the base or threshold water potential defined for a specific germination fraction g, and t_g is the time to radicle emergence of fraction g. If θ_H is a constant, then t_g must increase proportionately as ψ is reduced and approaches $\psi_b(g)$. This is illustrated graphically in Figure 1, where the time to radicle emergence (t_g) is inversely related to the ψ of the seed. According to Equation 1, the rectangles defined by any combination of t_g and ψ enclose equal areas (= θ_H). Thus, radicle emergence occurs when a given seed has accumulated the hydrotime units characteristic of that seed lot. This will take longer at low ψ than at high ψ, but the relationship between $\psi - \psi_b$ *(g)* and t_g conforms to that shown for all seeds in the population.

Equation 1 can be rearranged in the following way to illustrate the relationship between germination rates, or the inverse of time to radicle emergence ($GR_g = 1/t_g$), and ψ:

$$GR_g = 1/t_g = (\psi - \psi_b(g)) / \theta_H . \qquad (2)$$

Thus, a plot of GR_g versus ψ gives a straight line with a slope of $1/\theta_H$ and an intercept on the ψ axis equal to $\psi_b(g)$ (Fig. 2A). If this relationship is plotted for different germination percentages (e.g. Gummerson, 1986), a series of parallel lines are developed with a common slope $(1/\theta_H)$ but different intercepts $(\psi_b(g))$. Experimentally, it has been found that in most cases the $\psi_b(g)$ values vary among seeds in the population in a normal or Gaussian distribution (e.g. Gummerson, 1986; Bradford, 1990; Dahal and Bradford, 1990). One can therefore plot the relative frequency of a given ψ_b value in the seed population as a normal bell curve, which can be defined by its mean $(\psi_b(50))$ and standard deviation $(\sigma_{\psi b})$ (Fig. 2B). The term $\psi_b(g)$ represents this distribution of threshold values within the seed population.

How do $\psi_b(g)$ values relate to the times to radicle emergence? According to Equation 1, $\psi - \psi_b(g)$ (referred to as $\Delta\psi(g)$) multiplied by the time to germination of percentage g (t_g) is a constant (θ_H). In Figure 3B, $\Delta\psi(g)$ is indicated by the lengths of the horizontal arrows between the seed ψ (0 MPa in this case) and the $\psi_b(g)$ value of different germination fractions. In Figure 1A, the times to germination of different percentages are similarly indicated by horizontal arrows. The arrows connecting panels A and B in Figure 3 show that specific threshold values corresponding to the 10th, 50th, or 90th percentiles are directly related to the times of radicle emergence for those same percentages. The products of the lengths of the $\Delta\psi(g)$ arrows in panel B multiplied by the lengths of the t_g arrows in panel A for each value of g are all identical and equal to θ_H. Thus, as the seed ψ approaches the $\psi_b(g)$ value of a particular seed, the time to radicle emergence lengthens to maintain a constant total hydrotime. This is illustrated in Figure 3C, where the seed ψ has been reduced to –0.25 MPa for a seed population with a $\psi_b(50)$ of –0.5 MPa. For the median seed, the time to radicle emergence will be exactly doubled, since the value of $\Delta\psi(50)$ has been reduced by half (Fig. 3A). However, the time to 10% radicle emergence will be less than doubled, while the time to 90% will be more than tripled, since a given reduction in seed ψ has differential effects on $\Delta\psi(g)$ depending upon the specific threshold values. As shown in Figure 3C, $\Delta\psi(90)$ is now quite small, and a correspondingly long time will be required for radicle emergence of the 90th percentile of seeds (Fig. 3A). In addition, the maximum germination percentage attained will be less than 100%, since a portion of the $\psi_b(g)$ distribution extends above –0.25 MPa (hatched area under curve in Fig. 3C). This represents seeds whose germination thresholds are higher than –0.25 MPa, and which therefore will not ever germinate at that water potential. If the ψ was reduced further, the final germination percentage would decrease (more of the threshold distribution would exceed the ψ value), and the times to germination of the remaining seeds would increase such that the total hydrotime (i.e. $\Delta\psi(g) \times t_g$) remained constant. In this way, the hydrotime model simultaneously accounts for both the timing and the final germination percentages of seed populations in response to changes in ψ.

An additional advantage of the hydrotime model is that germination time courses at different ψ can be normalized on a common time scale, just as

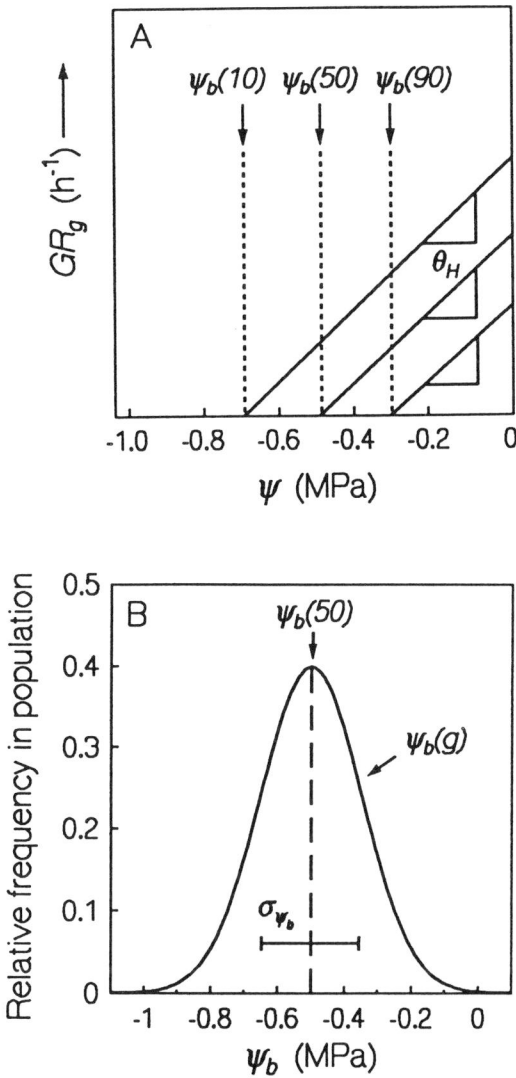

Figure 2. **A.** Germination rates ($GR_g = 1/t_g$) as a function of ψ for different germination fractions (g). Germination rates for a given fraction increase linearly with ψ above the threshold value ($\psi_b(g)$), which varies among seed fractions. The slope, however, is constant for all fractions and is equal to $1/\theta_H$. **B.** A normal distribution of ψ_b values, characterized by the mean ($\psi_b(50)$) and standard deviation ($\sigma_{\psi b}$). The relative frequency of occurrence of a particular ψ_b value is indicated by the bell-shaped curve, which is symbolized by $\psi_b(g)$

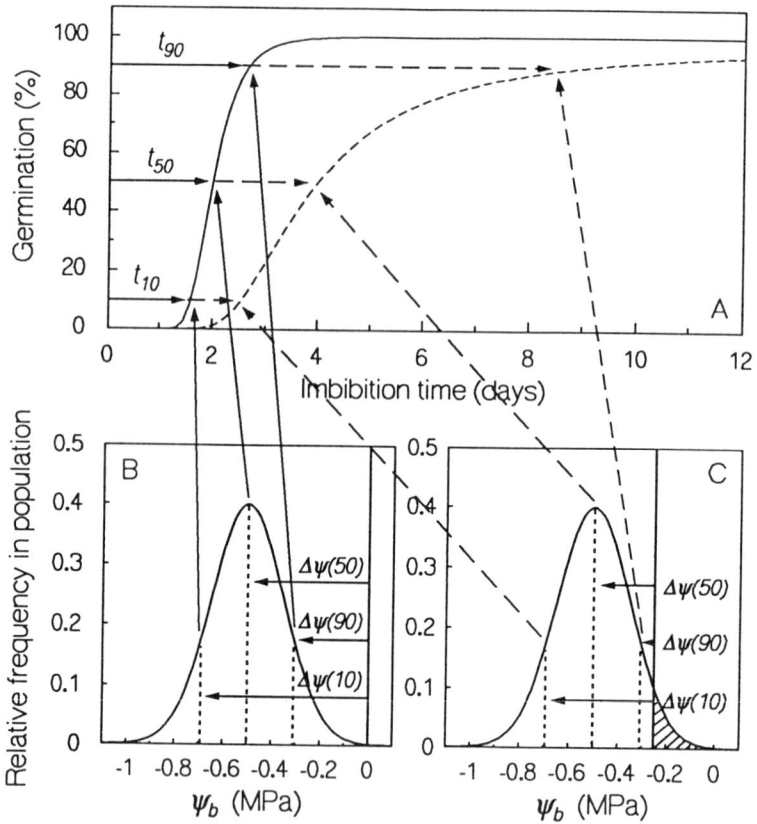

Figure 3. **A.** Germination time courses for a seed population at 0 MPa (solid curve) or at –0.25 MPa (dashed curve). The times to germination of 10, 50, and 90% of the seeds are indicated by the horizontal arrows in each case. **B.** The $\psi_b(g)$ distribution on which the time course at 0 MPa in panel A is based. The vertical line at 0 MPa indicates the ψ of the seed. The horizontal arrows indicate the values of $\psi - \psi_b(g)$, symbolized as $\Delta\psi(g)$ for the 10th, 50th and 90th percentiles. The arrows connecting panels A and B illustrate when particular seed fractions in the $\psi_b(g)$ distribution would germinate. In each case, the product of $\Delta\psi(g)$ times t_g is a constant equal to θ_H. **C.** The $\psi_b(g)$ distribution on which the time course at –0.25 MPa in panel A is based. The $\psi_b(g)$ distribution is identical to that in panel B, but the vertical line now indicates that the seed ψ is –0.25 MPa. Other details as for panel B. Note the greater effect on the timing of the last seeds to germinate compared to the effect on the first seeds to germinate, resulting in increased skewness in the time course curve. The hatched area under the $\psi_b(g)$ distribution curve in panel C represents seeds whose germination thresholds exceed the seed ψ, and which therefore will not complete radicle emergence at this ψ. Note that the same effect as occurred when the seed ψ was reduced by 0.25 MPa could be achieved by shifting the $\psi_b(g)$ distribution 0.25 MPa more positive

biological processes at a range of temperatures can be normalized on a common thermal time scale using degree-days. Using the parameters from the hydrotime model, germination time courses at any ψ can be normalized to the time course in water for that seed population. The relationship between the time to germination in water $(t_g(0))$ and the time to germination at any other ψ $(t_g(\psi))$ has been derived previously (Bradford, 1990) as:

$$t_g(0) / t_g(\psi) = 1 - (\psi / \psi_b(g)) . \qquad (3)$$

When $\psi = 0$ MPa, the right side of Equation 3 will equal 1, and $t_g(0) = t_g(\psi)$, as expected. When $\psi = \psi_b(g)$, then the right side of Equation 3 will equal zero, or $t_g(\psi)$ essentially becomes infinite. This is exactly what is meant by ψ_b: the ψ at which germination does not occur (or takes infinitely long). At any ψ between 0 MPa and $\psi_b(g)$, the time to germination at that ψ is increased relative to that in water by the factor $1 - (\psi / \psi_b(g))$. Thus, once $\psi_b(g)$ has been characterized for a seed lot (i.e. its mean and standard deviation are known), the germination time course at any ψ can be calculated relative to the time course in water for those same seeds. Alternatively, rearranging Equation 1,

$$t_g(0) = t_g(\psi) [1 - (\psi / \psi_b(g))] , \qquad (4)$$

shows that the time to germination at any ψ can be corrected back to that in water by multiplying by the bracketed factor. Conceptually, the ability to normalize germination time courses on a hydrotime basis indicates that seeds at a reduced ψ are progressing toward radicle emergence at the same pace as those in water, *given the amount of hydrotime they are accumulating per unit actual time.* Just as germination takes longer at a lower temperature, it also takes longer as ψ approaches the $\psi_b(g)$ threshold. Referring to Figure 3A, one can visualize that the curve for –0.25 MPa would result if the time axis for the curve at 0 MPa was simply stretched in inverse proportion to $\Delta\psi(g)$. As $\Delta\psi(g)$ decreased, the axis-stretching would increase proportionately, and this would result in the germination curve shown. Space does not permit further illustration of this normalizing function here, but a number of examples have been published previously (Bradford, 1990; Dahal and Bradford, 1990; 1994; Bradford and Somasco, 1994). The normalized time courses give a direct visual indication of the degree of fit of the hydrotime model to the data, and can identify data that deviate from expectations. The latter often indicates an alteration in physiology, as distinct from the delay anticipated simply from the lower rate of hydrotime accumulation at reduced ψ (e.g. Ni and Bradford, 1992; Dahal and Bradford, 1994).

Hydrotime and Dormancy

Seed dormancy is a complex phenomenon, and many factors undoubtedly contribute to the germination patterns and environmental responses observed among diverse plant species (Fenner, 1992; Kigel and Galili, 1995). However, as illustrated above, the hydrotime model can reproduce the types of germination time courses that are almost universally encountered as seeds enter or leave dormancy. That is, there is generally an increase in both final germination percentage and speed of germination as a seed population loses dormancy. In the hydrotime model, this pattern is an automatic consequence of the variation in threshold values among individual seeds and the relationship of the threshold distribution to the current seed ψ. It is significant to note that as far as the model is concerned, there is no distinction between a constant threshold distribution and a change in ψ, or a constant ψ with a change in the threshold distribution. For example, in comparing Figures 3B and 3C, exactly the same germination time course would result if the mean of the threshold distribution had been increased by 0.25 MPa as occurred when the ψ was decreased by 0.25 MPa. The model is sensitive only to $\Delta\psi(g)$, and not to the absolute values of either ψ or $\psi_b(g)$. If we visualize progressively shifting the $\psi_b(g)$ distribution to higher values, as the $\psi_b(g)$ values of part of the population exceed 0 MPa, those seeds would no longer be able to complete radicle emergence on water. This is at least a functional definition of dormancy, that seeds do not germinate on water when given otherwise suitable conditions. In a colloquial sense, we can consider that water is not 'wet' enough for those seeds to complete radicle emergence. We can therefore hypothesize that seed dormancy might be related to $\psi_b(g)$ values exceeding 0 MPa, and that as dormancy is broken by the appropriate environmental or hormonal signals, the $\psi_b(g)$ values of the entire population might shift toward progressively lower values. This would result in a series of germination time courses that become simultaneously more rapid and more complete, as is the case for the vast majority of dormancy data in the literature (Gordon, 1973).

This hypothesis has been tested experimentally for only a few cases thus far. Detailed studies of gibberellin- and abscisic acid-deficient tomato genotypes demonstrated that the effect of these hormones was to shift $\psi_b(g)$ distributions to more negative values when germination is promoted (i.e. in the presence of GA or absence of ABA) and to more positive values when germination is inhibited (in the presence of ABA or absence of GA) (Ni and Bradford, 1993). In the case of lettuce seed thermodormancy (or thermoinhibition), as the temperature increases toward the upper temperature limit for germination, the $\psi_b(g)$ distribution shifts progressively to more positive values, eventually reducing germination percentages (Bradford and Somasco, 1994). Ethylene is able to increase the upper temperature limit of lettuce seed germination, and it appears to act by maintaining more negative $\psi_b(g)$ values as temperatures increase (Dutta and Bradford, 1994). It is also possible to model on a theoretical basis germination time courses for a wide range of seed dormancy situations. Data from Perino and Côme (1977) can be used as an illustration of how the hydrotime model can accommodate time courses of germination as dormancy is alleviated. Apple seeds germinated poorly if imbibed directly at 30°C, but if first

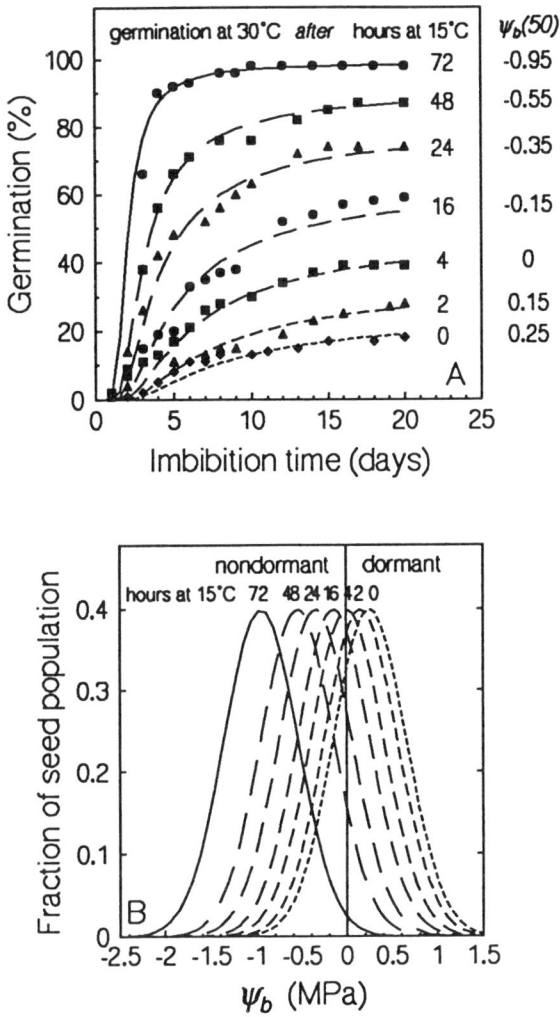

Figure 4. **A.** Escape of apple seeds from thermoinhibition at 30°C after increasing durations of preimbibition at 15°C. The symbols are the actual data taken from Perino and Côme (1977). The smooth curves are the time courses predicted by the hydrotime model based upon the $\psi_b(g)$ distributions shown in panel B. The $\psi_b(50)$ values (MPa) used to generate the predicted curves are indicated to the right of panel A. **B.** Theoretical $\psi_b(g)$ distributions used to generate the germination time courses in panel A. In each case, only the mean threshold ($\psi_b(50)$) was changed, with the hydrotime constant (θ_H) and the standard deviation ($\sigma_{\psi b}$) of the distribution remaining constant. The progressive negative shift in the $\psi_b(g)$ distribution can account for the entire pattern of germination responses, including both the rates and final percentages

imbibed at 15°C for increasing durations, they became capable of germinating more rapidly and to a higher percentage when transferred to 30°C (Fig. 4A). The symbols in Figure 4A are taken from from Perino and Côme (1977), but the solid lines are generated from the hydrotime model under the assumption that the effect of preincubation at 15°C is to progressively shift the $\psi_b(g)$ distribution to more negative values (Fig. 4B). That is, it is assumed that the standard deviation (width) of the $\psi_b(g)$ distribution is constant, and that the hydrotime constant (θ_H) also does not change. Simply shifting the mean of the $\psi_b(g)$ distribution to progressively lower values is sufficient to generate the complete family of time courses which match closely to the original data (Fig. 4A). Similar illustrative examples and additional discussion have been presented elsewhere (Bradford, 1995; 1996). While such theoretical modelling does not prove that $\psi_b(g)$ distributions actually change as predicted, the fit to the data is certainly suggestive, and the hypothesis is readily testable by measuring the ψ sensitivity of germination of seed populations in various dormancy states.

An Ecological Interpretation of the Hydrotime Model

If the hydrotime model provides an accurate description of germination responses to dormancy-inducing and dormancy-breaking factors, what are the implications for seed ecology? In many environments, seasonal or erratic rainfall is the primary determinant of seedling survival, often associated with a dry afterripening requirement or high temperature sensitivity (Kigel, 1995). In other cases, avoiding low temperatures and the danger of frost may be more important, requiring seeds to delay germination until after experiencing a cold period (Fenner, 1995; Egley, 1995; Benech-Arnold and Sánchez, 1995). According to the hydrotime model (or its extension to hydrothermal time; Gummerson, 1986; Dahal and Bradford, 1994), a major physiological response to these environmental signals is a shift in the $\psi_b(g)$ distribution of the seed population to more negative values. This has the effect of increasing the probability that a given seed will initiate and complete germination while sufficient water is available in its immediate environment. The linkage between water potential thresholds and germination rates is important, since it may be critical for seeds to germinate rapidly in environments where soil surfaces are alternately wet and dry (Allen *et al.*, 1993). As the $\psi_b(g)$ distribution shifts to more negative values, the rates of germination of all the seeds are increased. On the other hand, it has been pointed out that from an ecological perspective, slow germination is as effective as dormancy in preventing germination at unfavorable times in a fluctuating environment (Meyer and Monsen, 1991), as seeds are generally capable of tolerating dehydration if radicle emergence has not or has only recently occurred (Finch-Savage and McKee, 1989; Bruggink and van der Toorn, this volume). Maintaining a wide distribution of $\psi_b(g)$ values within the seed population assures that while some seeds will capitalize rapidly on favorable conditions, others will be more conservative, committing to radicle emergence only after a much longer period or not at all. Finch-Savage and Phelps (1993) and Allen *et al.* (this volume) have demonstrated convincingly that successive flushes of seedling emergence in the field following rainfall events can be described on the basis of water potential and temperature thresholds that allow fractions of the total seed population to germinate within specific time windows. Similarly, seasonal patterns of dormancy cycling controlled by temperature result in an increasing or decreasing likelihood that germination will be completed within a given hydrothermal period (Bouwmeester and Karssen, 1992; Benech-Arnold and Sánchez, 1995).

There is an ecological rationality to having various environmental signals influence germination capacity via effects on $\psi_b(g)$. As seasonal and environmental requirements are met which indicate to the seed that an opportune time to germinate is approaching (e.g. afterripening, stratification, light, nutrients, etc.), a negative shift in $\psi_b(g)$ will result in an increase in the fraction of seeds capable of germination and in the overall speed of germination. However, since this increased capability is based upon the water potential thresholds, the seed population will still remain highly sensitive to the current local water availability. Even if the $\psi_b(g)$ distribution shifts to quite low values in a physiological sense (–1 to –2 MPa), a relatively small decline in soil ψ can still have a dramatic effect on germination timing and percentage (c.f. Fig. 3). Thus, seasonal or environmental effects on the capacity for germination may act through physiological shifts in the $\psi_b(g)$ distribution, with the variation in $\psi_b(g)$ providing differential sensitivity to local conditions to ensure that there are both opportunistic and conservative individuals within the population. By responding to environmental factors via modification of their sensitivity to ψ, seed populations can achieve both long-term integration of their environmental (and evolutionary) history and regulation of their progress toward germination based upon current water availability.

References

Allen, P. S., Debaene, S. B. G. and Meyer, S. E. 1993. In: *Fourth International Workshop on Seeds: Basic and Applied Aspects of Seed Biology*, vol. 3, pp. 387–392 (eds. D. Côme and F. Corbineau). Paris: Association pour la Formation Professionelle de l'Interprofession Semences.

Benech-Arnold, R. L. and Sánchez, R. A. 1995. In: *Seed Development and Germination*, pp. 545–566 (eds. J. Kigel and G. Galili). New York: Marcel Dekker, Inc.

Bouwmeester, H. J. and Karssen, C. M. 1992. *Oecologia* 90: 88–94.

Bradford, K. J. 1990. *Plant Physiology* 94: 840–849.

Bradford, K. J. 1995. In: *Seed Development and Germination*, pp. 351–396 (eds. J. Kigel and G. Galili). New York: Marcel Dekker, Inc.

Bradford, K. J. 1996. In: *Plant Dormancy: Physiology, Biochemistry and Molecular Biology*, in press (ed. G. A. Lang). Wallingford, Oxon, U. K.: CAB International.

Bradford, K. J. and Somasco, O. A. 1994. *Seed Science Research* 4: 1–10.

Dahal, P. and Bradford, K. J. 1990. *Journal of Experimental Botany* 41: 1441–1453.

Dahal, P. and Bradford, K. J. 1994. *Seed Science Research* 4: 71–80.

Dutta, S. and Bradford, K. J. 1994. *Seed Science Research* 4: 11–18.

Egley, G. H. 1995. In: *Seed Development and Germination*, pp. 529–543 (eds. J. Kigel and G. Galili). New York: Marcel Dekker, Inc.

Fenner, M. 1992. *Seeds: The Ecology of Regeneration in Plant Communities*. Wallingford, Oxon, UK: CAB International.

Fenner, M. 1995. In: *Seed Development and Germination*, pp. 507–528 (eds. J. Kigel and G. Galili). New York: Marcel Dekker, Inc.

Finch-Savage, W. E. and McKee, J. M. T. 1989. *Annals of Applied Biology* 114: 587–595.

Finch-Savage, W. E. and Phelps, K. 1993. *Journal of Experimental Botany* 44: 407–414.

Gordon, A. G. 1973. In: *Seed Ecology*, pp. 391-409 (ed. W. Heydecker). London: Butterworths.

Gummerson, R. J. 1986. *Journal of Experimental Botany* 37: 729–741.

Kigel, J. 1995. In: *Seed Development and Germination*, pp. 645–699 (eds. J. Kigel and G. Galili). New York: Marcel Dekker, Inc.

Kigel, J. and Galili, G. 1995. *Seed Development and Germination*. New York: Marcel Dekker, Inc.

Meyer, S. E. and Monsen, S. B. 1991. *Ecology* 72: 739–742.

Ni, B.-R. and Bradford, K. J. 1992. *Plant Physiology* 98: 1057–1068.

Ni, B.-R. and Bradford, K. J. 1993. *Plant Physiology* 101: 607–617.

Perino, C. and Côme, D. 1977. *Physiologie Végétale* 15: 469–474.

40. A Statistical Perspective on Threshold Type Germination Models

K. PHELPS and W.E. FINCH-SAVAGE

Horticulture Research International, Wellesbourne, Warwick CV35 9EF, UK

Abstract

A need for empirical germination models that would give reliable predictions under field conditions led us to look critically at the statistical assumptions made in the concepts of thermal and hydrothermal time. Doubts were raised about the unquestioning description of rate temperature relationships by straight lines and the subsequent extrapolation of base temperatures from them. An alternative method where base temperatures are estimated from final percentage germination is suggested. This removes the linearity constraint from the rate temperature relationship and should have better predictive properties at low temperatures. An advantage of the method is that it predicts both the timing and percentage of seeds which will germinate in a given time interval. Thus it provides a unifying hypothesis which may aid interpretation of results from alternating temperature studies. Generalization of the method to include water potential is described and the methods are illustrated with several data sets.

Introduction

Studies on the prediction of seedling emergence patterns (Finch-Savage and Phelps, 1993) identified a need for germination models that could be applied under field conditions. Traditionally prediction of the timing of seed germination, as with many other biological processes, has been based on thermal time requirements estimated from constant temperature experiments. More recently the concept of hydrothermal time (Gummerson, 1986) has been proposed to incorporate the effect of water availability. Central to the concepts of both thermal time and hydrothermal time is the idea that each seed has a base temperature and water potential below which the germination process cannot proceed. Estimates of base temperature have been used to compare the performance of different cultivars (Scott *et al.*, 1984) and the effect of applying different treatments. Gummerson (1986) has suggested that differences in germination time amongst sugar beet seeds are due to variability in base water potential rather than base temperature. Bradford (1990) observed that base water potentials must vary within seed populations because final germination percentages are lowered as water potential is reduced.

Use of thermal and hydrothermal time for predicting the timing of germination under field conditions with fluctuating temperature and water potential provides a severe test of the models. In particular, the low temperature

R.H. Ellis, M. Black, A.J. Murdoch, T.D. Hong (eds.), Basic and Applied Aspects of Seed Biology, pp. 361–368.
© *1997 Kluwer Academic Publishers, Dordrecht. Printed in Great Britain.*

behaviour of models is important when early season crops are sown as temperatures are well below the optimum for germination. This led us to look more critically at the assumptions made in thermal and hydrothermal time and to suggest some alternative assumptions which are more acceptable statistically and may be more plausible biologically.

Examples of a Different Approach to the Concept of Thermal Time

Cabbage (*Brassica oleracea*) seeds were placed to germinate on moist absorbent filter paper under conditions of constant temperatures between 1°C and 35°C. Germination was recorded at regular intervals.

Figure 1 shows some of the resulting germination curves. To estimate thermal time requirements, the time to 50% germination (t_{50}) of viable seeds was estimated from all the curves and the reciprocal ($1/t_{50}$), representing the rate of germination of the 50th percentile, was plotted against the appropriate temperatures (Fig. 2).

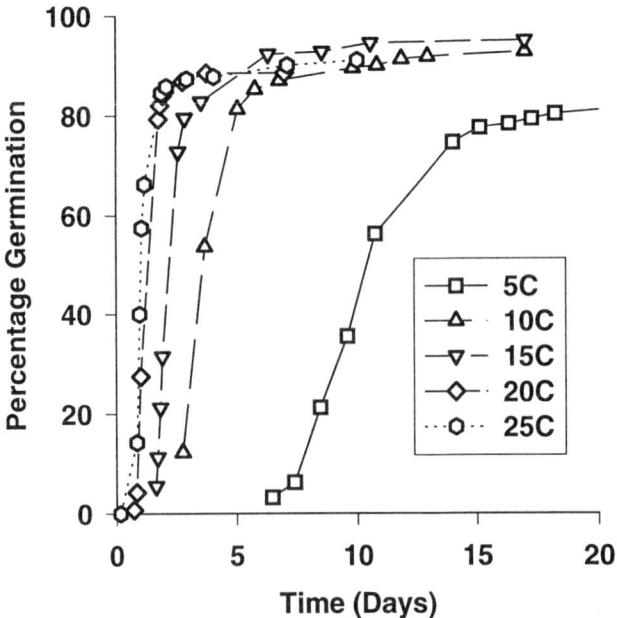

Figure 1. Cabbage. Distribution of germination times at constant temperatures

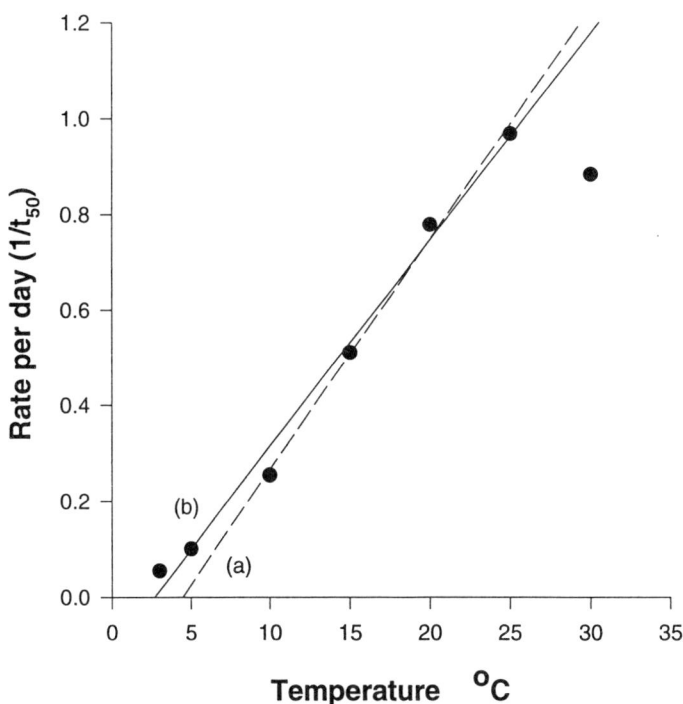

Figure 2. Cabbage. Germination rate $(1/t_{50})$ at all temperatures. Regression lines are for (a) all suboptimal temperatures (b) range 5°C to 25°C

Linear regression was used to find the equation of this line for the suboptimal $(< 30°C)$ range:

$$r = b_0 + b_1 T$$

where r is rate, T is temperature and b_0 $(= -0.0128)$, b_1 $(= 0.0164)$ are parameters.

From this equation we can calculate the base temperature $T_b = -b_0/b_1$ and the day degree (i.e. time above T_b) requirements $= 1/b_1(T-T_b)$.

The base temperature is a very poor estimate in the statistical sense. There are two reasons for this: 1) the line does not fit the data perfectly 2) it is an extrapolation. The line will never be a perfect fit because of the error involved in estimating the t_{50} from the germination curves but the extrapolated base temperature is very sensitive to the exact line fitted. For instance, if the 3°C and 5°C temperatures had not been included in the experiment a different regression line would have been obtained and the estimate of base temperature would have changed from 4.5°C to 2.7°C. This difference could be important in early season

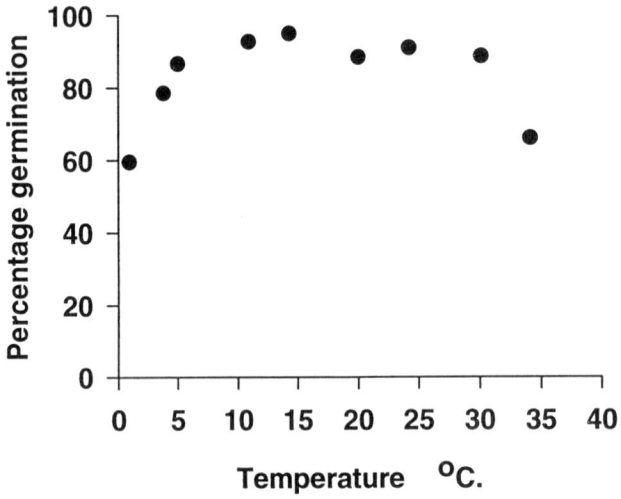

Figure 3. Cabbage. Final percentage germination at all temperatures

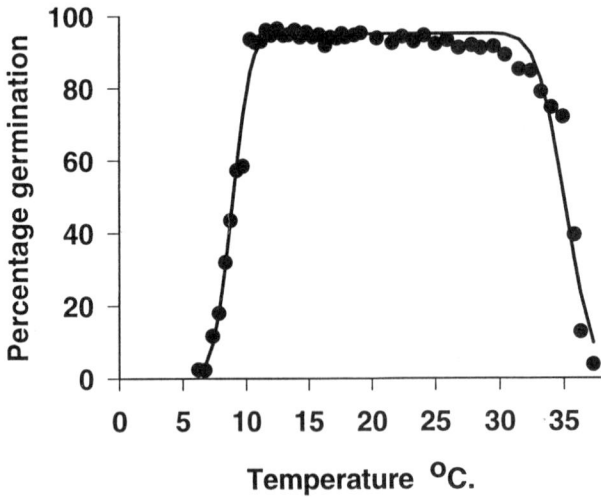

Figure 4. Tomato (Labouriau and Osborn, 1984). Final percentage germination. Fitted line is probit line adjusted for initial non-viability

sowings when temperatures are generally low.

One means of avoiding extrapolation would be to measure the base temperature directly. This would necessitate testing in a sequence of constant temperatures to find the highest temperature at which seeds failed to germinate. However the cabbage data (Fig. 3) show the flaws in this method. There is no one temperature at which the process stops. In cabbage, percentage germination is progressively reduced at both low and high temperatures.

A more complete example of this phenomenon (Fig. 4) was reported by Labouriau and Osborn (1984).

Figures 3 and 4 suggest that there is a range of base temperatures among the seeds. Furthermore the distribution of these bases can be inferred from the percentages germinated at each temperature (Murdoch *et al.*, 1989) using the technique of probit analysis (Finney, 1971).

Figure 3 suggests that more than half of the seeds have a base temperature less than 1°C but bases for some seeds may be as high as 5°C or 6°C. This suggests that the earlier approach is flawed and that we should make predictions for several subpopulations with different bases. This approach would also remove the constraint that the rate temperature relationship must be linear in order for a base to be estimated. Thus we may prefer to describe the data in Figure 2 by a curve where the rate at low temperatures gets close to zero but never actually reaches it. Use of curves rather than straight lines to describe rate temperature relationships is well-established in the entomological literature (e.g. Worner, 1992).

Although prediction under fluctuating temperatures using a curve is not quite as simple as calculating day degrees, it is not a major operation. Formally, we integrate the rate equation starting from the moment of sowing. Germination is assumed to occur when the integrand reaches a value of 1. In practice we use estimates of temperatures at hourly intervals and sum the rates (per hour) predicted by the curve for that temperature. The rates are accumulated until they reach a total of 1. To incorporate the base threshold the rate is set to 0 when the temperature falls below the base. If there is a range of bases, the calculation can either be made for several subpopulations each with a different base or by sampling bases at random from a normal distribution.

Application to Hydrothermal Time

Mathematically hydrothermal time (Gummerson, 1986) is an extension of thermal time. It can be expressed in the form:

$$r(T,\psi) = b(\psi - \psi_b)(T - T_b)$$

where $r(T,\psi)$ is the rate of germination at temperature T, potential ψ. It is based on the assumption that, at any constant temperature, there is a linear relationship between rate and water potential above a base water potential (ψ_b).

Table 1. Onion. Maximum percentage germination (from 200 seeds)

Water potential (MPa)	Temperature (°C)				
	5	10	15	20	25
0	91.5	89.0	90.5	96.0	88.0
-0.2	93.0	80.0	87.0	88.5	57.0
-0.4	81.5	81.0	88.5	87.5	8.5
-0.6	48.5	86.0	82.5	71.0	4.5
-0.8	11.5	76.0	49.0	29.0	1.0
-1	1.5	27.5	20.0	3.5	1.0

Table 2. Rate of germination $(100/t_{50})$ for regimes where more than 50% germinated. Figures in parentheses show predictions from the hydrothermal time model fitted at suboptimal temperatures

Water potential (MPa)	Temperature (°C)				
	5	10	15	20	25
0	8.0 (7.9)	14.3 (15.8)	25.0 (23.8)	33.3 (31.8)	28.6
-0.2	6.9 (6.1)	11.1 (12.3)	16.7 (18.5)	25.0 (24.6)	15.4
-0.4	4.3 (4.3)	9.1 (8.7)	12.5 (13.1)	15.4 (17.5)	–
-0.6	3.2 (2.6)	7.7 (5.2)	10 (7.8)	9.5 (10.4)	–

Conversely, for any constant water potential, there is a linear relationship between rate and temperature above a base, T_b.

The data from a further onion experiment involving a range of temperatures and water potentials suggests that the use of explicit bases could be extended to the study of water availability.

Table 1 shows the final percent germination obtained in each treatment regime. At each temperature percentage germination decreased as water potential decreased. However, the pattern of decrease was different at each temperature. As the temperature moved away from an optimal value between 10°C and 15°C the germination process became more sensitive to water availability: the water potential at which 50 percent of seeds germinated ranged from approximately –0.2 MPa at 25°C to below –0.8 MPa at 10°C. Conversely as the water potential decreased from 0 MPa to –0.8 MPa the optimal temperature for germination dropped from 20°C to 10°C.

The lack of germination made it difficult to measure germination rates from many of the temperature/water potential combinations. Table 2 shows the measurable rates along with a hydrothermal time relationship fitted to the suboptimal temperatures. Although this model describes the main features of the data, there are regions of systematic error and there is some indication that the temperature optimum decreases as water availability decreases. It may be preferable, therefore, to use curves to describe the rates. The use of separate thresholds removes any constraint on the type of curve that may be used.

Discussion

The approach taken here is more rigorous statistically than the traditional approach because it avoids over-interpretation of the extrapolated base temperature. It achieves this by considering bases explicitly rather than as components of the germination rates. It seems likely that the same principles apply to both ceiling and base thresholds. Explicit consideration of the bases has useful repercussions. The shape of the rate–temperature relationship is not constrained to be linear; any type of curve can be used, formulated if necessary to encompass the superoptimal range. If the aim of an experiment is merely to measure bases, regular recording of germination may be unnecessary. Use of a probit model to identify thresholds was recommended by Murdoch *et al.* (1989) but they found it difficult to interpret a 'unit of dose of alternating temperatures'. We suggest that a temperature threshold is interpreted as a switch: germination either proceeds or stops according to the value of the ambient temperature relative to the threshold. In temperature regimes which alternate round a threshold this will have the effect of increasing the germination time and producing a germination rate which is apparently slower than the rate observed at constant temperatures. Consideration of thresholds and use of nonlinear curves could explain some of the discrepancies noted at low temperatures by Ellis and Barrett (1994) in their studies of alternating temperatures and rates of germination in lentil.

The proposed method has an additional useful property. As well as predicting the rate of germination it predicts the number of seeds which will germinate during a given time period and temperature regime. Use of explicit bases also improves the assessment of the frequency distribution of times to germination. The study of frequency distributions and the conditions which must be met for thermal and hydrothermal time calculations to be valid will be discussed in a paper to be published elsewhere.

Acknowledgements

We thank the Ministry of Agriculture, Fisheries and Food for funding this work and J.R.A. Steckel for assistance with data and graphics.

References

Bradford, K.J. 1990. *Plant Physiology* 94: 840–9.
Ellis, R.H. and Barrett, S. 1994 *Annals of Botany* 74: 519–524.
Finch-Savage, W.E. and Phelps, K. 1993. *Journal of Experimental Botany* 44: 407–14.
Finney, D.J. 1971. *Probit Analysis (3rd. Edition)*. Cambridge University Press.
Gummerson, R.J. 1986. *Journal of Experimental Botany* 37: 729–41.
Labouriau, L. and Osborn, J.H. 1984. *Journal of Thermal Biology* 9: 285–94.
Murdoch, A.J., Roberts, E.H. and Goedert, C.O. 1989. *Annals of Botany* 63: 97–111.
Scott, S.J., Jones, R.A. and Williams, W.A. 1984. *Crop Science* 24: 1192–9.
Worner, S.P. 1992. *Environmental Entomology* 21: 689–99.

41. Accumulation of Seed Vigour During Development and Maturation

D.M. TEKRONY and D.B. EGLI

Department of Agronomy, University of Kentucky, Lexington, KY 40549–0091, USA

Abstract

Physiological maturity (PM, maximum accumulation of dry seed weight) represents maximum yield in grain crops and is proposed to represent maximum seed
quality (germination and vigour) for planting purposes of all crops. We investigated the relationship of seed maturation to seed quality for five agronomic and
four horticultural crops in several field environments. Annual evaluations of
several soyabean (*Glycine max* (L.) Merr.) cultivars were made from 1982 to 1992,
while one single cross hybrid of maize (*Zea mays* L.) was tested from 1985 to 1994.
All other crops were evaluated for two production years, except wheat (*Triticum
aestivum* L. em Thell.) with one year. High seed viability occurred first (early in
development) for all crops and was followed by maximum standard germination at
or before PM in all crops except tomato (*Lycopersicon esculentum* Mill.) and
pepper (*Capsicum spp.*). Maximum seed vigour occurred at or slightly before PM
in crops harvested as dry seed, but after PM for crops with seed harvested from
fleshy fruits (tomato, pepper). When soyabean and maize plants were stressed by
defoliation, drought or high temperatures during seed development, seed maturation was accelerated, but maximum seed vigour occurred at or before PM. Thus,
the maximum expression of seed vigour was closely related to the occurrence of
PM in all crops harvested as dry seed.

Abbreviations

Physiological maturity, PM; Days after pollination, DAP; Harvest maturity, HM;
Seed dry weight, SDW; Black layer, BL; Relative humidity, RH; Accelerated
ageing, AA; Tetrazolium chloride, TZ

Introduction

Plant reproductive development begins with formation of the flower bud and
progresses through anthesis, fruit development and accumulation of storage
materials in the seed. Reproductive development ends at PM, when the seed
reaches its maximum dry weight. Seed maturation is a rather nebulous term
which can describe several or all stages of seed development. We will refer to two
stages of maturation, PM and HM. Physiological maturity was first defined by
Shaw and Loomis (1950) as the stage in seed development when the seed
reaches its maximum dry seed weight and yield. This same stage was also
termed 'relative maturity' by Aldrich (1943), 'morphological maturity' by

R.H. Ellis, M. Black, A.J. Murdoch, T.D. Hong (eds.), Basic and Applied Aspects of Seed Biology, pp. 369–384.
© *1997 Kluwer Academic Publishers, Dordrecht. Printed in Great Britain.*

Anderson (1955) and more recently 'mass maturity' by Ellis and Pieta Filho (1992). Since the moisture of the seed is too great for mechanical harvest and threshing at PM, further desiccation must occur before direct harvesting is possible. Harvest maturity is defined as the first time the seed moisture declines to a harvestable level in those crops harvested as dry seeds and/or fruits.

Physiological maturity can be determined on an individual seed or plant community basis. Individual seeds in the plant community usually reach PM at different times making definitions of PM on a plant community basis less precise. For example, an individual soyabean seed reaches PM when the seed is completely yellow at a seed moisture concentration of ~ 550 g kg^{-1}, whereas reproductive growth stage R7 is a commonly accepted indicator of PM in a soyabean plant community (TeKrony *et al.*, 1979; TeKrony *et al.*, 1981). Although the average seed moisture at R7 is approximately the same as an individual seed at PM, the individual seeds in the community range from green to yellow (650 to 400 g kg^{-1} seed moisture, respectively) due to the indeterminate reproductive development of soyabean. The variation in the timing of seed maturation in other crops depends on the reproductive growth characteristics of each species. Our studies of seed maturity have always been based on the individual seed.

The measurement of PM usually requires repetitious measurement of seed dry weight during the seed filling period which has often been hampered by sampling variation in seed weight, making it difficult to determine the exact date of PM. For this reason an interactive regression analysis procedure has been used to estimate the time of PM as the intersect of two fitted lines during seed development (Pieta Filho and Ellis, 1991). Seed moisture has also been used to estimate PM; however the usefulness of this technique varies among species. Probably the most accurate method of determining PM is by measurement of ^{14}C assimilate uptake by the developing seed. In our studies with maize and soyabean, PM was preceded by a sharp decline in ^{14}C assimilate uptake by the seed, with final ^{14}C levels near zero at PM (TeKrony *et al.*, 1979; Hunter *et al.*, 1991). Changes in seed colour or other visual changes in the seed or fruit structure are also usually good indicators of PM. For example, the decline in ^{14}C assimilate uptake was closely associated with the change in colour from green to yellow for soyabean seed and the development of a brown layer in the placental–chalazal region of the maize seed. We have observed similar visual changes in seed or fruit colour associated with PM in our seed development studies with other agronomic and horticultural crops.

It is widely accepted that PM represents the end of the seed filling period and the maximum yield for any crop harvested as dry seed. The lingering question is, does PM also represent maximum seed quality for planting purposes? Harrington (1972) proposed that maximum seed quality was attained at PM, after which deterioration was initiated and seed germination and vigour declined, with the rate of decline dependent upon the storage environment. This hypothesis was supported by research for two decades, until a recent challenge by Ellis and associates who concluded that maximum seed quality does not occur until some time after PM (Ellis and Pieta Filho, 1992; Rao *et al.*, 1991; Pieta Filho and Ellis, 1991; Zanakis *et al.*, 1994). They concluded that maturing seeds of several dicot and monocot species do not attain maximum ability to survive storage (potential longevity) until sometime after PM. Thus, the objective of our research was to determine the relationship between the stage of seed development, maturation and seed quality (germination and vigor). To accomplish this we have evaluated numerous genotypes for several agronomic and horticultural crops across many production environments.

Experimental Methods

All seeds were produced in the field near Lexington, KY, USA (agronomic crops) or Fresno, CA, USA (horticultural crops). The species, cultivars and production years are listed in Table 1. Seeds of maize and soyabean were

Table 1. Crop species, cultivars and production years evaluated to determine the relationship between seed development and seed quality

Crop	Genus species	Cultivars (hybrids)	Years
Canola	*Brassica napus* L.	Lindora 00	1989–1990
Maize	*Zea mays* L.	F$_1$ single crosses:	
		B73 × Mo17	1985–1987
			1990–1994
		Mo17 × FR375	1988
		B73 × V922	1989
Lettuce[a]	*Latuca sativa* L.	Salinas	1987–1988
Onion[a]	*Allium cepa* L.		1987–1988
Pepper[a]	*Capsicum* spp.	Emerald giant	1987–1988
Soybean	*Glycine max* (L.) Merr	De Soto	1981–1983
			1985–1987
		Century	1990–1991
		Flyer, Linford	1991–1992
		Stafford, Pennyrile	
Tobacco	*Nicotiana tabacum* L.	TN86	1991–1992
Tomato[a]	*Lycopersicon esculentum* Mill	Jackpot	1987–1988
Wheat	*Triticum aestivum* L.em. Thell	Auburn, Double crop Ky 84-94-1, TAM 105	1986

[a]Produced at Fresno, CA, USA

produced for eight and six years, respectively, using the same genotypes, while all other species were produced for two years (except wheat with only one year). A weather station located near the field experiments provided data to relate the environmental conditions during seed development and maturation to seed quality for each production year and location. In all experiments the evaluations of seed development were made on an individual seed rather than a plant community basis. Thus, either individual flowers were tagged (self pollinated crops) or controlled pollinations were made (hybrid maize and onion) to allow harvest of seed at specific DAP. Hand harvests of seed were usually initiated early in seed development ($\sim 30\%$ of final SDW) and continued until past PM. In some crops where seeds are normally harvested dry, harvest continued past HM. Following harvest for those species where seeds are normally harvested dry, the seeds were dried either in the fruit (soyabean, canola, tobacco) or attached to the inflorescence (maize, lettuce, wheat, onion). Seeds were removed from the fruit and dried separately only for those species where seeds are produced in fleshy fruits (tomato, pepper). All seeds were dried in cloth bags or paper envelopes at $\sim 30°C$ with forced air until the seed moisture concentration (fresh weight basis) declined to < 140 g kg^{-1}. The seeds were hand threshed, sealed in zip-lock plastic bags and stored at 10°C until tested for quality within one to three months after harvest.

At each harvest a random sample of seeds was removed from the fruit or inflorescence and seed moisture was measured by drying at 105°C for 48 or 72 h. In most experiments fresh seeds were also planted immediately (within 6 h) for fresh seed germination following the testing guidelines of the Association of Official Seed Analysts (AOSA, 1993). All other evaluations of seed quality were made on 100 to 400 dry seeds using the standard germination test (AOSA, 1993) and several vigour tests. Seed vigour was measured using the tests most commonly used for each species by the seed industry. These included stress vigour tests (accelerated ageing, cold test), biochemical tests (electrical conductivity, tetrazolium chloride), seedling growth rate tests as described by AOSA (1983) and field emergence tests. In 1993 and 1994 a controlled deterioration test was conducted on B73 × Mo17 maize seeds at 40°C and 160 g kg^{-1} seed moisture to measure potential seed longevity (Ki) as described by Pieta Filho and Ellis (1991).

Results

We previously reported the changes in seed dry weight and moisture concentration occurring during seed development and maturation in soyabean, wheat and maize (Miles *et al.*, 1988; Rasyad *et al.*, 1990; Hunter *et al.*, 1991). Similar evaluations were made for each crop in these experiments with the date of PM determined from dry seed weight data as described by Crookston and Hill (1978). Thus, the results presented here will relate primarily to the relationship between the stage of seed development and maturation and seed quality.

A typical example of the seed quality changes occurring during seed development for F1 maize seed (B73 × Mo17) across two years (1986, 1987) for undefoliated and defoliated plants is shown in Figure 1. In both years seed dry weight increased at a constant rate until PM (data not shown); however PM occurred five days earlier in 1987 because of high temperature and drought stress during seed development. The stress in 1987 also reduced the final dry seed weight at PM, which occurred at a higher seed moisture concentration than in 1986 (Hunter *et al.,* 1991). Physiological maturity occurred earlier in both years when plants were defoliated at ~20 DAP. Hunter *et al.* (1991) concluded that black layer stage 4 provided a reliable indicator of PM regardless of the stress occurring during seed filling. In these and all other experiments with maize (Table 1), 100% seed viability and standard germination occurred early in seed development (~20% final DSW). Seed vigour, as measured by the cold test, was at low levels early in seed development, but reached maximum levels at or before PM in all field environments (Fig. 1B). Seedling growth rate for both shoots (Fig. 1D) and roots (data not shown) also reached maximum levels at or before PM as did seedling emergence in the field (Fig. 1A). Reduced membrane integrity was reflected by higher conductivity levels in immature seeds; however conductivity declined to low levels well before PM in all production environments (Fig. 1C). Thus, regardless of the vigour test or the level of stress during seed development, maximum seed vigour occurred at or before PM in 1986 and 1987 (Fig. 1).

Since the cold test is the universally accepted vigour test for maize, cold test results were related to black layer formation during seed development across eight production years for B73 × Mo17 (Fig. 2). A significant quadratic relationship showed low seed vigour levels early in development (BL 1, 2) and maximum vigour response slightly before PM (BL stage 4). Maximum seed vigour was maintained through complete black layer formation (BL 5), when the average seed moisture was 250 g kg^{-1}. Maize seeds in commercial production are frequently harvested before BL 5. Similar quadratic relationships were shown between black layer formation and both cold test and conductivity with maximum seed vigour occurring before PM across a wide range of maize genotypes (inbreds B73, Mo17; double, single and 3-way cross F1 hybrids and open pollinated F2 hybrids) and five production years (TeKrony and Hunter, 1995).

Recent reports for wheat, barley (*Hordeum vulgare* L.), soyabean and pearl millet (*Pennisetum glaucum*) showed that maximum potential longevity (Ki), another measure of seed vigour, was not achieved until 3 to 21 d after PM (Ellis and Pieta Filho, 1991; Rao *et al.,* 1991; Pieta Filho and Ellis, 1991; Zanakis *et al.,* 1994). Thus, in 1993 and 1994 the potential longevity of F1 maize seed (B73 × Mo17) was determined at each harvest during seed development. The 1994 results (Fig. 3) show that maximum levels of Ki occurred 3 d before PM and were maintained for 23 d. The changes in potential longevity during development were similar to the cold test in 1994 (Fig. 3) and in 1993 (data not shown).

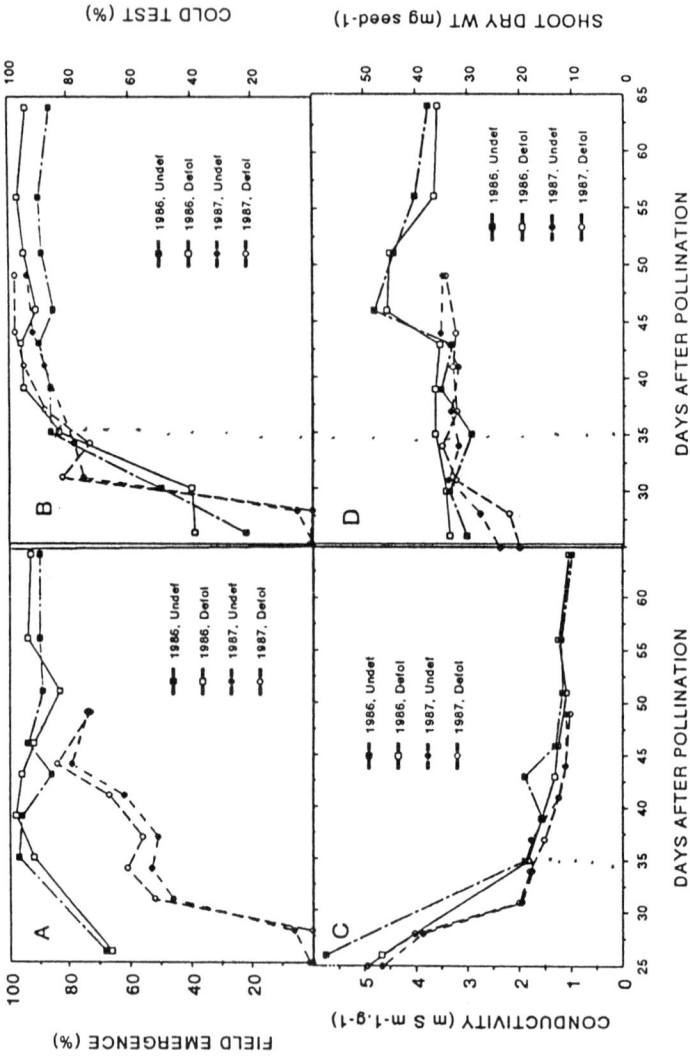

Figure 1. Changes in field emergence (A), cold test (B), conductivity (C) and shoot dry weight (D) for B73 × Mo17 maize seeds harvested from undefoliated and defoliated plants at various stages of seed development in 1986 and 1987. Physiological maturity occurred in 1986 at 51 (undefoliated) and 46 DAP (defoliated) and in 1987 at 44 (undefoliated) and 41 DAP (defoliated)

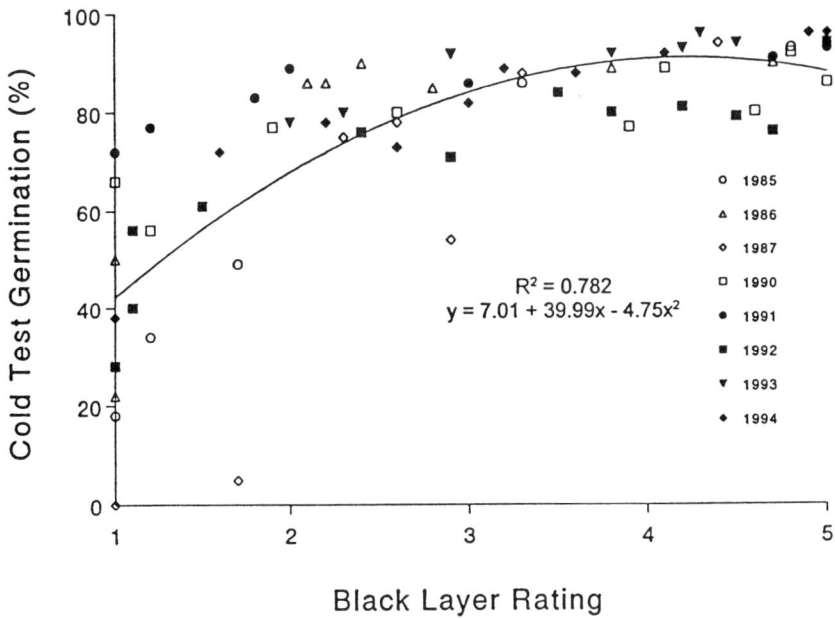

Figure 2. Relationship between cold test germination and black layer maturity for B73 × Mo17 maize seeds harvested at various stages of maturity from 1985 to 1994. Physiological maturity occurred at black layer stage 4

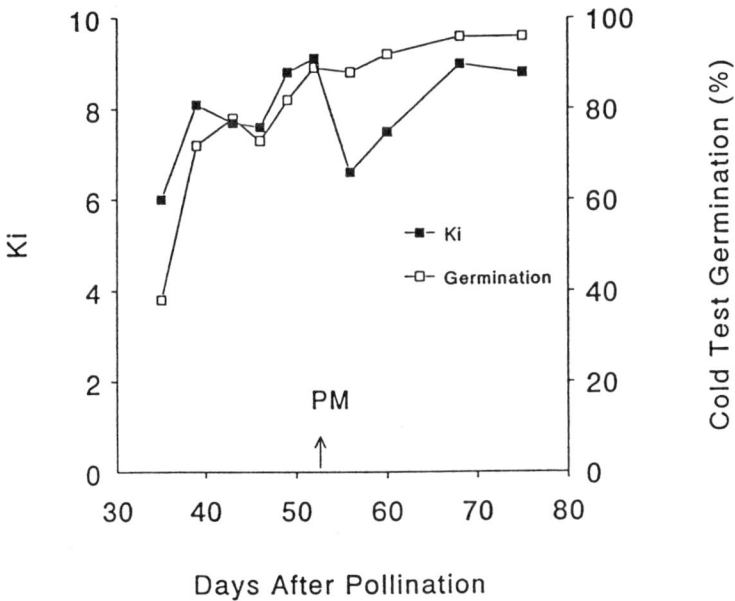

Figure 3. Change in potential longevity (Ki) and cold test germination at various stages of seed development and maturation for B73 × Mo17 maize seeds produced in 1994. Each symbol represents the mean of two samples from different replications. Date of physiological maturity indicated by PM

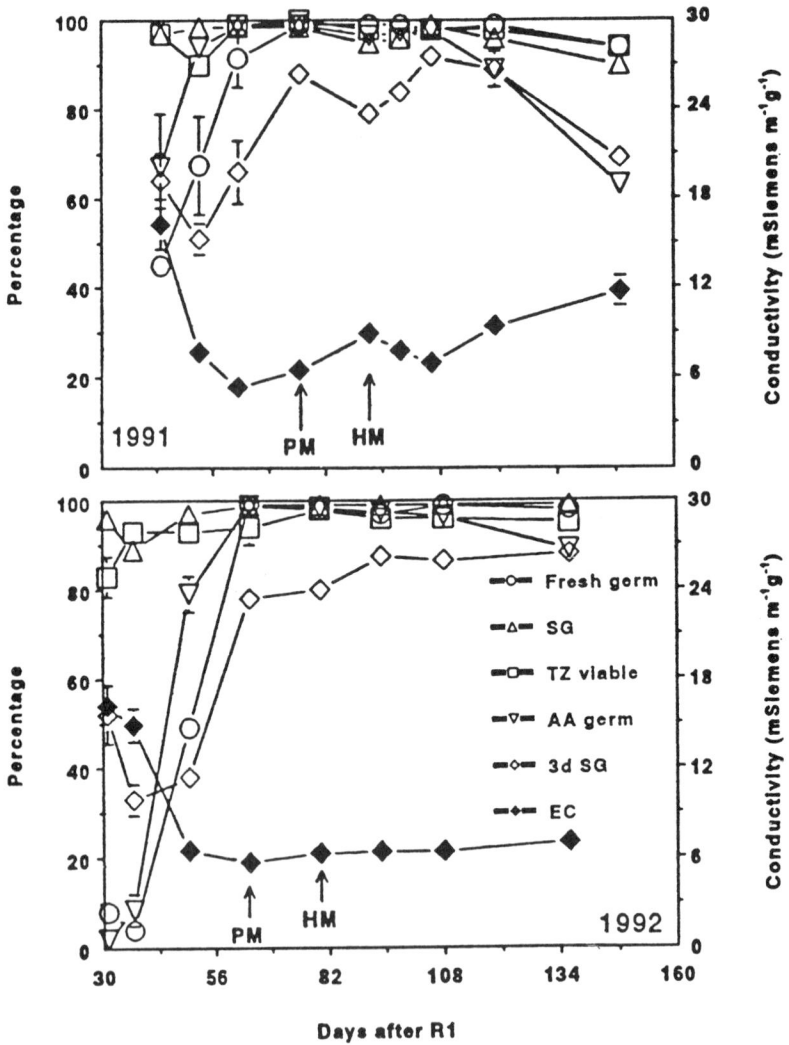

Figure 4. Seed quality for Pennyrile soyabean at various stages of seed development and maturation (R1 indicates first flowering) in 1991 and 1992. Dates of physiological maturity (PM) and harvest maturity (HM) indicated by arrow. SG, standard germination; TZ, tetrazolium chloride; AA, accelerated ageing; 3d, 3 day count of SG; EC, electrical conductivity

Typical changes in seed quality during soyabean seed development are shown for the cultivar Pennyrile in 1991 and 1992 (Fig. 4). The early harvests were made when the green seed first filled the pod locule ($\sim 25\%$ final SDW) and continued for 8 weeks past HM. The ability of fresh seeds to germinate was low for green, immature seeds but increased to $>95\%$ by PM. Maximum seed

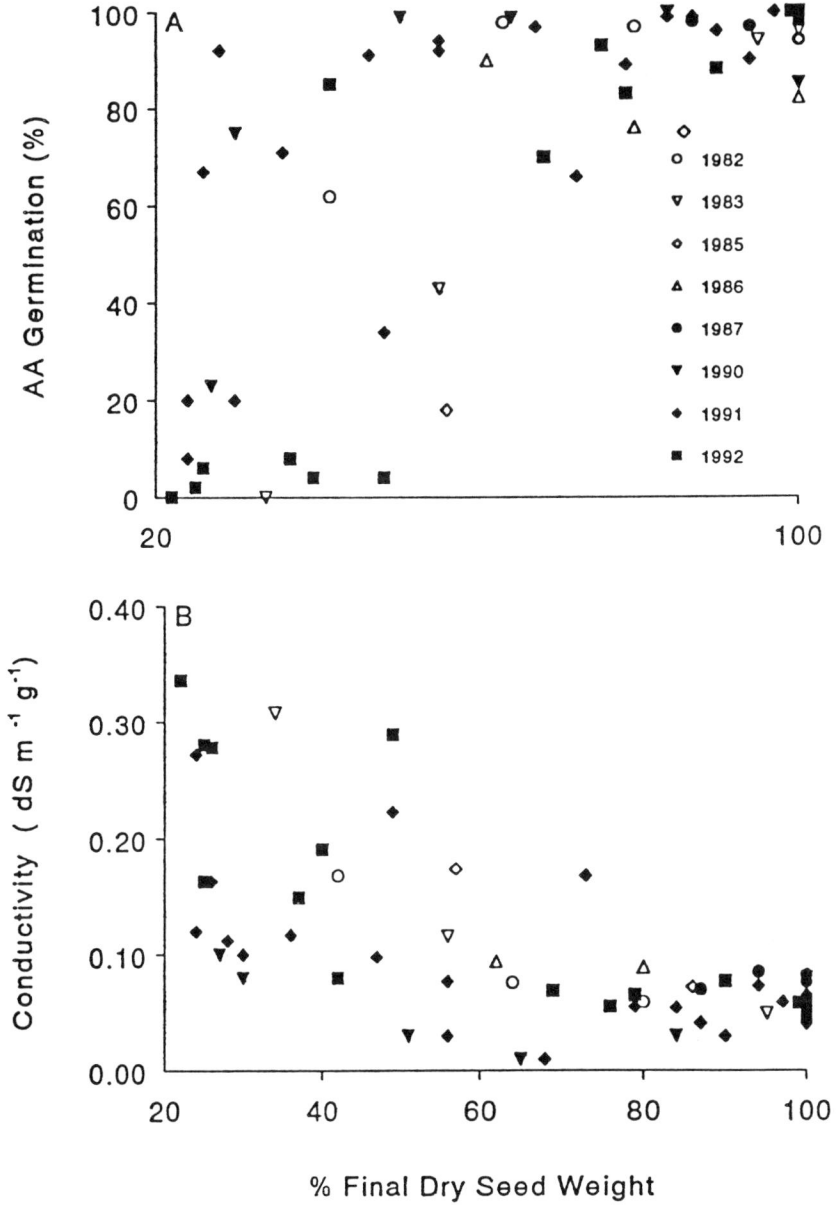

Figure 5. Relationship between accelerated ageing germination (AA) and conductivity and seed maturity at various stages of dry seed weight accumulation for seeds of several soyabean cultivars (Table 1) from 1982 to 1992. Physiological maturity occurred at 100% dry seed weight

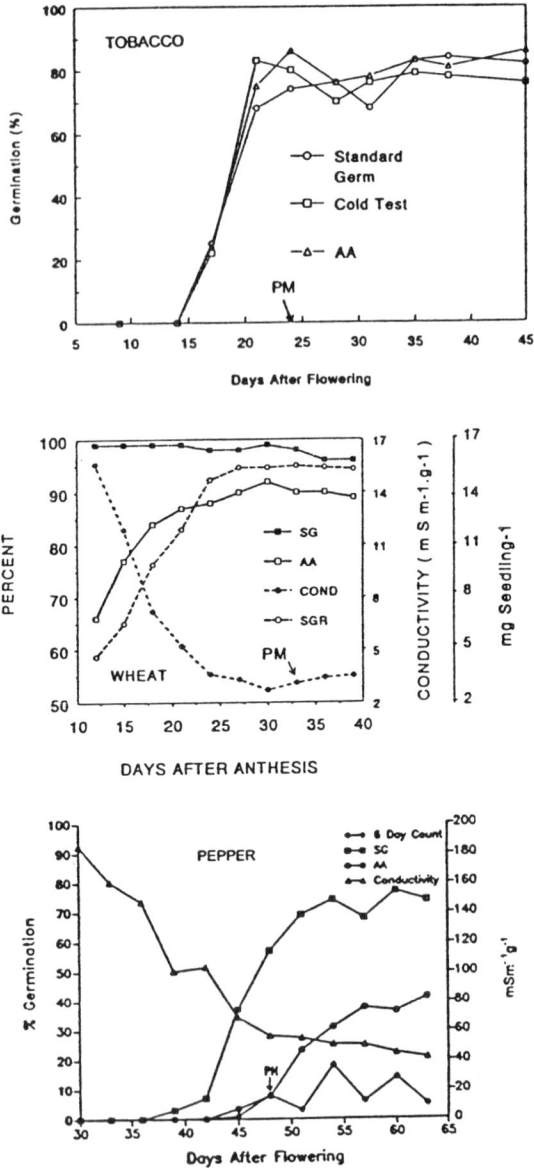

Figure 6. The changes in seed quality occurring at various stages in seed development and maturation for tobacco (1991), lettuce (1987), wheat (1986), canola (1990), pepper (1987) and tomato (1987). Date of physiological maturity (PM) indicated by arrow. Seed quality measurements; standard germination (SG), accelerated ageing (AA), conductivity (cond.), seedling growth rate (SGR), first count of standard germination test (4, 5, 6 day)

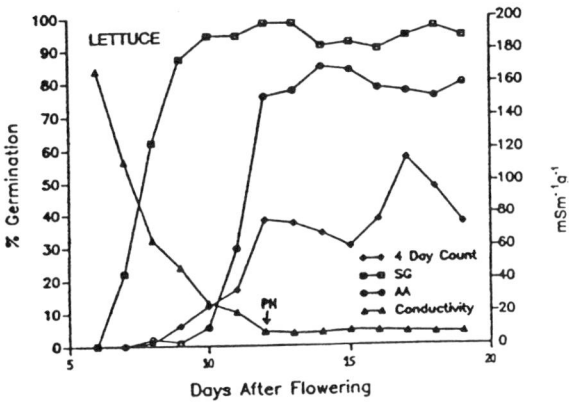

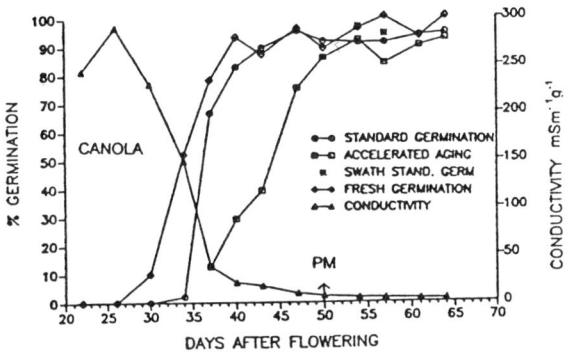

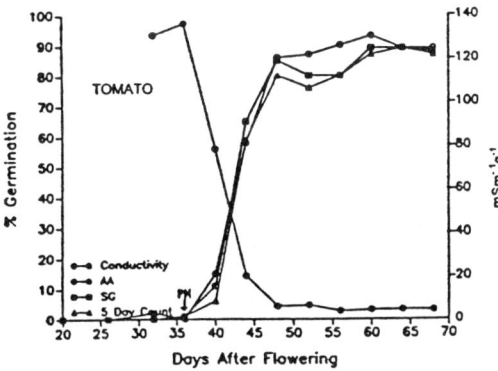

Figure 6. (cont.)

Table 2. Average and range in days before (–) or after (+) PM to maximum seed vigour

Crop	Genus species	Mean (days)	Range (days)
Canola	*Brassica napus* L.	–2	–4 to 0
Maize	*Zea mays* L.		
Inbred		0	–8 to +5
Single cross		–5	–10 to 0
Double cross		–10	–15 to –5
Lettuce	*Latuca sativa.* L	0	0
Onion	*Allium cepa* L.	–4	–6 to +9
Pepper	*Capsicum* spp.	+6	+4 to +9
Soybean	*Glycine max* (L.) Merr.	–7	–15 to 0
Tobacco	*Nicotiana tobacum* L.	0	0
Tomato	*Lycopersion esulentum* Mill.	+10	+7 to +12
Wheat	*Triticum aestivum* L. em. Thell	–6	–12 to 0

viability (TZ) and standard germination of dried seeds occurred in the immature seed at the first or second harvest. Seed vigour, as measured by the AA test, was low for immature seed, but increased to maximum levels at or before PM and maintained those levels through HM and 2 weeks post HM. The conductivity test also documented low levels of vigour for immature seeds and reached high seed vigour (low conductivity) before PM. When seed vigour was measured by the 3-day count of the standard germination test, changes in vigour during seed development were similar to AA or conductivity. Similar trends in seed quality were shown for 'Stafford', 'Linford' and 'Flyer' in 1991 and 1992 (data not shown) across the same harvest intervals and stages of seed development (Puteh, 1993).

The changes in seed vigour (AA and conductivity) which occurred for Pennyrile (Fig. 4) are very similar to that observed for eight years across several soyabean cultivars (Fig. 5). Although there was a wide range in seed vigour early in seed development, maximum levels consistently occurred by the time the seeds reached 90 to 100% (PM) of maximum dry seed weight. Similar to maize these trends are maintained even though plants were stressed by high temperatures, drought or defoliation (Vieira *et al.,* 1992) during seed development.

The changes in seed quality during development for a diverse group of crop species (tobacco, canola, lettuce, onion and wheat) whose seeds are harvested dry were similar to those shown for maize and soyabean. The relationship between seed development and seed quality for lettuce, canola, wheat and tobacco for one production year showed maximum expression of seed vigour at or before PM (Fig. 6). Similar trends were maintained in the second year for each species (data not shown). The only exception to the occurrence of maximum seed vigour at or before PM was for two species (tomato, pepper) whose seeds are produced in fleshy fruits (Fig. 6). Although the accumulation of dry seed weight followed normal trends for both species (data not shown), seed germination and vigour were low at PM. Maximum seed vigour (AA and conductivity) did not occur until ~ 10 and 6 d after PM in tomato and pepper, respectively.

Discussion

These results for nine crop species in many environments indicate that maximum seed quality (germination and vigour) occurs at or before PM for those species with seeds harvested as dry seed, but after PM for those species with seeds developing in fleshy fruits (Table 2). For the seven species harvested as dry seed it is clear that maximum levels of seed germination and vigour occur at or before PM. This relationship was maintained for maize and soyabean in a range of environmental conditions including high temperatures, drought and when seed filling was prematurely terminated by defoliation. Maximum seed quality occurred before PM in crop species producing small seeds (tobacco, canola, lettuce), large seeds (wheat, maize, soyabean) and a variety of fruit structures. These results clearly support the hypothesis proposed by Harrington (1972) that maximum seed quality is achieved at the end of the seed filling period (i.e. PM). Our results for those species harvested as dry seed are in conflict with recent reports by Ellis and associates for similar species which suggest that maximum potential seed longevity (their measure of seed vigour) occurs after PM (Ellis and Pieta Filho, 1992; Zanakis *et al.*, 1994; Pieta Filho and Ellis, 1991; Rao *et al.*, 1991; Ellis and Hong, 1994). There is general agreement however, between our results and those of Demir and Ellis (1992a; 1992b; 1993) that maximum seed quality does not occur until after PM for those species where seeds develop in fleshy fruits. This conclusion is also supported by studies of seed development in another fleshy fruited species, muskmelon (*Cucumis melo* L.), by Welbaum and Bradford (1989).

The contrasting conclusions from our investigations and those of Ellis and associates for those species harvested as dry seeds may relate to: (1) the harvest, drying and handling of high moisture and/or immature seeds, (2) whether seed harvests were made on an individual seed or plant community basis and (3) the methods used to measure seed vigour.

In the early stages of seed development the seed moisture concentration is high, ranging from 450 to 750 g kg^{-1}. Even at PM the average seed moisture of the species investigated exceeded 300 g kg^{-1} and was as high as 550 (soyabean) and 660 (onion) g kg^{-1}. It is not possible to measure seed vigour of such high moisture seeds. Thus, the seeds must be dried before testing. In our experiments we always dried the seeds attached to or enclosed in the detached fruit or inflorescence structures, while Ellis and associates dried seeds after removal from these structures. In our studies drying was done at 30°C, $\sim 50\%$ RH and seeds dried quite rapidly (7 d or less). Although we are uncertain how fast the seeds dried in their studies, the naked seeds may have dried slightly faster at the lower temperatures (15–20°C) and relative humidity (~15%) that they used. Thus, the seeds in our experiment may have continued maturation for a short time after harvest while attached to the fruit or inflorescence structure, which may partially explain the higher seed quality at an earlier stage of development. In preliminary tests with soyabean, we dried immature seeds removed from the pod and observed severe cracking of the testa, wrinkling of the seeds and much lower germination than for seeds dried in the pod. Zanakis *et al.* (1994) reported that enforced desiccation of naked seeds was fatal to most immature, green soyabean seeds. Thus, it seems possible that the rapid drying of naked seeds may explain the consistently lower quality in the early harvests of the Ellis experiments. Miles (1985) compared drying in several environments (seeds in pods) and observed maximum seed vigour at PM when seeds were dried normally (air or 30°C, 50% RH) and in the most severe drying conditions (30°C, 25% RH; 40°C, 50% RH). For this reason, we have dried all seeds either attached to the inflorescence or fruit at ~30°C, 50% RH. We also feel this drying technique is similar to what is presently done by commercial seed producers of maize, onion, tobacco, lettuce and wheat (if windrowed prior to threshing) seeds.

Because of seed to seed differences in maturation that can occur on a plant or within an inflorescence, we have consistently determined the time of PM on an individual seed basis. Thus, when evaluations are made for seed quality nearly all seeds tested were at the same stage of seed development when harvested. In some experiments conducted by Ellis and associates seeds were removed from the entire plant (soyabean, rice) or inflorescence (barley and wheat) which may have resulted in some immature seeds in the early samples and lower seed quality. This may provide an additional explanation for quality differences at PM between our experiments.

Many methods have been used for estimating seed vigour of dried seeds at various stages of seed development and maturation. We have attempted to use a combination of those seed vigour tests described in the Seed Vigour Testing Handbooks of ISTA (1995) and AOSA (1983) and most commonly used by the seed industry for each crop species. Ellis and Pieta Filho (1992) have proposed that this single or multiple test approach may provide misleading information because of small sample size (200 to 400 seeds) for each sample date. They concluded that seed quality is best determined by measuring the potential longevity (Ki) of seed in a controlled storage environment (40°C, 160 g kg^{-1}), which requires several germination tests after different storage periods for each sampling date. Our evaluations of maize seed (1993, 1994) show that maximum expression of seed vigour occurred at approximately the same stage of seed development when measured as potential longevity (Ki) or the cold test (Fig. 4). Although Ki is proposed to be a more accurate estimate of seed vigour, we got the same results with Ki and doubt that the differences reported in the maximum attainment of seed vigour can be related to seed testing methods. We conclude that the differences may result more from differences in harvest and drying techniques than from the method of measurement of seed vigour.

Recent seed maturation studies with wheat and barley (Ellis and Filho, 1992) conclude that maximum potential longevity does not occur until 3 to 21 d after PM. Since this contradicts the Harrington hypothesis that maximum seed quality is attained at PM, Ellis and co-workers have recommended that the term PM is misleading and should be replaced by a more appropriate term, mass maturity. It is doubtful that Shaw and Loomis (1950) assumed any implications to seed quality when this definition was proposed, even though it includes the word 'physiological'. Harrington (1972) and others have assumed that this term means maximum physiological seed quality for planting purposes, which appears to be correct for those crop species harvested as dry seed. Thus, the term mass maturity is probably not needed and the term PM can continue to be used.

In summary we conclude that PM represents a useful term to describe the maximum accumulation of dry seed weight and yield potential and should be maintained in the literature. Our seed development results with seven crop species which are commonly harvested as dry seed, show that maximum seed viability and germination was attained well before PM. The attainment of maximum viability was followed by the attainment of maximum vigour which occurred at or slightly before PM across a wide range of genotypes and production environments. For those species where seed develops in fleshy fruits (tomato, pepper), maximum seed germination and vigour occurred after PM. We speculate that the differences between our results and those of Ellis and associates regarding the time of maximum seed vigour during seed development relate primarily to differences in harvest and drying techniques and not seed vigour.

Acknowledgements

We recognize the research contributions of former and present graduate assistants: D. F. Miles, J.L. Hunter, S. E. Trawatha, A. Rasyad, J. Ferguson-Spears, A. Puteh, J. Woltz, University of Kentucky and J. Steiner, Professor and D. Roh, graduate assistant, Fresno State University who contributed to these studies.

References

Aldrich, S.R. 1943. *Journal of the American Society of Agronomy* 35: 667–680.
Anderson, S.R. 1955. *Agronomy Journal* 47: 483–487.
AOSA (Association of Official Seed Analysts). 1983. *Handbook No. 32*, AOSA, Lincoln, NE, USA.
AOSA (Association of Official Seed Analysts). 1993. *Journal of Seed Technology* 16: 1–113.
Crookston, R.K. and Hill, D.S. 1978. *Crop Science* 18: 867–870.
Demir, I and Ellis, R.H. 1992a. *Annals of Applied Biology* 121: 385–399.
Demir, I and Ellis, R.H. 1992b. *Seed Science Research* 2: 81–87.

Demir, I. and Ellis, R.H. 1993. *Seed Science Research* 3: 247–257.

Ellis, R.H. and Hong, T.D. 1994. *Annals of Botany* 73: 501–506.

Ellis, R.H. and Pieta Filho, C. 1992. *Seed Science Research* 2: 9–15.

Harrington, J.F. 1972. *Seed Biology*, Vol III, pp. 145–245 (ed. T.T. Kozlowski). New York: Academic Press.

Hunter, J.L., TeKrony, D.M., Miles, D.F. and Egli, D.B. 1991. *Crop Science* 31: 1309–1313.

ISTA (International Seed Testing Association). 1995. *Handbook of Vigour Test Methods* (3rd Edition), ISTA, Zurich, Switzerland. (In Press).

Miles, D.F. 1985. PhD Thesis. University of Kentucky, USA.

Miles, D.F., TeKrony, D.M. and Egli, D.B. 1988. *Crop Science* 28:700–704.

Pieta Fihlo, C. and Ellis, R.H. 1991. *Seed Science Research* 1: 163–177.

Puteh, A. 1993. PhD Thesis. University of Kentucky, USA.

Rao, N.K., Rao, S.A., Mengesha, M.H. and Ellis, R.H. 1991. *Annals of Applied Biology* 119: 97–103.

Rasyad, D.A., Van Sanford, D.A. and TeKrony, D.M. 1990. *Seed Science and Technology* 18:259–267.

Shaw, R.H. and Loomis, W.E. 1950. *Plant Physiology* 25: 225–244.

TeKrony, D.M., Egli, D.B., Balles, J., Pfeiffer, T. and Fellows, R.J. 1979. *Agronomy Journal* 71: 771–775.

TeKrony, D.M., Egli, D.B. and Henson, G. 1981. *Agronomy Journal* 73: 553–556.

TeKrony, D.M. and Hunter, J.L. 1995. *Crop Science* 35: 857–862.

Vieira, R.D., TeKrony, D.M. and Egli, D.B. 1992. *Crop Science* 32: 471–475.

Welbaum, G.E. and Bradford, K.J. 1989. *Journal of Experimental Botany* 40: 1355–1362.

Zanakis, G.N., Ellis, R.H. and Summerfield, R.J. 1994. *Experimental Agriculture* 30: 157–170.

42. Translation of Ribosomal Protein mRNAs in Maize Axes

E. SÁNCHEZ DE JIMÉNEZ, E. BELTRÁN and A. ORTIZ LÓPEZ

Departamento de Bioquímica, Facultad de Química, Universidad Nacional Autonoma de México, México, D.F. 04510

Abstract

The objective of this research is to study the mechanism(s) regulating ribosomal protein (rp) mRNA translation during maize germination. Northern analysis of RNA from non-germinated maize embryonic axes revealed rp-transcripts within the stored mRNA pool. S_4 and S_6 rp transcripts were stored as mature mRNAs while others seemed to be as pre-mRNAs. Preferential translation of rp-mRNAs was observed early in germination, as indicated by *in vivo* pulse labelling experiments with ^{35}S-methionine. Mobilization of S_6-transcript from ribonucleo-protein particles to polysomes was followed during germination. The effects of different hormones and environmental stresses were tested on this process. Recruitment of S_6 rp-transcript was found to be affected by insulin and stress. Inhibitors of either protein-phosphatases or kinases altered this process suggesting that it could be regulated by phosphorylation/dephosphorylation of protein(s) related to the translational apparatus. Insulin caused an increase in phosphoryla-tion of the ribosome integrated S_6 protein. These data suggest that S_6 phosphor-ylation might be involved in the mechanism responsible for rp-mRNA recruitment into polysomes. Consequently, in maize axes translation of rp-mRNA might be regulated by a signal transduction mechanism similar to other eukaryotic cells.

Introduction

Relevance of protein synthesis early on seed germination has been widely demonstrated (Bewley and Black, 1994). Quiescent embryonic axes undergo extended changes during imbibition, many of which activate the translational apparatus, inducing mobilization of stored mRNAs from RNP particles to polysomes (Gallie, 1993).

Lately, increasing information has been accumulated about the relevance of gene expression in eukaryotes regulated at the level of translation (Hershey, 1991). Different mechanisms for translational regulation have been reported in animal and other eukaryotes, such as selective mRNA polyadenylation (Sallés *et al.*, 1994; Hake and Richter, 1994), *trans*-acting factors (Bernstein and Ross, 1989) and phosphorylation and dephosphorylation of proteins of the transla-tional apparatus by different growth factors (Thomas *et al.*, 1982; Jefferies *et al.*, 1994). In plants, comparatively much less information on this matter has been described. Nevertheless, several reports indicate translational control during late embryogenesis through germination of seeds (Gallau and Dure, 1981;

R.H. Ellis, M. Black, A.J. Murdoch, T.D. Hong (eds.), Basic and Applied Aspects of Seed Biology, pp. 385–394.
© *1997 Kluwer Academic Publishers, Dordrecht. Printed in Great Britain.*

Kermode, 1990). In maize, we also have observed selective translation of stored mRNAs during germination (Sánchez de Jiménez and Aguilar, 1984). Further, translation of mRNAs codifying for abundant storage proteins in alfalfa embryos (Pramanik *et al.*, 1992) and oat seeds (Boyer *et al.*, 1992) have been found to be precisely regulated by development. Deeper understanding of this phenomenon comes from results showing selective mechanisms of translational control, such as repression of specific mRNA by an antisense RNA regulated by phytohormones (Rogers, 1988) or unfavourable codon usage by some mRNAs leading to differential translational rates (Boyer *et al.*, 1992). To gain further knowledge of this process, identification of the genetic information represented in stored mRNAs is contemplated as a main requirement and useful tool.

With this view, research has been conducted in our lab to get a better understanding of the mechanism regulating rp-mRNA translation. To this end, identification was first pursued of some rp-mRNA within the set of stored mRNAs and then search for effectors controlling rp-mRNA recruitment into polysomes.

Materials and Methods

Biological Material

Maize embryonic axes from *Zea mays* L. seeds, var. Chalqueño were dissected, disinfected and incubated for different periods of time in Murashige and Skoog medium in the dark at 25°C under sterile conditions. To study the effects of different effectors (cytokinins, auxins, insulin) and inhibitors (okadaic acid), the embryonic axes were incubated in their presence in the last 3 h of imbibition. For some experiments embryonic axes were exposed to stress, either heat or anoxia shock. Heat shock (45°C) was applied in the last hour of imbibition and for the anoxia treatment the embryonic axes were exposed to vacuum for the last 20 min.

In vivo *Pulse Labelling with* [35]*S-methionine*

In order to study the translation of stored mRNAs in maize embryonic axes, transcription was inhibited by imbibing 700 mg of embryonic axes in 1500 µl of α-amanitin (20 µg/ml). Inhibition of RNA synthesis was confirmed by testing ^{14}C-uridine incorporation into total and poly A^+ RNAs. These axes were incubated and *in vivo* labelled with 400 µCi of ^{35}S-methionine at the end of different periods of germination (3, 6 and 24 h) for 3h. The axes were homogenized in extraction buffer (20 mM Tris-Cl, pH 7.8, 5 mM $MgCl_2$, 20 mM KCl, 1 mM PMSF, 5 mM NaF, 0.5% β mercaptoethanol, 1% triton X-100 and 0.25 M sucrose) and centrifuged at 27 000*g* for 30 min. Ribosomes and postribosomal supernatant were separated using a 0.5 M sucrose pad and centrifuging at 280 000*g* for 4 h. [35]S -methionine incorporation into proteins of both fractions was determined.

In vivo *Phosphorylation of Ribosomal Proteins*

Seven hundred and fifty mg of embryonic axes was incubated for 24 h, and *in vivo* labelled with 500 µCi of ^{32}P-orthophosphate during the last 3 h of incubation. The ribosomal proteins were isolated as described before (Pérez-Méndez *et al.*, 1993) followed by 15% acrylamide gel electrophoresis using the Laemmli buffer system. The gels were stained with 0.2% Coomassie Brilliant Blue R-250, destained, and immersed in 0.1 M sodium salicylate solution (pH 5.7). Autoradiographs of the dried gels were obtained at –70°C using X-Omat film.

Isolation of Polysomal and Non-polysomal Fractions and RNA Extraction

Ribosomes from embryonic axes, incubated either for 6 or 24 h, were isolated and resolved into non-polysomal (NP) and polysomal (P) fractions using a 1.5 M sucrose pad and centrifuged at 200 000g for 3 h (Beltrán-Peña *et al.*, 1995). RNA of each fraction was extracted with phenol:chloroform:isoamyl alcohol (25:24:1). The phases were separated by centrifugation at 12 000g for 30 min and the RNA was precipitated from the aqueous phase with the addition of 0.1 volume of 3.0 M sodium acetate (pH 5.2) and 2.5 volumes of ice cold ethanol.

Total RNA was isolated using the extraction buffer (8 M guanidine Cl, 20 mM MES, pH 7.0, 20 mM EDTA and 50 mM β-mercaptoethanol) following the methodology described by Beltrán-Peña *et al.* (1995). The RNAs were analyzed by Northern blot as described below.

Northern Blot Analysis

Thirty µg of RNA was size fractionated in formaldehyde 1% agarose gels. RNA was transferred to Hybond N+ membranes, UV-cross linked and prehybridized in the hybridization buffer (7% SDS, 0.25 M Na_2HPO_4 pH 7.4, 1 mM EDTA pH 8, 1% BSA) for 2 h at 55°C. The denatured radiolabelled probe was diluted into the buffer (50 ng DNA/ml buffer) and hybridized overnight. The membranes were washed twice for 15 min at room temperature in 0.1% SDS, 2X SSC. Then the filters were washed once more for 15 min at room temperature in 0.1% SSC, 0.1% SDS. The probes used for this study were: (a) cDNA clone for ribosomal protein S_4, pDP1398 from chicken (donated by Andrew R. Zinn), (b) cDNA clone for ribosomal protein S_6, pS6-5 from rat (donated by Ira Wool), (c) genomic L_{16}, pHIS3CEN3 from rat (donated by Mohanish Deshmukh) and (d) cDNA clone for ribosomal protein L_3, pcDrpL3 from *Dictyostelium* (donated by Laura Steel). The probes were radiolabelled by random primer following the manufacturer's instructions (Dupont's Random Primer Extension Kit). The specific activity of the DNA probes was routinely about 10^7 cpm/µg.

Results

Identification of rp-transcripts within the Pool of Stored mRNAs

The first step was to confirm the presence of rp-mRNAs among the stored mRNAs. Identification of some of these rp-mRNAs was performed using RNA isolated from dry axes by Northern blot analysis, using heterologous rp-probes. The Northern blot autoradiograph of stored RNA showed a band of 0.9 kb for S_4 and 1.9 kb for S_6. In contrast for the LS rps, larger transcripts were observed (9.0 for L_3 and 11.9 kb for L_{16}) (Fig. 1). These results suggest that the former mRNAs were stored as mature mRNAs while the latter might be as pre-mature forms since they are much larger than the corresponding transcripts already reported for the corresponding animal mRNAs.

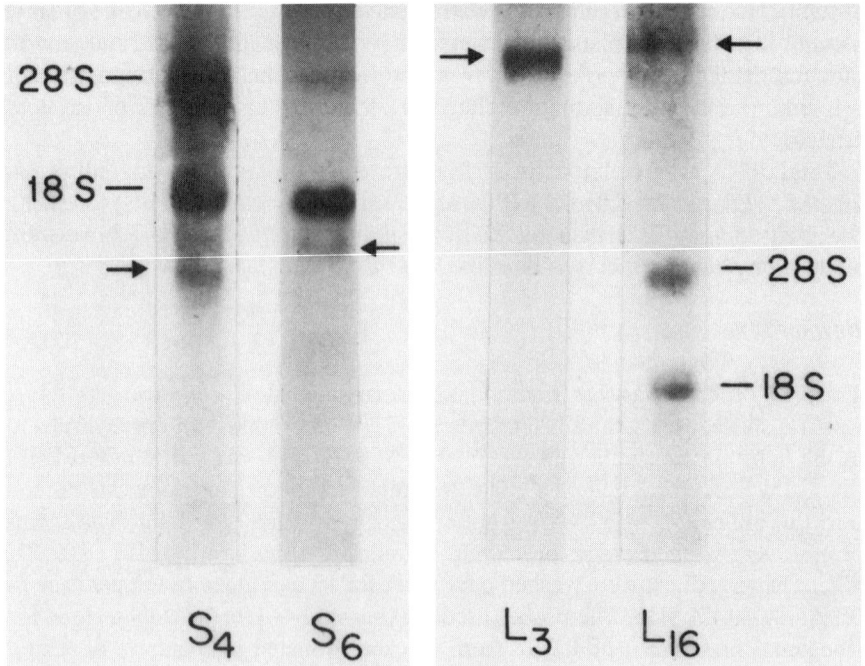

Figure 1. Northern blot analysis of total RNA isolated from dry embryonic axes. RNA samples from embryonic axes of dry seeds were electrophoresed and hybridized with cDNA probes for S_4 and S_6 rps of the small ribosomal subunit and with probes for L_3 and L_{16} rps of the large subunit. The arrows indicate the position of the respective transcript bands

Preferential Translation of rp-transcripts

In order to detect the major translation period for rps, α-amanitin-treated maize axes were incubated for different periods and fed during the last 3 h with ^{35}S-methionine. Incorporation of the labelled amino acid into proteins was measured in both cytoplasmic and ribosomal proteins. Cytoplasmic proteins showed approximately the same ^{35}S-methionine incorporation during 3, 6 and 24 h of incubation, while a preferential translation of ribosomal proteins was observed. This was particularly noticeable after 24 h of incubation (Fig. 2).

Mobilization of S6 mRNA into Polysomes

The process for rp-mRNA mobilization into polysomes was studied, polysomal (P) and non-polysomal (NP) fractions were isolated from embryonic axes incubated in sterile nutrient medium for either 6 or 24 h. Total RNA was extracted from each fraction and analyzed by Northern blot using a cDNA probe for S_6 as a marker of the process. Results indicated the presence of larger amounts of S_6 mRNA in the NP fraction at 0 h (Fig. 1) and 6 h than at 24 h, when most of this mRNA was present on the P fraction (Fig. 3). Based on these

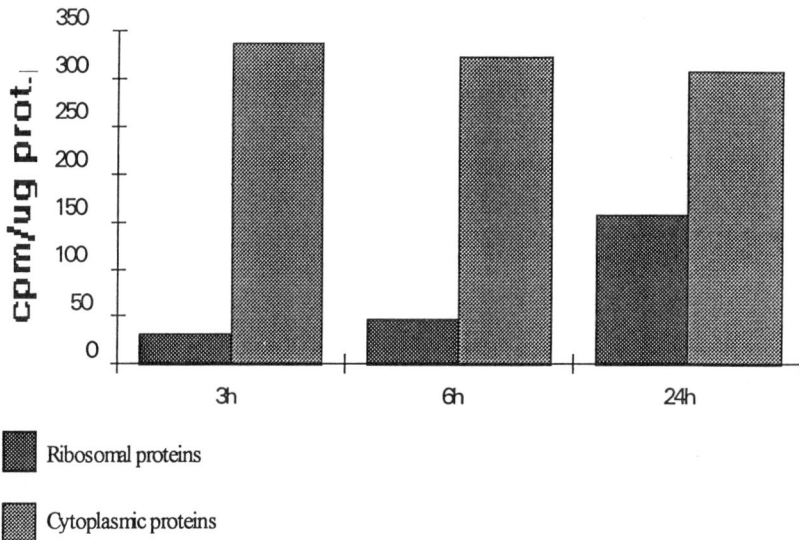

Ribosomal proteins

Cytoplasmic proteins

Figure 2. ^{35}S-methionine incorporation into proteins. Maize embryonic axes treated with α-amanitin were pulse-labelled with ^{35}S methionine (500 mCi) for the last 3 h of incubation. The cytoplasm and ribosomal proteins were isolated, TCA precipitated and ^{35}S-incorporation was measured in a scintillation counter using Bray's reagent

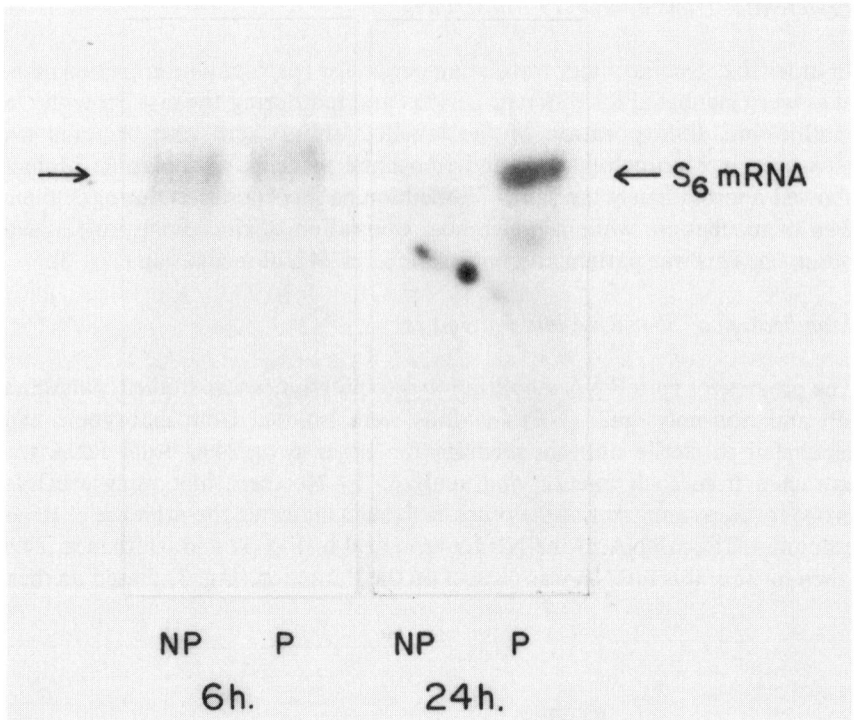

Figure 3. Mobilization of S_6-mRNA during germination. Axes incubated for 6 and 24 h were homogenized and ribosomes resolved in non-polysomal (NP) and polysomal (P) fractions. Total RNA was isolated from each fraction and electrophoresed for Northern blot analysis using S_6 cDNA as a probe. The arrows indicated the position of the S_6 mRNA

data, the following experiments were made on P fractions of 24 h incubated axes to study the effect of growth factors on S_6 mRNA recruitment for translation.

Effect of Growth Factors of S_6 mRNA Recruitment into Polysomes

As a first approach to elucidate the mechanism that regulates mRNA recruitment into polysomes, 24 h-incubated axes were stimulated for the last 3 h with either one of the following phytohormones: auxins – either native (IAA) or synthetic (2, 4-D, Dicamba) – or cytokinins such as zeatin, BAP, 2i-P and kinetin. Insulin, the most widely acting peptide growth factor in eukaryotic organisms, was also tested in the system. Results indicated similar S_6 mRNA levels in the polysomes of all the cytokinin-treated axes as the control (data not shown). Auxins, on the other hand, caused a slight increase of S_6 mRNA recruitment into polysomes, particularly the native auxin IAA. However,

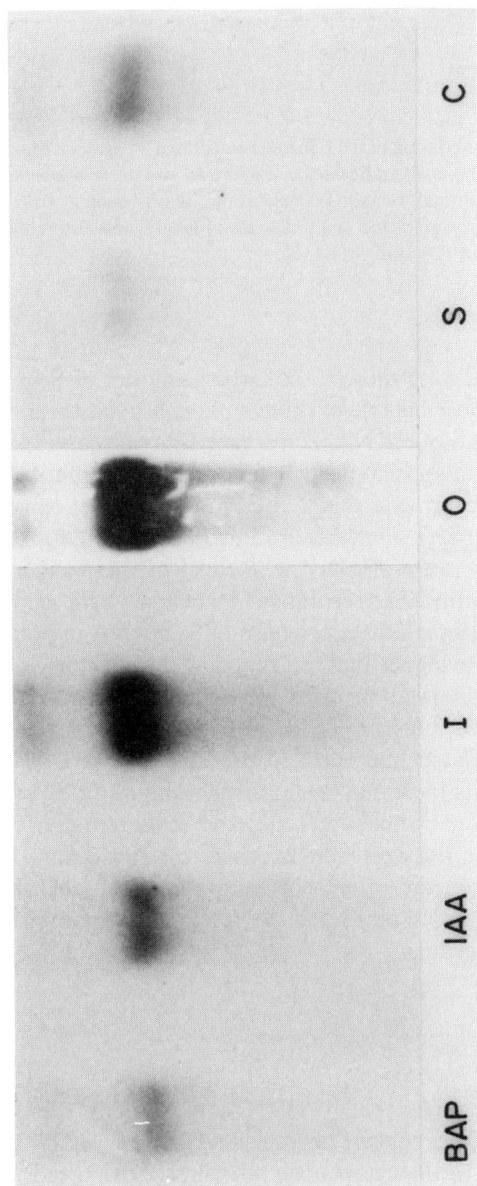

Figure 4. Effect of growth factors and environmental stress on S6-mRNA recruitment to polysomes. Maize embryonic axes incubated for 24 h were exposed for the last 2 h of incubation to either BAP (50 μM), IAA (20 mM), insulin (I, 200 μU), okadaic acid (O, 0.1 μM), heat or anoxia shock (stress: S), or water (C). RNA from the P fraction of each sample was isolated and used for Northern analysis as described in Figure 3. Note: Heat and anoxia shock gave similar results

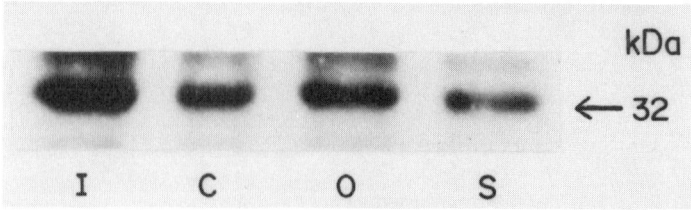

Figure 5. Phosphorylation of the ribosome-integrated S6 protein. Maize axes were incubated for 24 h under the conditions indicated in Figure 4, and *in vivo*-labelled with ^{32}P-orthophosphate (500 μCi) for the final 90 min of incubation. The ribosomal fraction was isolated from each axes group and rps extracted and analyzed by PAGE and autoradiography. Insulin (I), Control (C), Okadaic acid (O), and Stress (S)

insulin-treated axes showed the largest amount of S_6 mRNA in the P fraction (Fig. 4). The possible role of protein phosphorylation and dephosphorylation on the mechanisms of mRNA recruitment into polysomes was tested by addition of okadaic acid, a specific protein phosphatase inhibitor. Environmental stresses such as heat or anoxia shock, are known to affect rp phosphorylation (Scharf and Nover, 1982; Bailey-Serrés and Freeling, 1990); therefore these environmental factors were also tested. Results of these experiments showed a large increase of S_6 mRNA recruitment by okadaic acid while heat or anoxia shock significantly decreased the presence of S_6 mRNA in polysomes (Fig. 4).

To provide evidence that the action of the effectors is related to S_6 ribosomal protein phosphorylation, axes were incubated under the same conditions as before but in the presence of ^{32}P-orthophosphate. Ribosomes from these tissues were then isolated and their rp proteins analysed by PAGE and autoradiography. Results indicated that phosphorylation of S_6 protein was increased by insulin and okadaic acid application while inhibition of S_6 phosphorylation was observed when the axes were exposed to stress (anoxia or heat shock, Fig. 5). These data indicate direct correlation between the amount of S_6-transcript recruitment to polysomes and the level of phosphorylation on the ribosomal-integrated S_6 protein.

Discussion

The results indicate that rp-mRNAs are among the set of stored mRNA in maize axes. The rps are synthesized at very early stages of germination, preferentially to the cytoplasmic proteins (Figs. 1 and 2). Our data, and other previous reports (Gallie, 1993), suggest there might be specific regulatory mechanism(s) responsible for translational control during germination.

Selective mRNA recruitment into polysomes by action of growth factors has extensively been documented in eukaryotic organisms other than plants. These reports have demonstrated that within the 5'UTR (untranslatable region) of rp mRNAs there is a specific pyrimidine-rich consensus sequence, which seems to correspond to signals for selective mobilization of these mRNAs into polysomes (Mariottini and Amaldi, 1990; Levy *et al.*, 1991). Moreover, evidence has been presented regarding insulin and other growth factors as effectors for this process, probably through phosphorylation of the S_6 rp in the ribosome (Jefferies *et al.*, 1994); this has been interpreted as a stimulatory path for G1 cells to enter S phase (Thomas *et al.*, 1988). The data presented here shows a clear correlation between rp-mRNA recruitment into polysomes and the level of the S_6 phosphorylation (Fig. 4). Considering that quiescent cells of embryonic axes reinitiate metabolic activity and would move from G1 to S phase at the end of the germination period (Baiza *et al.*, 1989), and the high evolutionary conservation of the ribosomal proteins, these data might indicate that in plant cells rp-mRNAs could be regulated by a similar mechanism as in animals. The natural effector for this path, however, remains to be identified, although the observed strong effect of insulin suggests that plants might contain peptide growth factors similar to those present in animal cells. Addition of exogenous amounts of auxins to the axes also gave a positive result for S_6 mRNA recruitment into polysomes; nevertheless this signal was much lower than the one found for insulin (Fig. 4). Since auxins are known to promote alteration of the S_6 rp-phosphorylation in maize axes (Pérez *et al.*, 1990), this result might be interpreted as possible cross talk between the two transduction pathways, the one generated by auxins with the one induced by insulin.

In higher plants, and particularly in maize, a nucleotide sequence coding for a receptor-like protein kinase has been described (Walker, 1994). The putative polypeptide coded by this sequence, however, seems to correspond to a serine/threonine protein kinase, instead of tyrosine kinase as described for the growth factor-receptor in animal tissues. At present, no ligands for this receptor have been reported for plants. In our hands, an insulin-like peptide has been detected by RIA analysis in maize tissues (data not shown). Furthermore, inhibitors of the insulin signal transduction path such as wortmannin, an inhibitor of the phosphoinositide-3-kinase, an enzyme presumably a participant at an early step of the cascade of reactions (Powis *et al.*, 1994; Thelen *et al.*, 1994), and rapamycin, an inhibitor of the p70[s6k] kinase, the enzyme responsible for S_6 phosphorylation (Petritsch *et al.*, 1995) showed a strong inhibition of S_6 recruitment into the polysomal fraction of maize axes (data not shown).

Taking together all the above, we conclude that S_6 mRNA in maize axes is subjected to translational regulation during germination. The mechanism responsible for this process might be similar to the one described for other eukaryotic cells which seems to involve a signal transduction path probably initiated by an insulin-like peptide. Further work is currently being performed in our lab to confirm this hypothesis.

Acknowledgements

The authors are grateful to the donors of the rp-probes used in this research and to Andrea Sanjuan and Alfonso Suarez for technical assistance. This research was supported by Direccion General de Apoyo al Personal Académico (DGAPA), UNAM grant No. IN 200793.

References

Bailey-Serres, J. and Freeling, M. 1990. *Plant Physiology* 94: 1237–1243.
Baiza, A., Aguilar, R. and Sánchez de Jiménez, E. 1986. *Physiologia Plantarium* 69: 259–264.
Beltrán-Peña, E., Ortíz-López, A. and Sánchez de Jiménez, E. 1995. *Plant Molecular Biology* 28: 327–336.
Bernstein P. and Ross, J. 1989. *Trends in Biochemical Sciences* 14: 373–377.
Bewley, J. D. and Black, M. 1994. *Seeds. Physiology of Development and Germination*, Second Edition, pp. 445. New York, London: Plenum Press.
Boyer, S. K., Shotwell, M. A. and Larkins, B. A. 1992. *Journal of Biological Chemistry* 267: 17449–17457.
Gallau, G. A. and Dure, L. III. 1981. *Biochemistry* 20: 4169–4178.
Gallie, R. D. 1993. *Annual Review of Plant Physiology and Molecular Biology* 44: 77–105.
Hake, L. E. and Richter, J. D. 1994. *Cell* 79: 617–627.
Hershey, J. W. 1991. *Annual Review of Biochemistry* 60: 717–755.
Jefferies, H. B. J., Reinhard, C., Kozma, S. C. and Thomas, G. 1994. *Proceedings of the National Academy of Sciences* 91: 4441–4445.
Kermode, A. R. 1990. *Critical Reviews of Plant Science* 9: 155–195.
Levy, S., Avni, D., Hariharan, N., Perry, R.P. and Meyuhas, O. 1991. *Proceedings National Academy of Science of the USA* 88: 3319–3323
Marottini, P. and Amaldi, F. 1990. *Molecular and Cell Biology* 10: 816–822.
Pérez, L., Aguilar, R., Mendez, A. and Sánchez de Jiménez, E. 1990. *Plant Physiology*. 94: 1270–1275.
Pérez Méndez, A., Aguilar, R., Briones, E. and Sánchez de Jiménez, E. 1993. *Plant Science* 94: 71–79.
Petritsch, C., Woscholski, R., Edelmann, M. L. H., Parker, J. P. and Ballou, M. L. 1995. *European Journal of Biochemistry* 230: 431–438.
Powis, G., Bomjouklian, R., Berggren, M. M., Gallegos, A., Abraham, R., Ashendel, C., Zalkovo, L., Matter, F. W., Dodger, J., Grindey, G. and Vlahos, J. C. 1994. *Cancer Research* 54: 2419–2423.
Pramanik, S. K., Krochko, O. J. E. and Bewley, J. D. 1992. *Plant Physiology* 99: 1590–1596.
Rogers, J. C. 1988. *Plant Molecular Biology* 11: 125–138.
Sallés, F. J., Lieberfarb, M. E., Wreden, C., Gergen, J. P. and Strickland, S. 1994. *Science* 266: 1996–1999.
Sánchez de Jiménez, E. and Aguilar, R. 1984. *Plant Physiology* 75: 231–234.
Scharf, K. D. and Nover, L. 1982. *Cell* 30: 427–437.
Thelen, M., Wymann, P. M. and Langer, H. 1994. *Proceedings of the National Academy of Sciences USA* 91: 4960–4964.
Thomas, G., Martínez-Pérez, J., Siegmann, M. and Otto, A. M. 1982. *Cell* 30: 235–242.
Walker, J. C. 1994. *Plant Molecular Biology* 26: 1599–1609.

43. Cell Cycle Analysis in Dormant and Germinating Tomato Seeds

S.P.C. GROOT[1], R.D. DE CASTRO[1,2], Y. LIU[1,3] and R.J. BINO[1]

[1]*Center for Plant Breeding and Reproduction Research (CPRO-DLO), Department of Reproduction Technology, P.O. Box 16, 6700 AA Wageningen, The Netherlands;* [2]*Universidade Federal de Lavras (UFLA), Departamento de Fitotecnia, Laboratorio de Sementes, Cx. Postal 37, Lavras, MG, CEP 37200-000, Brasil;* [3]*Department of Horticulture, Hunan Agricultural College, Changsha 410128, P.R. China*

Abstract

Cell cycle activity is a prerequisite for growth and development. In germinating seeds, cell cycle activity has been shown to precede visible germination. The paper reviews cell cycle research with dormant and germinating tomato seeds. Nuclear DNA replication, as envisioned by flow cytometry, starts in the embryonic root tip and precedes visible germination. This DNA replication also occurs during priming of seeds and in the presence of abscisic acid, despite the inhibition of germination due to both these treatments. In dormant and gibberellin deficient tomato seeds DNA replication is blocked. Immunological detection of ß-tubulin levels, showed that accumulation of this cytoskeletal protein is correlated with the onset of nuclear DNA replication. The studies with dormant and gibberellin deficient tomato seeds give strong indications for the occurrence of a G_0 cell cycle phase in plants. The synchronization of the cell cycle in dry seeds, makes it an ideal material to study the regulation of growth and the cell cycle in plants. On the other hand, the analysis of cell cycle in germinating seeds will offer opportunities to get a more detailed knowledge of pre-germination seed development and its relation to seed quality.

Growth and Cell Cycle Activity

Seed germination is the start of the growth of a new plant. Growth can be described as the synergy between cell division, cell enlargement and cell differentiation, ultimately leading to complex organisms, small as the molecular model plant *Arabidopsis*, or large as the giant sequoia. The regulation of cell division in time and plane is of basic importance to multicellular organisms.

The sequence of processes occurring during cell division is often referred to as the cell cycle, which can be distinguished in the chromosome cycle and the parallel cytoplasmatic cycle (Alberts *et al.*, 1989). In the chromosome cycle, DNA synthesis, in which the nuclear DNA is duplicated, alternates with mitosis, in which the duplicate copies of the genome are separated. In the cytoplasmic cycle, cell growth, in which the many other components of the cell double in quantity, alternates with cytokinesis, in which the cell as a whole divides in two. Within the cell cycle four different phases can be distinguished. In

R.H. Ellis, M. Black, A.J. Murdoch, T.D. Hong (eds.), Basic and Applied Aspects of Seed Biology, pp. 395–402.
© *1997 Kluwer Academic Publishers, Dordrecht. Printed in Great Britain.*

higher plants and animals, DNA replication occurs during a discrete interval of the interphase, known as the S phase (S standing for synthesis). Since histones are also synthesized during this period, the S phase can be considered a time of chromosomal as well as DNA replication. Before and after the S period, interphase cells engage in growth and metabolic activity but do not engage in chromosome replication. These two phases are called the G_1 and G_2 (G standing for gap) phases of the interphase. The G_2 is, normally, followed by chromosome separation during cell division or mitosis (M phase). In diploid somatic cells in the G_1 phase, the nuclei contain 2C DNA levels, C being defined as the DNA content of generative nuclei. At the end of the S phase and in the G_2 phase these nuclei contain 4C DNA levels. With cell cycle research in animal systems, a fifth cell cycle phase has been defined, G_0, in which the cells, with a 2C DNA content, are in a quiescent stage. The metabolic activity of G_0 animal cells is distinctive from that of G_1 cells, with the production of phase specific proteins. A promoting factor is needed to bring the cells from the G_0 into the G_1 phase. With plants, intensive cell cycle research has only been initiated relatively recently, and as yet, no clear evidence has been presented for the existence of a G_0 phase in higher plants (D. Inzé personal communications).

Cell Cycle Activity in Seeds

One of the specific features of seeds is that they often can be stored dry and they can survive in the soil, without being damaged by alternating periods of dry or wet conditions. During evolution dormancy mechanisms have developed that in the natural situation prevent seeds from germinating in unfavourable seasons. However, once the seed has germinated, normally, there is no way back and the seedling will die upon drying. Till recently, little was known about cell cycle activities at the moment of germination and in the period preceding it.

Although in dry seeds the water content is too low for progression of the cell cycle, it was poorly known in which phase of the cell cycle the cells of the different seed organs were arrested. Nuclear replication stages in seeds had been identified using autoradiography and Feulgen staining, and it was found that embryos of some species (e.g. *Triticum durum, Zea mays, Vicia faba*) contained both 2C and 4C nuclei, while others (e.g. *Pinus pinea, Allium cepa, Lactuca sativa*) solely comprised 2C nuclei (Bewley and Black, 1978; Deltour, 1985). Apparently in a number of species the quiescent embryo arrests the cell cycle in G_0 or G_1, while in others species high amounts of embryonic cells become arrested in the late S or the G_2 phase during seed maturation. Conger and Carabia (1976) found with *Zea mays* embryos, that the proportion of 2C and 4C nuclei was approximately 1:1 in the root and 5:1 in the shoot tissue. According to the same authors, however, both roots and hypocotyls of dormant embryos of *Festuca arundinacea* and *Dactylis glomerata* predominantly contained 2C nuclei (Conger and Carabia, 1978). Only after imbibition and during germination did they observe an increase in the amount of 4C nuclei, indicating that during seed germination the cells enter the S or G_2 phase.

Flow cytometry with the use of DNA-specific fluorescent dyes, has made it possible to determine the amount of DNA in large numbers of nuclei within minutes. We have used this technique to quantify the amounts of nuclear DNA with high accuracy in various seed species and seed parts (Bino *et al.*, 1992; 1993). In this paper we review results obtained on cell cycle research with tomato seeds, as a model species, in relation to dormancy, germination, priming, artificial ageing and the plant hormones gibberellins and abscisic acid.

The Mature Dry Tomato Seed

In the mature and dry tomato seed, the embryo radicle tip contains predominantly nuclei with 2C DNA levels and between 5 and 11% of the nuclei in this tissue has 4C DNA levels (Bino *et al.*, 1992; Bino *et al.*, 1993; Liu *et al.*, 1994). The remaining embryo tissue, contains almost exclusively nuclei with 2C DNA levels, only some traces are observed from nuclei with 4C DNA. Analysis of the endosperm tissue showed that 75% of the nuclei contained 6C and 25% 12C DNA levels. From these figures it was concluded that during tomato seed maturation most nuclei in the embryo were arrested in G_0 or G_1 state of the cell cycle. In the triploid endosperm, apparently, endoreduplication of the nuclei had occurred prior to the desiccation of the seed.

Seed Germination

After 1 day of imbibition of the tomato seeds in water, the relative amount of nuclei with 4C DNA levels starts to increase. In the radicle tip 4C:2C ratios gradually rise towards 2.5 when seeds start to exhibit visible germination i.e. protrusion of the radicle (Bino *et al.*, 1992; Liu *et al.*, 1994). Also in the remaining embryo tissue a small progression towards 4C DNA levels is observed, but, here the 4C:2C ratio does not exceed 0.2 at the moment of germination. In the radicle tip, where most nuclei exhibit 4C DNA levels at the moment of germination, apparently the cell cycle has progressed toward the late S or the G_2 phase.

Dormancy

Normally tomato seeds do not possess dormancy. However, under certain cultivation conditions, freshly harvested seeds may be dormant to a large extent, and this dormancy can be broken by red light, a cold treatment, exogenous gibberellins or a period of dry storage (Groot and Karssen, 1992). Flow cytometric analysis of freshly harvested dormant seeds, which had been imbibed for a week in darkness, showed that nuclear DNA replication was blocked, since the 4C DNA level did not increase above that observed in the dry seeds (Y. Liu and S.P.C. Groot, unpublished results). Secondary dormancy can be induced with tomato, by irradiation of imbibed seeds with far-red light. Also in far-red light irradiated seeds the replication of nuclear DNA was blocked (Fig. 1). Breakage of the secondary dormancy of these seeds by white light, induced an increase in 4C DNA levels in the radicle tip, which preceded the actual germination.

Gibberellins

Studies with the gibberellin deficient *gib-1* mutant has provided important information on the role of this group of hormones in tomato seed development and germination (Groot, 1987). Gibberellins seem not to be essential for seed development, since mature *gib-1* seeds are indistinguishable from wild-type seeds, except for an average smaller size (Groot *et al.*, 1987b). The *gib-1* seeds, however, fail to germinate in water. It was shown that endogenous gibberellins, produced by the embryo, are needed to induce the enzymatic hydrolysis of the thick endosperm cell walls, thereby releasing the mechanical restraint for radicle protrusion (Groot and Karssen, 1987a; Groot *et al.*, 1988).

In mature *gib-1* seeds the 4C:2C ratios of the nuclei are comparable as in mature wild-type seeds (Liu *et al.*, 1994). However, there is no replication of DNA upon imbibition of the *gib-1* seeds in water. This blockage can be relieved by application of GA_{4+7}. The increase in 4C signal of wild-type seeds in GA_{4+7} is comparable to that observed with incubation in water.

The lack of nuclear DNA replication in imbibed dormant and in gibberellin deficient *gib-1* mutant seeds indicates that also in plants a G_0 phase exists. In these non-germinating seeds the embryonic cells, which are metabolically active, apparently need some kind of promoting factor to progress in the cell cycle. It seems reasonable to consider these embryonic cells as being in the G_0 phase of the cell cycle, corresponding to the situation with quiescent G_0 animal cells. To progress from the G_0 towards the G_1 phase, dormancy has to be broken, or exogenous gibberellins to be applied. Whether gibberellins are the promoting factor themselves, or act indirectly, remains to be elucidated. This reasoning doesn't exclude that embryonic cells in a dry non-dormant condition may also be arrested in the G_0 phase. However, in these cells the progression towards the G_1 phase may occur fast due to the presence or rapid production of the promoting factor, which is absent in dormant or water imbibed *gib-1* seeds.

Abscisic Acid

For tomato abscisic acid (ABA) deficient mutants have also been isolated. One of these mutant, the *sit^w* mutant, has been used to study the role of ABA in seed development and germination (Groot, 1987; Hilhorst, 1995). The *sit^w* mutant is

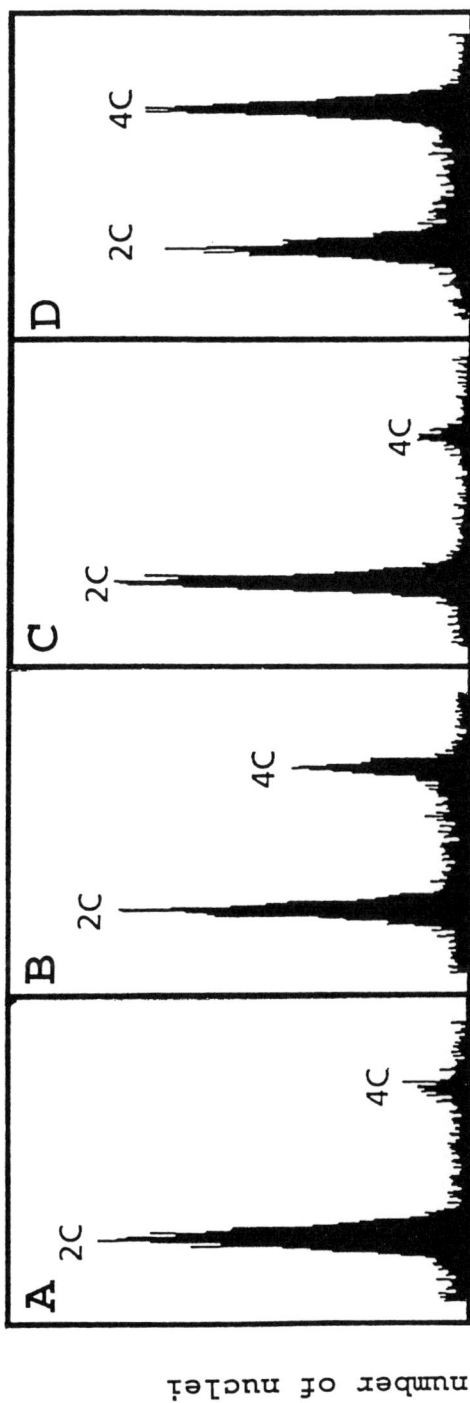

Figure 1. Histograms of flow cytometric analysis of nuclei from tomato embryo root tips from far-red irradiated and control seeds. **A** Nuclei from a dry seed; **B** Control seed imbibed 2 days in darkness; **C** Seeds irradiated 1 day with far-red light (5 min/h), followed by 1 day darkness (dormant seeds); **D** Far-red treated seeds (1 day) followed by 6 days darkness and 2 days white light (dormancy broken)

not completely devoid of ABA; developing mutant seeds contain about 3% of the ABA level of wild-type seeds (Groot *et al.* 1991). This strong reduction in ABA content has no significant effect on seed development, although the seeds germinate more readily compared to wild-type seeds (Groot and Karssen, 1992).

In the root tip from mature sit^w seeds the 4C:2C ratio of isolated nuclei is higher compared to those from wild-type seeds (Liu *et al.*, 1994). This suggests that in sit^w seeds, either, during maturation cell cycle activity is less efficiently arrested in the G_0 or G_1 state, or that germination processes had already started in the fruit before extraction of the seeds. The latter is in agreement with the observed viviparous germination of sit^w seeds in overripe fruits (Groot and Karssen, 1992). Upon imbibition of sit^w seeds in water, the nuclei in the radicle tip progress towards 4C DNA levels, as with the wild-type.

Imbibition of sit^w or wild-type seeds in 5 µM ABA inhibits the germination of both genotypes; however, nuclear DNA replication still commences (Liu *et al.*, 1994).

Osmotic Priming and Ageing

Priming of tomato seeds may be used to obtain seed lots with an accelerated and more uniform germination. Often priming is performed by incubating the seeds for days or even weeks in osmotic solutions and re-drying them after the treatment. The osmotic potential of the priming solutions is chosen such that the water content of the seeds is high enough for some pre-germination events to occur, but too low for actual germination.

We have also studied cell cycle activity during priming in 1.0 MPa PEG 6000, which prevents visible germination (Bino *et al.*, 1992). It was found that nuclear DNA replication in the radicle tip cells, still occurred during priming, although at a reduced rate. When the priming treatment was performed for a prolonged period of 14 days, most of these nuclei had progressed to 4C DNA levels. Similar results were also obtained by Baker and Bradford (1995), who showed that the rate of nuclear DNA replication was dependent on the osmotic potential of the priming solution.

A general disadvantage of priming is that the seeds become more sensitive to deteriorating conditions, such as high temperatures and high relative humidities, during subsequent storage and transport. It is very interesting to note, that this increased sensitivity is correlated with the 4C:2C ratios of the radicle tip nuclei (Baker and Bradford, 1995). A phenomenon that had previously also been observed with the evolutionary and structurally related pepper (*Capsicum annuum*) seeds (Saracco *et al.*, 1995).

β-Tubulin Accumulation

The cell cycle involves many regulatory and structural proteins. One of these proteins is β-tubulin, a component of the microtubular cytoskeleton. Studies using antibodies against tubulin have shown that the progression through the cell cycle is associated with changes in the specific organization of the microtubules (Hussey *et al.*, 1990; Traas *et al.*, 1992). In maize roots, it has been demonstrated that the progression of the cell cycle through the G_1 phase is dependent on the turnover of the microtubular cytoskeleton (Baluška and Barlow, 1993). However, the relation between synthesis of tubulins and cell cycle activity is not yet fully elucidated. At our laboratory, we are also studying β-tubulin levels during seed development and germination. With the use of commercially available anti-β-tubulin antibodies, at least three different β-tubulin isotypes can be detected on two-dimensional Western blots of radicle tips (de Castro *et al.*, 1995). All isotypes had a similar molecular weight of about 55 kDa. In the initial experiments no β-tubulin signal could be detected in extracts from radicle tips derived from dry seeds, whereas a weak signal was observed after one day's imbibition, a strong signal after two days', and a further accumulation after three days' imbibition, when germination had started (de Castro *et al.*, 1995). But, further development of the techniques has also enabled detection of a β-tubulin signal in the non-imbibed situation and a start of β-tubulin accumulation even prior to the onset of nuclear DNA replication (de Castro, unpublished results).

Analysis of β-tubulin levels in ABA treated tomato seeds, seeds from the *gib-1* mutant and during osmopriming, showed a clear correlation with the occurrence or absence of nuclear DNA replication (de Castro and coworkers, unpublished results). As with nuclear DNA replication, β-tubulin accumulation was strongest in the embryonic root tip compared to the other seed tissues. Far-red light treated secondary dormant tomato seeds do not exhibit an accumulation of β-tubulin, in contrast to non-irradiated control seeds (de Castro and Groot, unpublished results). Relief of dormancy by white light removed this blockage and β-tubulin accumulation was induced in the seeds.

Ageing of tomato seeds may result in a reduced rate of germination. Analysis of β-tubulin levels during imbibition of artificially aged tomato seeds, showed a clear correlation between rate of germination and rate of β-tubulin accumulation (de Castro *et al.*, 1995).

Conclusions

In dry tomato seeds a high proportion of the cells is synchronized in the G_0 or the G_1 phase of the cell cycle. After imbibition of the seeds the cell cycle is activated in the embryonic root tip, first noticeable by an accumulation of β-tubulin and later by nuclear DNA replication. The progression of cell cycle activity can be manipulated, by dormancy induction, priming and plant hormones as abscisic acid and gibberellins. This makes seeds ideal material to study the regulation of the cell cycle in plants. On the other hand, analysis of the cell cycle in germinating seeds will offer opportunities to get a more detailed knowledge of pre-germination seed development and its relation to seed quality.

References

Alberts, B., Bray, D., Lewis, J., Raff, M., Roberts, K. and Watson, J.D. 1989. *Molecular Biology of the Cell*, Second edition, pp. 1219. New York: Garland Publishing Inc.

Baker, E.H. and Bradford, K.J. 1995. *Proceedings of the Fourth National Symposium on Stand Establishment of Horticultural Crops, Monterey, California*, pp.69–76.

Baluška, F. and Barlow, P.W. 1993. *European Journal of Cell Biology* 61: 160–167.

Bewley, J.D and Black, M. 1978. *Physiology and Biochemistry of Seeds*, Vol 1, 306 pp. Berlin: Springer Verlag.

Bino, R.J, de Vries, J.N., Kraak, H.L. and van Pijlen, J.G. 1992. *Annals of Botany* 69: 231–236.

Bino, R.J., Lanteri, S., Verhoeven, H.A. and Kraak, H.L. 1993. *Annals of Botany* 72: 181–187.

Conger, B.V and Carabia, J.V. 1976. *Environmental and Experimental Botany* 16: 171–175.

Conger, B.V and Carabia, J.V. 1978. *Environmental and Experimental Botany* 18: 55–59.

de Castro, R.D., Zheng, X., Bergervoet, J.H.W., de Vos, C.H. and Bino, R.J. 1995. *Plant Physiology* 109: 499–504.

Deltour, R. 1985. *Seed Science Research* 1:37–44.

Groot, S.P.C. 1987. Ph.D. Thesis. Agricultural University Wageningen, N.L.

Groot, S.P.C. and Karssen, C.M. 1987a. *Planta* 171: 525–531.

Groot, S.P.C. and Karssen, C.M. 1987b. *Physiologia Plantarum* 71: 184–190.

Groot, S.P.C. and Karssen, C.M. 1992. *Plant Physiology* 99: 952–958.

Groot, S.P.C., van Yperen, I.I. and Karssen, C.M. 1991. *Physiologia Plantarum* 81: 73–78.

Groot, S.P.C., Kieliszewska-Rokicka, B., Vermeer, E. and Karssen, C.M. 1988. *Planta* 174: 500–504.

Hilhorst, H.W.M. 1995. *Seed Science Research* 5: 61–73.

Hussey, P.J., Lloyd, C.W. and Gull, K. 1988. *Journal of Biological Chemistry* 263:5474–5479.

Liu, Y., Bergervoet, J.H.W., de Vos, C.H., Hilhorst, H.W.M., Kraak, H.L., Karssen, C.M. and Bino, R.J. 1994. *Planta* 194:368–373.

Saracco, F., Bino, R.J., Bergervoet J.H.W. and Lanteri, S. 1995. *Seed Science Research* 5: 25–29.

Traas, J.A., Beven, A.F., Doonan, J.H., Cordewener, J. and Shaw, P.J. 1992. *Plant Journal* 723–732.

44. Regulation of an α-Type DNA Polymerase Activity During Maize Germination

J.M. VAZQUEZ-RAMOS, P. COELLO and E. GARCIA

Departamento de Bioquímica, Facultad de Química, UNAM. Ave. Universidad y Copilco, México D.F. 04510, México

Abstract

Three DNA polymerases can be separated by fractionation of crude protein extracts from maize embryonic axes through DEAE-cellulose. One of them, DNA polymerase 2 (DNA pol 2) has been purified 5000-fold and characterized. According to biochemical, immunological and physiological criteria, DNA pol 2 is an α-type enzyme. In support of this conclusion, we have found a DNA primase closely associated to DNA pol 2 along the entire purification procedure. With the help of antibodies developed against maize DNA pol 2, we have been able to determine its putative subunit composition. This enzyme is the only one, among DNA polymerases in embryonic axes, to increase in activity as the germination process is established. Since there is no increase in the amount of the different proteins that form DNA pol 2 during germination, the higher activity presented might be due to enzyme modification. We have found that DNA pol 2 is a phosphoprotein and that phosphorylation is a cyclic event during the 0–48 h period of germination. The implication of this is discussed. Finally, we show that pol 2 is a labile enzyme in seed deterioration and that it may be responsible, at least partially, for the decrease in vigour/viability of deteriorated seeds.

Introduction

Seeds in the dry state perform almost no detectable metabolism, but this soon restarts after seed imbibition in the form of protein synthesis, energy production, carbohydrate degradation, ribosomal assembly, membrane repair and/or reassembly, etc. A low but consistent DNA synthesis also takes place at these early hours, both in cotyledons and in embryos and evidence indicates that this synthesis is located both in organelles and in nuclei (Galli, 1982; Vázquez-Ramos, 1992); in the latter, DNA synthesis seems to be of repair type (Zaraín *et al.*, 1987). Nuclear DNA replication is a late event during germination and it is present in only a low proportion of the cells in the different meristematic embryo tissues (Baiza *et al.*, 1989). Not surprisingly, the number of cells at mitosis found in meristems at later times is also very low.

The time schedule in maize seeds for the events concerned with the duration of the cell cycle has been reported: most of the cells in dry embryonic axes are in the G1 state and S-phase begins by 15 h of imbibition; mitotic figures appear between 24 and 32 h of imbibition in the different meristems (Baiza *et al.*, 1989).

R.H. Ellis, M. Black, A.J. Murdoch, T.D. Hong (eds.), Basic and Applied Aspects of Seed Biology, pp. 403–412.
© *1997 Kluwer Academic Publishers, Dordrecht. Printed in Great Britain.*

DNA polymerases are central figures for the advancement of the S phase. In eukaryotes (except plants), 3 different replicative nuclear DNA polymerases (α, δ and ε) and one with repair function (β) have been reported (Kornberg and Baker, 1992); mitochondrial enzymes (γ) have also been described (Kornberg and Baker, 1992).

The knowledge of plant DNA polymerases lags far behind; the number of distinct polymerases present in plant cells or their physiological roles are unknown. There is some evidence of α-type (Litvak and Castroviejo, 1987; Bryant *et al.*, 1992) and δ-type (Richard *et al.*, 1991) enzymes and also of a β-type (Castroviejo *et al.*, 1990). Mitochondrial and chloroplastic enzymes have also been reported (McKown and Tewari, 1984; Heinhorst *et al.*, 1990).

Our work with DNA metabolism and the cell cycle during maize germination has led us to study maize DNA polymerases. In the past we have reported the purification and characterization of a DNA polymerase that could be classified as an α-type or replicative enzyme, since it has similar Mg, KCl, pH and temperature requirements, and also similar K_m for deoxyribonucleotides; although the response to specific polymerase α inhibitors is confusing, the enzyme was recognized by an antibody developed against an α-type calf thymus DNA polymerase (Coello *et al.*, 1992).

This enzyme has been further characterized; it was discriminated from other maize polymerases and its regulation during the process of germination was studied; these data are presented below.

Results and Discussion

Characterization of DNA Polymerase 2

By fractionating crude protein extracts from maize axes (imbibed for 24 h) through DEAE-cellulose, 3 peaks of DNA polymerase activity are obtained. Polymerases are named after their order of elution as 1, 2 and 3 (Coello *et al.*, 1992). Enzymes 1 and 3 have been partially purified and their characterization is under way. DNA polymerase 2 was purified 5000 fold (Coello *et al.*, 1992) and studied further because it presented the highest specific activity, which gradually increased during germination (see below). As stated before, the purified enzyme presented characteristics of an α-type, replicative polymerase.

Two important distinctive features of an α-type enzyme were determined for DNA pol 2: its processivity and the presence of an accompanying DNA primase activity.

The processivity of a DNA polymerase, which defines the mean length of oligonucleotides synthesized after a single primer binding event, was determined for DNA pol 2. It was found that the processivity value, under optimal *in vitro* conditions of ions, pH, concentration of deoxynucleotides and temperature was of about 18 nucleotides added per event (Coello and Vázquez-Ramos, 1995b), nonwithstanding the reaction time (Fig. 1). This value is similar to those

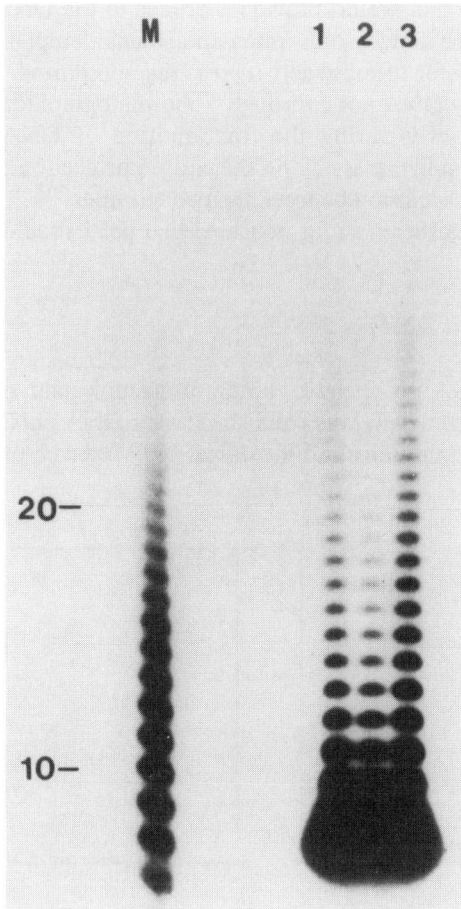

Figure 1. Processivity of DNA pol 2. DNA pol 2 was purified 2000-fold by passing protein extracts through DEAE-cellulose and Heparin-Sepharose. The assay for processivity contained, among other things, polydA/oligodT as template/primer, Mg^{++}, ^{32}P-ATP and DNA pol 2. The reaction took place for 15 (lane 1), 30 (lane 2) or 60 min (lane 3). M = ^{32}P-labelled oligonucleotide ladder

reported for animal α-type polymerases (Kornberg and Baker, 1992).

Every DNA polymerase α isolated to date is a multisubunit enzyme or holoenzyme, and two of the composing subunits constitute the DNA primase activity (Kornberg and Baker, 1992). We then studied whether a DNA primase activity copurified with DNA pol 2. To measure this activity, use was made of single-stranded templates, ribo- and deoxyribonucleotides (one of these radio-labelled) and a partially purified fraction of DNA pol 2. No primer was added and therefore any DNA synthesis taking place would be the result of the action of a DNA primase that would provide a primer to the DNA polymerase. We could detect primase activity only when an artificial template such as poly-dT was used. The assay for primase activity was then optimized. The next step was to extensively purify DNA pol 2 through 4 chromatographic steps and look for DNA primase activity during the fractionation of DNA pol 2. Primase copurified with the polymerase along the entire purification procedure (Fig. 2) indicating a close association between the two enzymes.

All the evidence gathered so far indicated that pol 2 should be considered an α-type enzyme.

An Antibody Against DNA pol 2

Highly purified DNA pol 2 elutes rapidly from molecular sieves suggesting a high molecular weight. This was confirmed when DNA pol 2 was electrophoresed under native conditions and identified, by Western blot, by the antibody

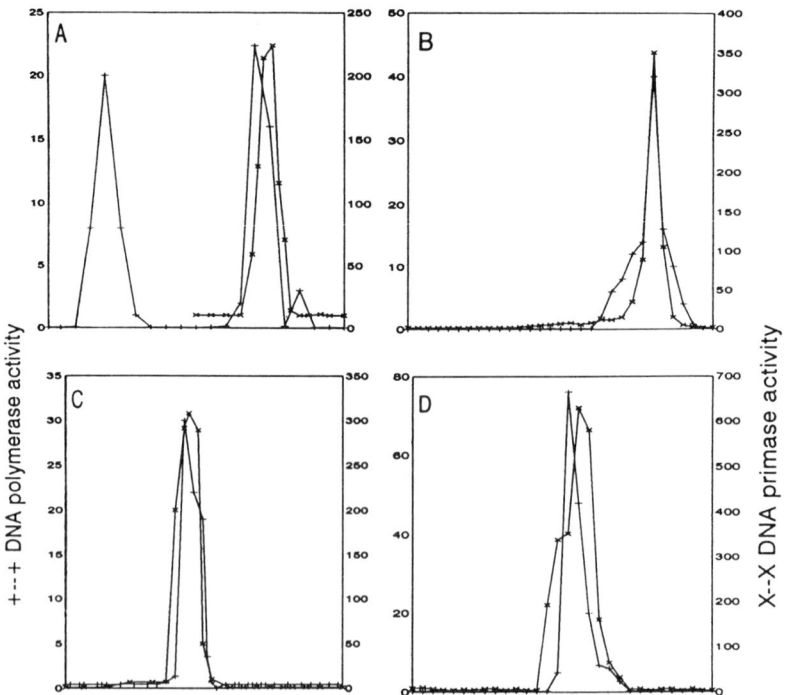

Figure 2. Copurification of DNA pol 2 and DNA primase. Crude protein extracts from 24 h imbibed maize axes were fractionated through DEAE-cellulose (A), Heparin-Sepharose (B), Superdex 200 (C) and Mono Q HR 5/5 (D). DNA polymerase 2 activity was the marker to follow and DNA primase activity was tested in all fractions

against the calf thymus α-enzyme; it was an enzyme with a molecular weight above 250 kDa. This property helped us to develop an homologous antibody against the maize enzyme, cutting the corresponding protein band from native gels and injecting it into rabbits. In Western blots of gels run under denaturing conditions, the homologous antibody recognized 7 protein bands of 90, 83, 70, 60, 55, 45 and 24 kDa. The 90 kDa band was shown, by activity gels, to be the catalytic subunit (Coello *et al.*, 1994). Every DNA polymerase α studied to date consists of polypeptides of around 160, 70, 60 and 50 kDa, the last 2 composing the DNA primase (Kornberg and Baker, 1992). DNA pol 2 has polypeptides of 70, 60, 55 and 45 kDa; some of them might constitute the primase activity.

Regulation of DNA pol 2 During Maize Germination

The antibody against DNA pol 2 was used to discriminate between the 3 DNA polymerases separated by DEAE-cellulose. Only DNA pol 2 but not the 1 or 3, was both immunodetected and immunoinhibited by the antibody, indicating that the 3 enzymes are not immunologically (serologically) related (Coello *et al.*, 1994).

DNA pol 2 is interesting from the physiological point of view since it is the only polymerase whose activity increases several fold as maize germination advances, reaching a peak of activity at a time when cells are fully engaged in DNA replication (Coello *et al.*, 1992; Coello and Vázquez-Ramos, 1995a).

The question arising was whether this increase in activity was due to synthesis of new enzyme. The homologous antibody developed against DNA pol 2 was used to answer this question. Proteins were extracted from maize axes imbibed for different times and, by Western blotting, the amount of either the native enzyme (holoenzyme) or that of the different protein subunits was evaluated. The result indicated that neither the amount of the holoenzyme (Fig. 3), nor the amount of the different protein subunits (Fig. 6A) changed during the period between 0 to 48 h of germination (Coello and Vázquez-Ramos, 1995a). This suggests that the increase in activity might be due to enzyme modification.

Since protein phosphorylation/dephosphorylation is a general eukaryotic mechanism for modifying enzymic activity, and human DNA polymerase α is a phosphoprotein (Wong *et al.*, 1986; Nasheuer *et al.*, 1991), we considered the possibility that DNA pol 2 was also a phosphate-modified enzyme. We found that indeed, DNA pol 2 is phosphorylated, but not all the time. By 3 h-pulse labelling maize axes with ^{32}P, and immunoprecipitating the resulting ^{32}P-labelled protein with the homologous antibody, we could determine the potential for DNA pol 2 phosphorylation at different times during germination. It was found that DNA pol 2 was not phosphorylated between 0–3 h of imbibition, the different subunits were phosphorylated between 11–14 h, again not phosphorylated between 21 and 24 h and re-phosphorylated between 45 and 48 h of germination. These results suggested that there were precise times at which (a) the kinases responsible for phosphorylating DNA pol 2 were present, or (b) DNA pol 2 was accessible to the kinases. Although we still do not know

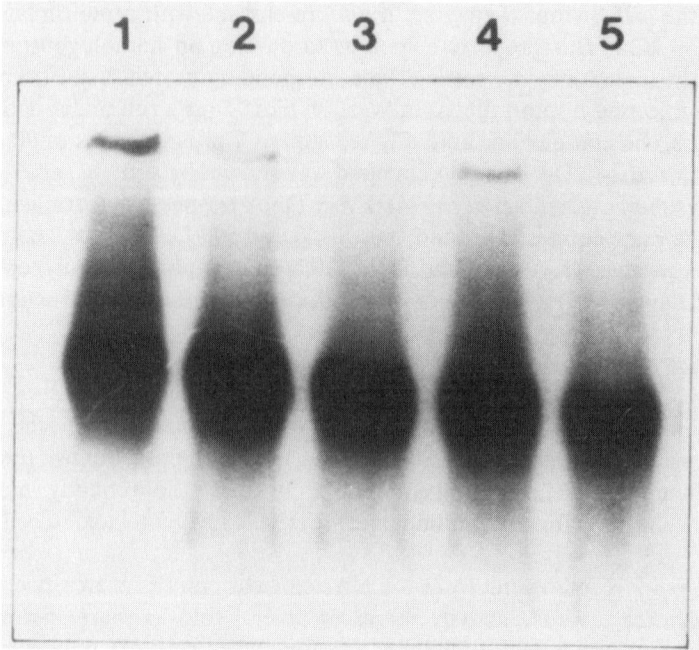

Figure 3. Changes in the amount of DNA pol 2 during germination. Proteins in crude extracts from maize axes imbibed for the indicated time periods were separated by native PAGE, the proteins were transferred to nitrocellulose filters and developed by the Western blot technique using the antibody against DNA pol 2. Lanes: 1, 0 h; 2, 3 h; 3, 14 h; 4, 24 h; 5, 48 h

the answer, we found that there was very little potential to phosphorylate the different subunits of DNA pol 2 between 0–11 h of germination, demonstrated by an experiment in which axes were 2 h, ^{32}P-pulse labelled at different times during this period and the proteins from crude extracts were immunoprecipitated using the homologous antibody. Very little label was present in the different protein subunits despite the time of film exposure (13 days, Fig. 4). On the contrary, the enzyme was labelled in the 11–14 h period and remained phosphorylated after that time (Coello and Vázquez-Ramos, 1995a); this was demonstrated after a pulse-chase experiment in which axes were ^{32}P-labelled in the period 11–14 h of germination, the label was washed out and the axes then re-incubated and samples taken every 2 h to examine by immunoprecipitation if the DNA pol 2 subunits had lost the label. The answer was that the label remained up to 24 h of germination (film exposure time was 2 days, Fig. 5).

All these results suggest that DNA pol 2 is modified by phosphorylation at a time very close to the time at which the S phase begins in meristematic cells and that the enzyme in those cells remains phosphorylated perhaps throughout the whole S phase. Since there is another period at which, by pulse labeling,

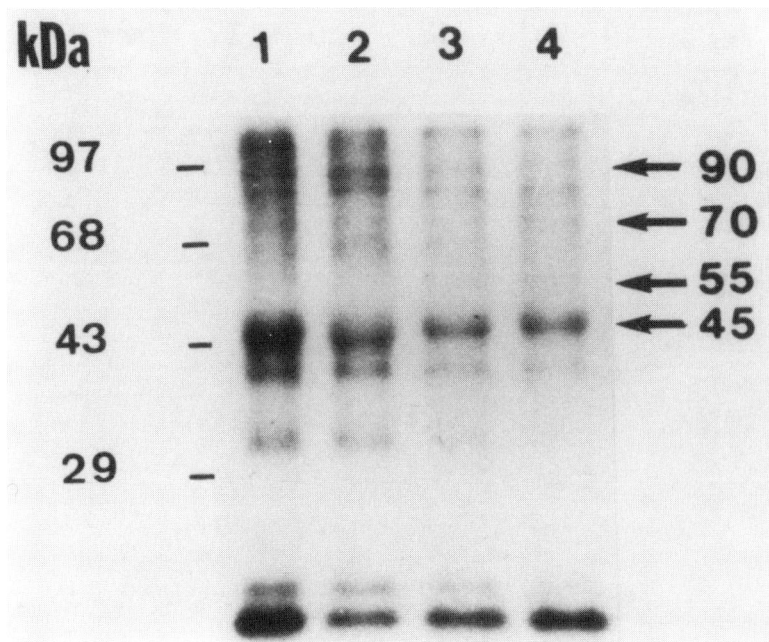

Figure 4. Radioactive labelling of DNA pol 2 during early germination. Embryonic axes were labelled with ^{32}P during the 3–5 h (lane 1), 5–7 h (lane 2), 7–9 h (lane 3) and 9–11 h (lane 4) of germination. Proteins in crude extracts were immunoprecipitated with the antibody against DNA pol 2, separated by SDS-PAGE and gels were processed for autoradiography. The film exposure time was 13 days

phosphate is incorporated to the DNA pol 2 (45–48 h), it means that at some point after 24 h of germination, DNA pol 2 is dephosphorylated to be rephosphorylated again later on. This cyclic addition of phosphate might represent a form of regulation of enzyme activity/assembly.

Our efforts to modify DNA pol 2 activity by *in vitro* phosphorylation–dephosphorylation of the enzyme did not produce reliable results. The effect of phosphorylation on enzyme processivity was also tested and we found that the 24 h DNA pol 2 did respond to phosphorylation, although marginally. Therefore, we still do not know what the effect of *in vivo* enzyme phosphorylation may be. We do know, however, that the increase in DNA pol 2 activity as germination proceeds is not due to *de novo* synthesis of the enzyme (Coello and Vázquez-Ramos, 1995a). We are still working on this topic.

Replicative DNA polymerases are fundamental to every living cell. Since DNA replication is part of the germination process, then replicative poly-

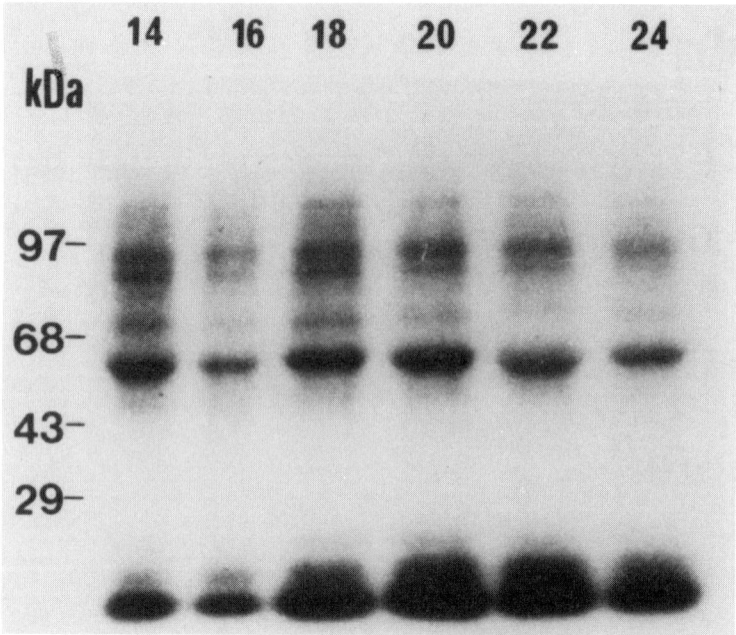

Figure 5. Pulse-chase of the radioactive label in DNA pol 2 after the 11–14 h period of germination. Embryonic axes were labelled with ^{32}P during the 11–14 h period and then radioactive label was thoroughly washed. Axes were reincubated and samples were taken every 2 h; proteins in crude extracts from these axes were immunoprecipitated using the homologous antibody, separated by SDS-PAGE and gels were processed for autoradiography. Numbers above the lanes represent the time of germination at which samples were removed. The film exposure time was 2 days (compare with Figure 4)

merases will also be of the highest importance. A consequence of loss of vigour/ viability in maize seeds is a decrease in the capacity of cells in axes to synthesize DNA; DNA polymerase activity is also greatly reduced (Vázquez-Ramos *et al.*, 1988). We have been able to show that in those seeds in which viability has been reduced to less than 40%, DNA pol 2 activity is very much reduced. Furthermore, this reduction in activity is due to the disappearance of the different subunits that compose the enzyme (Fig. 6B). Therefore, DNA pol 2 is labile in seed deterioration and its loss is probably one of the causes of the reduction of general DNA synthesis during seed germination and, possibly, of loss of seed vigour/viability.

Our future efforts will be devoted to determining the effect of phosphorylation on DNA pol 2 (activity, processsivity, assembly), and which amino acid residues

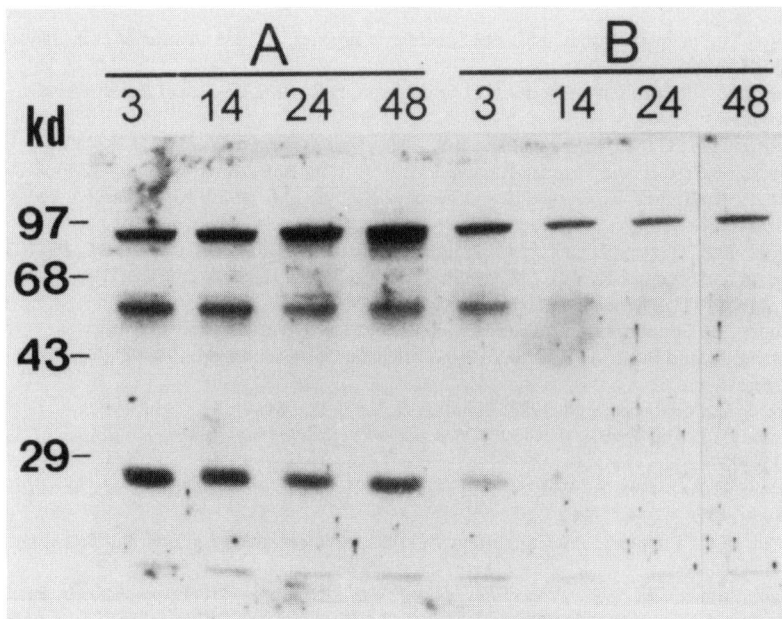

Figure 6. Amount of DNA pol 2 subunits during germination and the effect of seed deterioration. Proteins in crude extracts either from control (A) or deteriorated (B) seed axes imbibed for 3, 14, 24 and 48 h (A and B) were separated by SDS-PAGE, processed as for Western blotting and developed using the antibody against DNA pol 2

are phosphorylated; also, which polypeptides contain the DNA primase activity. It is in our scope to pursue the cloning of the gene for the protein that possesses the catalytic activity (90 kDa protein) and in this way to know its expression along germination and possible regulatory mechanisms.

Acknowledgements

We acknowledge receiving a grant from OAS-Plant Biotechnology.

412 J.M. Vazquez-Ramos, P. Coello and E. Garcia

References

Baiza, A.M., Vázquez-Ramos, J.M. and Sánchez de Jiménez, E. 1989. *Journal of Plant Physiology* 135: 416–421.
Bryant, J.A., Fitchett, P.N., Hughes, S.G. and Sibson, D.R. 1992. *Journal of Experimental Botany* 43: 31–40.
Castroviejo, M., Gatius, M.T. and Litvak, S. 1990. *Plant Molecular Biology* 15: 383–397.
Coello, P., García, E. and Vázquez-Ramos, J.M. 1994. *Canadian Journal of Botany* 72(6): 818–822.
Coello, P., Rodríguez, R., García, E. and Vázquez-Ramos, J.M. 1992. *Plant Molecular Biology* 20: 1159–1168.
Coello, P. and Vázquez-Ramos, J.M. 1995a. *European Journal of Biochemistry* 231(1): 99–103.
Coello, P. and Vázquez-Ramos, J.M. 1995b. *Plant Physiology* 109: 645–650.
Galli, M.G. 1982. *Physiologia Plantarum* 56: 245–250.
Heinhorst, S., Cannon, G. and Weissbach, A. 1990. *Plant Physiology* 92: 939–945.
Kornberg, A. and Baker, T. 1992. *DNA Replication*, Second Edition, pp. 198–199. New York: WH Freeman and Co.
Litvak, S. and Castroviejo, M. 1987. *Mutation Research* 181: 81–91.
McKown, R.L. and Tewari, K.K. 1984. *Proceedings of the National Academy of Sciences USA* 81: 2354–2358.
Nasheuer, H.P., Moore, A., Wahl, F. and Wang, T.S.F. 1991. *Journal of Biological Chemistry* 266: 7893–7903.
Richard, M.C., Litvak, S. and Castroviejo, M. 1991. *Archives of Biochemistry and Biophysics* 287: 141–150.
Vázquez-Ramos, J.M. 1992. *Proceedings of the Fourth International Workshop on Seeds. Basic and Applied Aspects of Seed Biology* pp. 317–322 (Ed. D. Come and F. Corbineau). Paris: ASFIS.
Vázquez-Ramos, J.M., López, S., Vázquez, E. and Murillo, E. 1988. *Journal of Plant Physiology* 133: 600–604.
Wong, S.W., Paborsky, L.R., Fisher, P.A., Wang, T.S.F. and Korn, D. 1986. *Journal of Biological Chemistry* 261: 7958–7968.
Zaraín, M., Bernal-Lugo, I. and Vázquez-Ramos, J.M. 1987. *Mutation Research* 181: 103–110.

45. DNA Integrity and Synthesis in Relation to Seed Vigour in Sugar Beet

M. REDFEARN[1], N.A. CLARKE[1], D.J. OSBORNE[2], P. HALMER[3] and T.H. THOMAS[1]

[1]IACR-Broom's Barn, Higham, Bury St. Edmunds, Suffolk, IP28 6NP; [2]Oxford Research Unit, The Open University, Foxcombe Hall, Oxford, OX1 5HR; [3]Germain's (UK) Ltd., Hansa Rd., Hardwick Industrial Estate, King's Lynn, Norfolk, PE30 4LG, UK

Abstract

DNA repair, including the ligation of DNA breaks, is one of the earliest events during rehydration from the dry state in seeds. The γ-irradiation of dry embryos produces DNA damage detectable on alkaline gels. Repair has been shown to occur in sugar-beet embryos following a two-hour period of imbibition. The effect of vigour on DNA repair following irradiation was investigated using two sugar-beet growers lots.

Improvements in seed vigour following the primed advancement treatment have been demonstrated through accelerated seedling emergence in the field. Earlier studies have shown that the advanced seed had a higher ratio of extractable high molecular weight RNA to high molecular weight DNA than untreated seed. Using Feulgen staining and image analysis, the proportions of 2C and 4C nuclei were calculated. Nuclei from the advanced seed showed a higher proportion of 2C-4C intermediates, 4C and 8C DNA contents than untreated seed indicating that the advancing treatment had progressed the onset of DNA replicative synthesis.

Introduction

Decreasing vigour and viability affects DNA integrity in two ways. Firstly, DNA strand breaks (and other damage) occur during the loss of viability in rye (Cheah and Osborne, 1978) and maize (Vázquez-Ramos et al., 1988). This is shown by greater DNA fragmentation when DNA is extracted, denatured and run on alkaline agarose gels (Elder et al., 1987). Secondly, the DNA repair capacity declines with decreasing viability. DNA ligase function is decreased in rye (Elder et al., 1987) and maize (Vázquez et al., 1991). A decrease in DNA polymerase activity is also seen in deteriorated maize (Vázquez-Ramos et al., 1988) and barley (Yamaguchi et al., 1978). High vigour/viability embryos possess DNA with very little damage and effective repair systems. To demonstrate that high vigour/viability embryos have good repair systems, the DNA is artificially damaged using γ-irradiation. This damage is visualized on an alkaline gel where there is a movement of the DNA profile to a lower mean molecular weight in comparison with DNA extracted from untreated embryos. Following a period of imbibition, fragments are converted back to higher molecular weight by DNA repair. This has been shown in rye (Elder et al.,

R.H. Ellis, M. Black, A.J. Murdoch, T.D. Hong (eds.), Basic and Applied Aspects of Seed Biology, pp. 413–420.
© 1997 Kluwer Academic Publishers, Dordrecht. Printed in Great Britain.

1987) and it is thought that DNA repair occurs by a patch, with the incorrect nucleotide(s), being excised and replaced and then ligated (Osborne, 1983). The aim of this work was to determine if DNA repair was detectable in sugar-beet embryos using this method and if so, how vigour affects the capacity of embryos to repair artificially induced DNA damage.

DNA repair-type synthesis occurs during osmotic priming (Ashraf and Bray, 1993) followed by DNA replication in the embryonic meristem. The effect of the seed treatment, advancing, on the sugar-beet DNA content was investigated. The prolonged advancement treatment was developed by Durrant and Mash (1992) and involves imbibing the seeds for 8 h and then keeping the seeds at 124% moisture content for 90 h. Advanced seeds show accelerated seedling emergence in the field (Thomas *et al.*, 1993) and in the cold stress test (Redfearn, unpublished results) compared to seeds which have been thiram-steeped for 6 h. Advanced seeds also have a larger ratio of extractable high molecular weight RNA to DNA compared to thiram-steeped seeds or untreated seeds (Redfearn *et al.*, in press). In this study, the effect of the advancement treatment on the DNA status was investigated using Feulgen staining and image analysis. The proportion of nuclei with each DNA content was calculated to determine the stage in the cell cycle that the nuclei were held following the advancement treatment

Materials and Methods

The commercial sugar-beet 'seed' is botanically termed the fruit while the 'true seed' refers to the embryo plus its perisperm and integuments (Richard *et al.*, 1989). Growers lots from two cultivars, A and B, were analysed. A growers lot refers to seeds from a cultivar produced by one farmer. The embryos were extracted from untreated sugar-beet seeds of one growers lot of cultivar A for the initial DNA repair experiment. To compare the DNA repair of seedlots with different vigour levels, two growers lots from cultivar B were selected based on assessments made previously in the field and in laboratory tests. For the DNA content determination, embryos were extracted from the advanced or untreated seeds (growers lot from cultivar A). The true seed was extracted from the seed coat and then imbibed in deionized water for 1 h at room temperature to soften the surrounding tissue. The embryo was removed using tweezers and rinsed in deionized water. The embryos were then placed in fixing solution (1:1 acetic acid:ethanol) overnight for the DNA content experiment or dried back to their original weight overnight at 24°C for the DNA repair work.

DNA Repair

The embryos were left untreated (control), γ-irradiated or γ-irradiated then imbibed. Dry embryos were exposed to γ-irradiation of 100 Kr for 105 min (1000 Grays) from a ^{137}Cs-source. The embryos were imbibed on 5.5 cm

Whatman 1 filter paper discs soaked in 1 ml of 1.5% sucrose solution for 2 h at 24°C. The embryos were dried back to their original dry weight at 24°C. All the dry embryos were stored in the refrigerator. For the extraction of DNA, ten embryos per microcentrifuge tube were used.

The method of DNA extraction and alkaline gel electrophoresis was adapted from Elder *et al.* (1987). The embryos in each 0.5 ml microcentrifuge tube were frozen in liquid nitrogen before grinding in the homogenizing medium (0.15 M NaCl, 0.1 M EDTA, 0.5% SDS in 50 mM Tris–HCl (pH 7.8) solution) using a metal crusher. The samples were centrifuged to remove particulate matter and incubated with proteinase K (from *Tritirachium album*) at 37°C for 3 h. The DNA was precipitated in ethanol overnight at –20°C. Following centrifugation, the DNA pellets were dissolved in TE buffer (pH 7.6) and incubated with RNase A (from bovine pancreas) for 2 h at 37°C. Alkaline loading buffer was added to each sample before denaturing at 80°C.

The alkaline gel was pre-run in an alkaline buffer (30 mM NaOH and 1 mM EDTA (pH 8)) for 3×30 min periods (changing the buffer after each run of 30 min) in the fridge at 50 V. Once the samples were loaded, the gel was run in fresh alkaline buffer at 24 mA (12 V) in the fridge overnight. The gel was neutralized and stained using ethidium bromide. To remove unbound ethidium bromide, the gel was rinsed in water. The gels were photographed using a Polaroid MP 4 land camera and a Flowgen transilluminator. The intensity of the ethidium bromide-stained DNA was quantified using an image analyser (Aequitas 1A, Dynamic Data Links Ltd, Cambridge).

DNA Content

The Feulgen stain was prepared using fuchsin basic stain. The embryos were stained using the conventional Feulgen staining method (De Tomasi, 1936) and the root tips mounted on slides.

The intensity of staining of the nuclei was quantified using image analysis. Each slide was magnified ($\times 400$) under a microscope (Reichert Diastar) and the image transferred to the computer using a video camera (Toshiba) focused on the microscope eye-piece. The image analysis programme (Aequitas IA) was set-up with specific brightness and contrast levels. A scale on a glass slide was used to calibrate the microscope and image analyser. The average intensity and detected area of the stained nucleus and the background intensity were measured for 100 nuclei (25 nuclei per slide) for each treatment.

The intensity readings were determined by calculating the difference between the background readings and the average intensity readings of each nucleus. The DNA content per nucleus (in arbitrary units) was calculated by multiplying the intensity of the stained nucleus by the detected area.

Results and Discussion

DNA Repair

Figure 1 shows the fragmentation profile of the DNA samples run on an alkaline gel. The higher molecular weight DNA is found closest to the origin and the molecular weight decreases progressively down the gel. γ-Irradiation of the embryos increases the fragmentation of the DNA which is seen by an increase in smaller molecular weight DNA compared to unirradiated embryos. The fragmentation profile of the control embryos has the highest intensity reading at approximately 20 units from the origin whereas the irradiated embryos have a DNA profile which peaks at approximately 60 units. If the irradiated embryos are imbibed for two hours, then the DNA profile moves to a higher molecular weight, with the maximum intensity at around 20 units from the origin.

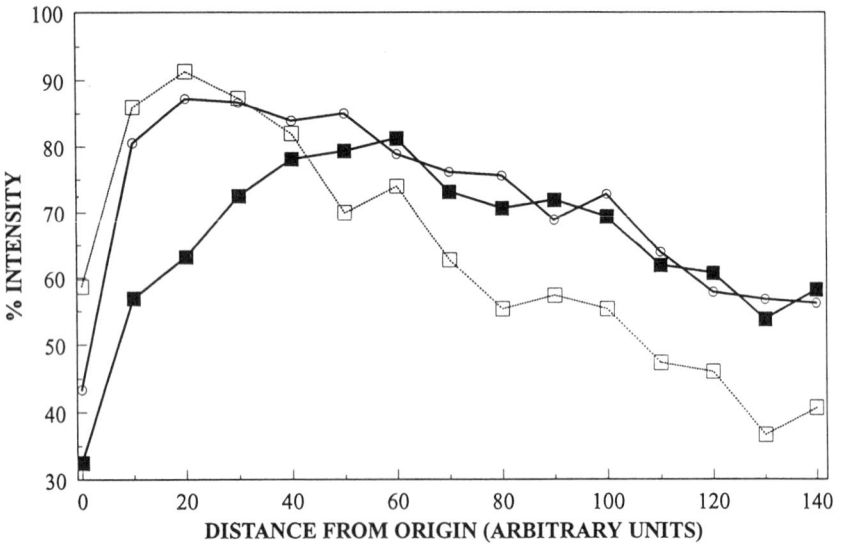

Figure 1. The DNA fragmentation profiles of the control (□), γ-irradiated (■) and γ-irradiated then imbibed (o) embryos run on an alkaline gel

The intensity of ethidium bromide-stained DNA was calculated at four positions down the alkaline gel for each lane. In Table 1, segment A includes the measurements nearest to the origin which corresponds to the highest molecular weight DNA fragments. The smaller DNA fragments migrate further down the gel so that segment D includes the smallest molecular weight

Table 1. The percentage of the total intensity of ethidium bromide-stained DNA in each segment of the alkaline gel. Segment A is found at the top of the gel nearest the origin; subsequent segments are positioned down the gel. Each value in the table is the mean of 10 readings and the percentage of the total intensity is calculated for each mean

Segment	Control (%)	γ-Irradiated (%)	γ-Irradiated–imbibed (%)
A	19.2	10.9	19.1
B	30.8	27.7	26.9
C	30.8	34.1	28.8
D	19.2	27.3	25.2
Total	100	100	100

fragments. The percentage of the total intensity readings for each segment was calculated. The DNA from the control embryos had nearly 20% of the total fragments in the highest molecular weight segment A whereas the irradiated embryos had just over 10%. The irradiated embryos which were subsequently imbibed for 2 h had 19% of the total fragments in the highest molecular weight segment. This demonstrates that the 2 h imbibition period following irradiation facilitates a joining of the smaller molecular weight fragments to produce more higher molecular weight fragments; this process being DNA repair.

There were no consistent differences found between the extent of DNA repair of high and low vigour embryos, during imbibition following the irradiation treatment. The seed used was commercially available and therefore of high quality so that the differences between high and low vigour were small. In this experiment, imbibition, which facilitates DNA repair, took place at an optimum temperature of 24°C. Earlier work has shown that there are smaller differences between the performance of the high and low vigour seeds when they are germinated at 20°C than when they are germinated under stressful conditions such as at 9°C (Redfearn, unpublished data). Therefore the extent of DNA repair under stressful conditions of artificially damaged DNA may well be greater for the high vigour seeds compared to the low vigour seeds. We have yet to compare the DNA repair of different seedlots under conditions such as low temperatures.

DNA Content

The percentage of root tip nuclei with each DNA content was calculated for untreated and the advanced seed. The 1C value corresponds to the DNA content of the unreplicated haploid chromosome complement (Bennett and Smith, 1976). At the completion of maturation in most plant embryos, the cycle arrests in the G1 phase with a 2C DNA content (Baker and Bradford, 1995).

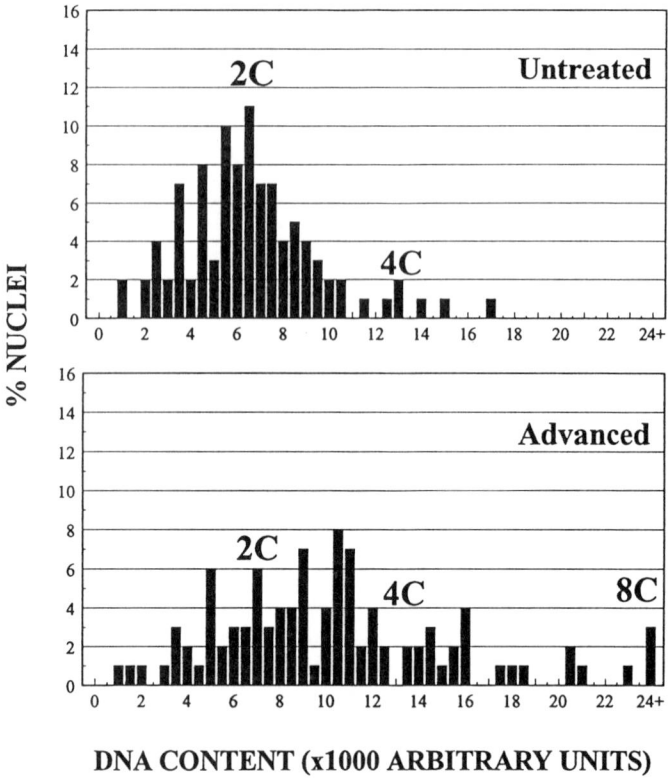

Figure 2. The effect of the primed advancement treatment on the DNA contents of cultivar A root tip cells

This 2C peak is clearly seen in the untreated root tip nuclei in Figure 2. The nuclei from the advanced seeds show a higher proportion of DNA contents larger than 2C in comparison with untreated seeds which mostly have nuclei with a 2C DNA content. Table 2 shows the decrease in the percentage of 2C nuclei when the seed is advanced. The percentage of nuclei with the 4C DNA content rises from 5% to 20% following the advancement treatment. There is a small increase in 8C nuclei from 0% in untreated seed to 3% in advanced seed. This indicates that the advancement treatment facilitates DNA replication which arrests at the first post S-phase because no cell division was seen in the advanced root tip cells. These results are similar to those determined using flow cytometry in pepper (Lanteri *et al.*, 1993, 1994) and tomato root tip cells where osmopriming increased the percentage of 4C nuclei (Bino *et al.*, 1992; Lanteri *et al.*, 1994; Baker and Bradford, 1995). There was no evidence of cell division

Table 2. The percentage of root tip nuclei from advanced or untreated embryos with DNA contents of 2C, 4C or 8C calculated from Figure 2

	DNA content		
Seed treatment	2C	4C	8C
Untreated	93[a]	5[c]	0
Advanced	70[b]	20[d]	3

[a] and [b] are significantly ($p < 0.05$) different
[c] and [d] are significantly ($p < 0.05$) different

occurring in the advanced root tip cells. This was also seen in leek during priming where DNA synthesis occurred without cell division (Bray *et al.*, 1989).

The primed advancement treatment advances seed in the field and in the cold stress germination test in addition to increasing the percentage of nuclei with larger DNA contents than 2C. Lanteri *et al.* (1994) found that the advancement of germination in the laboratory, following osmopriming (the reduction in mean time to germination) was significantly ($p50.01$) correlated with the percentage of 4C nuclei. Therefore, the acceleration in rate of germination by a seed treatment, such as advancement or priming, facilitates DNA replication without cell division. In addition, the drying back stage following the advancement treatment evidently did not reduce seed performance so that the first post S-phase, at which the sugar-beet embryos arrested, is not desiccation sensitive.

Conclusion

DNA repair was detected in sugar-beet embryos during a 2 h DNA content imbibition period following damage by γ-irradiation. There were no consistent differences between the DNA repair capability of high and low vigour seeds. This may be due to the seeds analysed being of similar high quality and the repair taking place under optimum conditions.

Advanced seeds show improved performance in the field and in the cold stress germination test. The 2 h imbibition period, followed by drying, increased the percentage of nuclei with larger DNA contents (2C–4C intermediates, 4C and 8C) without cell division occurring. Therefore, the advancement treatment facilitated nuclear DNA replication which terminated at the post S-phase. The post S-phase in sugar-beet embryos is not desiccation sensitive because drying, following the advancement treatment, did not cause a decrease in seed performance.

Acknowledgements

The authors would like to acknowledge Germain's (U.K.) Ltd. for the donation of the seeds and for carrying out the primed advancement treatment on the seeds. This work was funded by Germain's (U.K.) Ltd.

References

Ashraf, M. and Bray, C.M. 1993. *Seed Science Research* 3: 15–23.

Baker, E.H. and Bradford, K.J. 1995. In: *Proceedings of the Fourth National Symposium on Stand Establishment of Horticultural Crops*, pp.69–76.

Bennett, M.D. and Smith, J.B. 1976. *The Philosophical Transactions of the Royal Society of London* B274: 227–274.

Bino, R.J., de Vries, J.N., Kraak, H.L. and van Pijlen, J.G. 1992. *Annals of Botany* 69: 231–236.

Bray, C.M., Davison, P.A., Ashraf, M. and Taylor, R.M. 1989. *Annals of Botany* 63: 185–193.

Cheah, K.S.E. and Osborne, D.J. 1978. *Nature* 272: 593–599.

De Tomasi, J.A. 1936. *Stain Technology* 11: 137–144.

Durrant, M.J. and Mash, S.J. 1992. *Annals of Applied Biology* 120: 151–159.

Elder, R.H., Dell'Aquila, A., Mezzina, M., Sarasin, A. and Osborne, O.J. 1987. *Mutation Research* 181: 61–71.

Lanteri, S., Kraak, H.L., Ric de Vos, C.H. and Bino, R.J. 1993. *Physiologia Plantarum* 89: 433–440.

Lanteri, S., Saracco, F., Kraak, H.L. and Bino, R.J. 1994. *Seed Science Research* 4: 81–87.

Osborne, D.J. 1983. *Canadian Journal of Botany* 61: 3568–3577.

Redfearn, M., Clarke, N.A., Halmer, P., Osborne, D.J. and Thomas, T.H. In: *Proceedings of the 58th Congress of the Institut International Recherches de Betteravières*, Beaune.

Richard, G., Raymond, P., Corbineau, F. and Pradet, A. 1989. *Seed Science and Technology* 17: 485–497.

Thomas, T.H., Jaggard, K.W., Durrant, M.J., Mash, S.J. and Armstrong, M.J. 1993. In: *Proceedings of the 56th Winter Congress of the Institut International Recherches de Betteravières*, Brussels, pp. 437–448.

Vazquez, E., Montiel, F. and Vázquez-Ramos, J.M. 1991. *Seed Science Research* 1: 269–273.

Vazquez-Ramos, J.M., Lopez, S., Vazquez, E. and Murillo, E. 1988. *Journal of Plant Physiology* 133: 600–604.

Yamaguchi, H., Naito, T. and Tatara, A. 1978. *Japanese Journal of Genetics* 53: 133–135.

46. Ethanol, a Respiratory By-product: An Indicator of Seed Quality

P.K. KATAKI and A.G. TAYLOR

Department of Horticultural Sciences, New York State Agricultural Experiment Station, Cornell University, Geneva, NY 14456, USA

Abstract

Ethanol production by seeds was developed as a rapid biochemical test, and found to be a sensitive indicator of seed quality. Seeds of maize, soyabean, cotton, rice and lettuce were aged at 75% relative humidity and 45°C to obtain five aged samples for each species including a nonaged control. Ageing reduced the percentage germination and the coefficient of the rate of germination (CRG). Samples of differentially aged seed lots were subjected to aerobic and anaerobic conditions. Maize and soyabean seeds were imbibed in distilled water while seeds of cotton, rice and lettuce were imbibed with 50 mM glucose in 5 mM potassium phosphate adjusted to a pH of 5.6. Cotton seeds were ground and particles from 3.35 to 4.76 mm were used, while lettuce was scarified. Ethanol was analysed using immobilized enzyme electrode technology and verified by gas chromatography. Anaerobiosis induced greater amounts of ethanol production compared to aerobic treatments. In general, the ethanol production increased with ageing under aerobic conditions while the reverse trend was measured under anaerobic conditions. The anaerobic to aerobic ethanol ratio (ANA ethanol ratio) was calculated and was correlated with the CRG.

Introduction

Biochemical compounds accumulating in aged seeds during the early stages of germination are potential tools for developing rapid seed quality tests. Ethanol production by germinating seed is one such compound which has been studied in relation to seed quality. In germinating seed, pyruvate formed via glycolysis is utilized by the tri-carboxylic acid (TCA) cycle and oxidative phosphorylation to generate ATP required to sustain metabolic processes in plants (Goodwin and Mercer, 1988). However, the lack of oxygen under anaerobic conditions diverts pyruvate from the TCA cycle and oxidative phosphorylation to form lactic acid and with the acidification of the cytosol to form ethanol (Davies *et al.*, 1974; Roberts *et al.*, 1984).

Deteriorating seed lots have been shown to produce higher levels of ethanol compared to good quality seed during germination under aerobic conditions (Woodstock and Taylorson, 1981; Gorecki *et al.*, 1985; Amable and Obendorf, 1986). During seed deterioration, mitochondrial membranes are presumed to be damaged thereby disrupting oxidative phosphorylation (Priestley, 1986). Consequently, pyruvate formed via glycolysis will be utilized by the fermentation

R.H. Ellis, M. Black, A.J. Murdoch, T.D. Hong (eds.), Basic and Applied Aspects of Seed Biology, pp. 421–427.
© *1997 Kluwer Academic Publishers, Dordrecht. Printed in Great Britain.*

pathway to produce ethanol. Under anaerobic conditions, accumulation of ethanol by good quality seed should be higher compared to aged seeds due to their more efficient pathway and a greater demand for ATP. Collectively, these studies suggest that differential ethanol production from aged seed lots could be used as a seed quality test, but a test has not been established. The present study is therefore an attempt to take advantage of this differential ethanol producing capability of seed as a means of establishing a rapid vigour test by analyzing the total ethanol from imbibing seeds of soyabean, maize, rice, lettuce and cotton.

Materials and Methods

Seed Source, Ageing and Germination Testing

Seed of soyabean (*Glycine max* (L.) Merrill, cultivar 'Hardin'), cotton (*Gossypium hirsutum* L., cultivar 'GS 510'), rice (*Oryza sativa* L., cultivar 'Lemont'), maize (*Zea mays* L., cultivar 'Jubilee') and lettuce (*Lactuca sativa* L., cultivar 'Salinas') were initially equilibrated at 75% relative humidity (RH) in Plexiglas chambers above a mixture of glycerol–water (Forney and Brandl, 1992). This procedure resulted in a seed moisture content of 12.9, 12.6, 11.5, 9.7, and 8.1% (fresh weight basis) for maize, rice, soyabean, cotton and lettuce, respectively as determined by the oven method (ISTA, 1985). Immediately upon equilibration, seeds were sealed in plastic covered aluminium pouches and placed in a water bath at 45°C for 4 to 32 days (d). At each sampling time, pouches were opened, seeds stored in plastic ziploc bags at 30% RH and 5°C. Five quality seed lots, including the nonaged control seeds, for each of the five crop species were analysed for germination, coefficient of the rate of germination (CRG) and ethanol production.

For each treatment, four replications of 25 seeds were placed on moistened paper towels and germinated following conditions described in *The Rules for Testing Seeds* (AOSA, 1993). Seeds with visible radicle protrusion were considered germinated. The coefficient of the rate of germination (CRG) was calculated as follows (Bewley and Black, 1994):

$$CRG = \{\Sigma n \ / \ (\Sigma \ t \times n)\} \times 100$$

where n = number of seeds completing germination on day t.

Ethanol Production Under Aerobic and Anaerobic Conditions for Maize and Soyabeans

Aerobic and anaerobic treatments were performed in 25°C incubators and all analysis was based on four replications for each treatment. To create an aerobic condition yet allowing seeds to imbibe water, a 115 ml analytical test filter funnel (Nalge Company, Rochester, NY) with a perforated removable base and membrane filter (0.45 micron, 47 mm) was used. The funnels were filled with half the total amount of 62–88 micron glass microbeads (Class 4A, size 1723, 2.42 density, Cataphote Inc., Jackson, MS) which provided an inert medium for the seeds. Seeds were placed on the surface and covered with the remaining amount of glass beads. The glass bead to seed ratio was 10:1 on a weight basis. This apparatus allowed for removal of excess water by vacuuming from the moistened glass beads without disturbing the seeds. At the sampling time (24 h), distilled water was poured into each analytical funnel and vacuumed off to collect the leachate for analysis, after which the seeds were removed immediately, homogenized (using an Ultra-Turrax, Janke & Kunkel Inc., Cincinnati, Ohio) in a known amount of ice cold distilled water and centrifuged. The supernatant from the homogenized sample and the leachate sample were immediately stored at –80°C until analysis.

For the anaerobic treatment, the seed moisture content was increased by placing seeds on a screen above water for 24 h to reduce imbibition damage. Seeds were then placed in plastic vials into which a known volume of distilled water was poured and purged with a steady flow of nitrogen gas for 5 min. At the sampling time (24 h) the leachate was poured into separate tubes and sealed for analysis while the seeds were homogenized as described in the aerobic treatment.

Ethanol Production Under Aerobic and Anaerobic Conditions for Cotton, Lettuce and Rice

Seeds of cotton were gently ground in an electric coffee mill and the partially ground seed lots were sized through a 4.76 and 3.35 mm sieve (U.S. standard sieve series, Arthur H. Thomas Company, Philadelphia, PA, USA). Lettuce seeds were also gently scarified in the coffee mill, which removed the seed coverings. Intact seeds of rice were used for the aerobic and anaerobic treatments. A solution of 50 mM glucose in 5 mM potassium phosphate adjusted to pH 5.6 was used in this study as the imbibing medium for both the aerobic and anaerobically treated seeds. Four replications for each of the aged seed lots were subjected to aerobic and anaerobic conditions as described above and samples were collected at 24 h imbibition time for cotton and rice, and 12 h for lettuce.

Analysis of Ethanol, Lactate, Methanol and Acetaldehyde

The leachate and homogenate samples were analysed with a Biochemistry Analyser (model 2700, Yellow Spring Instrument Inc., Yellow Springs, Ohio). This unit employs membrane bound immobilized enzymes which catalyzes the reaction of substrate to produce H_2O_2 which is then oxidized to produce a signal current. Ethanol and lactate were determined simultaneously from the same sample with alcohol and lactate oxidase membranes, respectively.

The ethanol membrane was sensitive to methanol and there was a concern that the presence of methanol may confound ethanol measurements. It has been shown that during seed development and maturation large amounts of methanol are produced; however, a low amount was produced during pre-radicle emergence stage during seed germination (Obendorf *et al.*, 1990). Therefore, the results from the YSI unit at 24 h sampling time was compared to results from a Gas Chromatograph (GC), H.P. 5890 series II for each sample. The GC method has been described by Kataki (1995).

Analysis of Microbial Contamination

To assess the extent of microbial contamination during ageing, three replications of 25 seeds each, for all ageing treatments were plated on petri dishes containing 2% malt agar, 8% bacto agar, and 10% sodium chloride (adapted from Harman, 1972). After plating, the plates were checked daily for the number of contaminated seeds for 7 d. To minimize microbial growth during aerobic and anaerobic treatments, three replications of each ageing treatment for the aerobic and anaerobic conditions were set up as previously described. Distilled water containing 100 ppm streptomycin and 0.2% thiram (Gustafson Inc., Dallas, TX) were used as the test solution and the results from this set up were compared to the results of the experimental technique using distilled water only.

Results

Duration of ageing at 45°C and 75% RH generated different quality seed lots as evidenced by the decline in germination and CRG for all the crop species (Table 1).

The ethanol was determined from leachate and seed homogenate and values were combined to calculate the total ethanol produced from seeds. The trends for all species revealed that as the duration of seed ageing increased there was a general decrease in ethanol production under anaerobic conditions, but an increase in ethanol during the early phases of ageing under aerobic conditions. The ethanol production ratio under anaerobic to aerobic (ANA) conditions revealed a decline with ageing period. The ANA was found to be linearly correlated with CRG for all species.

Methanol and acetaldehyde were only detected in trace amounts from the samples by gas chromatography (Kataki, 1995). Lactate was detected in all samples, but not related to seed quality. The effect of microbial contamination was minimal both after ageing and during the 24 h hydration period.

Discussion

Many physiological and biochemical changes occur during seed ageing and this

Table 1. Percent germination, coefficient of the rate of germination, total ethanol production under anaerobic and aerobic conditions and the ANA ethanol ratio

Days aged	Percent germination	CRG	Total ethanol production (mg/g seed)		ANA ethanol ratio
			Anaerobic	Aerobic	
Maize					
0	99	55	5.42	0.40	13
2	99	45	5.50	0.63	8.8
4	98	38	4.34	0.57	7.6
11	17	28	2.84	0.64	4.4
18	0	–	1.05	0.32	3.2
Soyabean					
0	100	54	6.09	0.52	12
2	99	50	5.95	1.07	5.6
4	99	35	5.99	3.47	1.7
9	31	29	3.92	2.57	1.5
12	0	–	3.01	1.20	2.5
Cotton					
0	100	50	9.98	0.13	78
2	99	38	9.50	0.23	42
4	96	39	8.05	0.64	13
11	88	36	5.39	0.49	11
21	55	35	4.33	1.72	2.5
32	0	–	2.44	1.84	1.3
Lettuce					
0	100	47	6.11	0.36	17
2	100	42	3.62	0.47	7.6
4	98	34	2.59	0.60	4.3
11	93	28	0.90	0.65	1.4
21	18	25	0.84	0.57	1.5
32	0	–	0.85	0.51	1.6
Rice					
0	97	18	2.98	0.06	50
2	93	16	1.92	0.07	29
4	92	15	1.92	0.14	14
11	35	14	1.31	0.09	14
21	0	–	0.97	0.08	12

subject has been reviewed by Priestley (1986). Impaired energy synthesis mechanisms and reduced respiration capacity are early events during seed ageing (AOSA, 1983). Therefore, direct or indirect methods to measure mitochondrial efficiency should provide a biochemical basis for seed quality. Woodstock and Taylorson (1981) and Reedy and Knapp (1990) have suggested that the production of ethanol, a by-product of respiration, during the early phases of germination may be used as a seed vigour test. However, a reliable test has not been developed.

The ANA ethanol ratio has been shown to be a sensitive index of seed quality for all five species. The ANA ethanol ratio was based on the findings that high quality seeds had the greatest ethanol production under anaerobic conditions, and that ethanol production decreased with seed ageing, while under aerobic conditions, these trends were generally reversed. Thus a small change in seed quality resulted in a large change in the measured ANA ethanol ratio.

It was found that ANA was correlated with CRG using intact seeds of soyabean and maize by using water as the imbibing medium. Chemical modification of the imbibing media, 50 mM glucose in 5 mM potassium phosphate adjusted to pH 5.6 (Hole *et al.*, 1992) was needed to increase ethanol production in the other three species. Glucose acted as a substrate for glycolysis and an acidic pH favours the production of ethanol in comparison with lactate (Davies *et al.*, 1974). The second variable which was manipulated was to alter the physical integrity of the seeds. In the case of cotton, grinding increased the surface to volume ratio, while in lettuce only the outer coverings were abraded.

In conclusion, it appears that seeds that naturally produce high levels of ethanol even under aerobic conditions such as maize and soyabean do not require a modification of the chemical medium. This interpretation is consistent with earlier work that has shown that corn, soyabean and pea seeds produced large quantities of ethanol while many other species including rice and lettuce produced only small amounts of ethanol (Raymond *et al.*, 1985). We therefore suggest the use of a solution containing glucose with a favourable pH for crop species that produce low amounts of ethanol under aerobic conditions (e.g. rice and lettuce). In addition, for crop species with relatively impermeable seed coats to solute diffusion, scarification of the seed coat can enhance ethanol production. The combination of buffered glucose solution and scarification of lettuce seeds increased the sensitivity of ANA ethanol ratio as a rapid seed quality indicator in comparison to the use of intact seeds imbibed in water. It is also suggested that due to the different capacities of various crop species to produce ethanol under aerobic and anaerobic conditions, experimentally derived minimal acceptable ANA ethanol ratios should be developed for different species. Seed lots with ratios above a pre-determined level may be categorized as highly viable and vigorous.

Acknowledgement

This project was partially supported from a grant from the American Seed Research Foundation.

References

Amable, R. A. and Obendorf, R. L. 1986. *Journal of Experimental Botany* 37: 1364–1375.

AOSA (Association of Official Seed Analysts). 1983. *Seed Vigor Testing Handbook.* Contribution no. 32.

AOSA (Association of Official Seed Analysts). 1993. *Journal of Seed Technology* 16 (3): 1–113.

Bewley, J. D. and Black, M. 1994. *Seeds: Physiology of Development and Germination.* Second Edition, pp. 445. New York, London: Plenum Press.

Davies, D. D., Grego, S. and Kenworthy, P. 1974. *Planta* 118: 297–310.

Forney, C. F. and Brandl, D. G. 1992. *Horticultural Technology* 2: 52–54.

Goodwin, T. W. and Mercer, E. I. 1988. *Introduction to Plant Biochemistry.* Second Edition, pp. 677. Oxford: Pergamon Press.

Gorecki, R. J., Harman, G. E. and Mattick, L. R. 1985. *Canadian Journal of Botany* 63: 1035–1039.

Harman, G. E. 1972. *Phytopathology* 62: 206–208.

Hole, D. J., Cobb, B. G., Hole, P. S. and Drew, M. C. 1992. *Plant Physiology* 99: 213–218.

ISTA (International Seed Testing Association). 1985. *Seed Science and Technology* 13: 338–495.

Kataki, P. K. 1995. Ph.D. Thesis. Cornell University, NY.

Obendorf, R. L., Koch, J. L., Gorecki, R. J., Amable, R. A. and Aveni, M.T. 1990. *Journal of Experimental Botany* 37: 1364–1375.

Priestley, D. A. 1986. *Seed Aging: Implications for Seed Storage and Persistence in Soil,* pp. 304. Ithaca, London: Cornell University Press.

Raymond P., Al Ani, A. and Pradet, A. 1985. *Plant Physiology* 79: 879–887.

Reedy, M. E. and Knapp, A. D. 1990. *Journal of Seed Technology* 14: 74–82.

Roberts, J. K. M., Callis, J., Jardetzky, O., Walbot, V. and Freeling, M. 1984. *Proceedings of the National Academy of Sciences* USA 81: 6029–6033.

Woodstock, L. W. and Taylorson, R. B. 1981. *Plant Physiology* 67: 424–428.

47. Semipermeable Layer in Seeds

A.G. TAYLOR, M.M. BERESNIEWICZ and M.C. GOFFINET
Department of Horticultural Science, New York State Agricultural Experiment Station, Cornell University, Geneva, NY, 14456, USA

Abstract

Semipermeability may be defined as the ability of seed coverings to allow water uptake and gas exchange, while solute diffusion is restricted or prevented. The presence and location of a semipermeable layer was studied in seed coats of cabbage, leek, onion, tomato, and pepper. Morphological studies did not reveal a semipermeable layer in cabbage seed coats, and all subsequent research was performed on seeds of the other four species. Electron microscopy studies revealed that the semipermeable layer is located at the innermost layer of the seed coat just next to the endosperm. Ultrastructurally, the layer was similar for the four species, typically amorphous, highly compact, but easily distinguished from the remainder of the seed coat and endosperm tissue. The layer was permeable to water while inhibiting uptake of lanthanum salts. Histochemical analysis revealed that the semipermeable layer in seed coats of leek and onion was composed primarily of cutin, while in tomato and pepper the layer was composed of suberin.

Introduction

The seed coat provides the morphological structure that separates the embryo and other nutritive tissue from the environment. In those species in which the seed coat allows water uptake, it is generally assumed that the seed coat is permeable to solutes. However, seed coats may provide a physical and/or chemical barrier to diffusion of compounds both in and out of the seeds. The term semipermeable was used in the early twentieth century to describe the condition in which the seed coverings allow water uptake and gas exchange, while solute diffusion is restricted or prevented (cited by Beresniewicz *et al.*, 1995b). The composition of the semipermeable layer has been studied by histochemical staining and reports vary with species. The composition has been postulated to be cutin, suberin or cellulose and the layer has been described morphologically as a cuticularized or cutinized membrane or suberized layer impregnated with fats (cited by Beresniewicz *et al.*, 1995a).

Early investigators provided insight into the existence, location and probable composition of the semipermeable layer; however, studies were limited by the research tools of the time. This research reports on the use of electron microscopy and the use of lanthanum, a non-radioactive label, to investigate the presence of a semipermeable layer in seed coats of two species each in the Liliaceae and the Solanaceae. Histochemical methods were developed to distinguish between cutin and suberin based on enzymatic or chemical hydrolysis prior to staining.

R.H. Ellis, M. Black, A.J. Murdoch, T.D. Hong (eds.), Basic and Applied Aspects of Seed Biology, pp. 429–436.
© *1997 Kluwer Academic Publishers, Dordrecht. Printed in Great Britain.*

Materials and Methods

Five crop seeds were studied: cabbage (*Brassica oleracea* L. var. *capitata*, cv. Hybrid Condor), leek (*Allium porrum* L., cv. Pancho), onion (*Allium cepa* L., cv. Texas Early Grano), tomato (*Lycopersicon esculentum* L., cv. Peelmech), and pepper (*Capsicum annuum* L., cv. California Wonder). Non-germinable seed samples were obtained by ageing seeds of cabbage, leek, onion, tomato and pepper at 45°C and 90% relative humidity for 10, 4, 10, 22 and 11 days (d), respectively (Taylor *et al.*, 1995).

Lanthanum Uptake

The structure and localization of the semipermeable layer from non-germinable seeds was examined by a microprobe and with a transmission electron microscope (TEM). Seeds of all species were first presoaked in aerated water for 4 h, then individual seeds were each transferred to a well of a 96-well microtiter plate containing 100 µl of water per well and soaked for an additional 24 h at 20°C. The leachate was transferred to new microtiter plates and ninhydrin reagent was added to each well. Plates were then incubated for 45 min at 95°C for colour development to test for amino acid leakage (Taylor *et al.*, 1995). Individual seeds that were found not to leak were placed in vials containing 2% lanthanum nitrate and vials were placed on a rotating wheel for 24 h.

Microprobe Studies

Seeds that had been perfused with lanthanum were cut into halves with a razor blade, freeze-dried, mounted on carbon stubs, sputter coated with carbon and viewed with a JEOL 733 electron microprobe in the back-scattered electron imaging mode at 15 kV. The appearance of lanthanum (white areas) was determined with the Tracor Northern 5500 energy dispersive X-ray analysis (EDX) system at a 40° take-off angle. Globoid crystals located in seed samples were analysed for 100 s. Lanthanum amounts were based on the L-series.

TEM and (EDX)

Seeds that had been perfused with lanthanum were prepared for histology. From each seed, a segment of testa and subjacent endosperm tissue was carefully removed with a razor blade to include approximately one cubic millimeter. Segments were chemically fixed for 1 h in 3% glutaraldehyde in Sorensen's phosphate buffer, pH 7.2. After rinsing in buffer twice, segments were post-fixed for 1 h in 1% aqueous osmium tetroxide (light microscopy studies only) and again rinsed twice in buffer. Specimens were then dehydrated at 30 min per step in a graded ethanol series (15%, 30%, 50%, 70%, 95%, 100%, and again 100%). Specimen vials were kept in ice water throughout fixation and dehydration.

After dehydration, specimens were vacuum infiltrated and embedded in LR White resin (London Resin Co., Surrey, England). Sections for TEM were cut with a diamond knife at approximately 90 nm thickness and viewed, without post-staining, using a JEOL 100SX TEM operated at 100 kv. For EDX, sections approximately 100 nm thick were prepared in the same way as previously described for TEM and were placed on fine-mesh copper grids. Sections were viewed, without post-staining the tissue, using a JEM 200CX TEM operated at 200 kv in scanning transmission electron microscope (STEM) bright field and annular dark field modes. The location of the perfused lanthanum was determined with the Tracor Northern EDX system operated at a 72° take-off angle. Selected globoid crystals in seed samples were analysed for elemental composition for 100 s.

Light Microscopy and Histochemical Analysis

Light microscopy was used to observe the presence and localization of various histochemical stains in seed coats of leek, onion, tomato and pepper. Sections 1.5 μm-thick were prepared as described for TEM, except without soaking to test for amino acid leakage and uptake of lanthanum. Sections of the semipermeable layers were viewed with a light microscope at 400 × – 1000 ×. The following histochemical stains were used to test for the presence of the compounds in parentheses: aniline blue (callose), periodic acid and Schiff's reagent (polysaccharides), ruthenium red (pectin), phloroglucinol (lignin), Sudan III and IV (lipids, cutin, suberin).

Fluorescence microscopy was used to observe seed tissues treated with several fluorescent compounds and stains. Tissues were frozen in M-1 Embedding Matrix (Lipshaw Manufacturing Co., Detroit, Michigan) at –20°C and 5 μm-thick sections were cut with a Microtome-Cryostat (International Equipment Co., Needham Heights, Massachusetts). Sections were flattened on glass slides, stained and observed for epifluorescence with a Zeiss Photomicroscope at a magnification of 400 × – 1000 ×, using filter set A (filter excitation 365 nm, dichroic filter 395 nm, emission filter 397 nm) or set B (filter excitation 400–440 nm, dichroic filter 460 nm, emission filter 470 nm). The following fluorescent histochemical stains were used to test for the presence of the compounds in parentheses: calcofluor with filter set A (cellulose), fluorol yellow with filter set A (lipids), natural fluorescence with filter set A (polyphenolic substances), Sudan III and IV with filter set B (cutin and suberin), berberine-crystal violet with filter set B (suberin and phenolics).

No histochemical stain is able to clearly differentiate between suberin and cutin, therefore protocols were developed to distinguish between these substances. Suberin is composed of two components, one is lignin-like, the other cutin (cited by Beresniewicz *et al.*, 1995a). Methods were developed first to hydrolyze cutin or the cutin portion of suberin, and then apply the histochemical stains. Cutin was hydrolyzed by either cutinase prepared from *Fusarium solani* f. sp. *pisi* or by NaOH. Potato (*Solanum tuberosum*) tuber peel with visible suberized wounds and cucumber (*Cucumis sativus*) fruit peel were used as known sources of suberin and cutin, respectively. After hydrolysis, seed samples were freeze-sectioned with the cryostat or free-hand sectioned and sections then stained with phloroglucinol, berberine/crystal violet, or Sudan (as previously described).

Results

All non-germinable cabbage seeds exhibited amino acid leakage after a 24-h period, indicating that the seed coat was not a barrier to solute diffusion. Preliminary electron microscopy studies also did not reveal a semipermeable layer within the seed coat of this species. Therefore, all subsequent research was performed on the other four species.

General trends were observed with respect to deposition of lanthanum in leek, onion, tomato and pepper. Electron microprobe micrographs taken in the back-scattered electron imaging mode revealed that lanthanum (observed as the white band) was deposited as a continuous area in the inner region of the seed coat for leek (Fig. 1), onion (Fig. 3), tomato (Fig. 5) and pepper (Fig. 7). Elemental analysis of the white band observed with the electron microprobe revealed lanthanum in relatively large amounts in all species, thus confirming the accumulation of this heavy metal.

TEM micrographs revealed black crystalline precipitates within the inner seed coat structure after lanthanum perfusion of seeds of leek (Fig. 2), onion (Fig. 4), tomato (Fig. 6) and pepper (Fig. 8). Precipitates were analysed and found to be lanthanum crystals, which accumulated through all seed coat layers but the innermost in these four crops. The innermost layer not penetrated by lanthanum resembled an amorphous membrane surrounding the endosperm in all four species (Figs. 2, 4, 6 and 8). Elemental analysis of this amorphous semipermeable layer revealed trace amounts of lanthanum in leek and onion, while no lanthanum was found in this layer in tomato and pepper.

Histochemical staining of seed coat cross sections revealed positive staining for all four species with Sudan III and IV in both visible and UV light, and also with fluorol yellow (data not shown). Hydrolysis of cutin or the cutin portion of suberin by cutinase or by NaOH was performed on potato tuber peel, cucumber fruit peel, and all four seed samples. Cucumber fruit peel revealed no staining or intermittent staining with Sudan III and IV, phloroglucinol or berberine/crystal violet, while the potato tuber exhibited medium to strong staining with the same stains (Table 1).

Leek and onion revealed similar staining as found in cucumber fruit peel, indicating that cutin was the primary material found in the semipermeable layer of those species. Staining patterns of the semipermeable layer in tomato and pepper were similar to those of the potato peel, indicating that suberin was the primary constituent in the semipermeable layer of those species.

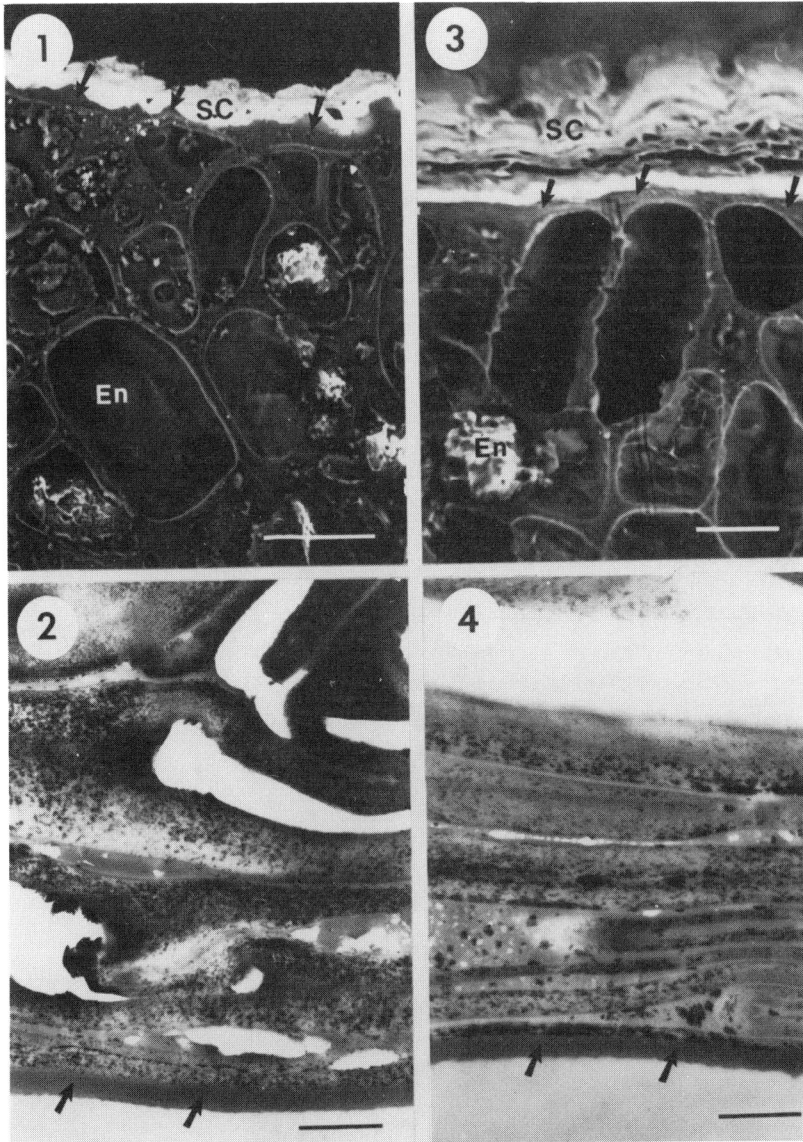

Figures 1–4. Anatomy of leek and onion seed coats. Longitudinal section through mature seed showing the seed coat and endosperm viewed by microprobe (Fig. 1, leek; Fig. 3, onion). Note region of lanthanum deposition (white band) subtended by semipermeable layer (arrows). Bar = 50 μm. Seed coat viewed by TEM showing semipermeable layer as a gray band (arrows) (Fig. 2, leek; Fig. 4, onion). Bar = 1 μm. Abbreviations: En, endosperm; SC, seed coat

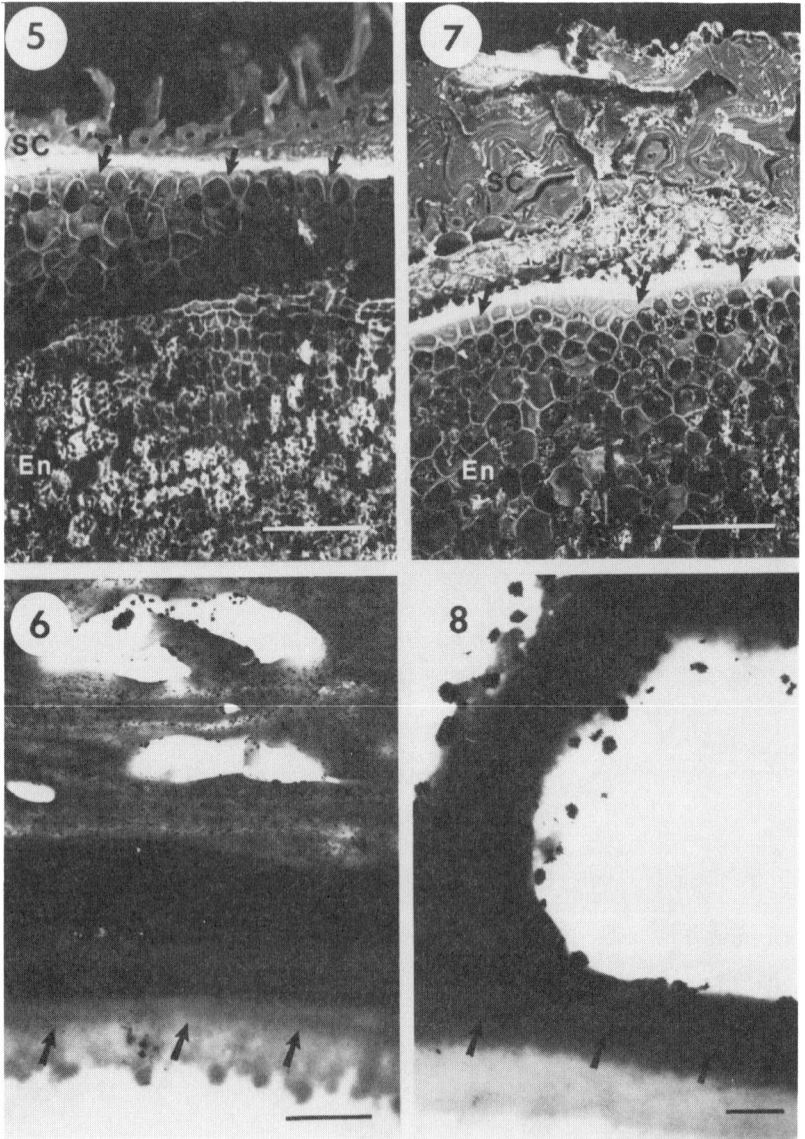

Figures 5–8. Anatomy of tomato and pepper seed coats. Longitudinal section through mature seed showing the seed coat and endosperm viewed by microprobe (Fig. 5, tomato; Fig. 7, pepper). Note region of lanthanum deposition (white band) subtended by semipermeable layer (arrows). Bar = 100 μm. Seed coat viewed by TEM showing semipermeable layer as a gray band (arrows) (Fig. 6, tomato; Fig. 8, pepper). Bar = 1 μm. Abbreviations: En, endosperm; SC, seed coat

Table 1. Histochemical staining of cucumber fruit peel, potato tuber peel, and the semipermeable layer (SPL) of leek, onion, tomato and pepper seed coats

Crop	Plant part	Hydrolysis treatment	Sudan III & IV[a]	Phloro-glucinol[b]	Berberine/ crystal violet[c]
Cucumber	Fruit peel	Cutinase	–	+/–	+/–
		NaOH	–	–	+/–
Potato	Tuber peel	Cutinase	+++	+++	+++
		NaOH	++	+++	+++
Leek	SPL	Cutinase	–	+/–	–
		NaOH	+/–	+/–	–
Onion	SPL	Cutinase	–	+/–	–
		NaOH	+/–	+/–	–
Tomato	SPL	Cutinase	++	++++	++++
		NaOH	+++	+++	+
Pepper	SPL	Cutinase	++	+++	++++
		NaOH	++	++	++

[a] Viewed with UV light, sudan stains for lipids, cutin and suberin.

[b] Viewed with white light, phloroglucinol stains for lignin.

[c] Viewed with UV light, berberine/crystal violet stains for phenolics.

Key to staining: – no staining, +/– intermittent staining, + faint staining, ++ slight staining, +++ medium staining, ++++ strong staining

Discussion

Investigations have been made on the leakage of compounds from seeds in association with seed ageing and the seed's uptake of vital stains in cabbage, leek, onion, tomato and pepper. In general, the leakage rate of amino acids and sugars was greatest from non-germinable seeds of cabbage, and less so from leek and onion (Lee *et al.*, 1995; Taylor *et al.*, 1995). Tomato and pepper were found to leak only small amounts of amino acids under the same test conditions (Taylor *et al.*, 1995). Further studies focused on the uptake and staining of tetrazolium chloride (TZ) from non-aged seeds. Beresniewicz *et al.* (1995b) found that the percentage of intact seeds of cabbage, leek, onion, tomato and pepper showing some degree of staining by TZ was 100, 74, 22, 8 and 7%, respectively. Leek and onion seeds have been shown to be susceptible to mechanical damage resulting in seed coat cracks. Seed coat cracks enhance both solute uptake and leakage compared to intact seeds (Beresniewicz *et al.*, 1995c). Collectively from these data, cabbage was found to have permeable seed coats, tomato and pepper relatively impermeable seed coats, while onion and leek seed coats were intermediate in their responses.

A semipermeable layer was found in the seed coats of leek, onion, tomato and pepper as determined by the deposition of lanthanum in electron microscopy studies. Ultrastructurally, the layer was similar for the four species in being typically amorphous, highly compact, but easily distinguished from the remainder of the seed coat and endosperm tissues. This layer also appears to be responsible for limiting leakage of amino acids and sugars and the uptake of tetrazolium salts. The phenomenon of semipermeability appears to be widespread in seeds of higher plants. A survey of over 500 species including 40 plant families revealed the presence of a semipermeable layer, except in certain species belonging to the Cruciferae (Brassicaceae), Leguminosae (Fabaceae) and Cistaceae (cited by Beresniewicz *et al.*, 1995b).

The chemical nature of the semipermeable layer was elucidated by either enzymatic or chemical hydrolysis of the seed coats followed by a series of histochemical stains. Based on these methods, the semipermeable layer in leek and onion was found to be composed of cutin, while this layer in tomato and pepper was composed of suberin. Sudan allowed the lipid component of cutin or suberin to fluoresce with UV light, phloroglucinol detected the lignin portion of suberin under white light after the hydrolysis procedure, and berberine/crystal violet allowed fluorescence of the phenolic components of suberin. Hydrolysis of cutin would reveal no or little staining with Sudan III and IV, since the lipids would be removed by the treatment. Staining of suberin by phloroglucinol was enhanced after hydrolysis, as the cutin portion was removed, thus exposing the lignin component of the polymer. Staining by berberine/crystal violet would resemble the staining as found for phloroglucinol, strong staining of suberin due to the phenolics of the lignin portion, with little staining of cutin. These hydrolysis methods may be used in conjunction with traditional histochemical methods to detect the presence of cutin or suberin in seeds and other plant tissues.

Acknowledgement

This project was partially funded from a grant from Germains, UK.

References

Beresniewicz, M. M., Taylor, A. G., Goffinet, M. C. and Koeller, W. D. 1995a. *Seed Science and Technology* 23: 135–145.

Beresniewicz, M. M., Taylor, A. G., Goffinet, M. C. and Terhune, B. T. 1995b. *Seed Science and Technology* 23: 123–134.

Beresniewicz, M. M., Taylor, A. G., Goffinet, M. C. and Terhune, B. T. 1995c. *Plant Varieties and Seeds* 8 (2): 87–95.

Lee, S. S., Taylor, A. G., Beresniewicz, M. M. and Paine, D. H. 1995. *Plant Varieties and Seeds* 8 (2): 81–86.

Taylor, A. G., Lee, S. S., Beresniewicz, M. M., and Paine, D. H. 1995. *Seed Science and Technology* 23: 113–122.

48. Stress, Protein Biosynthesis and Loss of Vigour and Viability in Cereal Seed

C.M. BRAY

School of Biological Sciences, University of Manchester, 3.614 Stopford Building, Oxford Road, Manchester M13 9PT, UK

Abstract

Specific lesions have been identified in components of the protein synthesizing system of cereal embryos from seed lots of reduced vigour. These lesions have been implicated in reduced levels of protein synthesis in wheat embryos when germination takes place under both optimal and stress conditions. Lesions appear in polyA$^+$ RNA (mRNA) metabolism, ribosome integrity but not ribosomal RNA breakdown, and aminoacyl tRNA synthesis. The rate of breakdown of stored mRNA and accumulation of *de novo* synthesized mRNA transcripts are both less rapid in low-vigour embryos and represent major lesions affecting the rapidity and successful completion of the germination process. Additional lesions have been demonstrated in the endosperms of cereal seeds as viability is lost particularly in the responses, both quantitatively and qualitatively, of the levels and pattern of proteins synthesized when seeds are subjected to stress conditions, e.g. heat shock, during germination. The transport properties of scutellar epithelial membranes also change during loss of seed viability thereby affecting the transport of mobilized endosperm reserves to the cereal embryo. These lesions are discussed in the context of effects on germination performance and, ultimately, seedling establishment.

Introduction

The importance of the cultivation of wheat as a major food source for *Homo sapiens* has long been recognized and studies of the germination process have contributed information which has assisted in the attainment of increased yields of grain. Studies in recent years have demonstrated that seed lots of equal high percentage germination may vary in their ability to produce crops and differences in seedling establishment and resultant decreased grain yields can often be attributed to variations in seed vigour (Stormonth, 1978). Whereas a seed is considered viable when it possesses the capacity to germinate normally under the optimal conditions of the standard germination test (Roberts, 1972), the concept of seed vigour is a much more nebulous concept and is often manifest in reduced rates of seedling emergence and establishment only when seeds are required to germinate in the non-ideal conditions found in the field. The stresses encountered during field germination can vary widely and include water stress (osmotic stress), flooding, temperature stress (heat, cold), and the many physical stresses associated with varying soil conditions during and after sowing.

R.H. Ellis, M. Black, A.J. Murdoch, T.D. Hong (eds.), Basic and Applied Aspects of Seed Biology, pp. 437–449.
© *1997 Kluwer Academic Publishers, Dordrecht. Printed in Great Britain.*

The cereal embryo, and in particular the wheat embryo, has been the subject of intense studies over the years with respect to the molecular events occurring during imbibition and the onset of germination. These investigations have demonstrated that RNA and protein biosynthesis commence within minutes of the imbibition of water by cells of the dehydrated embryo (Spiegel and Marcus, 1975). The successful reactivation of the protein synthesizing system of the quiescent cereal embryo and maintenance of the integrity of its translational apparatus are essential if germination is to be completed successfully with subsequent emergence of the radicle. The first few hours following completion of water uptake by the wheat embryo is a period of rapid change within the embryo as the pattern of protein synthesis shifts from that of a 'maturation mode' operative whilst the grain was on the mother plant to that of a 'germination mode' compatible with the physiological processes involved in development of the immature tissues found within the embryo into fully functional roots, shoots and leaves of the young seedling. The messenger RNA population within the embryo of the mature wheat grain undergoes an almost complete turnover during the first 5–6 h following imbibition at 20°C (Smith and Bray, 1982) and a shift in the patterns of proteins synthesized over this time period ensues (Lane, 1991 and references therein).

Any lesions which might arise in the protein synthesizing system of the cereal seed in either embryo, scutellum or aleurone tissue, might be expected to have a deleterious effect on the germination capacity of the seed. Studies in our laboratory over the past several years have been aimed at identifying the specific lesions which appear in the components of the protein synthesising system of cereal seeds of reduced vigour and viability and assessing the relative importance of these lesions in the process of loss of seed vigour and viability. This report describes the results of some of these investigations.

Results and Discussion

Early studies using pulse-labelling techniques to investigate protein biosynthetic capacity in wheat embryos isolated from seed lots of high viability but differing in vigour demonstrated that the rate of protein biosynthesis in isolated embryos during the early hours of imbibition at an optimal temperature (20°C) was related to the vigour of the seed lots (Blowers et al., 1980). When wheat embryos are germinated at sub-optimal temperatures (e.g. 10°C), then the differences between rates of protein biosynthesis in embryos from seed lots of different vigour ratings are amplified and differences persist at later germination stages than are seen when germination occurs at 20°C. This relationship between the rate of protein biosynthesis in germinating cereal embryos and the vigour rating of the seed lot holds true for seeds from a wide range of seed lots and from different varieties (Table 1). Wheat embryos which remain attached to the endosperm for the duration of the pulse labelling studies also synthesize protein at rates indicative of the vigour rating of the seed lot indicating that the

Table 1. Protein biosynthesis and vigour ratings in wheat embryos germinated at 10°C

		[^{14}C]amino acid incorporation expressed as % high vigour, high viability incorporation levels (set to 100%)	
Viability (% germination)	Vigour rating	5–6 h	23–24 h
var. Hobbit samples			
89	92	100	ND
92	90	83	100
84	88	83	100
88	72	73	62
85	69	30	47
64	55	10	13
var. Atou samples			
97	91	100	100
95	67	45	53
83	16	15	12

differences observed in protein biosynthetic rates are physiological differences and are not introduced artificially by the mass isolation technique (Johnston and Stern, 1957) employed to isolate embryos.

Further studies have attempted to identify the nature of the lesions which contribute to the reduced rates of protein biosynthesis in embryos from seed lots of reduced vigour. Lesions have been identified in low-vigour embryos in several components essential to the integrity of the protein synthesizing system of wheat embryos. These include amino acyl tRNA synthetase activities, ribosome integrity (but not gross ribosomal RNA degradation) possibly involving ribosomal proteins required in binding amino acyl tRNA to the ribosome (Blowers *et al.*, 1985), purine nucleotide and nucleotide sugar levels (Standard *et al.*, 1983) and polyadenylated RNA (mRNA, polyA$^+$ RNA) levels (Smith and Bray, 1984). Of these lesions, those involved in mRNA biosynthesis and degradation, i.e. mRNA turnover, appear most relevant to loss of germinative capacity and have been the subject of more intense investigations over recent years.

When wheat embryos from three seed lots of different vigour ratings and one seed lot of low viability were germinated at 20°C differences in the levels of polyA$^+$ RNA in the embryos could be observed over the first 24 h of imbibition (Fig. 1). High and low-vigour seed lots could be distinguished by their different patterns of polyA$^+$ RNA turnover during early germination but it was only at about the 5 h germination period at 20°C that the level of polyA$^+$ RNA in the embryo was a reflection of the vigour (and viability) rating of the seed lot

(Fig. 1). The reason for this is that the polyA$^+$ RNA profiles (Fig. 1) are composite curves arising from two opposing phenomena i.e. degradation of the stored mRNA present in the quiescent embryo and the transcription of new polyA$^+$ RNA species following the onset of germination. Use of the fungal toxin a-amanitin, a potent inhibitor of RNA polymerase II and hence polyA$^+$ RNA biosynthesis, in the imbibition medium has permitted the dissection of these two components of the composite mRNA turnover profile in wheat embryos during germination (Smith and Bray, 1982; Rushton and Bray, 1987) the results of which are summarised in Figure 2.

Wheat embryos from seed lots of different vigour ratings imbibed in the presence of 12 mM a-amanitin were still capable of degrading the polyA$^+$ RNA stored in the quiescent embryo (Fig. 2A). The rate at which this stored polyA$^+$

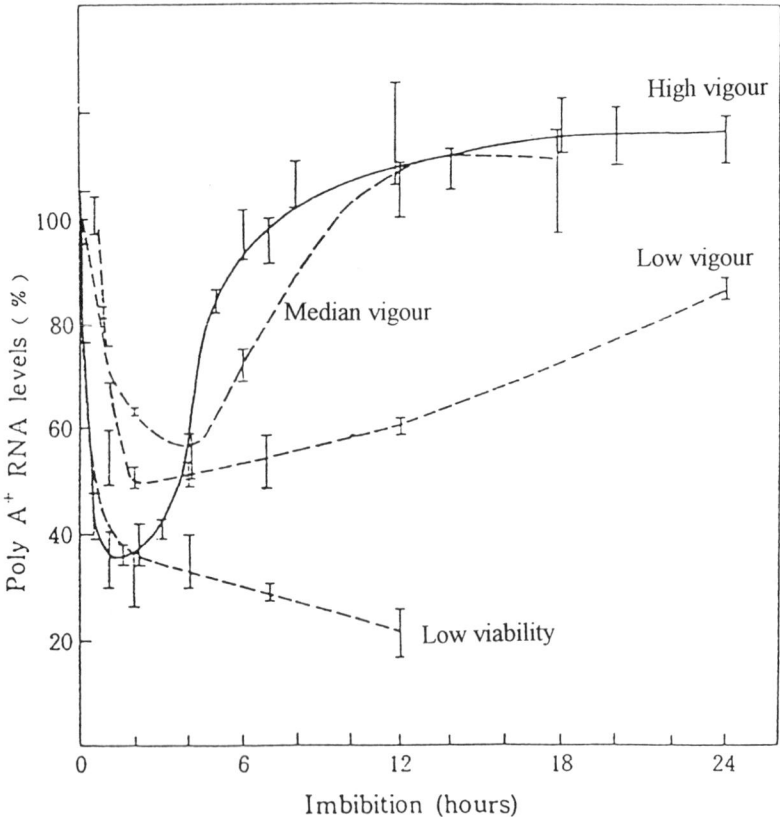

Figure 1. Polyadenylated RNA levels in germinating wheat embryos. Results represent average of six independent determinations

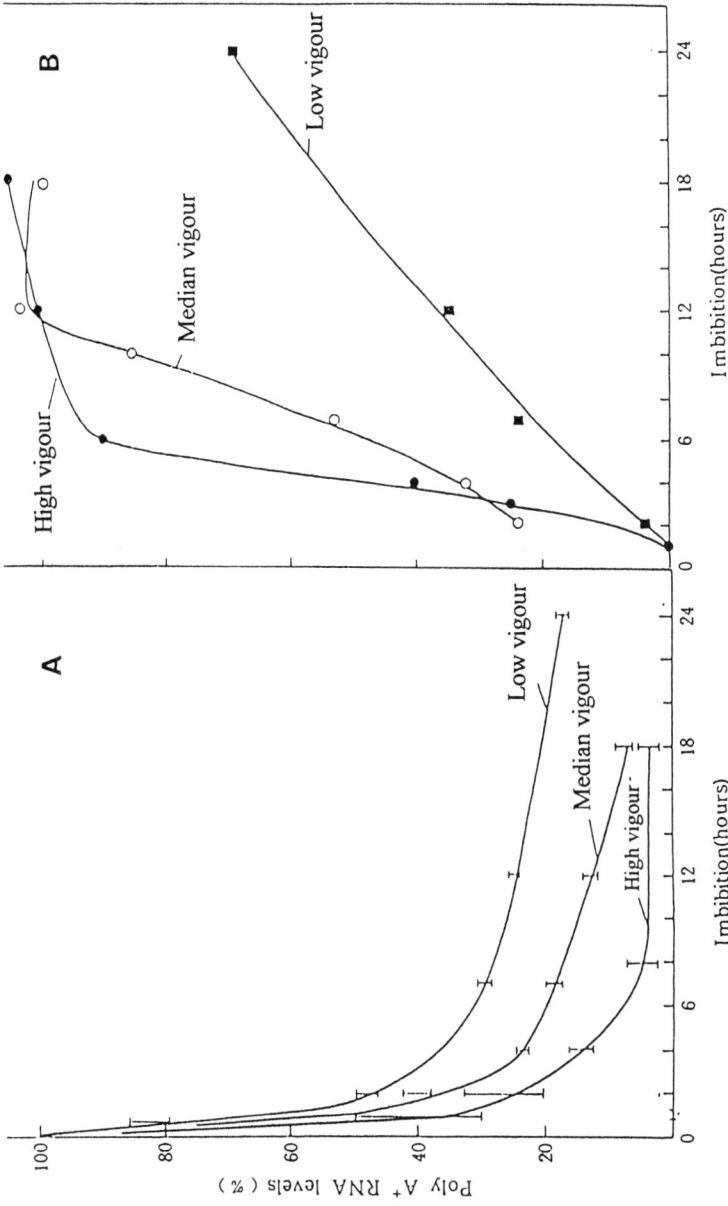

Figure 2. (A) Stored mRNA levels and (B) *de novo* synthesized polyA⁺ RNA transcript levels in wheat embryos during germination. Results represent the average of six independent determinations

Table 2. Changes in *de novo* and stored poly A⁺ RNA levels in wheat embryos during germination

Germination time (h)	Relative proportion (%) stored or *de novo* synthesized poly A⁺ RNA in embryo					
	High vigour		Medium vigour		Low vigour	
	De novo	Stored	*De novo*	Stored	*De novo*	Stored
4	74	26	57	43	23	77
7	94	6	73	27	43	57
12	95	5	88	12	59	41
18	96	4	92	8	73	27

RNA was degraded declined with loss of seed vigour and the time taken for 50% of the stored polyadenylated RNA to be degraded was 0.7 h, 1.2 h and 2h in high, medium and low-vigour embryos respectively. Degradation of stored polyA⁺ RNA commenced immediately upon imbibition of water by the quiescent embryos suggesting that the quiescent wheat embryo possesses all of the necessary enzymes for this rapid degradation and that there is no requirement for *de novo* synthesis of these enzymes during germination. This degradation of mRNA transcripts within the wheat embryo during germination must be highly selective since specific mRNA transcripts (e.g. those for the Em and Ec protein) contained within the stored mRNA population are degraded whilst simultaneously other mRNA species e.g. ribosomal protein transcripts and *de novo* synthesized transcripts for germin and histone proteins are either maintained at a constant level or accumulate in the wheat embryo during germination (Lane, 1991).

By subtraction of the residual levels of stored polyA⁺ RNA (Fig. 2A) from total polyA⁺ RNA levels (Fig. 1), the levels of newly transcribed mRNA species in wheat embryos during germination can be estimated (Fig. 2B). The changing pattern of the ratio of conserved: *de novo* synthesized polyA⁺ RNA species over the 24 h germination period studied was characteristic of the vigour rating of the seed lot from which embryos were isolated. These ratios are quantified at four different time points in Table 2 and demonstrate the very rapidly decreasing contribution of stored polyA⁺ RNA to the potential mRNA pool of high-vigour wheat embryos after the initial 4 h imbibition period. Increased imbibition times are required before a similar situation occurs in low-vigour seed, e.g. at 12 h imbibition only 5% of the mRNA population of high-vigour wheat embryos is comprised of stored mRNA species whilst 40% of the low-vigour embryo mRNA population still comprises stored mRNA transcripts. There was also little difference between the polyA⁺ RNA levels in low viability (6%) wheat embryos at various germination stages (Fig. 1) and the polyA$^+$ RNA levels found in low-vigour wheat embryos imbibed in a-amanitin (Fig. 2A). This suggests that the inability of a wheat embryo to germinate is associated with an inability to transcribe new mRNA species.

Collectively these results indicate that the production of *de novo* mRNA transcripts during the hours following water imbibition is an essential pre-requisite for successful completion of the germination process and that the ability of the wheat embryo to accumulate new mRNA species has a direct bearing on the rapidity with which the seed will eventually germinate. However it must also be remembered that at least some of the mRNA transcripts contained within the stored mRNA population of the quiescent wheat embryo could still have an important role to play in the re-initiation of protein biosynthesis in the first 2–3 h of imbibition. However, since up to 75% of the stored polyA$^+$ RNA population of the quiescent wheat embryo can be degraded during the first 2 h of imbibition without any decline in protein biosynthetic activity of the embryo (Rushton and Bray, 1987) then this would suggest that only a small fraction of the stored mRNA pool is required to fulfil this role.

Commencement of the growth phase of germination in wheat embryos coincides with the synthesis of specific proteins not found in the mature quiescent embryo (Grzelczak *et al.*, 1982). Their synthesis requires *de novo* mRNA transcription following imbibition since the mRNAs for these proteins are absent from the stored mRNA pool of the quiescent embryo. Changes in the turnover patterns of polyA$^+$ RNA in wheat embryos during loss of vigour would be reflected in lower levels of 'growth phase proteins' mRNA transcripts in low-vigour embryos compared to high-vigour embryos at comparable early germination stages (5–10 h onwards). As a consequence, inefficient transcription or translation of growth phase genes or their mRNA transcripts could be a major contributory factor which would explain, at least in part, the slower rates of germination of lower-vigour seed lots. In addition, altered ratios of stored and *de novo* synthesized mRNA transcripts at later germination stages in low-vigour embryos could affect, both qualitatively and quantitatively, the pattern of proteins synthesized during early germination and so affect the rate of germination and subsequent seedling establishment.

Since seeds without a viable embryo will not germinate, then it is not too surprising to find that most biochemical studies on loss of seed viability have concentrated on embryo material whereas less effort has been directed to the investigation of those changes occurring in seed structures outside the embryo. Whilst deleterious changes occurring in the scutellum, aleurone and endosperm tissues in cereals may not affect germinability *per se* they will have significant effects on seedling growth and vigour by affecting storage reserve mobilisation and transport to the growing embryo (Aspinall and Paleg, 1971; Petruzzelli and Taranto, 1989). Cereal seeds offer an ideal system in which to study the expression of the effects of ageing on embryo and non-embryo tissues of seeds due to the ease of physical separation of the embryo from the non-embryonic tissues comprising scutellum and endosperm.

α-Amylase synthesis and secretion by aleurone cells surrounding the starchy endosperm is a major prerequisite to initiate the mobilization of the carbohydrate reserves contained in cereal endosperm for use by the growing embryo. Wheat seed which are losing viability accumulate lesions in aleurone cells which result in decreased levels of synthesis and release of α-amylase during the first few days of germination (Aspinall and Paleg, 1971; Petruzzelli and Taranto, 1989) and an increased threshold at which gibberellic acid elicits a response in aleurone cells (Aspinall and Paleg, 1971). These studies were performed on mixed populations of seed samples which contained both viable and non-viable seeds. Other studies on wheat seed lots in which viable seeds had been separated from non-viable seeds at the three-day post-imbibition stage have demonstrated that loss of viability was accompanied by a failure to increase endosperm α-amylase levels in non-viable seeds above a basal level at least up to 12 days post imbibition (Fig. 3) during which time the endosperm of viable seeds exhibits at least a 6-fold increase in α-amylase levels (Livesley and Bray, 1991). Surprisingly, this reduced capacity for α-amylase synthesis and secretion was not the result of a parallel decrease in levels of protein synthesis in aleurone layers of non-viable seeds since *in vivo* rates of protein synthesis were at least as high in aleurone layers from non-viable seeds

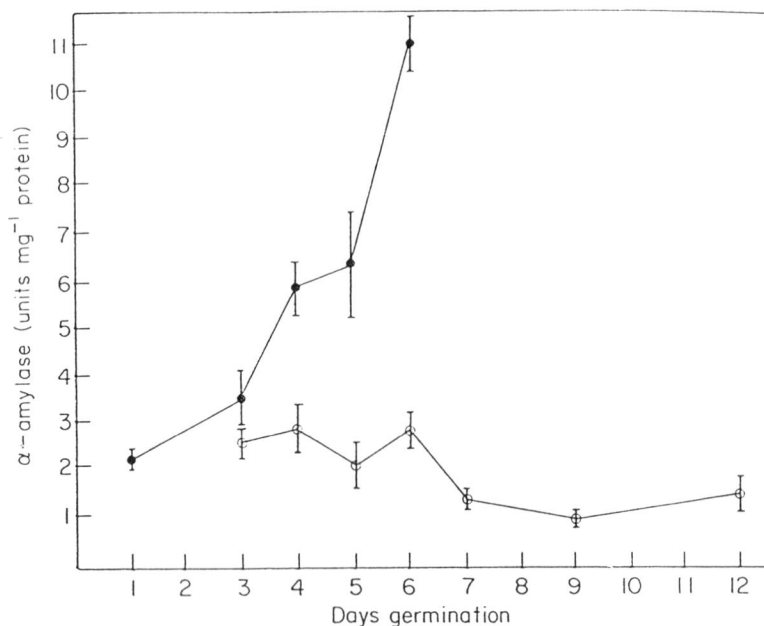

Figure 3. α-Amylase activity development profile in germinating wheat seeds.
●—● viable seeds; o—o non-viable seeds. Error bars represent s.d. of three replicate samples

as in viable seeds at similar post-imbibition stages (Livesley and Bray, 1991). However, the aleurone layers of these non-viable seeds appeared to be non-responsive to exogenously applied gibberellic acid (Livesley and Bray, 1991). Analysis of the α-amylase isoenzyme patterns of the wheat endosperm during germination and loss of viability demonstrated that the inability to increase α-amylase levels in non-viable endosperm tissue was due to a reduced synthesis of high pI isoform α-amylases and a failure to commence synthesis of the low pI isoforms (Livesley and Bray, 1991). These low pI isoforms have been demon-strated to accompany the large increase in total α-amylase activity seen at this time (Sargeant and Walker, 1978).

One of the more accurate indicators of seed vigour is the accelerated ageing test which measures the seed's ability to withstand an applied stress, namely to survive long periods of incubation at high temperature and humidity. Low-vigour seeds exhibit a reduced ability to survive and germinate following exposure to such stress conditions and this is in contrast to the behaviour of high-vigour seeds incubated under an identical stress. The conditions of the accelerated ageing test mimic those of a sub-lethal heat shock regime, albeit for a more prolonged period of time. Heat shock induces the synthesis of a number of so-called heat-shock or heat-stress proteins (hsps) in both prokaryotic and eukaryotic organisms, including plants, which serve to protect the organism and permit survival under such adverse conditions. Helm *et al.* (1989) demonstrated that the synthesis of heat-shock proteins and their mRNAs is a normal part of wheat embryo development. They also showed that imbibing wheat embryos were capable of expressing an enhanced heat-shock response and that this response was related to seed vigour. Some heat-shock proteins function as molecular chaperones, these proteins being 'slow' ATPases which aid protein folding by binding and stabilising unfolded or partially folded proteins prevent-ing these proteins being degraded and directly facilitating their folding.

Aleurone cells of the wheat seed also respond to heat shock by exhibiting typical quantitative and qualitative changes in protein synthesis patterns. Their *in vivo* rate of protein synthesis is reduced in response to the heat-shock and gross changes in the pattern of proteins synthesized are observed with the initiation of synthesis of specific classes of proteins characteristic of the heat shock response (Livesley and Bray, 1993). A brief sub-lethal heat shock (4 h at 42°C) abolished α-amylase synthesis in aleurone layers from both viable and non-viable 3-d imbibed wheat seeds but only aleurone layers from viable seeds were capable of recommencing substantial α-amylase biosynthesis during a recovery period at 20°C following the heat shock (Livesley and Bray, 1993). Qualitative differences in the pattern of protein synthesis in response to heat-shock and recovery from heat shock were also demonstrated between aleurone layers from 3-d-imbibed aleurone layers from viable and non-viable wheat seeds. These studies demonstrated that aleurone layers from non-viable seeds, although metabolically still active and capable of high rates of protein synthesis, were impaired in their response(s) to stress during imbibition in comparison to aleurone layers from viable seed.

The suppression of α-amylase synthesis in aleurone layers in response to heat shock has also been demonstrated in barley seeds (Belanger *et al.*, 1986). This suppression of α-amylase biosynthesis was mediated via a rapid destruction of otherwise stable α-amylase mRNAs and not by translational control as reported for several other systems. The degradation of α-amylase mRNA coincided with the rapid destruction of endoplasmic reticulum (E.R.) within aleurone cells. This disappearance of the E.R. was suggested to be the cause of the destabilization and eventual selective degradation of the α-amylase mRNA normally present in the polysomal particles associated with the rough E.R. in aleurone cells during germination (Belanger *et al.*, 1986).

Supporting evidence for lesions arising in the protein synthesizing and secretory systems in cereal endosperm aleurone cells has been provided by studies on protein disulphide isomerase (PDI) activity in wheat endosperm during loss of viability. PDI is the enzyme which catalyses the formation of the correct disulphide bonds in proteins during folding into their correct 3-D configuration following synthesis (Freedman *et al.*, 1984) such bonds often being found in secretory and cell surface proteins as a result of post translational modifications of proteins in the lumen of the E.R. Microsome preparations from aleurone layers from germinating viable wheat seeds exhibit higher levels of PDI activity than equivalent preparations from aleurone layers of non-viable seeds at the same post-imbibition time points (Livesley *et al.*, 1992). The low PDI activity in the aleurone layers of non-viable seeds, which still retain metabolic activity and significant levels of protein biosynthesis, may be a reflection of the lack of proliferation of the E.R. in these aleurone layers of non-viable wheat seeds (Livesley and Bray, 1991). Reduced levels of PDI in aleurone cells of non-viable seeds will have an adverse effect on protein folding and secretion of active hydrolytic enzymes by these aleurone layers with consequential detrimental effects on endosperm reserve mobilization.

Transport of mobilized reserves from the endosperm of cereal seed to the growing embryo is mediated via the scutellum which synthesizes a battery of membrane transport proteins during the early stages of germination to permit the transport of simple peptides, amino acids and sugars across the scutellum into the embryo. The barley scutellum appears to be the earliest structure within the seed to initiate protein synthesis following imbibition with almost all protein synthesis within the seed at 4.5 h imbibition being accounted for by this tissue (Stoddart *et al.*, 1973). One of the earliest detectable changes in the barley scutellum during germination is the synthesis of plasma membrane transport proteins and in particular that of the scutellar peptide transporter (Fig. 4).

This transporter is involved in the transport of the initial products of endosperm reserve protein digestion, i.e. small peptides, into the scutellar epithelial cells where they are degraded into amino acids for onward transport to the embryo. Thus the scutellum acts as a specialized absorptive and transport tissue during germination and the maintenance of this function is crucial for successful redistribution of the endosperm reserves to the embryo to support growth processes during germination and seedling establishment. Peptide

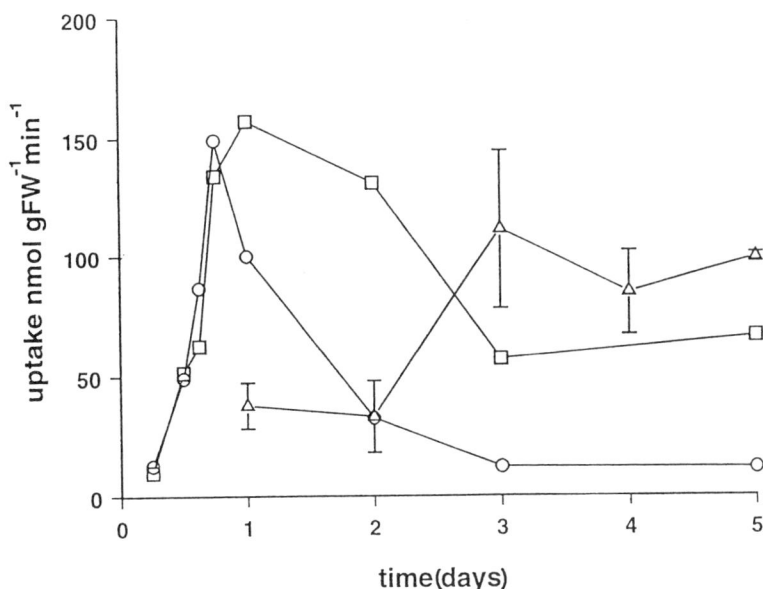

Figure 4. Peptide and amino acid transport activity development in germinating barley scutellar tissue. Δ—Δ amino acid transport activity; O—O peptide transport activity in scutellar tissue from embryos isolated from grain at 5 h imbibition stage; □—□ peptide transport activity in scutellar tissue in intact grains

transport activity develops in the scutellar epithelial cells within the first few hours of germination and precedes amino acid transport activity development during germination (Fig. 4, W.A. Waterworth and C. West, unpublished results). This observation reflects the timing of the appearance of the end products of reserve protein degradation within the endosperm of germinating cereals with small peptides appearing before amino acids. Non-viable barley seeds do not synthesize the scutellar peptide transporter (Table 3) and are thus incapable of furnishing essential supplies of amino acid nitrogen to the embryo to support the rapid growth phase in the hours following completion of water imbibition. Reserve mobilization deficiencies, which may include scutellar transport lesions, have been suggested previously as being a potential contributory factor to loss of cereal seed viability (Bhattacharya and Sen-Mandi, 1985).

A controlled coordinated response by the cells of the dehydrated orthodox seed to the imbibition of water is required from the reconstruction of cellular structures and reinitiation of the metabolic processes required to initiate and complete the germination process and then, ultimately, successfully establish a healthy vigorous young seedling. In the field these processes are often undertaken under the challenges of a hostile environment. Seeds of high vigour can

Table 3. Peptide transport activity in viable and non-viable barley scutellum

Source of scutellum	Peptide transport activity (nmoles peptide transported/g fresh wt/min)		
	1 day	3 days	5 days
Viable barley seeds	127	213	28.3
Non-viable barley seeds	25.0	13.8	18.0

resist and respond effectively to applied stresses, e.g. temperature and osmotic stress, and still germinate rapidly. As seeds lose vigour their response to applied stresses appears to become uncoordinated and has been shown to produce both quantitative and qualitative changes in patterns of protein synthesis in cereal embryos. In particular the ability to respond to imbibition by synthesizing 'growth phase' associated mRNAs and their proteins and simultaneously degrading stored mRNAs associated with cereal embryo development as rapidly as possible following water imbibition may be a major factor affecting cereal seed germinative vigour and ultimately viability.

Additional lesions appearing in the scutellum and aleurone layer of cereal seed may be more relevant to the post germinative phases of growth involved in seedling establishment but these are of equal importance to the grower. Again, loss of cereal seed vigour can be associated with a reduced or disorganized response to stress in these terminally differentiated tissues. It is of interest to note that the physiological response to stress has been shown to be a major player in determining metazoan lifespan. Genes have been identified in the nematode *C. elegans* in which mutations result in increased stress resistance (heat shock, free radicals) and increased lifespan. The identification of such genes demonstrated to affect lifespan and ageing in *C. elegans* and the association of long-lived mutants with increased thermotolerance is not too far removed from the scenario linking the cereal embryo response to heat shock with cereal seed vigour and loss of viability. Could homologous genes to those found in *C. elegans* affect the lifespan of seeds? If so, their identification and exploitation in the areas of seed quality and germplasm conservation could be of great significance.

Acknowledgements

Financial support from the BBSRC, SERC and the Rank Prize Funds is acknowledged. The author also acknowledges the significant contribution from the many collaborators whose research is cited in the text.

References

Aspinall, D and Paleg, L.G. 1971. *Journal of Experimental Botany* 18: 40–78.

Belanger, F.C., Brodl, M.R. and Ho, T.H.D. 1986. *Proceedings of the National Academy of Sciences (USA)* 83: 1354–1358.

Bhattacharyya, S. and Sen-Mandi, S. 1985. *Annals of Botany* 56: 475–479.

Blowers, L.E., Stormonth, D.A. and Bray, C.M. 1980. *Planta* 150: 19–25.

Freedman, R.B., Brockway, B.E. and Lambert, N. 1984. *Biochemical Society Transactions* 12: 929–932.

Grzelczak, Z.F., Sattolo, M.H., Hanley-Bowdoin, L.K., Kennedy, T.D. and Lane, B.G. 1982. *Canadian Journal of Biochemistry* 60: 388–397.

Helm, K.W., Petersen, N.S. and Abernethy, R.H. 1989. *Plant Physiology* 90: 598–605.

Johnston, F.B. and Stern, H. 1957. *Nature* 179: 160–161.

Lane, B.G. 1991. *Federation of American Societies for Experimental Biology Journal* 5: 2893–2901.

Livesley, M.A. and Bray, C.M. 1991. *Annals of Botany* 68: 69–73.

Livesley, M.A. and Bray, C.M. 1993. *Seed Science Research* 3: 179–186.

Livesley, M.A., Bulleid, N.J. and Bray, C.M. 1992. *Seed Science Research* 2: 97–103.

Petruzzelli, L. and Taranto, G. 1989. *Physiologia Plantarum* 76: 289–294.

Roberts, E.H. 1972. In: *The Viability of Seeds* pp 14–58 (ed. E.H. Roberts). London: Chapman and Hall Ltd.

Rushton, P.R. and Bray, C.M. 1987. *Plant Science* 51: 51–59.

Sargeant, J.G. and Walker, T.S. 1978. *Stärke* 30: 160–163.

Smith, C.A.D. and Bray, C.M. 1982. *Planta* 176: 413–421.

Smith, C.A.D. and Bray, C.M. 1984. *Plant Science Letters* 34: 335–343.

Spiegel, S. and Marcus, A. 1975. *Nature* (London) 256: 228–230.

Standard, S.A., Perret, D. and Bray, C.M. 1983. *Journal of Experimental Botany* 34: 1047–1054.

Stoddart, J.L., Thomas, H. and Robertson, A. 1973. *Planta* 112: 309–321.

Stormonth, D.A. 1978. In: *Development in the Business and Practice of Cereal Seed Trading and Technology*, pp 49–76 (ed. P.R. Hayward). London: The Gavin Press.

49. Priming-induced Nuclear Replication Activity in Pepper (*Capsicum annuum* L.) Seeds. Effect on Germination and Storability

S. LANTERI[1], P. BELLETTI[1], C. MARZACH[1], E. NADA[1], L. QUAGLIOTTI[1] and R.J. BINO[2]

[1]*DI.VA.P.R.A., Plant Breeding and Seed Production, via P. Giuria 15, 10126 Turin, Italy;* [2]*CPRO-DLO, PO Box 16, 6700 AA Wageningen, The Netherlands*

Abstract

Using flow cytometry, we observed that priming treatments in PEG solutions might induce DNA replication in the embryo root tips of pepper seeds. Under the same osmotic condition the amount of induced DNA synthesis was proportional to the length of the treatment and its effectiveness in improving seed performance, as measured by the reduction in mean germination time (MGT). However, different osmotic treatments might exert the same effect on MGT while inducing a different amount of nuclei to enter the synthetic phase. The activation and progress in DNA synthesis during priming, therefore, appear to be strongly influenced by the osmotic conditions and length of the treatment.

Osmoconditioning of controlled deteriorated seeds exerted various effects on seed germination, depending on seed deterioration. Under the same treatment, the amount of priming-induced DNA synthesis was lower than in unaged seeds or it was not induced at all.

Osmotic treatments considerably lowered seed tolerance to adverse storage conditions as compared with untreated seeds. However, seeds in which DNA replication was induced by priming were more sensitive to controlled deterioration than seeds in which priming did not induce nuclei to enter the synthetic phase.

Introduction

Several methods are used to precondition seeds in order to accelerate the rate of germination and to improve seedling uniformity. The method most commonly used is osmopriming, which consists of the incubation of seeds in an osmoticum dissolved in water, for a specific period of time and at a certain temperature. After priming, seeds are re-dessicated to allow storage and handling. Incubation in an osmotic solution prevents germination by restraining processes which are directly involved in radicle protrusion, such as cell wall extension. Other physiological and biochemical reactions may proceed, albeit often to a reduced rate. The improved seed performance after priming has been explained by the completion of DNA repair mechanisms (Osborne, 1983) and by a more favourable metabolic balance of primed seeds at the start of germination in water (Dell'Aquila *et al.*, 1978). During priming changes in protein patterns

R.H. Ellis, M. Black, A.J. Murdoch, T.D. Hong (eds.), Basic and Applied Aspects of Seed Biology, pp. 451–459.
© *1997 Kluwer Academic Publishers, Dordrecht. Printed in Great Britain.*

have also been observed (Davison and Bray, 1991).

In this paper we report the effect of priming on nuclear replication and germination of pepper seeds unaged and controlled deteriorated. Furthermore, we report the effect of priming on pepper seeds' storability. Our aim was to gain a better understanding of induction of nuclear replication and restoration of seed quality following priming treatments.

Materials and Methods

Seed and Seed Treatment

Pepper (*Capsicum annuum* L.) seeds of a commercial lot were obtained from SAIS (Societa Agricola Italiana Sementi, Cesena, Italy). They were submitted to 3 controlled deterioration treatments as described below. Seeds were kept in a cabinet at 75% RH and 20°C for 2 days (d), then transferred to laminate foil packets, sealed and kept at 45°C for 4–6 or 8 d.

Deteriorated seeds, as well as unaged seeds, were osmoprimed using PEG 6000 solutions at an osmotic potential of –1.1 or –1.5 MPa by placing them on filter paper saturated with the solutions at 20°C in the dark. All seeds were primed for 6, 10 or 14 d. After priming, the seeds were washed with running tap water to remove the osmotic agent and dried back to the initial seed moisture content (7.5% on a dry weight basis) in a drum with ventilated air (Koopman, 1963), above a saturated solution of $CaCl_2$ with 32% RH at 20°C. The moisture content was estimated using the oven method described by ISTA (1993).

Two samples of unaged pepper seeds were osmoprimed for 10 d at –1.1 MPa and for 6 d at –1.5 MPa. Subsequently they were subjected to controlled deterioration treatments performed as previously described and kept at 45°C for 24, 32, 48, 96, 120, 144 and 240 h.

Germination Tests

Germination tests were carried out according to the ISTA guidelines for germination under laboratory conditions (1993). A solution of 0.2% KNO_3 was used to saturate the germination substrate.

To determine mean germination time (MGT) seeds were counted daily for 14 d and were considered germinated when radicles had emerged at least 1 mm. The MGT was evaluated according to the formula $\Sigma nd/N$ (n = number of germinated seeds on each day; d = number of days from the beginning of the test; N = total number of germinated seeds). To determine germination percentage the number of normal seedlings was counted after 6, 10 and 14 d. Germination percentages were transformed to arcsin square roots before statistical analysis. Data on germination percentage (GP) and MGT were analysed by Tukey's HSD test.

Table 1. Effect of controlled deterioration at 45°C and osmopriming in PEG solutions on germination percentage (GP) and mean germination time (MGT) of pepper seeds. Data are means $\pm$ SE of four replicates of 100 seeds. Within a column, means with the same letter are not significantly different ($p < 0.01$; Tukey's HSD test)

Treatment	GP (%)	MGT (d)
Control	89.5$\pm$1.2 a	5.0$\pm$0.3 d
Artificial ageing for 4 d	74.9$\pm$1.5 b	6.9$\pm$0.0 e
Artificial ageing for 6 d	33.0$\pm$1.6 c	9.0$\pm$0.3 f
Artificial ageing for 8 d	24.9$\pm$1.4 d	9.1$\pm$0.4 f
Priming at –1.1 MPa for 6 d	91.5$\pm$1.3 a	2.6$\pm$0.3 b
Priming at –1.1 MPa for 10 d	86.8$\pm$1.1 a	2.2$\pm$0.2 a
Priming at –1.1 MPa for 14 d	86.9$\pm$1.5 a	2.0$\pm$0.1 a
Priming at –1.5 MPa for 6 d	91.5$\pm$1.7 a	3.4$\pm$0.2 c
Priming at –1.5 MPa for 10 d	92.9$\pm$1.8 a	3.3$\pm$0.1 c
Priming at –1.5 MPa for 14 d	88.8$\pm$1.8 a	2.6$\pm$0.1 b

Preparation of Nuclear Samples

Samples of nuclei for flow cytometry were prepared as reported by Lanteri *et al.* (1993). One mm of the distal part of the embryo was used for analysis. To detect DNA, 10 mg/l of propidium iodide (PI)) was added to the isolation buffer (Saxena and King, 1989). For each sample at least 20 seeds were used and the determination was made in duplicate.

Fluorescence was measured using a FACScan flow cytometer (Becton & Dickinson, Mountain View, CA, USA) equipped with a 488 nm light source (argon laser). Two filters were used to collect the red fluorescence due to PI staining of the DNA, one transmitting at 585 nm with a bandwidth of 420 nm (FL2), the other transmitting above 620 nm (FL3). FL2 and FL3 were registered on a logarithmic scale. Forward and side light scatter were measured simultaneously. The flow rate was set at about 200 nuclei/s and at least 10^4 nuclei were analyzed for each sample. Debris was excluded from analysis by appropriately raising FSC and FL3 thresholds to values selected experimentally. Data were recorded in a Hewlett Packard computer (HP 9000, model 300) using CellFit software (Becton & Dickinson).

Results

Effect of Ageing and Priming Treatments on Fresh Seeds

The GP of fresh pepper seeds was 89.5 and the MGT 5 d. The reduction in GP and the increase in MGT due to controlled ageing treatments were proportional to their duration (Table 1). The most severe ageing condition (45°C for 8 d) induced a decrease in GP to 24.9% and increased the MGT by 4.1 d. Priming decreased MGT without affecting GP (Table 1). The most effective priming treatments were those performed at the lower osmotic potential (–1.1 MPa) for 10 and 14 days, which induced a decrease in MGT of about 3 d. There was no germination during the osmotic treatments.

Effect of Priming Treatments on Artificially Aged Seeds

In seeds aged for 4 d, priming for 6 to 14 d at –1.1 or –1.5 MPa induced a reduction in MGT similar to that observed in unaged seeds (Figs. 1A,B); furthermore partial recovery of GP, proportional to length of treatment, was observed in both priming conditions (data not reported).

In seeds aged for 6 d, priming slightly increased GP and exerted a marked effect on MGT (Figs. 1A,B). After 10 or 14 d at –1.1 MPa, MGT was lowered to 2.9 and 2.7 d, respectively.

In seeds aged for the longest duration (8 d) priming was detrimental to seed germination. After 6 d at –1.1 MPa, GP was reduced to about 14% and after 14 d only 3.5% of seeds germinated (data not reported). A similar pattern was observed at –1.5 MPA.

Effect of Priming and Ageing on Nuclear Replication

Flow cytometry analysis of nuclear DNA content in the embryo root tips of dry, fully matured pepper seeds, revealed only one peak at channel 100 (Fig. 2), corresponding to diploid pre-replication chromosomal configuration (G1).

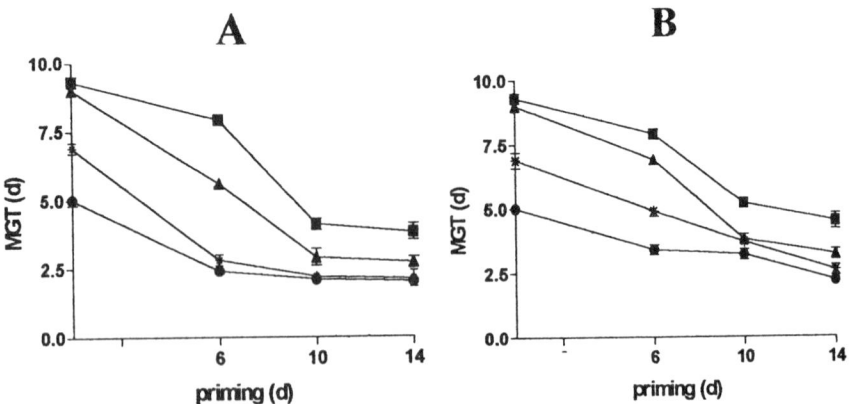

Figure 1. Mean germination time (MGT) of pepper seeds unaged (●) and artificially aged at 45°C for 4 (✕), 6 (▲) and 8 (■) days after priming in PEG solutions for 6, 10, 14 days (d) at the osmotic potentials of –1.1 (A) and –1.5 (B) MPa. Standard errors are indicated where they exceed the size of the symbols

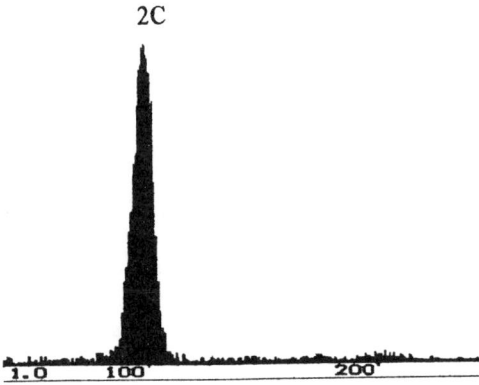

Figure 2. Flow cytometric histogram of nuclei from the embryo root tips of pepper seeds showing one peak at the 2C DNA content (channel 100)

Priming did not induce nuclear replication in the embryo root tips of unaged seeds only when performed for 6 d at –1.1 MPa. The number of nuclei induced in 4C was correlated to treatment duration (Figs. 3A,B). Under the same treatment duration there was a higher percentage of nuclei in 4C after priming at the lower osmotic potential. DNA profiles observed in radicle meristems of seeds primed for 14 d are reported in Figures 4 A,B.

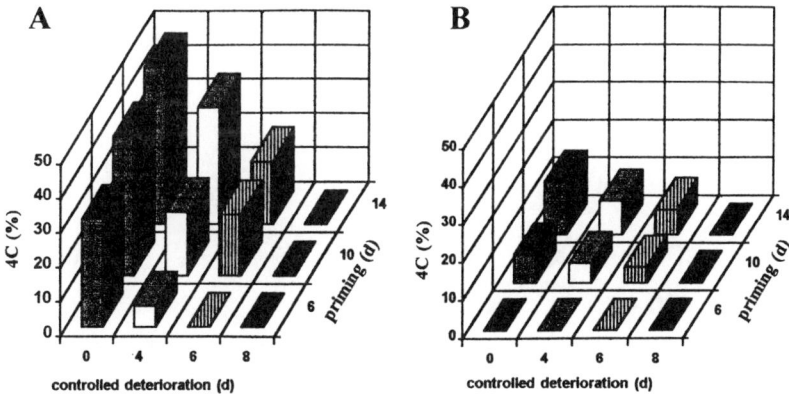

Figure 3. Percentage of 4C nuclei in the embryo root tips of pepper seeds unaged and artificially aged at 45°C for 4, 6 or 8 d, after priming at the osmotic potential of –1.1 (A) and –1.5 (B) MPa, for 6, 10 or 14 d

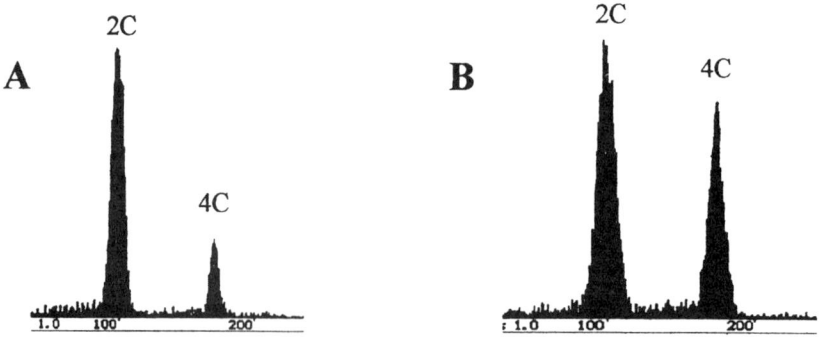

Figure 4. Flow cytometric histograms of nuclei from the embryo root tips of pepper seeds primed in PEG solution for 14 days at the osmotic potential of –1.5 MPa (A) and –1.1 MPa (B). The peak at channel 100 corresponds to the pre-replicative DNA content (2C), while the peak at channel 190 corresponds to a post-replicative DNA content (4C)

In aged seeds a marked decrease in nuclear replication was observed after osmotic treatment (Figs. 3A,B). In embryo root tips of seeds aged 8 d neither priming condition induced nuclei to enter the synthetic phase. Priming at –1.1 MPa always induced DNA synthesis on seeds aged for 4 d but, in seeds aged for 6 d, only when performed for the duration of 10 and 14 d (Fig. 3A). Priming at the osmotic potential of –1.5 MPa induced nuclear replication activity only on 4 and 6 d aged seeds when performed for 10 and 14 d (Fig. 3B).

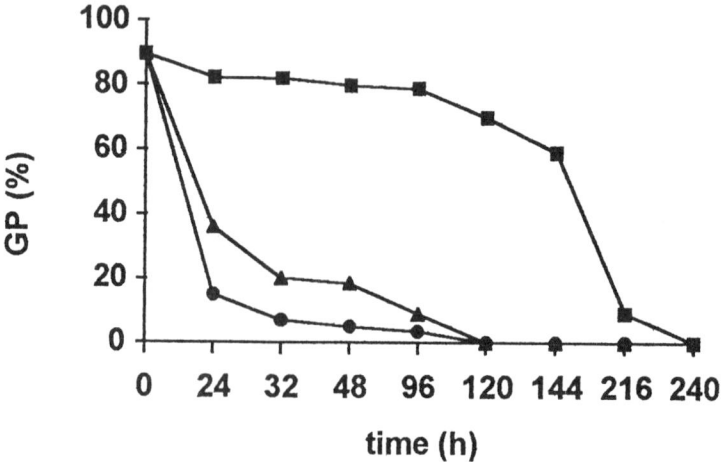

Figure 5. Effect of controlled deterioration at 45°C on viability of pepper seeds unprimed (■), and primed in PEG solutions at –1.5 MPa for 6 d (▲) and –1.1 MPa for 10 d (●)

Effect of Priming Treatments on Seed Storability

Germination performance of primed seeds was adversely affected by controlled deterioration treatments. Ageing at 45°C reduced the GP of seeds primed at −1.5 and −1.1 MPa respectively to 32% and 11.5% after only 24 h, while unprimed seeds still produced 82% normal seedlings (Fig. 5). In primed seeds a further decrease in the percentage of germination was observed after 32–96 h of ageing and after 120 h no normal seedlings were produced. Unprimed seeds lost their germinability only after 240 h of ageing (Fig. 5).

Discussion

Osmotic preconditioning prevents seed germination since the water content is kept below the threshold necessary for radicle growth (Bradford, 1990). Notwithstanding, during priming, DNA, RNA and proteins are synthesized (Ashraf and Bray, 1993; Bray *et al.*, 1993; Cruz Garcia *et al.*, 1995). Our results show that priming also induced nuclear replication activity in the embryo root meristems of *Capsicum annuum.* The rate and amount of DNA synthesis depended on the osmotic potential of the solution and the duration of priming. Activation of DNA synthesis, therefore, appears to be a late event which follows the activation of other molecular processes.

Unless performed at −1.5 MPa for 6 d, priming always induced DNA replication in root tips of unaged pepper seeds. The amount of DNA synthesis was higher after priming at the lower osmotic potential and was correlated to priming duration. However, there was not always a direct link between the amount of DNA synthesis and seed performance; e.g. the longest priming treatment at −1.5 MPa was as effective on seed performance as the shortest treatment at −1.1 MPa, which induced larger numbers of nuclei to enter G2. Both priming at −1.1 MPa for 6 d and at −1.5 MPa for 14 d, in fact, reduced MGT by 3 d but induced respectively about 30 and 12% of nuclei to enter G2.

The effect of controlled deterioration on seed performance, as measured by the reduction in GP and the increase in MGT, was proportional with its duration. The effect on germination of subsequent priming depended on the osmotic potential and its length as well as the degree of seed deterioration. Osmoconditioning of 4 d aged seeds completely reversed the effect of ageing and even caused enhancement of MGT up to the level of unaged osmoprimed seeds. In more deteriorated seeds (6 d ageing) priming induced a less marked effect although a partial recovery of seed performance and viability was observed. Differently, in strongly deteriorated seeds (8 d ageing), priming induced a further deterioration by reducing the GP from about 25.9 to 5%. This result accords with the findings of Roberts (1981) and Pandey (1989), who reported that severe damage developing in the phase preceding seed death is irreversible.

The amount of priming-induced DNA synthesis in controlled deteriorated seeds was lower than in unaged seeds. This indicates that decline in seed viability is preceded by slowing down of metabolism related to germination, including DNA replication. When observed, the amount of DNA synthesis induced by a specific priming condition, was directly correlated to the restoration of seed vigour.

Interestingly in 4 d aged seeds, in spite of reduced DNA synthesis and the apparent lower metabolic advancement, the positive effect on germination performance of 14 d priming treatment at −1.1 MPa was analogous to that observed in unaged seeds. Recovery of seed performance after priming of controlled deteriorated seeds was also observed by Coolbear *et al.* (1990) and Fujikura and Karssen (1992). This result, together with the previously described results obtained in unaged seeds, shows that the length of a priming treatment greatly influences seed performance.

Osmotic treatments on unaged seeds at −1.1 MPa for 10 d and at −1.5 MPa for 6 d considerably lowered seed tolerance to adverse storage conditions as compared with untreated seeds. In our pepper seed lot, therefore, the enhanced germination performance of primed seeds accompanies an enhanced susceptibility to ageing. This contradicts results previously obtained by Georghiou *et al.* (1987), which observed an increased longevity of pepper seeds subjected to osmoconditioning prior to storage, but agrees with results obtained in tomato by Argerich *et al.* (1989) and in lettuce by Tarquis and Bradford (1992).

Although a drastic drop in viability of primed seeds was observed after only 24 h of storage at 45°C, seeds primed at −1.1 MPa, containing about 40% of nuclei in G2 were more sensitive to ageing than seeds primed at −1.5 MPa, in which no nuclei were induced to enter the synthetic phase. This could be a consequence of a higher DNA content, which is a more vulnerable target for mutation inducing factors. Alternatively the increased sensitivity could be a consequence of their more advanced progress in germinative events, making seeds less resistant to deteriorative factors imposed during storage. On the whole it appears that the nucleic acid DNA content of the embryo only plays an additive role in influencing seed storability.

From our previous and present results it appears that DNA synthesis is affected by the water potential and length of priming as well as the degree of seed deterioration. As reported by Coolbear *et al.* (1990) nucleic acid synthesis is a pre-requisite for germination but does not appear the only metabolic factor influencing seed germination. For instance, activation of DNA repair, whose efficiency is presumably proportional to the length of the preconditioning treatment, is of extreme importance. Hence, before DNA synthesis can start, both DNA repair mechanisms and cell cycle related processes must first be activated. During priming other processes, not necessarily leading to DNA replication, may also be important in improving seed performance. These processes could be activated at lower osmotic potentials than those required for DNA replication, but could enable seed to improve its efficiency in synthesising DNA during the first hours of germination. Lanteri *et al.* (1993) observed that priming treatments which do not induce DNA replication both accelerate and increase the amount of DNA synthesis during the period of imbibition which precedes visible germination.

References

Argerich, C.A., Bradford, K.J. and Tarquis, A.M. 1989. *Journal of Experimental Botany* 40: 593–598.

Ashraf, M. and Bray, C.M. 1993. *Seed Science Research* 3: 15–23.

Bradford, K.J. 1990. *Plant Physiology* 94: 840–849.

Bray, C.M., Davison, P.A., Ashraf, M. and Taylor, R.M. 1989. *Annals of Botany* 63: 185–193.

Coolbear, P., Slater, R.J. and Bryant, J.A. 1990. *Annals of Botany* 65: 187–195.

Cruz García, F., Jimenez, L.F. and Vázquez-Ramos, J.M. 1995. *Seed Science Research* 5: 15–33.

Davison, P.A. and Bray, C.M. 1991. *Seed Science Research* 1: 29–35.

Dell'Aquila, A., Savino, G. and De Leo, P. 1978. *Plant and Cell Physiology* 19: 348–354.

Fujikura, Y. and Karssen, C. 1992. *Seed Science Research* 2: 23–31.

Georghiou, K., Thanos, C.A. and Passam, H.C. 1987. *Annals of Botany* 60: 279–285.

ISTA. 1993. *Seed Science and Technology* 21, Supplement, Rules 1993.

Koopman, M.J.F. 1963. *Proceedings of the International Seed Testing Associations* 28: 853–860.

Lanteri, S., Kraak, L., De Vos, C.H.R. and Bino, R.J. 1993. *Physiologia Plantarum* 89: 433–440.

Osborne, D.J. 1983. *Canadian Journal of Botany* 61: 3568–3577.

Pandey, D.K. 1989. *Seed Science and Technology* 17: 391–397.

Roberts, E.H. 1981. *Seed Science and Technology* 9: 359–372.

Saxena, P.K. and King, J. 1988. In: *Biotechnology in Agricultural and Forestry. Plant Protoplast and Genetic Engineering*, pp 328–342 (ed. Y.P.S Bajaj). New York: Springer-Verlag.

Tarquis, A.M. and Bradford, K.J. 1992. *Journal of Experimental Botany* 43: 307–317.

50. Induction of Desiccation Tolerance in Germinated *Impatiens* Seeds Enables Their Practical Use

T. BRUGGINK and P. VAN DER TOORN

S&G Sandoz Seeds, P.O. Box 26, 1600 AA Enkhuizen, The Netherlands

Abstract

Selection of rapidly germinating seeds increased the percentage of usable plants obtained from a seed lot of *Impatiens*. Without further treatment such seeds were susceptible to desiccation damage during sowing. Exposure of germinated seeds to a mild water stress for several days, by incubation in PEG solution, induced desiccation tolerance. Best results were obtained after 7 days (d) incubation in PEG solution of -1.0 MPa. Seeds treated in this way and subsequently desiccated produced nearly 100% usable plants and had a shelf life of 6 months.

Introduction

Selection of seeds based on their germination rate gives the opportunity to obtain the most vigorous seeds from a batch. This was shown for cauliflower, leek and onion by Finch-Savage (1986) and for carrot by Finch-Savage and McQuistan (1988). In all these crops the most rapidly germinating seeds gave rise to the highest percentage of usable plants.

However, the resulting product of such a selection consists of germinated seeds which have a short shelf life and are prone to desiccation damage because seeds generally lose their tolerance to desiccation at the moment of germination (Bewley and Black, 1982; Leopold, 1990; Hong and Ellis, 1992). Several methods were developed in the past to utilize germinated seeds commercially. The fluid drilling technique (Gray, 1981; 1984), in which seeds are sown in a gel or directly from water into cellular trays, requires specialized equipment. The technique of low moisture content germinated seeds (LMCG) (Finch-Savage and McKee, 1988), in which germinated seeds are dried to a moisture content at which they just survive, has the risk of desiccation damage during handling of the seeds. Both methods have not become accepted in the market.

Our finding that desiccation tolerance can be reinduced after seeds have germinated (Bruggink and van der Toorn, 1995) opens new possibilities for the use of germinated seeds. In the present paper some aspects of selection of germinated seeds and of the induction of desiccation tolerance after germination are studied for the bedding plant *Impatiens walleriana*.

R.H. Ellis, M. Black, A.J. Murdoch, T.D. Hong (eds.), Basic and Applied Aspects of Seed Biology, pp. 461–467.
© *1997 Kluwer Academic Publishers, Dordrecht. Printed in Great Britain.*

Material and Methods

Seeds of *Impatiens walleriana* 'Impulse red' were germinated in aerated water at 20°C. Germinated seeds were selected at the moment that radicle protrusion was visible.

To induce desiccation tolerance germinated seeds were incubated on blotter paper moistened with a PEG 8000 solution at a temperature of 8°C for different periods of time. Several concentrations of PEG were used resulting in different potentials of the solution as calculated according to Michel (1983). At the end of incubation seeds were rinsed in tap water, centrifuged inside nylon gauze bags to remove free water and either dried back slightly to a moisture content of 45% (fresh weight basis) or they were desiccated to a moisture content of 5%. Drying took place by exposing seeds to air of 40% relative humidity and 20°C. Seed moisture content was determined by oven drying at 130°C for 2 h.

Germinated seeds with a moisture content of around 45% were stored at 8°C in Petri dishes which allowed minimal air exchange but prevented further desiccation. Germinated seeds with a moisture content of around 5% were stored in sealed aluminium bags at a temperature of 8°C.

Survival of desiccated germinated seeds was determined from their ability to produce a well developed root system after a specified number of days of reimbibition on blotting paper at 25°C.

The percentage usable plants was determined by sowing replicates of 100 seeds in a peat soil mix. The number of well developed seedlings was determined after 14 d at 20°C.

For a simulation of the effect of conditions during sowing on germinated seeds, seeds were spread out in a 5 mm layer in a tray which was placed in an environment of 30% relative humidity and a temperature of 20°C. Seeds were carefully stirred every 20 min. At regular intervals samples were taken to measure seed moisture content and percentage usable plants.

Sucrose content of whole seeds was determined on replicate samples of 25 seeds as described previously (Bruggink and van der Toorn, 1995).

Results

Seeds with the highest germination rate showed the highest percentage of usable plants (Table 1). Seeds with lower germination rates showed progressively lower percentages of usable plants. Seeds which had not yet germinated after 112 h of incubation resulted in only 64% usable plants.

Germinated seeds which are not tolerant to desiccation may easily suffer damage during sowing (Fig. 1). When seeds were exposed to simulated sowing conditions for more than 1 h the resulting percentage usable plants decreased. Moisture content of seeds had decreased from the initial 45 to 20% during that period and after 2 h it was reduced to 8%. In comparison, seeds with induced desiccation tolerance showed a similar decrease in moisture content but without

Table 1. Effect of time of germination of seeds of *Impatiens walleriana* 'Impulse red' on the subsequently obtained percentage usable plants. The middle column shows the size of the fraction which germinated in the specified period. Seeds were germinated at a temperature of 20°C until first germination occurred. After that temperature was 20°C in daytime (8 h) and 15°C at night (16 h)

Time of germination (h)	Size of fraction (%)	Usable plants (%)
<72	14.3	96
72–80	26.9	95
80–96	24.9	91
96–112	16.6	83
>112	17.3	64
Control seeds	100.0	84

adverse effects on later emergence.

Germinated *Impatiens* seeds were incubated in PEG solutions of different potentials, ranging from 0 to –3 MPa; incubation lasted for 1, 2 or 7 d at a temperature of 8°C. Subsequently seeds were desiccated and their ability to form secondary roots on reimbibition was measured. Seeds which were desiccated immediately after germination without further treatment gave survival percen-

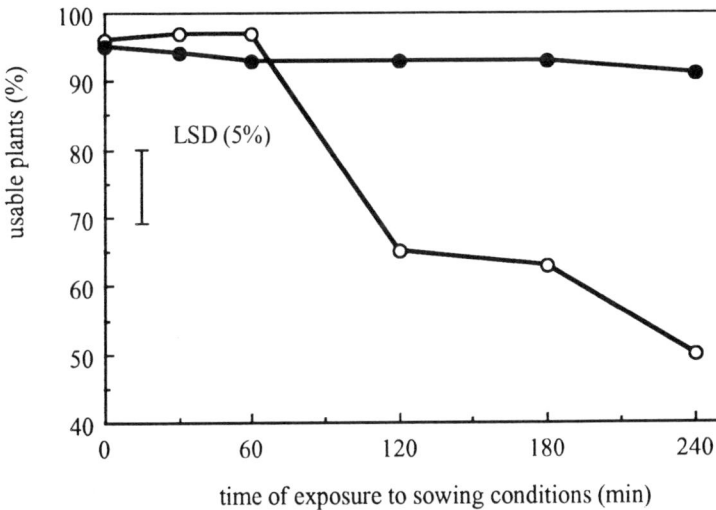

Figure 1. Effect of duration of simulated sowing conditions (20°C, 30% RH) on the percentage usable plants obtained from germinated seeds of *Impatiens*, with (closed circles) or without (open circles) induced desiccation tolerance. The vertical bar shows the least significant difference at the 5% level

tages of less than 10% (not shown). Seeds which were desiccated after 1, 2 or 7 d incubation in PEG gave survival percentages of 90% or higher as measured by root formation after 7 d of reimbibition (not shown). A longer incubation period, however, resulted in more rapid formation of secondary roots (Fig. 2). The differences in survival and rate of secondary root formation between the

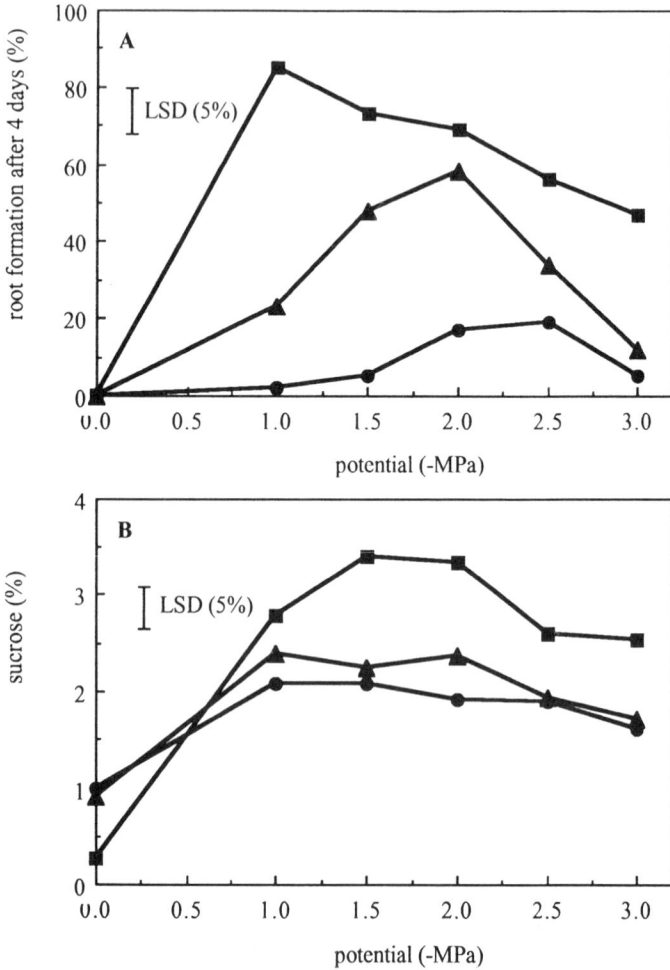

Figure 2. Effect of incubation in PEG solutions of different potential at 8°C on germinated seeds of *Impatiens*. Incubations lasted for 1 (circles), 2 (triangles) or 7 d (squares). After incubation seeds were desiccated. The Figures show the percentage of desiccated seeds which showed growth of secondary roots after 4 days of reimbibition on blotter paper at 25°C (A), and the sucrose concentration of the germinated seeds at the end of the incubation period in PEG solution (B). A potential of 0 represents incubation in water. The vertical bars show the least significant difference at the 5% level

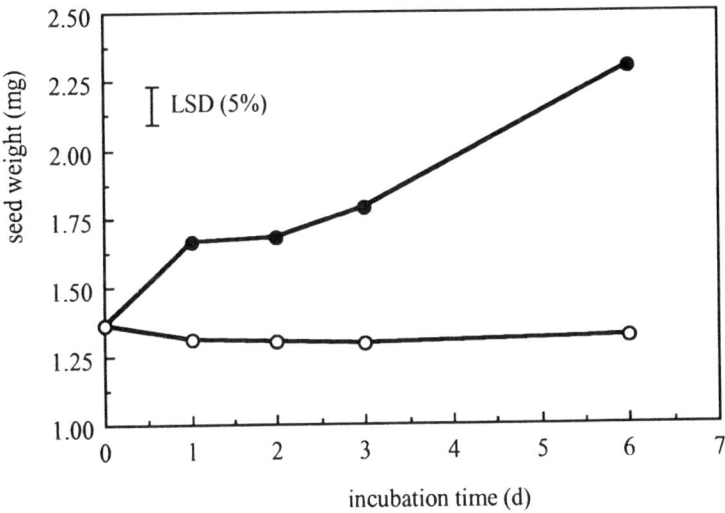

Figure 3. Average weight of germinated *Impatiens* seeds during incubation in water (closed circles) or in PEG solution of –1.5 MPa (open circles), both at a temperature of 8°C. The vertical bar shows the least significant difference at the 5% level

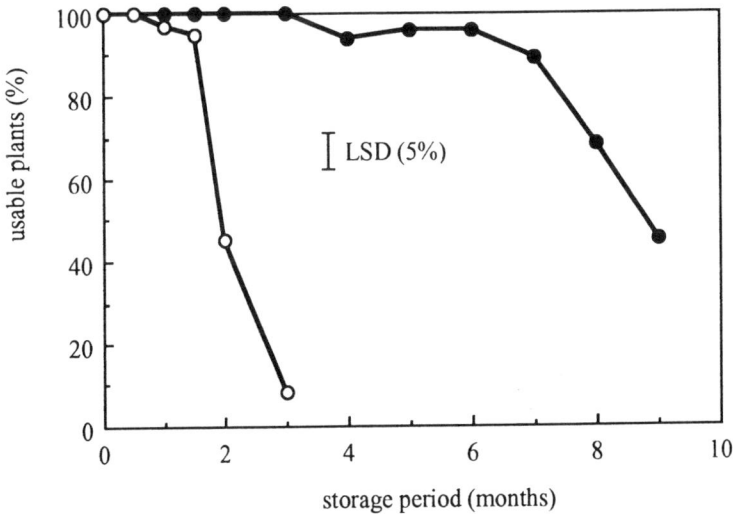

Figure 4. Effect of storage of germinated *Impatiens* on percentage usable plants. Seeds were either stored at a seed moisture content of 45% (open circles) or at a seed moisture content of 5% (closed circles). Seeds were stored at 8°C and sown in soil after different durations of storage. The vertical bar shows the least significant difference at the 5% level

different potentials of PEG solutions were small. Incubation in water was not effective. In all cases, except for incubation in water, sucrose concentration increased (Fig. 2). Sucrose concentration of the seeds before incubation was 1.0% on a dry weight basis (not shown). A longer incubation period resulted in more sucrose build-up, again with only minor differences between the different PEG concentrations. Incubation in water resulted in a decrease in sucrose content with increased incubation time.

Figure 3 shows that incubation in PEG (–1.5 MPa, 8°C) prevented further growth of germinated *Impatiens* seeds, whereas at the same temperature in water seeds almost doubled in fresh weight in 7 d.

The shelf life of non-induced non-desiccated seeds was compared to that of seeds in which desiccation tolerance was induced and which were subsequently desiccated to a moisture content of 5%. All seeds were stored at 8°C. Initially biweekly and later on monthly a germination test was carried out on soil (Fig. 4). Performance of the non-desiccated seeds showed a sharp drop after around 6 weeks of storage whereas the desiccated seeds remained at a fairly constant level for 6 months.

Discussion

Selection of rapid germinating *Impatiens* seeds resulted in higher percentages of usable plants compared to control seeds. This result is similar to that found for a number of vegetable seeds (Finch-Savage, 1986; Finch-Savage and McQuistan, 1988). This indicates that it is probably a general principle that seeds which are able to germinate rapidly have a higher chance of producing a normal seedling.

Germinated *Impatiens* seeds were sensitive to desiccation as shown in Figure 1. Exposure to air for more than 1 h, as may easily occur during handling of seeds in practice, resulted in a sharp drop in performance of the seeds. This drop did not occur when desiccation tolerance had previously been induced in the seeds. The percentage usable plants obtained from non induced seeds, remained at around 50%, even for seeds which had been dried for 4 h. Storage of such seeds, however, will rapidly lead to a complete loss of viability (unpublished results).

Induction of desiccation tolerance occurred when seeds were exposed to such conditions that radicle growth was prevented, as was the case with incubation in PEG solution (Fig. 3). One day incubation in PEG solution already gave high survival percentages but longer incubation periods resulted in faster formation of roots. This indicates that seeds suffer less damage from desiccation when they were incubated for a longer period previously. In parallel with increased desiccation tolerance the sucrose content of the seeds increased from 1% before incubation to around 2% after a 1 or 2 d incubation and to over 3% after a 7 d incubation. This supports the hypothesis that an increased sucrose content may be necessary to confer desiccation tolerance (Koster and Leopold, 1988; Blackman *et al.*, 1992; Bruggink and van der Toorn, 1995). However, it is unlikely that sucrose is the only factor necessary for desiccation tolerance

(Leprince *et al.*, 1993; Ooms *et al.*, 1994). Synthesis of heat soluble proteins seems also to be important in the case of germinated seeds (Bruggink and van der Toorn, 1995). Nevertheless, sucrose concentration can, at least in the case of *Impatiens*, be used as a measure of the degree of desiccation tolerance which is induced. The increase in sucrose concentration might be caused by the fact that radicle growth has ceased (Fig. 3) while metabolic processes, like breakdown of reserves into sugars, continued.

When germinated seeds were stored with a high moisture content (45%) their shelf life was limited to around 6 weeks. Probably seeds deteriorated because of continued respiration and the occurrence of fungi on some of the seeds. When such factors are eliminated by desiccation to a moisture content similar to that of untreated seeds (5%) shelf life was extended to around 6 months. This is still shorter than the shelf life of untreated seeds, indicating that, like in primed seeds (Tarquis and Bradford, 1992), the events associated with germination of seeds have a negative effect on shelf life.

Conclusion

Selection of seeds showing rapid germination can for *Impatiens* be used as a method to increase the percentage usable plants. When desiccation tolerance is induced in germinated seeds their handling is facilitated and by subsequent desiccation their shelf life may be increased to around 6 months. Because of the easy induction of desiccation tolerance, germinated *Impatiens* seeds could be used as a model system for studies on the mechanisms of desiccation tolerance.

References

Bewley, J.D. and Black, M. 1982. *Physiology and Biochemistry of Seeds in Relation to Germination II.* Berlin: Springer-Verlag.
Blackman, S.A., Obendorf, R.L. and Leopold, A.C. 1992. *Plant Physiology* 100: 225–230.
Bruggink, T. and van der Toorn, P. 1995. *Seed Science Research* 5: 1–4.
Finch-Savage, W.E. 1986. *Annals of Applied Biology* 108: 441–444.
Finch-Savage, W.E. and McKee, J.M.T. 1988. *Annals of Applied Biology* 113: 415–424.
Finch-Savage, W.E. and McQuistan, C.I. 1988. *Journal of Agricultural Science, Cambridge* 110: 93–99.
Gray, D. 1981. *Horticultural Reviews* 3: 1–27.
Gray, D. 1984. *Aspects of Applied Biology 7. Crop establishment: biological requirements and engineering solutions*: 153–172.
Hong, T.D. and Ellis, R.H. 1992. *Journal of Experimental Botany* 43: 239–247.
Koster, K.L. and Leopold, A.C. 1988. *Plant Physiology* 88: 829–832.
Leopold, A.C. 1990. In: *Stress in Plants: Adaptation and Acclimation Mechanisms* (eds. Alscher, R.G. and Cumming, J.R.). New York: Wiley-Liss.
Leprince, O., Hendry, G.A.F. and McKersie, B.D. 1993. *Seed Science Research* 3: 231–246.
Michel, B.E. 1983. *Plant Physiology* 72: 66–70.
Ooms, J.J.J., Wilmer, J.A. and Karssen, C.M. 1994. *Physiologia Plantarum* 90: 431–436.
Tarquis, A.M. and Bradford, K.J. 1992. *Journal of Experimental Botany* 43: 307–317.

51. Osmotic Relations and Cell Wall Acidification as the Prerequisites of the Start of Elongation in the Seed Axial Organs

O.V. ANTIPOVA

Institute of Plant Physiology, Russian Academy of Sciences, Botanical Street 35, 127276 Moscow, Russia

Abstract

In *Vicia faba minor* seeds, early germination occurs by cell elongation until the axis is 1 cm in length. It is a good model to study some processes which prepare the axial organs for elongation. These processes develop in the water content range from 60% (fr wt), at which physical water absorption is complete, to 72–73%, at which the radicles protrude. The first process is the additional accumulation of osmotically active solutes. This accumulation follows from (1) measurements of osmotic potential in cell sap from axial organs and hypocotyls, (2) estimation of main osmotic components, mostly sugars and K^+, and (3) examination of vacuole: cytoplasm area ratio. Accumulation of osmotic solutes provides further water inflow into the vacuole necessary for the cell elongation. The second process is acidification of cell wall resulting from H^+-extrusion due to H^+-ATPase activation. This follows from: (1) axes begin to acidify the ambient solution 2 h prior to radicle emergence; (2) acidification is inhibited by diethylstylbestrol and stimulated by fusicoccin; (3) fusicoccin stimulates radicle emergence, water uptake and cell elongation as the acid buffer did while vanadate and diethylstylbestrol inhibited radicle emergence. Thus, acidification of cell walls providing wall loosening and acid growth, in combination with active water flow into the cells, allows them to start the elongation.

Introduction

Growth initiation in the axial organs represents the key process in germination, providing for further development of the plant organism of the next generation. The primary and obligatory process in the growth of axial organs is cell elongation (Obroucheva, 1992). Cell division begins either simultaneously with cell elongation or later. It is cell elongation that provides rapid growth of embryo axis and its immediate contact with soil water. For this reason, the initiation of growth in the axial organs of germinating seeds should be considered in terms of the initiation of cell elongation.

The initiation of growth in germinating seeds is usually considered as the culmination of metabolic activation in the course of seed imbibition. However, such activation occurs not only in the axial organs but in the endosperm or cotyledons as well. The goal of this work was to find out and characterize the specific metabolic process(es) preparing for cell elongation in the axial organs.

R.H. Ellis, M. Black, A.J. Murdoch, T.D. Hong (eds.), Basic and Applied Aspects of Seed Biology, pp. 469–477.
© *1997 Kluwer Academic Publishers, Dordrecht. Printed in Great Britain.*

Materials and Methods

The most suitable plant material was the seeds of *Vicia faba minor* L. (broad beans, cv. Aushra) with hypogeal germination. Their radicle emergence results from the elongation of hypocotyl cells, then the elongation of basal radicle cells begins, and only 20 h after the start of elongation, mitotic activity in the root meristem commences (Obroucheva and Antipova, 1985).

The seeds imbibed on pleated filter paper in distilled water inside glass-covered trays at 27°C in a dark thermostat. Early germination was recorded as the time of radicle protrusion in 50% of seeds.

The water potential was measured by compensation method; the osmotic potential was measured cryoscopically with a Video-Osmolab-4000 (Germany). Ions were measured with ion-selective electrodes with an Orion Research Ionalyzer-90 (USA). The total sugars were determined by the phenolic method; sucrose and fructose were measured using the Roe method. The amount of glucose was calculated by subtracting the sucrose and fructose amounts from the total sugars. Cell and vacuole areas were measured on electron micrographs. pH was measured for 30 min with a pH-meter, the measuring cell of which contained 100 axial organs in 10 ml 1 mM KCl. Details of the methods used have been provided elsewhere (Obroucheva et al., 1993; Obroucheva and Antipova, 1994).

Results

As the cell elongation needs the water inflow, water relations in embryo axes called for close inspection. The hydration curve of axial organs (Fig. 1) can be described as a rapid rise up to 60% water content (wc, % fr wt), then a lag-phase corresponding to slow wc increase to 72% (visible germination) followed by gradually increased wc due to vacuolation of elongating cells. The hydration curve of the axial organs in dead seeds (killed by heating at 80°C for a week) shows that water absorption proceeds only to 60% wc. Thus 60% wc is the upper limit for axis hydration by purely physical forces, mainly by matrix forces. Therefore, the hydration of axial organs above 60% needs the physiological mechanisms providing further water uptake.

To characterize the contribution of matrix and osmotic forces to water uptake by axial organs, the water and osmotic potentials were compared within the 60–72% wc range (Table 1). The osmotic forces dominate in the axis hydration at lag-phase, when the water absorption by physical forces was completed.

The amount of osmotic solutes was directly measured in the cell sap from axial organs, the sugars and K^+-ions being known as the main osmotic constituents in plants.

Figure 2 demonstrates that the axial organs contained at early imbibition significant amounts of ions, namely potassium, nitrate, and chloride, accumulated as early as during seed maturation on the maternal plant. From 68% wc

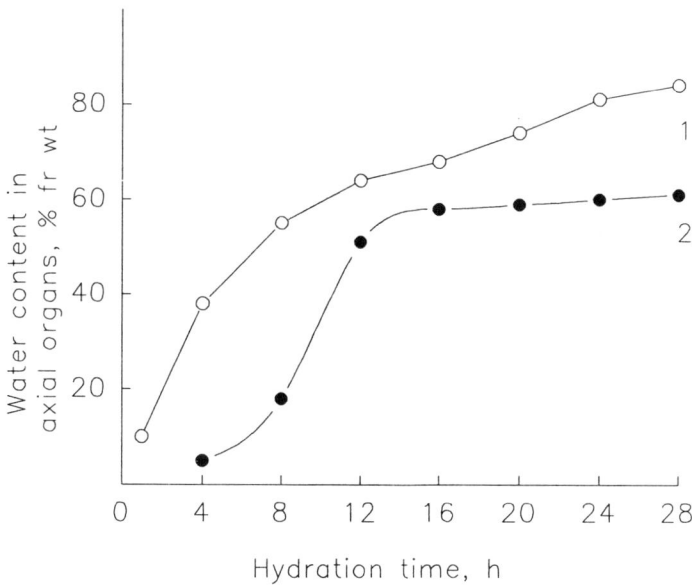

Figure 1. Water absorption by axial organs of intact (1) and dead (2) seeds

on, potassium became a dominant ion. Fructose and sucrose were the main sugars at 60% wc. As the wc increased, the amount of sucrose did not alter, but fructose disappeared while glucose accumulated and became a dominant sugar. The contribution of amino acids to the development of osmotic forces was less than 15 % while the effect of cations (Ca, NH_4) is negligible (data not shown).

Figure 3 shows the relative contributions of cations, anions, and sugars to the osmotic pressure in imbibing axial organs. At 60–65% wc, the ions played a much more important role than the sugars. However, from 68% wc on (prior to growth initiation), the ions and the sugars equally contributed to osmotic potential, and potassium became the major ion.

Table 1. Ratio of osmotic to water potential in imbibing seeds

Water content in axial organs, % fr wt	Water potential MPa	Osmotic potential	
		MPa	% of water potential
60	−11.4	−2.5	22
65	−4.7	−1.7	36
68	−2.4	−1.6	67

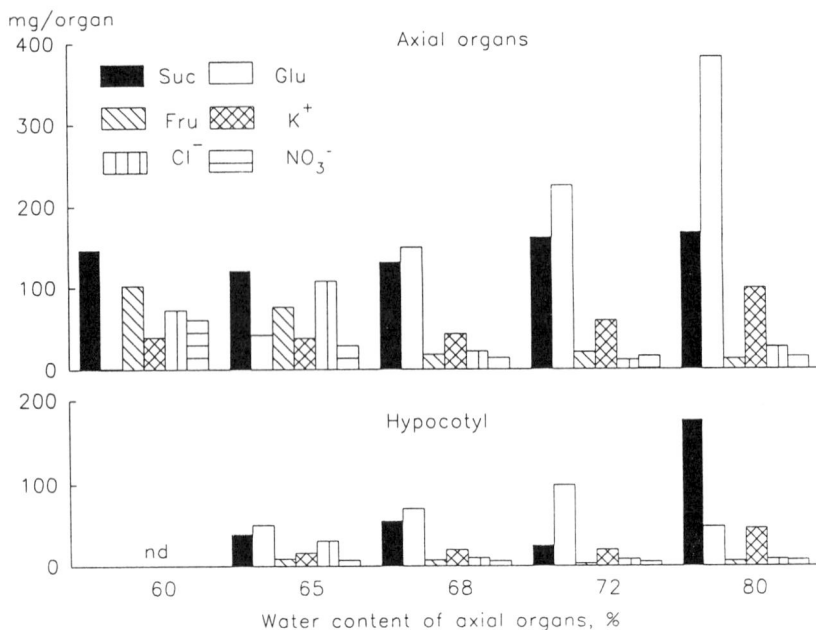

Figure 2. Amount of osmotic solutes in cell sap from axial organs of imbibing seeds. Suc: sucrose; Glu:glucose; Fru: fructose

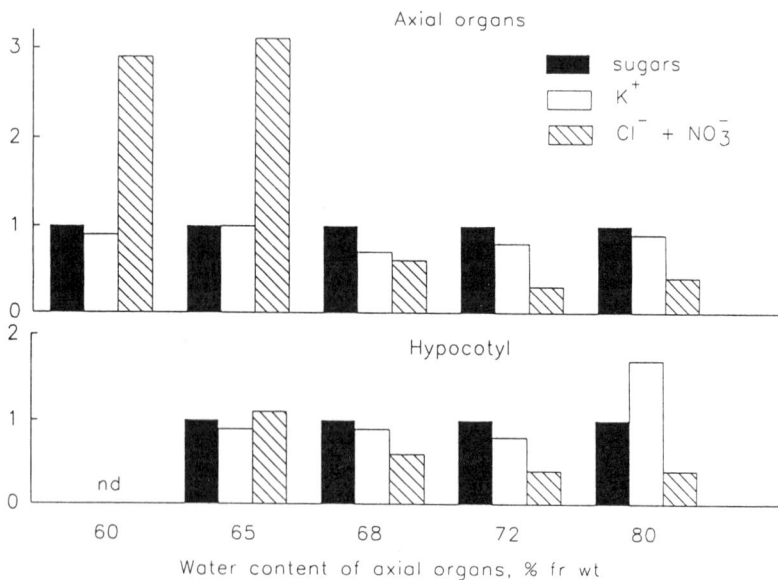

Figure 3. Relative contribution of main osmotic components to the osmotic pressure of cell sap from axial organs of imbibing seeds. The osmotic pressure produced by the sugars at each water content was taken to be one

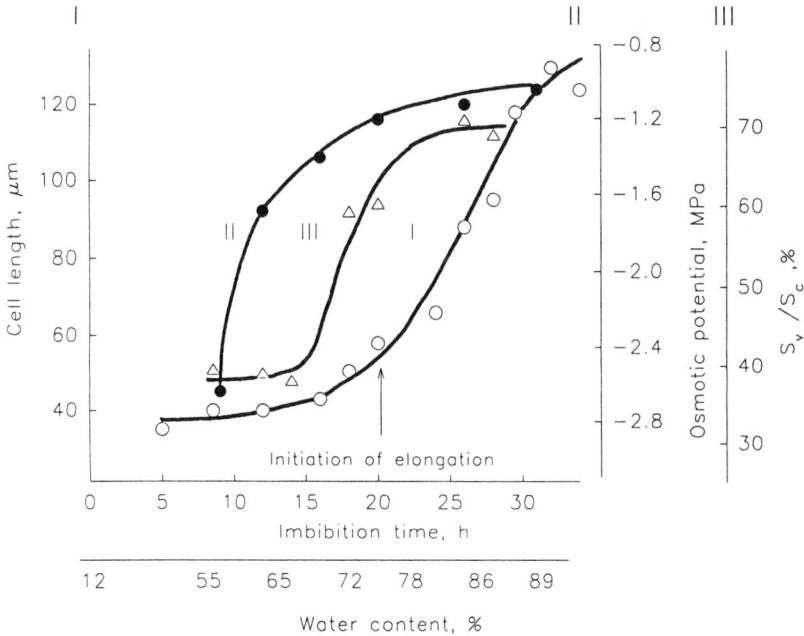

Figure 4. Cell elongation (I), osmotic potential of cell sap (II), and cell vacuolation (III) in hypocotyls in the course of cell preparation for elongation. S_v and S_c are the areas of vacuoles and cells on the sections, respectively

472Therefore, in axial organs imbibed to 65% wc and higher, the accumulation of solutes, mainly glucose and potassium, occurred in amounts adequate to provide additional water absorption by the cells, a prerequisite for the initiation of cell elongation.

Accumulation of solutes in the axial organs is directly related to changes in the cell vacuolar system, a compartment that is involved in osmotic water transport into the cells. The vacuolation was followed in the cortical cells of hypocotyl and compared with the dynamics of osmotic potential in hypocotyl and cell transition to elongation.

During imbibition, the small vacuoles were restored from the protein bodies, accompanied by the hydrolysis of stored proteins. The small vacuoles then fused into larger groups, sometimes with the remaining cytoplasm inside them. Their further fusion resulted in the formation of a large central vacuole. At 55–65% wc in the hypocotyl, the protein bodies occupied about 40% of the cell area. Prior to growth initiation, the vacuoles formed from the protein bodies already occupied 62–63% of the cell area as a result of their fusion (Fig. 4). During further hydration of the hypocotyl, the vacuolar system enlarged more slowly, covering almost 70% of the cell area. The vacuolation of cells in the axial organs follows the increase in osmotic potential and precedes the cell elongation.

In summary, the results clearly indicate the leading role of osmotically-driven water inflow due to accumulation of glucose and K^+-ions derived from stored starch and phytin during their degradation in imbibing axes. This accumulation of osmotic solutes in the axial organs is evident at 68% wc, that is prior to radicle emergence and well before the import of metabolites from the cotyledons. Therfore, the axial organs themselves produce the sufficient amounts of osmotically active substances to maintain the water transport into their cells. The accumulation of solutes is followed by development of vacuolar system as a cell compartment closely related to osmoregulated water transport to the cell prior to their elongation.

Nevertheless, the accumulation of solutes is not sufficient for elongation to begin because cell elongation requires cell wall loosening, i.e. an enhanced ability of wall constituents to slide apart under increasing cell turgor. Hence, cell wall loosening is a process necessary for the cell enlargement. Cell wall loosening can be induced by its acidification, a phenomenon of so-called acid growth (Kutschera, 1994). To test the contribution of wall acidification to the initiation of cell elongation, the acidification of ambient solution by the axial organs was estimated.

Table 2 shows no acidification of the ambient solution by the axial organs imbibed to a wc of 60–65%. From a wc of 68% on, the axial organs markedly lowered the pH of external solution. Thus, the cells of axial organs exuded the protons into the ambient solution. The most likely source of the protons is proton-ATPase located in the plasmalemma. Its operation can be tested by using diethylstilbestrol, its inhibitor (Balke and Hodges, 1979), and fusicoccin, its stimulator (Marre, 1979). Fusicoccin sharply enhanced the acidification exerted by axial organs, the effect being prominent only in those axial organs that imbibed to a wc of 68% or more. No acidification of the ambient solution or fusicoccin action occurred in the axial organs, the wc of which was lower (60

Table 2. Acidification of ambient solution by axial organs estimated by the change in pH (Δ pH). Embryos were hydrated in 5×10^{-4} M diethylstilbestrol or water for periods shown. Fusicoccin (2×10^{-5} M) was then added and the pH was measured again after 40 min

Hydration time (h)	Water content in axial organs, % fresh weight	Acidification of medium after hydration in		Additional acidification by fusicoccin Δ pH
		Diethylstilbestrol	Water	
9	60	0	0.01	0
12	65	0	0.09	0.02
16	68	0.05	0.34	0.15
20	75	0.31	0.65	0.21
45	85	0.25	0.59	0.20

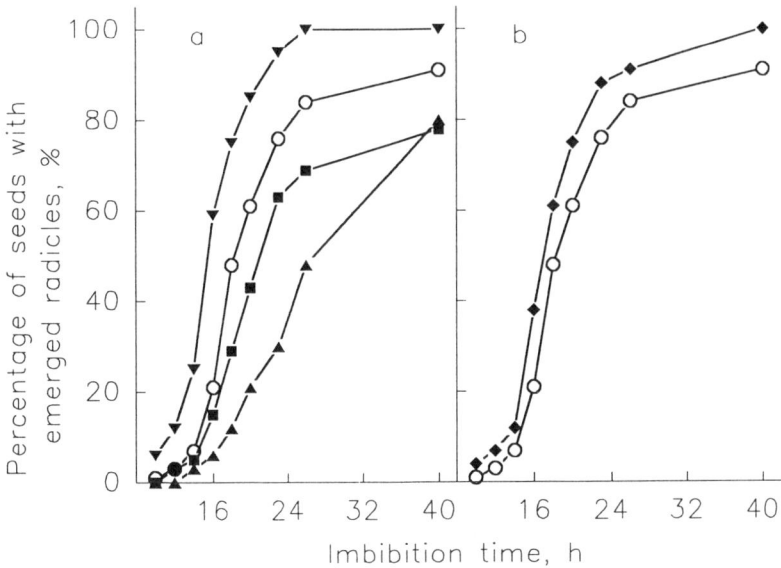

Figure 5. Effect of (a) fusicoccin, and inhibitors of H$^+$-ATPase and (b) acid buffer or IAA, on radicle emergence in broad bean seeds. ◆, 1 mM phosphate–citrate buffer, pH 3.9 or 5 × 10^{-5} M IAA; ▼, 5 × 10^{-5} M fusicoccin; ■, 5 × 10^{-4} M diethylstilbestrol; ▲, 5 × 10^{-3} M orthovanadate; ○, water. Note the curves for IAA and the buffer solution in (b) are the same

and 65%) as the table shows. Diethylstilbestrol also showed no effect at these wc levels. At a higher wc, this inhibitor diminished the degree of acidification almost to twofold. Thus, at a wc of 68% in the bean axial organs, H$^+$-ATPase begins to operate at the plasmalemma, providing for the proton extrusion into the cell walls.

To decide whether or not the initiation of the proton-ATPase functioning was related to the initiation of cell elongation, the action of fusicoccin, diethylstilbestrol, and orthovanadate, another inhibitor of the plasmalemma H$^+$-ATPase (Jacobs and Taiz, 1980), on the rate of radicle protrusion was studied (Fig. 5). Fusicoccin stimulated early germination even more actively than the acid buffer, indole-3-acetic acid (10^{-5} M) did not affect the initiation of cell elongation at early germination, whereas the inhibitors of the proton–ATPase retarded it. Hence, the cell elongation began earlier in the presence of H$^+$-ATPase activator and was delayed by its inhibitors. These data indicate that the acid growth could be a mechanism for preparing and initiating cell elongation because the cells of axial organs are capable not only of acidifying the ambient medium, but also of performing a more rapid elongation in such a medium. The experiments with fusicoccin, an activator of H$^+$-ATPase, and with diethylstilbestrol and orthovanadate, inhibitors of the enzyme, suggest that both processes, i.e. the acidification of ambient solution and the initiation of cell elongation, depend on the operation of this enzyme.

Discussion

The results described can be interpreted as evidence for acid growth as a mechanism for preparing and initiating cell elongation because the cells of axial organs are capable not only of acidifying the ambient medium but of initiating a more rapid elongation.

Therefore, in imbibing broad bean axes a second physiological mechanism contributing to the initiation of cell elongation begins to operate at 68% water content. This mechanism includes an activation of H^+-ATPase located in plasmalemma that results in proton extrusion from the cytoplasm through the plasmalemma into the cell walls. Acidification of cell walls destabilizes the chemical bonds between the cell wall polymers (Hohl *et al.*, 1991), activates both the enzymes partially degrading the hemicelluloses (Fry, 1989; Keller and Taylor, 1989) and extensin, the protein capable of cleaving the cellulose interfibrillar bonds (McQueen-Mason and Cosgrove, 1995). As a result, the extensibility of cell wall rises, thus providing the possibility of cell enlargement.

The preparation for cell elongation in the axial organs needs two special physiological mechanisms, the operation of which starts when the metabolic processes in the axial organs are fully activated. The first physiological mechanism is the accumulation of solutes, mainly of sugars and K^+-ions, due to degradation of the reserves; it provides further water transport into the axial organs mainly by osmotic forces. The second physiological process constitutes the acidification of cell walls by the protons extruded from the cytoplasm with the aid of ATPase situated in the plasmalemma. It provides the cell wall loosening resulting in their higher extensibility. Owing to enhanced water inflow to the vacuoles, the turgor pressure on the cell walls increases; as the cell wall components became capable of loosening, they slide apart making the volume enlargement possible. In this way the cell elongation commences being supported later on by building in the newly synthesized cell wall constituents.

Acknowledgement

This work was supported by the Russian Foundation for Fundamental Research, grant no. 95-04-11024.

References

Balke, N.E. and Hodges, T.R. 1979. *Plant Physiology* 63: 48–52.

Fry, S.C. 1989. *Physiologia Plantarum* 75: 532–536.

Hohl, M., Hung, Y.N. and Schopfer, P. 1991. *Plant Physiology* 95: 1012–1018.

Jacobs, M. and Taiz, L. 1980. *Proceedings of National Academy of Sciences USA* 77: 7242–7246.

Keller, C.P. and Taylor, J.E.P. 1989. *Canadian Journal of Botany* 67: 2944–2952.

Kutschera, U. 1994. *New Phytologist* 126: 549–569.

Marre, E. 1979. *Annual Review of Plant Physiology* 30: 273–288.

McQueen-Mason, C.J. and Cosgrove, T.J. 1995. *Plant Physiology* 107: 87–100.

Obroucheva, N.V. 1992. In: *Root Ecology and Its Practical Application. Proceedings of the Third Symposium of International Society of Root Research*, pp.13–16 (eds. L. Kutschera, M. Sobotik and E. Hubl). Klagenfurt: Verein zur Wurzelforschung.

Obroucheva, N.V. and Antipova, O.V. 1985. *Soviet Plant Physiology* 32: 932–941.

Obroucheva, N.V. and Antipova, O.V. 1994. *Russian Journal of Plant Physiology* 41: 391–395.

Obroucheva, N.V., Antipova, O.V. and Ivanova, I.M. 1993. *Russian Journal of Plant Physiology* 40: 641–646.

52. Ultrastructure of Solid Matrix-Primed Endospermic and Nonendospermic Seeds

A. DAWIDOWICZ-GRZEGORZEWSKA

Institute of Experimental Botany, University of Warsaw, ul. Krakowskie Przedmieœecie 26/28, 00–927 Warsaw, Poland

Abstract

It was shown that solid matrix priming (SMP) with Micro-Cel E of carrot (endospermic) and cucumber (nonendospermic) seeds improved their germination performance by shortening the mean germination time, lowering the spread and increasing the tolerance to low temperatures. The cytological studies performed on these seeds allowed us to conclude that developmental advancement in seed metabolism occurred during the studied period of SMP. So far, other ultrastructural data concerning cellular events taking place during SMP are not available. It was found that an advancement in the catabolic phase of germination *sensu stricto* was taking place in the radicles of embryos of both types of seeds, where degradation of storage protein and lipid bodies, followed by starch accumulation was stated. In the endosperm tissue of the carrot seed, the catabolic changes were restricted to the part enclosing the radicle, where degradation (only partial) of storage protein bodies and complete degradation of storage cell wall components were noted. In contrast, the data showed that the breakdown of the storage materials in the endosperm (weakening of endosperm) represented a secondary effect of SMP, which was probably controlled and stimulated by the metabolically advanced embryo.

Introduction

It has been reported that priming with solid matrix carrier Micro-Cel E synchronizes and accelerates germination of several seeds (including carrot and cucumber) more effectively than osmoconditioning (OC) in PEG solutions (Khan *et al.*, 1992a,b). Despite the wide acceptance of priming as a seed quality improving technique, its physiological basis is not clearly understood. It is known that priming allows the progress of the prolonged phase II of water uptake (prior to radicle protrusion, germination *sensu stricto*), but prevents the start of phase III, the radicle protrusion phase (Karssen *et al.*, 1991). It has also been reported that seeds are metabolically active during priming in osmotic solutions, evidence has been provided for syntheses of DNA, RNA and proteins (Bray *et al.*, 1989; Davison and Bray, 1991; Bray, 1995) as well as for activity of some enzymes involved in the PPP pathway, glycolysis, proteolysis and lipid metabolism (Smith and Cobb, 1991). Moreover, it has also been observed that endosperm-containing small seeds show particularly beneficial responses to the OC treatments (Bradford, 1990; Van der Toorn, 1989; Nonogaki *et al.*, 1992)

R.H. Ellis, M. Black, A.J. Murdoch, T.D. Hong (eds.), Basic and Applied Aspects of Seed Biology, pp. 479–487.
© *1997 Kluwer Academic Publishers, Dordrecht. Printed in Great Britain.*

thus suggesting that the decrease in the mechanical restraint of the endosperm (i.e. weakening) due to the hydrolysis of storage cell wall components might explain the beneficial effects of priming.

In the light of the hitherto obtained data two explanations of the physiological mechanisms involved in seed priming can be proposed: (i) repair of cellular and subcellular damage that could accumulate during seed development and storage (*reviewed*, Bray, 1995); (ii) advancement of metabolic events preparing for radicle protrusion (Heydecker and Coolbear, 1977; Welbaum and Bradford, 1991).

The object of this study has been to relate the cytological data (based on the light and transmission electron microscopy observations) to (ii) of the above hypotheses and answer the following questions:

- Are changes related to the degradation of storage materials taking place in the embryonic axis (of carrot and cucumber) and the endosperm (of carrot) seeds during SMP in Micro-Cel E?

- Does the weakness of endosperm constitute a primary and main effect of priming in endosperm-containing seeds?

There are preliminary data for secondary (S) umbel carrot seeds (Dawido-wicz-Grzegorzewska and Maguire, 1993) which show that due to the SMP treatment in Micro-Cel E, some developmental advance in both the embryonic axis and the endosperm is attained. Moreover, S umbel seed embryos attain the dimensions and developmental stage comparable to primary (P) ones; this explains uniformity of the early post-germinative growth matriconditioned seeds. In the present study, in order to answer the above questions, the primary umbel carrot seeds, which are the most developed at harvest time, have been investigated; the results are compared to those obtained for endospermless seeds of cucumber.

Material and Methods

Plant Material

Carrot (*Daucus carota* L. cv. Nantes) and cucumber (*Cucumis sativus* L. cv. Mieszko) seeds were supplied in 1993 and 1994 by Experimental Seed Station (PNOS) in Ozarów Mazowiecki near Warsaw. In the case of carrot seeds, only P umbel seeds were studied. All seeds were stored in paper bags, at ambient temperature. The seed of carrot contains the living, thick-walled endosperm which occupies the bulk of the seed and the small, axillary linear embryo, separated from the endosperm by a narrow, acellular space. The seed of cucumber is devoid of endosperm, so the embryo is directly enclosed by a seed coat. At harvest time, all studied seeds were morphologically mature, as indicated by the presence of well differentiated cotyledons and embryonic axis.

Matriconditioning

Seeds were matriconditioned with Micro-Cel E (Manville, Filtration and Minrals) mixed with water at appropriate ratios, each time determined experimentally for every lot of seeds. The ratios (by weight) of seeds:carrier:water were as follows: for carrot – 1.0:0.5:1.9 and for cucumber – 1.0:0.32:1.0. The seeds were mixed thoroughly with water and solid carrier in 100 ml glass jars, not firmly closed. Jars were transferred to +15°C in continuous light for 6 days (d). The moisture equilibrium water content of seeds at the end of SMP corresponded to $\psi = -1.0$ MPa to -1.5 MPa. Seeds matriconditioned for 0, 2, 4 and 6 d were rinsed with water and samples of 15 seeds were taken after each selected period of SMP for cytological examination.

Electron (TEM) and Light (LM) Microscopy

Small blocks of tissues (about 2 mm), containing embryonic radicle and surrounding endosperm from carrot seeds and/or radicle with the lower part of the hypocotyl from cucumber seeds were processed for TEM studies. They were fixed in 3% glutaraldehyde in 0.05 M PIPES (piperazine N-N-bis-ethanol sulphonic acid) buffer and postfixed in 1% OsO_4 in 0.05 M cacodylate buffer (pH 7.2), washed in cacodylate buffer, dehydrated in graded ethanol-propylene oxide series and embedded in Spurr's resin. The blocks of tissues containing longitudinal halves of the whole carrot and cucumber seeds (without seed coats) were processed separately for LM studies, following a similar but significantly prolonged procedure. Ultrathin sections were stained with water-saturated uranyl acetate and 0.4% lead citrate on uncoated nickel grids, and semi-thin ones with 1% toluidine blue with 1% azure II in 1% borax solution. Ultrathin sections were viewed with Hitachi 600 TEM at 60 or 100 kV and semi-thin ones with Optiphot 2 (Nikon) LM.

Results and Discussion

Ultrastructure of Seeds Before Matriconditioning

Embryo

Three primary tissues: protoderm, procambium and ground parenchyma were present in the quiescent embryos of carrot and cucumber seeds at harvest time. All embryo cells (including cotyledons) had thin primary cell walls which enclosed the protoplast having characteristics typical for mature dormant or quiescent seeds (Dawidowicz-Grzegorzewska, 1989; Dawidowicz-Grzegorzewska and Beranger-Novat, 1989; Dawidowicz-Grzegorzewska and Podstolski, 1992). Storage protein and lipid bodies occupied most of the cell's volume (Figs. 1 and 2). Protein bodies varied in diameter from 0.3 to 3.0 μm, depending on the

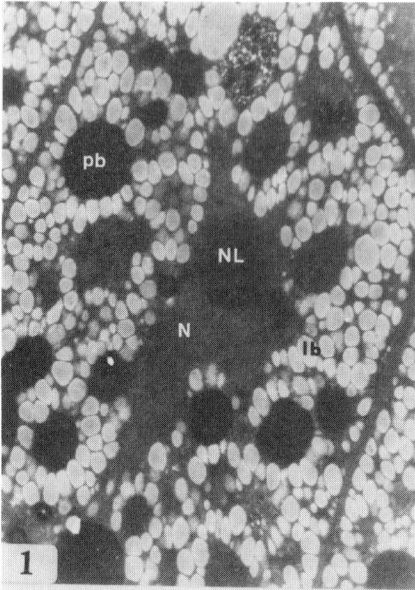
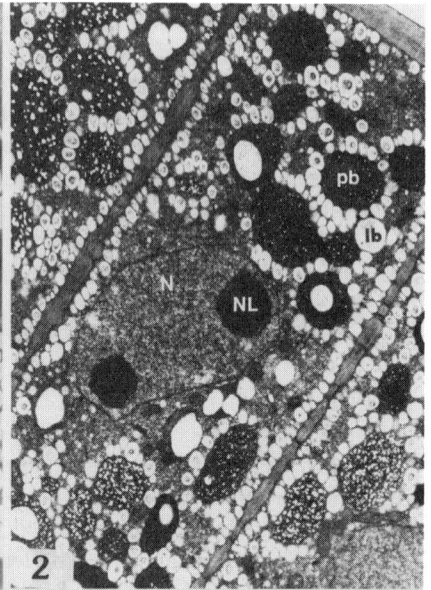
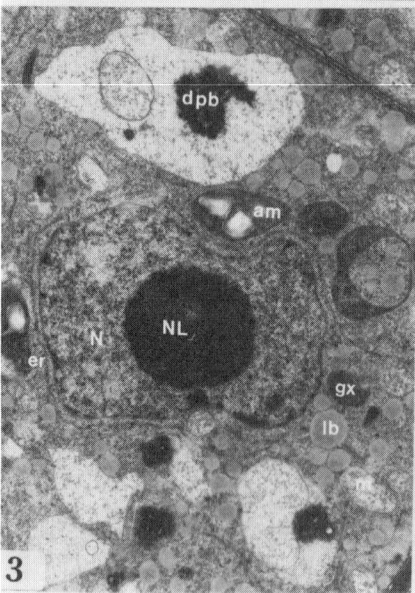
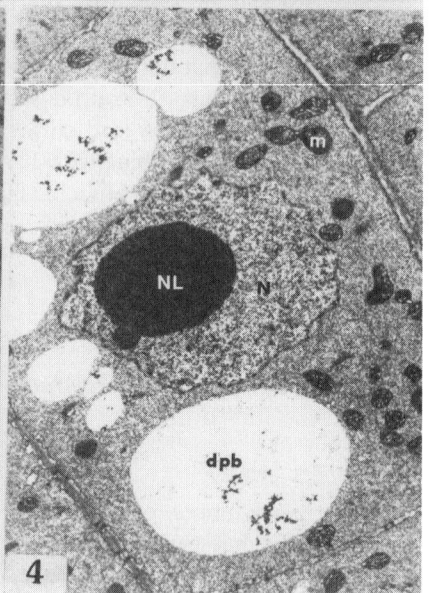

tissue. The smallest were contained in protoderm and procambial tissues, whereas the larger ones were found in the ground parenchyma of cotyledons. The protein bodies contained none, one or several globoid inclusions with phytin, and occasionally, in carrot seeds small calcium oxalate crystals were also found. Such morphology of protein bodies suggests that they attained the final, mature stage of development during embryogenesis. Lipid bodies were very abundant in both kinds of seeds. They were arranged in shells around the protein bodies and along the walls. Starch was negligible and present only occasionally in few amyloplasts. The nucleus was irregularly shaped and contained condensed chromatin with prominent chromocentres in cucumber seeds; nucleoli in both seeds were small and contained only the dense, fibrillar component. The protoplast was poor in organelles, such as mitochondria, ER and Golgi membranes – all these structures had reduced morphology, typically for a dormant tissue.

Endosperm

The carrot seed embryo was surrounded by a thick and viable endosperm (up to 15 cells layered), which was considerably thinner in the micropylar end around the root cap and adjacent part of the radicle (Fig. 5). Unlike the embryo, the endosperm stored carbohydrates, presumably β-mannans (Halmer, 1985) in the form of thickened primary cell walls (Fig. 7), but not starch. Similarly as in the embryo, the most abundant organelles were storage protein and lipid bodies, but both were larger than those in the embryo (protein bodies up to 8 μ in diameter). Moreover, protein bodies contained alternatively two types of inclusions: crystal-containing and phytin-containing globoids or prominent, druse-shaped calcium oxalate crystals as the form of calcium storage (Spitzer and Lott, 1982). Mitochondria and proplastids were scarce and had a very reduced fine structure. Neither ER, Golgi structures nor ribosomes were found, the last in contrast to the dormant embryonal tissue where monosomes were always present.

Figures 1 and 2. The ground meristem cells in the embryonic axis from unconditioned carrot (1) and cucumber (2) seeds. Note the presence of numerous storage protein (*pb*) and lipid (*lb*) bodies; some protein bodies contain globoids. Nucleus (*N*) contains one or two small nucleoli (*NL*). Magnification 6000 ×

Figures 3 and 4. The ground meristem cells in the embryonic axis after 4 d of SMP in the carrot (3) and after 6 d of SMP in the cucumber (4) seeds. Note the presence of the several degraded protein bodies (*dpb*) bearing electron-dense (dark) remnants of storage material and of less numerous (3), or absent lipid bodies (4). Note also the presence of mitochondria (*m*), glyoxysomes (*gx*), profiles of ER (*er*) and amyloplasts (*am*) with starch. Nucleus and nucleolus are enlarged, chromatin diffused (compare with Figs 1 and 2). Magnification 6000 ×

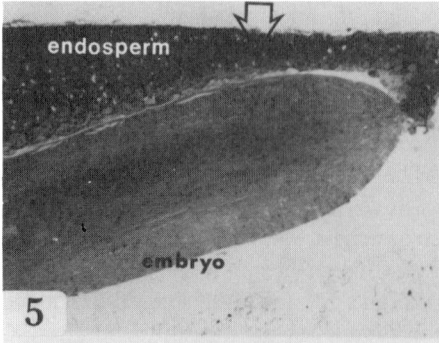
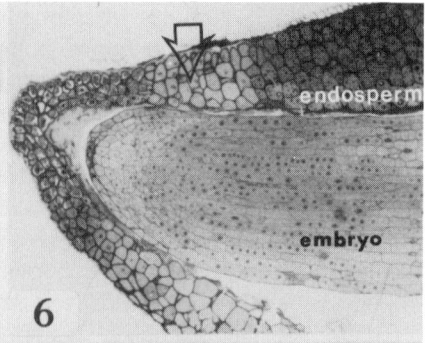
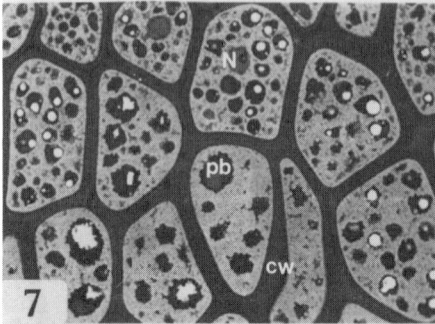
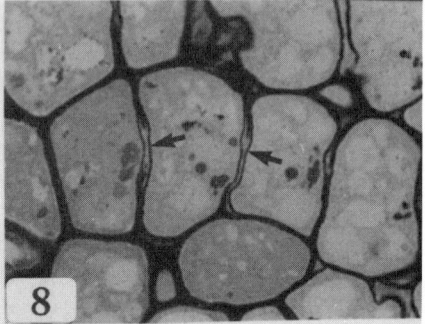
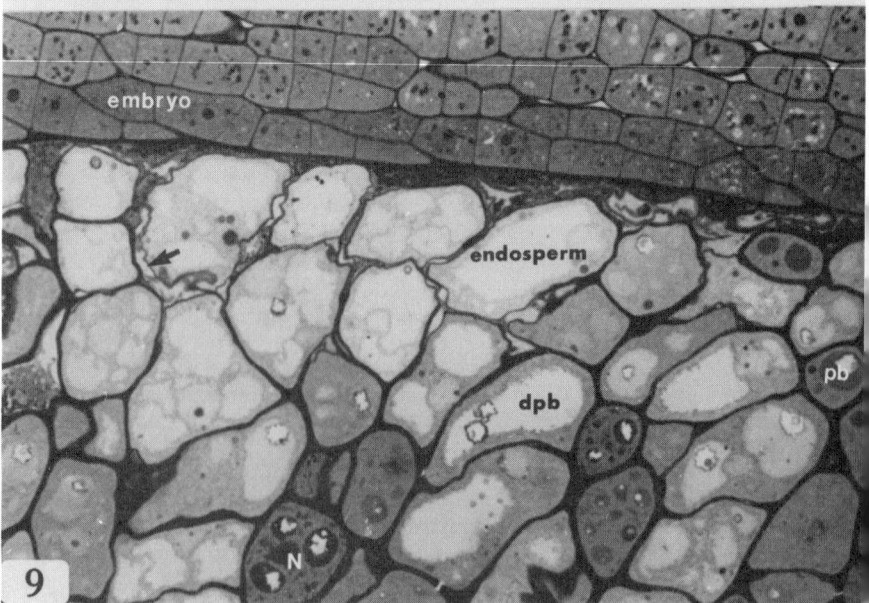

Ultrastructure of Seeds During and After Matriconditioning

Embryo

During 6 d of SMP, the fine structural changes related to the mobilization of storage materials occurred similarly in both carrot and cucumber seeds. They were limited to the radicle, adjacent part of the lower hypocotyl and root cap cells from the 2nd day of SMP in the protoderm and progressed gradually during the whole treatment inwards to the ground meristem and basipetally towards cotyledons. It was observed that both types of storage organelles (bearing proteins and lipids) were undergoing degradation. Protein bodies were firstly converted into several small vacuoles containing some flocculent, residual storage proteins. Gradually, these vacuoles fused or disintegrated within the protoplast (Figs. 3 and 4). As a rule, mobilization of storage lipids followed that of proteins: the number of lipid bodies decreased, whereas concomitantly glyoxysomes proliferated and starch grains appeared in the amyloplasts. Simultaneously, it was observed that rough ER, cytosolic polysomes, mitochondria, Golgi structures and amyloplasts were intensively developing and proliferating. Important changes were also observed in the fine structure of the nucleus: the size of nucleus and nucleolus increased, chromatin dispersed, and the granular component of nucleolus appeared. All these observations suggest that transcriptional, translational and biosynthetic activities related to the catabolism and interconversions of storage materials have been restored during the SMP period. The observed changes are taking place in the embryos of both the endosperm-containing and nonendospermic seeds and are similar to those reported for the catabolic phase of germination *sensu stricto* in apple seeds (Dawidowicz-Grzegorzewska, 1989).

Figure 5. The embryonic axis enclosed by the endosperm (half of the endosperm is removed) in the unconditioned carrot seed. Both tissues are stained dark with toluidine blue owing to the presence of storage proteins. An arrow shows the part magnified in Fig. 7. Magnification 85 ×

Figure 6. The part of the embryonic axis enclosed by the endosperm after 4-d period of SMP in the carrot seed. Note the weakening of endosperm arround the radicle tip, which is clear, as well as the embryonic axis (both devoid of storage proteins). An arrow shows the part magnified in Fig. 8. Magnification 85 ×

Figures 7 and 8. Details of the endosperm from the unconditioned carrot seed (Fig. 7), containing thick walls (*cw*), protein bodies (*pb*, with unstained globoid and calcium oxalate inclusions) between numerous (grey background) lipid bodies; the same tissue from 4 d conditioned (Fig. 8) carrot seed ; note the presence of the thin, hydrolyzed cell walls devoid of the middle lamellae (arrows). Both magnification 1000 ×

Figure 9. 6 d of SMP of carrot seed; the endosperm cells neighbouring the embryonic axis are more weakened than on the opposite side; note also that partly degraded protein bodies (*dpb*) are still enclosed by the lipid bodies (grey background) and the starch is not present unlike in the embryo (double arrows). Magnification 1000 ×

Endosperm

The fine structural changes taking place in the endosperm of P umbel carrot seeds during 6 d of SMP were limited to its micropylar end which surrounds the radicle and the root cap. They started on the 3rd or 4th day of SMP, that is, when the neighbouring embryonic tissues were already 'activated'. They involved the partial degradation of protein bodies, where matrix was solubilized, whereas globoid and oxalate crystal inclusions remained unchanged until the end of priming. Such partly degraded bodies sometimes fused, but, as a rule, were never disintegrated and remained enclosed by a limiting membrane. The next observed step was the hydrolysis of the thick storage primary walls and of middle lamellae until the point of cell separation (Figs. 8 and 9). The number and the fine structure of reserve lipid bodies were not changed, unlike those in the embryo. The rest of the cytoplasmic organelles, and the nucleus, retained characteristic features typical for dormant or quiescent cells. The structural features of the observed storage protein and cell wall hydrolysis suggest their autolytic character due to the activity of enzymes already pre-existing in dormant endosperm cells, as in the date palm seeds (Chandra Sekhar and DeMason, 1990). The timing and localization of changes described here for carrot seed endosperm suggest very strongly stimulation coming from the embryo: they started and were much more advanced in the close vicinity of the radicle. The embryo-controlled weakening of endosperm in fully hydrated tomato and celery seeds was previously shown by Groot et al. (1989) and Van der Toorn (1989). Finally, after the whole period of SMP the carrot seed endosperm was 'weakened' only in its micropylar end, whereas the remaining part (more than 90% of tissue) was not changed (Fig. 6). In conclusion, some catabolic changes, which advance the seeds germinative metabolism, were also taking place in the endosperm tissue but, in the light of the presented data, seem to be of secondary character as induced and controlled by the embryonic axis.

Conclusions

The most known benefits of seed matriconditioning, such as enhanced speed and synchrony of germination and increased tolerance to low temperatures, in the light of the presented data can be interpreted as follows:

– during solid matrix priming with Micro-Cel E some developmental advance in both the embryonic axis and endosperm is attained;

– ultrastructural observations performed during the 6-d period of SMP indicate that the breakdown and interconversions of storage materials, specific to the catabolic phase of germination *sensu stricto* are taking place, both in the axis and in the micropylar part of the radicle-enclosing endosperm;

- weakening of endosperm, one result of SMP, constitutes a secondary effect, which is stimulated by the embryonic axis, primarily activated during the treatment;

- the proliferation of various cellular membranes and organelles which occurrs in the embryonic axis during the SMP period can be regarded as responsible for the increased tolerance of primed seeds to low temperatures.

Acknowledgements

This work was supported by a Grant No. 6 P 20400405 from the State Committee for Scientific Research (KBN).

References

Bradford, K.J. 1990. *Plant Physiology* 94: 840–849.
Bray, C.M., Davison, P.A., Ashraf, M. and Taylor, R.M. 1989. *Annals of Botany* 63: 185–193.
Bray, C.M. 1995. In: *Seed Development and Germination* , pp. 767–789 (eds. J. Kigel and G. Galili). New York, Basel, Hong Kong: Marcell Dekker. Inc.
Chandra Sekhar, K.N. and DeMason, D.A. 1990. *Planta* 181: 53–61.
Davison, P.A. and Bray, C.M. 1991. *Seed Science Research* 1: 29–35.
Dawidowicz-Grzegorzewska, A. 1989. *Journal of Plant Physiology* 135: 43–51.
Dawidowicz-Grzegorzewska, A. and Beranger-Novat, N. 1989. *Journal of Experimental Botany* 40: 913–918.
Dawidowicz-Grzegorzewska, A. and Podstolski, A. 1992. *Annals of Botany* 69: 39–46.
Dawidowicz-Grzegorzewska, A. and Maguire, J.D. 1993. In: *Basic and Applied Aspects of Seed Biology, Proceedings of the Fourth International Workshop on Seeds*, pp. 1039–1045 (ed. D. Côme and F. Corbineau). Paris: Universite P. et M. Curie Press.
Groot, S.P.C., Kieliszewska-Rokicka, B., Vermeer, E. and Karssen, C.M. 1988. *Planta* 174: 500–504.
Halmer, P. 1985. *Physiologie Vegetale* 23: 107–125.
Heydecker, W. and Coolbear, P. 1977. *Seed Science and Technology* 5: 33–425.
Karssen, C.M, Haigh, A., Van der Thorn, P. and Weges, R. 1991. In: *Recent Advances in the Development and Germination of Seeds*, pp. 269–279 (ed. R.B. Taylorson). New York: Plenum Press.
Khan, A.A., Abawi, G.S. and Maguire, J.D. 1992a. *Crop Science* 32: 231–237.
Khan, A.A., Maguire, J.D., Abawi, G.S. and Ilyas, S. 1992b. *Journal of American Society of Horticultural Sciences* 117: 41–47.
Nonogaki, H., Matsushima, H. and Morohashi, Y. 1992. *Physiologia Plantarum* 85: 167– 172.
Smith, P.T. and Cobb, B.G. 1991. *Physiologia Plantarum* 82: 433–439.
Spitzer, E. and Lott, N.A. 1982. *Canadian Journal of Botany* 60: 1381–1391 and 1399–1403.
Van der Toorn, P. 1989. PhD Thesis. Agricultural University of Wageningen, The Netherlands.
Welbaum, G.E. and Bradford, K.J. 1991. *Journal of Experimental Botany* 42: 393–399.

e Molecular Mechanism of Seed Deterioration
tion to the Accumulation of Protein–
lehyde Adducts

SHI, A. KAMATAKI and M. ZHANG

_ooratory of Environmental Biology, Botanical Garden, Faculty of Science, Tohoku University, North Campus, Kawauchi, Aobaku, 980-77 Sendai, Japan

Abstract

Many dry seeds evolve various volatile compounds during storage, of which carbonyl compounds, especially acetaldehyde (Ald), are the most active in causing their deterioration. Although the longevity of seeds is generally prolonged with decreasing RH during storage, the contents of Ald were higher at 44% RH than at 75% RH in short-lived seeds, such as lettuce and cocklebur, compared with long-lived pea seeds. The production and accumulation of Ald in seeds occurred even at $-3.5°C$, in which the operation of mitochondrial respiration was undetectable but seed vigour was gradually lost. In cocklebur, Ald was contained in both axial and cotyledonary tissues, which content increased as the storage period was prolonged, and the loss of seed vigour occurred in both tissues. Moreover, the contents of proteins soluble in phosphate-buffered saline (PBS, pH 7.4) declined as their viabilities were lost. Thus, we presumed that the endogenously accumulated Ald in the seeds may cause seed ageing by deteriorating seed proteins. Hence, a sensitive competitive ELISA method, which was improved by us, was used for the quantification of endogenous Ald–protein adducts (APA) in dry seeds. As a result, the seeds treated with exogenous Ald rapidly lost their viabilities with increasing RH, accompanied by the accumulation of APAs. However, this adduct was detected even in the PBS-insoluble but urea-soluble proteins from lettuce seeds exposed to Ald, and in naturally aged seeds of lettuce and cocklebur that had been stored at low RHs. Thus, the modification of functional proteins in seeds by Ald seems to be the most important primary trigger of seed deterioration under dry storage conditions.

Introduction

Orthodox seeds are capable of maintaining their vigour for a long term under low RH and temperature, but gradually deteriorate and finally lose their viability. However, the actual mechanisms of seed deterioration under such conditions are still unknown. We have found that many dry seeds evolve various volatile compounds during storage and the evolved volatiles accumulate within the container for seed storage, thus causing seed deterioration (Zhang et al., 1993; 1994; 1995a). Fifty-nine volatiles derived from seeds were identified using GC–MS (Zhang et al., 1995b), of which carbonyl compounds, such as Ald and 3-methylbutanal, were especially effective in causing seed deterioration (Zhang

R.H. Ellis, M. Black, A.J. Murdoch, T.D. Hong (eds.), Basic and Applied Aspects of Seed Biology, pp. 489–498.
© *1997 Kluwer Academic Publishers, Dordrecht. Printed in Great Britain.*

et al., 1994). Ald is produced by all seed species tested, though 3-methylbutanal is not detected in pea, and it is contained more abundantly within the short-lived oil or protein seeds, such as lettuce, carrot and soyabean, than within the long-lived starchy seeds, such as rice and pea, even when they were stored at – 3.5°C (Zhang *et al.*, 1995b). The important point is that the contents of Ald were higher at lower RHs rather than at a high RH of 75% (Zhang *et al.*, 1995b). These facts suggest that the slow deterioration of orthodox seeds under low RHs may be involved in the accumulation of toxic Ald. On the other hand, there are many cases in which the contents of soluble proteins decrease as seeds deteriorate (see Priestley, 1986). The present study was carried out to examine the probable involvement of endogenously produced Ald in the slow ageing of seeds proceeding even under extremely low RHs and in the naturally occurring ageing in severely dehydrated seeds through the modification of functional proteins by non-enzymatic reaction.

Materials and Methods

Seeds of lettuce (*Lactuca sativa* L. cv. Aochirimen), soyabean (*Glycine max* Merr. cv. Miyagishirome), carrot (*Daucus carota* L. var. *sativus* Hoffm cv. Kuroda-5-sun), rice (*Oryza sativa* L. cv. Sasanishiki) and pea (*Pisum sativum* L. cv. Alaska) were used. All seeds were stored at −3.5°C until use and showed germination of 90% or more. However, only cocklebur (*Xanthium pennsylvanicum* Wallr.) seeds were stored at 8°C. RHs for storage were controlled using a saturated solution of LiCl for 12% RH, K_2CO_3 for 44% RH and NaCl for 75% RH, which were adopted to make sorption zones 1, 2 and 3, respectively (Zhang *et al.*, 1994).

Seed storage, assay of germinability, application of Ald to seeds, determination of volatile contents in seeds using GC and identification of volatiles by GC–MS were performed as described in previous papers (Zhang *et al.*, 1994; 1995a). Soluble proteins of seeds were extracted using phosphate-buffered saline (PBS, pH 7.4) or CAPS (pH 10.5). In some case, the PBS-insoluble parts of the proteins were further extracted by 8% urea solution including 2% NaCl. The amounts of proteins were determined by Lowry's method (1951) or Bradford's method (1976). For immunoassay, only the PBS or urea soluble fraction was used after being dialysed against the same PBS and condensed.

For quantification of endogenously formed APA, however, a sensitive competitive enzyme-linked immunosorbent assay (ELISA) method was newly developed (Zhang *et al.*, submitted), while APA formed in the presence of exogenous acetaldehyde was quantified by means of the direct ELISA (Perata *et al.*, 1992). For both ELISA, polyclonal antibody against acetaldehyde (1 mM) bovine serum albumin was obtained from rabbits (Israel *et al.*, 1986). In the direct ELISA, immunoplate wells were coated with seed proteins (100 µg/ml) in phosphate buffered saline (PBS; pH 7.4) for 2 h. The wells were then washed with water and blocked with 0.5% gelatin for 1 h. After being washed with PBS containing 0.05% Tween-20 (PBST), the second antibody (goat anti-rabbit immunoglobulin bound to alkaline phosphatase diluted in PBS and 0.5% gelatin) was added in the wells. After 1 h, the wells were twice rinsed with PBST then washed again with water twice, and an alkaline phosphatase substrate, p-nitrophenyl phosphate, was added to each well. The reaction was stopped by addition of 2.5 M KOH and A_{405} was photometrically assayed. In the solid competitive ELISA, however, the procedure of the direct ELISA was mainly improved in a process in which the immunoplate was pre-coated by Ald–gelatin adducts which were prepared at 1 mM Ald and 10 mM $NaCNBH_3$ and then added to a mixture of the sample solution and rabbit antibody before binding the second antibody (Zhang *et al.*, submitted).

Results and Discussion

Figure 1 shows the existence of 11 kinds of aldehydic volatile compounds within carrot seeds stored under 44% RH at 23°C for 8 months. As shown previously (Zhang *et al.*, 1994) Ald was the most toxic volatile which accelerated the deterioration of many kinds of seeds. The contents of Ald were then compared among some seed species stored at –3.5°C or 23°C for 8 months under different RHs (Fig. 2). The Ald contents in long-lived pea seeds increased with decreasing RH, while those in short-lived lettuce and cocklebur seeds were maximal at 44% RH. This trend was almost equal to that at –3.5°C. Of more importance was the fact that the Ald contents were far higher in the short-lived fatty seeds than in the long-lived starchy seeds.

According to Vertucci and Leopold (1984, 1987), O_2 uptake in soyabean seeds becomes obvious in sorption zone 3, but it was not detected in sorption zone 2. Similar results were obtained with soyabean seeds by Zhang *et al.* (1995c), who pointed out that the mitochondrial activity of the stored seeds could be deduced from the differences in alcoholic fermentation between the seeds stored with or without HCN gas. Therefore, in order to examine indirectly the development of mitochondrial respiration in seeds during a storage period at different RHs and temperatures the ethanol evolution was assayed after 8 months' storage with or without 1 mM HCN gas (Table 1). When seeds were stored at –3.5°C, the ethanol evolution increased with increasing RH, in which any effect of HCN was not detected. At 23°C in the absence of HCN, however, the ethanol evolution was not detected at 75% RH in both starchy and fatty seeds and was maximal at 44% RH. Similarly, any effect of HCN gas on the ethanol evolution was not found at both 12 and 44% RHs, while at 75% RH in sorption zone 3, the application of HCN caused tremendous increases in ethanol evolution in all seed species tested (Table 1).

These facts suggest that if the duration is relatively short, mitochondrial activity hardly develops during storage at –3.5°C regardless of seed species, while if they were stored at 23°C in sorption zone 3, HCN-sensitive mitochondrial respiration would develop. However, even at 23°C, the mitochondrial

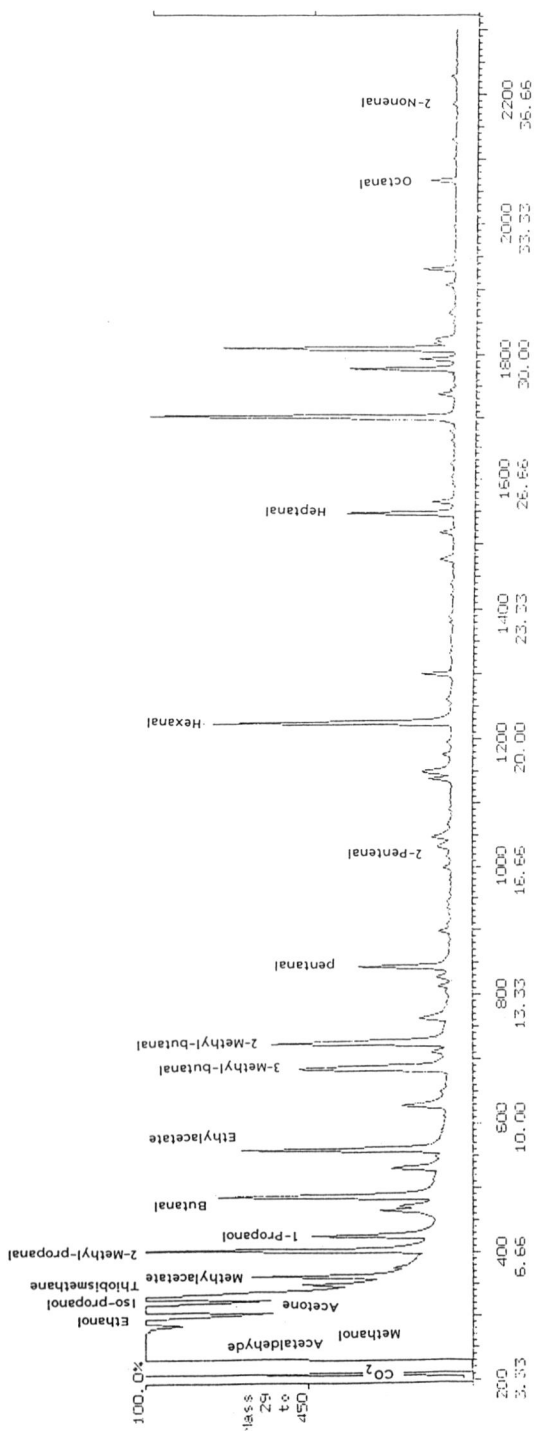

Figure 1. Chromatogram of aldehydic volatiles from carrot seeds stored at 44% RH and 23°C

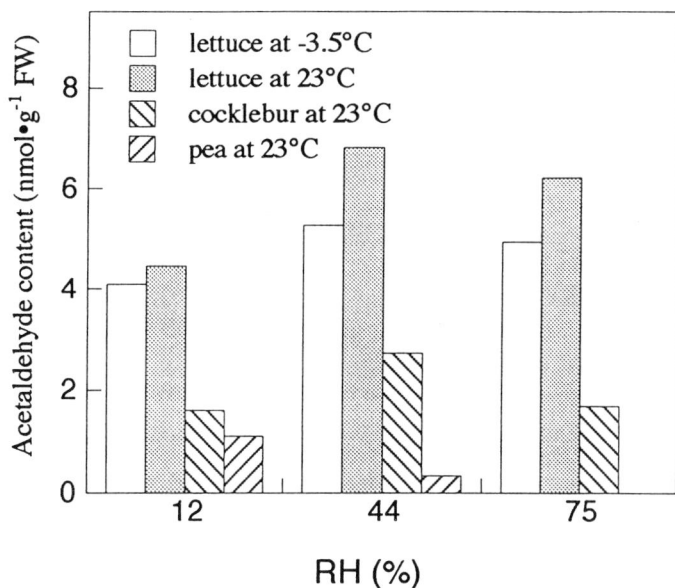

Figure 2. The contents of Ald in seeds stored at –3.5° or 23°C for 8 months under different RHs

Table 1. Effect of HCN on ethanol evolution by dry seeds under different RHs at –3.5°C and 23°C during a storage period with or without 0.3 mM HCN gas for 8 months

| Seed | HCN | \multicolumn{3}{c}{Ethanol evolution (pmol.g^{-1}Fw.day^{-1})} | | | | |
| | | \multicolumn{3}{c}{–3.5°C} | \multicolumn{3}{c}{23°C} | |
		12% RH	44% RH	75% RH	12% RH	44% RH	75% RH
Rice	–	0	0.7	1.0	0.6	1.1	0
	+	0	0.9	1.0	0.7	1.0	50.1
Pea	–	0	0.4	0.9	0.8	2.2	0
	+	0	0.4	1.0	0.7	2.1	59.1
Cocklebur	–	0.3	0.9	1.8	1.2	1.9	0
	+	0.3	0.9	1.7	1.3	1.7	9.4

respiration could not operate if the seeds were stored at either sorption zone 1 or 2. This possibility has been supported by the higher consumption of CoA and acetyl CoA at 75% RH than at 53% RH in both axial and cotyledonary cocklebur seed tissues (Zhang *et al.*, 1995c). From these facts, it was suggested that the seed deterioration occurring slowly at lower temperatures and RHs may proceed through a mechanism different from that occurring at higher temperatures and RHs. The production and accumulation of Ald in seeds stored under lower RHs and temperatures (Fig. 2) suggest its probable involvement in causing the seed deterioration via the modification of proteins.

Nevertheless, the longer the duration of storage, the higher was the content of acetaldehyde in cocklebur seeds (Fig. 3). The content was the highest in the seeds which had completely lost their growth potentials. Interestingly, the content of ethanol was highest in the seeds just before their death, as seen in soyabean seeds stored at 75% RH for 3 weeks (Zhang *et al.*, 1995b,c). That is, the inability of development of mitochondrial respiration in aged seeds elevates the activity of fermentation indirectly, but the declining supply of acetyl CoA (Zhang *et al.*, 1995c) and the gradual inactivities of alcohol and acetaldehyde dehydrogenases (MacLeod, 1952) by further prolonging the storage period reduced the content of ethanol and reversely increased that of acetaldehyde in dead seeds (Fig. 3). The higher content of Ald in this stage, however, may result not only from its accumulation as an intermediate of fermentation but also from the oxygenation of unsaturated fatty acids. These facts suggest that the probable modification of seed proteins by carbonyl compounds would continue even after seeds had completely lost their viability. Moreover, Table 2 shows that the

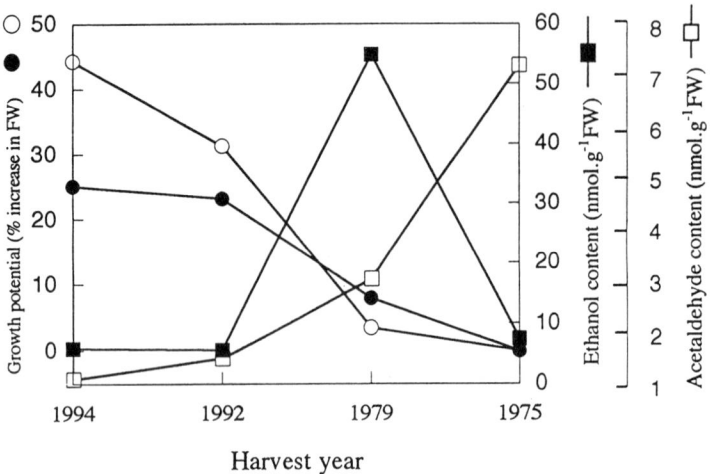

Figure 3. Growth potentials of axial (○) and cotyledonary (●) tissues of cocklebur seeds harvested in different years and the contents of ethanol (■) and Ald (□)

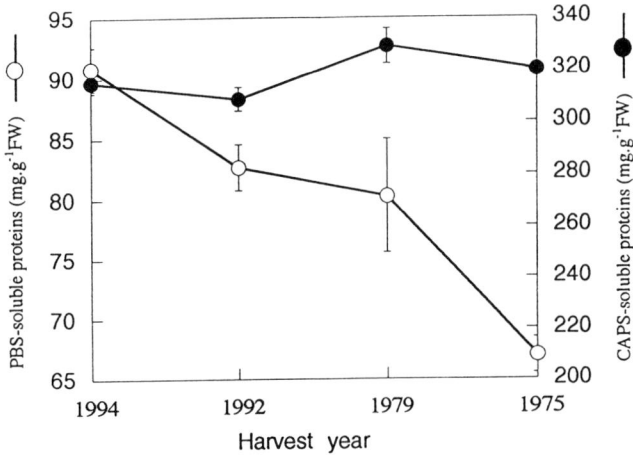

Figure 4. The contents of PBS- or CAPS-soluble proteins from cocklebur seeds harvested at different years and stored at 8°C

deterioration of seeds proceeds similarly in both axial and cotyledonary tissues.

For some time, there has been a hypothesis that seed ageing is associated with the denaturation of seed proteins, as shown by solubility (see Priestley, 1986), although it has not been regarded as important by most researchers. The decrease in solubility has been explained to arise from interactions of proteins with lipid peroxidation products, from sulphhydryl oxidation proteins leading to their malfunction, and from the Maillard reaction between reducing sugars and proteins. However, the detailed elucidation of the mechanism by which seed proteins are denatured remains unknown.

Figure 4 clearly shows that the content of PBS-soluble proteins decreases with increasing duration of storage in cocklebur seeds as well as in many other seeds, which suggests that the decreased PBS-solubility of seed proteins during natural storage may be associated with their viability. However, the contents of proteins soluble in alkaline CAPS (pH 10.5) buffer showed little difference between non-viable and viable seeds (Fig. 4). Therefore, the lower solubility in PBS of proteins from aged seeds may arise from their acidification, which is probably due to the reduction in number of amino bases in polypeptides. Also these results are in support of our assumption that the trapping of amino bases in proteins with endogenously evolved aldehydic compounds via Shiff's reaction would lead to the loss of seed vigour during the storage period.

In order to explore this possibility, lettuce seeds were stored with or without exogenous Ald under different RHs for only 1 month and then extracted for PBS-soluble proteins. The residues were again extracted for urea-soluble proteins, in which the use of CAPS was avoided for convenience for immu-

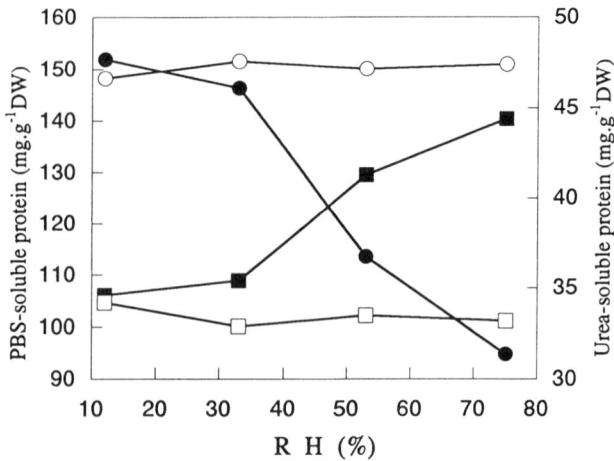

Figure 5. PBS-soluble (○, ●) and urea-soluble (□, ■) proteins from lettuce seeds stored at 23°C under different RHs for 30 d with (●, ■) or without (○, □) 2 mM gaseous acetaldehyde

noassay (Fig. 5). In view of the reason that the duration of storage was only 1 month, the contents of both the PBS-soluble and insoluble but urea-soluble proteins were almost constant regardless of RH when the seeds were not exposed to exogenous Ald. In contrast, in the seeds subjected to exogenous Ald, the contents of the PBS-soluble proteins decreased expectedly with increasing RH, while those of the urea-soluble ones increased slightly with RH, although most of the urea-soluble proteins had been removed by centrifugation of the precipitate which was formed by dialysis against PBS for application to immunoassay. These results clearly show that the denaturation of seed proteins during the storage period is due to modification by aldehydic compounds produced by them, probably to the formation of APA, and finally leading to aggregation.

Figure 6 shows the viabilities and APA contents in the same samples as those used for Figure 5. Corresponding to the decrease in the PBS-soluble proteins as shown in Figure 5, only in the lettuce seeds subjected to exogenous Ald during the storage period, was the viability rapidly lost as the RH increased. In contrast to this, the amounts of APA in both the PBS-soluble (Zhang *et al.,* submitted) and urea-soluble proteins increased with increasing RH as expected only when the seeds were exposed to exogenous Ald. These amounts were so abundant that we could easily detect them by means of the direct ELISA. However, APA from the control lettuce seeds which were stored at 75% RH was hardly detected by this methodology.

An important problem is whether or not the non-enzymatic modification of seed proteins with exogenous Ald can occur between their proteins and

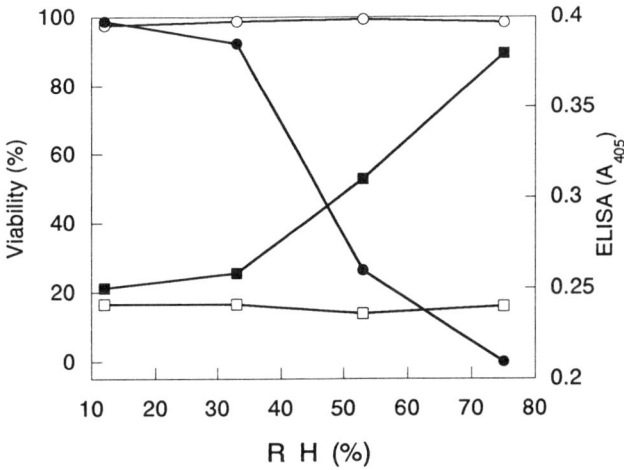

Figure 6. Changes in viability (O, ●) and APA contents (□, ■) in urea-soluble proteins shown as A_{405} in the same samples as those in Figure 5: without (O, □) or with (●, ■) 2 mM acetaldehyde

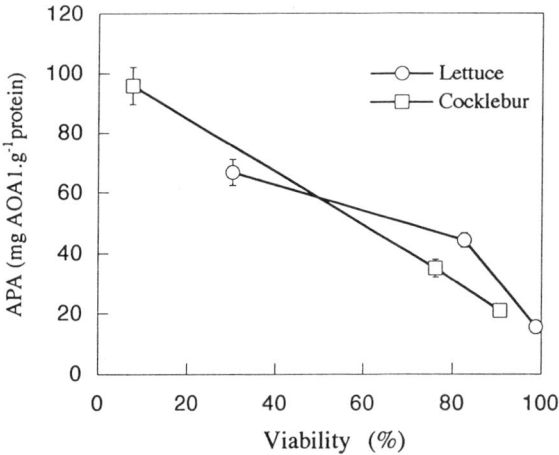

Figure 7. Interrelation between the loss of viability and the accumulation of PBS-soluble APA in lettuce and cocklebur seeds

endogenously evolved Ald under the natural storage conditions which have generally been adopted in seed banks. The APA which would naturally be produced under the low concentrations of endogenous Ald could not be detected by the direct immunoassay. Therefore, in order to detect the endogenously formed APA, we adopted the newly devised solid competitive ELISA (Zhang *et al.*, submitted) in an experiment shown in Figure 7.

As shown in Figure 7, the amounts of naturally produced APA in both cocklebur and lettuce seeds increased as the duration after their harvest was prolonged and their viabilities were lost. It is thus evident that the seed deterioration occurring during storage would be accompanied by the modification of seed proteins with aldehydic compounds produced by them, probably leading to the decrease in solubility through trapping of amino bases in proteins as shown in the following process.

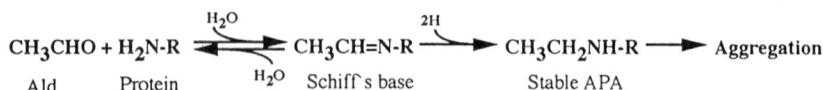

$$CH_3CHO + H_2N\text{-}R \underset{H_2O}{\overset{H_2O}{\rightleftharpoons}} CH_3CH{=}N\text{-}R \xrightarrow{2H} CH_3CH_2NH\text{-}R \longrightarrow \text{Aggregation}$$

Ald   Protein   Schiff's base   Stable APA

In conclusion, the rapid deterioration of seeds in sorption zone 3 would be due to both the consumption of co-enzymes and the formation of APA (Fig. 6). However, it has been supposed that the metabolic activities would proceed only under moisture contents above sorption zone 2 (Priestley, 1986). Nevertheless, the production of Ald occurred even at 12% RH giving sorption zone 1 (Fig. 3), which implies that APA could be formed even in sorption zone 1 if the supply of H^+ for stabilization of APA was ensured. This suggests that the slow loss of seed viability occurring during storage under lower RHs is mainly due to the non-enzymatic denaturation of functional proteins via Shiff's reaction with aldehydic compounds.

References

Bradford, M.M. 1976. *Analytical Biochemistry* 72: 248–254.

Israel, Y., Hurwitz, E., Niemela, O. and Arnon, R. 1986. In: *Proceedings of the National Academy of Science*, USA 83: 7923–7927.

Lowry, O.H., Rosebrough, N.J., Farr, A.L. and Randall, R.J. 1951. *Journal of Biological Chemistry* 193: 265–275.

MacLeod, A.M. 1952. *Transactions of the Botanical Society*, Edinburgh 36: 18–33.

Perata, P., Vernieri, P., Armellini, D., Bugnoli, M., Tognoni, F. and Alpi, A. 1992. *Plant Physiology* 98: 913–918.

Priestley, D.A. 1986. *Seed Aging: Implications for Seed Storage and Persistence in the Soil* pp.156–158. Ithaca and London: Comstock Publishing Assoc.

Vertucci, C.W. and Leopold, A.C. 1984. *Plant Physiology*,75: 114–117.

Vertucci, C.W. and Leopold, A.C. 1987. *Plant Physiology* 84: 1038–1043.

Zhang, M., Lui,Y.,Torii, I., Sasaki, H. and Esashi,Y. 1993. *Seed Science and Technology* 231: 359–373.

Zhang, M., Maeda, Y., Furihata, Y., Nakamaru, Y. and Esashi, Y. 1994. *Seed Science Research* 4: 49–56.

Zhang, M., Nakamaru, Y., Tsuda, S., Nagashima, T. and Esashi, Y. 1995a. *Plant and Cell Physiology* 36: 157–164.

Zhang, M., Yajima, H., Umezawa, Y., Nakagawa, Y. and Esashi, Y. 1995b. *Seed Science and Technology* 23: 59–68.

Zhang, M., Yoshiyama, M., Nagashima, T., Nakagawa, Y. and Esashi, Y. 1995c. *Plant and Cell Physiology* 36: 1189–1195.

Zhang, M., Nagata, S., Miyazawa, K., Kikuchi, H. and Esashi, Y. *Plant Physiology* (in submission).

54. Structural and Biochemical Changes in the Plasma Membrane from Dry and Imbibed Embryos

O. GARCIA-RUBIO[1], S. SANCHEZ-NIETO[1], C. ENRIQUEZ-ARREDONDO[1], J. SEPULVEDA[2], A. CARBALLO[3] and M. GAVILANES-RUIZ[1]

[1]Depto. de Bioquímica, Fac. de Química, Conj. E, UNAM, Cd. Universitaria, 04510; [2]Instituto de Fisiología Celular, UNAM, Cd. Universitaria, 04510, México, D.F.; [3]Colegio de Posgraduados, Montecillo, Edo. de México, México

Abstract

It is generally assumed that the plasma membrane suffers adjustments in structure and function during the onset of germination due to water penetration into the seed. However, there are no biochemical data supporting this possibility. In this work, leakiness, respiration and water uptake were measured. In addition, plasma membranes were purified from dry and hydrated embryos and morphology and the protein content were analyzed. Data revealed differences that can be related to the surface charge and protein abundance in the membranes from dry and imbibed embryos. Such parameters can be associated with changes in the physiological activities of the embryo that are membrane-dependent.

Introduction

Electrolyte leakage which occurs at the start of imbibition of dry seeds has been related to an impairment of the cell membrane integrity (Simon and Raja-Harun, 1972; Powell and Matthews, 1978). It was proposed that cell membranes could lose their native structure due to the gradual dehydration that takes place during the seed's natural desiccation, or to the rapid inrush of water during the first minutes of imbibition (Crowe et al., 1989; Bruni and Leopold, 1991). It was initially suggested that membrane disruption was due to the formation of non-bilayer structures such as inverted hexagonal phases. However, further investigations have precluded this possibility, and have established the presence of a bilayer structure regardless of the extent of dehydration (Seewaldt et al., 1981; Priestley and De Kruiff, 1982; Crowe and Crowe, 1986).

The objective of this work is to investigate the biochemical changes that could be associated with the physical and physiological transitions during imbibition, especially in the plasma membrane. Biochemical differences in purified plasma membrane fractions from dry and imbibed maize embryos were observed and can help to explain the dynamics of membrane components in the onset of germination.

R.H. Ellis, M. Black, A.J. Murdoch, T.D. Hong (eds.), Basic and Applied Aspects of Seed Biology, pp. 499–505.
© 1997 Kluwer Academic Publishers, Dordrecht. Printed in Great Britain.

Materials and Methods

Embryos were manually dissected from maize dry seeds, hybrid A62 (Colegio de Posgraduados, Montecillo, Edo. de México) and imbibed in water at 30°C for different periods of time under sterile conditions.

Water content was expressed on a dry weight basis using 30 embryos for the different periods of imbibition.

Twenty dry or imbibed embryos were immersed in distilled water, and leakiness was monitored by measuring water conductivity every five minutes.

Oxygen consumption was measured in both dry and imbibed embryos. A set of 3 embryos was placed in 3 ml of 50 mM KH_2PO_4 pH 7 and the oxygen uptake was determined using a Clark electrode.

Crude microsomal membranes were obtained from homogenized embryos as described by Sánchez-Nieto et al. (1992). Plasma membranes were purified with the aqueous two-polymer partitioning system according to Larsson et al. (1987), using a 6.8% PEG and 6.8% Dextran mixture for dry embryos and a 6.7% similar mixture for imbibed embryos. Two succesive partitionings were used to obtain the final purified plasma membrane fraction (U_2).

H^+-ATPase activity was used as a marker for the plasma membrane according to Faraday and Spanswick (1992). ATP hydrolysis (vanadate-sensitive), was measured as described by Sánchez-Nieto et al. (1992).

Protein was measured according to Lowry et al. (1951) as modified by Peterson (1972).

Electron microscopy images were obtained from membranes fixed in 3% glutaraldehyde and postfixed in a cold 2% OsO_4 solution. Samples were dehydrated and embedded in propylene oxide and Epon 812 during 48 h. Sections were stained for 30 min with 2% uranyl acetate and a lead solution and viewed in a JEOL 12EX-11 electron microscope.

Results and Discussion

The time course of some physiological activities during early imbibition was studied. Figure 1A shows that the rate of leakiness is higher the shorter the imbibition time is. The extent of leakiness, after 60 min, was different between the dry and 5 h imbibed embryos. Water uptake occurred in a biphasic manner: an initial rapid phase was observed in the first hour, while a slower phase occurred between 1 and 25 h of imbibition (Fig. 1B). Embryo respiration also showed an initial rapid phase followed by a slower phase (Fig. 1C), similar to the pattern described by Ehrenshaft and Brambl (1990). These data confirm the existence of some metabolic paths in the membrane, such as oxidative phosphorylation, which can start to function in the isolated embryo early in germination (Pradet, 1982), although the membranes are not totally sealed. Therefore, the occurrence of these events is simultaneous.

In order to find possible differences between the plasma membranes from dry

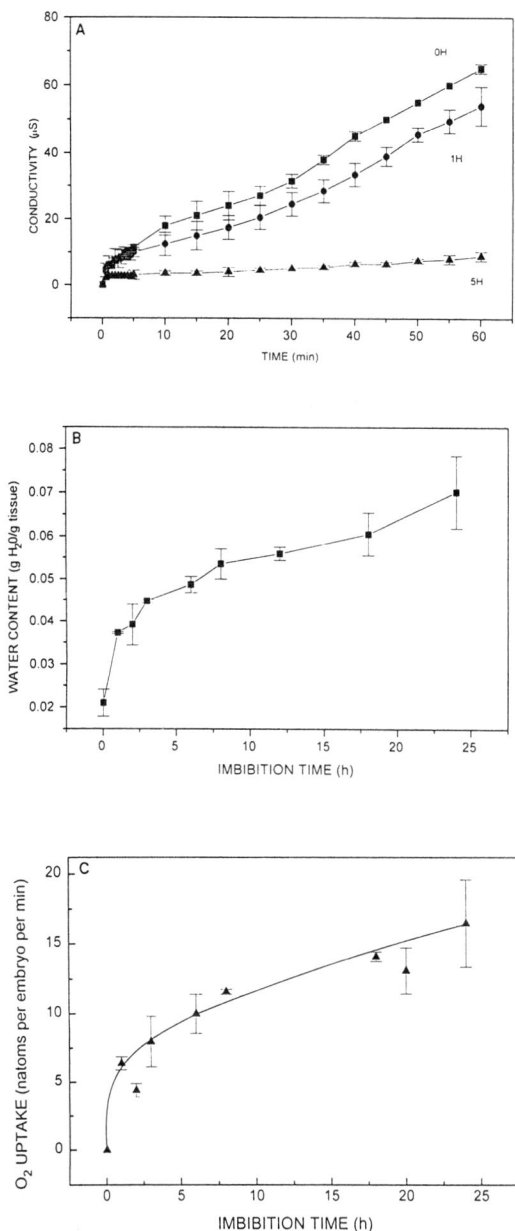

Figure 1. Physiological parameters of maize embryos during imbibition. A. Time course of electrolyte leakage from dry and imbibed maize embryos. Batches of twenty dry (■), 1 h (●) and 5 h (▲) imbibed embryos were placed in distilled water and conductivity measured at the indicated times. B. Water uptake of embryos. Groups of thirty embryos were imbibed the times shown and water content was measured as described under Materials and Methods. C. Respiration of dry and imbibed embryos. Three maize embryos were imbibed the indicated times and then oxygen uptake was measured

Table 1. Requirement of polymer concentrations to purify plasma membranes from dry and hydrated embryos*

PEG, Dextran (%)	ATPase Activity T_0	(%) T_1
6.7	39.1 ± 3.8	87.6 ± 4.1
6.8	89.0 ± 3.1	67.9 ± 4.0

*Dry (T_0) or 1 h imbibed (T_1) embryos were used to obtain purified plasma membranes by the procedure of aqueous two-polymer partitioning system, with the respective polymer concentrations as indicated. 10 mg of membrane protein were used to quantitate the activity of the H^+-ATPase from plasma membrane (vanadate-sensitive ATP hydrolysis)

and imbibed embryos that could account for the above behaviour, a subcellular fraction was purified obtaining an 88 to 89% plasma membrane enrichment (Table 1). Since different polymer concentrations were employed to obtain the same level of purification, plasma membranes from these two different sources have different surface charges, probably due to the membrane insertion of protein and phospholipid synthesized *de novo*. McDonnell *et al.* (1982) have reported phospholipid biosynthesis during the first 24 h of germination.

The electron microscopy analysis of the whole crude membranes (Figs. 2 A,B), shows that membranes obtained from hydrated embryos had a high level of differentiation. Membrane vesicles from RER are plainly differentiated after 1 h of imbibition. This is an important observation, since RER is the biosynthesis site of the membrane components. It must be mentioned that it has been reported that formation of polyribosomes and protein synthesis are already found at 45 min of imbibition (Spiegel and Marcus, 1975). In contrast, purified plasma membrane fractions (U2) from dry or hydrated embryos exhibited very similar, homogenous vesicles, as has been found in other plant tissues (Johansson *et al.*, 1995). Nevertheless, it can not be ruled out that the morphology of the dry membranes could have been acquired upon contact with the aqueous environment during the tissue homogenization. In the latter case, this could be a reflection of the response of the membrane to the water penetration of the seed. The presence of small, loose bilayer pieces (Figs. 2C,D) was observed, although it is possible that this fragmentation is an artifact or part of the injury factors affecting the membrane systems (Bewley, 1986).

When the protein content of the purified plasma membranes from dry or imbibed embryos was compared (Fig. 3), it was observed that there was a transient increase in the protein enrichment of the membrane, with the maximum observed at 2 h of imbibition. However, the total amount of protein obtained in the respective homogenates was always a constant value. The physiological significance of this peak is unknown and may be the result of the balance of membrane protein synthesis and degradation.

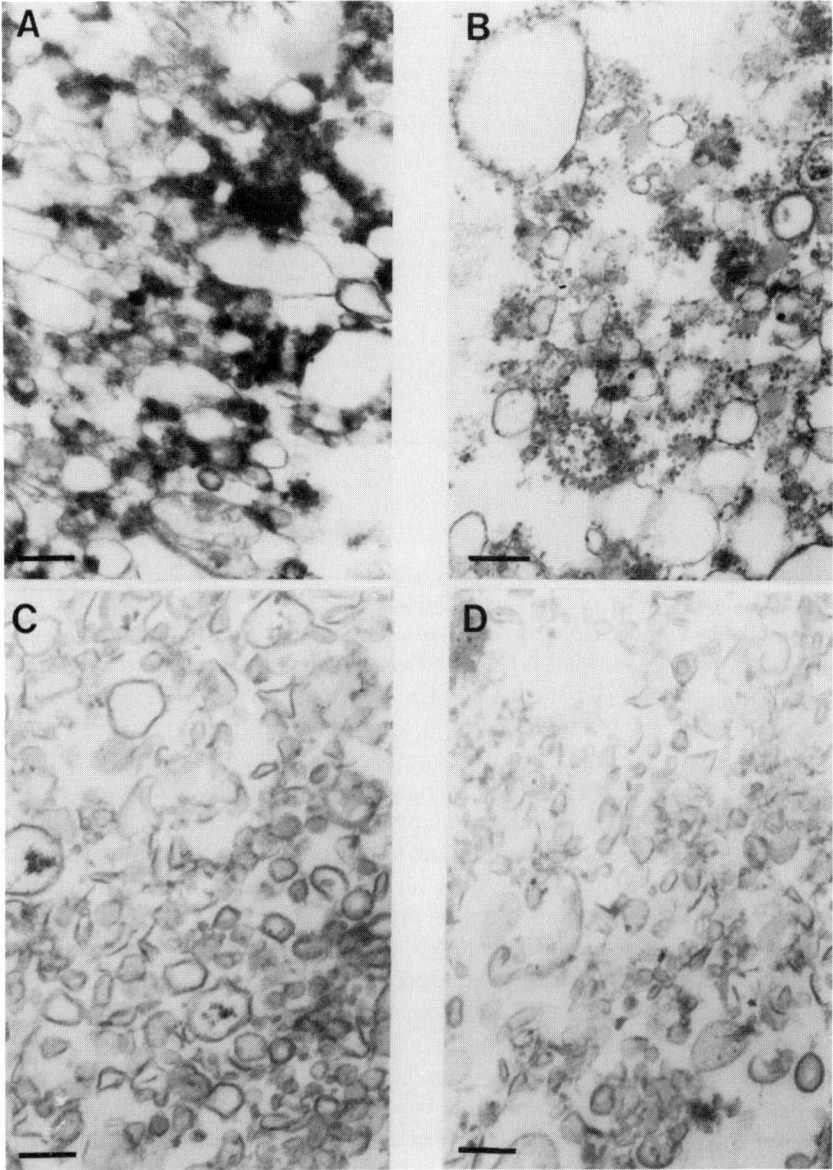

Figure 2. Electron micrographs of total membranes (A,B) and purified plasma membrane (C,D), from dry (A,C) and imbibed embryos (B,D). Membrane samples were processed as described under Materials and Methods. Bar length, 200 nm

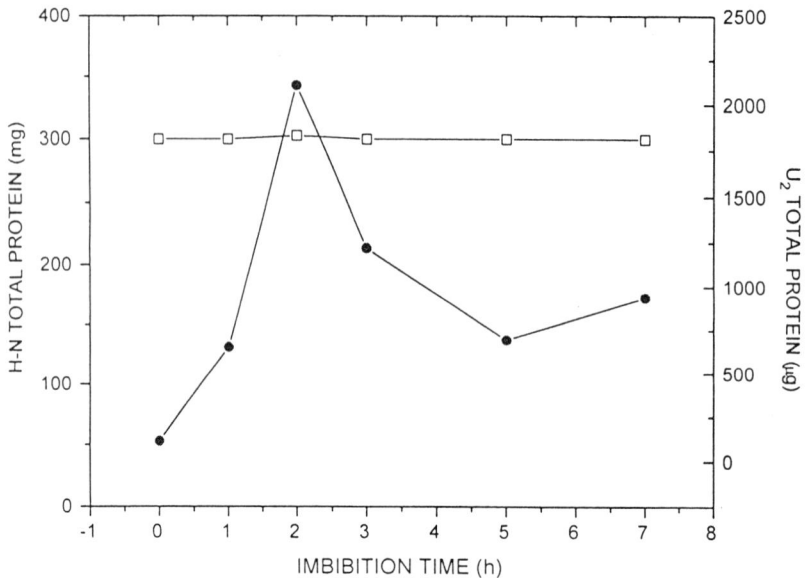

Figure 3. Total protein obtained from homogenates minus the nuclear fraction and from purified plasma membranes from dry and imbibed embryos. Lots of 8 g of embryos imbibed for the indicated times were used to obtain the homogenates without nucleus (H-N, □) and the plasma membranes (U2, ●) as described under Materials and Methods and protein was quantified

Our results suggest that in some membranes as RER, the transition from the dry to the hydrated state is large enough to be discerned in their morphology. In the case of the plasma membrane, the differences are evident only at the biochemical level, as in the cases of the amount of protein and the surface charge. These changes can be related to the physiological activities of the membrane, such as permeability and specific transport.

Acknowledgements

The authors are grateful to Drs. Marietta Tuena de Gómez-Puyou and Irma Bernal-Lugo for encouragement and helpful discussions during this work, and to Drs. Adriana Ortiz-López, Diego González-Halphen and Javier Plasencia de la Parra for the critical reading of the manuscript. We also thank Dr. Georges Dreyfus for allowing the use of the electron microscopy facilities at the Instituto de Fisiología Celular, UNAM. This project has been financed by grants IN206691 from DGAPA, UNAM, and 4836-N9406 from CONACyT, México.

References

Bewley, J. D. 1986. In: *Physiology of Seed Deterioration* pp. 27–45. Madison: Crop Sci. Soc. Amer. Spec. Publ. No. 11.

Bruni, F. and Leopold, A. C. 1991. *Physiologia Plantarum* 81: 359–366.

Crowe, J. H., Crowe, L. M., Hoekstra, F. A. and Wistrom, C. A. 1989. In: *Seed Moisture* pp. 1–14. (eds. P.C. Stanwood and M. B. McDonald) Madison: Crop. Sci. Soc. Amer. Publ. No. 14.

Crowe, L. M. and Crowe, J. H. 1986. In: *Membranes Metabolism and Dry Organisms* pp. 210–230 (ed. A. Carl Leopold) Ithaca, London: Comstock Publishing Associates.

Ehrenshaft, E. and Brambl, R. 1990. *Plant Physiology* 93: 295–304.

Faraday, C. D. and Spanswick, R. M. 1992. *Journal of Experimental Botanay* 43: 1583–1590.

Johansson, F., Olbe, M., Sommarin, M. and Larsson, C. 1995. *Plant Journal* 7: 165–173.

Larsson, C., Widell, S. and Kjellbom, P. 1987. In: *Methods in Enzymology* pp. 558–568 (eds. L. Packer and R. Douce) San Diego: Academic Press.

Lowry, O. H., Rosebrough, N. J., Farr, A. J. and Randall, R. F. 1951. *Journal of Biological Chemistry* 193: 265–275.

McDonnell, E. M., Francis, G. P., Mirbahar, R. B., Tomos, A. D. and Laidman, D. L. 1982. *Journal of Experimental Botany* 33: 631–642.

Sánchez-Nieto, S., Rodríguez-Sotres, R., González-Romo, P., Bernal-Lugo, I. and Gavilanes-Ruíz, M. 1992. *Seed Science Research* 2: 105–111.

Seewaldt, V., Priestley, D. A., Leopold, A. C., Feigenson, G. W. and Goodsaid-Zalduondo, F. 1981. *Planta* 152: 19–23.

Simon, E. W. and Raja-Harun, R. 1972. *Journal of Experimental Botany* 23: 1076–1085.

Spiegel, S., Marcus, A. 1975. *Nature* 256: 228–230.

Peterson, G. L. 1972. *Analytical Biochemistry* 83: 346–356.

Powell, A. A. and Matthews, S. 1978. *Journal of Experimental Botany* 29: 1215–1229.

Pradet, A. 1982. In: *The Physiology and Biochemistry of Seed Development, Dormancy and Germination* pp. 347–369 (ed. A. A. Khan). New York, Amsterdam: Elsevier Biomedical Press.

Priestley, D. A. and De Kruijff, B. 1982. *Plant Physiology* 70: 1075–1078.

55. Comparison of the Storage Potential of Soyabean (*Glycine max*) Cultivars with Different Rates of Water Uptake

D.A. HAHALIS[1] and M.L. SMITH[2]

[1]IERM, School of Agriculture, University of Edinburgh; [2]Scottish Agricultural College, West Mains Road, Edinburgh, EH9 3JG, UK

Abstract

Five soyabean cultivars were grown under the same conditions at the Agricultural Research Station, Mingora, Pakistan. All cultivars were subjected to accelerated ageing conditions (41°C, 100% R.H.) for between one and four days (d). Large differences in the percentage of normal seedlings were revealed after ageing. In addition, seedling vigour parameters such as shoot and root fresh and dry weight reduced with increased ageing periods. In all cultivars, root growth was more sensitive to accelerated ageing than shoot growth. The cultivar Pioneer-9581 had the lowest water uptake in contrast to all others tested, due to a large proportion of hard seeds. This cultivar also showed the slowest moisture uptake when subjected to ageing test, had the lowest decline of the percentage of normal seedlings and total fresh weight per normal seedling. From these results for Pioneer-9581, water uptake could be effectively regulated by the seed coat but this was not the case for the other cultivars tested. The variation in the structure of the soyabean seed coat was discussed in relation to imbibition and establishment as well as storage potential.

Introduction

In soyabeans (*Glycine max* L. Merr.), the seed coat is highly permeable; thus seeds can absorb moisture easily (Ragus, 1987). As a result, seeds tend to be more susceptible to weathering in the field (Burchett *et al.*, 1985; TeKrony *et al.*, 1980) as well as to humid tropical environment under open storage conditions (Ellis *et al.*, 1982). However, seeds that absorb water more slowly could have a greater initial viability. High viability is also an important factor affecting the storability of seeds (Roberts, 1986). Seeds that absorb moisture slowly might also be expected to have a better storability under open storage conditions in the tropics due to the logarithmic relationship between seed moisture content and longevity in soyabeans (Ellis *et al.*, 1982).

It has also been shown that a lower rate of moisture absorption due to a large proportion of hard seeds resulted in prolonged storage potential in soyabeans (Potts *et al.*, 1978; Hartwig and Potts, 1987), long beans, *Vigna sesquipedalis* (Abdullah *et al.*, 1992) and cotton, *Gossypium hirsutum* (Christiansen *et al.*, 1960; Patil and Andrews, 1985). The objective of this study was to compare the

R.H. Ellis, M. Black, A.J. Murdoch, T.D. Hong (eds.), Basic and Applied Aspects of Seed Biology, pp. 507–513.
© *1997 Kluwer Academic Publishers, Dordrecht. Printed in Great Britain.*

storage potential and seedling growth of a cultivar with a high proportion of hard seeds to others with few hard seeds. All cultivars were grown at the same site and could be reasonably expected to be of similar quality.

Materials and Methods

Five seed lots of soyabean were grown under the same conditions at the Agricultural Research Station, Mingora, Pakistan. All cultivars matured approximately 135 d after sowing, had similar seed weights and all subsequently were stored in a cold room at 4°C until testing (Table 1).

Table 1. Days of maturity after sowing and seed weights of the cultivars

Cultivars	Days of maturity	100 seed weights (g)
Forrest	134	20.1
Bay	133	20.3
HSC-591	134	16.7
Hartwig	140	19.4
Pioneer-9581	137	16.7

The percentage of normal seedlings was determined for 4 replications of 50 seeds placed 2 cm apart as a single row on a double layer of moistened paper towel and covered with another single layer of towel. Seeds were kept inside polythene bags at 25°C for 8 d when germination was assessed according to International Seed Testing Association rules (1993). Seeds that remained hard after the 8th d were carefully scarified and subjected to another germination test. The total percentage of normal seedlings was expressed as the sum of the two tests.

Seed moisture content was determined following grinding and drying for 16 h at 105°C (International Seed Testing Association, 1993). Seed moisture content was expressed as a percentage of the fresh weight of seed.

The rate of water uptake was measured for 20 individually weighed seeds. Each single seed was placed into a single compartment of a compartmentalized petri dish all filled to capacity (5 ml) with distilled water. This was left at room temperature (around 20°C). After 6 h each seed was removed, blotted dry and weighed. Hard seeds were those that showed no seed coat wrinkling during that time and the number of those was expressed as a percentage.

Accelerated ageing was performed at 41 ± 1°C and 100% relative humidity (R.H.) by placing samples of surface sterilized seeds (10 s, 5% sodium hypochlorite) on a single layer of muslin cloth placed on top of a metal sieve for support. The sieve was placed inside a desiccator containing 4–5 cm of distilled water. The desiccator was sealed with petroleum jelly and placed in a water bath running at 41°C.

Dry weight was obtained after drying in an oven at 70°C until a constant weight was achieved. Shoot and root fresh and dry weight were recorded in all normal seedlings only for the first germination test with unscratched seeds. Results were expressed as mg per normal seedling.

Results

Large differences in the percentage of normal seedlings between the cultivars were shown after ageing. There were no differences between the unaged controls in all cultivars for the percentage of normal seedlings (Fig. 1). The cultivar Pioneer-9581 was resistant to ageing showing no reduction in the percentage of

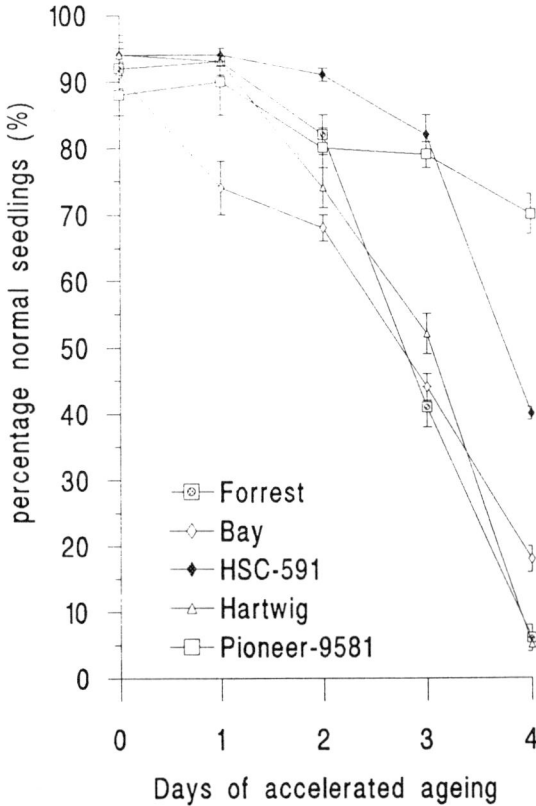

Figure 1. The effect of accelerated ageing on the percentage of normal seedlings. SE is shown as a bar

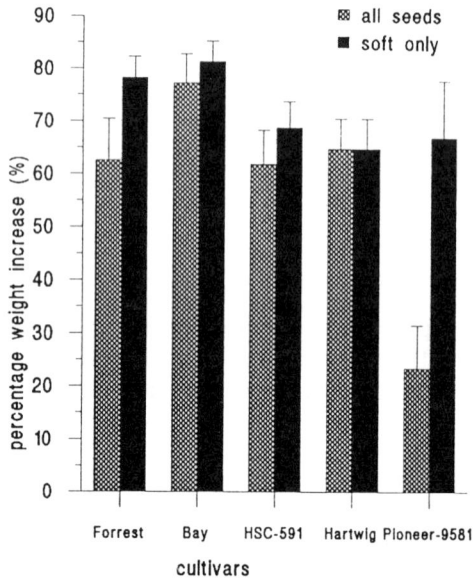

Figure 2. The percentage weight increase after 6 h immersion in water. SE is shown as a bar

normal seedlings after 1 d of ageing and only an 18% drop after 4 d in comparison with the unaged control ($p < 0.01$) (Fig. 1). Whereas the cultivar Bay was sensitive to ageing and showed an 18% drop after 1 d of ageing ($p < 0.001$) and 73% after 4 d ($p < 0.001$) (Fig. 1).

The large proportion of hard seeds (65%) in Pioneer-9581 resulted in a lower rate of water uptake (Fig. 2). In the other cultivars, the proportion of hard seeds was small with no effect on the rate of water uptake (Fig. 2). When hard seeds were excluded from the calculation, the differences in the rates of imbibition were not statistically significant across all cultivars (Fig. 2).

There were differences in the moisture uptake between cultivars (Fig. 3). Pioneer-9581 had a slow rate of absorption reaching 25% m.c. after 4 d of ageing (Fig. 3). In contrast Bay had a higher rate of absorption reaching 33% m.c. after 4 d of ageing (Fig. 3). As ageing period increased, greater differences in the moisture uptake were shown between Pioneer-9581 and the other cultivars (Fig. 3).

Shoot growth (expressed as fresh and dry weight in milligrams) declined with increased ageing in most cultivars except that the rate of decline was much less for Pioneer-9581 (Figs. 4a,b). Shoot fresh weight per normal seedling was different for all unaged controls. Pioneer-9581 showed the minimum value (294 mg/normal seedling) whereas Bay showed the maximum value (689 mg/normal seedling) ($p < 0.001$) (Fig. 4a). After 4 d of ageing, the shoot fresh weight per

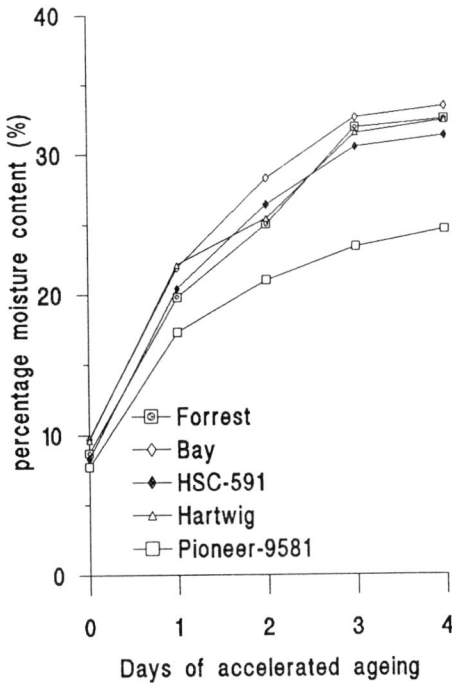

Figure 3. The seed moisture content during the accelerated ageing test. SE is smaller than the symbols

normal seedling in Pioneer-9581 reduced by 24% in comparison to the unaged control ($p<0.01$) (Fig. 4a). However, the other cultivars had a more pronounced decline after 4 d of ageing (Fig. 4a). This ranged from 67% in Hartwig ($p<0.001$) to 43% in HSC-591 ($p<0.001$) (Fig. 4a). Shoot dry weight per normal seedling followed the same pattern of reduction as the shoot fresh weight (Fig. 4b).

Root growth was more sensitive to accelerated ageing than shoot growth and all cultivars showed a rapid fall after 1 d of ageing. After 4 d of ageing, the decline ranged from 72% in Forrest in comparison to the unaged control ($p<0.001$) to 65% in Pioneer-9581 ($p<0.001$) (Fig. 5a). Shoot dry weight per normal seedling reduced in the same way as shoot fresh weight (Fig. 5b).

Discussion

The initial seed quality across all cultivars was expected to be similar as the cultivars were grown under the same conditions, matured at about the same

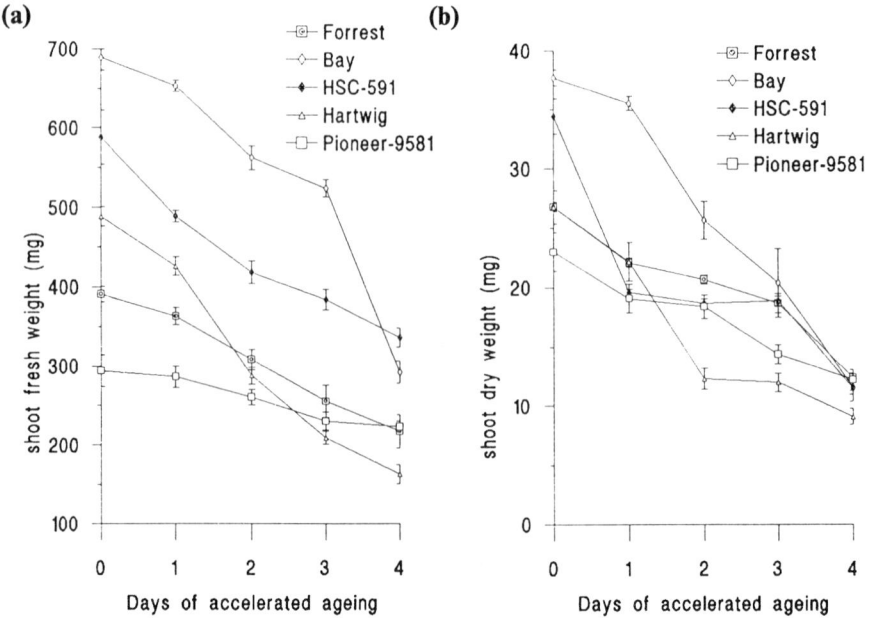

Figure 4. (a) The shoot fresh weight per normal seedling and (b) the shoot dry weight per normal seedling after ageing for different periods of time. SE is shown as a bar

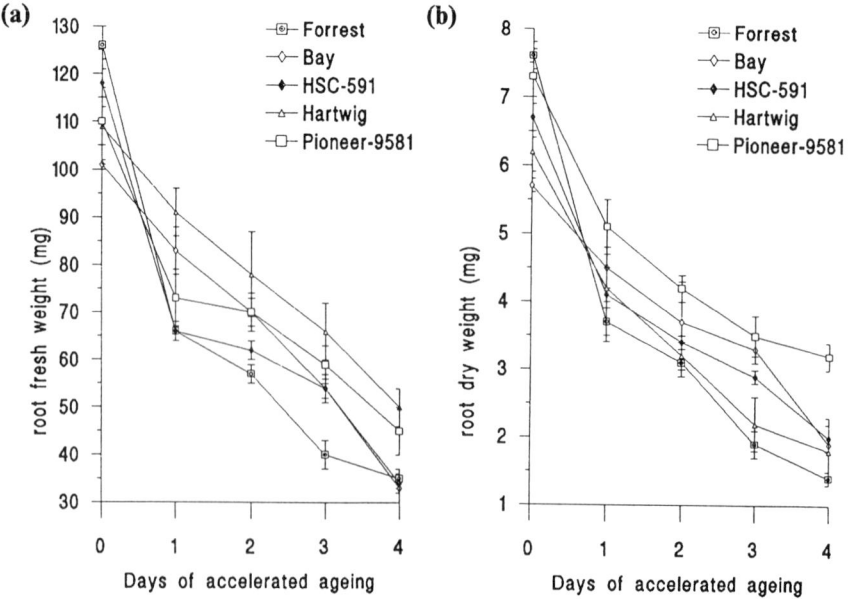

Figure 5. (a) The root fresh weight per normal seedling and (b) the root dry weight per normal seedling after ageing for different periods of time. SE is shown as a bar

time, had similar seed weights and were stored in the same cold room before testing (Table 1).

Although the unaged controls of all cultivars were shown to have the same high percentage of normal seedlings, large differences were observed in the seedling growth between the cultivars. In Pioneer-9581, the speed of germination was undoubtedly slow due to the high proportion of hard seeds and therefore a lower seedling growth was expected. However, for the other cultivars, these differences could not be attributed to variation in the rate of imbibition but may have reflected genetic differences between the cultivars.

The large proportion of hard seeds in the cultivar Pioneer-9581 was found to prolong storage potential due to the lower rate of moisture absorption during the accelerated ageing test. Similar results have been reported by Potts *et al.* (1978). The efficacy of hard seeds to prolong storage potential could be related to amounts of hard seeds in the seed lot and difference in the rate of moisture absorption between hard and soft seeds during ageing.

Seedlings produced from hard seeds had a smaller shoot growth reduction when subjected to ageing. In contrast, root growth significantly reduced after only 1 d of ageing in all cultivars. Bulat (1963) found that in Leguminosae the radicle tip was the most sensitive area to ageing.

It appears from this study that seed lots of soyabeans that possess a large proportion of hard seeds could prolong storage potential, prevent much of the decline in shoot growth but not root growth in aged seeds.

Acknowledgements

D. Hahalis gratefully acknowledges the support of a grant from the Greek State Scholarship Foundation. Also, we are much indebted Dr. Ehsanullah Khan for kindly supplying the seed samples.

References

Abdullah, W.D., Powell, A.A. and Matthews, S. 1992. *Seed Sciece and Technology* 20: 141–147.
Burchett, C.A., Schapaugh, W.T. Jr., Overley, C.B. and Walter, T.L. 1985. *Crop Science* 25: 655–660.
Bulat, R. 1963. *Proceedings of International Seed Testing Association* 28: 713–751.
Christiansen, M.N., Moore, R.P. and Rhyne, C.L. 1960. *Agronomy Journal* 52: 81–84.
Ellis, R.H., Osei-Bonsu, K. and Roberts, E.H. 1982. *Annals of Botany* 50: 69–82.
Hartwig, E.E. and Potts, H.C. 1987. *Crop Science* 27: 506–508.
ISTA (International Seed Testing Association). 1993. *Seed Science and Technology* 21: Supplement.
Patil, V.N. and Andrews, C.H. 1985. *Seed Science and Technology* 13: 193–199.
Potts, H.C., Duangpatra, J., Hairston, W.G. and Delouche, J.C. 1978. *Crop Science* 18: 221–224.
Ragus, L.N. 1987. *Seed Science and Technology* 15: 285–296.
Roberts, E.H. 1986. In: *Physiology of Seed Deterioration*, pp. 101–123 (eds M.B. McDonald and C.J. Nelson). Madison: Crop Science Society of America.
TeKrony, D.M., Egli, D.B. and Phillips, A.D. 1980. *Agronomy Journal* 72: 749–753.

56. Changes in Sucrose, Cyclitols and their Galactosyl Derivatives with Seed Ageing

P.K. KATAKI[1,2], M. HORBOWICZ[1,3], A.G. TAYLOR[2] and R.L. OBENDORF[1]

[1]*Seed Biology, Department of Soil, Crop and Atmospheric Sciences, 619 Bradfield Hall, Cornell University, Ithaca, New York 14853–1901;* [2]*Department of Horticultural Sciences, New York State Agricultural Experiment Station, Geneva, New York 14456, USA;* [3]*Research Institute of Vegetable Crops, 96-100 Skierniewice, Poland*

Abstract

Soluble carbohydrates have been studied in relation to seed desiccation tolerance and storability. In this study the changes in sucrose, cyclitols and their galactosyl derivatives were related to seed ageing of maize, rice, soyabean, cotton and lettuce. Seeds were aged at 75% relative humidity and 45°C. Ageing reduced percent germination and the coefficient of the rate of germination for all crop species studied. A decline in raffinose in maize axis and rice embryo tissues correlated with the decrease in percent germination. A similar correlation between the raffinose series of oligosaccharides and ageing was not found in seeds of soyabean, cotton and lettuce possibly due to changes in the carbohydrate pools during ageing. Sorbitol, an acyclic polyol, increased significantly ($r > 0.80$**) from undetectable or trace amounts to high levels as a function of seed ageing for all five species.

Introduction

Long term storability of seed has been associated with desiccation tolerance (Koster, 1991; Leprince *et al.*, 1993). Nonreducing sugars, including sucrose and the raffinose series of oligosaccharides, are one of multiple factors required for seed desiccation tolerance and storability (Leprince *et al.*, 1990; Bernal-Lugo and Leopold, 1992; Blackman *et al.*, 1992). Horbowicz and Obendorf (1994) suggested that accumulated galactosyl cyclitols, such as galactinol or fagopyritols, function in the same role as raffinose and stachyose in facilitating desiccation tolerance and storability.

Because seed desiccation tolerance and storability are associated with accumulation of non-reducing carbohydrates, our objective was to analyze sucrose, cyclitols, and their galactosyl derivatives in maize, rice, soyabean, cotton and lettuce seeds during ageing to identify a common marker for detection of seed deterioration within and among crop species. Based on tentative identification, sorbitol (D-glucitol), an acyclic polyol found in numerous plant species, accumulates during rapid ageing of seeds for the five crop species analysed. It is suggested that sorbitol may serve as a marker for deterioration in rapidly aged seeds.

R.H. Ellis, M. Black, A.J. Murdoch, T.D. Hong (eds.), Basic and Applied Aspects of Seed Biology, pp. 515–522.
© *1997 Kluwer Academic Publishers, Dordrecht. Printed in Great Britain.*

Materials and Methods

Commercial stocks of high quality maize (*Zea mays* L., cultivar 'Jubilee'), rice (*Oryza sativa L., cultivar* 'Lemont'), soyabean (*Glycine max* [L.] Merrill, cultivar 'Hardin'), cotton (*Gossypium hirsutum* L., cultivar 'GS 510'), and lettuce (*Lactuca sativa* L., cultivar 'Salinas') seeds were obtained in 1991 and stored at 5°C and 30% relative humidity (RH) for 2 years. For seed ageing, seeds were equilibrated at 75% RH in plexiglass chambers above a mixture of glycerol:water (specific gravity, 1.151) (Forney and Brandl, 1992). This procedure resulted in seed moisture concentrations of 12.9, 12.6, 11.5, 9.7, and 8.1% of fresh weight for maize, rice, soyabean, cotton and lettuce, respectively. Immediately upon equilibration, seeds were sealed in plastic covered aluminum pouches and placed in a water bath at 45°C for 4 to 32 days. At each sampling time, pouches were opened, and seeds were stored in plastic Ziploc polyethylene bags at 30% RH and 5°C. Seed lots of five different qualities, including the nonaged control seeds, for each of the five crop species were analysed for germination, abnormal seedlings, coefficient of the rate of germination (CRG), and soluble carbohydrates.

Germination of four replications of 25 seeds followed the Rules for Testing Seeds (AOSA, 1993). Seeds with visible radicle protrusion were considered germinated. Seedlings with stubby roots, missing cotyledons (soyabean, cotton and lettuce), stem or root necrosis were classified as abnormal. The CRG was calculated as described in Bewley and Black (1994): CRG = $[\Sigma n / \Sigma(tn)] \times 100$, where n = number of seeds completing germination on day t.

Soluble carbohydrates were extracted and analysed from three replications of the axis, scutellum and endosperm of five maize seeds, embryo and endosperm of 10 rice seeds, axis of five soyabean seeds, axis and cotyledons of five cotton seeds, and 10 whole lettuce seeds. Tissues were extracted in ethanol:water (1:1, v/v) with phenyl α-D-glucoside as internal standard. Extracts were heated, centrifuged, evaporated and dried overnight over P_2O_5. Residues were derivatized with trimethylsilylimidazol:pyridine (1:1, v/v) and analyzed by high resolution gas chromatography on a DB-1 capillary column with FID detection (Horbowicz and Obendorf, 1994). Derivatized carbohydrates were identified by retention time in comparison to authenic standards and confirmed by analysis of hydrolysis products, GC-MS and/or ^{1}H-NMR. Sorbitol, digalactosyl glycerol and digalactosyl *myo*-inositol were tentatively identified.

Results

Duration of ageing at 45°C and 75% RH resulted in different quality seed samples of each species, as evidenced by the decline in germination and CRG, and an initial increase in percent abnormal seedlings (Figs. 1A,D, 2A,D, 3A). Soluble carbohydrates had similar profiles during seed ageing for axis or embryo tissues and for scutellum or endosperm tissues (Kataki, 1995). Therefore, only results for axis or embryo tissues are presented.

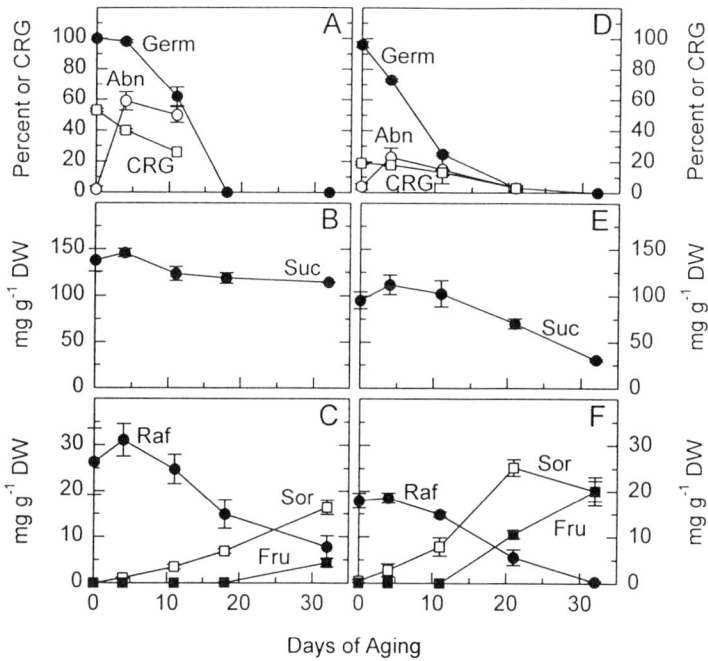

Figure 1. (A–C) Maize. (D–F) Rice. (A,D) Percent germination, percent abnormal seedlings, and coefficient of the rate of germination (CRG) of seeds, (B,E) sucrose, and (C,F) raffinose, sorbitol, and fructose plus glucose concentrations in maize axes and rice embryos as a function of days of seed ageing. Vertical bars represent s.e. of the mean

Maize axis tissues contained 17% of their dry weight as soluble sugars, oligosaccharides and polyols. Sucrose (82% of total) and raffinose (16% of total) were the major components (Figs. 1B,C) with small amounts of stachyose, *myo*-inositol, galactinol and digalactosyl glycerol. Sorbitol appeared in axis tissues and increased markedly with duration of ageing (Fig. 1C); fructose and lesser amounts of glucose increased after loss of viability. Galactinol, stachyose and digalactosyl glycerol declined with advanced ageing in maize axes without a corresponding increase in galactose. Ageing increased *myo*-inositol and decreased sucrose slightly (Fig. 1B), but there was no change in total soluble carbohydrates.

Rice embryos contained 11% of their dry weight as total water soluble carbohydrates. Sucrose (83% of total) and raffinose (16%) were prominent in rice embryos. The loss of sucrose (Fig. 1E) and raffinose (Fig. 1F) in rice embryos was significant at 21 and 32 d of ageing and associated with loss of viability (Fig. 1D). Sorbitol increased markedly during ageing, and fructose (mainly) plus glucose continued to increase after loss of viability (Fig. 1F). Galactose was not detected.

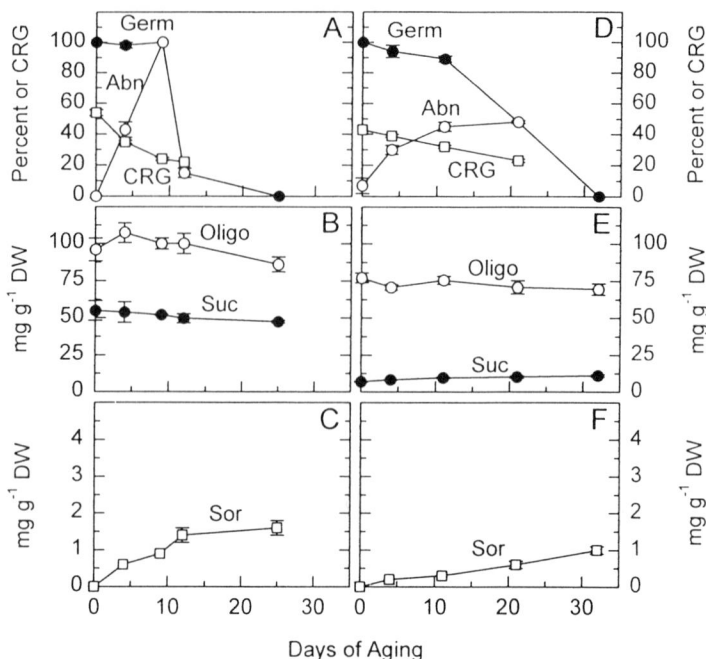

Figure 2. (A–C) Soyabean. (D–F) Cotton. (A,D) Percent germination, percent abnormal seedlings, and coefficient of the rate of germination (CRG) of seeds, (B,E) oligosaccharide and sucrose, and (C,F) sorbitol concentrations in axes of soyabean and cotton as a function of days of seed ageing. Vertical bars represent s.e. of the mean

Soyabean axes contained 17% of their dry matter as soluble carbohydrates, including stachyose (46% of total), sucrose (32%), raffinose (7%), galactopinitol A (3%), galactopinitol B (3%) and verbascose (2%). The remaining 7% included D-pinitol, D-*chiro*-inositol, *myo*-inositol, fagopyritol B$_1$, galactinol, digalactosyl glycerol, ciceritol, digalactosyl *myo*-inositol and a few unknowns. Galactopinitol A, galactopinitol B, fagopyritol B$_1$ and galactinol in soyabean axes decreased after 12 d of ageing with associated loss of viability (Fig. 2A). *myo*-Inositol, D-*chiro*-inositol, raffinose and verbascose increased slightly in aged seeds, but total oligosaccharides (83 to 73% stachyose) changed little. Small amounts of stachyose appeared to be converted to raffinose and verbascose. Sucrose (Fig. 2B), D-pinitol, digalactosyl glycerol, ciceritol and stachyose remained unchanged. Sorbitol increased from 0 to 1.6 mg g^{-1} dry weight after 25 d of ageing (Fig. 2C), but the change was less dramatic than in maize and rice. Absence of glucose, fructose and galactose may be the cause of the small accumulation of sorbitol in aged soyabean.

The carbohydrate composition in the cotton axis was complex. Total soluble

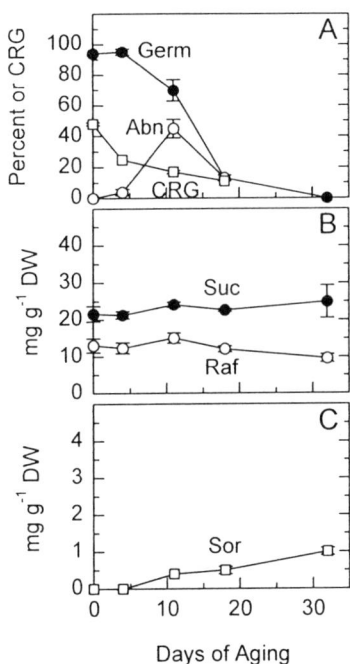

Figure 3. (A–C) Lettuce. (A) Percent germination, percent abnormal seedlings, and coefficient of the rate of germination (CRG) of seeds, (B) sucrose and raffinose, and (C) sorbitol concentrations in whole seeds as a function of days of seed ageing. Vertical bars represent s.e. of the mean

carbohydrates were 10% of the dry weight. Raffinose (65% of total) was much higher than sucrose (7%) and stachyose (8%) (Fig. 2E). *myo*-Inositol, galactinol, digalactosyl glycerol, digalactosyl *myo*-inositol and a few unknowns were also detected. Sorbitol in the cotton axis increased from nondetectable in nonaged seeds to 1 mg g^{-1} dry weight of aged axis tissue (Fig. 2F), whereas fructose, glucose and galactose were not detected. *myo*-Inositol and sucrose increased gradually, but galactinol, digalactosyl glycerol, digalactosyl *myo*-inositol, raffinose and stachyose either remained unchanged or increased slightly. The ratio of sucrose:oligosaccharide remained unchanged during ageing.

Lettuce seeds had 4% of their seed dry weight as soluble carbohydrates, with sucrose (60% of total) and raffinose (36%) as the major components (Fig. 3B). Lettuce seeds had increased sorbitol (Fig. 3C) and *myo*-inositol with seed ageing (Fig. 3A). D-*chiro*-Inositol, sucrose, galactinol, and raffinose were unchanged.

During seed ageing, changes in soluble carbohydrates other than sorbitol were species specific. In seeds of all five crops, a significant (1% level) increase in sorbitol was observed during ageing.

Discussion

Fructose and glucose were detected only in highly aged seeds of maize and rice seeds but not in soyabean, cotton or lettuce. Free galactose was not detected in any seed. Presence of the reducing sugars is not desirable because they are thought to promote deterioration through formation of Amadori products and Maillard reactions (Wettlaufer and Leopold, 1991).

Sucrose, a non-reducing sugar, is normally present in high concentration in seeds. Sucrose concentration decreased significantly during ageing in the axis of maize and embryo of rice, but declined less in soyabean. Sucrose remained unchanged during ageing of lettuce seeds, but in cotton axes, sucrose increased slightly, probably from degradation of raffinose.

Raffinose and stachyose inhibit crystallization of sucrose and therefore may aid in the protective role of sucrose by formation of a stable glassy state (Koster, 1991). This vitrification of sugars around fluid phase phospholipids lowers their fluid-to-gel phase transition temperature during desiccation (Koster *et al.,* 1994). No consistent trend among the five crop species studied was found for changes in sucrose, raffinose, stachyose and verbascose during seed deterioration. In maize and rice, the reduction in sucrose and raffinose correlated with the decline in seed germinability. Sucrose synthase (EC 2.4.1.13), encoded by the *Sus1* gene, converts sucrose to UDP-glucose and fructose, and *Sus1* is expressed at elevated sucrose levels (Lin and Varner, 1991) in embryo tissues of maize (Chen and Chourey, 1989) and other cereals (Shaw *et al.,* 1994). This would be consistent with the loss of sucrose, specific accumulation of fructose, and conversion of fructose to sorbitol by sorbitol dehydrogenase during seed ageing of maize and rice at 45°C. It is likely that accumulation of reducing sugars is minimized by conversion of fructose and glucose to sorbitol. Seeds that did not accumulate fructose and glucose (soyabean, cotton and lettuce) accumulated small amounts of sorbitol during ageing.

Seeds of some species accumulate galactosyl derivatives of *myo*-inositol, D-pinitol and/or D-*chiro*-inositol that also have been proposed to confer desiccation tolerance and long term storability (Horbowicz and Obendorf, 1994). For example, desiccation tolerant buckwheat (*Fagopyrum esculentum* Moench) seeds accumulate fagopyritols (galactosyl derivatives of D-*chiro*-inositol) but only traces of raffinose and stachyose. This led Horbowicz and Obendorf (1994) to suggest that accumulation of galactosyl cyclitols probably contributes to seed desiccation tolerance and storability.

Galactinol, the galactosyl donor in the biosynthesis of raffinose, stachyose and verbascose (Dey, 1990), is present in small concentrations in mature seeds. Galactinol pools were most prominent in soyabean, cotton and lettuce. Galactinol decreased during seed ageing of soyabean and cotton. In soyabean a small amount of stachyose, the predominant oligosaccharide, appeared to be converted to raffinose plus verbascose. In cotton and lettuce, small amounts of raffinose, the predominant oligosaccharide, appeared to be converted to sucrose plus stachyose during ageing, thereby contributing to the small increase in sucrose during ageing. Thus, the decline in raffinose correlated with seed ageing in rice and maize but not in soyabean, cotton and lettuce. Previous studies showed that a decline in raffinose was correlated with deterioration in maize seeds (Bernal-Lugo and Leopold, 1992), but in soyabean seeds deterioration occurred with high levels of stachyose (Sun and Leopold, 1993).

Cyclitols are effective scavengers of hydroxyl radicals *in vitro*, especially the O-methyl cyclitols such as D-pinitol (Orthen *et al.*, 1994). *myo*-Inositol, an effective hydroxyl radical scavenger *in vivo* (Smirnoff and Cumbes, 1989), remained unchanged during ageing of maize and lettuce, but a gradual increase was detected in rice, soyabean and cotton. D-*chiro*-Inositol, D-pinitol and their galactosyl derivatives were detected in soyabean, but the concentrations remained stable during seed ageing.

Sorbitol is the only compound which stands out as a common marker among all five crop species. Consistently, sorbitol was negatively correlated with percent germination and CRG for all species and seed parts, and therefore correlates positively with seed ageing. Sorbitol increased from undetectable levels for the nonaged control to 1 to 20 mg g^{-1} dry weight in aged seeds. Consistent with the current evidence that sorbitol accumulates in desiccated seeds during ageing, sorbitol accumulates in response to salt and water stress in transgenic plants (Jensen and Bohnert, 1993). There is strong evidence to support the existence of a sorbitol synthesis pathway in plants. Shaw and Dickinson (1984) detected sorbitol in developing maize kernels, but its concentration decreased rapidly during seed maturation. The enzymes ketose reductase (EC 1.1.1.14) that converts fructose to sorbitol and aldose reductase (EC 1.1.1.21) that converts glucose to sorbitol have been detected in axes of germinating soyabean seeds (Kuo *et al.*, 1990). Bartels *et al.* (1991) isolated a clone (pG 22-69) which encoded a putative gene product of 35 kDa with a high structural homology to mammalian genes that encode an NADPH dependent aldose reductase. Ooms *et al.* (1993) detected a seed maturation protein in *Arabidopsis* that is antigenically similar to a protein with homology to aldose reductase. Thus, aldose reductase enzyme or mRNA could be assayed as a diagnostic test for seed ageing.

The rapid ageing treatment is a highly stressful environment for seeds. The high relative humidity of 75% and temperature of 45°C may have created a stressful environment for the release of fructose, glucose and/or galactose with conversion to and accumulation of sorbitol in rapidly aged seeds. The increase in sorbitol was large during rapid ageing of rice and maize that also released fructose and glucose, but comparatively moderate in lettuce, cotton and soyabean that released only trace amounts of glucose and fructose. Galactose did not accumulate in any of the treatments indicating a possible conversion of galactose to glucose and excess glucose to sorbitol. This response was consistent in axis, embryo, scutellum, endosperm and cotyledon tissues. Sorbitol accumulation probably requires a combination of high humidity and high temperature plus the release of reducing sugars.

Though it is very tempting to propose accumulation of sorbitol as a good marker for seed deterioration, the phenomenon should be analysed in seeds aged at lower water activities and temperature. Sorbitol accumulation in naturally aged seeds needs to be verified.

Acknowledgements

This research was conducted as part of Western Regional Research Project W-168 and funded in part by grants from The American Seed Research Foundation, The Kosciuszko Foundation, and Pioneer Hi-Bred International, Inc.

References

AOSA, 1993. *Journal of Seed Technology* 16: 1–13.
Bartels, D., Engelhardt, K., Roncarati, R., Schneider, K., Rotter, M. and Salamini, F. 1991. *European Molecular Biology Organization Journal* 10: 1037–1043.
Bernal-Lugo, I. and Leopold, A.C. 1992. *Plant Physiology* 98: 1207–1210.
Bewley, J.D. and Black, M. 1994. *Seeds: Physiology of Development and Germination, Second Edition*. New York: Plenum Press.
Blackman, S.A., Obendorf, R.L. and Leopold, A.C. 1992. *Plant Physiology* 100: 225–230.
Chen, Y.C. and Chourey, P.S. 1989. *Theoretical and Applied Genetics* 78: 553–559.
Dey, P.M. 1990. In: *Methods in Plant Biochemistry Volume 2 Carbohydrates*, pp. 189–218 (ed. P.M. Dey). New York: Academic Press.
Forney, C.F. and Brandl, D.G. 1992. *HortTechnology* 2: 53–54.
Horbowicz, M. and Obendorf, R.L. 1994. *Seed Science Research* 4: 385–405.
Jensen, R.G. and Bohnert, H.J. 1993. *Abstracts of Papers American Chemical Society* 206: 159.
Kataki, P.K. 1995. Ph.D. Thesis, Cornell University, Ithaca, New York, USA.
Koster, K.L. 1991. *Plant Physiology* 96: 302–304.
Koster, K.L., Webb, M.S., Bryant, G. and Lynch, D.V. 1994. *Biochimica et Biophysica Acta* 1193: 143–150.
Kuo, T.M., Doehlert, D.C. and Crawford, C.G. 1990. *Plant Physiology* 93: 1514–1520.
Leprince, O., Bronchart, R. and Deltour, R. 1990. *Plant Cell and Environment* 13: 539–546.
Leprince, O., Hendry, G.A.F. and McKersie, B.D. 1993. *Seed Science Research* 3: 231–246.
Lin, L.-S. and Varner, J.E. 1991. *Plant Physiology* 96: 159–165.
Ooms, J.J.J., Léon-Kloosterziel, K.M., Bartels, D., Koornneef, M. and Karssen, C.M. 1993. *Plant Physiology* 102: 1185–1191.
Orthen, B., Popp, M. and Smirnoff, N. 1994. *Proceedings of the Royal Society of Edinburgh Section B (Biological Sciences)* 102: 269–272.
Shaw, J.R., Ferl, R.J., Baier, J., St.Clair, D., Carson, C., McCarty, D.R. and Hannah, L.C. 1994. *Plant Physiology* 106: 1659–1665.
Shaw, J.R. and Dickinson, D.B. 1984. *Plant Physiology* 75: 207–211.
Smirnoff, N. and Cumbes, Q.J. 1989. *Phytochemistry* 28: 1057–1060.
Sun, W.Q. and Leopold, A.C. 1993. *Physiologia Plantarum* 89: 767–774.
Wettlaufer, S.H. and Leopold, A.C. 1991. *Plant Physiology* 97: 165–169.

57. The Role of Methyl Jasmonate in Germination of *Amaranthus caudatus* L. Seeds

J. KEPCZYNSKI and B. BIALECKA

Department of Plant Physiology, University of Szczecin, Felczaka 3a, 71-412 Szczecin, Poland

Abstract

Non-dormant *Amaranthus caudatus* seeds germinated easily at temperatures from 20 to 35°C; however the germination was partially inhibited at 40°C. Application of methyl jasmonate (Me-Ja) inhibited seed germination. The inhibition caused by 3×10^{-4} M Me-Ja was completely reversed by ethephon at temperatures between 20 to 40°C. Inhibition due to 10^{-3} M Me-Ja was largely overcome by ethephon at 24°C. ACC was able to reverse the inhibitory effect of 3×10^{-4} M Me-Ja, but it did not affect germination in the presence of 10^{-3} M Me-Ja. Me-Ja 10^{-3} M inhibited both ethylene production and synthesis of ACC.

Introduction

Jasmonic acid (Ja) and methyl jasmonate (Me-Ja) are widespread in the plant kingdom (Parthier, 1991). These compounds promote senescence, abscission, tuber formation, tendril coiling and fruit ripening or inhibit callus growth, root growth and photosynthesis. Ja stimulated the germination of dormant embryos of *Acer tataricum, A. platanoides* and *Malus domestica* (Berestetzky *et al.*, 1991; Daletskaya and Sembdner, 1989; Ranjan and Lewak, 1992), but inhibited the germination of embryos isolated from non-dormant seeds of *Acer tataricum* (Berestetzky *et al.*, 1991), seeds of *Avena sativa* and *Triticum durum* (Daletskaya and Sembdner, 1989). Me-Ja was found to inhibit the germination of *Helianthus annuus* and *Amaranthus caudatus* seeds (Corbineau *et al.*, 1988; Kepczynski and Bialecka, 1994).

This study was undertaken in order to obtain more detailed information about the mechanism by which Me-Ja affects germination of *Amaranthus caudatus* seeds. We were interested in determining 1) the effect of Me-Ja on the germination of *Amaranthus caudatus* seeds at temperatures between 20°C and 40°C, and 2) the nature of any interaction between Me-Ja and ethephon or the precursor of ethylene biosynthesis, ACC. In order to find out whether the effect of Me-Ja is related to ethylene biosynthesis, ethylene production and content of ACC was determined.

R.H. Ellis, M. Black, A.J. Murdoch, T.D. Hong (eds.), Basic and Applied Aspects of Seed Biology, pp. 523–529.
© *1997 Kluwer Academic Publishers, Dordrecht. Printed in Great Britain.*

Material and Methods

Plant Material

Commercially available seeds of *Amaranthus caudatus* cv. *atropurpureus* harvested in 1982 were stored dry at $-20°C$. These non-dormant seeds were used in all experiments.

Germination

Seeds were incubated in 5 cm Petri dishes on filter paper moistened with 1.5 ml of distilled water or the same volume of aqueous solutions of Me-Ja, ethephon and ACC at pH 6.6. Five replications with 50 seeds in each test were used. Experiments were performed in the dark at temperatures from 20 to 40°C. All manipulations were performed under green light. Seeds with a radicle about 2 mm long were considered to have germinated.

Ethylene Determination

Seeds (0.1 g) were placed in 25 ml flasks with 1 ml water or Me-Ja solution. The flasks were sealed with rubber stoppers and incubated at 24°C in darkness. Ethylene content in the gas phase of the flask was determined by withdrawing a 1 ml sample with a gas-tight syringe and injecting into a Hewlett-Packard 5890 equipped with a flame ionization detector and stainless column with Poropak Q 80/100 mesh. Ethylene concentrations were determined after 1, 2 and 3 days of incubation.

ACC Determination

Seeds (0.3 g) which were incubated in water or 10^{-3} M Me-Ja solution for 1 or 3 d at 24°C in darkness were homogenized with cold 80% ethanol. After extracting two times the combined ethanol extracts were evaporated to dryness in vacuo and redissolved in 5.5 ml of water. The ACC content was determined according to the method of Lizada and Yang (1979).

Results and Discussion

Seeds of *Amaranthus caudatus* germinated easily in a wide range of temperatures from 20 to 35°C (Fig. 1). The highest temperature used, 40°C, retarded seed germination. Me-Ja had a strong inhibitory effect on germination (Fig. 2). The effect of 10^{-4} M jasmonate on seed germination was dependent on temperature markedly inhibiting germination at 20 and 40°C, though not 25 to 35°C (Fig. 2). It was observed that 10^{-3} M jasmonate almost completely inhibited seed germination at temperatures between 20 and 40°C not only at

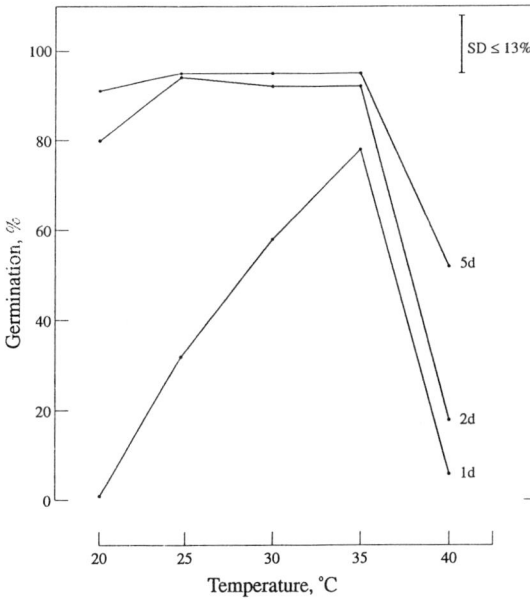

Figure 1. The influence of temperature on germination of *Amaranthus caudatus* seeds after 1, 2 and 5 d of incubation

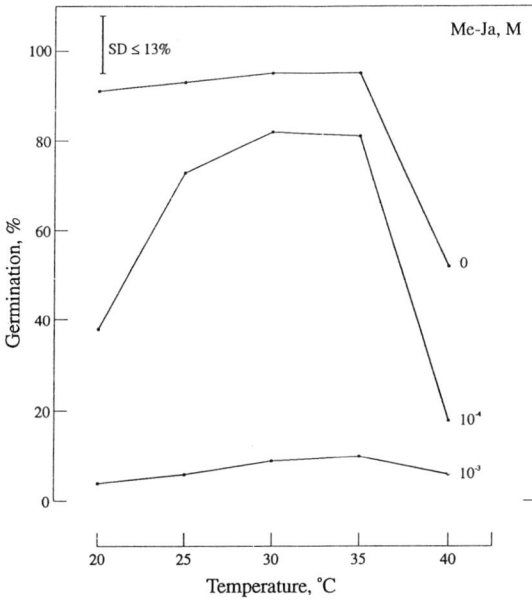

Figure 2. The influence of methyl jasmonate on germination of *Amaranthus caudatus* seeds at different temperatures after 5 d of incubation

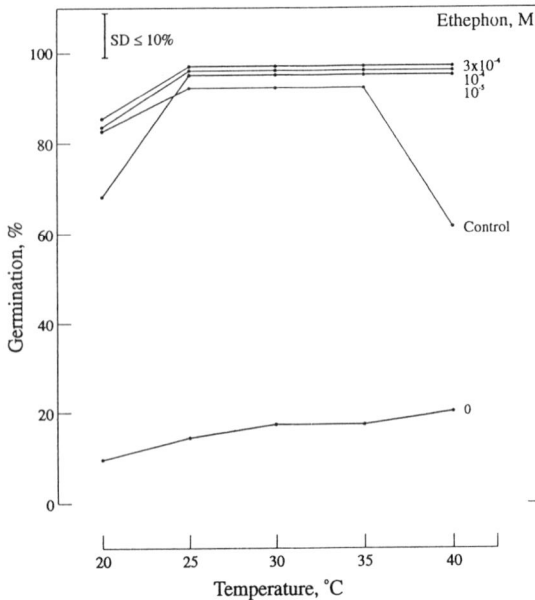

Figure 3. The influence of ethephon on the germination of *Amaranthus caudatus* seeds in 3×10^{-4} M methyl jasmonate at different temperatures after 5 d of incubation. The control line shows germination in water

24°C as was shown previously (Kepczynski and Bialecka, 1994). The results presented are in agreement with earlier observations that Ja or Me-Ja is an inhibitor of germination of non-dormant embryos of *Acer tataricum* (isolated from stratified seeds) (Berestetzky *et al.*, 1991) and seeds of *Avena sativa, Triticum durum, Helianthus annuus* and *Amaranthus caudatus* (Daletskaya and Sembdner, 1989; Kepczynski and Bialecka, 1994).

Previous studies showed that ethephon, an ethylene releasing compound, was highly effective in reversing inhibition of *A. caudatus* seed germination caused by ABA (Kepczynski, 1986), which shows physiological similarities to Me-Ja (Staswick, 1995). An inhibitor of ethylene biosynthesis, AVG, increased inhibition of seed germination due to ABA. In view of such results the next experiment on the interaction between Me-Ja and ethephon was examined (Fig. 3). Almost complete inhibition of seed germination in the presence of 3×10^{-4} M Me-Ja was fully antagonized by $10^{-5} - 3 \times 10^{-4}$ M ethephon at temperatures from 20 to 40°C. Figure 4 shows that the inhibitory effect of 10^{-3} M of Me-Ja at 24°C was largely overcome by ethephon. It was observed in the next experiment that ACC, a precursor of ethylene biosynthesis, was also effective in reversing the inhibition although to a lesser extent than ethephon (Fig. 5). The inhibition by 3×10^{-4} M Me-Ja was almost completely antagonized by 10^{-3} M ACC. The effect of the highest concentration of Me-Ja was not

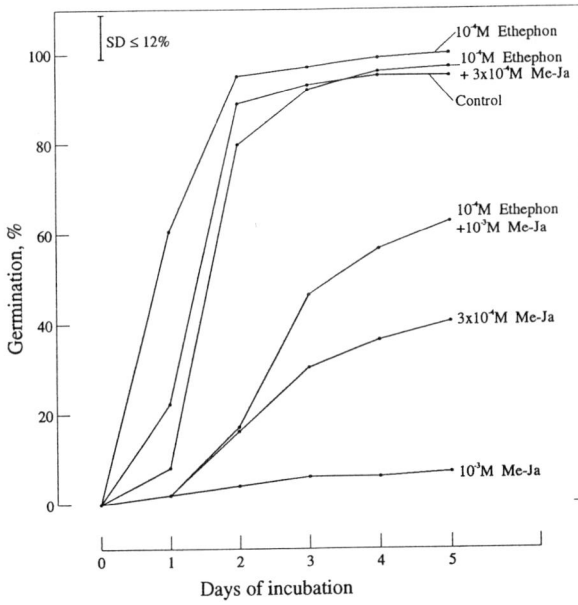

Figure 4. The effects of ethephon and methyl jasmonate on the germination of *Amaranthus caudatus* seeds at 24°C

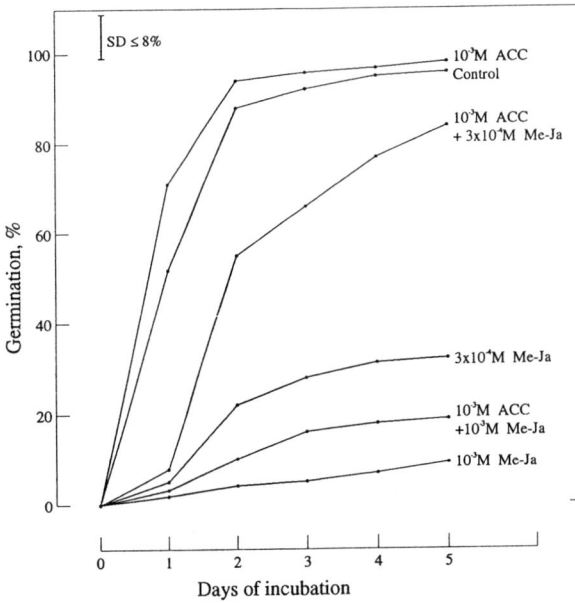

Figure 5. The effects of ACC and methyl jasmonate on the germination of *Amaranthus caudatus* seeds at 24°C

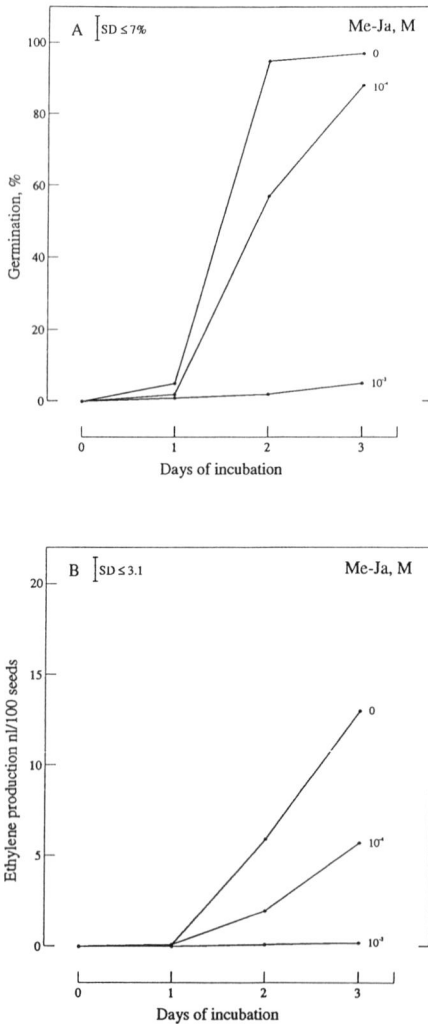

Figure 6. The influence of methyl jasmonate on germination (A) and ethylene production (B) by *Amaranthus caudatus* seeds at 24°C. The flasks were closed during 1, 2 or 3 d of incubation

reversed by ACC. The results presented may suggest that the inhibitory effect of Me-Ja on seed germination relates to the inhibition of ethylene production. As expected 10^{-3} M Me-Ja inhibited both seed germination and ethylene production (Fig. 6). Me-Ja at this concentration also reduced ACC synthesis (Fig. 7).

The results presented indicate that methyl jasmonate, inhibiting seed germination, increases the requirement for ethylene and may suggest the involvement of ethylene biosynthesis in the response of seeds to Me-Ja.

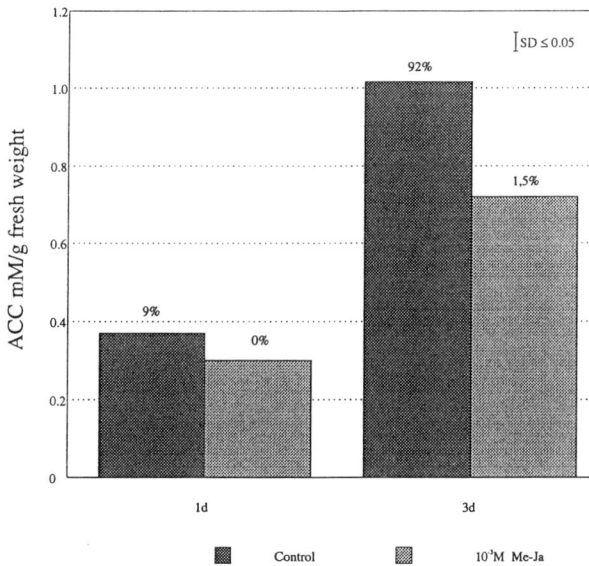

Figure 7. The influence of methyl jasmonate on ACC production by *Amaranthus caudatus* seeds after 1 or 3 d of incubation at 24°C. Germination percentages are shown above each bar

References

Berestetzky, V., Dathe, W., Daletskaya, T., Musatenko, L. and Sembdner, G. 1991. *Biochemie und Physiologie der Pflanzen* 187: 13–19.

Corbineau, F., Rudnicki, R.M. and Côme, D. 1988. *Plant Growth Regulation* 7: 157–169.

Daletskaya, T. V. and Sembdner, G. 1989. *Fisiologiya Rastenii* (Russ.) 36: 1118–1123.

Kepczynski, J. 1986. *Physiologia Plantarum* 67: 588–591.

Kepczynski, J. and Bialecka, B. 1994. *Plant Growth Regulation* 14: 211–216.

Lizada, M.C.C., Yang, S.F. 1979. *Analytical Biochemistry* 100: 140–145.

Parthier, B., Bruckner, C., Dathe, W., Hause, B., Herrmann, G., Knofel, H.D., Kramell, R., Lehmann, J., Miersch, O., Reinbothe, S., Sembdner, G., Wasternack, C. and Zur Nieden, U. 1991. In: *Progress in Plant Growth Regulation. Proceedings of the 14th International Conference on Plant Growth Substances*, pp. 276–286 (eds C.M. Karssen, L.C. Van Loon and D. Vreugdenhil). Dordrecht: Kluwer Academic Publishers.

Ranjan, R. and Lewak, S. 1992. *Physiologia Plantarum* 86: 335–339.

Staswick, P.E. 1995. In: *Plant Hormones Physiology, Biochemistry and Molecular Biology*, pp. 179–187 (ed. P.J. Davis). Dordrecht: Kluwer Academic Publishers.

58. Biochemical Events after Priming and Osmoconditioning of Seeds

K.-H. KOEHLER, B. VOIGT, H. SPITTLER and M. SCHELENZ

Botanisches Institut der E.-M.-Arndt-Universität, Grimmer Str. 88, D-17487 Greifswald, Germany

Abstract

Priming of white cabbage seeds (cv. Baltikol), followed by drying periods of different lengths, accelerates germination and increases germination percentage after a second imbibition. The desiccation tolerant phase seems to be extended over the onset of DNA synthesis in the radicles. Even seeds with visible radicle and split seed coat are desiccation tolerant. This could be important for the fluid drilling of the seeds. The longer the time of the first imbibition, the more intensive are protein, RNA and DNA syntheses. Pretreated seeds during the second imbibition period reach the high level of synthesis of these macromolecules observed in seeds imbibed only once. Prerequisites for germination processes are manifested during the priming which are responsible for the promotive effect of the treatment. Experiments with osmoconditioned (imbibed in mannitol) corn cockle seeds show a promotive effect on protein, RNA and especially DNA synthesis. Two-dimensional separation of soluble protein, however, did not show dramatic changes in the protein pattern of osmoconditioned seeds.

Introduction

In contrast to weeds and wild plants the seeds of cultivated plants are deliberately produced, carefully harvested and stored, and finally sown into a well prepared seed bed. Presowing treatments such as priming and osmoconditioning are used to improve germination behaviour, which is based on endogenous physiological and ecological 'memories' of the seed, especially under suboptimal natural conditions. What are the biochemical processes underlying these presowing treatments and how are these processes consciously influenced?

Materials and Methods

Commercial seeds of white cabbage (*Brassica oleracea* var. *capitata* L., cv. Baltikol) were used for the priming experiments in which seeds were soaked in water for a definite time, then redried for 24 h at a temperature of 30°C and stored at 8°C, before using them for a second imbibition.

The osmoconditioning experiments using mannitol solutions of different molarity were done with *Agrostemma githago* L. seeds harvested in the

R.H. Ellis, M. Black, A.J. Murdoch, T.D. Hong (eds.), Basic and Applied Aspects of Seed Biology, pp. 531–536.

Botanical Garden in Greifswald.

Incorporation experiments were carried out according to Hecker and Koehler (1979) and Voigt (1988). Isolated embryos were incubated in radio-active labelled precursor solutions for 2 h at 20°C in the dark.

Two-dimensional electrophoresis was done according to O'Farrell (1975) modified by Voigt (1988).

Results and Discussion

A hindrance to the use of the positive effects of priming for commercial application is the variability of results described in the literature. This variability is based on differences among species, varieties, and sometimes seed lots as well as on the treatment (temperature used for drying, time of imbibition etc.).

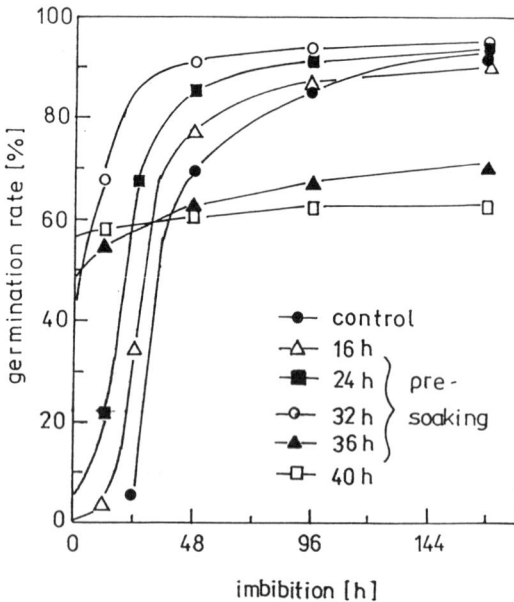

Figure 1. Germination rates of white cabbage seeds imbibed at 25°C in the dark after presoaking treatments of 16, 24, 32, 36 and 40 h

White cabbage seeds (cv. Baltikol) were imbibed for the times indicated in Figure 1 and redried at 30°C for 24 h. The seeds were then stored for another 48 h at 8°C. The moisture content of the seeds was between 15 and 18%, corresponding to the initial moisture content of the seeds before the experiment.

Figure 1 shows that the time until radicle protrusion decreases with increasing

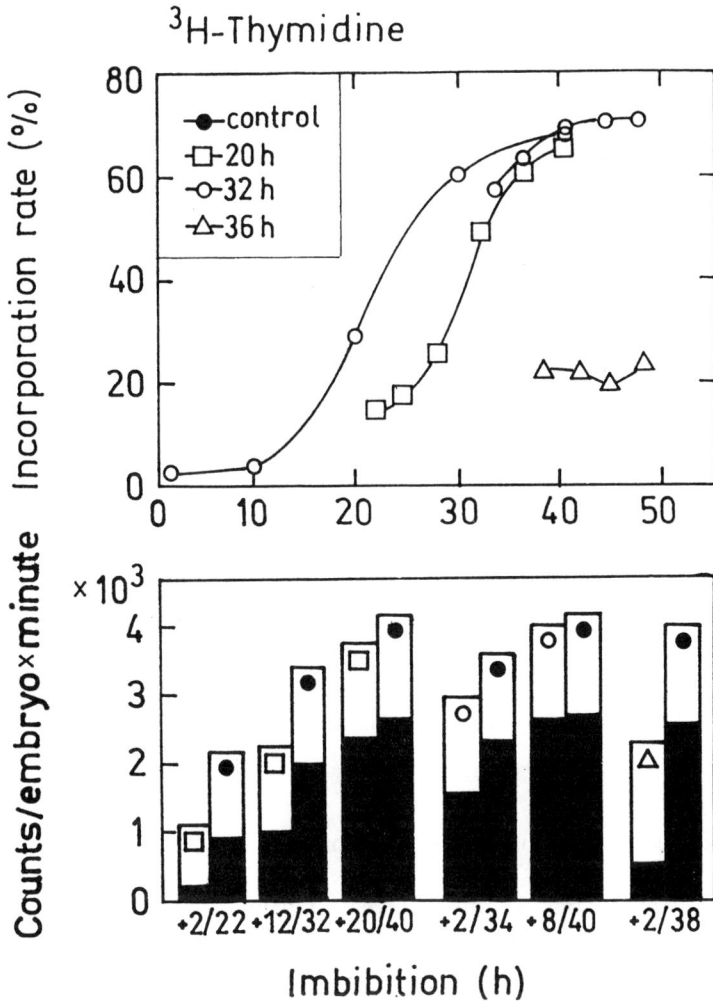

Figure 2. DNA synthesis, measured as incorporation rate of H-thymidine, in white cabbage seeds after different times of priming. The absolute values for incorporation (black) and uptake (white) are shown in comparison to the control

presoaking time. An optimum is reached after 32 h of presoaking, after which approximately 45% of the seeds have a split seed coat and the radicles are visible (similar to rape, see Schopfer and Plachy, 1985). This is in contrast to other species, like corn cockle (Koehler *et al.*, 1984). After 36 h of presoaking, however, a clear deterioration of the seeds is observed. The germination rate after 36 and 40 h of presoaking is approximately 30–40% lower than the control.

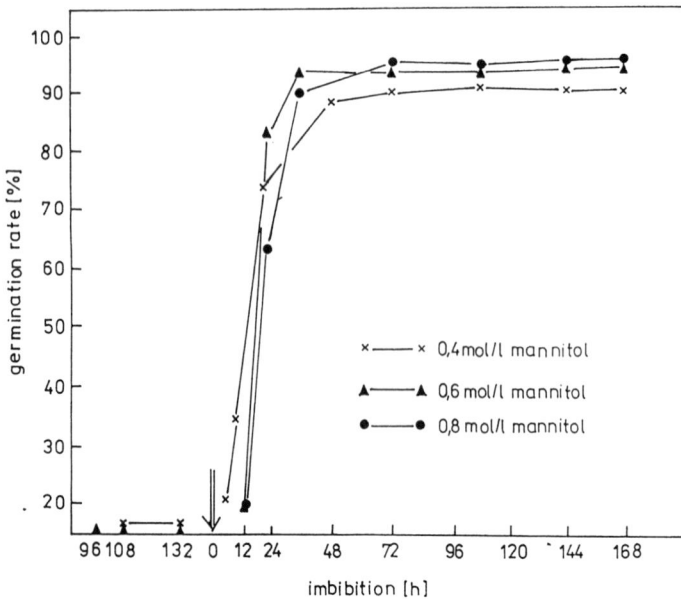

Figure 3. Germination of corn cockle seeds osmoconditioned in mannitol solutions of different molarity. Transfer to water after 132 h treatment in mannitol

In this connection it is of interest how long the desiccation tolerant period remains.

Figure 1 confirms that the time of presoaking and the redrying treatment are decisive for the promotive effect, i.e. acceleration of germination and a higher uniformity of germination (Bradford, 1986). The germination rate, however, cannot be increased by this treatment, which was found for a seed lot with diminished germinability (data not shown).

The incorporation rates for precursors of protein and RNA synthesis (data not shown) point to a fast recommencement of these syntheses during the second imbibition period. DNA synthesis shows only little reduction after redrying the seeds (Fig. 2). If the seeds are presoaked for 20 h and then redried, DNA synthesis during the second imbibition reaches within 2 h the level observed during the first imbibition after 17 h. If the seeds are presoaked for 32 h, DNA synthesis starts in the second imbibition at the level reached in seeds imbibed for 26 h. A decrease of DNA synthesis, however, is detectable after 36 h of presoaking. Though, for a short time even after radicle protrusion, redrying does not severely disturb DNA synthesis in the second imbibition period, i.e. the desiccation tolerant period in cabbage seeds seems to be extended over the onset of DNA synthesis (in contrast to *Agrostemma githago*, Koehler *et al.*, 1984).

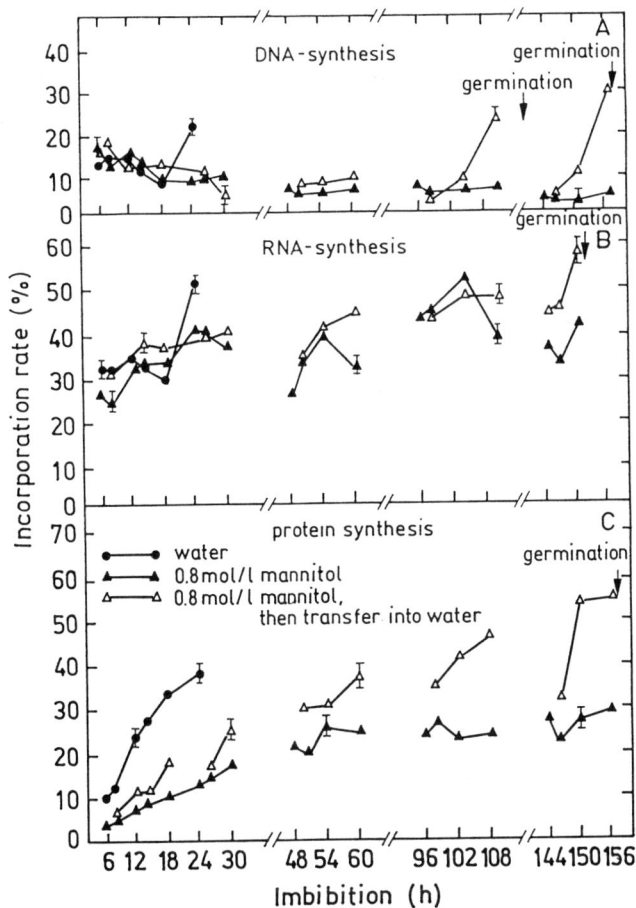

Figure 4. Protein, RNA and DNA synthesis of corn cockle seeds, osmoconditioned for different times, after transfer to water

Figure 3 shows the germination behaviour of corn cockle seeds during osmoconditioning in mannitol solutions with different water potential for 132 h. In water (not shown) radicle protrusion occurs first after about 20 h and after 72 h reaches approximately 98%. There was no germination during the osmoconditioning in the mannitol solutions, except in the concentration of 0.4 mol/l mannitol where 16% of the seeds were germinated after 132 h. The water potential of the mannitol solutions used in our experiments was too low to allow sufficient water uptake by the embryo to exceed the necessary threshold for radicle emergence. Transfer of the osmoconditioned seeds to water results in a very rapid and synchronous germination. The lower the mannitol concentration, the more rapid the germination after transfer to water, i.e. the necessary threshold of the water potential for germination is reached after different times of soaking. During the osmoconditioning treatment some prerequisites for macromolecule synthesis were developed (Fig. 4). The longer the osmoconditioning treatment, the more intense was macromolecular synthesis after transfer to water.

In preliminary experiments on protein patterns in *Agrostemma* embryos we could not find any proteins induced after osmoconditioning and transfer to water.

It can be concluded that priming and osmoconditioning influence germination by changing the water potential of the seeds as well as by allowing the development of prerequisites for pregerminative biochemical processes, among them the synthesis of macromolecules (Koehler *et al.*, 1992).

References

Bradford, K.J. 1986. *Horticultural Science* 21: 1105–1112.

Hecker, M. and Koehler, K.-H. 1979. *Developmental Biology* 69: 270–280.

Koehler, K.-H., Schelenz, M. and Mueller, U. 1984. In: *Ertragserhoehung und Qualitaetsverbesserung in der Saatgutproduktion 1. Kongress- und Tagungsberichte der Martin Luther Universitaet Halle-Wittenberg*: 24–37.

Koehler, K.-H., Schmerder, B. and Sheikhany, H. 1992. *Journal of Plant Physiology* 139: 528–532.

O'Farrell, P.H. 1975. *Journal of Biological Chemistry* 250: 4007–4021.

Schopfer, P. and Plachy, C. 1985. *Plant Physiology* 77: 676–686.

Voigt, B. 1988. Ph.D. Thesis. University of Greifswald, Germany.

59. Sinapine Leakage for Detection of Seed Quality in *Brassica*

P.C. LEE[1,2], A.G. TAYLOR[1] and D.H. PAINE[1]

[1]*Department of Horticultural Sciences, New York State Agricultural Experiment Station, Cornell University, Geneva, New York 14456;* [2]*Department of Soil, Crop and Atmospheric Sciences, Cornell University, Ithaca, New York 14853, USA*

Abstract

Sinapine leakage has been developed as a rapid seed viability test in *Brassica* and other Cruciferous seeds on a single seed basis, in which non-viable seeds leak fluorescent sinapine while viable seeds do not leak. Sinapine leakage was used to predict germination for 11 lots, and the prediction was significantly correlated with the actual germination. The major source of errors in this test is false-negatives (F–), i.e. those seeds that do not leak and do not germinate. The sinapine leakage index (SLI) was calculated to assess F– for any seed lot by dividing the number of non-germinable seeds that leaked by the total number of non-germinable seeds. The SLI was found to be related to seed coat permeability. Chemical analysis of seed coats was conducted on 4 lots with different SLI values. Cutin content was negatively correlated with SLI, and cutin may restrict the diffusion of sinapine throughout the seed coat. Sinapine leakage provided the basis for the development of a novel seed coating system to detect and upgrade seed quality. Seeds were first imbibed and then coated with a cellulose filler to adsorb sinapine leakage. The coated seeds are then dried and sorted by UV colour sorting equipment to upgrade seed quality by removal of fluorescent coated seeds.

Introduction

Sinapine (3,5-dimethoxy-4-hydroxycinnamoylcholine) is the choline ester of sinapic acid (Tzagoloff, 1963; Austin and Wolff, 1968) and is present in seeds of *Brassica* and other genus of the Brassicaceae (Fenwick, 1979; Bouchereau *et al.*, 1991). Sinapine leakage was observed from non-viable seeds during imbibition (Taylor *et al.*, 1988) and has been shown to be a better predictor of cabbage (*B. oleracea* L. var. *capitata*) seed viability than electrolyte conductivity (Hill *et al.*, 1988). Sinapine leakage has been shown to be related to seed quality in cauliflower and broccoli (*B. oleracea* L. var. *botrytis* and *italica*, respectively) (Lee and Taylor, 1995). Hence, detecting sinapine leakage from non-germinable seeds could be used as a rapid seed viability test in *Brassica*.

The accuracy of using sinapine leakage as a method to detect seed germinability was studied by Lee and Taylor (1995). The non-leaking non-germinable seeds were called false negatives (F–) and became the major source of errors in predicting germination. The sinapine leakage index (SLI) was calculated to assess F– for any seed lot by dividing the number of non-germinable seeds that

R.H. Ellis, M. Black, A.J. Murdoch, T.D. Hong (eds.), Basic and Applied Aspects of Seed Biology, pp. 537–545.
© *1997 Kluwer Academic Publishers, Dordrecht. Printed in Great Britain.*

leaked by the total number of non-germinable seeds. Sinapine leakage was found to be related to seed coat permeability (Taylor *et al.,* 1993b), and seed lots with more permeable seed coats had higher SLI values (Lee and Taylor, 1995). The SLI of a seed lot was found to decrease during the early phases of ageing. This decline was attributed to non-viable tissues on the axis which would directly affect germinability without greatly contributing to leakage.

A novel seed coating system was developed to exploit sinapine leakage as a method to non-destructively detect and upgrade *Brassica* seed quality (Taylor *et al.,* 1990; 1994). The seed coating used filler materials that adsorb sinapine leakage from hydrated seeds, identifying non-viable seeds by fluorescence (Taylor *et al.,* 1991). Sorting was performed manually under UV light, and seeds were sorted into two fractions based on fluorescence. Sowing the non-fluorescent fraction resulted in a greater percentage of normal seedlings than the non-sorted seeds. Seed conditioning was performed by sorting fluorescent from non-fluorescent coated seeds with a UV colour sorter to demonstrate commercial scale sorting (Taylor *et al.,* 1993a).

In this work, the relation between actual germination and the germination predicted by sinapine leakage was studied. Chemical analysis was conducted on seed coat to identify the compound that influenced seed coat permeability to sinapine. Finally, seed coating technologies were used to integrate sinapine leakage for upgrading seed quality.

Materials and Methods

Seed Lots

Seeds of cabbage 'Condor', cauliflower 'Blue Diamond', hybrid broccoli 'Excelsior', and radish 'Champion' (*Raphanus sativus*) were obtained from Harris Moran Seed Co. Cabbage 'Rio Verde' and broccoli 'Commander' were supplied by Northrup King Co. Two cabbage varieties, 'Resistant Golden Acre' and 'Copenhagen Market Early' (CME), and two cauliflower varieties, 'Snowball X' and 'Snowball 76', were provided by Ferry-Morse Seed Co. Broccoli 'Packman' and 'Pirate', and cauliflower 'Paleface' were obtained from Germain's Inc., UK. Seed lots were obtained in different years and stored at 5°C and 30% relative humidity (RH) and tested for the percent germination following the Rules for Testing Seeds (AOSA, 1988).

Seed Ageing and Fluorescent Leakage

The seeds of 'CME' lot were equilibrated at 75% RH in a Plexiglas chamber at room temperature. The relative humidity was maintained by a solution of glycerol and water (Forney and Brandl, 1992) with 1.151 specific gravity. The water activity of the seeds was verified with a Protimeter dewpoint meter (Protimeter plc, Marlow, Bucks, UK). The moisture content corresponded to

9.4% as determined by a Sinar Datatec P25 moisture analyzer (Sinar Technology Ltd., Weybridge, Surrey, UK). The equilibrated seeds were then heat sealed in aluminium bags and submerged in a 45°C water bath for varying periods of time. At the end of the ageing periods, the seeds were dried under ambient conditions prior to further use. Four replications of 50 aged seeds for different ageing periods were placed on moistened white blotters in a 10.2×10.2 cm germination container. These containers were maintained at 20–30°C (AOSA, 1988). Seeds on blotters were visually examined for fluorescent leakage and marked under UV light every 24 h for 3 days (d). The percentage of seeds that exhibited radicle emergence (defined as germination) was recorded after 3 d.

Prediction of Germination

Four replications of 50 seeds for 11 lots were placed in germination containers and examined for the fluorescent leakage as described above. The predicted germination (PG) was calculated as: (%) $PG = [1 - (f/n)] \times 100$, where f is the number of seeds that exhibited fluorescent leakage, and n is the number of seeds tested.

SLI and Compositional Analysis on Seed Coats

The sinapine leakage index of each lot was calculated as: $SLI = x_f/x_t$, where x_t is the total number of non-germinable seeds, and x_f is the number of non-germinable seeds exhibiting fluorescent leakage. Compositional analysis on the seed coats was carried out by the Division of Nutritional Sciences, Department of Animal Science at Cornell University, Ithaca, NY. Seed coats of selected seed lots were removed from enough seeds to produce 10 g seed coats for each lot. Analyses for pectin, lignin, cellulose, hemicellulose and cutin were carried out following Van Soest *et al.* (1991). The tannin content was determined following Porter *et al.* (1986) using condensed tannins from *Desmodium ovalifolium* as the standard. Data were expressed as percentages.

Seed Coating and Sorting

Thirty grams of seed were soaked in aerated water at 25°C for 4 h. Seeds were then coated in a rotating coating pan using a mixture containing 30 g of micro crystalline cellulose 'Avicel ph101' (FMC Corp. Newark, Delaware) as the filler and 1.5 g of carboxy methyl cellulose 'CMC 7H' (Hercules, Inc., Wilmington, Delaware) as the binder and using distilled water as necessary. After coating, coated seeds were dried at 40°C for 1 h in a convection air oven. Four replications of 50 each were randomly selected for the 'non-sorted' fraction for germination. Sorting was then accomplished by hand under UV light. Those seeds with no fluorescent expression were selected as the 'sorted' fraction. The seed coatings were washed off with cold water prior to the germination test. Non-coated seeds were used as the control. Germination tests were performed on blotter paper at 25°C and the first count of radicle emergence was taken after 3 d. The final germination was recorded after 10 d and the percent normal seedlings determined.

Results

Sinapine leakage was observed from non-germinable seeds as a fluorescent halo surrounding the imbibed seed on cellulose blotters under UV light, and these observations are consistent with our earlier investigations. A sample of cabbage seeds was aged for periods of time to investigate the relationship between fluorescent leakage and germination. The percentage germination (radicle emergence) revealed a negative sigmoidal shaped curve with ageing, and there was a concomitant increase in the percentage of seeds that leaked sinapine with the loss of germination (Fig. 1).

The germination was predicted from the number of leaking seeds in the test sample. The predicted germination and the actual germination (normal + abnormal) of the 11 seed lots were significantly correlated at the 1% level ($r = 0.88**$) (Table 1). The difference between predicted and actual germination was less than 5 percentage points for eight of the 11 lots.

The sinapine leakage index, SLI, was determined for different seed lots and then four lots representing a range of SLI values from 0.2 to 1.0 were used for

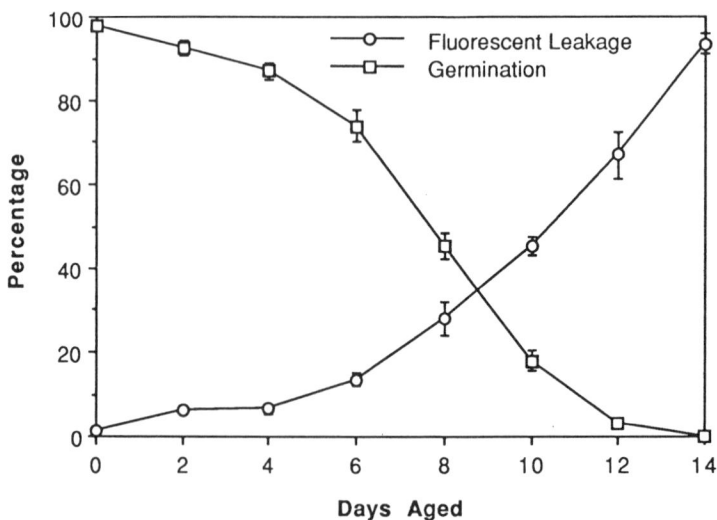

Figure 1. The effect of ageing on the percent fluorescent leakage and germination of cabbage 'CME'

Table 1. The predicted and actual germination of 11 lots

Crop	Variety	Germination (%)	
		Predicated	Actual
Cabbage	Hybrid condor	81	82
	Rio verde	91	86
	Resistant golden acre	77	79
	Copenhagen market early	99	92
Cauliflower	Snowball X	96	84
	Snowball 76	91	88
	Blue diamond	99	98
Broccoli	Commander	87	83
	Hybrid excelsior	83	80
	Packman	65	71
	Pirate	85	89

the analysis of seed coat composition (Table 2). A radish sample was used as this lot was found to have a SLI of 1.0. Pectin, a minor component, ranged from 4% to 6% and was correlated with SLI ($r = 0.98*$). Tannins, which included insoluble and soluble tannins, composed 5% to 10% of the seed coat. No significant correlation was shown between tannins and SLI, although insoluble tannins were negatively related with SLI. Cellulose and the crude lignin were the two

Table 2. Seed coat compositional analysis of four lots with different SLI values

Compound	SLI			
	0.2	0.74	0.91	1.0
Pectins	3.9	4.9	5.7	5.9
Tannins	8.8	8.9	4.9	10.3
Insoluble tannins	6.7	5.7	3.8	3.3
Soluble tannins	2.1	3.2	1.1	7.0
Hemicellulose	9.1	7.8	11.1	12.0
Cellulose	24.6	24.0	24.3	18.1
Crude lignin	31.6	28.1	20.2	21.1
Phenolic lignin	10.4	15.6	13.8	17.6
Cutin	21.2	12.5	6.4	3.5

Data expressed as a percent

major components of the seed coat, and no consistent trends were found between these two materials and SLI. Cutin, a component of the crude lignin, was found in high concentrations in seed lots with low SLI, while lots with high SLI values contained little cutin. Cutin content was negatively correlated with SLI ($r = -0.98*$) based on our survey of four lots.

A sequence of events that integrates sinapine leakage and coating technologies to upgrade *Brassica* seed quality is shown (Fig. 2). Leakage of sinapine from hydrated non-germinable seeds was adsorbed by the cellulose during coating. Coated seeds were dried, and fluorescent coatings were observed from non-

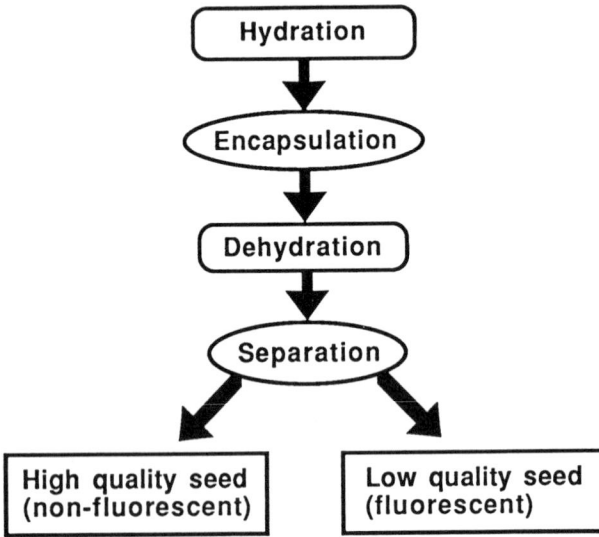

Figure 2. A sequence of events integrating sinapine leakage and coating technologies to upgrade *Brassica* seed quality

germinable seeds under UV light.

The effect of coating and sorting by sinapine leakage on germination of five lots was studied (Table 3). By removing the fluorescent fraction, the percentage of germinated seeds for varieties CME, Paleface and Pirate increased to more than 95%. The initial germination of Packman was very low; however, after sorting the percentage of germinated seeds and normal seedlings increased by 41 and 29 percentage points, respectively. In general, the percentage of germinated seeds after 3 d and normal seedlings for a lot were increased in the sorted fraction compared to the non-coated control or coated but non-sorted fraction.

Table 3. The effect of coating and sorting by sinapine leakage on germination of five varieties of *Brassica*

Variety	Radical emergence (%)			Normal seedlings (%)		
	Non-coated	Coated non-sorted	Coated sorted	Non-coated	Coated non-sorted	Coated sorted
CME	85 b	88 b	87 a	88 b	95 a	98 a
Paleface	79 b	80 b	96 a	71 b	66 b	84 a
Snowball 76	77 ab	72 b	86 a	81 b	75 c	86 a
Pirate	82 b	85 b	95 a	79 b	79 b	93 a
Packman	47 b	53 b	88 a	52 b	54 b	81 a

Mean separation for radicle emergence and normal seedlings within rows by LSD (5%)

Discussion

Sinapine is a reserve material formed during seed development on the mother plant and has been found to be the major aromatic choline ester in *Brassica* seeds (Tzagoloff, 1963; Bouchereau *et al.*, 1991). Sinapine has been found in all *Brassica* lots examined in our laboratory including various Plant Introduction lines from the USDA. Fluorescent leakage from *Brassica* seeds in association with non-viability was reported by Kugler (1952) and Zhang and Yan (1964). Taylor *et al.* (1988) verified that the major compound responsible for the observed fluorescence was sinapine. Based on our experience with different *Brassica* crops and varieties, the loss of germination was accompanied with an increase in the percentage of seeds that leaked sinapine as illustrated by a single lot of cabbage (Fig. 1). Therefore, sinapine leakage has the potential to be used as a rapid test to predict germination.

There are several advantages of using sinapine leakage as a viability test for *Brassica,* and these merits satisfy the criteria for seed testing suggested by Matthews (1981). First, the expression of fluorescent leakage was consistent throughout different crops and varieties regardless of the degree of seed ageing. Second, sinapine leakage occurs after several hours of imbibition and requires no pre-treatment. In contrast, the tetrazolium test generally requires pre-treatments such as pre-soaking and/or cutting. Third, sinapine leakage provides information on a single seed basis. Hence, this method provides a good way of estimating viability. Finally, the method is non-destructive, and combined with the fact that seeds are desiccation tolerant during the early phases of germination, sinapine leakage can be exploited with other seed technologies to upgrade seed quality. A system that integrates hydration, coating and conditioning (Fig. 2) results in a process that can be adapted for commercial seed conditioning.

The factors influencing the accuracy of the use of sinapine leakage are F– and false positives (F+) (Lee, 1994; Lee and Taylor, 1995). Germination could be over estimated due to the presence of F– seeds or be under estimated by the F+ seeds. Seedlings from the F+ seeds were generally found to be abnormal or died later in the test, and the presence of F+ seeds was generally low (Lee and Taylor, 1995). The major source of errors was the F– seeds, and our data indirectly illustrates the effect of F– seeds on germination prediction and upgrading after sorting. The leakage test over estimated germination for Snowball X and Rio Verde in comparison with the actual germination (Table 1). Both lots had a high percentage of F– seeds, and the SLI values were low, 0.2 and 0.4, respectively (data not shown). Upgrading of germination was measured in all lots in the seed coating study; however, F– seeds would reduce the efficacy of this enhancement.

The partial answer for the presence of F– seeds may be found in the permeability of the seed coat to diffusion of sinapine. Sinapine leakage was found to be retarded by the seed coat, and seed coat cracks increased the rate of leakage (Taylor *et al.*, 1993b). A negative correlation was found between SLI of 13 lots with seed coat permeability as measured by the T50 (time for 50% of heat killed seed to leak) (Lee and Taylor, 1995). A semi-permeable layer has been found in seed coats of leek (*Allium porrum* L.), onion (*Allium cepa* L.), tomato (*Lycopersicon esculentum* L.) and pepper (*Capsicum annuum* L.); however, a semi-permeable layer was not found in the seed coat of cabbage (Beresniewicz *et al.*, 1995a). The major compounds responsible for the semi-permeability were cutin in leek and onion or suberin in tomato and pepper (Beresniewicz *et al.*, 1995b). The chemical analysis of the seed coats in this study revealed that lower SLI lots had higher cutin contents. Therefore, cutin in the seed coat may restrict the leakage of sinapine from the embryo to the environment. Further studies are needed on seed coat morphology and histochemistry in *Brassica*.

In conclusion, sinapine provides a convenient marker due to its detection by fluorescence. Sinapine leakage can be used to estimate seed viability as a rapid laboratory test and to upgrade seed quality on a commercial scale. Sinapine provides a model system to both assess and enhance seed germinability on a single seed basis.

Acknowledgement

This project was partially funded from a grant from Germain's Inc., UK.

References

AOSA (Association of Official Seed Analysts). 1988. *Journal of Seed Technology* 12: 1–122.
Austin, F. L. and Wolff, I. A. 1968. *Journal of Agricultural and Food Chemistry* 16: 132–135.

Beresniewicz, M. M., Taylor, A. G., Goffinet, M. C. and Koeller, W. D. 1995a. *Seed Science and Technology* 23: 135–145.

Beresniewicz, M. M., Taylor, A. G., Goffinet, M. C. and Terhune, B. T. 1995b. *Seed Science and Technology* 23: 123–134.

Bouchereau, A., Hamelin, J., Lamour, I., Renard, M. and Larher, F. 1991. *Phytochemistry* 30: 1873–1882.

Fenwick, G. R. 1979. *Journal of the Science of Food and Agriculture* 30: 661–663.

Forney, C. F. and Brandl, D.G. 1992. *Horticultural Technology* 2: 52–54.

Hill, H. J., Taylor, A. G. and Huang, X. L. 1988. *Journal of Experimental Botany* 39: 1439–1448.

Kugler, I. 1952. *Naturwissenschaften* 39: 213.

Lee, P. C. 1994. M.S. Thesis. Cornell University, NY, U.S.A.

Lee, P. C. and Taylor, A. G. 1995. *Plant Varieties and Seeds* 8: 17–28.

Matthews, S. 1981. *Seed Science and Technology* 9: 543–551.

Porter, L. J., Hrstich, L. N. and Chan, B. G. 1986. *Phytochemistry* 25: 223–230.

Taylor, A. G., Churchill, D. B., Lee S. S., Bilsland, D. M. and Cooper, T. M. 1993a. *Journal of the American Society for Horticultural Science* 118: 551–556.

Taylor, A. G., Hill, H. J., Huang, X. L. and Min, T. G. 1990 and 1994. U.S. patent no. 4,975,364, and European Patent Convention no. 364,952.

Taylor, A. G., Huang, X. L. and Hill, H. J. 1988. *Journal of Experimental Botany* 39: 1433–1438.

Taylor, A. G., Min, T. G. and Mallaber, C. A. 1991. *Seed Science and Technology* 19: 423–434.

Taylor, A. G., Paine, D. H. and Paine, C. A. 1993b. *Journal of the American Society for Horticultural Science* 118: 546–550.

Tzagoloff, A. 1963. *Plant Physiology* 38: 202–206.

Van Soest, P. J., Robertson, J. B. and Lewis, B. A. 1991. *Journal of Dairy Science* 74: 3583–3597.

Zhang, G. H. and Yan, Q. S. 1964. *Bulletin of Plant Physiology* 3: 21–25.

60. Temperature and the Rate of Germination of Dormant Seeds of *Chenopodium album*

A.J. MURDOCH and E.H. ROBERTS

Department of Agriculture, The University of Reading, Earley Gate, P.O. Box 236, Reading RG6 6AT, UK

Abstract

Rate of germination is often used as a measure of seed vigour. But germination may be retarded not only by low seed quality but also by dormancy.

The effect of dormancy on the rate of germination of seeds of *Chenopodium album* was examined at a range of constant and alternating temperatures on a temperature gradient plate. Although alternating temperatures overcame much of the dormancy in seeds of *C. album*, the mean rate of germination ($\Sigma[n]/\Sigma[Dn]$ where n is the number of seeds germinating on day D) was largely a function of mean temperature. The effects on rate of germination of both thermoperiod (time spent at the warmer temperature in diurnal alternating-temperature cycles) and temperature amplitude (daily maximum minus daily minimum) at sub-optimal temperatures (at which the maximum daily temperature was less than 25°C) were statistically significant. The amount of variation explained was small however, so that the rate of germination was relatively independent of the final germination percentage at a given mean temperature. At mean temperatures over 25°C, the rate of germination decreased with increase in mean temperature.

Introduction

The rate of germination of an individual seed cannot be measured, but it is possible to time how long it takes to complete the process of germination to an endpoint such as radicle emergence. The rate of progress towards germination can then be calculated by taking the reciprocal of this period.

Using this approach to quantify the developmental physiology of plants, it is commonly found that the rate of development is a linear function of temperature (Roberts, 1988). With respect to the germination of non-dormant seeds at sub-optimal constant temperatures, there is a positive linear relation of the rate of germination to temperature between the base temperature, T_b, (at which the rate is zero since no seeds germinate), and the optimum temperature, T_o, (at which, by definition, seeds germinate most rapidly). At supra-optimal constant temperatures, there is another, but negative, linear relationship between the optimum temperature and the ceiling temperature, T_c at which the rate is again zero (Roberts, 1988).

The effect of fluctuating temperatures can be evaluated by quantifying the rate of germination as a function of the mean temperature of the treatment. Garcia-Huidobro *et al.* (1982) showed that wide temperature fluctuations may result in

R.H. Ellis, M. Black, A.J. Murdoch, T.D. Hong (eds.), Basic and Applied Aspects of Seed Biology, pp. 547–553.
© *1997 Kluwer Academic Publishers, Dordrecht. Printed in Great Britain.*

increases in the germination rate of non-dormant seeds above that exhibited by seeds germinated at constant temperature. By contrast, Ellis and Barrett (1994) showed that non-dormant seeds of *Lens culinaris* Medikus (lentil) behaved consistently with the hypothesis that rate of germination responds instantaneously to the current temperature and that alternating temperatures did not increase the rate of germination. The only exception was in alternations which included sub-zero temperatures where germination was delayed.

This paper examines the rate of germination of dormant seeds. In such seeds, final germination may vary greatly between constant and fluctuating temperatures at the same mean temperature depending on the extent to which the temperature treatment relieves dormancy. *Chenopodium album* is an ideal subject for such studies since the dormant seeds are very sensitive to fluctuating temperatures and germination at sub-optimal mean temperatures varies with the amplitude of temperature alternations (Murdoch *et al.*, 1989). It is, therefore, possible to investigate whether the rate of germination at a given mean temperature varies with the relief of dormancy caused by the temperature treatment. If the rate were to vary in this way, the periods prescribed for germination tests on dormant seeds might have to take account of whether constant or alternating temperatures are used.

Materials and Methods

Seeds

Seeds of *Chenopodium album* L. were collected from a natural infestation on arable land at Sonning-on-Thames, Berkshire, UK during a dry afternoon in September, 1974. After sieving, the fruits were dried for 10 d on a shaded glasshouse bench. The perianth of *C. album* was then rubbed off gently and the seeds were cleaned using a seed blower. The seeds were then stored in closed glass containers at 12.5% moisture content at –20°C for about 10 years before use. This storage environment helped to preserve seed dormancy and viability.

Germination Tests

Germination of *C. album* was tested for a period of 24 d on a two-dimensional temperature gradient plate (Murdoch *et al.*, 1989) employing gradients of 3–40°C in each direction, which were applied daily for 8 and 16 h, respectively. The working area of the surface of the plate was subdivided by a polystyrene matrix which produced a systematic arrangement of 156 alternating and 13 constant temperature treatments. One hundred seeds were germinated in each treatment on top of Whatman 3MM chromatography paper which was moistened with 0.01 M potassium nitrate and continually exposed to diffuse laboratory light. The minimum criterion of germination was 2 mm root or shoot growth. Germinated seeds were counted and removed frequently in order to construct germination progress curves.

Analysis of Data

The mean time to germination, *g*, for the seeds in each treatment was calculated as

$$g = [\Sigma(D.n)]/\Sigma n$$

where, *n*, is the number of seeds germinating *D* days after the commencement of imbibition. The reciprocal of the mean time gives the mean rate of progress towards germination, $1/g$.

Results

Germination of *Chenopodium album* seeds along the constant temperature gradient (0°C amplitude) diagonal of the temperature gradient plate increased with increase of temperature to about 25°C, the optimal constant temperature for germination, and then decreased with further increase in temperature. Alternating temperatures caused significant loss of dormancy in regimes with a mean temperature lower than 25°C (Murdoch *et al.*, 1989). This loss of dormancy depended on the amplitude of alternation and the thermoperiod (defined as the time spent at the warmer temperature in consecutive 24 h temperature cycles). Examination of the germination progress curves of sub-optimal treatments at mean temperatures of 9 or 15°C shows great differences in loss of dormancy (final germination percentage) between treatments (Figs. 1,2).

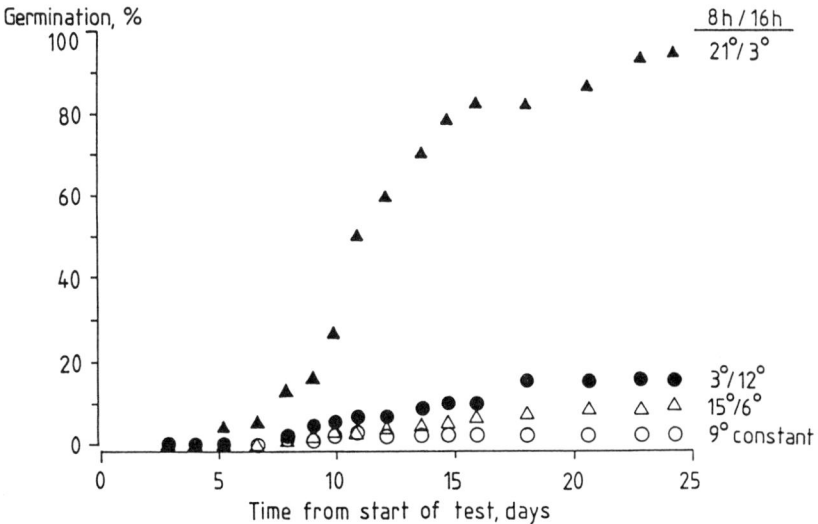

Figure 1. Progress of germination of *Chenopodium album* at a mean temperature of 9°C. Single replicates of c. 100 seeds were germinated at (○) 9°C constant, or at alternating temperatures (8 h/16 h) of (△) 15°/6°C; (●) 3°/12°C; (▲) 21°/3°C

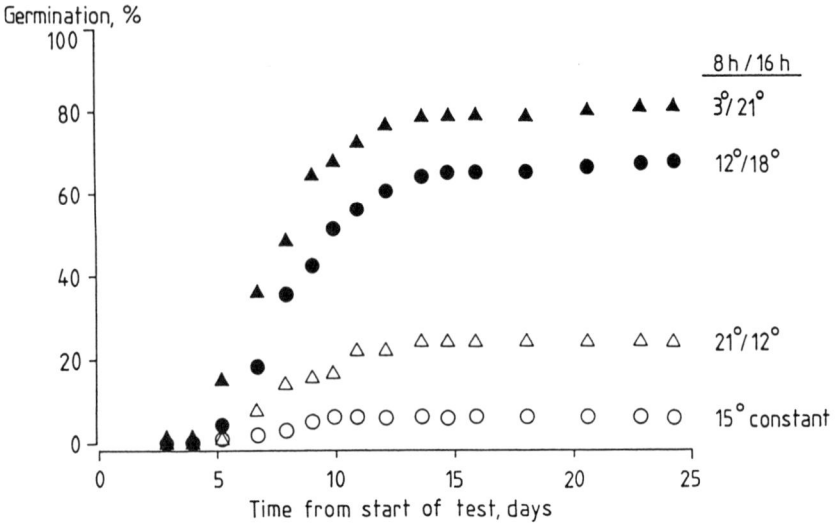

Figure 2. Progress of germination of *Chenopodium album* at a mean temperature of 15°C. Single replicates of c. 100 seeds were germinated at (○) 15°C constant, or at alternating temperatures (8 h/16 h) of (△) 21°/12°C; (●) 12°/18°C; (▲) 3°/21°C

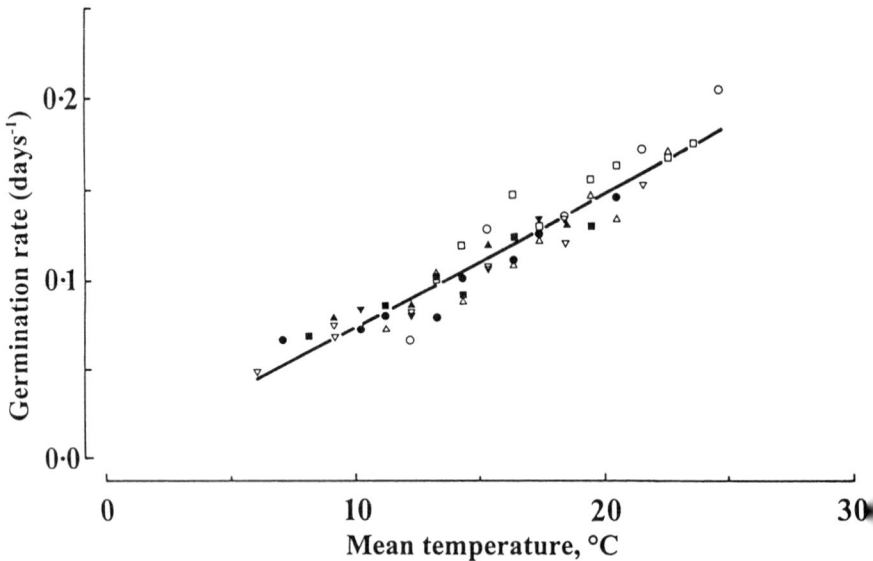

Figure 3. Mean rate of germination of *Chenopodium album* seeds at sub-optimal temperatures. Amplitudes (°C) of alternating temperatures are as follows: (○) 0°; (□) 3.1°; (△) 6.1°; (▽) 9.2°; (●) 12.3°; (■) 15.4°; (▲) 18.5°; (▼) 21.6°. The common line, $1/g = -0.000737 + 0.00746T$, was fitted according to the analysis (b) in Table 1

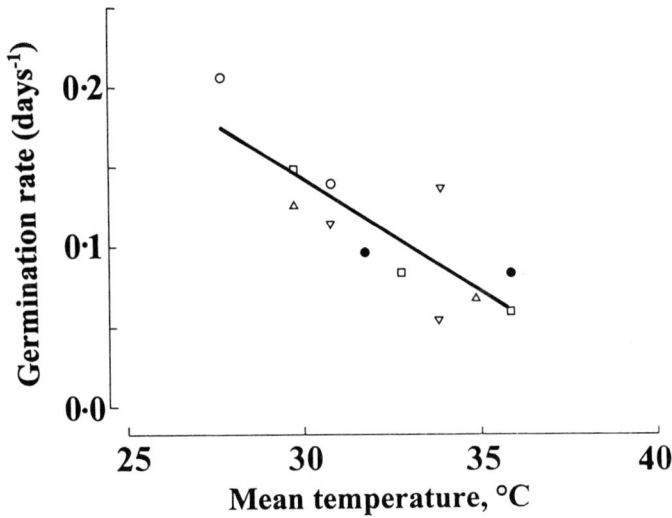

Figure 4. Mean rate of germination of *Chenopodium album* seeds at supra-optimal temperatures. Other details as Figure 3. The common line is $1/g = 0.562 - 0.0140T$

Table 1. Analysis of variance of mean rates of germination of *Chenopodium album* seeds germinated in sub-optimal temperature treatments (at which the maximum daily temperature was $\leqslant 24.6°C$)

Model	Degrees of freedom	Sum of squares	F-ratio
Total	50	0.06160	
(a) Separate lines for each amplitude and thermoperiod	23	0.00164	
(b) Common lines for all amplitudes and thermoperiods	49	0.00564	
Difference (b–a)	26	0.00400	2.15*
(c) Linear effect of amplitude which is the same at each thermoperiod	47	0.00421	
Difference (c–a)	24	0.00257	1.50[NS]
Difference (c–b)	2	0.00143	7.97*

*Significance: $p < 0.05$; [NS] = not significant

Mean rates of germination of all sub-optimal treatments (where $T_{max} < T_o$) were nevertheless largely a function of mean temperature which accounts for over 90% of the variation (Fig. 3; Table 1). However, a simple mean temperature model differs significantly from the general model in which separate lines for each amplitude at each thermoperiod are fitted to rate of germination as a function of mean temperature. Including linear terms in amplitude (model c, Table 1), significantly improves the goodness of fit, and the residual sum of squares is not significantly different from the general model.

At supra-optimal temperatures (where $T_{max} \geqslant T_o$), germination rate decreased with mean temperature, apparently linearly (Fig. 4). However, there is a wide scatter of points and any prediction of the time taken for seeds to germinate at supra-optimal temperatures is likely to be imprecise.

Discussion

Although the amount of data is limited, it is interesting both with a view to predicting field emergence and when prescribing temperature regimes for seed testing, that the mean rate of germination in sub-optimal treatments largely depended on the mean temperature and not on the efficacy of alternating temperatures to relieve dormancy. Although these results superficially support those of Ellis and Barrett (1994), a conflict with those of Garcia-Huidobro *et al.* (1982) as discussed in the introduction is not necessarily true. In this analysis, we have not considered trans-optimal treatments (i.e. alternating-temperature treatments in which the lower temperature is below the optimum and the upper temperature above it), and we report variable results at supra-optimal temperatures. At sub-optimal temperatures, the small effect of amplitude does not appreciably alter the positive linear relationship between the mean rate of germination and mean temperature.

If they were found to be general, the implications of these observations for seed testing are important: the duration of the germination tests at sub-optimal temperatures should depend on mean temperature and not on the loss of dormancy achieved. Clearly, the aim is to relieve as much dormancy as possible, and for this purpose it was shown previously (Murdoch *et al.*, 1989) that an alternating temperature of wide amplitude and cool mean temperature is appropriate; but having assured maximum loss of dormancy in this way, the duration of the test is not dependent on success in relieving dormancy and should be inversely related to the mean temperature chosen. Biologically, this conclusion is surprising since loss of dormancy due to alternating temperatures is not instantaneous: seeds either need a pretreatment before exposure to a temperature change (Totterdell and Roberts, 1979; 1981) or loss of dormancy progresses as a function of the number of temperature cycles received (Totterdell and Roberts, 1980; Probert *et al.*, 1987). Moreover, there is considerable seed-to-seed variation in the period of pretreatment or in the number of cycles needed. The interpretation of the results is nevertheless straightforward: although the absolute number of seeds losing dormancy varied with mean temperature, amplitude and thermoperiod, the relative rate of loss of dormancy largely depended on mean temperature (and probably also, the number of diurnal temperature cycles experienced).

Even when they germinated it is probable that most seeds in the population initially possessed some dormancy which was lost during the test, so that in this experiment a delay in germination between dormant and non-dormant seeds might not be detected. Hence the results presented here for a given seed lot do not imply that the rate of germination is independent of seed dormancy. A further important qualification is that the rate of germination does vary at given mean temperatures where different dormancy breaking agents are active. For example, the rate of germination of *Panicum maximum* tends to be slower in water compared to potassium nitrate (Goedert, 1984).

References

Ellis, R.H. and Barrett, S. 1994. *Annals of Botany* 74: 519–524.

Garcia-Huidobro, J., Monteith, J.L. and Squire, G.R. 1982. *Journal of Experimental Botany* 33: 297–302.

Goedert, C.O. 1984. Ph.D. Thesis, The University of Reading, U.K.

Murdoch, A.J., Roberts, E.H. and Goerdert, C.O. 1989. *Annals of Botany* 63: 97–111.

Probert, R.J., Gajjar, K.H. and Haslam I.K. 1987. *Journal of Experimental Botany* 38: 1012–1025.

Roberts, E.H. 1988. In: *Plants and Temperature. Symposia of the Society of Experimental Biology*, (Eds S.P. Long and F.I. Woodward). Company of Biologists Ltd., Cambridge.

Totterdell, S. and Roberts, E.H. 1979. *Plant Cell and Environment* 2: 131–137.

Totterdell, S. and Roberts, E.H. 1980. *Plant Cell and Environment* 3: 3–12.

Totterdell, S. and Roberts, E.H. 1981. *Plant Cell and Environment* 4: 75–80.

61. Hydration up to Threshold Levels as the Triggering Agent of the Processes Preparing Germination in Quiescent Seeds

N.V. OBROUCHEVA

Institute of Plant Physiology, Russian Academy of Sciences, Botanical Street 35, 127276 Moscow, Russia

Abstract

Water operates as a triggering agent in quiescent seeds: in the course of seed hydration to threshold levels the metabolic systems are activated step by step culminating in the initiation of cell elongation. Two experimental approaches were used: (1) close inspection of a particular process in dependence of water content and (2) incubation of imbibed seeds under the conditions preventing their further hydration (humid air or polyethylene glycol 6000). In the experiments with pea seeds the following threshold hydration levels (% fr wt) were shown: 45% – starch degradation, 45% – final activation of respiration, 52–55% degradation of reserve proteins. With the axial organs of broad bean seeds, the accumulation of osmotic solutes was evident at 65–68% and activation of H^+-ATPase in the plasmalemma resulting in acid growth at 68–70%. By close inspection of literature threshold hydration levels were found for other main metabolic systems. A general scheme of early germination is presented: hydration up to threshold water content levels triggers the main metabolic systems in the axial organs step by step, thus performing the first (metabolic activation) and the second (preparation for cell elongation) stages of early germination. In the cotyledons only metabolic activation occurs triggered by water inflow.

Introduction

As activation of metabolism and visible germination occur only in hydrated seeds, water is routinely considered as a prerequisite to the germination. However, its role far exceeds its availability during seed germination. Here, water entering the seed is considered as an agent triggering not only various metabolic systems and the processes preparing for cell elongation in the axial organs, but the early germination itself.

Triggering of any system implies its sharp transition from inactive to active state. For this reason, the triggering role of water can be revealed by close inspection of activation of any process as a function of water content (wc). If any system does not function within a certain water content range, but then is abruptly activated by a small rise in tissue hydration level, it can be said with assurance that the system is triggered by water content increase to a threshold level.

R.H. Ellis, M. Black, A.J. Murdoch, T.D. Hong (eds.), Basic and Applied Aspects of Seed Biology, pp. 555–562.
© *1997 Kluwer Academic Publishers, Dordrecht. Printed in Great Britain.*

The triggering role of water can be confirmed if the seed wc are maintained at constant levels for a long period of time by seed incubation either in humid air or in PEG solution (polyethylene glycol, an osmoticum). At wc below the threshold level, the seeds are unable to begin the process under study in spite of sufficiently long incubation, but at higher wc the seed has already started the process and continued it.

Various metabolic processes in germinating seeds are considered in this paper in terms of their triggering by seed hydration to threshold levels.

Results

A clear example of the triggering role of tissue hydration is starch degradation studied in the cotyledons and axial organs of imbibing pea seeds (Obroucheva *et al.*, 1988). Figure 1 shows the decrease in starch content as a function of tissue wc (% fr wt).

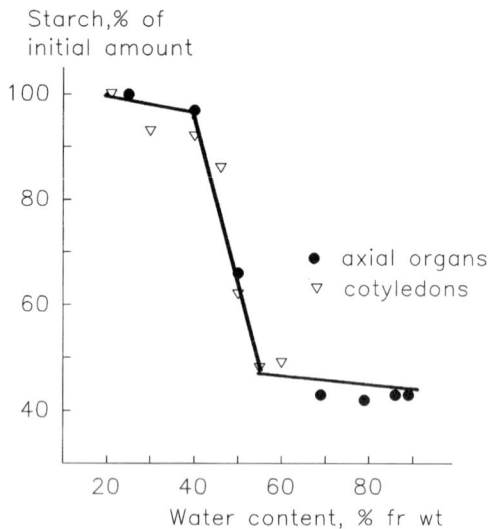

Figure 1. Starch degradation in germinating pea seeds as a function of their water content

A drop of starch amount occurred at roughly 45% wc; starch degradation started at the same water content in both cotyledons and axial organs. To be certain that it is the hydration up to 45% water content that triggers the starch degradation, pea seeds imbibed to 20, 30, 40, 50 or 60% wc were kept in a humid chamber for up to one week.

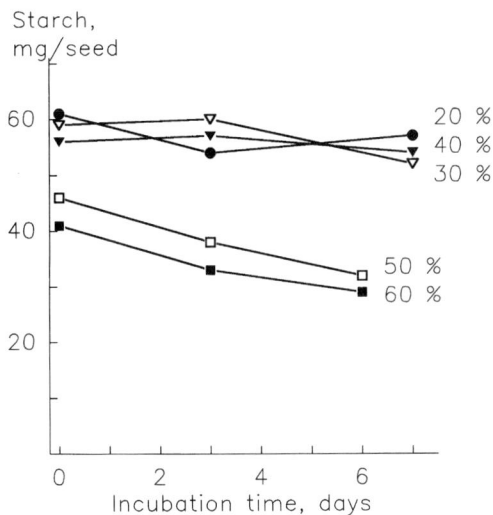

Figure 2. Starch content (mg/seed) of germinating pea seeds maintaining their water contents in humid air

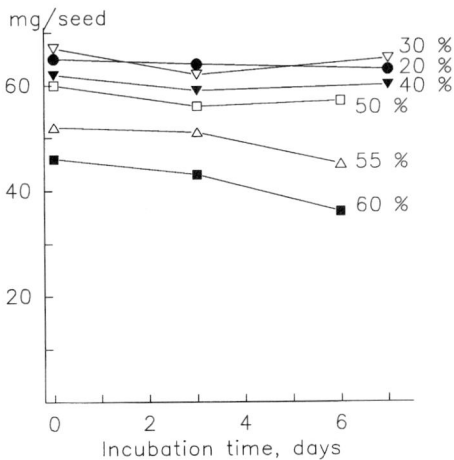

Figure 3. Protein content (mg/seed) of germinating pea seeds maintaining their water contents in humid air

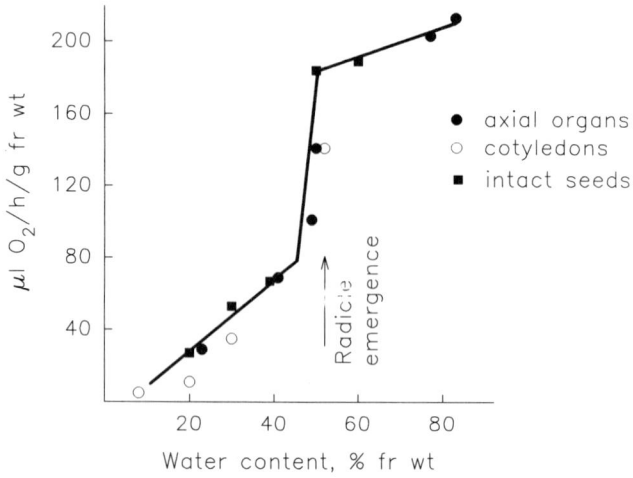

Figure 4. Respiration rate in germinating pea seeds as a function of their water content

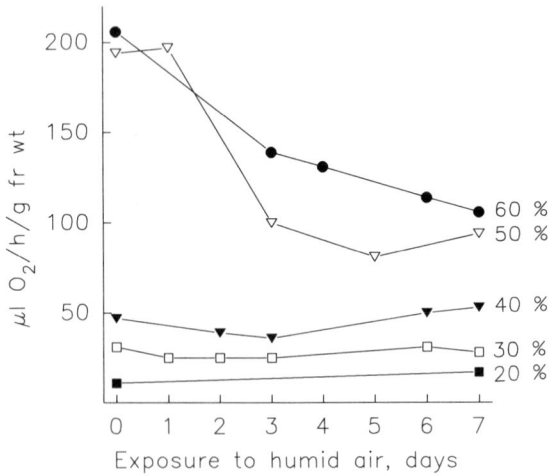

Figure 5. Respiration rate in germinating pea seeds maintaining their water contents in humid air

Their water content did not change significantly. The seeds with 20, 30 and 40% wc did not start starch hydrolysis at all (Fig. 2), although 7 days (d) was a sufficiently long time to begin it. The seeds imbibed to 50 or 60% wc, which had already begun the starch degradation, proceeded with it (Fig. 2). Therefore, 45% wc appears to be threshold hydration level at which starch degradation commences.

The same experimental design was applied to protein degradation. Pea seeds imbibed to 20, 30, 40, 50, 55 or 60% wc were incubated in a humid chamber for 7 d.

The seeds with 20–50% wc were unable to start proteolysis throughout this long period (Fig. 3). The seeds at 55 and 60% wc had already initiated the proteolysis and continued it (Fig. 3). Hence, 52–53% wc can be proposed as the threshold water content at which the protein degradation is triggered in these pea seeds.

The initial activation of respiration was described at 23% wc by Vertucci (1989) in the soyabean seeds. In the pea seeds the respiration was studied at higher wc (Obroucheva and Kovadlo, 1985). Like in soyabean seeds, initially the respiration rate slowly increased in pea seeds, cotyledons and axes as the imbibition proceeded but within 40–50% wc range it sharply rose to a constant level typical of vegetative tissues (Fig. 4).

To be confident that final activation of respiration is triggered by the rise in hydration level from 40 to 50% wc, the aforementioned experimental design was repeated. The pea seeds, which maintained their wc at 20–40%, continued to respire at the same rate, without any enhancement of respiration.

In the seeds imbibed previously to 50 or 60% wc the respiration gradually declined due to substrate exhaustion (Fig. 5). The fact that the seeds at 40% wc did not activate their respiration during a week, can be interpreted as inadequate hydration. Therefore, 45% would appear to be a threshold level for final activation of respiration in pea.

The radicle protrusion occurring by cell elongation is also triggered by hydration of axial organs to 72%. The seeds imbibed to 60, 65 and 68% wc and kept in PEG-6000 to prevent their further hydration, were unable to germinate for 30 h, although the seeds in water accomplished it in a few hours (Table 1).

Table 1. Radicle protrusion in broad bean seeds kept at constant water contents. To maintain water contents at 60, 65 and 68%, PEG 6000 concentrations were 30, 20 and 15%, respectively

Water content in axial organs, % ft wt	Germination time (h)	
	Seeds in water	Seeds in PEG
60±1.5	9	>30
65±1.0	6	>30
68±1.0	3	>30

Discussion

Taken together these observations clearly indicated that the increase in water content to threshold levels triggered the metabolic processes and growth initiation. With this in mind, the literature was screened to select the experiments, in which the development of various processes in the course of imbibition was followed in dependence of seed hydration, and to find the wc level, at which each process began.

The data collected (Fig. 6) are arranged along typical hydration curves of axial organs and cotyledons in such a way that the threshold wc levels for each process are indicated by arrows. The schematic representation of this hypothesis was originally published in 1993.

Seed hydration, as a triggering agent, acts over a period from the very beginning of imbibition till radicle emergence by cell elongation. The initiation of cell divisions is not triggered by wc increase but occurs in fully-hydrated axial organs, like other processes beginning after radicle emergence.

The activation of metabolic systems occurs stepwise in the course of seed hydration. The first to be activated are the respiration and amino acid metabolism at 20–23% wc. Apparently their activation is due to sufficient hydration of enzyme molecules providing their conformational readiness to interact with the substrate molecules which can be delivered at this level of cytoplasm hydration.

Next are triggered the metabolic systems operating inside the more complexly organized structures which need higher wc (45–50%) to become active. The initiation of transcriptional activity of genes requires not only activation of corresponding enzymes but triggering of a gene regulatory ensemble as well. For the beginning of protein synthesis, the delivery of tRNAs and mRNAs to the ribosomes and proper conformation of ribosomal proteins organized into subunits are necessary. Final activation of respiration is provided by incorporation of soluble subunits of succinate dehydrogenase and cytochrome oxidase into the inner membranes of mitochondria (Asahi and Maeshima, 1983). The beginning of starch and protein degradation needs water penetration into starch grains and protein bodies, the solubilization of densely packed substrates and the conformational changes inside hydrolytic enzyme molecules.

The aforementioned processes begin to operate automatically in the course of seed hydration because water, if available, enters the seeds without hindrance up to 55–60% wc. Such water inflow is due to dry seed behaviour as a capillary-porous body, water being bound to various hydrophilic biopolymers. This level of hydration known as water absorption by pure physical forces, is estimated by the wc reached by dead seeds. Therefore, the metabolic activation is triggered unimpeded step by step by water entering and completed in the cotyledons by radicle emergence and in the axial organs by 55–60% wc.

After metabolic activation the second stage of germination, i.e. preparation for elongation, occurs only in the axial organs and includes accumulation of osmotic solutes, vacuolation and cell wall acidification by activation of

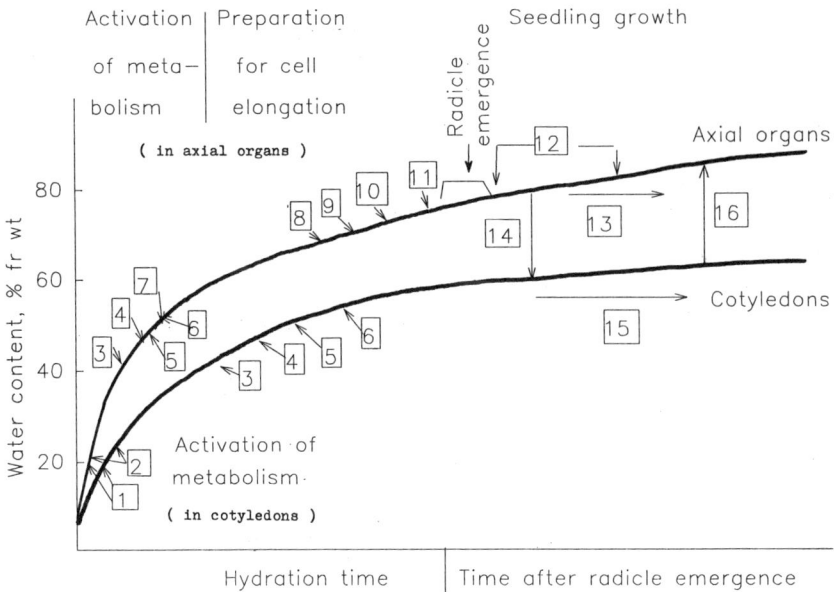

Figure 6. Three stages of seed germination and triggering role of water inflow

1. Primary activation of respiration, lettuce (Eldan and Mayer, 1972), soyabean (Vertucci, 1989)
2. Onset of amino acid metabolism, wheat (Linko and Milner, 1959)
3. Onset of mRNA synthesis, rape (Comai and Harada, 1990), maize (Dommes and Van de Walle, 1983)
4. Final activation of respiration, pea (Obroucheva and Kovadlo, 1985)
5. Onset of starch degradation, pea (Obroucheva *et al.*, 1988)
6. Onset of reserve protein degradation, pea (Obroucheva *et al.*, 1988)
7. Activation of protein synthesis, wheat (Marcus *et al.*, 1966)
8. Accumulation of osmotic solutes, broad bean (Obroucheva *et al.*, 1993)
9. Enlargement of vacuole, broad bean (Obroucheva *et al.*, 1993)
10. Activation of H^+-ATPase in plasmalemma, broad bean (Obroucheva and Antipova, 1994)
11. Elongation start in the axial organs, broad bean (Obroucheva, 1993)
12. Beginning of mitotic activity
13. Onset of de novo biosynthesis of hormones and polyamines
14. Induction of hydrolases by the axial organs
15. Degradation of major reserves
16. Import of degradation products into the axial organ

plasmalemma H^+-ATPase (Obroucheva and Antipova, 1989). The osmotic solutes, mainly glucose and potassium ions, are accumulated at 65–68% wc and drive the further water inflow. Acidification of cell walls resulting in their enhanced extensibility starts at 68–70% wc, being also triggered by the hydration.

Until radicle emergence, the abscissa presents the hydration time, necessary for seed to imbibe up to a certain wc level. It differs from the imbibition time because it does not depend on the conditions of water uptake (temperature, area of seed contact with water, permeability of seed coat etc.), but takes proper account only to the wc reached.

After radicle emergence, in fully-hydrated tissues, the time duration is measured as always. This third stage of germination involves the growth processes resulting in seedling emergence, *de novo* synthesis of various phytohormones, including a hormonal stimulus transported to the cotyledons; it induces there a mobilization of the bulk of reserves, the degradation products being translocated to the axial organs to meet their growth and energy demands.

Figure 6 shows that different seeds initiate the same metabolic processes at the same wc levels, for example, the activation of amino acid metabolism or mRNA synthesis. Therefore the triggering of various systems by the hydration to their specific threshold levels is a common seed strategy. It permits the seeds to behave adequately to their environment.

Acknowledgements

This work was supported by the Russian Foundation for Fundamental Research, grant no. 95-04-11024.

References

Asahi, T. and Maeshima, M. 1983. In: *The New Frontiers in Plant Biochemistry*, pp. 119–132. The Hague: Nijhoff/Dr.Junk Publishers.

Comai, L. and Harada, Y.J. 1990. *Proceedings of National Academy of Sciences USA* 87: 2671–2674.

Dommes, J. and Van de Walle, C. 1983. *Plant Physiology* 73: 484–487.

Eldan, M. and Mayer, A.M. 1972. *Physiologia Plantarum* 26: 67–72.

Linko, P. and Milner, M. 1959. *Plant Physiology* 34: 392–396.

Marcus, A., Feeley, J. and Volcane, T. 1966. *Plant Physiology* 41: 1167–1172.

Obroucheva N.V. 1993. In: *Basic and Applied Aspects of Seed Biology. Proceedings of the Fourth International Workshop on Seeds*, pp. 275–281 (eds. D. Côme and F. Corbineau). Paris: University P. et M. Curie.

Obroucheva, N.V. and Antipova, O.V. 1989. In: *Structural and Functional Aspects of Transport in Roots*, pp. 41–44 (eds. B. C. Loughman and O. Gasparicova). Dordrecht: Kluwer Academic Publishers.

Obroucheva, N.V. and Antipova, O.V. 1994. *Russian Journal of Plant Physiology* 41: 391–395.

Obroucheva, N.V., Antipova, O.V. and Ivanova, I.M. 1993. *Russian Journal of Plant Physiology* 40: 641–646.

Obroucheva, N.V. and Kovadlo, L.S. 1985. *Soviet Plant Physiology* 32: 753–761.

Obroucheva, N.V., Kovadlo, L.S. and Prokofiev, A.A. 1988. *Soviet Plant Physiology* 35: 322–328.

Vertucci, C.W. 1989. *Physiologia Plantarum* 77: 172–176.

62. A Cysteine Endopeptidase (SH-EP) in Germinated *Vigna mungo* Seeds: Post-translational Processing and Intracellular Transport

T. OKAMOTO and T. MINAMIKAWA

Department of Biology, Tokyo Metropolitan University, Minami-osawa, Hachioji, Tokyo, 192-03 Japan

Abstract

A plant cysteine endopeptidase, designated SH-EP, is a major protease occurring in cotyledons of *Vigna mungo* seedlings, and acts to degrade seed globulin stored in protein bodies (protein storage vacuoles). SH-EP is synthesized on membrane-bound ribosomes as a 43-kDa intermediate through the cotranslational cleavage of a signal sequence, and the intermediate is processed further to the 33-kDa mature enzyme via 39- and 36-kDa intermediates. (a) *N-terminal processing* – Experiments of *in vitro* processing of the SH-EP intermediates revealed that two processing enzymes, VmPE-1 and VmPE-2 (*V. mungo* processing enzymes 1 and 2), are involved in the processing. VmPE-1 was purified from the day-3 cotyledons. The enzyme is the same type of protease as asparaginyl endopeptidases in terms of the primary structure and substrate specificity. (b) *C-terminal processing* – The amino acid sequence of SH-EP deduced from the cDNA contains C-terminus with Lys-Asp-Glu-Leu (KDEL) tail, which is known as a retention signal for endoplasmic reticulum (ER) while mature SH-EP is localized in protein bodies. The analysis for C-terminal amino acid residues of SH-EP indicated that the C-terminal propeptide of 10 amino acid residues containing the KDEL tail is processed to form mature SH-EP.

Introduction

Most proteases are synthesized as higher molecular weight precursors, and the precursors are post-translationally processed to mature forms (Kassell and Kay, 1973; Mach *et al.*, 1994). It has been reported that the prosequence of such protease precursors functions in masking the active site (James and Sielecki, 1986), intracellular transport (Chrispeels and Raikhel, 1992) and correct folding of proteases (Vernet *et al.*, 1991). Plant cysteine proteases are also synthesized as precursors which have the large N-terminal prosequence, and those precursors are cleaved to the mature enzymes (Mitsuhashi and Minamikawa, 1989; Holwerda *et al.*, 1990; Koehler and Ho, 1990).

In cotyledons of germinated *V. mungo* seeds, a cysteine endopeptidase, termed SH-EP, plays a major role in the degradation of seed storage protein (Mitsuhashi *et al.*, 1986). SH-EP is synthesized on membrane-bound polysomes as a large inactive 45-kDa precursor, which is cotranslationally processed to a 43-kDa intermediate through the cleavage of a 2-kDa signal sequence. The 43-kDa

R.H. Ellis, M. Black, A.J. Murdoch, T.D. Hong (eds.), Basic and Applied Aspects of Seed Biology, pp. 563–568.
© *1997 Kluwer Academic Publishers, Dordrecht. Printed in Great Britain.*

intermediate of SH-EP is processed further to 33-kDa mature enzyme via 39- and 36-kDa intermediates (Mitsuhashi and Minamikawa, 1989). In addition, a KDEL sequence which is known as the retention signal for ER exists in the C-terminus of the amino acid sequence of SH-EP deduced from the cDNA, while mature SH-EP localizes in protein bodies. Here we report the mechanism of the N- and C-terminal processing of SH-EP.

Materials and Methods

V. mungo seeds were germinated on layers of wet filter paper at 27°C in darkness, and cotyledons of the day-3 seedlings were collected. Purification of SH-EP and immunological methods were carried out as described by Mitsuhashi and Minamikawa (1989). VmPE-1 was purified according to Okamoto and Minamikawa (1995). In vitro processing was as described by Okamoto et al. (1994). Analysis of amino acid composition and sequence, and mass spectroscopy were carried out according to Matsuoka et al. (1994).

Results and Discussion

N-terminal Processing of SH-EP

When extracts from day-3 cotyledons of V. mungo were incubated at 27°C, the 43-kDa intermediate of SH-EP was observed to be processed to the 33-kDa mature enzyme through 39- and 36-kDa intermediates (Fig. 1A). Experiments with protease inhibitors suggested that at least two enzymes, designated VmPE-1 and VmPE-2, participate in this post-translational processing (Okamoto et al., 1994). VmPE-1 is involved in the processing step from the 43-kDa intermediate to the 36-kDa intermediate and VmPE-2 from the 36-kDa intermediate to the 33-kDa mature form (Fig. 1B). VmPE-1 was purified from the day-3 cotyledons. This protease has a molecular mass of 33 kDa as estimated by SDS-PAGE (Fig. 1C).

Jackbean asparaginyl endopeptidase (legumain) has the strict substrate specificity toward the carboxyl side of Asn residues (Abe et al., 1993). The N-terminal amino acid sequence of VmPE-1 (Fig. 1D) and the immunoreactivity of VmPE-1 to the antibody against the jackbean asparaginyl endopeptidase (Fig. 1C) indicated that VmPE-1 is homologous to the asparaginyl endopeptidase in terms of the primary structure. In addition, a fluorometric assay of VmPE-1 using synthetic peptides as the substrate revealed that the protease has substrate specificity to Asn residue.

The hydropathy plot of the prosequence of SH-EP shows that two Asn residues at positions 72 and 74 occur in a hydrophilic region of the prosequence or on the molecular surface (Fig. 1E). These Asn residues on the molecular surface may be the sites that Asn-specific proteases act to cleave (Hara-

A

kDa
43 —
39
36 —
33 —

0   1   2   3   4   6   8   12
Incubation time (h)

B

mRNA
|— translation on membrane-bound ribosomes
45 kDa polypeptide
|———————— Signal peptidase
43 kDa polypeptide
39 kDa polypeptide |— VmPE-1
36 kDa polypeptide
|———————— VmPE-2
33 kDa polypeptide (mature form)

C

1   2   3
kDa
94-
67-
43-
30-   ◄33 kDa
20.1-
14.4-

D

	1	5	10
VmPE-1	D E	– GT RWAVL	I
Jackbean Asn EPase	D E V	GT RWAVL	A
Castor bean VPE	D Q L	GT RWAVL	A
Vetch proteinase B	F E	– GT RWA I L	L

E

Hydropathy index

●: Asn residue

3
2
1
0
-1
-2
-3

0   20   40   60   80   100   120
Number of amino acid residues

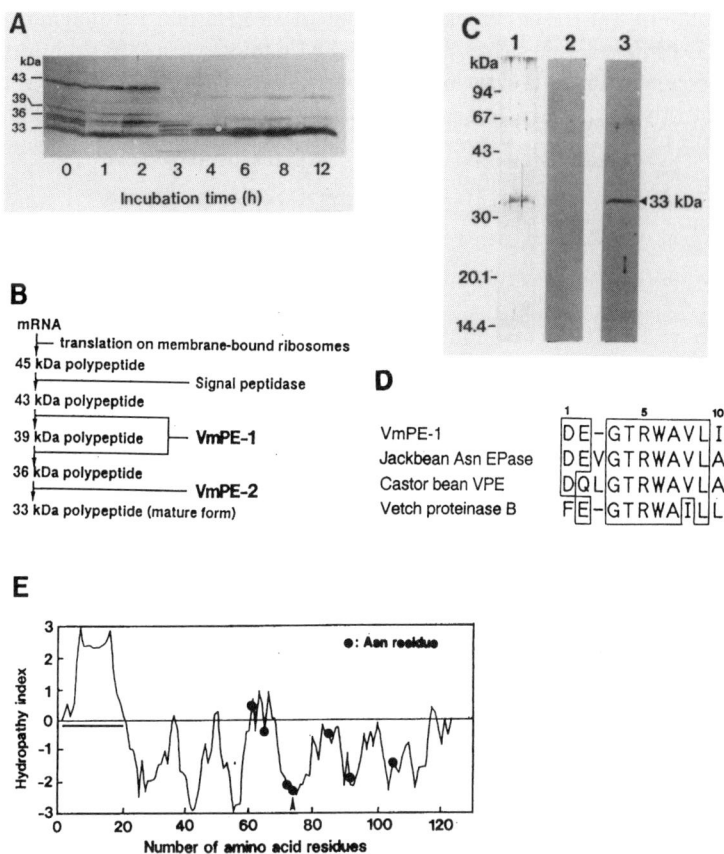

Figure 1. Post-translational processing of SH-EP in cotyledons of *Vigna mungo* seedlings and characterization of a processing enzyme, VmPE-1. **A**, Time course of post-translational processing of SH-EP in extracts from the cotyledons. An extract from the cotyledons of day-3 seedlings was incubated for 0 to 12 h at 27°C, and analyzed by SDS-PAGE/immunoblotting using an antiserum raised against SH-EP. **B**, A postulated post-translational processing of SH-EP. The C-terminal propeptide consisting of 10 amino acid residues has been shown to be cleaved during maturation of SH-EP (Okamoto *et al.*, 1994). However, this processing step is not included in the panel, since the step has not been positioned yet. The molecular masses indicated were estimated from the mobility on SDS-PAGE. **C**, SDS-PAGE and immunoblotting of purified VmPE-1. SDS-PAGE and silver-staining of purified VmPE-1 (lane 1), immunoblotting with the antiserum against SH-EP (lane 2) and with the antibody against the jackbean asparaginyl endopeptidase (lane 3). **D**, The N-terminal amino acid sequence of VmPE-1 and its similarities to other proteases. The N-terminal 10 residues of VmPE-1 were compared with those of the jackbean asparaginyl endopeptidase (Takeda *et al.*, 1994), the vacuolar processing enzyme of castor bean (VPE; Hara-Nishimura *et al.*, 1993) and proteinase B from vetch seeds (Becker *et al.*, 1995). **E**, The hydropathy plot of N-terminal prosequence of SH-EP. The amino acid sequence of SH-EP was cited from Akasofu *et al.* (1989). The arrow indicates the Asn residue present in the ERFNIN motif (Karrer *et al.*, 1993). The bar shows the putative signal peptide region

MAMKKLLWVVLSLSLVLGVANSFDFHEKDLESEESLWDLYERWRSH   46

HTVSRSLGEKHKRFNVFKANVMHVHNTNKMDKPYKLKLNKFADMTN   92

HEFRSTYAGSKVNHHKMFRGSQHGSGTFMYEKVGS̈V̈PASVDWRKKG   138

A̲V̲TDVKDQGQCGSCWAFSTIVAVEGINQIKTNKLVSLSEQELVDCD   184

KEENQGCNGGLME̲S̲A̲F̲E̲F̲IKQKGGITTESNYPYTAQEGTCDESKVN   230

DLAVSIDGHENVPVNDENALLKAVANQPVSVAIDAGGSDFQFYSEG   276

VFTGDCNTĎLNHGVAIVGYGTTVDGTNYWIVRNSWGPEWGEQGYIR   322

M̲Q̲R̲Ň̲I̲S̲K̲K̲E̲G̲L̲CGIAMMA̲S̲Y̲P̲I̲K̲Ň̲S̲S̲D̲Ň̲P̲T̲*GSLSSPKDEL,   362

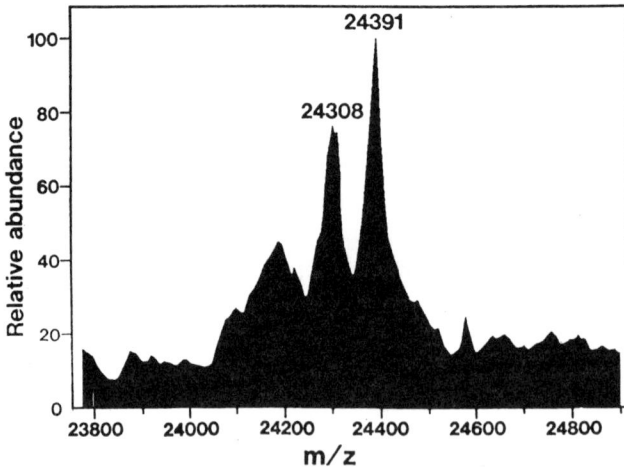

Figure 2. Amino acid sequence and mass spectrum of SH-EP. The amino acid sequence of SH-EP deduced from the nucleotide sequence of SH-EP cDNA (Akasofu *et al.*, 1989). The amino acid sequence corresponding to the peptides obtained by CNBr cleavage of pyridylethylated SH-EP is underlined. Dots and asterisk indicate N-terminal and C-terminal amino acid residues of 33 kDa mature SH-EP, respectively. The mass spectrum of purified SH-EP was reconstituted from the data recorded with electrospray mass spectrometry

Nishimura *et al.*, 1993). The Asn residue at position 74 is present in the ERFNIN motif which is strictly conserved in the N-terminal prosequence of cysteine proteases other than cathepsin B-like proteases (Karrer *et al.*, 1993). A number of cysteine proteases occurring in storage organs of growing seedlings are known to have similarities (40 to 60%) to SH-EP with respect to the amino acid sequence of the mature regions (Yamauchi *et al.*, 1992), and they have a conserved Asn residue in the ERFNIN motif of the prosequence (Karrer *et al.*, 1993). Among these plant cysteine proteases, EP-B and aleurain, both from barley seedlings, are known to be subjected to multi-fashioned post-translational processing similar to SH-EP (Koehler and Ho, 1990; Holwerda *et al.*, 1990). From these joint observations, we postulate that VmPE-1 and related processing enzymes occur ubiquitiously in germinated seeds and function in post-translational processing of cysteine proteases that act to degrade seed storage proteins. The postulated processing of cysteine proteases in plant seedlings mediated by VmPE-1 and related Asn-specific proteases may be different from the autocatalytic processing observed for the enzymes such as cathepsin B (Mach *et al.*, 1994), cathepsin L (Smith and Gottesman, 1989) and papain (Vernet *et al.*, 1991).

C-terminal Processing of SH-EP

The C-terminal tetrapeptide KDEL has been shown to be a signal for retention in the ER lumen of mammalian and plant cells (Warren, 1987; Chrispeels, 1991). Lee *et al.* (1993) identified a cDNA clone from *Arabidopsis thaliana* as a homologue to the ERD2 gene family, whose product is the receptor for the ER retention signal, and showed that this plant gene was able to complement a lethal phenotype of the yeast *erd2* deletion mutant. These facts are apparently inconsistent with the evidence that SH-EP acts to degrade storage globulin which has been deposited in protein bodies.

The direct N-terminal sequence analysis of the mature SH-EP and the amino acid composition and N-terminal sequence analysis of peptide fragments of pyridylethylated SH-EP cleaved by CNBr indicated that Thr-352 might be the C-terminus of mature SH-EP (Fig. 2). The mass spectrum of purified SH-EP indicated the presence of two molecular species with masses, 24,308 and 24,391 (Fig. 2). These mass values are accounted for by the presence of two polypeptide chains, one extending from Val-128 to Thr-352 and the other from Ser-127 to Thr-352. These results indicated that the mature SH-EP has Thr-352 at the C-terminus and that the C-terminal decapeptide including a KDEL tail of the precursor protein is removed by post-translational processing during the maturation of SH-EP (Fig. 2).

References

Abe, Y., Shirane, K., Yokosawa, H., Matsushita, H., Mitta, M., Kato, I. and Ishii, S. 1993. *Journal of Biological Chemistry* 268: 3525–3529.

Akasofu, H., Yamauchi, D., Mitsuhashi, W. and Minamikawa, T. 1989. *Nucleic Acids Research* 17: 6733.

Becker, C., Shutov, A. D., Nong, V. H., Senyuk, V. I., Jung, R., Horstmann, C., Fischer, J., Nielsen, N. C. and Müntz, K. 1995. *European Journal of Biochemistry* 228: 456–462.

Chrispeels, M. J. 1991. *Annual Review of Plant Physiology and Plant Molecular Biology* 42: 21–53.

Chrispeels, M. J. and Raikhel, N. V. 1992. *Cell* 68: 613–618.

Hara-Nishimura, I., Takeuchi, Y. and Nishimura, M. 1993. *Plant Cell* 5: 1651–1659.

Holwerda, B. C., Galvin, N. J., Baranski, T. J. and Rogers, J. C. 1990. *Plant Cell* 2: 1091–1106.

James, M. N. and Sielecki, A. R. 1986. *Nature* 319: 33–38.

Karrer, K. M., Peiffer, S. L. and DiTomas, M. E. 1993. *Proceedings of the National Academy of Science USA* 90: 3063–3067.

Kassel, B. and Kay, J. 1973. *Science* 180: 1022–1027.

Koehler, S. M. and Ho, T-H. D. 1990. *Plant Cell* 2: 769–783.

Lee, H., Gal, S., Newman, T.C. and Raikhel, N.V. 1993. *Proceedings of the National Academy of Sciences USA* 90: 11433–11437.

Mach, L., Mort, J. S. and Glössl, J. 1994. *Journal of Biological Chemistry* 269: 13030–13035.

Matsuoka, K., Seta, K., Yamakawa, Y., Okuyama, T., Shinoda, T. and Isobe, T. 1994. *Biochemical Journal* 298. 435–442.

Mitsuhashi, W., Koshiba, T. and Minamikawa, T. 1986. *Plant Physiology* 80: 628–634.

Mitsuhashi, W. and Minamikawa, T. 1989. *Plant Physiology* 89: 274–279.

Okamoto, T. and Minamikawa, T. 1995. *European Journal of Biochemistry* 231: 300–305.

Okamoto, T., Nakayama, H., Seta, K., Isobe, T. and Minamikawa, T. 1994. *FEBS Letters* 351: 31–34.

Smith, S. M. and Gottesman, M. M. 1989. *Journal of Biological Chemistry* 264: 20487–20495.

Takeda, O., Miura, Y., Mitta, M., Matsushita, H., Kato, I., Abe, Y., Yokosawa, H. and Ishii, S. 1994. *Journal of Biochemistry* 116: 541–546.

Vernet, T., Khouri, H. E., Laflamme, P., Tessier, D. C., Musil, R., Gour-Salin, B. J., Storer, A. C. and Thomas, D. Y. 1991. *Journal of Biological Chemistry* 266: 541–546.

Warren, G. 1987. *Nature* 327: 17–18.

Yamauchi, D., Akasofu, H. and Minamikawa, T. 1992. *Plant and Cell Physiology* 33: 789–797.

63. Alleviation of Chilling Injury by Ethephon in Pea Seeds

L. PETRUZZELLI[1] and F. HARREN[2]

[1]Ist. Germoplasma, C.N.R., via G. Amendola 165\A, 70126 Bari, Italy; [2]Dept. Molecular Laser Physics, University of Nijmegen, Toernooiveld, 6525 ED Nijmegen, The Netherlands

Abstract

Exposing *Pisum sativum* seeds to 5°C at the beginning of imbibition markedly delayed germination and ethylene production. The presence of ethephon (ETH) improved tolerance to chilling injury, whereas L-α-(2-aminoethoxyvinyl) glycine hydrochloride (AVG) and, more evidently, 2,5-norbornadiene (NBD) tended to enhance chilling injury. Both these harmful effects were counteracted by adding ETH. On the other hand, cobalt, though slightly inhibiting ethylene synthesis, was effective in protecting against chilling injury. When testing other metals (silver, nickel, copper, lead, mercury, zinc and potassium), the greatest beneficial effect on germination and ethylene production was observed with zinc. Also, a reduced K^+ efflux occurred in the presence of cobalt and zinc ions and, to a lesser level, of ETH and nickel. Thus, it seems likely that ethylene is partially required for prevention of chilling injury and the effects of Zn, Co and, to a small extent, of Ni salts may not relate to their anti-ethylene properties.

Introduction

Low temperatures between 0 and 10°C are of major agricultural importance, since they markedly affect plant growth and development with consequent productivity reduction. At cellular and molecular levels, the effects induced by chilling stress are numerous and involve membrane organization and functions, respiration and sugar metabolism (Levitt, 1980).

In the case of germinating seeds, the imbibitional damage under chilling conditions has been well documented and has been generally ascribed to faulty membrane reorganization caused by the synergistic effects of low moisture content and chilling in the physical state of membrane lipids (Dogras *et al.*, 1977; Leopold and Musgrave, 1979). However, results in contrast with this interpretation have been reported (Hobbs and Obendorf, 1972; Bochicchio *et al.*, 1991). Furthermore, little is known about the metabolism of seeds damaged by chilling and whether the physiological changes which occur during chilling can be repaired after transfer to a warmer temperature. In this regard, even pea seeds, that are considered insensitive to chilling, show profound changes on their cellular membranes when briefly exposed to an initial temperature of 5°C (Di Nola and Mayer, 1985; 1986; 1987).

On the other hand, it is well known that a stimulation of ethylene production

R.H. Ellis, M. Black, A.J. Murdoch, T.D. Hong (eds.), Basic and Applied Aspects of Seed Biology, pp. 569–576.
© *1997 Kluwer Academic Publishers, Dordrecht. Printed in Great Britain.*

is generally induced by chilling stress in plant organs or tissues of a number of plants upon the transfer to warmer temperatures (Wang and Adams, 1982; Chen and Patterson, 1985; Gay et al., 1991). Although Wang (1987) did not observe any reduction of chilling injury by application of aminooxyacetic acid, an inhibitor of ACC synthesis, it remains quite unclear whether ethylene increase may contribute to the manifestation of chilling injury or is simply a response to the chilling stress.

The present paper was initiated to evaluate whether an initial chilling exposure could influence the capability of *Pisum sativum* seeds to synthesise ethylene during the subsequent germination at 20°C. With the aim of verifying the involvement of ethylene in chilling injury we have examined the effects on chilled seeds of ethephon, a well known ethylene releasing compound, and of antagonists of ethylene production or action. Aggravation by AVG and NBD was not paralleled by the effects of $CoCl_2$ according to Small et al. (1991), and this finding led us to extend the study to other heavy metals.

Materials and Methods

Seeds of pea (*P. sativum* L., cv. Frijaune) purchased from a commercial source were used as experimental material. Prior to chilling and germination, seeds were surface-sterilized in 0.5% sodium hypochlorite for 5 min followed by thorough rinsing with sterile distilled water. For cold treatment, samples of 100 seeds were incubated at 5°C for 8 or 16 h on one sheet of Whatman No. 1 filter paper moistened with 45 ml double-distilled water or treatment solution in 150 mm diameter glass Petri dishes. After chilling, seeds were rinsed with distilled water and incubated for a further 72 h at 20°C on filter paper moistened with 20 ml of distilled water in four 150 mm diameter Petri dishes each containing 25 seeds. To apply NBD, Petri dishes containing seeds were placed inside 4.5 l airtight glass desiccators. Liquid NBD was applied to an extra 5 cm Petri dish where it completely evaporated within few minutes. The desiccators were incubated at 5°C and, after chilling, flushed with a stream of air for 30 min before transferring seeds to 20°C. For all germination experiments, each treatment had three replicates and experiments were repeated twice with similar results.

Ethylene production was measured using laser-driven photo-acoustic equipment already described (Petruzzelli et al., 1994). When ethylene quantification was performed by gas-chromatography, 50 chilled seeds after incubation for 20 and 30 h at 20°C were aseptically transferred to 100 ml flasks containing sterile distilled water. The flasks were sealed with silicone-rubber stoppers and incubated in darkness at 25°C with shaking at 50 cycles/min. After 4 h, 2 ml samples were removed from the flasks and injected into a Hewelett-Packard 5890 gas chromatograph fitted with a hydrogen flame ionization detector and a 183×0.32-cm stainless steel column containing 80/100 Poropack Q. Ethylene concentrations were evaluated from calibration curves provided by standard ethylene mixtures and were corrected for ethylene background in empty flasks.

For the measurement of K^+ efflux, 50 chilled or unchilled seeds after 20 h of imbibition at 20°C, were incubated in 10 ml sterile distilled water into 100 ml Erlenmeyer flasks. After incubation at 25°C with shaking at 50 cycles/min, the potassium content of the leachates was determined electrochemically from 1:4 dilution of the leachate with a potassium electrode (Ingold, type 152213000) using a Radiometer ION83.

All results represent the mean of at least four replicates. For laser-driven PAC ethylene detection, representative data are shown.

Results

Increasing the duration of 5°C exposure at the start of water imbibition progressively delayed subsequent germination of pea seeds at 20°C (Fig. 1A).

At the same time, while ethylene evolution by control peas increased rapidly after the initial 15 h after seeding, in seeds exposed to 5°C for 8 h ethylene evolution was markedly reduced. Additional exposure for a total of 16 h to 5°C resulted in a further decline in ethylene synthesis upon transfer to 20°C (Fig. 1B).

To verify the role of ethylene in chilling injury, we applied ethephon to peas during the 16 h chilling exposure. As shown in Figure 2, the presence of 10 mM ethephon reduced chilling injury. The presence of AVG, an inhibitor of pyridoxal-phosphate-mediated reactions as the conversion of S-adenosyl-methionine to 1-aminocyclopropane-1-carboxylic acid in ethylene biosynthesis (Yu and Yang, 1979), induced a further slight reduction in germination percentage whereas the application of NBD, a powerful competitive inhibitor of ethylene action (Sisler and Yang, 1983), resulted in a marked aggravation of chilling injury. Also, application of ethephon in combination with AVG or NBD counteracted the negative effects of inhibitors.

When we tested Ag, Ni and Co salts, well known antagonists of ethylene production or action (Beyer, 1976; Lau and Yang, 1976), a great beneficial effect was observed with $CoCl_2$ while the other two salts did not aggravate but slightly tended to reduce the injury (Table 1). Among the other metals, a complete prevention of chilling injury was induced by zinc.

The capability of Zn^{2+} to alleviate chilling injury was associated with a great increase of early ethylene evolution (Table 2). Also, Ni^{2+} slightly increased while $CoCl_2$ decreased ethylene production. No significant effect was induced by the other ions (data not shown). Furthermore, almost no ethylene was detected in presence of AVG.

The protection from chilling injury induced by zinc and cobalt ions was parallel to their capability to avoid completely the increase in potassium leakage produced by chilling (Fig. 3). The lowest rate of K^+ leakage was from peas exposed to Zn^{2+} while the highest rate of K^+ was from seeds exposed to AVG.

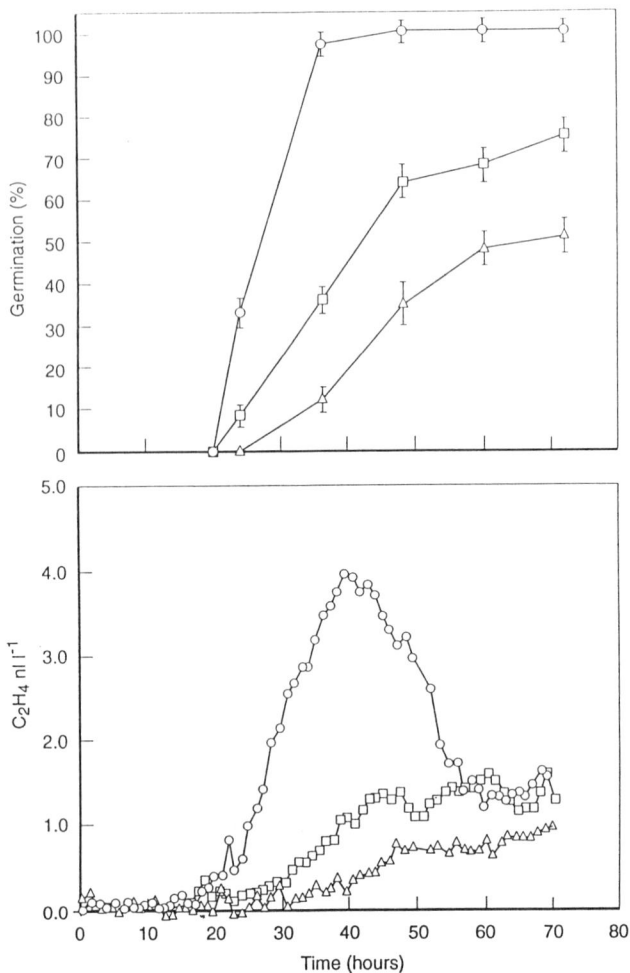

Figure 1. Effect of 8 and 16 h exposure to 5°C on (A) subsequent germination and (B) ethylene evolution at 20°C of *P. sativum* seeds: (○) unchilled control seeds, (□) 8 h chilled seeds and (△) 16 h chilled seeds. The vertical bars denote SE

Discussion

The present results show that the germinability of peas is delayed when seeds are first imbibed at 5°C and then transferred to 20°C (Fig. 1A). However, unlike other plant tissues (Abeles *et al.*, 1992), pea seeds exposed to chilling produce markedly less ethylene than unchilled seeds upon their transfer to warmer

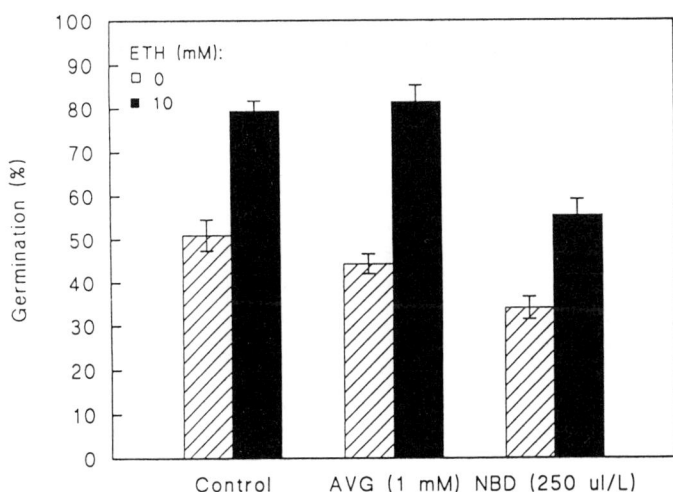

Figure 2. The effect of NBD and AVG applied without (cross-hatched) or with (solid) 10 mM ethephon during the 16 h chilling period on subsequent germination of peas at 20°C. Germination was measured at 72 h. The vertical bars denote SE

Table 1. Effect of exposure to various metals during the 16 h chilling on subsequent germination after 72 h at 20°C. Means ± SE

Chilling solutions	Concentrations giving highest germination (mM)	Germination percent (%)
Chilled control	–	51.0 ± 3.6
CoCl$_2$	5.0	84.2 ± 2.0
AgNO$_3$	0.1	57.6 ± 2.7
NiSO$_4$.6H$_2$O	1.0	60.2 ± 2.4
HgCl$_2$	0.01	52.9 ± 2.0
PbCl$_2$	0.01	51.9 ± 2.0
CuCl$_2$.2H$_2$O	0.01	54.3 ± 4.2
KCl	10.0	55.1 ± 3.0
ZnCl$_2$	10.0	94.0 ± 2.2

temperatures. Also, exogenous ethylene reduces the chilling injury and, on the contrary, NBD and also AVG, to a low extent, lead to increased chilling sensitivity (Fig. 2). These findings seem to indicate that in germinating seeds ethylene is required to partially prevent chilling injury. However, this suggestion appears to be not supported by the increased tolerance to chilling observed in presence of cobalt (Table 1). A beneficial effect by cobalt ions has been also

Table 2. Changes in ethylene production in peas exposed to AVG, Co^{2+}, Ni^{2+} and Zn^{2+} during 16 h chilling period. Ethylene production was determined 20 and 30 h after transfer to 20°C by gas-chromatography. Each value denotes mean $\pm$ SE

| Incubation medium | C_2H_4 production (pl seed^{-1} h^{-1}) | |
	20 h	30 h
Chilled control	11.2 ± 3.2	33.2 ± 6.4
$CoCl_2$, 5 mM	3.0 ± 1.0	23.2 ± 3.8
AVG 1 mM	0	1.5 ± 0.4
$NiSO_4.6H_2O$ 1 mM	12.4 ± 1.0	58.0 ± 2.8
$ZnCl_2$ 10 mM	33.0 ± 5.4	132.5 ± 17.8

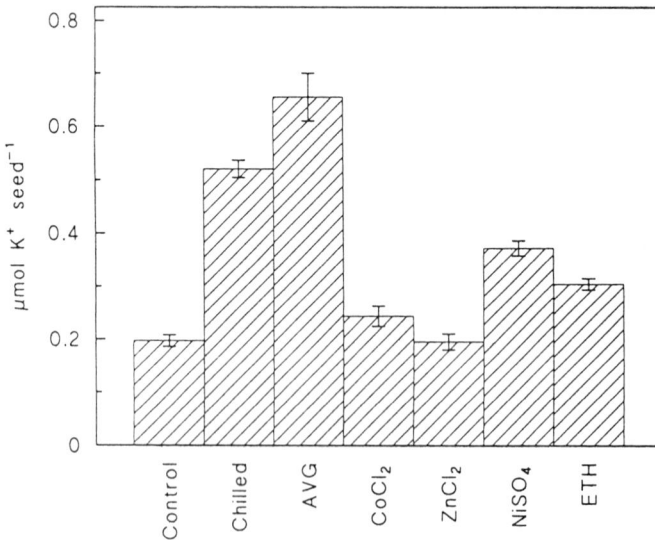

Figure 3. Changes in potassium efflux by pea seeds at 20 h after transfer from 5 to 20C. Data are means +SE of experiments performed with 50 seeds. ETH: ethephon

reported in bean seeds for soaking injury (Small *et al.*, 1991). Although Ag and Ni salts are less effective with chilled peas than with soaked beans, in agreement with the above authors we found that zinc prevents the injury while potassium is totally inadequate. Also, zinc and, to a lesser extent, nickel stimulate ethylene production (Table 2).

Thus, it may be deduced that the effect of these heavy metals is probably not related to their anti-ethylene properties as suggested by Small *et al.* (1991). The beneficial effects of Co, Zn and, to a extent, Ni salts may be due to their ability to prevent some deleterious reaction in embryonic cells.

This possibility seems to be supported by the data of potassium efflux which reflects membrane integrity (Mullett and Considine, 1980) and plays an important role during the first hours of germination (Fig. 3; Cocucci and Cocucci, 1977). In fact, the effect of zinc and cobalt and also, to a lesser extent, of ethephon and nickel to reduce K^+ efflux capacity corresponds to the effects of these compounds on seed germination. This correlation is supported by the greater K^+ efflux induced by AVG. It may therefore be assumed that metal ions induce a protective effect on chilled seed membranes and membrane integrity is necessary for the recovery from chilling injury, thus confirming the great number of observations correlating ethylene production with cell membrane state (Etani and Yoshida, 1987; Kende, 1993).

As zinc, cobalt and, to a lesser extent, nickel can affect ion fluxes (Cocucci and Morgutti, 1986) and a close interaction between proton pump activity and ACC-oxidase has been recently reported (Malerba *et al.*, 1995), we are currently investigating the possibility that they act through changes in ionic composition of embryonic cells.

Acknowledgements

This research was supported by the National Research Council of Italy, Special Project RAISA, Sub-project N 2 Paper No 2530. The competent assistance of Dr N. Aurisano for potassium analysis is also gratefully acknowledged.

References

Abeles, P., Morgan, W. and Saltveit, Jr M. E. 1992. *Ethylene in Plant Biology.* New York: Academic Press.
Beyer, E. M. 1976. *Plant Physiology* 58: 268–271.
Bochicchio, A., Coradeschi, M.A., Zienna, P., Bertolini, M. and Vezzana, C. 1991. *Seed Science Research* 1: 85–90.
Chen, Y. Z. and Patterson, B. D. 1985. *Australian Journal of Plant Physiology* 12: 377–385.
Cocucci, S. M. and Cocucci, M. 1977. *Plant Science Letters* 10: 85–95.
Cocucci, S. M. and Morgutti, S. 1986. *Physiologia Plantarum* 68: 497–501.
Di Nola, L. and Mayer, A. M. 1985. *Phytochemistry* 24: 2549–2554.
Di Nola, L. and Mayer, A. M. 1986. *Phytochemistry* 25: 2255–2259.
Di Nola, L. and Mayer, A. M. 1987. *Phytochemistry* 25: 1591–1593.
Dogras, C. C., Dilley, D. R. and Herner, R. C. 1977. *Plant Physiology* 60: 897–902.
Etani, S. and Yoshida, S. 1987. *Plant and Cell Physiology* 28: 83–91.
Gay, C., Corbineau, F. and Come, D. 1991. *Environmental and Experimental Botany* 31: 193–200.
Hobbs, P. R. and Obendorf, R. L. 1972. *Crop Science* 12: 664–667.
Kende, H. 1993. *Annual Review of Plant Physiology and Plant Molecular Biology* 44: 283–307.

Lau, O. L. and Yang, S. F. 1976. *Plant Physiology* 58: 114–117.

Leopold, A. C. and Musgrave, E. M. 1979. *Plant Physiology* 64: 702–705.

Levitt, J. 1980. *Responses of Plants to Environmental Stresses. I. Chilling, Freezing and High Temperature Stresses*, Second Edition, pp. 23–64. Orlando: Academic Press.

Malerba, M., Crosti, P. and Bianchetti, R. 1995. *Journal of Plant Physiology* 145: 711–716.

Mullett, J. H. and Considine, J. A. 1980. *Journal Experimental Botany* 31: 151–162.

Petruzzelli, L., Harren, F. and Reuss, J. 1994. *Environmental and Experimental Botany* 34: 55–61.

Sisler, E. C. and Yang, S. F. 1983. *Plant Physiology* 72-S: 40.

Small, J. G. C., Botha, F. C., Pretorious, J. C. and Hoffman, E. 1991. *Journal of Experimental Botany* 42: 277–280.

Wang, C. Y. and. Adams, D. O. 1982. *Plant Physiology* 69: 424–427.

Wang, C. Y. 1987. *Physiologia Plantarum* 69: 253–257.

Yu, S. B. and Yang, S. F. 1979. *Plant Physiology* 64: 1074–1077.

64. Flow Cytometric Analysis of Sugar-Beet Seeds Different in Vigour

E. SLIWINSKA

Department of Genetics and Plant Breeding, University of Technology and Agriculture, 85-796 Bydgoszcz, Poland

Abstract

The nuclear replication stages in different tissues of dry mature sugar-beet seed were determined using flow cytometry. The following samples were analysed: (a) whole seed ball, (b) true seed removed from the pericarp, (c) true seed without radicle, (d) radicle. In all the cases most cells were arrested in G_1/G_0 phase of cell cycle. Nevertheless, the presence of cells arrested in G_2 phase was observed. The highest G_2/G_1 ratio was found in the radicle. The seed lots of three sugar-beet varieties, PN Mono 1, Jastra (both $2 \times$) and Maria ($3 \times$), different in vigour (low, medium and high vigour for each variety) were investigated in order to study a correlation between vigour and G_2/G_1 ratio in the radicle. A significantly higher G_2/G_1 ratio for seed of low vigour was obtained suggesting that flow cytometry can be used for identification of the low vigour seed lots which need special treatment in order to improve their quality.

Introduction

Flow cytometry (FCM) is commonly used in sugar-beet breeding and seed production for control of ploidy level (De Laat *et al.*, 1987; Brown *et al.*, 1991; Sliwinska and Steen, 1995; Demilly *et al.*, 1995). FCM involves rapid analysis of the degree of fluorescence of nuclei isolated from plant tissue. Since the degree of fluorescence is linearly related to the DNA content, flow cytometry can be used for accurate measures of the ploidy level as well as of the cell cycle activity (Galbraith, 1984).

Different intact plant tissues, such as leaf, root, stem, petal or anther can be analysed (Galbraith *et al.*, 1983; Galbraith, 1984). Bino *et al.* (1992, 1993), and Lanteri *et al.* (1992) showed that FCM can also be used for determination of nuclear replication stages in seed tissues of some species. They found that most embryonic cells of tomato and pepper dry seed were arrested in G_1 phase of the cell cycle. Only upon imbibition of the seed in water or osmotic solution was an augmentation of the 4C signals observed, indicating that the cells had entered the synthetic phase of nuclear division. Flow cytometry of pepper seed appeared to be helpful to determine the effectiveness of priming by a direct count of the number of G_2 nuclei (Lanteri *et al.*, 1993). Demilly *et al.* (1995) showed the usefulness of flow cytometry for seed testing, especially for ploidy determination of dry seeds and for seed species identification. They also demonstrated that it

R.H. Ellis, M. Black, A.J. Murdoch, T.D. Hong (eds.), Basic and Applied Aspects of Seed Biology, pp. 577–584.

was possible to test ploidy homogeneity of sugar-beet seed lots by seed embryo analyses.

Besides the ploidy, the quality, i.e. germination capacity and vigour, is also tested in seed production. The laboratory germination test usually follows the ISTA rules (ISTA, 1985). However, there is no standard vigour testing method (Fiala, 1989). Many different vigour tests for sugar-beet seed have been reported: moisture stress (Perry, 1978; Lovato and Cagalli, 1992), cold tests (Akeson and Widner, 1980; Kraak et al., 1984; Lovato and Cagalli, 1992), packed-sand test (Akeson and Widner, 1980), conductivity and accelerated ageing test (Kraak et al., 1984) and the root length vigour test (Sadowski, 1991). However, further investigations are still desired (Fiala, 1989).

The present study aims at adapting flow cytometry to sugar-beet seed testing. The nuclear replication stages in different tissues of dry mature sugar-beet seed have been determined. Seed lots of three sugar-beet varieties, disparate in vigour, have been investigated in order to study a correlation between G_2/G_1 ratio in radicle and germination characteristics such as germination capacity and root length vigour index (RLVI).

Material and Methods

Experiment 1

Commercial sugar-beet seeds of diploid variety PN Mono 1 and triploid variety Adonis were investigated. The following samples were analysed in 30 replications for each variety:

a)   whole seed ball (with pericarp),
b)   true seed (with no pericarp),
c)   true seed without radicle,
d)   radicle.

A simple blender for removing true seed from the pericarp was used.

The samples were prepared for flow cytometric analysis according to Galbraith et al. (1983), with some minor modifications. Seed tissues were chopped with a sharp razor blade in a Petri dish containing 0.5 ml nucleus isolation buffer (Tampon ploidie, Chemunex, Moisons-Alfort, France) supplemented by 4′,6-diamidino-2-phenylindole (DAPI; Colorant ploidie, Chemunex, Moisons-Alfort, France) with 1000:1 ratio. After choppings the suspension was passed through a 30 μm mesh nylon filter and analysed after about half an hour. Preparations were kept on ice.

DNA content was measured in a flow cytometer Partec CA II (Partec, Munster, Germany) equipped with an HBO 100W/2 lamp (Osram, Germany), KG1, B38, UG1 filters, CG435 barrier filter, TK420 dichromic mirror, and 40 × 0.8 quartz objective. All analyses were performed using logarithmic amplification. In each sample 2000 to 5000 nuclei were analysed at the rate of 40–100 nuclei/s. Computer programme Partec DPAC V2.0 has been used for data analysis. The weighted mean of 30 replications was calculated to determine the share of particular nuclear replication stages.

Experiment 2

Seeds of three sugar-beet varieties, PN Mono 1 (2×), Jastra (2×), and Maria (3×), representing deliveries from several growers, were used in this study. Three seed lots of each variety, different in vigour, were investigated.

Laboratory germination test was performed for 14 days (d) in pleated filter paper, at 20°C (ISTA, 1985) and at 65% substrate moisture (Jassem *et al.*, 1993). The germination capacity after 4 and 14 d as well as the root length vigour index (the percentage of seeds with primary root over 15 mm in length after 96 h; Sadowski, 1991) were determined. Three 100-seed replicates per seed lot were tested.

Radicles isolated from true seeds (two radicles for each sample) were used in flow cytometry. The samples were prepared and analysed in the same way as in experiment 1, in 30 replications for each seed lot.

A Student's *t*-test for independent samples was performed for the G_2/G_1 ratios in particular varieties. After angular transformation of germination test results, the analysis of correlation was made.

Results and Discussion

Experiment 1

Flow cytometric analysis of different sugar-beet seed tissues revealed four peaks (Fig. 1). The coefficient of variation (CV) for the G_1/G_0 peak ranged from 2.82% to 6.06%. Most of the cells were arrested in the G_1/G_0 phase of cell cycle (the peak 2C for a diploid variety and 3C for a triploid one). The next three peaks showed DNA amounts 1.5, 2 and 3 times higher than the biggest one did, respectively. They corresponded with DNA levels 3C, 4C and 6C for a diploid variety and 4C, 6C and 8C for a triploid one.

The third peak represented embryo cells arrested in G_2 phase whose share in total nuclear DNA content was relatively highest in the radicle (Table 1). It confirmed the results of Bino *et al.* (1992, 1993) who ascertained that the cells in G_2 phase are primarily located in the root-tip region of the embryo. In consequence, the G_2/G_1 ratio in radicle was also higher than that in other tissues. Therefore, the radicle seems to be most convenient for cell cycle investigations.

The second peak might correspond with S phase of embryo cells and/or with endosperm cells enclosing the lower part of the radicle (Artschwager, 1927; Bennett and Esau, 1936). The presence of 6C (2× variety) and 8C (3× variety)

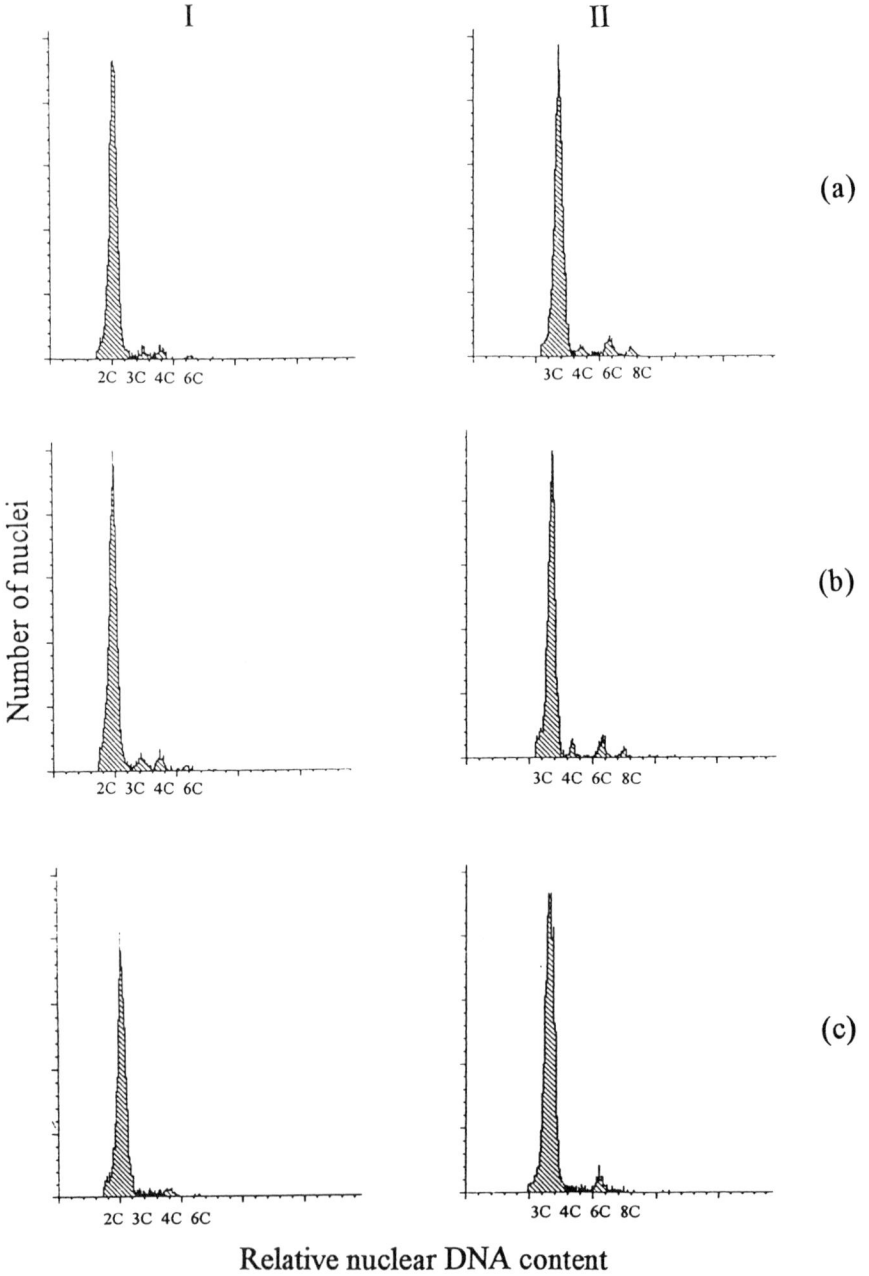

Figure 1. Flow cytometric analysis of nuclei from diploid (I) and triploid (II) sugar-beet seed. (a) whole seed ball, (b) true seed removed from the pericarp, (c) true seed without radicle, (d) radicle

peaks is probably caused by endoreplication (Marciniak, 1991; Bino *et al.*, 1992; 1993). Only the limited amount of 3C/6C (for diploid) and 4C/8C (for triploid) cells was found in true seed without radicle, whereas the radicle contained a much higher number of such cells (Table 1). This suggests that they originated from the endosperm. In addition to the peaks mentioned, the 12C, 16C and even 24C peaks were observed in some samples (data not shown), which confirmed the presence of endoreplicated cells in sugar-beet seed tissues.

The similarity between the distribution of nuclear replication stages in the whole seed ball and in the isolated true seed (Table 1) confirmed the suggestion that the cells of pericarp in mature sugar-beet seed ball contain no nuclei. Preparing samples for flow cytometry requires careful chopping of the tissue, which is much more troublesome for the whole seed ball than for the true seed removed from the pericarp. Therefore the latter is more convenient for the seed flow cytometric analysis.

Experiment 2

The results of laboratory germination test are shown in Table 2. The seed lots of particular varieties represented low, medium and high vigour, measured as a root length vigour index (RLVI).

Flow cytometric analysis made for each variety proved that seeds of low vigour show the highest G_2/G_1 ratio in radicle, i.e. the share of the cells arrested in G_2 stage of the cell cycle is relatively higher than in more vigorous seed (Table 2). Significant negative correlation between the G_2/G_1 ratio and RLVI was

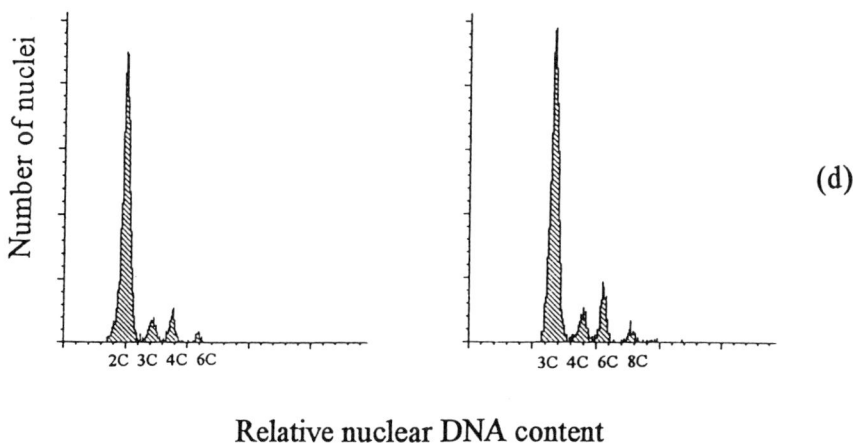

Relative nuclear DNA content

Figure 1 (cont).

Table 1. The share of nuclei with different DNA content expressed as C values in tissues of dry mature sugar-beet seeds of two varieties

Seed tissue	Relative nuclear DNA content (peak's area %)					G_2/G_1 ratio
	2C	3C	4C	6C	8C	
PN Mono 1 (2×)						
Whole seed ball	89.41	3.38	5.90	1.31		0.066
True seed	85.86	4.23	7.12	2.79		0.083
True seed without radicle	92.46	1.05	5.56	0.93		0.060
Radicle	82.36	5.93	9.60	2.11		0.116
Adonis (3×)						
Whole seed ball		94.45	2.22	2.56	0.77	0.027
True seed		93.44	2.06	3.75	0.75	0.040
True seed without radicle		97.67	0.11	2.15	0.07	0.022
Radicle		83.75	4.58	8.61	3.06	0.103

Table 2. Results of flow cytometric analysis and laboratory test for different seed lots of three sugar-beet varieties

Variety	G_2/G_1 ratio	Root length vigour index (%)	Germination capacity after 4 days (%)	Germination capacity after 14 days (%)
PN Mono 1 (2×)	0.066 a[a]	20	66	84
	0.052 a	45	88	92
	0.040 b	52	87	92
Jastra (2×)	0.144 a	20	53	76
	0.125 a	31	60	77
	0.102 b	48	85	85
Maria (3×)	0.101 a	2	40	88
	0.075 b	39	73	79
	0.068 b	55	80	88

[a]Values for a particular variety followed by the same letter are not significantly different at $p = 0.05$ (Student's t-test for independent samples)

Table 3. Correlation between G_2/G_1 ratio in radicle and laboratory test parameters for the seed of three sugar-beet varieties

Variety	Laboratory test parameters	G_2/G_1 ratio
PN Mono 1 (2×)	Root length vigour index	−0.960*
	Germination capacity after 4 days	−0.863
	Germination capacity after 14 days	−0.887
Jastra (2×)	Root length vigour index	−0.999**
	Germination capacity after 4 days	−0.961*
	Germination capacity after 14 days	−0.930*
Maria (3×)	Root length vigour index	−1.000**
	Germination capacity after 4 days	−1.000**
	Germination capacity after 14 days	0.315

*Significant at $p = 0.05$
**Significant at $p = 0.01$ (Student's t-test)

found (Table 3). This correlation confirmed the suggestion of Bino *et al.* (1992, 1993) that the arrest of cell cycle activity in G_1 may have a physiological significance. Cells at G_2 seem to be more sensitive to stress conditions (Deltour, 1985) and to factors affecting nuclear division and chromosome morphology (Sybenga, 1972). It is possible that the external conditions during maturation and storage of seeds as well as delayed seed harvest increase the share of G_2 cells (Bino *et al.*, 1993).

The present study suggests that a low proportion of G_2 cells in sugar-beet seeds suggests that fast and uniform germination can be expected. It is proposed that flow cytometric analysis of sugar-beet seed could be used in seed production for identification of low vigour seed deliveries which should be eliminated or specially treated (washed, primed) before blending and marketing.

Acknowledgements

The author would like to thank Deleplanque et CIE, France, for affording possibilities for flow cytometric analyses, Mr. Arnauld Godefroy for technical assistance, Prof. M. Olszewska and Prof. M. Jassem for scientific consultations.

References

Akeson, W. R. and Widner, J. N. 1980. *Crop Science* 20: 641–644.

Artschwager, E. 1927. *Journal of Agricultural Research* 34/1: 1–25.

Bennett, C. W. and Esau, K. 1936. *Journal of Agricultural Research* 53/8: 595–620.

Bino, R. J., De Vries, J. N., Kraak, H. L. and Van Pijlen, J. G. 1992. *Annals of Botany* 69: 231–236.

Bino, R. J., Lanteri, S., Verhoeven, H. A. and Kraak, H. L. 1993. *Annals of Botany* 72: 181–187.

Brown, S. C., Devaux, P., Marie, D., Bergounioux, C. and Petit, P. X. 1991. *Le Technoscope de Biofutur* 47: 2–14.

De Laat, A. M. M., Gohde, W. and Vogelzang, M. J. D. C. 1987. *Plant Breeding* 99: 303–307.

Deltour, R. 1985. *Journal of Cell Science* 75: 43–83.

Demilly, D., Lamaire M., Billy, B. and Ducournau, S. 1995. *24th ISTA Congress, Seed Symposium, Copenhagen, Denmark, June 7–16.*

Fiala, F.Y. 1989. *Seed Science and Technology* 17: 153–167.

Galbraith, D. W. 1984. *Cell Culture and Somatic Cell Genetics of Plants* 1: 765–777.

Galbraith, D. W., Harkins, K. R., Maddox, J. M., Ayres, N. M., Sharma, D. P. and Firoozabady, E. 1983. *Science* 220: 1049–1051.

ISTA (International Seed Testing Association). 1985. *Seed Science and Technology* 13, Annex 5.

Jassem, M., Sliwinska, E. and Zornow, A. 1993. *Seed Science and Technology* 21: 203–211.

Kraak, H.L., Vos, J., Perry, D. A. and Bekendam, J. 1984. *Seed Science and Technology* 12: 731–745.

Lanteri, S., Bino, R. J. and Kraak, H. L. 1992. *Capsicum Newsletter* 4: 249–253.

Lanteri, S., Kraak, H. L., De Vos, C. H. R., and Bino, R. J. 1993. *Physiologia Plantarum* 89: 433–440.

Lovato, A. and Cagalli, S. 1992. *Seed Science and Technology* 21: 61–67.

Marciniak, K. 1991. *Acta Societatis Botanicorum Poloniae* 60/3–4: 273–284.

Perry, D. A. 1978. *Seed Science and Technology* 6: 159–181.

Sadowski, H. 1991. *Biuletyn Instytutu Hodowli i Aklimatyzacji Roslin* 177: 71–82.

Sybenga, J. 1972. *General Cytogenetics*: 34–56. New York: Elsevier.

Sliwinska, E. and Steen P. 1995. *Journal of Applied Genetics* 36(2): 111–118.

65. Effect of Elevated Temperatures during Grain Development on Seed Quality of Barley (*Hordeum vulgare* L.)

I.S.K. SYANKWILIMBA, M.P. COCHRANE and C.M. DUFFUS

Crop Science and Technology Department, The Scottish Agricultural College (SAC, Edinburgh) West Mains Road, Edinburgh EH9 3JG, Scotland, UK

Abstract

In order to investigate the effects of elevated temperatures during grain development on seed quality, an experiment was carried out using two controlled environment rooms. Plants of barley (*Hordeum vulgare* L. cv. Blenheim) were subjected to different temperature regimes (30°C as a maximum) during grain development and, after harvest, the seeds were tested for vigour. It was found that elevated temperatures imposed during the early stages of grain development caused a reduction in grain dry weight. Seed lots produced under elevated temperature regimes were also found to differ in germination characteristics and seedling growth. Root numbers, length of seminal roots and seedling dry matter showed strong positive correlations with grain dry weight. There was no statistically significant correlation between the length of the first leaf (plumule length) and grain dry weight.

Introduction

Cereal crops grown in many parts of the tropics are routinely subjected to temperatures above 30°C. Such elevated temperatures have been shown to exert adverse effects on yield and on grain quality. Little is known, however, about the specific effects of elevated temperatures during grain development on the quality of the grain in relation to seed vigour. Experiments conducted under field conditions and in controlled environments showed that grains grown at high temperatures had lower grain weight than those grown at lower temperatures (MacLeod and Duffus, 1988; Tester *et al.*, 1991; Savin *et al.*, 1994; Jenner, 1994). Since elevated temperatures during grain development cause a reduction in grain weight, it is possible that they may also affect seed quality.

An understanding of the effect of elevated temperatures during grain development of barley may be important for the production of high quality seed in tropical countries such as Zambia. Such information may also be useful to plant breeders involved in the development of cultivars better adapted to various temperature conditions.

R.H. Ellis, M. Black, A.J. Murdoch, T.D. Hong (eds.), Basic and Applied Aspects of Seed Biology, pp. 585–592.
© *1997 Kluwer Academic Publishers, Dordrecht. Printed in Great Britain.*

Materials and Methods

Plants of cv. Blenheim were grown, six plants per pot (20 cm diameter), in peat - based compost in a glasshouse. Prior to anthesis, plants were transferred to two identical growth rooms and were kept at 18°C and 16 h day length at a light intensity of 140 $\mu E\,m^{-2}\,s^{-1}$ at ear level. As soon as 50% of the ears had anthesed (4 May, 1994), plants were subjected to different temperature regimes as indicated below:

Temperature Regimes

1. 18°C to harvest-ripeness.
2. 18°C for 25 days (d) and then 30°C to harvest-ripeness.
3. 30°C for 10 d and then 18°C to harvest-ripeness.
4. 30°C to harvest-ripeness.

Day one of the temperature treatment was 5 May. Ears were harvested 960°C days after anthesis, hand-threshed, and the grains were divided into two seed lots: date 1 – ears anthesed between 23 April and 4 May; date 2 – ears anthesed 5–14 May. Germination tests were carried out in 90 mm diameter Petri dishes containing two filter papers to which 5 ml, 8 ml or 10 ml of distilled water was added. Chitted seeds were removed each day and the germination percentage was determined after 72 h.

Seedling growth analysis was carried out using paper towelling. Grains were weighed individually before they were glued to the paper towelling with a non-toxic adhesive. There were four replicates of 25 grains each. The papers were sprayed with distilled water, rolled loosely and then placed in an upright position in a wire basket. The baskets were enclosed in black polythene bags to minimize water loss and to prevent any light reaching the grains. They were incubated at 5°C for 4 d to break any dormancy in the grains, and then they were transferred to an incubator at 18°C. Seedling measurements were taken after 7 d. Roots and plumules were cut off at grain level and they and the remainder grains (grains after the removal of plumules and roots) were dried to constant weight in an oven at 70°C.

Results

When tested using 5 ml of water per Petri dish, grains grown in all temperature regimes had a percentage germination >94% indicating uniformly high viability (Fig. 1). However, grains which had experienced elevated temperatures during grain development had a higher percentage germination at 8 ml and 10ml of water than grains which had experienced a low temperature throughout grain development. Grains grown in the low temperature regime had a higher grain dry weight than those grown in elevated temperature regimes (Fig. 2).

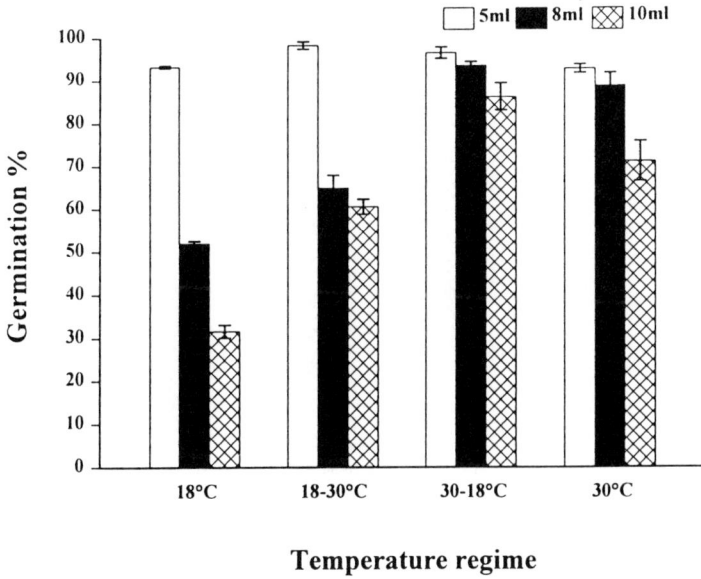

Figure 1. Germination in 3 water levels (5 ml, 8 ml and 10 ml) of grain of barley cv. Blenheim grown under different temperature regimes during grain development. Error bars SEM; n = 3. Seeds for these germination tests were obtained from the seed lots harvested from ears that anthesed between 23 April and 4 May (date 1)

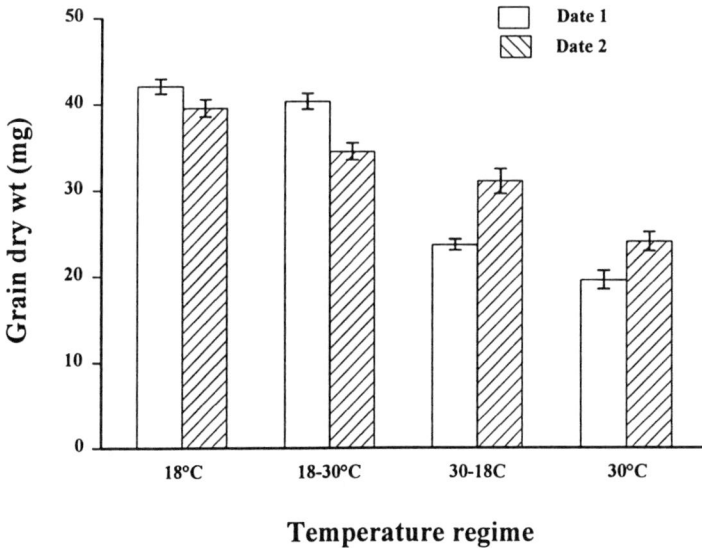

Figure 2. Grain dry weight of cv. Blenheim grown under different temperature regimes. Error bars SEM; n = 4. Date 1 represents ears anthesed between 23 April and 4 May; Date 2 represents ears anthesed 5–14 May

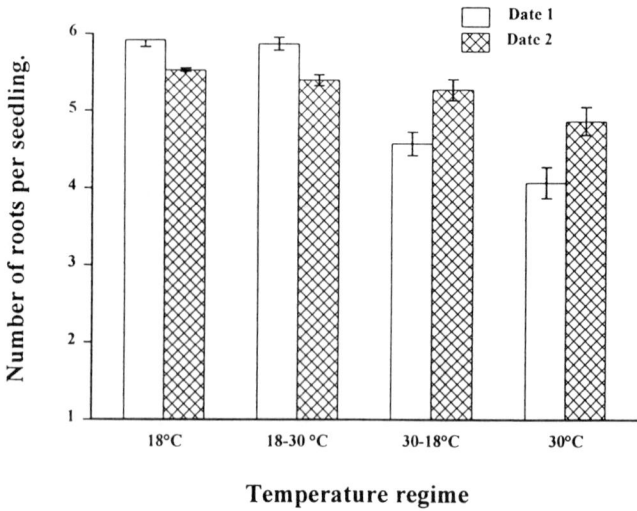

Figure 3. Number of roots in 7 day old seedlings grown from grains of cv. Blenheim. Dates as in Figure 2. Error bars SEM; n = 4

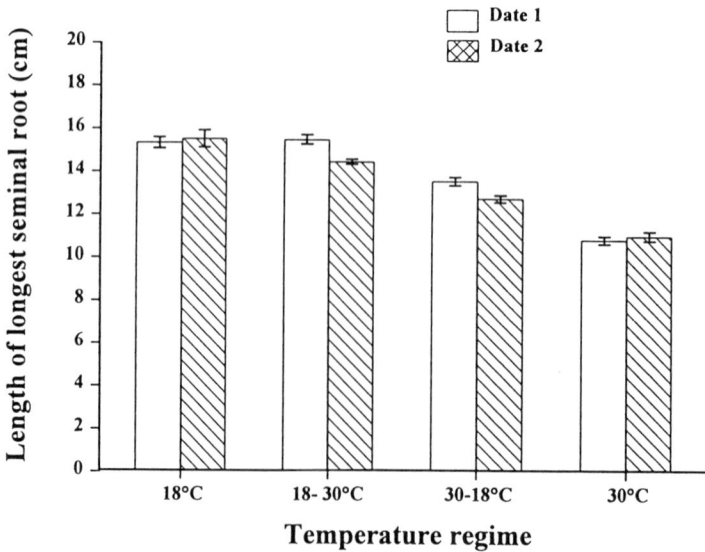

Figure 4. Root length of the longest seminal root (cm) of 7 day old seedlings grown from grains of cv. Blenheim. Dates as in Figure 2. Error bars SEM; n = 4

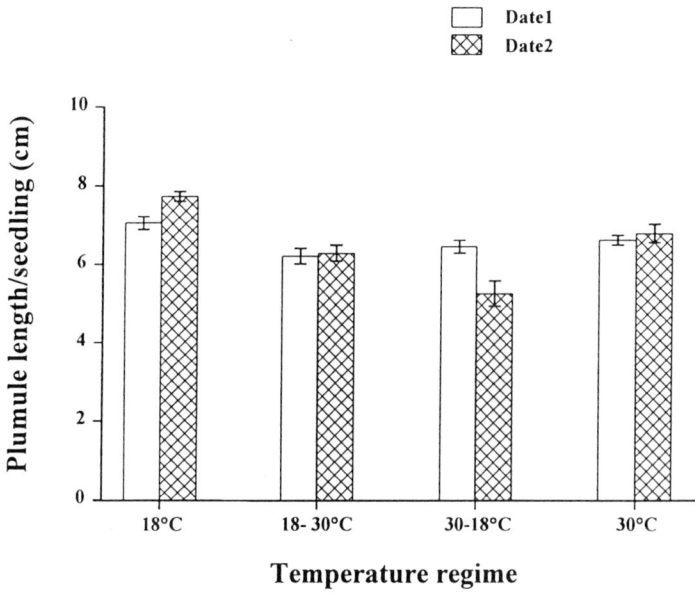

Figure 5. Plumule length (cm) of 7 day old seedlings grown from grains of cv. Blenheim. Dates as in Figure 2. Error bars SEM; n = 4

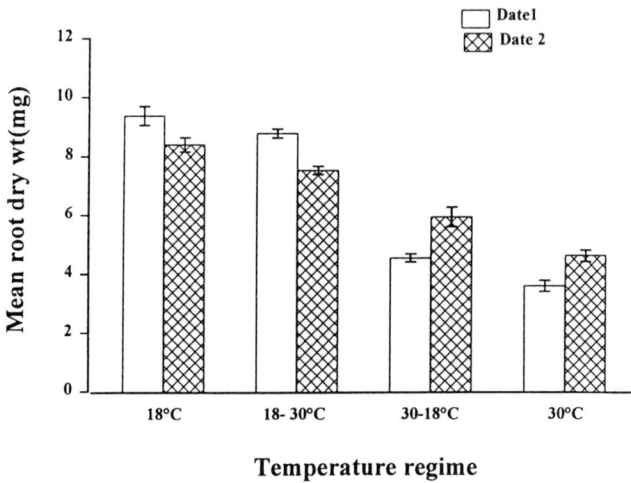

Figure 6. Root dry weight (mg) of 7 day old seedlings grown from grains of cv. Blenheim. Dates as in Figure 2. Error bars SEM; n = 4

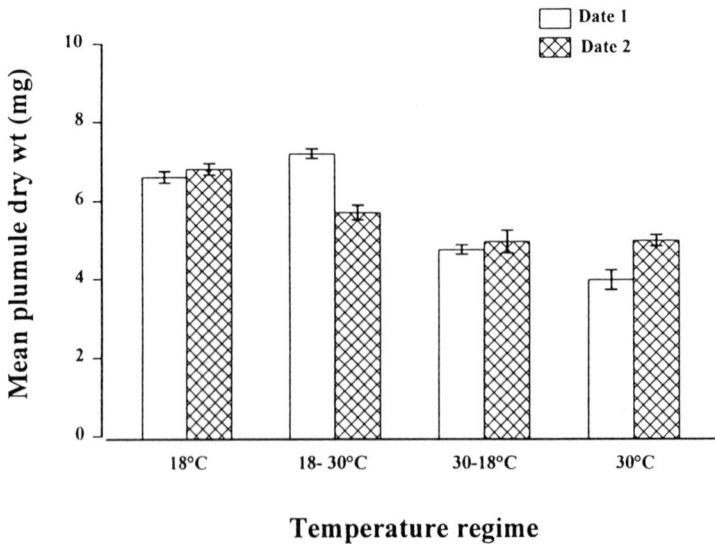

Figure 7. Plumule dry weight (mg) of 7 day old seedlings grown from grains of cv. Blenheim. Dates as in Figure 2. Error bars SEM; $n = 4$

Table 1. Correlation coefficients between seedling growth characteristics

	Grain dry wt	Root number	Root length	Plumule length	Root dry wt	Plumule dry wt
Grain dry wt						
Root number	0.925					
Root length	0.869	0.791				
Plumule length	0.163	0.010	0.209			
Root dry wt	0.986	0.919	0.897	0.195		
Plumule dry wt	0.937	0.881	0.844	0.277	0.929	
Remainder grain dry wt	0.954	0.855	0.803	0.226	0.948	0.899

Seedlings from grains grown in elevated temperatures had fewer roots and had seminal roots which were shorter than those of seedlings from grains which had experienced low temperature throughout development (Figs. 3 and 4). Plumule length was similar in the seedlings grown from all the seed lots (Fig. 5). The mean dry weights of the roots and plumules of seedlings from grains grown at 18°C were higher than those of seedlings from grains grown at 30°C (Figs. 6 and 7). Root numbers, length of seminal roots and seedling dry matter showed strong positive correlations with grain dry weight. There was no statistically significant correlation between the length of the first leaf (plumule length) and grain dry weight (Table 1). The root dry weight : shoot dry weight ratios for seedlings from grains grown at 18°C were 1.42 and 1.23 for dates 1 and 2 respectively, and, for seedlings from grains grown at 30°C, 0.90 and 0.92 for dates 1 and 2 respectively. In all seed lots, over 31% of the dry weight of the grains remained after 7 d of seedling growth (data not shown).

Discussion

The results of germination tests using 10 ml of water showed that low percentage germination was observed in grain grown at 18°C from anthesis until harvest-ripeness and in grain that experienced low temperature during the early stages of grain development and then reached harvest-ripeness at an elevated temperature. Kelly and Briggs (1992) attributed low percentage germination under excess water to the presence of microbial activity that competed for oxygen. The seed lots which were grown in temperature regimes 1 and 3 were harvested from the same growth room and so were likely to have had similar microfloras on the surface of their grains, but nevertheless, they had very different germinabilities. Microorganisms located inside the pericarp may have influenced germination, but it is perhaps more likely that the low percentage germination obtained in grains grown in temperature regimes 1 and 2 was due to low temperature-induced dormancy which was not expressed when 5ml water was used in the germination test. Grains from plants which were in an elevated temperature regime (30°C) for the whole period of grain development, had a lower mean dry weight than the grains grown in the other temperature regimes. This observation supports the findings of Nicholls (1982), Wardlaw *et al.* (1989), Jenner (1991), and Savin *et al.* (1994) who reported that single grain weight was less for grains grown at high temperatures than for those grown at lower temperatures.

The seedling growth analysis revealed that elevated temperatures at early stages of grain development may subsequently affect seedling growth characteristics. Data indicated that grains from ears grown at a low temperature from anthesis until harvest-ripeness produced strong seedlings with more roots, longer seminal roots and more dry matter than the grains that were exposed to elevated temperature for part or all of the grain development period. There were strong positive correlations between seedling characteristics and grain dry weight, but there was no statistically significant correlation between grain dry weight and the length of the plumule. It is interesting to note that plumule length in this seedling growth test has been used as an indicator of seed vigour (Perry, 1977). In this regard, the number of roots per seedling and the length of the longest seminal roots may be better indicators of vigour in adverse environmental conditions in the tropics where seedlings that have a long extensive root system may have a better chance of establishment. The reason being that, under drought conditions, long roots may be able to grow into lower soil layers and thus continue drawing water from the soil to sustain seedling growth.

The root dry weight/shoot dry weight ratios indicate that seedlings from grains grown at low temperature and those from grains grown at elevated temperatures differed in their partitioning of carbon between root and shoot. Seedling dry weight was lowest in the seed lots grown at 30°C, but even in these seed lots, 31.8% percent of grain dry weight remained after 7 d seedling growth in the dark, and so it would appear that the relatively poor seedling growth in these seed lots was not due to lack of food reserves. Differences in growth may have been due to differences in embryo size, or in the efficiency of mobilization and utilization of food reserves.

Acknowledgement

This research project was funded by the Zambian Government through the Zambian Agricultural Research and Extension Project (ZAREP) from African Development Bank and World Bank funds.

References

Jenner, C. F. 1991. *Australian Journal of Plant Physiology* 18: 165–177.
Jenner, C. F. 1994. In: *Proceedings of 44th Australian Cereal Chemistry Conference,* pp. 69–71. Ballarat, Australia.
Kelly, L. and Briggs, D. E. 1992. *Journal of the Institute of Brewing* 98: 395–400.
MacLeod, L. C and Duffus, C. M. 1988. *Australian Journal of Plant Physiology* 15: 367–375.
Nicholls, P. B. 1982. *Australian Journal of Plant Physiology* 9: 373–383.
Perry, D. A. 1977. *Seed Science and Technology* 5: 709–719.
Savin, R., Stone, P. J and Nicolas, M. E. 1994. In: *Proceedings of 44th Australian Cereal Chemistry Conference*, pp. 58–59. Ballarat, Australia.
Tester, R. F. South, J. B., Morrison, W. R. and Ellis, R. P. 1991. *Journal of Cereal Science* 13: 113–127.
Wardlaw, I. F., Dawson, I. A. and Manibi, P. 1989. *Australian Journal of Agricultural* Research 40: 15–24.

66. Relationship between Standard Germination, Accelerated Ageing Germination and Field Emergence in Soyabean

D.M. TEKRONY and D.B. EGLI

Department of Agronomy, University of Kentucky, Lexington, KY 40546-0091, USA

Abstract

Twenty six field emergence experiments were conducted over a 10 year period (two to four planting dates per year) to evaluate the effect of seed bed conditions on the relationship between standard germination (SG), accelerated ageing germination (AA) and field emergence in soyabean (*Glycine max* L. Merrill). Seed bed conditions were characterized by the field emergence index (FEI = mean field emergence/mean standard germination × 100). The ability of the laboratory tests to predict field emergence was evaluated using the prediction accuracy (proportion of the seed lots in each treatment with a specified quality level that had a field emergence above a minimum level). The FEI varied from 108 to 44 across the 26 experiments and the prediction accuracy varied from 0 to 100%. The prediction accuracy was high for both SG and AA in ideal field conditions and decreased as soil stress increased. The AA test had a higher prediction accuracy than SG in moderate stress conditions. Lowering the minimum acceptable field emergence from 80 to 60% improved the prediction accuracy for SG and AA, but only seed lots with AA ⩾ 80% had acceptable prediction accuracy over a wide range in seed bed conditions.

Introduction

The combination of planting rate and the proportion of the planted seeds that emerge determine the plant population. Emergence varies widely and is influenced by the germination and vigour of the planting seed and the conditions in the seed bed. Soil temperature, water and oxygen levels, micro-organisms and soil structure affect the ability of the seed to germinate and the seedling to emerge from the soil (Burris, 1976; Powell, 1988).

Seed quality is measured and reported to provide an indication of expected emergence when the seed is planted in the field. However, it has been difficult to establish consistent relationships between measures of quality and field emergence. Significant correlations of standard germination or various measures of seed vigour and field emergence have been reported for soyabean (Edje and Burris, 1971; TeKrony and Egli, 1977; Johnson and Wax, 1978; Kulik and Yaklich, 1982). The relationships were not consistent across experiments and could not predict potential performance. Combining results from several tests into a vigour index (TeKrony and Egli, 1977) or a regression model (Luedders

R.H. Ellis, M. Black, A.J. Murdoch, T.D. Hong (eds.), Basic and Applied Aspects of Seed Biology, pp. 593–600.
© *1997 Kluwer Academic Publishers, Dordrecht. Printed in Great Britain.*

and Burris, 1979) did little to improve prediction accuracy. These inconsistent relationships between quality and emergence are probably due to variation in seed bed conditions. It may be unrealistic to expect tests conducted in controlled conditions to relate to performance in the wide range of soil conditions that may be encountered in the field.

Egli and TeKrony (1995) included seed bed conditions in their evaluation of the relationship between standard germination, accelerated ageing germination and field emergence. Both tests accurately predicted emergence in non-stress field environments but the ability to predict emergence decreased as soil stress increased. They found that accelerated ageing germination was a better predictor of field emergence than standard germination in less than ideal field conditions. Thus, the relationship between laboratory quality tests and field emergence was dependent upon seed bed conditions.

Our objective in this research was to extend our previous work (Egli and TeKrony, 1995) evaluating the relationship between seed quality and field emergence in a range of seed bed conditions by including consideration of the minimum field emergence needed to produce an adequate plant population.

Materials and Methods

Field emergence experiments were conducted near Lexington, KY over 10 years (Table 1, Egli and TeKrony, 1995). Some seed lots were harvested the previous autumn and some had been in warehouse or controlled environment storage for 18 to 30 months. Standard germination (AOSA, 1983) and accelerated ageing germination (AOSA, 1993) were measured before planting the first emergence experiment each year.

There were one to four planting dates each year to provide a range in seed bed conditions (Table 1). One hundred seeds were sown in a single 6 m row at a depth of 2.5 to 5.0 cm. There were three or four replications in a randomized complete block design. Final emergence counts were taken after emergence had stopped.

The field emergence index (FEI, Eq. 1) was calculated for each field emergence experiment as an index of seed bed conditions (Egli and TeKrony, 1995).

$$\text{FEI} = \frac{\text{Mean emergence}}{\text{Mean standard germination}} \times 100 \qquad [1]$$

An FEI of 100 represented ideal seed bed conditions and seed bed stress increased as the index decreased below 100.

The performance of seed lots was assessed by determining if the emergence was equal to or above a minimum acceptable emergence. A quality test accurately predicted the performance of a seed lot if the emergence of the lot was above the minimum. Prediction accuracy was computed for each experiment as the proportion of the seed lots having a defined quality level (e.g.

SG ≥ 80%) that had emergence above the minimum level (Egli and TeKrony, 1995). For example, if there were 10 seedlots in an experiment with standard germination equal to or above 80% (the defined quality level) and 5 of these seed lots had a field emergence equal to or above 80% (the assumed minimum field emergence), the prediction accuracy would be 50% (5/10). We evaluated minimum field emergence levels of 60, 70 and 80%.

The prediction accuracy concept evaluates predictive ability of a test relative to the seed lot producing an acceptable plant population instead of trying to predict the exact level of emergence.

Results

Average standard germination was above 80% for all years except 1985 (Table 1) and the germination of most seed lots in each experiment was above 80% (Egli and TeKrony, 1995). The average accelerated ageing germination was much lower and the range within each experiment was usually from 90% or greater to near zero (Egli and TeKrony, 1995). Mean field emergence ranged from 39 to 86% (Table 1) and the FEI varied from 108 to 44, reflecting the wide range in seed bed conditions across the 26 experiments.

Table 1. Mean seed quality and field emergence of the seed lots used in the 26 field emergence experiments, 1980 to 1993

| | | | | Field emergence experiment | | | | | |
| | | Germination | | 1 | | 2 | | 3 | |
Year	Seed lots no.	Standard (%)	Accelderated ageing (%)	PD*	EM (%)	PD	EM (%)	PD	EM (%)
1980**	40	82	62	4/7	54	4/21	57	5/1	64
1981	52	90	42	4/8	64	6/11	79	–	–
1984	29	89	63	5/12	72	7/10	67	–	–
1985	12	77	50	4/16	64	6/29	83	–	–
1987	16	91	77	4/29	79	5/14	86	6/15	85
1988	33	91	61	5/4	54+	5/11	81	7/29	79
1989	17	88	60	4/25	39	5/4	45	6/8	58
1991	15	95	82	4/17	64	6/5	71	7/2	75
1992	38	94	62	4/20	64	6/3	52	7/2	70
1993	20	90	41	6/2	59	–	–	–	–

*PD, planting date; EM, emergence

**A fourth planting was made in 1980 on July 12, Mean emergence was 77%

+, 17, only 17 seed lots were included in the first planting

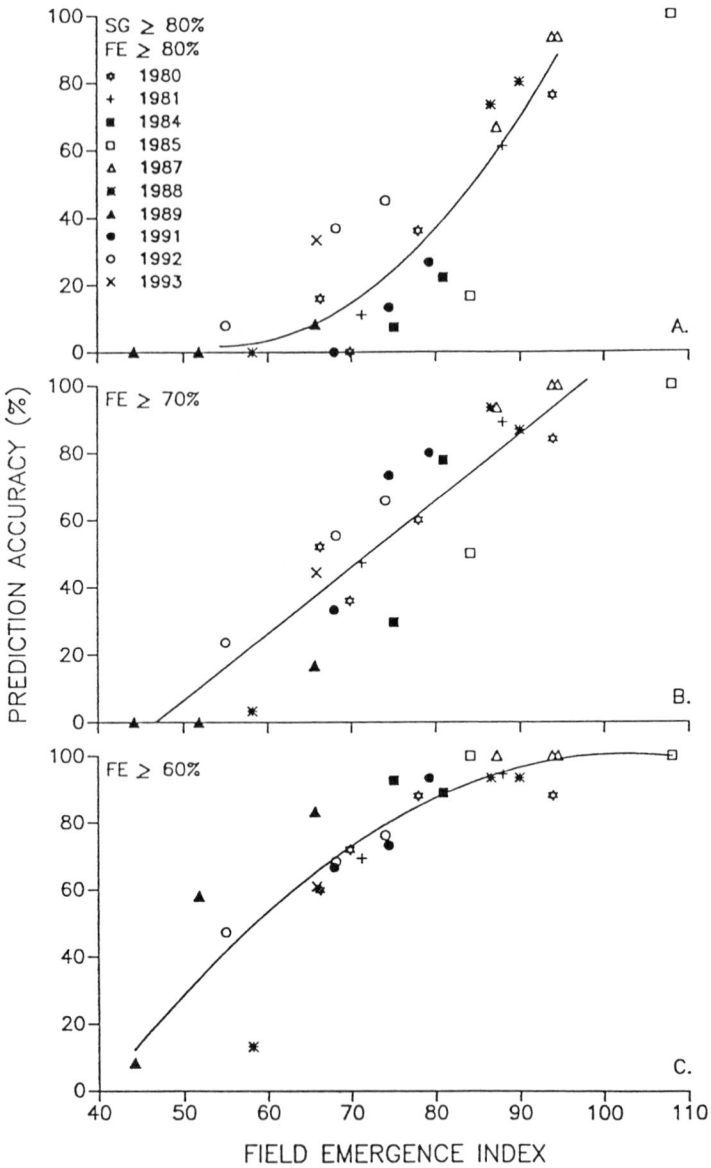

Figure 1. The relationship between prediction accuracy and field emergence index for soyabean seed lots with standard germination of 80% or more and a minimum field emergence of 80% (A) (from Egli and TeKrony, 1995), 70% (B), or 60% (C). The regression equations relating prediction accuracy (PA) to field emergence index (FEI) are (n = 26) (A) PA = 162.23 – 5.88 FEI+0.0539 FEI^2, R^2 = 0.80, (B) PA = 92.374+1.977 FEI, r^2 = 0.82, (C) PA = –173.60+5.354 FEI – 0.026 FEI^2, R^2 = 0.81

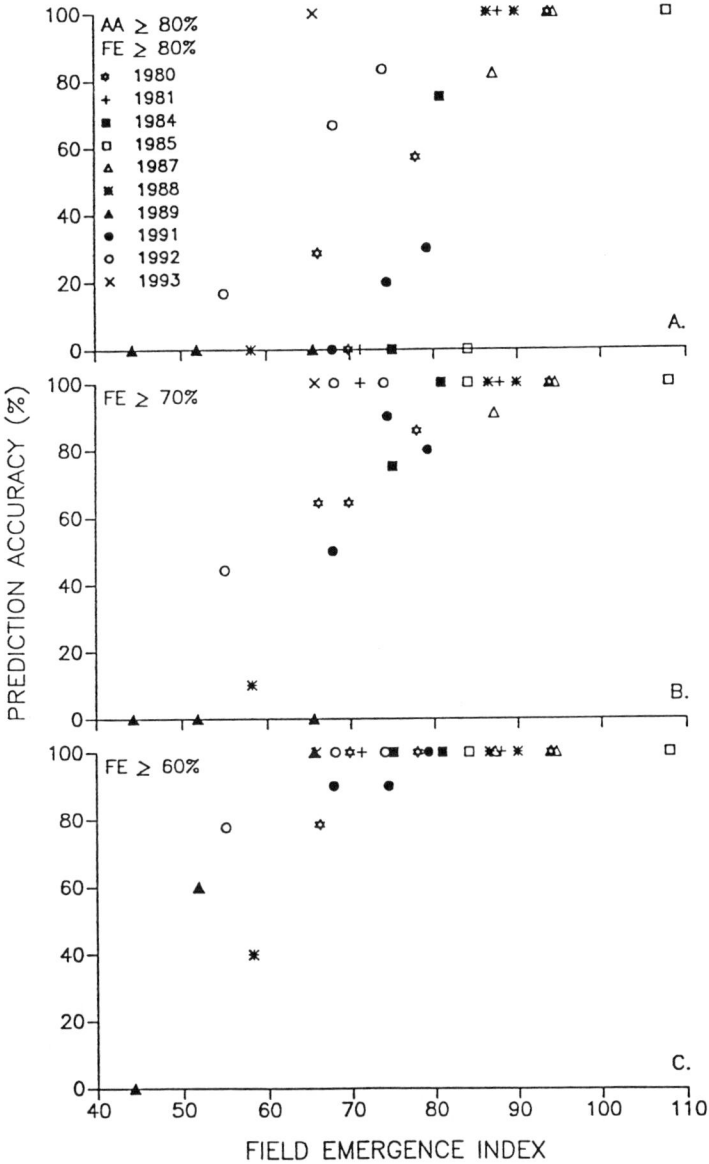

Figure 2. The relationship between prediction accuracy and field emergence index for soyabean seed lots with an accelerated ageing of 80% or greater and minimum field emergence of 80 (A), 70 (B) or 60% (C)

The prediction accuracy for seed lots with a standard germination of 80% or greater was 100% for minimum field emergence levels of 60, 70, and 80% in ideal field conditions (FEI $\geqslant$ 100, Fig. 1). As the FEI decreased (seed bed stress increased), the prediction accuracy decreased, but the decrease occurred faster for a minimum field emergence of 80% (Fig. 1A) than for minimum field emergence levels of 70 (Fig. 1B) or 60% (Fig. 1C). Most of the seed lots with 80% or greater germination produced satisfactory stands at an FEI of 80 if only 60% emergence was required, but less than half of the seed lots exhibited satisfactory performance if 80% emergence was required. The prediction accuracy decreased rapidly in all cases as the FEI decreased below 80.

Selecting seed lots based on the accelerated ageing test (AA $\geqslant$ 80%) provided higher prediction accuracy at lower FEI (Fig. 2) than standard germination. Prediction accuracy remained high at lower FEI when the minimum field emergence was 60 (Fig. 2C) compared with 80%. Prediction accuracy remained above 80% until the FEI approached 85 for a minimum field emergence of 80%, and a FEI of 80 for a minimum field emergence of 70%. However, the prediction accuracy was above 80% until the FEI reached 65 for a minimum field emergence of 60%.

Discussion

Prediction accuracy is a measure of the ability of quality tests to identify seed lots that produce satisfactory plant populations in the field. A satisfactory plant population will be adequate for maximum yield and is defined by the planting rate and a selected level of field emergence. Ideally, the prediction accuracy would equal 100%, meaning that all seed lots with the selected quality level produced satisfactory populations. For seed lots with a standard germination $\geqslant$ 80%, this ideal level of prediction occurred only when FEI $\geqslant$ 100 (Fig. 1) Thus, standard germination accurately predicted performance only in ideal field conditions as reported previously (TeKrony *et al.*, 1987; Egli and TeKrony, 1995). Prediction accuracy for a vigour test, accelerated ageing, was higher than for standard germination (Fig. 2) in less than ideal field conditions (FEI $<$ 100), which is consistent with the widely held opinion that vigour tests provide better estimates of field performance in stress conditions (TeKrony *et al.*, 1987). Under severe stress (FEI $<$ 60), neither of the two tests were able to predict performance.

The practical interpretation of our results obviously depends on knowing the range in FEI that will normally occur in producers fields. We found no relationship between soil temperature and FEI (Egli and TeKrony, 1995); consequently we could not use environmental data to estimate FEI. However, our data from 26 field experiments provide an estimate of the FEI that might occur in Kentucky. Excluding the unrealistically-early April planting dates, there were 18 experiments with planting dates in May, June and July. The average FEI for these experiments was 78 and, more importantly, the range was from 52 to 108. The FEI of seven of the experiments (39%) was between 60 and 80. Thus, soyabean in Kentucky will frequently be planted in seed beds where the FEI is between 60 and 80 and it would be helpful to have a quality test that had a high prediction accuracy for these seed beds.

Standard germination and accelerated ageing did not produce a high prediction accuracy for FEI between 60 and 80 when the minimum field emergence level was 80% (Fig. 1, Egli and TeKrony, 1995). Raising the quality level to 90% improved prediction accuracy but the prediction accuracy's were still not high for FEI of 60 to 80 (Egli and TeKrony, 1995).

Another factor, not previously considered, that influences the prediction accuracy is the minimum field emergence required to produce a plant population that is adequate for maximum yield. Lowering the minimum emergence to 70 or 60% resulted in higher prediction accuracy at lower FEI for both standard germination and accelerated ageing germination (Figs. 1 and 2). The prediction accuracy for seed lots with an accelerated ageing germination of 80% or greater and 60% emergence was near 100% until the FEI reached 65. By comparison, seed lots with a standard germination of 80% or greater had a prediction accuracy of only 50% in the same conditions. Clearly, prediction accuracy can be improved by lowering the minimum field emergence. In fact, we were able to get high prediction accuracy over the range of FEI likely to be encountered in normal soyabean production (100 – 65) by using high vigour seed and lowering the minimum field emergence to 60%. Recommended seeding rates in Kentucky are well above the populations needed for maximum yield (Herbek and Bitzer, 1988) and 60% emergence will provide adequate plant populations.

Conclusions

At planting time the objective of a soyabean producer is to obtain a plant population that is adequate for maximum yield. Planting seed quality, seed bed conditions and planting rate combine to determine emergence and population. Seed quality and seed bed conditions were important in determining emergence in our experiments. Seed lots selected on the basis of standard germination produced adequate populations only in ideal field conditions. Lowering the minimum field emergence needed to produce an adequate population did not improve the performance of the standard germination test to adequate levels. To be sure of getting adequate populations across the range in seed bed conditions likely to occur in normal soyabean production it was necessary to lower the minimum emergence level to 60% and use seed lots with an accelerated ageing germination of 80% or greater.

The use of high vigour seed lots and recommended planting rates ensured adequate plant populations over a wide range in field conditions. However, at extremely low FEI's none of the seed lots, regardless of the vigour level, produced adequate emergence.

References

Association of Official Seed Analysts. 1983. *Seed Vigour Testing Handbook.* Publication No. 32, AOSA, USA.

Association of Official Seed Analysts. 1993. *Journal of Seed Technology* 16: 1–113.

Burris, J.S. 1976. *Journal of Seed Technology* 1: 158–74.

Edje, O.T. and Burris, J.S. 1971. *Agronomy Journal* 63:536–538.

Egli, D.B. and TeKrony, D.M. 1995. *Seed Science and Technology* 23: 595–607.

Herbek, J.H. and Bitzer, M.J. 1988. *Soybean Production in Kentucky Part III. Planting Practices and Double Cropping.* Lexington, KY: Univ. of KY.

Johnson, R.R. and Wax, L.M. 1978. *Agronomy Journal* 70: 273–278.

Kulik, M.M. and Yaklich, R.W. 1982. *Crop Science* 22: 766–770.

Luedders, V.D. and. Buries, J.S. 1979. *Agronomy Journal* 71: 877–879.

Powell, A.A. 1988. *Advances in Research and Technology of Seeds* 11: 29–61.

TeKrony, D.M. and Egli, D.B. 1977. *Crop Science* 17: 573–577.

TeKrony, D.M., Egli, D.B. and White, G.M. 1987. In: *Soybeans: Improvement, Production and Uses,* 2nd Edition (ed. J.R. Wilcox). Madison, WI.: ASA-CSSA-SSSA.

67. Gene Expression During Maize Seed Germination

J.J. ZUÑIGA, F. CRUZ, A. GOMEZ, M.P. SANCHEZ and
J.M. VAZQUEZ-RAMOS
Departamento de Bioquímica, Facultad de Química, UNAM. Avenida Universidad y Copilco, México 04510, D.F., México

Abstract

Two experimental approaches have been used to study changes in gene expression during maize germination: the germination of osmoprimed seeds and the stimulation of germination by cytokinins. Both have in common that DNA replication and the cell cycle take place several hours earlier than it would happen during germination of non-treated seeds. Moreover, in the case of cytokinin-stimulated seeds, the hormonal stimulatory effect disappears if embryo axes are also treated with RNA synthesis inhibitors, indicating that transcription is necessary. We have been studying the changes in gene expression that occur when osmoprimed seeds are put to germinate or when seeds are treated with cytokinins. The methodology used is Differential Display. Although we have not been able to find many differences, several messages seem to appear as a result of accelerating germination by either of the two treatments. The corresponding cDNAs to the messages have been partially sequenced and the sequence compared in data banks.

Introduction

The proliferation and differentiation events ocurring during seed embryogenesis give place to a mature embryo body. However, cellular proliferation stops at a certain growing state, before the seed dehydration and reserve accumulation processes start, and remain in this state until appropriate germination conditions appear (Bewley and Black, 1994). Thus, metabolic events during early germination play a key role in reactivating cellular proliferation – which has been arrested during the quiescent period – a pre-requisite for new seedling development.

In our group, we define germination as all the metabolic events that prepare cells in embryonic axes to reinitiate cell division; cell proliferation would then lead to seedling establishment. Several metabolic events take place during early germination, including protein, RNA and DNA synthesis, in addition to the initiation of the cell cycle (Bewley and Black, 1994). However, there is no evidence of the existence of a control of gene expression that would regulate the transition from quiescence to a proliferative metabolism; i.e. the genetic 'switch off' of genes coding, among others, for reserve and dehydration protection proteins, and the 'switch on' of germination-specific genes.

We are interested in the evaluation of changes in gene expression, directly associated with germination specific processes. For this, we have followed two

R.H. Ellis, M. Black, A.J. Murdoch, T.D. Hong (eds.), Basic and Applied Aspects of Seed Biology, pp. 601–610.
© *1997 Kluwer Academic Publishers, Dordrecht. Printed in Great Britain.*

experimental approaches: cytokinin stimulation of maize embryonic axes germination and maize seed osmopriming.

Cytokinins have been reported as stimulators of seed germination (Van Staden, 1983). Benzyl adenine (BA), a synthetic cytokinin, stimulates DNA metabolism during maize germination by stimulating DNA synthesis, probably of both repair and replicative types (Zaraín *et al.*, 1987; Vázquez-Ramos and Reyes, 1990). As a consequence, the S phase of the cell cycle is activated hours before it could happen if BA was not exogenously added; by 24 h of germination, BA treated axes show 3 times more mitotic figures than non BA-treated axes (Reyes *et al.*, 1991). During the early hours of germination (0–6 h), BA also stimulates enzymatic activities like protein kinases and polyADP-ribosyl polymerase (Zúñiga *et al.*, 1995). The biochemical response of cells to BA is very fast. Interestingly, DNA metabolism stimulated by BA is α-amanitin repressible, indicating that RNA synthesis is a requisite for such stimulation (Vázquez-Ramos and Reyes, 1990). Thus, a possibility existed that the quick response of cells to BA was mediated not only by enzymic activation but by changes in gene expression.

Seed osmopriming is a process by which germination is also accelerated. During osmopriming, seeds are partially hydrated due to a restrictive water availability; germinative metabolism probably occurs, but no radicle protrusion is observed. No DNA replication seems to take place at this stage, which can last days (Bray *et al.*, 1989). When the optimal water potential is restored, germination proceeds faster and in a uniform fashion, and DNA synthesis is accelerated (Bray *et al.*, 1989).

At this point, an analogy can be found between the effects of BA on seeds and seed osmopriming: both stimulate DNA metabolism. Since the evidence suggests that DNA synthesis stimulation (by BA) is, at least partially, due to transcriptional activation, and there is no evidence of the occurrence of replicative DNA synthesis during seed osmopriming, we have devoted our efforts to find genes whose expression was activated/accelerated by either BA or release from osmopriming, that could represent germination specific markers.

Material and Methods

Maize Embryonic Axes Imbibition and RNA Extraction

Maize (*Zea mays,* var. chalqueño) embryonic axes were hand dissected, disinfected with 0.5% NaClO and incubated at 25°C between two sterile discs of Whatman paper No. 1 for different time periods with either water or BA (10^{-6} M).

Total RNA was prepared using Tryzol (GIBCO BRL, Life Technologies) according to Chomczinski (1993). Purity and integrity of total RNA was measured spectrophotometrically and by denaturing agarose gel electrophoresis. The final RNA preparation was treated with 0.2 U/µl of *RNase*-free D*Nase* I (Boehringer-Mannheim).

Osmopriming

Conditions for establishing maize osmopriming have been reported by Cruz-García *et al.* (1995). Briefly, we used polyethylene glycol (PEG-8000) to produce an osmotic potential of –1.7 MPa, according to Michel and Kaufmann (1973). Maize seeds were incubated in Petri dishes containing two discs of sterile Whatman No. 1 paper and 40 ml of PEG-8000. The plates were incubated for 10 days (d). Germination rate and uniformity of osmopriming seeds were evaluated following the cumulative germination after 24, 36, 48, and 60 h of imbibition at 25°C.

mRNA Differential Display

The mRNA differential display (DD) technique was followed according to Liang and Pardee (1992), using the following commercial primers obtained from Operon Technologies, Alameda, Cal. Anchor primers: 1 mM of either oligo-dT$_{11}$CC, oligo-dT$_{11}$AA, oligo-dT$_{11}$GA or oligo-dT$_{11}$CG. PCR amplification to produce double strand cDNA was made in the presence of (α^{35}S)dATP with a second 10mer with arbitrary sequence: 5′-CTGGGCAACT-3′ (OPM-06), 5′-GTCTTGCGGA-3′ (OPM-09), 5′-GACCTACCAC-3′ (OPM-15), or 5′-AGGTCTTGGG-3′ (OPM-20) according to the manufacturer. Reaction samples were separated on a 6% DNA sequencing gel and exposed, as indicated.

Northern Blot Analysis

Total RNA (up to 60 mg) was heat denatured, subjected to electrophoresis on agarose gels with 2.2 M formaldehyde at 50 V for 4–5 h, and transfered to nylon membranes. The blots were UV crosslinked, prehybridized at 65°C in 250 mM Na$_2$HPO$_4$/NaH$_2$PO$_4$ pH 6.8, 1 mM EDTA, 7% SDS, 1% albumin, and hybridized to random primer labelled cDNA probes. After hybridization, the blots were washed twice at 65°C for 15 min with washing solution containing 100 mM Na$_2$HPO$_4$/NaH$_2$PO$_4$ pH 6.8, 1 mM EDTA, 0.1% SDS, and exposed to X-O-Mat film at –70°C.

DNA Sequencing

The sequencing assays for the DD isolated bands were made with the Perkin Elmer *AmpliCycle* Sequencing kit, according to the manufacturer. The thermal cycling conditions were as follows: one initial step at 95°C, 2 min, followed by 25 cycles at 94°C, 1 min→65°C, 1 min→72°C, 1 min. Reaction samples were separated on a 6% DNA sequencing gel and exposed, as indicated.

Results and Discussion

DNA synthesis stimulation by BA is α-amanitin repressible, but that is not the case for DNA polymerase activity (Vázquez-Ramos and Reyes, 1990), which remains at about the same level; other factors, BA inducible, should be responsible for such stimulation. An interesting possibility is that of protein phosphorylation. Protein kinase activity is considerably enhanced during early germination by BA and this stimulation is also α-amanitin repressible (Table 1). BA may be promoting the activation of DNA polymerases through the induction of protein kinase(s) that would stimulate their activity. We have recently reported that DNA polymerase 2, a replicative enzyme, is a phospho-protein that is modified in a cyclic fashion during germination (Coello and Vázquez-Ramos, 1995). We are currently trying to relate the effect of BA on DNA metabolism with phosphorylation of DNA pol 2.

Another interesting effect of BA on DNA metabolism during germination is that if it is applied in short pulses, at 0–3 or 3–6 h after initial imbibition, the curve of stimulation is different. Addition in the 0–3 h period produces DNA synthesis stimulation only during the first 9–12 h of germinaton, and the synthesis decreases to equal control levels (Fig. 1). On the other hand, addition of BA during the 3–6 h period of imbibition causes a stimulatory effect similar to that produced by continuous addition of the hormone (Fig. 1).

These results suggest that BA promotes 2 kinds of DNA synthesis and also that the 3–6 h pulse causes the stabilization of the effect, which is only transitory if BA is added only for the three initial hours. Indirectly, they also suggest that something new appears during the 3–6 h BA pulse that causes both stimulation and stabilization.

Maize osmopriming has a beneficial effect on germination of seeds, which can

Table 1. BA-stimulation of protein kinase (PK) and polyADP-ribosyl polymerase (PARP) activities during early maize germination

Imbibition time (h)	PK (cpm/μg protein)		PARP (cpm/μg protein)	
	–BA	+BA	–BA	+BA
3	10.67±0.53	14.64±2.7 (37%)[a]	3.16±0.8	7.14±1.3 (139%)
(α-amanitin)	4.60±1.6	6.29±2.8	–	–
6	10.73±0.86	19.53±2.1 (82%)	6.56±0.6	7.68±1.6 (17%)
(α-amanitin)	5.04±1.8	8.78±3.1	–	–

[a]Numbers in brackets indicate % of stimulation over control sample

SD, n = 3 independent experiments

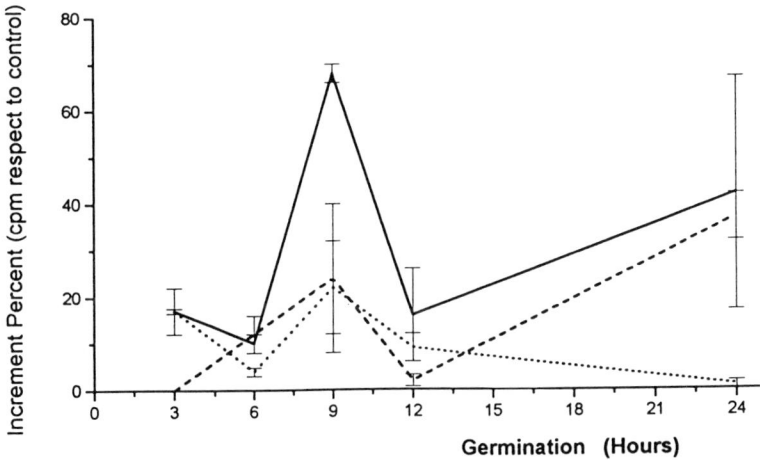

Figure 1. Effect of BA pulses on stimulation of DNA synthesis during maize germination. Maize embryonic axes received a 3 h pulse of BA (10^{-6} M) during the 0–3 h ($\cdots\cdots$) or 3–6 h (- - -) of germination, or BA was present all the time (——). DNA synthesis was followed as incorporation of [^{3}H]thymidine. Results are reported as % over control axes that received no BA. s.d. n = 5

Table 2. Incorporation of radioactive precursors into DNA, RNA and proteins during subsequent germination of osmoprimed maize embryos

Imbibition time (h)	DNA (cpm/µg DNA) Control	OP	RNA (cpm/µg RNA) Control	OP	Protein (cpm/µg prot) Control	OP
10	1910 ± 71.3	4800 ± 225	5220 ± 409	12500 ± 500	3100 ± 605	7190 ± 230
20	5320 ± 21.3	10625 ± 310	12800 ± 920	16900 ± 422	9940 ± 131	1220 ± 126

SD, n = 3 independent experiments

Table 3. Mitotic index in radicle meristems of embryo axes from osmoprimed maize seeds during subsequent germination

Imbibition time (h)	Secondary root Control	OP	Primary root Control	OP
10	0.0	0.15 ± 0.01	0.0	0.0
20	0.6 ± 0.1	1.8 ± 0.2	0.4 ± 0.09	1.5 ± 0.12
30	1.4 ± 0.25	3.2 ± 0.35	0.8 ± 0.1	26 ± 0.3

SD, n = 4 independent experiments

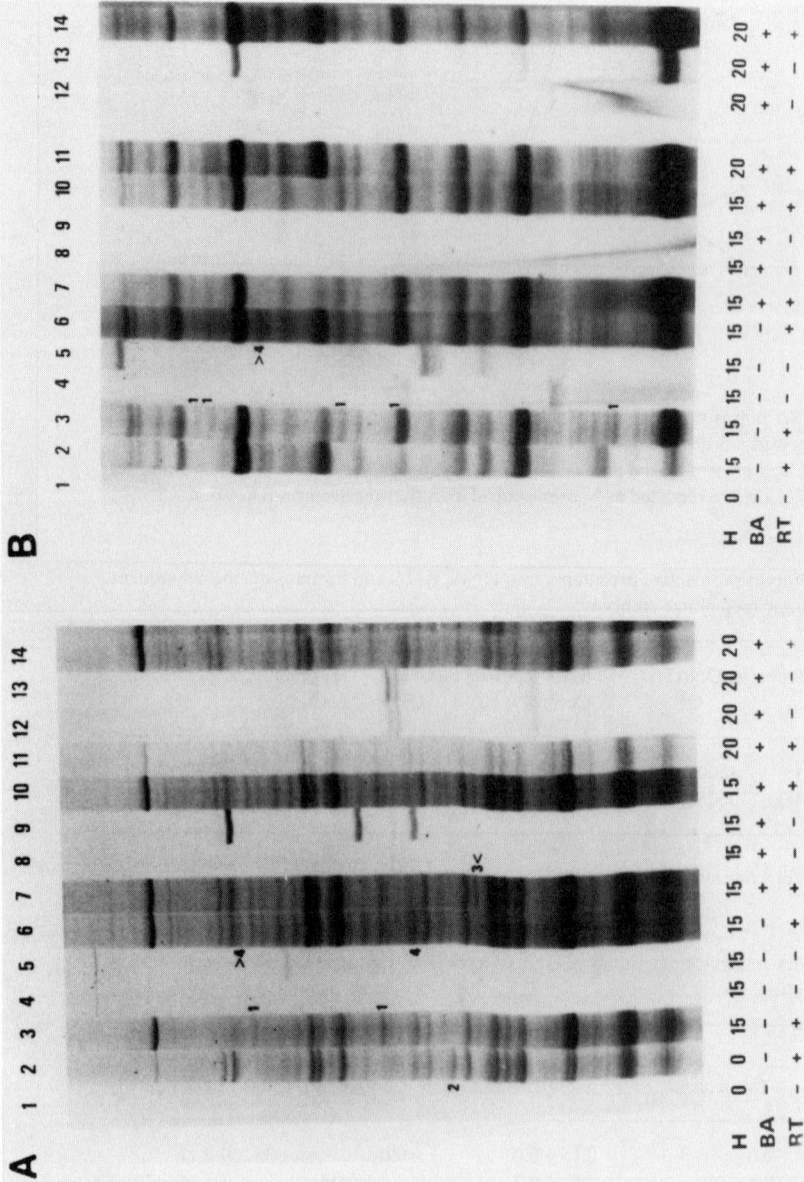

Figure 2. Differential display of RNA maize embryonic axes imbibed with/without BA for 0, 15 and 20 h. The following combination of anchor and random decamer primers was used: A, oligo-dT$_{11}$CC and OPM-6; B, oligo-dT$_{11}$GA and OPM-15; Numbers along the lanes indicates cDNAs that increase (1) or decrease (2) independently of BA treatment, and cDNAs that increase (3) or decrease (4) due to BA treatment. Arrow heads beside numbers indicate cDNAs that were cut from the gel, confirmed by Northern blot and sequenced. (RT−) = samples treated without *Taq DNA* polymerase. H = time in hours; BA = benzyl adenine

remain under the repressive conditions for weeks, and then, when the optimal water potential is restored, germinate faster and more uniformly than non-osmoprimed maize seeds. We have found no evidence that DNA replication or mitosis occur during maize osmopriming (Cruz *et al.*, 1995); however, when the germination conditions are reestablished, DNA replication and mitosis take place very quickly, many hours before they would appear in non-osmoprimed seeds (Tables 2 and 3). Again, a sudden reactivation of DNA metabolism and the cell cycle is the result of 'advancing' germination, this time caused by the osmopriming treatment.

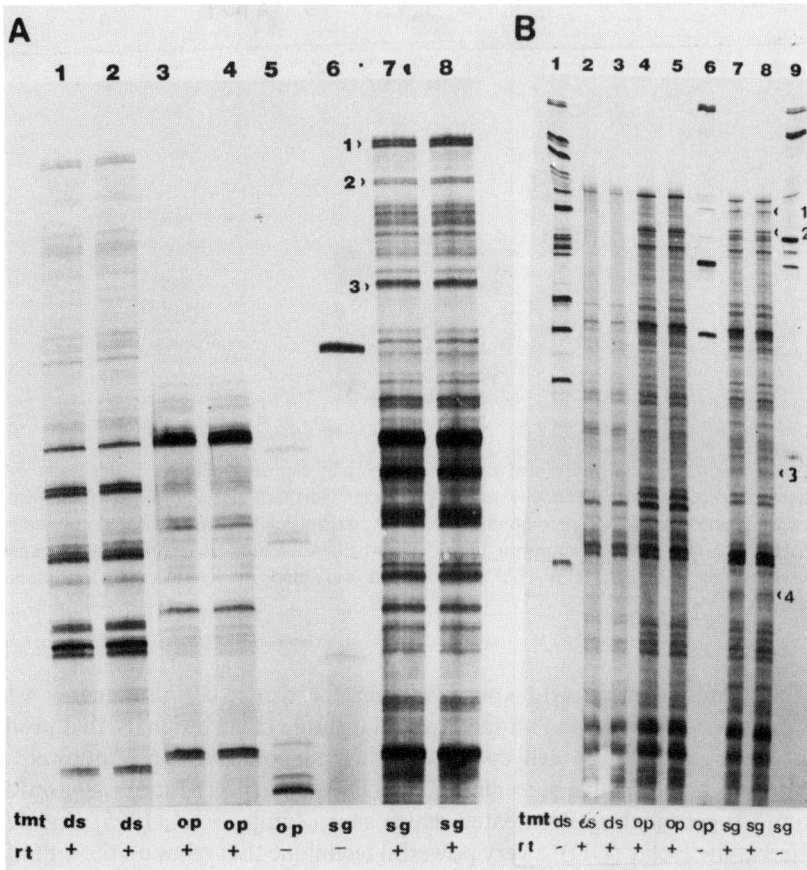

Figure 3. Differential display of RNA from osmoprimed germinated-seeds. The following combination of anchor and random decamer primers was used: A, oligo-dT$_{11}$CG and OPM-20; B, oligo-dT$_{11}$AA and OPM-9. Arrow heads beside numbers indicate cDNAs that appear at 24 h germination; ds = dry seed; op = 10 d osmoprimed seeds; sg = 24 h subsequent germinated seeds; RT = addition of reverse transcriptase

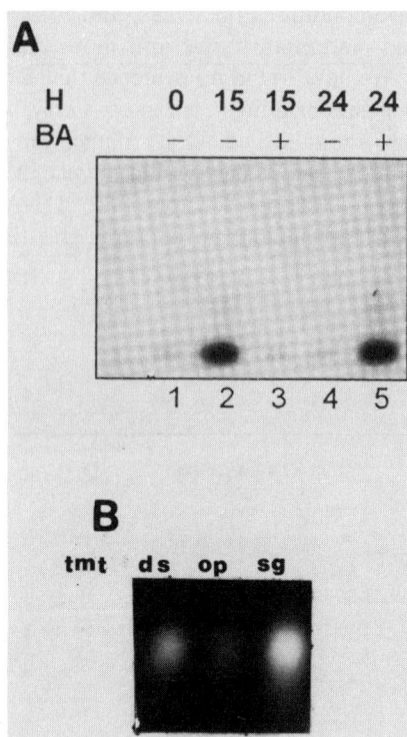

Figure 4. Expression of osmoprimed and BA-differentially expressed cDNAs. Bands isolated from differential display experiments was used as probes for Northern blot analysis to follow variation in gene expression of the corresponding genes. A, expression of BA-modulated transcript for BARM 15.3 (Fig. 2B); H = time in hours; BA = benzyl adenine. B, expression of transcript corresponding to SG20.1 (Fig. 3A). Experiments were done as described in Materials and Methods

Our main interest is to know whether there is a certain time at which germination-specific signals appear, and the nature of such signals, that produce the reinitiation of the cell cycle. For this purpose, we have followed two strategies: a subtractive screening of a cDNA library produced using mRNA from 5 h germinating BA-treated maize axes (Zúñiga *et al.*, 1995), and RNA differential display (DD), a very powerful technique that relies on the amplification by PCR of subpopulations of mRNA, extracted from germinating maize axes imbibed for different periods. Both BA-stimulated and osmoprimed-germinating maize axes have been used. The results of screening of a cDNA library will not be described here.

For DD, an oligomer of dT_{11}+ 2 more nucleotides in any combination (i.e. C-T, C-G, C-C, etc.) are used at the 3' end to produce single strand cDNAs; the second oligomer is any one of a series of 10mers (of random sequence) that will select a subpopulation of the synthesized cDNAs. Some results of DD are shown in Figure 2. Bands in Figure 2A (oligos: OPM-6, $dT_{11}CC$) follow a different behaviour: band 1 (BARM 6.1, 350 bp) is present at 15 h of germination only in control but not in BA-treated axes; bands 2 (BAEM 6.1, 300 bp) and 3 (BAEM 6.2, 250 bp) are messages that increase with time, but it is more evident in BA-treated axes. Bands 1 (BARM 15.2, 320 bp) and 2 (BARM 15.3, 200 bp) in Figure 2B (oligos: OPM-15, $dT_{11}GA$) represent messages that are present in 15 h germinated, non-BA treated axes (control) and only marginally in BA-treated axes (same time). However, both bands reappear later on, at 24 h in BA-treated axes, whereas both virtually disappear in control axes.

Figure 3A (oligos: OPM-20, $dT_{11}CG$) represents messages that accumulate in 24 h germinated, osmoprimed seed axes, present at low level in the dry seed (except band 3) but virtually absent in 10 d osmoprimed seeds. Reamplified messages correspond to cDNAs of 350 bp (band 1, SG20.1), 300 bp (band 2, SG20.2) and 250 bp (band 3, SG20.3). Figure 3B (oligos: OPM-9, $dt_{11}AA$) also shows messages that appear at 24 h germinated, osmoprimed seed axes, that seem not to be present in dry seeds or in 10 d osmoprimed seeds. Band 1 (SG9.1) correspond to a message of 700 bp; band 2 (SG9.2), 500 bp; band 3 (SG9.3), 300 bp and band 4 (SG9.4), 250 bp.

Northern blot analysis using some of these cDNAs as probes (Fig. 4) confirmed results obtained in differential display: the transcript for BARM 15.3 was present in 15 h germinated, non-BA treated axes (control) and only marginally in BA-treated axes (same time). However, the transcript reappeared at 24 h in BA-treated axes, whereas it virtually disappeared in control axes (Fig. 4A). A 150-base sequence corresponding to BARM 15.3 cDNA was compared in EMBL-EMNEW data bank and revealed 88% identity with a previously reported *Zea mays* clone 5c04g02 3' end, not identified, yet. The transcript for SG20.1 is present in dry seed; its level is low during osmopriming, but rises during subsequent germination. A 200-base sequence corresponding to SG20.1 was compared in EMBL-EMNEW data bank, showing 60% identity to *E. coli* RecA protein. A 250-base sequence corresponding to SG9.2 was also compared showing 65% *Glycine max* tubulin.

The results presented here indicate that while there is *de novo* transcription stimulated by either BA or osmopriming during maize germination, such expression must consist of scarcely represented messages in the mRNA population, that should code for tightly regulated control proteins.

In this sense, we are currently exploring other combinations of anchor and random primers in order to increase the possibility of finding regulated genes. We are also focusing on obtaining the full length cDNA of the genes which are transcribed during germination in order to know their identity and pattern of expression.

References

Bray, C. M., Davison, P. A., Ashraf, M. and Taylor, R. M. 1989. *Annals of Botany* 63: 185–193.

Bewley, J .D. and Black, M. 1994. *Seeds: Physiology of Development and Germination.* London: Plenum Press.

Chomczynski, P. 1993. *Biotechniques* 15: 3: 532–536.

Coello, P. and Vázquez-Ramos, J.M. 1995. *European Journal of Biochemistry* 231: 99–103.

Cruz-García, F., Jiménez, L. F. and Vázquez-Ramos, J. M. 1995. *Seed Science Research* 5: 15–23.

Liang, P. and Pardee, A. B. 1992. *Science* 257: 967–971.

Michel, B. and Kaufmann, M. K. 1973. *Plant Physiology* 51: 914–916.

Reyes-Jiménez, J., Jiménez-García, L. F., González, M. A. and Vázquez-Ramos, J. M. 1991. *Seed Science Research* 1: 113–117.

Van Staden, J. 1983. *Physiologia Plantarum* 55: 60–72.

Vázquez-Ramos, J. M. and Reyes, J. 1990. *Canadian Journal of Botany* 45: 649–653.

Zaraín, H. M., Bernal-Lugo, I. and Vázquez-Ramos, J. M. 1987. *Mutation Research* 181: 103–110.

Zúñiga-Aguilar, J. J., Gómez-Gutierrez, A., López-Villaseñor, I. and Vázquez-Ramos, J. M. 1995. *Seed Science Research* 5: 219–226.

68. Genotypic, Phenotypic and Opportunistic Germination Strategies of Some Common Desert Annuals Compared with Plants with Other Seed Dispersal and Germination Strategies

Y. GUTTERMAN

Jacob Blaustein Inst. for Desert Research and Dept. of Life Sciences, Ben-Gurion University of the Negev, Sede Boker Campus 84990, Israel

Abstract

In the Negev Desert with unpredictable distribution and low amounts of rain, *Spergularia diandra* is a common annual. Several factors may regulate its low germinability, such as genotypic and phenotypic factors during seed maturation, including seed position and plant age affecting seed coat colour, day length, temperature, and post maturation conditions. Seed dispersability may be influenced by the diversity of seed coat structure and relative weight within the three genotypes and the three phenotypes. The opportunistic strategy and low germination regulated also by light and temperature, and the quantitative long day response for flowering, allow this species to emerge and survive in very great numbers in large areas. It produces large numbers of tiny seeds which disperse after maturation and 'escape' from massive consumption.

In *Schismus arbicus*, another common plant in the Negev, the maternal and environmental influences during seed maturation, and the escape strategy of seed dispersal combined with an opportunistic strategy for germination and flowering are similar to *S. diandra*.

In contrast are plants that protect their seeds which are dispersed by rain and have a 'cautious' dispersal and germination strategy. An example is the rare *Blepharis* spp. which produces a few relatively large seeds. *Mesembryanthemum nodiflorum* is an example of a plant with intermediate strategies of seed dispersal and germination.

The Negev Desert: Rains and Plant Survival Strategies

The Negev Desert is part of the northern Saharo-Arabian Desert which receives small and unpredictable amounts of rain in winter. Summers are long with high temperatures and low relative humidity during the day, and there are about 190 nights with dew throughout the year. Fluctuations of precipitation from one year to another may be very great. In 1977/78 a total of 51.2 mm of rain fell on 23 days (d), distributed over 230 d. Two years later 168.9 mm of rain fell on 43 d distributed over 205 d. The average annual precipitation in this area was 102 ± 9 mm over the last 19 years. The distribution of rain through the season is also unpredictable. There are years during which the amount of rain that falls in 1 d

R.H. Ellis, M. Black, A.J. Murdoch, T.D. Hong (eds.), Basic and Applied Aspects of Seed Biology, pp. 611–622.
© *1997 Kluwer Academic Publishers, Dordrecht. Printed in Great Britain.*

may equal the whole rainfall of another year. On 25 January 1991, 51 mm of rain fell, which was similar to the total precipitation in 1977/78. Also typical to this desert is the massive seed consumption by ants, birds, rodents, etc. (Evenari *et al.*, 1982; Gutterman, 1993; 1994b).

Species of annual plants inhabiting the Negev Desert have developed at least two opposite strategies of survival (Fig. 1). One is the 'protection' and ombrohydrochoric strategy of seed dispersal, as in *Blepharis* spp. (Acanthaceae) which is a rare plant in limited areas of the deserts of Israel and the Sinai Peninsula. On the other hand there are the 'escape' strategy of seed dispersal and 'opportunistic' strategy of germination, as in *Schismus arabicus* Nees (Poaceae) and *Spergularia diandra* (Guss.) Heldr. et Sart. (Caryophyllaceae) which are very common annuals in the Negev and other areas of Israel. *Mesembryanthemum nodiflorum* L. (Aizoaceae) which once in several years covers very large areas, displays intermediate strategies of seed dispersal and germination. *Asteriscus hierochunticus* (Michon) Wikl. (= *A. pygmaeus*) (Asteraceae) achenes display an intermediate strategy of germination by which achenes require relatively long periods of wetting to engender germination. The achenes differ in their germinability according to their position in the capitulum (Zohary, 1966; Gutterman *et al.*, 1967; Gutterman, 1972; 1980/81; 1994a,c; 1996a,b; Feinbrun-Dothan, 1977; 1986; Feinbrun-Dothan and Danin, 1991; Gutterman and Evenari, 1994; Gutterman and Ginott, 1994) (Fig. 1).

The Escape Seed Dispersal and Opportunistic Seed Germination Strategy: *Spergularia diandra* and *Schismus arabicus*

Spergularia diandra produces large numbers of tiny dust-like seeds. Populations of the central Negev highlands near Sede Boker were found to consist of three plant genotypes with differing seed coat structure: hairy, partially-hairy or glabrous. When seeds have matured at the beginning of summer, a narrow opening appears at the top of the capsule. When the wind is strong enough the seeds are sucked out and dispersed. The hairy seeds, which weigh relatively less, are dispersed much further distances than the genotypes with partially-hairy or glabrous seed coats (Figs. 2a,b). However, when the seeds land on the soil surface, the glabrous seeds can be moved further along the soil surface than the hairy seeds. The seeds enter cracks in the soil crust, or shallow depressions, where they become covered by soil particles and thus escape from massive seed predation. Different genotypes may germinate from different microhabitats.

Spergularia diandra seedlings may emerge in depressions even after less than 10 mm of rain (Loria and Noy-Mier, 1979/80), which is not enough for seedling survival. However, even if a few small amounts of rain follow with a few days between each shower, seedlings may develop, flower and produce a large number of seeds. The later the seeds germinate in the season with rain, the longer the day length and the shorter the time until first flower appearance (Fig. 3). The low germinability of these seeds is also regulated by the genotype, as well

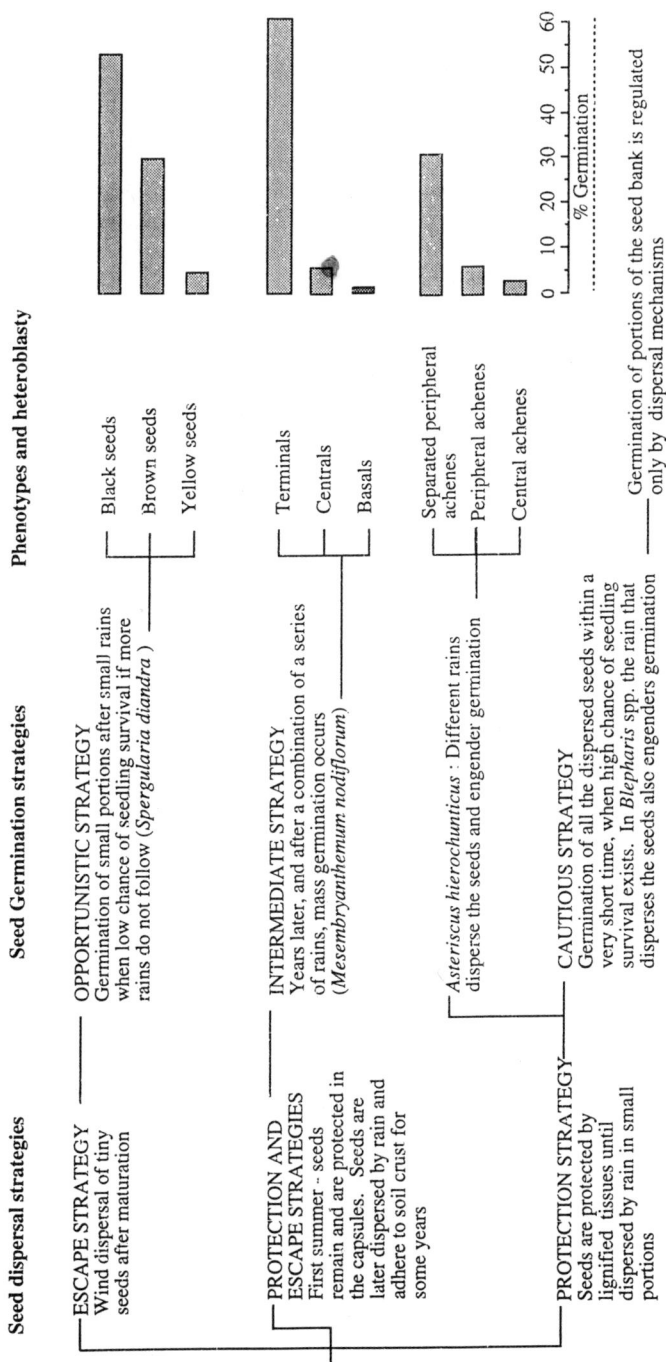

Figure 1. Summary of two extreme and one intermediate strategies of seed dispersal and germination found in *Spergularia diandra, Blepharis* spp., *Asteriscus hierochunticus* and *Mesembryanthemum nodiflorum*. The phenotypic heteroblasty of three of these species is shown

(a)   **(b)**

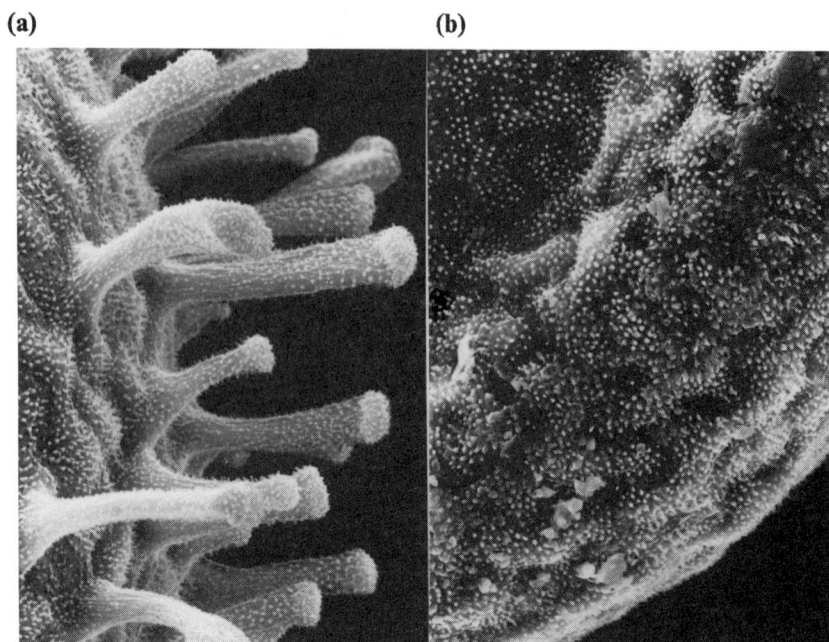

Figure 2. Spergularia diandra brown seeds magnified × 1000: (a) hairy; (b) glabrous

Table 1. Germination (% ± s.e.) of Spergularia diandra seeds harvested near Sede Boker on 27 June 1989 and wetted by water on 20 January 1994 at natural winter temperatures of 3–21°C in light and dark with daily short illuminations (5 min) (Gutterman, 1994b)

	Germination % ± s.e.			
	Days of wetting			
Treatment	4	7	9	17
Light	5.5 ± 1.7	37.0 ± 2.0	46.5 ± 4.6	47.5 ± 2.6
Dark	0	9.0 ± 1.0	25.0 ± 2.6	25.0 ± 2.7
Daily temp. °C	7–15	3–17	5–21	6–20

as phenotypic influences during maturation such as day length (Fig. 4), temperature, position on the plant, and plant age, which influence the seed coat colour (Fig. 1). During the period of wetting, light and temperature are very important factors which influence the low percentage of seeds of S. diandra which germinate from the seed bank. Even under optimal conditions of temperature and water only a small portion of the S. diandra seed population

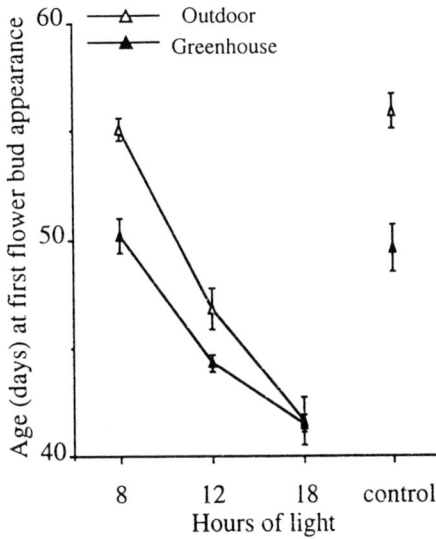

Figure 3. Effect of photoperiod on *Spergularia diandra* average plant age ($\pm$s.e.) at appearance of first flower bud. Average 9–10 plants in each treatment. Experiment started on 7 December 1993. Control in natural day length

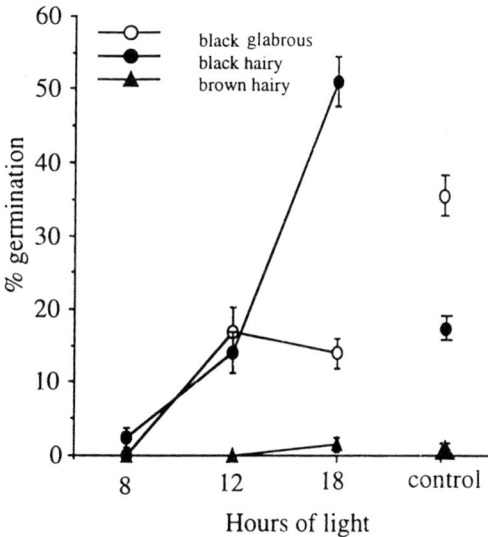

Figure 4. Comparison of germination (%$\pm$s.e.), after 30 d of wetting, of black and brown hairy, and black glabrous *Spergularia diandra* seeds matured under 8, 12 and 18 h day lengths outdoors and under natural day length and temperatures in the growing season (control). Seeds were collected between 7 and 9 March 1994 from mother plants that had developed from seeds harvested on 27 June 1989 near Sede Boerk. The germination experiment began on 10 March 1994. Control in natural day length

(a)

(b)

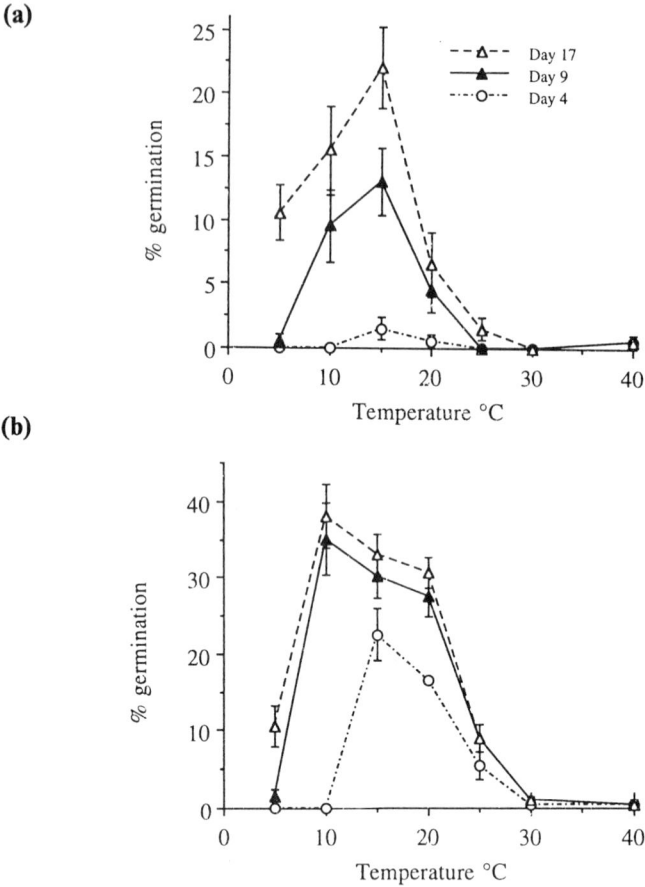

Figure 5. Germination (%±s.e.) after 4, 9 and 17 d of wetting *Spergularia diandra* seeds harvested on 27.6.1989 near Sede Boker and wetted by water on 5.9.1993 in constant temperatures of 5 to 40°C: (a) in dark with short illuminations; (b) in light

germinates. After 17 d of wetting up to 38% of the seeds germinated (Figs. 5a,b). In natural winter temperatures up to 48% germination occurred after 17 d of wetting in Petri dishes (Table 1) (Gutterman, 1994b). Usually after a rain of 10 to 15 mm in winter, the soil surface is wet enough for germination only during the first 2–4 d after the rainfall. Accordingly, the percentage of germination may be very low in the natural habitats, which reduces the risk of germination occurring after such a small rainfall.

Since at each germination opportunity only a very small percentage of the long living *S. diandra* seeds in the seed bank in the soil germinate, the length of dry storage also may have an influence on the amount and speed of germination. The older the seeds, at least up to 14 months, the earlier, faster and higher the

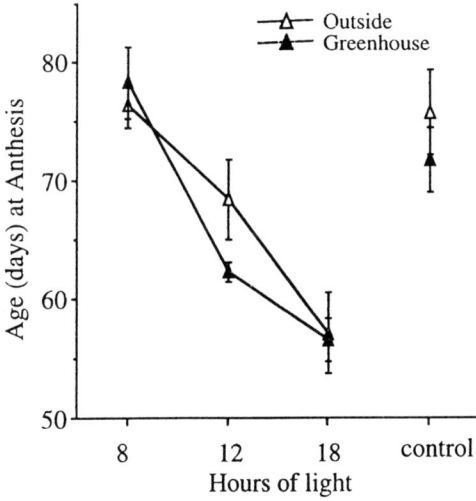

Figure 6. Effect of photoperiod outdoors and in greenhouse on average plant age (±s.e.) of *Schismus arabicus* at anthesis. Control in natural day length

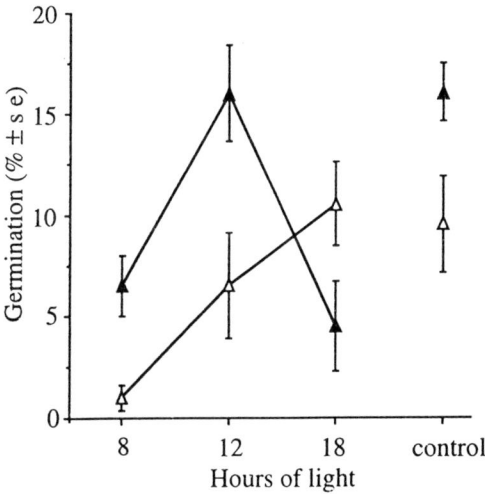

Figure 7. Day length outdoors and in greenhouse during maturation affecting *Schismus arabicus* seed germinability after 4 d of wetting. Control in natural day length

percentage germination (Gutterman, 1993; 1994a,b; 1996b.).

In *Schismus arabicus* all the very small, glassy, caryopses are similar and are dispersed by wind after maturation at the beginning of summer. They are larger than *Spergularia diandra* seeds and resemble grains of sand in colour and shape. In this way seeds 'escape' from massive seed consumption and become widespread after maturation.

Schismus arabicus caryopses germinate after less rainfall than *Spergularia diandra* and seedlings may survive for a relatively long period even after one fall of rain at high temperatures and low relative humidity (Gutterman and Evenari, 1994). This gives them a higher chance to survive until the next rain. Also in *Schismus arabicus*, as in *Spergularia diandra*, the later the seedling appearance in longer days, the shorter the time until first flower appearance (Fig. 6).

There are several environmental and maternal influences from the time of seed development and maturation until germination, which regulate the low germination percentage of the seed bank of this plant:

1) Day length and temperature during maturation: under each combination of conditions the germinability is different (Fig. 7).

2) Temperatures between maturation and the time of germination regulates after ripening. Germination will take place only after caryopses are stored for 70 d at 40°C or after one or two summers. The lower the temperature during dry storage the lower the percentage germination. Caryopses stored at −18°C do not germinate even after more than two years but do germinate if in addition they are exposed to at least 70 d at 40°C (Table 2) (Gutterman, 1994b).

3) Temperature and light during the time of wetting and germination are very important regulators of the time of radicle appearance, rate and final levels of germination (Figs. 8a,b) (Gutterman, 1996a).

4) In alternating winter temperatures of 18 h at 15°C and 6 h at 5°C large differences in germination levels occur in comparison with caryopses germinated in 5 to 15°C, in two sets of experiments using seeds harvested in 1991 and 1993. Large differences also occurred in alternating summer temperatures of 20 and 40°C. In 18 h at 20 and 6 h at 40°C germination was much higher than in 18 h at 40°C and 6 h at 20°C (Table 3) (Gutterman, 1996a).

The Protection Strategy of Seed Dispersal and Cautious Strategy of Germination

The protection strategy of seed dispersal and cautious strategy of germination have been found in *Blepharis* spp. The seeds are protected by the lignified sepals

Table 2. The influence of dry storage temperature on the germination percentage ($\pm$s.e.) of *Schismus arabicus* caryopses during 6 d of wetting at 20°C in light (L) or dark with short illuminations (5 min) during daily observations (D). 4×50 caryopses for each treatment in 50 mm diameter Petri dishes. (A) Freshly harvested mature *Schismus arabicus* caryopses collected from a natural population near Sede Boker on 18 April 1993; (B) Caryopses harvested on 16 May, 1993 and stored at constant temperatures of 40°C or –18°C for 70 days. (C) Caryopses harvested on 28 April 1991 and stored at room temperature of 10–27°C (Gutterman, 1994b)

	Dry storage conditions	Wetting at 20°C L or D	Germination % after 6 days of wetting
A	*Harvest 18 April 1993*		
	Freshly harvested	L	0.5
		D	0
B	*Harvest 16 May 1993*[a]		
	40°C	L	8.0 ± 1.8
	40°C	D	51.0 ± 2.6
	–18°C	L	0.5 ± 0.3
	–18°C	D	0
C	*Harvest 28 April 1991*[a]		
	10–27°C	L	28.5 ± 2.3
	10–27°C	D	43.5 ± 1.2

[a]This germination experiment started on 25 July 1993

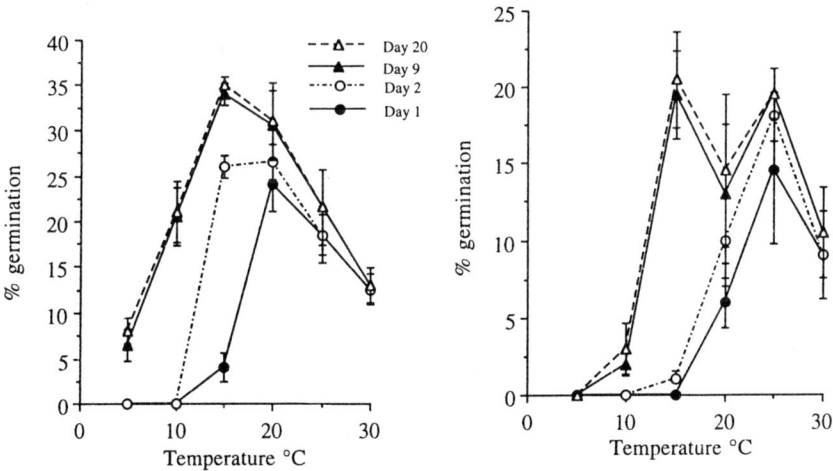

Figure 8. Germination ($\pm$s.e.) after 1, 2, 9 and 20 d of wetting of *Schismus arabicus* seeds harvested on 28 April 1991 and wetted by water on 4 May 1993 in constant temperatures of 5 to 30°C: (a) in dark with short illuminations; (b) in light

Table 3. The influence of alternating temperature during wetting on the germination percentage (±s.e.) of *Schismus arabicus* caryopses harvested on (A) 18 April 1993 and (B) 28 April 1991. The caryopses were stored at room temperatures and transferred to dry storage at 40°C for 60 days before the germination experiments which took place in summer 1993. Four replicates of 50 caryopses per Petri dish were used and transferred between constant winter temperatures of 5 and 15°C or constant summer temperatures of 20 and 40°C. All the treatments of this germination experiment were in dark with 5 min of light each 24 h. After four cycles of 24 h all the treatments continued at 15°C or 20°C, respectively (Gutterman, 1996a).

	% Germination (±s.e.)					
	A. Caryopses harvested on 18.4.93			B. Caryopses harvested on 28.4.91		
	At alternating temperatures		Constant temperatures	At alternating temperatures		Constant temperatures
Hours of wetting in temperatures	Day 1	Day 2	Day 8	Day 1	Day 2	Day 8
18h 5°C → 6 h 15°C	0	12.0±2.7	22.5	0	9.0±3.8	30.5
18 h 15°C → 6 h 5°C	24.0±3.5	29.5±2.5	30.0	32.0±2.9	40.0±3.5	43.0
5°C →	0	2.5±1.5	16.0	0	1.0±0.5	21.5
15°C →	12.5±2.8	16.0±2.4	16.0	36.5±5.1	44.5±6.0	44.5
18 h 20°C → 6 h 40°C	12.5±1.5	19.0±1.7	21.0±2.0	49.5±2.0	53.0	55.5±1.5
18 h 40°C → 6 h 20°C	0	0	11.0±1.9	0	0	19.5±1.2
20°C →	22.5±1.7	25.0±3.0	25.0±3.0	26.0±2.0	32.0	35.0±0.6
40°C →	0	0	0	0	0	0

and bracts and enclosed by the hard capsules. In this plant the regulation of germination is wholly dependent on the dispersal mechanisms. A few capsules will explode only after a long period of wetting by rain. The seeds that are released will germinate immediately in the same rainfall in a wide range of temperatures, from less than 8°C to 40°C, in light and dark (Gutterman, 1973; 1993; 1994b) (Fig. 1).

The achenes of *Asteriscus hierochunticus* are protected by the lignified capitulum and released, whorl by whorl, by rain. The achenes do not germinate immediately after dispersal and achenes from different whorls display differing germinability. The achenes separated from the peripheral whorl germinate to the highest level and those from the centre to the lowest (Gutterman, 1993; Gutterman and Ginott, 1994) (Fig. 1).

Protection and Escape Strategies of Seed Dispersal and Intermediate Strategies of Germination

Mesembryanthemum nodiflorum displays intermediate strategy between the two extremes described above. During the first summer after maturation the seeds remain on the parent plant and are protected in the capsules. The seeds are later dispersed by rain and adhere to the soil crust for several years. In this way the seeds are protected from seed consumption until they germinate. After several years, if a number of rains follow each other and dilute the salts on the soil surface, where the seeds are situated, mass germination occurs. These seeds can survive and germinate well for at least 22 years. According to their position in the capsule during maturation they differ in their germinability for many years. In most cases the seeds from the terminal capsules germinate to the highest level and those from the basals to the lowest (Gutterman, 1980/81; 1994a,c) (Fig. 1).

References

Evenari, M., Shanan, L. and Tadmor, N. 1982. *The Negev. The Challenge of a Desert*, Second edition, pp. 266. Cambridge, Mass: Harvard University Press.

Feinbrun-Dothan, N. 1977. *Flora Palaestina*, Part Three-Text. 469 pp. Jerusalem: Israel Academy of Sciences and Humanities.

Feinbrun-Dothan, N. 1986. *Flora Palaestina*, Part Four-Text. 487 pp. Jerusalem: Israel Academy of Sciences and Humanities.

Feinbrun-Dothan, N. and Danin, A. 1981. *Analytical Flora of Eretz-Israel.* pp. 1040. Cana, Jerusalem (Hebrew).

Gutterman, Y., Witztum, A. and Evenari, M. 1967. *Israel Journal of Botany* 16: 213–234.

Gutterman, Y. 1972. *Oecologia* 10: 145–149.

Gutterman, Y. 1973. In: *Seed Ecology,* pp. 59–80 (ed.W. Heydecker). London: Butterworths.

Gutterman, Y. 1980/81. *Israel Journal of Botany* 29: 93–97.

Gutterman, Y. 1993. *Seed Germination in Desert Plants. Adaptions of Desert Organisms,* pp 253. Berlin, Heidelberg, New York: Springer Verlag.

Gutterman, Y. 1994a. *Israel Journal of Plant Sciences* 42: 261–274.

Gutterman, Y. 1994b. *Botanical Review* 60: 373–425.

Gutterman, Y. 1994c. *Israel Journal of Plant Sciences* 42: 197–205.

Gutterman, Y. 1996a. *Journal of Arid Environments*. 32: (in press).

Gutterman, Y. 1996b. *Journal of Arid Environments* (in press).

Gutterman, Y. and the late Evenari, M. 1994. *Israel Journal of Plant Sciences* 42: 1–14.

Gutterman, Y. and Ginott, S. 1994. *Journal of Arid Environments* 26: 149–163.

Loria, M. and Noy-Meir, I. 1979/80. *Israel Journal of Botany* 28: 211–225.

Zohary, M. 1966. *Flora Palaestina*, Part One-Text. 367 pp. Jerusalem: Israel Academy of Sciences and Humanities.

69. Comparative Ecophysiology of Seed Germination Strategies in the Seven Pine Species Naturally Growing in Greece

A. SKORDILIS and C.A. THANOS

Department of Botany, University of Athens, Athens 15784, Greece

Abstract

Seven out of the eleven European pine species grow naturally in Greece. *Pinus halepensis, P. brutia* and *P. pinea* are low-altitude, typical Mediterranean species. The Greek populations of *P. nigra, P. heldreichii, P. sylvestris* and *P. peuce* grow in high altitudes and are located at the southern limits of their natural world distributions.

Seed germination data concerning temperature dependence as well as light and/or stratification requirements are presented and discussed in relation to the individual species characteristics. Fire resilient *P. halepensis* and *P. brutia* are characterized by quite low germination in the dark, throughout their optimal temperature range (10–20°C); in addition, their germination is photosensitive (white-light promoted and far-red inhibited). Prolonged stratification is always beneficial to *P. brutia* but detrimental to *P. halepensis* seeds. Germination of *P. pinea* seeds is also slow but indifferent to light conditions and restricted to a very narrow temperature range, around 20°C. *P. nigra* seeds are fast germinating over a wide temperature range and rather indifferent to light. Germination of the deeply dormant *P. heldreichii* seeds as well as of the less dormant ones of *P. sylvestris* requires light and/or stratification.

By combining germination data for each species with the particular seasons of seed dispersal and the climatic conditions of the respective habitats, timing schedules of seed germination and subsequent seedling emergence in nature are proposed.

Introduction

Despite the relatively small area of Greece, the majority of European pine species grow naturally in the country, in the form of mixed or unmixed forests and/or isolated stands. *Pinus halepensis* Miller and *P. brutia* Ten. are common, low-altitude pine species, covering extended areas around the Mediterranean basin. *P. brutia* largely replaces *P. halepensis* in the north-eastern part and there is a well defined spatial isolation in their natural distributions; the shortest distance (in northern Greece) is about 50 km (Panetsos, 1975). *P. pinea* L. also a typical Mediterranean pine, is found in lowland areas often on sandy ground near the sea. Its natural range is currently uncertain, since this species has been widely planted for centuries due to its edible seeds. *P. nigra* Arnold, is a sub-Mediterranean pine, which ranges widely through southern Europe. Its natural

R.H. Ellis, M. Black, A.J. Murdoch, T.D. Hong (eds.), Basic and Applied Aspects of Seed Biology, pp. 623–632.
© *1997 Kluwer Academic Publishers, Dordrecht. Printed in Great Britain.*

variability is reflected in its great taxonomic complexity at the subspecies and variety levels. *P. heldreichii* Christ *(P. leucodermis* Antoine) is a montane and subalpine pine confined to high elevations in the Balkan peninsula and in southern Italy, often replacing *P. nigra* at high altitudes. *P. sylvestris* L. is a widespread European Siberian pine; throughout its extensive distribution, this variable montane species has been divided into several subspecies and varieties. *P. peuce* Griseb. is a montane Balkan species surviving today only in a few humid mountain systems. The latter four species hardly reach the Mediterranean coast and their populations in Greece are located more or less at the southern limits of their natural world distributions.

Figure 1 shows the natural distribution of all pine species in Greece while a list of their main characteristics is presented in Table 1 (data from Pozzera, 1959; Little and Critchfield, 1966; Critchfield and Little, 1969; Krugman and Jenkinson, 1974; Panetsos, 1981; Greuter *et al.*, 1984; Strid, 1989; Ministry of Agriculture, Greece, 1992; Gaussen *et al.*, 1993).

Despite their economic and ecological significance, the ecophysiological diversity of pines is often not fully appreciated. Yet the variation among species

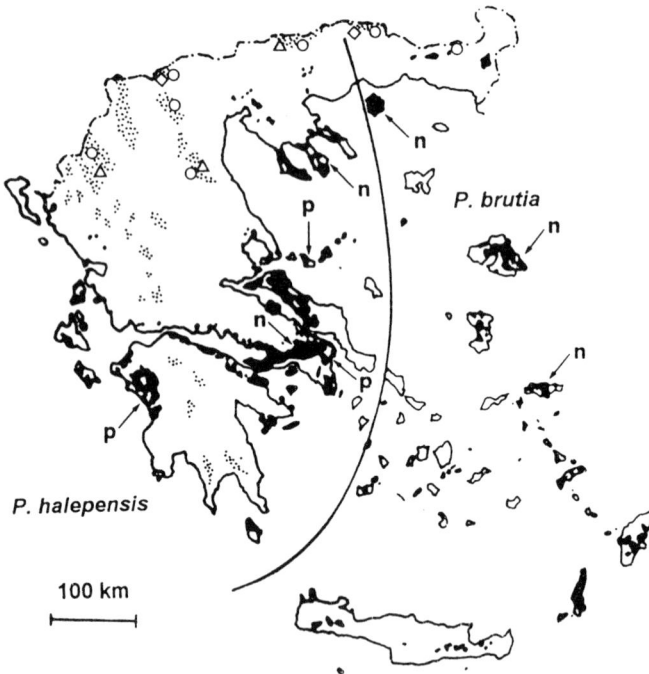

Figure 1. The geographical distribution of *Pinus brutia (■), P. halepensis* (■)*, P. nigra* (□ and **n**: isolated occurrence), *P. pinea* (**p**), *P. heldreichii* (Δ), *P. sylvestris* (○) and *P. peuce* (◇)

Table 1. A list of characteristics for the seven pine species naturally growing in Greece

	P. brutia	P. halepensis	P. heldreichii	P. nigra	P. peuce	P. pinea	P. sylvestris
Common name	East Mediterranean pine	Aleppo pine	Heldreich pine	Austrian pine	Balcan pine	Umbrella pine	Scotch pine
Natural distribution	East Mediterranean, Black Sea, Syria, Iran	Central and Western Mediterranean	Southern Balkans and Central Italy	Southern Europe and Western Asia	Southern Balkans (limited distribution)	Around Mediterranean basin and Black Sea	Europe and North Asia
Altitude (m)	0–850 (1200)	0–600	1300–2300	600–2150	~1700	0–150	1100–1800
Total cover in Greece (ha)	196 000	372 000	8300	282 000	10	200	21 000
Mature tree height (m)	15–20 (30)	10–20 (30)	20–30	40–50	20–30	15–25	20–40
Seed bearing age (y)	7–10	7–10	?	15–40	12–30	?	5–15
Seed bearing intervals (y)	1	1	?	2–5	3–4	3–6	4–6
Flowering dates	March–April	March–April	May–July	May–June	May–June	April–May	April–May
Cone ripening dates	April–May	April–May	Aug.–Sept.	Sept.–October	Sept.–October	Nov.–Dec.	Sept.–October
Seed production (years after flowering)	3	3	2	2	2	2.5	2
Seed dispersal dates	May–June (some remain closed)	May–July (many remain closed)	Sept.–October	March–April	Sept.–October	May–June	Dec.–March
Mean seed weight (mg)	35–90	~20	~30	~50	~30	~850	~10

in morphology and life history is quite large. Concerning seed germination, relatively little research has been carried out in the Mediterranean species *P. brutia, P. halepensis* and *P. pinea* (e.g. Magini, 1955; Calamassi *et al.*, 1980; 1984; Thanos and Skordilis, 1987; Skordilis and Thanos, 1995). On the other hand, seed germination data available for *P. sylvestris, P. nigra, P. heldreichii* and *P. peuce* refer to northern provenances. The regeneration of all these pines depends exclusively upon their seeds; therefore a study on their particular ecophysiological adaptations may contribute considerably to the conservation of these species as well as of their habitats in Greece.

Materials and Methods

All seeds were either hand collected by the authors or offered by the Forestry Division, Ministry of Agriculture, Greece. Seed provenances used in the present study were, respectively: *P. sylvestris* from Kozani, *P. heldreichii* from Konitsa, *P. nigra* from Kastoria, *P. pinea* from Attica, *P. peuce* from Mt Voras, *P. brutia* from Thasos, Samos, Rodos and Lasithi and *P. halepensis* from Halkidiki, K. Vourla, Attica, Istiaia and Steni.

Germination tests were performed with 5 replicates of 25 seeds per Petri dish (diameter 9 cm, lined with two discs of filter paper and moistened with 5 ml of distilled water). Experiments were carried out in plant growth chambers (W.C. Heraeus GmbH BK Model 5060 EL, Germany) where the temperature was kept constant within $\pm 0.5°C$. The germination experiments presented in Figure 3 were carried out on a temperature- and light-programmable growth cabinet, model GB48 (Conviron, Canada) equipped with a lamp canopy of 48 incandescent bulbs (Sylvania 50A19, 50 W, 227 V) and 10 fluorescent tubes (Sylvania Cool White FR96T12/CW/VHO-235/1). Temperature was kept constant while a photoperiod of 12 h was applied. The broad-band far-red (FR) light source, was obtained through a FR filter (one blue and 2 red plexiglas layers, Rohm: No. 627 and No. 501, respectively). Cold moist stratification was accomplished by maintaining imbibed seeds in the dark, at a temperature of $3 \pm 1°C$.

Germination was recorded every 1 or 2 days (d) and was considered complete when no additional seeds germinated. The criterion of germination was visible radicle protrusion; seeds exhibiting abnormal germination were excluded from germination counts. All germination percentages were based on the filled seed portion only; empty seeds determined by dissection tests, performed after completion of each experiment, were not counted. The mean percentage of unsound seeds was less than 5% for *P. heldreichii, P. peuce, P. pinea* and *P. sylvestris* seedlots, around 10% for all *P. halepensis* and *P. brutia* seedlots and around 25% for *P. nigra*. Data concerning *P. brutia* and *P. halepensis* seed germination, shown in Figures 2 and 3, represent mean germination values for all the seedlots used.

Results

Final dark germination, in the range of 5 to 25°C, was found to vary significantly between the different pine species (Fig. 2). These differences can be grouped and summarized as follows: *P. brutia* and *P. halepensis* germinated promptly in a rather wide temperature range, 10–20°C. On the basis of germination rates, 20°C could be considered as the optimal temperature. Germination of both these species could also take place at the marginal temperatures, 5 and 25°C, but to a significantly lower level; in addition, the germination rate at 5°C was extremely low, while at 25°C it was relatively fast. Similar results were obtained for *P. nigra,* although the maximum germination percentages were scored in the narrower range of 15–20°C. Once more, 20°C was the optimal germination temperature. In comparison with the two previous species, the temperature-dependence germination curve of *P. nigra* seems to have shifted towards higher temperatures.

Dark germination of *P. pinea* seeds was feasible only at 20°C; its inability to germinate at a wider range seems to be a characteristic of the species.

P. sylvestris and *P. heldreichii* exhibited a primary seed dormancy expressed

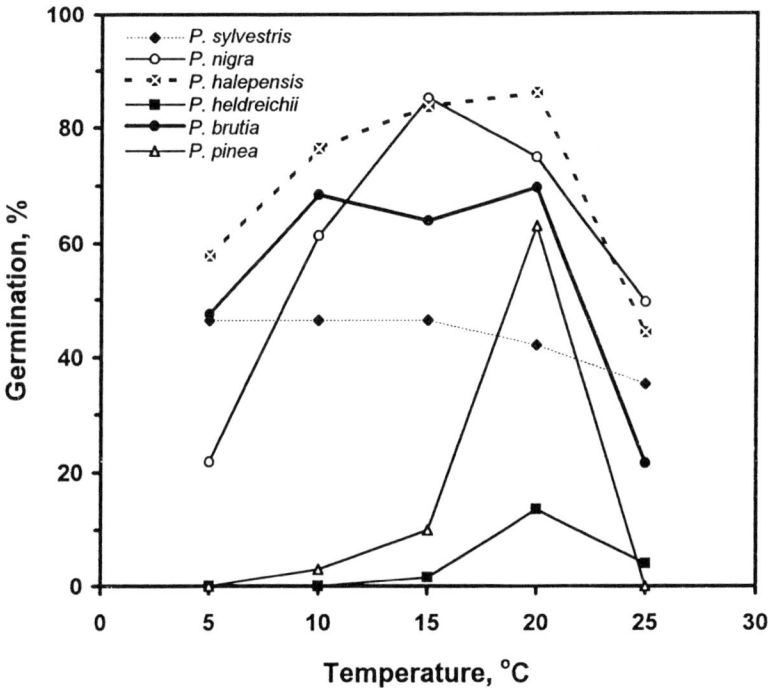

Figure 2. Final dark germination as a function of temperature

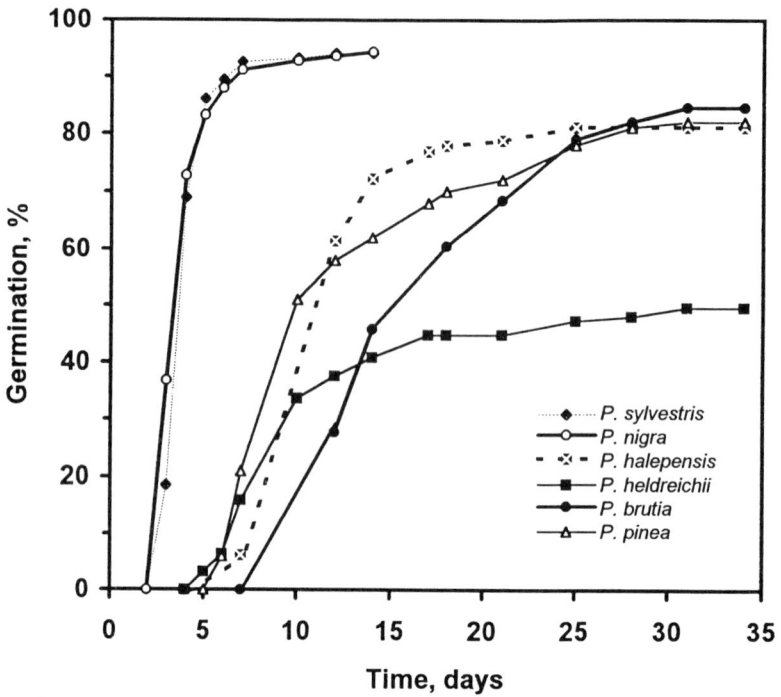

Figure 3. Time course of seed germination at 20°C in a 12 h white light photoperiod

by a restricted germinability throughout the temperature range. However, the degree of dormancy differed significantly among the two species. *P. sylvestris* final dark germination values were quite constant, around 40%, for all temperatures tested and germination rate was optimal at 20°C. *P. heldreichii* seeds, on the other hand, were deeply dormant since the highest germination percentage (obtained at 20°C) was less than 15%.

Germination time courses of all pine species are shown in Figure 3. Germination took place at the optimal temperature of 20°C and, in addition, a 12 h white light photoperiod was applied. Diurnal light resulted in a general increase of germination rate; in each species tested, germination curves were advanced (compared to dark controls, data not shown in the figure) by 1 to 4 d, depending on the species. Moreover, a significant promotion of final germination was observed in *P. sylvestris* and *P. heldreichii*. Primary dormancy was fully released by white light in *P. sylvestris* seeds but only partially in the more dormant ones of *P. heldreichii* (cf. Fig. 2).

Concerning germination rate, *P. nigra* and *P. sylvestris* are relatively fast germinators since their almost identical time course curves were completed

Table 2. The effects of light (WL: white light; FR: far-red) and prechilling (one month at 3°C on the final level (G) and rate (R) of germination at the optimum temperature (20°C), as compared to dark controls

	WL promotion		FR	Chilling promotion	
	G	R	inhibition	G	R
P. brutia	(+)	+	+	(+)	+
P. halepensis	–	+	+	(–)	+
P. heldreichii	+	+	–	+	+
P. nigra	–	+	–		+
P. peuce	–	–		+	+
P. pinea	–	+	–		
P. sylvestris	+	+	–	+	+

(+), In dormant seedlots of *P. brutia*, the final germination percentage is improved significantly by WL and/or chilling

(–), Prolonged prechilling treatments (exceeding 2 months) are detrimental to *P. halepensis* seeds

within a week. On the other hand, the germination of the four other species, even at these optimal conditions, was considerably slower.

The effects of light and prechilling on the final level and rate of germination at 20°C, as compared to dark controls are summarized in Table 2. Far-red light resulted in a total inhibition of *P. halepensis* and *P. brutia* seed germination, while no significant differences in the final germinability were observed in *P. nigra*, *P. pinea* and *P. sylvestris* seeds (detailed data not shown). The effect of prechilling prior to transfer to 20°C in the dark, was shown to deviate among the species tested (detailed data not shown), escalating from an increase of germination speed only (in *P. halepensis*, *P. brutia*, *P. nigra*) to a full release of dormancy in *P. sylvestris* and *P. heldreichii* seed germination.

P. peuce seeds were found to be extremely dormant. No germination was recorded at any temperature and light regime used. However, preliminary results indicate that a prolonged stratification (exceeding 4 months) is required for germination induction.

Discussion

A considerable array of differences was observed among the seven pine species concerning cone and seed characteristics, cone ripening and seed dispersal periods as well as germination behaviour (temperature dependence, germination speed, FR inhibition, chilling requirement). These differences may be attributed to varying ecophysiological strategies in regard to the temporal pattern of seedling emergence and establishment for each particular species.

According to the variants of this strategy, seed germination is timed either during autumn and early winter (in southern, mild and dry areas) or in spring (in regions with relatively cold and moist climatic conditions) or even throughout the wet season (in intermediate conditions), depending on the species.

The two typical, low-altitude Mediterranean, fire-resilient pines *P. halepensis* and *P. brutia* are distinguished by a high degree of cone serotiny (resulting in abundant canopy seed banks; Daskalakou and Thanos, 1994) as well as precocious reproduction. Laboratory germination is quite slow but feasible throughout the range of cool temperatures. Although short periods of stratification improve germination rates of both species, long term chilling of *P. halepensis* seeds was eventually proven detrimental (Skordilis and Thanos, 1995). The germination of the above species in the field is realized soon after the onset of the rainy season (mid October–November) and may continue to the end of the winter. The light sensitivity of both species (FR fully inhibits germination) indicates the 'invasive' potential of colonizing open habitats (pioneer species). It must be mentioned that populations of these species tend to survive as seeds through recurrent summer fires. Mature individuals have a low tolerance of fire but populations are fire-resilient through abundant seed reproduction and delayed seed release, similarly to the fire-resilient pines of North America (McCune, 1988). In addition, together with *P. pinea*, they are the only European pines where seeds are produced three years after cone initiation (two years for the other species, Table 1).

Our results concerning temperature dependence of *P. pinea* seed germination are in agreement with similar studies on other provenances (Magini, 1955). In contrast to the other pine species studied in the present work, the much heavier seeds of *P. pinea* are wingless; thus they are dispersed mainly under the canopy of the parent trees where seedling recruitment is eventually observed (Masetti and Mencuccini, 1991). Therefore, the indifference of *P. pinea* seed germination towards FR may be viewed as an adaptation to the below-canopy recruitment.

The deeply dormant seeds of *P. heldreichii* displayed an absolute stratification requirement, which could not be substituted by prolonged illumination. Analogous results were also shown by Borghetti *et al.* (1986, 1989) for Italian populations of this species. Under natural conditions, the presence of primary seed dormancy probably does not permit germination during autumn; thus germination takes place in spring, after seeds have experienced the low temperatures of winter, thus preventing young seedlings from being exposed to damaging freezing temperatures.

P. nigra seed germination was quite fast in a wide temperature range (also in agreement with populations of other origins; Paci, 1989), indicating that seed germination in nature will take place in spring, soon after dispersal. In contrast to the dry summer conditions prevailing in *P. halepensis* and *P. brutia* distribution areas, water availability in the mountainous habitats ensures the successful survival of the spring germinating *P. nigra* seedlings.

P. sylvestris seed dispersal, in nature, takes place at the end of winter. In this species, the promotive effect of stratification and light on the release from primary dormancy, exhibited by a large proportion of the seed population, could be considered an adaptive strategy, which either delays germination if seeds are covered by snow or enhances germination at an open, snow-free position, suitable for seedling survival.

P. peuce, occurring as a Eurasiatic relic in some mountains of the Balkan peninsula (Wilhelm, 1987), is taxonomically very different from all other pine species studied in the present work (Prus-Glowaski *et al.*, 1985). The observed difficulties in germinating these extremely dormant seeds are in agreement with results from other provenances (Djordjeva, 1967). Due to its very extended prechilling requirement, germination in nature may take place the second or even third year following dispersal.

Acknowledgements

This work was partly supported by the European Union (PROMETHEUS research project, Environment Research Programme, EV5V-CT94-0482, Climatology and Natural Hazards).

References

Borghetti, M., Verdramin, G.G., Giannini, R. and Schettino, A. 1989. *Acta Ecologica – Œcologia Plantarum* 10: 45–56.
Borghetti, M., Verdramin, G.G., Veneziano, A. and Giannini, R. 1986. *Canadian Journal of Forest Research* 16: 867–896.
Calamassi, R., Falusi, M. and Tocci, A. 1980. *Annali dell' Istituto Sperimentale per la Selvicoltura (Arezzo)* 11: 193–230.
Calamassi, R., Falusi, M. and Tocci, A. 1984. *Silvae Genetica* 33: 133–139.
Critchfield, W. and Little, E. 1969. *USDA Forest Service, Miscellaneous Publication* No. 991, 97 pp., Washington, D.C.
Daskalakou, E.N. and Thanos, C.A. 1994. In: *Proceedings of the 2nd International Conference on Forest Fire Research*, Coimbra, Portugal 21–24 November 1994, Vol. II, pp. 1079–1088. (ed. D.X. Viegas)
Djordjeva, M. 1967. *Bulletin Scientifique, Section A, Academie des Sciences RSF Yugoslavie* 12: 145.
Gaussen, H., Heywood, V.H. and Chater, A.O. 1993. In: *Flora Europaea, Vol. I*, pp. 40–44, (eds Tutin *et al.*). Cambridge: Cambridge University Press.
Greuter, W., Burdet, H.M. and Long, G. 1984. *Med-Checklist, Vol.1*, pp. 33–35. Conservatoire et Jardin Botaniques de le Ville de Geneve.
Krugman, S.L. and Jenkinson, J. 1974. In: *Seeds of Woody plants in the United States. USDA Forest Service, Handbook no 450*. pp 598–638 (ed. C.S. Schopmeyer).
Little, E. and Critchfield, W. 1969. *USDA Forest Service, Miscellaneous Publication* No. 1144, 51 pp., Washington, D.C.
Magini, E. 1955. *L' Italia Forestale e Montana* 10:106–124.
Masetti, C. and Mencuccini, M. 1991. *Ecologia Mediterranea* 17:103–118
McCune, B. 1988. *American Journal of Botany* 75: 353–368.
Ministry of Agriculture, Greece. 1992. *Results of the First National Census of Forests*. pp.134.
Paci, M. 1989. *L' Italia Forestale e Montana* 44: 411–423.
Panetsos, C.P. 1975. *Silvae Genetica* 24:163–168.

Panetsos, C.P. 1981. *Annales Forestales (Zagreb)* 9: 39–77.
Pozzera, G. 1959. *L'Italia Forestale e Montana* 14: 196–206.
Prus-Glowaski, W., Szweykowski, J. and Nowak, R. 1985. *Silvae Genetica* 34: 162–170.
Skordilis, A. and Thanos, C.A. 1995. *Seed Science Research* 5: 151–160.
Strid, A. 1989. *Mountain Flora of Greece. Vol I.* Cambridge: Cambridge University Press.
Thanos, C.A. and Skordilis, A. 1987. *Seed Science and Technology* 15: 163–174.
Wilhelm, K. 1987. *Plant Systematics and Evolution* 162: 133–163.

70. Germination of *Orobanche* Seeds: Some Aspects of Metabolism during Preconditioning

A.M. MAYER and N. BAR NUN

Department of Botany, The Hebrew University of Jerusalem, Jerusalem, 91904, Israel

Abstract

Response to stimulants, such as strigol or its analogues, of seeds of *Orobanche aegyptiaca* requires preconditioning in water for 6 days (d). Treatment with salicylic acid during preconditioning results in a sharp rise in respiration, which is insensitive to cyanide. When seeds become responsive to stimulant they no longer respond to salicylic acid and respiration is fully cyanide sensitive. During preconditioning uptake of thymidine by the seeds is not inhibited by aphidicolin. When they become responsive thymidine uptake is inhibited by aphidicolin. During preconditioning no major proteins are broken down, but protein synthesis as indicated by leucine incorporation can be detected after 8 h. During the preconditioning phase unusual, active metabolism occurs, which may be responsible for the transition from non-responsive to responsive seeds.

Introduction

The germination of the parasitic plants, reviewed by Joel *et al.* (1994), has received little attention compared to the extensive research into the physiology and biochemistry of seeds in general (Mayer and Poljakoff-Mayber, 1989; Come and Corbineau, 1993). The germination of these seeds is regulated in order to increase the likelihood of subsequent effective attachment to the host. Research has centered on the obligate root parasites of the genus *Orobanche* devoid of chlorophyll and the genus *Striga*, a hemi-parasite whose germination exhibits some adaptations which parallel those of *Orobanche*. These are economically the most important parasitic weeds, world wide. Seeds of both *Orobanche* and *Striga* undergo a clearly defined transition from a state in which they are unable to germinate and cannot respond to a germination stimulant to a physiological state in which they can germinate after exposure to a germination stimulant.

In order to become responsive to a germination stimulant the seeds must be placed in water for a period of time measured in days, a treatment termed preconditioning (Joel *et al.*, 1991; 1994). Exposure to the stimulant during this preconditioning period does not induce germination and may even inhibit subsequent germination. The transition is clear and well defined, but its physiology and biochemistry has hardly been studied.The seeds become committed to germination only after preconditioning and exposure to stimulant. Both *Striga* and *Orobanche* seeds may be dormant in the soil for long periods of time, tens of years (Lopez-Granados and Torres Garcia, 1993). While

R.H. Ellis, M. Black, A.J. Murdoch, T.D. Hong (eds.), Basic and Applied Aspects of Seed Biology, pp. 633–639.
© *1997 Kluwer Academic Publishers, Dordrecht. Printed in Great Britain.*

dormant they will not respond to preconditioning and to the addition of a stimulant. For *Striga*, the nature of the stimulant is more or less recognized and it is akin to strigol. For the genus *Orobanche*, the nature of the stimulant is still not known, although there are indications that it too might be related to strigol.

Early work on the respiration of *Striga* shows that its rate of respiration changes during preconditioning and after germination (Vallance, 1951). In *Striga* changes in respiratory pathways may occur during preconditioning (Giraudeau and Fer, personal communication).

Clearly changes in metabolic pathways and events take place which are well defined and can be experimentally probed. The changes in respiration, protein metabolism and nucleic acid metabolism are accompanied by the ability of the seeds to respond to a stimulant and to germinate.

Here we report on some of the changes in respiratory mechanisms in broomrape during preconditioning and the early steps of respiration and on some other metabolic events occurring during the preconditioning and early respiration.

Materials and Methods

Seeds of *Orobanche aegyptiaca* were collected near Kibbutz Usha and stored at room temperature in the laboratory. The seeds were disinfected, and germinated as described by Bar Nun and Mayer (1993), the stimulants used being GR7 or GR24, depending on the experiment. Oxygen uptake by the seeds was measured as described by Bar Nun and Mayer (1993). Inhibitors were applied to the seeds for 30 min immediately before measurement of oxygen uptake. Salicylic acid was given at a final concentration of 20 μM and KCN at 1 mM. Measurement of incorporation of thymidine into DNA and uridine into RNA and leucine into protein are described in Mayer and Bar Nun (1994). Total RNA was isolated by the method of Chang *et al.* (1993) and RNA was separated by agar gel electrophoresis in 1% agarose in Tris-acetate buffer 0.04 M containing 0.001 M EDTA for 45 min and stained with ethidium bromide.

Results and Discussion

Respiration

During the preconditioning period the respiration of *Orobanche* seeds shows an intial rise, followed by a decrease till the stage at which they become responsive to stimulant. When salicylic acid is applied to seeds, during the preconditioning period, their respiration increases dramatically during the first 5 d of preconditioning, but when they become receptive to stimulant, on d 7, their respiration no longer responds to salicylic acid (Fig. 1). Salicylic acid is known to engage the alternative, cyanide resistant pathway of germination (Kapulnik *et al.*,

1992). Hence it appears that during preconditioning this pathway can be activated. Following the onset of germination the seeds respond only to a limited extent to the addition of salicylic acid. We tested the response of both normal and salicylic acid induced respiration to cyanide. The addition of cyanide to the seeds in the absence or presence of salicylic acid is shown in Figure 2. The data in this figure clearly indicate that the respiration induced by salicylic acid is indeed cyanide resistant, during preconditioning. Most of the respiration during the germination phase is cyanide sensitive, whether salicylic acid was applied or not. Moreover, respiration of seeds which were not stimulated by a germination stimulant, GR24, was totally repressed by cyanide. Thus while they were insensitive to cyanide during preconditioning they became sensitive after preconditioning. These results clearly show that the repiratory pathway of the seeds changes during the preconditioning period.

Protein Synthesis and Nucleic Acid Metabolism

Extremely active protein synthesis or a turn-over occurs during preconditioning (Fig. 3). Proteins become labelled at this stage. Already after 8 h of imbibition we detected some 14 proteins which incorporated label, a pattern maintained throughout the preconditioning period. The uptake of label becomes much less evident in seeds which germinate for 12 d in absence of host. The seeds during

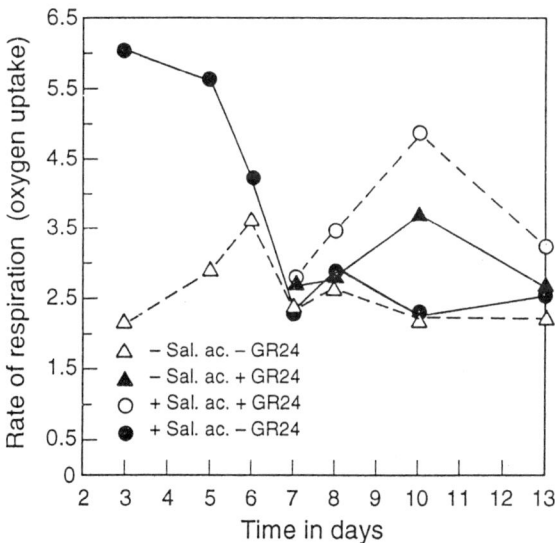

Figure 1. Oxygen uptake of *Orobanche* seeds, during preconditioning and germination, following treatment with salicylic acid (20 µM) in the absence or presence of germination stimulant GR24, given on day seven. (Oxygen uptake as µl O_2/min/100 mg seeds)

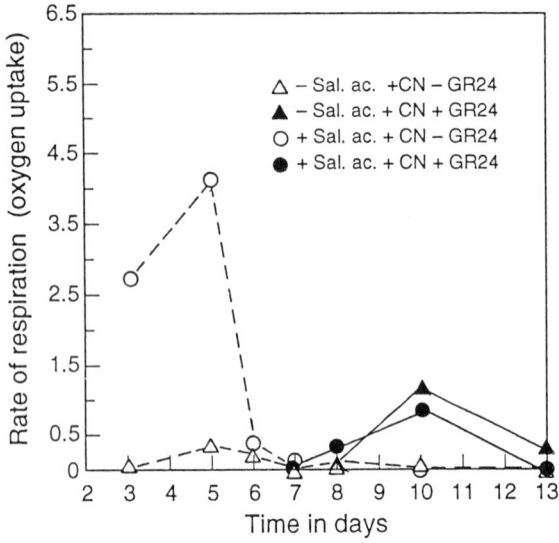

Figure 2. Oxygen uptake of *Orobanche* seeds, during preconditioning and germination, following treatment with salicylic acid (20 μM) and cyanide (1 mM) in the absence or presence of germination stimulant GR24, given on day seven. (Oxygen uptake as μl O_2/min/100 mg seeds)

Comparison of bands at different periods with those labelled after 8h imbibition followed by 4h radioactive leucine.

Figure 3. Schematic presentation of labelling of proteins with [^{14}C]leucine during preconditioning and germination. (Numbers on left are molecular weights, kDa)

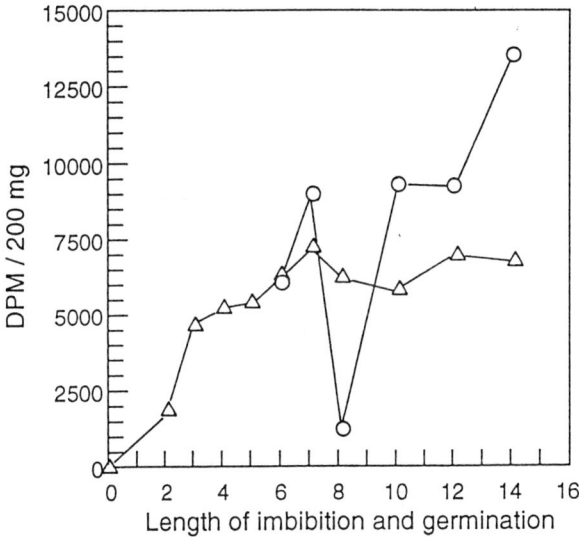

Figure 4. Pattern of thymidine incorporation into total DNA isolated from *Orobanche* seeds during preconditioning and germination. Seeds treated with water throughout, exposed for 2 h to labelled thymidine (Δ). Seeds treated for six days with water and then with 10 ppm of GR7, exposed for 2 h to labelled thymidine (O)

preconditioning also took up radioactive thymidine into total DNA (Fig. 4). However, just at the transition stage, this incorporation dropped dramatically, to 10% of its value at the stage of when they become responsive to stimulant. This drop was highly significant (replicates not shown here). When germination begins, thymidine uptake is rapidly resumed. It seemed possible that the thymidine uptake was due to different mechanisms during preconditioning and

Table 1. Effect of aphidicolin on thymidine incorporation by *Orabanche aegyptiaca* seeds. Results as dpm/200 mg fresh wt. in DNA extracted from the seeds after addition of labelled [3H]thymidine for 2 h

Treatment	Control[a]	Aphidicolin (300 μM[b])	% Inhibition
Imbibition 4 d	446 (±42)	344 (±49)	23 (n.s.)
Imbibition 6 d + 4 d germination[c]	1331 (±75.5)	246 (±94)	81.5

[a]Control seeds treated with 0.1% DMSO; [b]Aphidicolin dissolved in 0.1% DMSO; [c]GR7 added on d 6, to induce germination

dpm, disintegrations per min

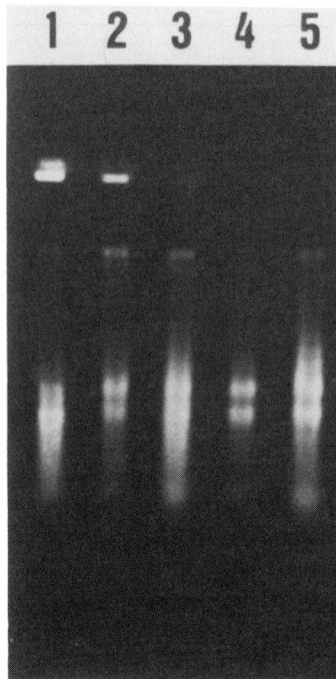

Figure 5. Agar gel electrophoresis of *Orobanche* RNA. Lane 1, dry seeds; Lane 2, 3 days imbibed; Lane 3, 6 days imbibed; Lane 4, 6 days imbibed + 2 days GR7; Lane 5, 6 days imbibed + 4 days GR7

germination. This is borne out by experiments using aphidicolin, which inhibits replicative DNA synthesis but not DNA repair (Table 1). It is very evident that during preconditioning thymidine is not inhibited, indicating that it is due to a repair mechanism of some kind. During germination, when of course the seedling root begins to elongate, thymidine incorporation is almost totally inhibited, indicating that at this stage DNA replication, as expected begins.

Distinct changes in the electrophoretic pattern of RNA isolated from the seeds can be discerned. Some very high molecular weight material which may be DNA is present in dry seeds and disappears at the end of the preconditioning period. The mobility of the major bands which represent ribosomal RNA also changes at about the onset of the response to stimulant (Fig. 5). Uridine incorporation into total RNA increases more or less linearly throughout preconditioning and germination and is much the same in seeds treated with stimulant or not, but is somewhat lower in the seeds not treated with stimulant (Mayer and Bar Nun, 1994, and unpublished).

The major phenolic compound present in the dry seeds is tricin, as identified by Professor J.B. Harborne. Its level falls considerably during preconditioning, and in germinating seeds caffeyl esters begin to appear, apparently close in structure to orobanchin. The function of tricin in the seeds is unclear, but the fact that its level falls during preconditioning may be significant in the transition process.

Clearly very marked changes occur in the seeds of *Orobanche* during preconditioning. They are particularly evident at the stage at which the seeds are becoming or have become sensitive to the externally-applied stimulant. It is too early to formulate detailed theories about which of these steps, if any are directly involved in the transition stage. Considerably more work, both at the biochemical and the molecular biology level is needed to clarify this.

References

Bar Nun, N. and Mayer, A.M. 1993. *Phytochemistry* 34: 39–45

Chang, S., Puryear, J. and Cairney, J. 1993. *Plant Molecular Biology Reporter* 11: 113–116.

Côme, D. and Corbineau, F. 1993. *Proceedings of the 4th International Workshop on Seeds – Basic and Applied Aspects of Seed Biology* (Volumes 1–3), pp. 1118, Paris: ASFIS.

Joel, D.M., Back, A., Kleifeld, Y. and Gepstein, S. 1991. In: *Progress in Orobanche Research,* pp.147–156 (eds. K. Wegman and L.J. Musselman). Tubingen, FRG: Eberhard-Karls University.

Joel, D.M., Steffens, J.C. and Matthews, D. 1994. In: *Seed Development and Germination,* pp.567–596 (eds. J. Kigel, M. Negbi and G. Gallili). New York: Marcel Dekker.

Kapulnik, Y., Yalpami, N. and Raskin, I. 1992. *Plant Physiology* 100 : 1921–1926.

Lopez Granados, F. and Garcia Torres, L. 1993. *Weed Research* 33: 319–327.

Mayer, A.M. and Poljakoff-Mayber, A. 1989. *The Germination of Seeds,* Fourth Edition, pp.270. Oxford: Pergamon Press.

Mayer, A.M. and Bar Nun, N. 1994. In: *Biology and Management of Orobanche, Proceedings of the 3rd International Workshop on Orobanche and Related Striga Research,* pp. 147–156 (eds. A.H. Pieterse, J.A.C. Verklij and S.J. ter Borg). Amsterdam, The Netherlands: Royal Tropical Institute.

Vallance, K.B. 1951. *Annals of Botany* 15: 109–128.

71. Ecophysiology of Seed Germination in Composites Inhabiting Fire-prone Mediterranean Ecosystems

M.A. DOUSSI and C.A. THANOS

Department of Botany, University of Athens, Athens 15784, Greece

Abstract

Within the framework of a research project concerning adaptive mechanisms of postfire regeneration in Mediterranean ecosystems, the ecophysiology of seed germination was studied in the following composites: *Dittrichia viscosa* (L.) Greuter, *Helichrysum stoechas* (L.) Moench ssp. *barrelieri* (Ten.) Nyman and *Phagnalon graecum* Boiss. & Heldr. All three species are common in the Mediterranean region; the former is a colonizer while the latter two are usually present in phrygana (low-shrub, fire-prone Mediterranean vegetation). As the result of their numerous, anemochorous dispersal units the three species are endowed with the potential to invade disturbed areas, in general, and burned ones, in particular.

Germination in *Dittrichia* was generally very low in the dark (0–10%). Moreover, an absolute light requirement was revealed and germination could be promoted even by green safelight. In *Helichrysum*, final dark germination reached its highest value (ca. 50%) at relatively low temperatures (10°C, 15°C). Light and nitrates (optimal concentration 20 mM) promoted germination dramatically above dark control. Nevertheless, full induction of germination required the presence of both these factors. *Phagnalon* germinated optimally (70–100%) over a broad temperature range (10–25°C), in the dark. However, light proved beneficial at suboptimal temperatures. In the three species studied, light and, in the case of *Helichrysum*, nitrate availability are particularly important for seed germination and eventual seedling recruitment in the postfire environment.

Introduction

Wildfires are an integral component of ecosystem structure and function in all three major vegetation types (pine forests, maquis and phrygana) which dominate the Mediterranean-type environment of Greece. Postfire regeneration is characterized by the remarkable resilience of established plants through resprouting from both epicormic meristems and below-ground organs and also through germination from a soil- or plant-stored seed bank. A potentially important additional contribution to postfire regeneration is the long-distance dispersal of seeds into recently burned areas.

One characteristic that allows a plant species to exploit newly disturbed sites is the production of large numbers of seeds which may be dispersed long distances. Apparent adaptations that would allow long-distance dispersal, i.e. small, light seeds with structures such as a pappus, are certainly prominent

R.H. Ellis, M. Black, A.J. Murdoch, T.D. Hong (eds.), Basic and Applied Aspects of Seed Biology, pp. 641–649.
© *1997 Kluwer Academic Publishers, Dordrecht. Printed in Great Britain.*

among pioneers in newly burned sites (Whelan, 1986). Many references are given by Kozlowski and Ahlgren (1974) to the well dispersed seeds that germinate soon after the fire. *Haplopappus tenuisectus* and *Gutierrezia sarothrae* (both composites) are postfire pioneer shrubs, disseminated by wind in the Sonoran Desert. In the Mediterranean maquis, several annuals may appear in large numbers during the first year after fire; these annuals are apparently opportunistic invaders and are not stimulated by fire *per se*. It has been proposed that the long-term presence of disturbed habitats around the Mediterranean produced pioneer, weedy genotypes that became highly successful invaders (Cody and Mooney, 1978).

Dittrichia viscosa is a densely glandular, viscid perennial, up to 130 cm tall with stems woody at the base. It is widely distributed in South Europe and grows on pine woods, dry stream-beds, ditches, cliffs, fallow fields, roadsides, waste places and along hillside paths. *Helichrysum stoechas* ssp. *barrelieri* is a woody-based perennial, 10–50 cm tall, growing on edges of pine forests and in phrygana or maquis on stony hillsides from Sicily to Turkey. *Phagnalon graecum* is a dwarf shrub up to 30 cm tall, growing in phrygana on dry, rocky ground throughout South-eastern Europe (Tutin *et al.*, 1964–1980; Burnie, 1995).

All three plant species mentioned above belong to the Compositae. Apart from certain general rules for laboratory germination (e.g. Ellis *et al.*, 1985; ISTA, 1993), there is practically no information concerning the taxa under study. In the broad context of studying the adaptive mechanisms of postfire regeneration in Mediterranean ecosystems, the ecophysiology of their seed germination was investigated.

Materials and Methods

Plant Material

Achenes of *Dittrichia viscosa* (L.) Greuter (formerly *Inula viscosa* (L.) Aiton) were collected in December 1993 from plants growing in the University Campus. Achenes of *Helichrysum stoechas* (L.) Moench ssp. *barrelieri* (Ten.) Nyman and *Phagnalon graecum* Boiss. & Heldr. were collected in June 1994 from a burned pine forest at Mt Parnes, Attica, five years after the fire. Nomenclature follows Tutin *et al.* (1964–1980). Achenes, hereafter called seeds for simplicity, were stored in moisture- and light-proof containers at room conditions. The average, air dry seed weight for the three species was 0.240, 0.030 and 0.046 mg, respectively. The percentage of unsound (empty and non-viable) seeds (revealed by dissection under a stereomicroscope) was considerable: 20, 50 and 30%, respectively. At the end of germination tests, non-viable seeds were found disintegrated, while those considered viable were still hard, with a firm embryo. Seeds were used after an initial storage of 6 months at room conditions; thereafter, no changes in germination characteristics were observed throughout the experimentation period.

Germination Conditions

Germination experiments were conducted in glass Petri dishes (7 cm in diameter) lined with two filter paper discs and moistened with 3 ml of distilled water or other imbibition medium. Criterion of germination was the visible protrusion of the radicle through the fruit wall. In one set of experiments, measurements were taken regularly (and germinated seeds were discarded after each count) while in another, germination was recorded only once, at the end of the incubation period. The former tests were considered finished when no additional seeds germinated. Germination was expressed as a percentage from five samples of 25 seeds each (except for the experiments on *H. stoechas* where 50 seeds were sown per dish). At the end of each experiment ungerminated seeds were dissected and inspected under a stereomicroscope; unsound seeds were counted out and all the results and graphs presented are consequently corrected for germinable seeds only. For germination experiments in darkness, seeds were incubated within light-proof, metal containers in controlled temperature cabinets (Model BK 5060 EL, W.C. Heraeus GmbH, Germany) where temperature was kept constant within $\pm 0.5°C$ of the value set.

Light Sources

The red and far-red broad-band irradiations were produced by a bank of tubes and the light was filtered through an appropriate combination of coloured Plexiglas sheets (each 3 mm thick, Röhm GmbH, Germany). Red (R) light ($11.0 \ \mu mol \ m^{-2} s^{-1}$) was produced by ten red fluorescent tubes (TL 20W/15, Philips) and one layer of red Plexiglas, 501. Far-red (FR) light ($18.2 \ \mu mol \ m^{-2} s^{-1}$) was obtained by twelve white incandescent tubes (Philinea 6276X60 W, Philips), filtered through three layers of Plexiglas (two blue, 627, and one red, 501). All manipulations of imbibed seeds were carried out under a dim green safelight ($0.05 \ \mu mol \ m^{-2} \ s^{-1}$; one green fluorescent tube F 15T8.G.6, 15 W Green-Photo, General Electric, USA, two Plexiglas sheets, one red-orange, 478, and one green, 700). Total flux density values refer to light in the visible range (400–800 nm) at the seed surface, calculated from the measurements taken with a spectroradiometer (ISCO SR, USA).

Statistical Analysis

Statistically significant differences were assessed using the two-sample *t*-test (for two means) and the Newman-Keuls multiple range test (for more than two means); percentage values were arcsine transformed prior to statistical tests (Zar, 1974).

Results

The germinability of *Dittrichia* seeds was very low (0–8%) throughout the temperature range 5–30°C when seeds remained in total darkness and germination was scored only once, at the end of the incubation period. When seeds were exposed to safelight (for a few min each time) during the routine inspection for germination, a dramatic induction of germination took place (Fig. 1). At 5 and 10°C no effect of safelight was obtained. Further study of this light-facilitated

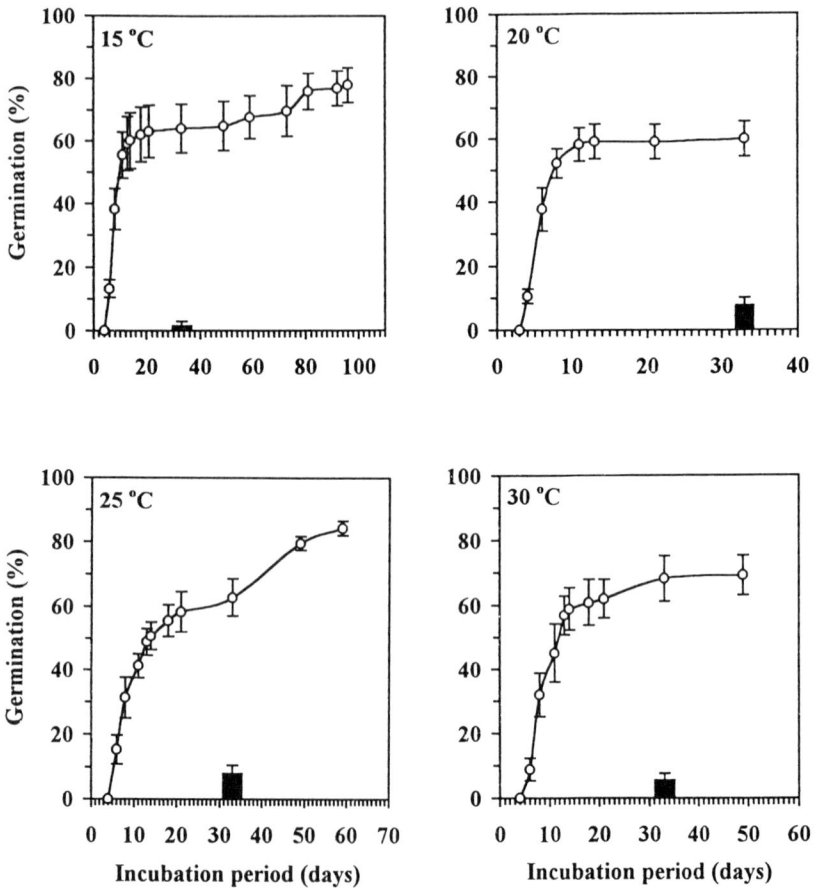

Figure 1. Time-courses of dark germination of *Dittrichia viscosa* seeds at various constant temperatures. Seeds were exposed for a few minutes to a dim green safelight while being inspected for germination. Bars represent final dark germination values when seeds were examined only once, at the end of a 33-day-long incubation period. Vertical lines indicate ±s.e.

Table 1. Mean (±s.e.) germinability of *Dittrichia viscosa* seeds at 20°C, in the dark. Light pulses were given 1 d after onset of imbibition. With the exception of safelight (seeds inspected twice weekly under dim green safelight), in all other cases germination was recorded only at the end of the incubation period (30 d). Statistically different ($p < 0.05$) means are followed by different letters

Treatment	Germination (%)
Darkness	8.3 ± 1.8^a
Safelight	60.0 ± 5.6^c
5 min R	69.6 ± 3.2^c
10 min FR	19.8 ± 3.4^b
5 min R + 10 min FR	26.0 ± 4.3^b
5 min R (in 3 consecutive d)	90.5 ± 1.7^d

promotion of germination revealed phytochrome mediation (Table 1). Maximum germinability (90%) was obtained with 5 min R per day (d) given for 3 consecutive days. Prechilling at 5°C induced only a slight enhancement of subsequent dark germination at 20°C; 2 and 4 weeks of chilling resulted in 25.9 ± 6.6 and $19.4 \pm 3.2\%$, respectively. Nitrates were ineffective and showed germination similar to that in water.

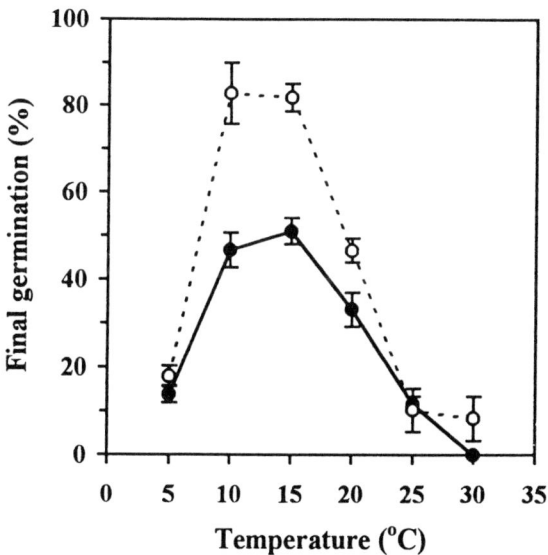

Figure 2. The germinability of *Helichrysum stoechas* ssp. *barrelieri* seeds as a function of temperature. Seeds were inspected for germination only once, at the end of the incubation period (●) or weekly under a dim green safelight (○). Vertical lines indicate ±s.e.

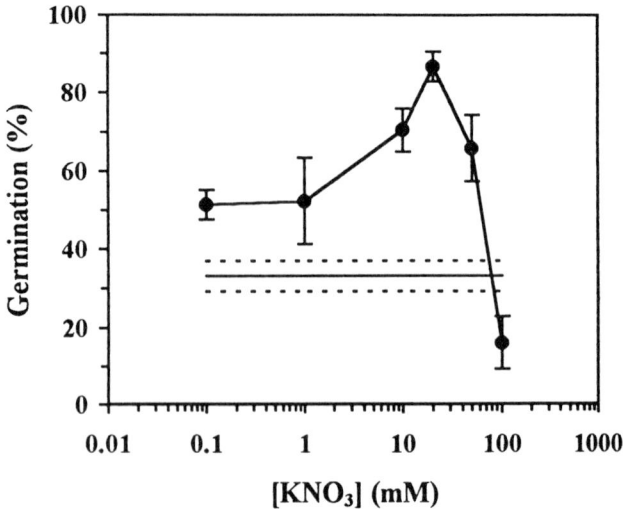

Figure 3. Final germination (at 20°C, in the dark) of *Helichrysum stoechas* ssp. *barrelieri* seeds as a function of potassium nitrate concentration. The horizontal line represents final germination in water (stippled lines: ±s.e.). Vertical lines: ±s.e.

Dark germination of *Helichrysum* was optimal at a rather narrow range of cool temperatures. A maximum final germination level of about 50% was obtained at 15 and 10°C while at 20°C the germination scored was 33%. A sharp decline of germinability was observed at more extreme temperatures (5, 25 and 30°C). However, the presence of green safelight, routinely used for recording germination in darkness, resulted in significant (10°C: 0.001

Table 2. Mean (±s.e.) germinability of *Helichrysum stoechas* spp. *barrelieri* seeds at 20°C, in the dark. Light pulses were given 1 d after onset of imbibition. With the exception of safelight (seeds inspected twice weekly under dim green safelight), in all other cases germination was recorded only at the end of the incubation period (30 d). Statistically different ($p < 0.05$) means are followed by different letters

Treatment	Germination (%)
Darkness	33.1 ± 3.9^a
Safelight	46.5 ± 2.7^b
5 min R	89.6 ± 2.7^c
20 mM KNO$_3$	86.5 ± 3.8^c
20 mM KNO$_3$ + 5 min R	96.4 ± 1.3^d

$<p<0.002$, 15°C: $p<0.001$, 20°C: $0.02<p<0.05$) enhancement of germinability over the dark control (Fig. 2). Moreover, dark germination was greatly enhanced in the presence of nitrates (Fig. 3). Maximum promotion of germination (87%) was obtained with a concentration of 20 mM KNO_3. A statistically significant enhancement of germinability was clearly observed at 10–50 mM; moreover, a considerable stimulation of germination was observed even at 0.1 mM although this result would need further confirmation. Finally, the concentration of 100 mM proved inhibitory. As shown in Table 2, although both light and nitrates resulted, individually, in a statistically significant promotion of germination, the combined presence of these factors was most inductive (final germination: 96%).

Dark germination of *Phagnalon* was optimal (73–96%) over the whole temperature range 10–25°C (maximum percentages at 20 and 15°C) with little or no germination at more extreme temperatures (30 and 5°C). Pulses of green safelight, given during the inspection of seeds for germinated ones, resulted in a significant (5°C: $0.005<p<0.01$, 10, 25 and 30°C: $p<0.001$) enhancement of germination over samples which had remained in total darkness throughout the incubation period (Fig. 4). No significant effect of nitrates was detected when tested at the various constant temperatures in the dark.

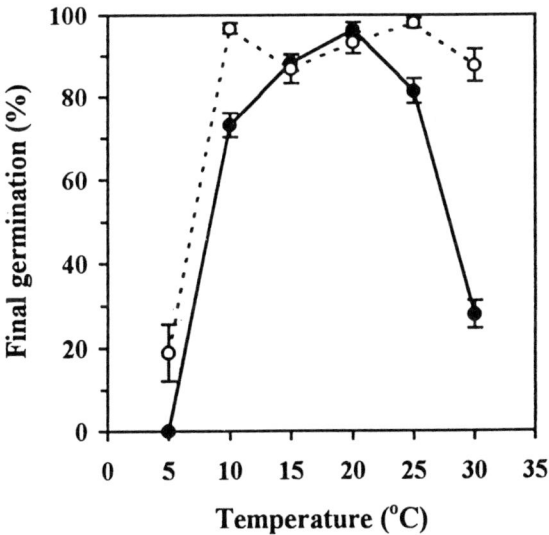

Figure 4. The germinability of *Phagnalon graecum* seeds as a function of temperature. Seeds were inspected for germination only once, at the end of the incubation period (●) or weekly under a dim green safelight (○). Vertical lines indicate ±s.e.

Discussion

In the present work seeds of *Dittrichia* were found to be deeply dormant. Very low germination (up to 8%) was scored in the dark; however, 90% germination was obtained with 5 min R per d given for 3 consecutive days. The small seeds of *Dittrichia* were clearly light-requiring. Even a few short exposures to a dim, green safelight strongly promoted germination, suggesting a very low threshold level of photosensitivity. On the other hand, prechilling at 5°C resulted in only a slight enhancement of subsequent dark germination, thus emphasizing the role of light as the principal agent of germination induction.

Seeds of *Helichrysum* exhibited a partial dormancy since only a portion (30–50%) germinated in the dark under the relatively cool, 'Mediterranean' temperature range, 10–20°C. Nitrates (optimal concentration 20 mM) resulted in a dramatic enhancement of germinability at 20°C, in the dark, thus establishing *Helichrysum* as the first instance of nitrate-promoted species in the postfire Mediterranean flora. The stimulatory effect of nitrates on seed germination of numerous plant species is, nowadays, well documented. In addition, both the American and the International Associations for seed testing have, since 1954, adopted officially the systematic use of 0.2% (about 20 mM) KNO_3 in their suggested germination protocols for many species (AOSA, 1981; ISTA, 1993). A curve of seed germination induction as a function of nitrate concentration similar to that obtained with *Helichrysum* has been observed in numerous cases (e.g. in the chaparral fire annual *Emmenanthe penduliflora*; Thanos and Rundel, 1995). Fire leads to a massive volatilisation of simple nitrogenous compounds, mainly nitrate and ammonium (e.g. DeBano *et al.*, 1979). On the other hand, several measurements in various ecosystems have shown a depletion of available nitrates in the presence of actively growing plants (e.g. Pons, 1989). Thus, a dormancy relief mechanism triggered by an appropriate level of nitrates would serve as a competition-avoiding, gap detector. As far as role of light is concerned, regular pulses of safelight were able to increase the final germination level in the range of optimal temperatures. The effect of a short exposure to R was of the same magnitude as that of the nitrates; moreover, a significantly increased effect (full germination induction, 96%) was obtained when both treatments were applied together.

In *Phagnalon*, dark germination was maximal (73–96%) in the range of 10–25°C (optimal temperatures: 20 and 15°C) while a considerable level (nearly 30%) was scored also at the warm 30°C. Although light was not required at the optimal temperatures it was found to significantly enhance germinability at the marginal ones. The 'aggressive' germination behaviour exhibited by *Phagnalon* is typical of an opportunist colonist with a capacity for expansion into disturbed land (not necessarily a burned one).

The results presented show that illumination of all three species was an important (to a different extent for each species) inducing agent of germination. This conclusion is in agreement with the well established fact that most light-requiring seeds are small-sized. In a survey of 69 species, many of the small-seeded species (<0.1mg) showed a light requirement while the majority of species with seed weights 1 mg or more were light-indifferent (Grime *et al.*, (1981). Several authors have already reported light stimulation within the Compositae (e.g. Mott, 1972; Atwater, 1980; Willis and Groves, 1991; Plummer and Bell, 1995). Light requirement of seed germination is considered as a potentially significant mechanism for gap detection and this suggestion could well also extend to burned areas (as shown previously with Mediterranean plants; Thanos and Skordilis, 1987; Thanos *et al.*, 1995). Changes of light quantity and, more importantly, quality can be encountered in the postfire environment as a result of soil disturbance and/or removal of the 'far-red-enrichment filters' of the canopy and the litter. The prolific seed production and effective dispersal coupled with the inductive role of light place the three species in the category of 'widely dispersed' colonizers of disturbed habitats (e.g. recently burned sites).

Acknowledgements

This work was fully supported by European Union (PROMETHEUS research project, Environment Research Programme, EV5V-CT94-0482, Climatology and Natural Hazards).

References

AOSA (Association of Official Seed Analysts). 1981. *Journal of Seed Technology* 6: 1–125.
Atwater, B. R. 1980. *Seed Science and Technology* 8: 523–573.
Burnie, D. 1995. *Wild Flowers of the Mediterranean*, pp. 320. London: Dorling Kindersley.
Cody, M. L. and Mooney, H. A. 1978. *Annual Review of Ecology and Systematics* 9: 265–321.
DeBano, L. F., Eberlein, G. E. and Dunn, P. H. 1979. *Soil Science Society of America Journal* 43: 504–509.
Ellis, R. H., Hong, T. D. and Roberts, E. H. 1985. *Handbook of Seed Technology for Genebanks. Volume II. Compendium of Specific Germination Information and Test Recommendations*. Rome: International Board for Plant Genetic Resources.
Grime, J. P., Mason, G., Curtis, A. V., Rodman, J., Band, S. R., Mowforth, M. A. G., Neal, A. M. and Shaw, S. 1981. *Journal of Ecology* 69: 1017–1059.
ISTA (International Seed Testing Association). 1993. *Seed Science and Technology* 21: Supplement.
Kozlowski, T. T. and Ahlgren, C. E. 1974. *Fire and Ecosystems*. New York: Academic Press.
Mott, J. J. 1972. *Journal of Ecology* 60: 293–304.
Plummer, J. A. and Bell, D. T. 1995. *Australian Journal of Botany* 43: 93–102.
Pons, T. L. 1989. *Annals of Botany* 63: 139–143.
Thanos, C. A., Kadis, C. C. and Skarou, F. 1995. *Seed Science Research* 5: 161–170.
Thanos, C. A. and Rundel, P. W. 1995. *Journal of Ecology* 83: 207–216.
Thanos, C. A. and Skordilis, A. 1987. *Seed Science and Technology* 15: 163–174.
Tutin, T. G., Heywood, V. H., Burges, N. A., Valentine, D. H., Walters, S. M. and Webb, D. A. 1964–1980. *Flora Europaea, Volumes 1–5*. Cambridge: Cambridge University Press.
Whelan, R. J. 1986. In: *Seed Dispersal*, pp. 237-271 (ed. D. R. Murray). Academic Press Australia.
Willis, A. J. and Groves, R. H. 1991. *Australian Journal of Botany* 39: 219–228.
Zar, J. H. 1974. *Biostatistical Analysis*. Englewood Cliffs, NJ: Prentice-Hall.

72. SEED: A Computer-assisted Learning Package on Seed Longevity

A.D. MADDEN[1], R.H. ELLIS[2] and S.B. HEATH[1]

[1]Centre for Computer-based Learning in Land Use and Environmental Sciences (CLUES), University of Aberdeen, Aberdeen AB9 2UB; [2]Department of Agriculture, The University of Reading, Earley Gate, P.O. Box 236, Reading RG6 6AT, UK

Abstract

SEED is a computer-assisted learning (CAL) package for use by students on university courses, and by seed technologists for professional updating in the workplace. It provides a supported learning environment of tutorial reinforced by multiple choice questions and exercises. This ensures that the package can be used without the support of a lecturer or trainer. CAL packages such as SEED use the flexibility and versatility of information technology, and in so doing, help to make teaching and learning more efficient, adaptable and enjoyable, thereby benefitting both students and lecturers.

Introduction

The seed viability equation (Ellis and Roberts, 1980a,b; Roberts and Ellis, 1989) quantifies the relationships between the longevity of seeds of orthodox species in air-dry storage and the temperature and moisture content of the seed storage environment. A sound understanding of these relationships is important to both seed technologists and to students of seed physiology, not least because it is necessary for practical advice on appropriate seed storage regimes to be quantitative. Unfortunately, these models are often seen as complex and difficult to understand by many students; particularly students in the biological sciences who may be intimidated by mathematical equations.

In recent years computer software has become considerably more versatile and usable. As a result, it has become possible to develop educationally valuable computer-assisted learning (CAL) courseware which is of use to students on university courses and for professional updating (Heath and Young, 1992). The exploitation of these developments has been made practical by the dramatic decline in the cost of computer hardware. The equipment necessary for CAL no longer represents a major capital investment, with the result that suitable computers are now common on university campuses and in a far greater range of workplaces. Through the use of CAL tutorials and simulation models students can learn at their own pace, and can benefit from those features of the computer, such as animations and graphics, which can be used to enliven and clarify the subject matter.

R.H. Ellis, M. Black, A.J. Murdoch, T.D. Hong (eds.), Basic and Applied Aspects of Seed Biology, pp. 651–655.
© *1997 Kluwer Academic Publishers, Dordrecht. Printed in Great Britain.*

SEED – a CAL Module on Seed Longevity

SEED was produced to take advantage of the increased availability of personal computers and of their potential as teaching aids.

It was developed by Ellis, Schofield, Tidball and Young (1995) with two aims:

1. to explain how environmental factors affect the longevity of orthodox seeds;

2. to describe the seed viability equation, and to explain its derivation.

It is designed to help overcome the difficulties encountered by biological scientists in understanding the seed viability equation.

The SEED tutorial was written with the authoring tool Asymetrix ToolBook. It explains:

– the seed survival curve, and shows how the probit transformations can be used to describe the seed survival curves simply (Fig. 1)

– the effects of storage environment on seed longevity

– variation in longevity due to differences between seed lots (Fig. 2)

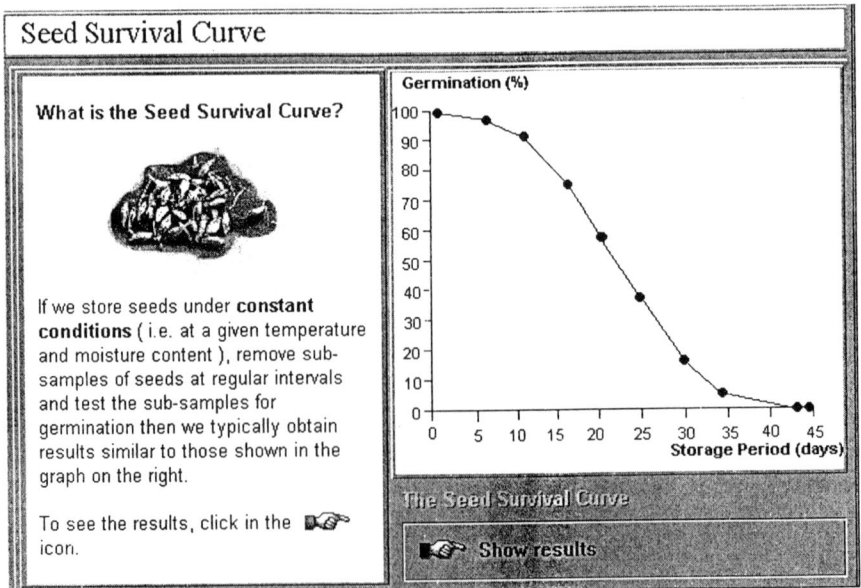

Figure 1. SEED describes the seed survival curve, and explains probit analysis

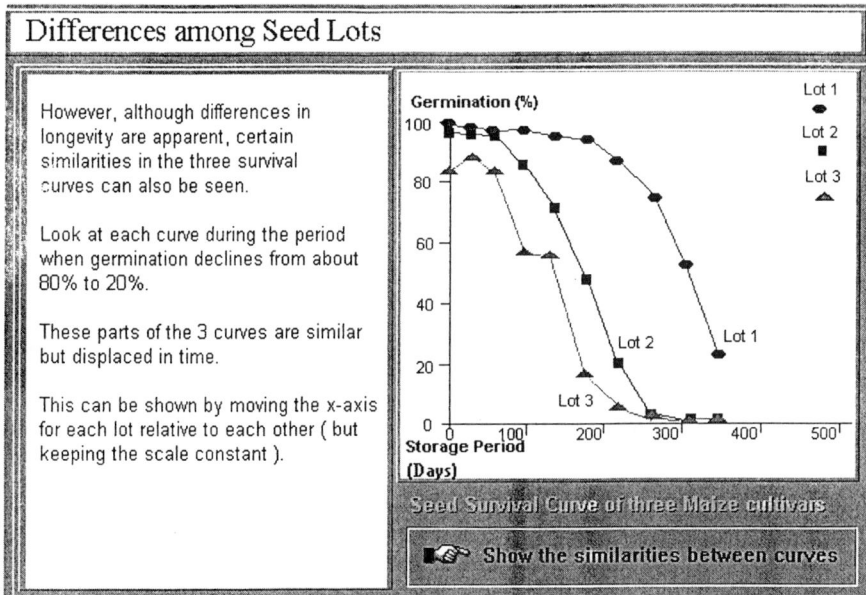

Figure 2. The effect of differences among seed lots on the seed survival curve is illustrated

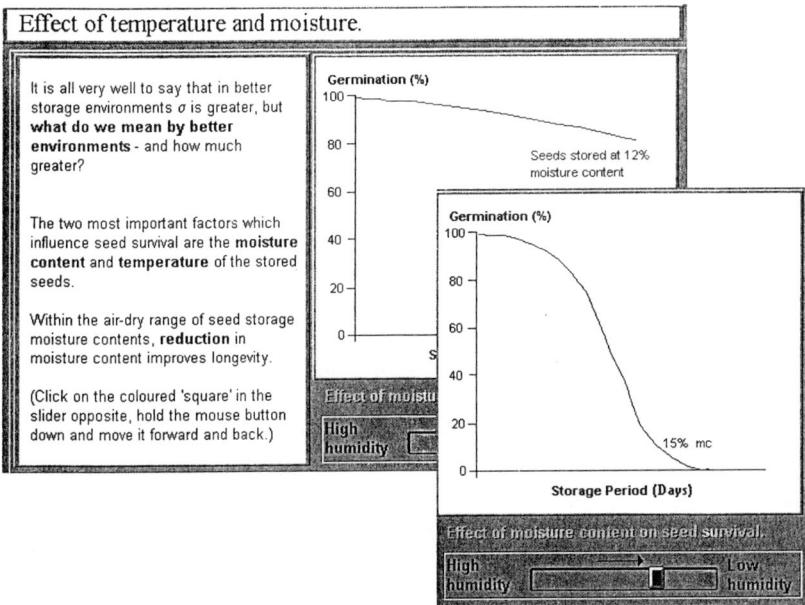

Figure 3. The effects of differences in temperature and moisture content on seed survival are shown

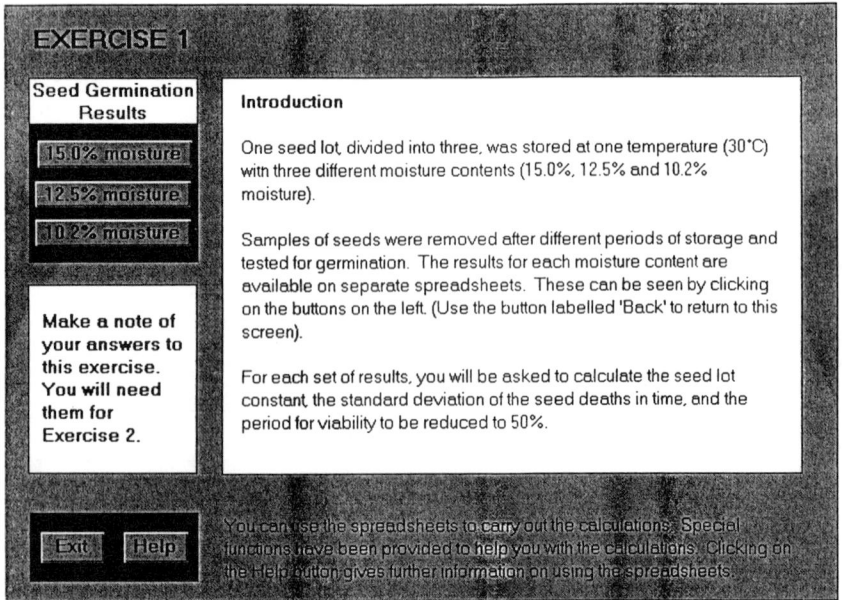

> **EXERCISE 1**
>
> **Seed Germination Results**
>
> 15.0% moisture
> 12.5% moisture
> 10.2% moisture
>
> **Make a note of your answers to this exercise. You will need them for Exercise 2.**
>
> **Introduction**
>
> One seed lot, divided into three, was stored at one temperature (30°C) with three different moisture contents (15.0%, 12.5% and 10.2% moisture).
>
> Samples of seeds were removed after different periods of storage and tested for germination. The results for each moisture content are available on separate spreadsheets. These can be seen by clicking on the buttons on the left. (Use the button labelled 'Back' to return to this screen).
>
> For each set of results, you will be asked to calculate the seed lot constant, the standard deviation of the seed deaths in time, and the period for viability to be reduced to 50%.
>
> Exit   Help
>
> You can use the spreadsheets to carry out the calculations. Special functions have been provided to help you with the calculations. Clicking on the Help button gives further information on using the spreadsheets.

Figure 4. SEED includes a series of exercises to enable users to test their understanding of the tutorial material

- the effects of differences in temperature and moisture content on longevity (Fig. 3)

- different measures of longevity

- the low moisture-content limit to the application of the seed viability equation

- the development of the seed viability equation.

The chapters in the tutorial are interspersed with exercises (Fig. 4) designed to ensure a full understanding of the subject matter. These were developed with the spreadsheet software Microsoft Excel.

Users of SEED can see the seed viability equation being developed step by step, and can observe the effect of changing parameters. The major benefit to students is that they are able to work through the tutorials and exercises at their own pace, rather than to a pace dictated by a lecturer. Unlike a traditional textbook, the courseware provides an animated display with many supporting graphics to illustrate the dynamic relationships involved. In addition, the multiple choice questions which form part of the package give students immediate feedback on their understanding. If students have difficulties with a topic, they can repeat the appropriate section as many times as they wish without feeling intimidated.

Acknowledgements

Further information and details of how to obtain a copy of SEED can be obtained from CLUES, William Guild Building, University of Aberdeen, AB9 2UB, UK or email: clues@aberdeen.ac.uk. SEED was developed under the Teaching and Learning Technology Programme (TLTP) jointly funded by the four UK higher education bodies, HEFCE, HEFCW, SHEFC and DENI.

References

Ellis, R.H. and Roberts, E.H. 1980a. *Annals of Botany* 45: 13–30.
Ellis, R.H. and Roberts, E.H. 1980b. *Annals of Botany* 45: 31–47.
Ellis, R.H., Schofield, J.L., Tidball, J. and Young, C.P.L. 1995. *Seed*, v 0.9. (ed. Heath S.B.). University of Aberdeen: MERTaL Courseware.
Heath, S.B. and Young, C.P.L. 1992. *Computers and Education* 19: 49–55.
Roberts, E.H. and Ellis, R.H. 1989. *Annals of Botany* 63: 39–42.

73. Free Radicals in Seeds – Moving the Debate Forward

G.A.F. HENDRY

NERC Unit of Comparative Plant Ecology, Dept. of Animal & Plant Sciences, The University, Sheffield S10 2TN, UK

Abstract

If we are to show that oxygen, in its various forms, plays a significant causal role in loss of viability in seeds, then it would be a considerable advance experimentally (and philosophically) if we were to pinpoint the sources of activated oxygen. This paper considers one possible fruitful line of research.

Introduction

Throughout biology there is a price to be paid for dependence on oxygen. In animals, including humans, oxygen plays a central destructive role in stressed and diseased states and in the slow but inexorable manifestation of ageing. In the vegetative autotrophic plant, however, the status of oxygen is enigmatic and unique. Simultaneously, the photosynthesizing plant may bear chloroplasts undergoing hyperoxia while the roots survive anoxia. Despite synthesizing a complete complement of molecular defences against oxidative attack, (the chloroplast being the richest source of anti-oxidants including (pro)vitamins A, C and E) potentially damaging forms of oxygen are generated during photosynthesis, photorespiration and mitochondrial respiration particularly following the imposition of stress. Destruction of DNA, pigments, protein and lipids following free radical formation is associated with uptake of transition metals (particularly Fe and Cu) into photosynthetic and non-photosynthetic vegetative tissue and the build-up of a now substantially characterized stable free radical based on a quinone. So much is now well established and introductory accounts are available for biological systems (Halliwell and Gutteridge, 1989) and more specifically in the context of plants (Hendry and Crawford, 1994).

In seeds and seedlings the opportunities for transition metal-catalyzed radical reactions are probably limited. Nevertheless, the build-up of an identical, stable free radical does occur in germinating seeds and seedlings, linked to the loss of vigour, loss of desiccation tolerance and mortality. Evidence as to the origin and location of these free radicals is now available (Hendry et al., 1992; Atherton et al., 1993). In germinating tissue, during imposed desiccation, free radical generation can be pinpointed to the developing mitochondrion (Leprince et al., 1994; 1995). In recalcitrant seeds, with significant rates of respiration, free radicals are generated in the embryonic axis following loss of water (Hendry et

R.H. Ellis, M. Black, A.J. Murdoch, T.D. Hong (eds.), Basic and Applied Aspects of Seed Biology, pp. 657–663.
© 1997 Kluwer Academic Publishers, Dordrecht. Printed in Great Britain.

al., 1992; Finch-Savage *et al.*, 1994). However, in orthodox seeds, with low rates of respiration, the role of free radicals in determining mortality is far from resolved. A review of the role of oxygen and other free radicals in seed biology has been provided (Hendry, 1993) and it is not intended to reiterate this subject here. Instead in this paper we attempt to show that advances in free radical research in vegetative plant biology may have distorted our view of seeds allowing us to overlook the rather special conditions applying to seed biology. If we can adjust our focus away from the vegetative plant to the seed we may be better placed to understand the role of free radicals in seed survival.

Oxygen and Oxidative Damage in the Vegetative Plant

In the recent literature a large number of environmental stresses have been described in which oxidative attack has been implicated in the vegetative plant including seedlings (Table 1).

In addition a number of natural events in the life cycle have also been shown to involve, directly or indirectly, destructive oxidative attack, some of which are listed in Table 2.

There is also a growing literature on the role of oxygen in the degeneration of plant products and the preservation of plant-derived foods (e.g. Aruoma and Halliwell, 1991).

Table 1. Examples of environmental stresses in which activated oxygen has been directly and indirectly implicated

---

Edaphic and atmospheric
    Ozone, sulphur dioxide, nitrogen dioxide
    Herbicides
    Heavy metals – Cu, Cd, Al, Fe, Ca
    Nutrient deficiencies – Ca, Mg, K
    UV-B

Biotic
    Fungal disease and the hypersensitive response
    Polymerization chemistry of defence compounds

Climatic
    Water stress (especially combined with high irradiances)
    Low temperatures (combined with high irradiances)
    High temperatures
    Post flooding re-aeration injury

(original references are available in Hendry, 1993, 1994)

Table 2. Free radicals in plant senescence

Full-term senescence of leaves
Ripening of fruits
Post-fertilization senescence of petals
Degeneration of ageing chloroplasts (or chloroplast membranes)
Senescence of root nodules

(original references available in Lesham, 1988; Hendry, 1993; Merzlyak and Hendry, 1994)

The Growth of Research into Oxygen and Free Radicals in Plants

It will be no surprise then to discover that the literature on oxygen and free radical processes in plants has bloomed in recent years – with an eight-fold expansion in the number of original refereed papers in mainstream publications in the past ten years (estimated from *Current Awareness in Biomedicine – Oxygen Radicals*). However, the number of papers which specifically deal with free radical reactions (under various guises) in seeds, including recalcitrant seeds and germinating seeds was no more than about 10 a year to 1993 and probably less than 5% of the total in 1995. There are indications that this will expand in the next few years. Nevertheless, the point cannot be made too strongly that the background and current thinking on free radicals in seed biology is greatly influenced if not dominated by research into the vegetative plant.

The Status of Free Radical Research into Seeds and Seed Longevity

The search for molecular explanations for loss of seed longevity has been linked in recent decades to age-related or stress-related changes in DNA conformation, in rRNA, in lipid structure and membrane function together with capacity to repair altered DNA or damaged lipids (e.g. Roberts, 1972; Parish and Leopold, 1978; Osborne, 1980; Ohlrogge and Kernan, 1982; Priestley, 1986; Wilson and McDonald, 1986). From this valuable research an informed debate has ensued on age-related events in seeds undergoing loss of vigour or viability. The more recent contribution of free radicals (and activated oxygen in particular) to this debate has been to push the possible causal mechanism of loss of vigour (or viability) one step closer to the initiating step.

If the biological tissue being studied were almost anything other than seeds (or spores) it would be a relatively easy logical jump to move from observations on DNA or lipid damage to oxygen and free radical attack. Just such a link has been established in studies of cancers, of ageing in mammalian tissues, of electron transport and membrane function in plant chloroplast and mitochondrion. The link between degradation of lipid-rich plant products and activated oxygen is also now well established. But seeds are different!

Seeing Free Radicals from the Viewpoint of the Seed

How Free Radicals are Generated in Biology

In most biological tissues (other than seeds or spores) it is broadly accepted that the various forms of activated oxygen are generated in one of four ways:

1. Through enzyme catalysis. Examples from plants include the activities of phenol oxidases, the several peroxidases, other oxidases and oxygenases, though with some of these the transfer of single electrons to dioxygen is more readily demonstrated *in vitro* than *in vivo*. In plants the generation of O_2 through the water-splitting reaction of photosynthesis and H_2O_2 by glycollate oxidase during photorespiration are particularly well documented.

2. Through the leakage of electrons to di-oxygen in the course of electron transfer. Further and more reactive forms of activated oxygen may arise enzymically or non-enzymically, the latter particularly through the reaction of iron (Fe^{3+}) with hydrogen peroxide. In mammalian tissues the electron transfer chain of the mitochondrion is considered to be the most important source of partially reduced forms of oxygen. In the vegetative plant, under conditions of high rates of photosynthesis the chloroplast will be the major source of activated oxygen. The well-studied cytochrome P_{450} systems of animal endoplasmic reticula are also important sources of leaked electrons though this may be a minor route in greening plants (Hendry, 1986).

3. Decline in (or absence of) protection afforded by constitutive anti-oxidants. Decline in protection by certain enzymes such as the peroxidases would theoretically allow for the build up of reactive superoxide. However, in this author's opinion there are no more than a handful of convincing examples in the plant literature where an insufficiency of protective enzyme activity can be shown by itself to result in free radical damage.

4. Autoxidation reactions where oxygen oxidizes biological molecules in the course of which oxygen itself is partially reduced to more reactive forms. Examples include oxidation of reduced forms of flavins, thiol amino acids and aldehydes. With the extremely large number of compounds of secondary metabolism present in plants it is probably impossible to put into perspective autoxidation reactions as a source of reactive oxygen. We know full well that autoxidation reactions do take place when plant tissues are being homogenized in the laboratory.

There is now a sizeable literature which suggests that all of these routes are functioning sources of free radicals in seedlings, in the later stages of germination following imbibition and (though the literature is smaller) in moist recalcitrant seeds. The problem is to demonstrate whether these four routes are present *and active* in dry-stored seeds.

How Free Radicals might be Generated in Relatively Dry Orthodox Seeds

We need to demonstrate or confirm the following.

1. In dry-stored seeds, are those enzymes, which in other tissues generate free radicals, active *in vivo?* Although some oxygenases have particularly low minimum water activities, it is likely that the actual activity of most of these enzymes in radical generation will be impeded by lack of adequate hydration.

2. Do dry-stored orthodox seeds maintain functioning (not just functional) electron transfer systems? This would include seeds (or carpel tissues) which contain green plastids or proplastids. Further, in seeds with a seed coat barrier, do the impediments to gaseous exchange as well as restrictions due to dehydration also limit the operation of electro-genic reactions (and electron-leakage)?

3. Is there an age-related depletion of anti-oxidants (or associated protective or re-cycling enzymes) in dry-stored orthodox seeds? Appreciable amounts of ascorbic acid, for example, can be detected in seeds from herbaria.

These are questions which need to be answered if we are to understand the role of oxygen in seed mortality. It may be that the answer to all of these questions is a qualified yes but that in quantity insignificant to account for loss of viability, even over several years. In the context of dry-stored seeds one fruitful line of research would be to explore the significance of autoxidation reactions.

Do Seeds Undergo Autoxidation?

Autoxidation is certainly significant in the degradation of foods derived from plants or plant products. Indeed the chemistry of repetitive radical chain reactions has been studied extensively in unsaturated lipid foodstuffs. Evidence of autoxidation in viable and intact seeds is there to be found. Sun and Leopold (1995), for example, have investigated the chemical (that is non-enzymic) reaction in soybean seed cotyledons and axes, under conditions of accelerated ageing. They found that elimination of protein carbonyls (a measure of oxidative stress) which had built up slowly in the cotyledons over several years was lowest in seeds with low vigour. They also recorded a rise in protein fluorescence with age probably due to the operation of the Maillard browning reaction (firmly associated with ageing in human tissues), through the formation of sugar aldehydes and products of lipid peroxidation. The authors rightly question whether or not these reactions could have occurred as an effect rather than cause of death and conclude that there is a causal correlation. Another example, again with aged soyabean seeds, has been provided by Khan *et al.* (1996) where the testa (but not the cotyledons) of dry seeds of soyabean underwent lipid peroxidation and accumulated a stable free radical, particularly when illuminated. As the moisture content of these testa was less than 10 mg/g dry wt. it is difficult to see this other than as autoxidation. These events were associated with a decline in viability. There are other examples in the seed literature where autoxidation is at least a plausible explanation. However, care needs to be taken before concluding that changes to seed tissues on exposure to atmospheric oxygen will be due, necessarily, to autoxidation. Not all browning reactions are autoxidations, browning of the testa of *Acer platanoides* for example, on exposure to air is readily suppressed by heating to 90°C for 5 min (Greggains and Hendry, unpublished).

However, if autoxidation does prove to be important in loss of viability in dry-stored seeds, it would be rewarding to explore the use of natural and some of the non-natural anti-oxidants now widely used to retard the oxidative decomposition of lipid-rich foods. The application of food science technology would be limited, however, by the need to keep the seed alive! The wider and perhaps more enduring benefit, however, might be to shed light on the otherwise unexplained molecular mechanisms which determine the differences in longevity that exist between seeds of different species, both in the laboratory and in the field.

References

Aruoma, O. I. and Halliwell, B. 1991. *Free Radicals and Food Additives.* London: Taylor and Francis.

Atherton, N. M., Hendry, G. A. F., Mobius, K., Rohrer, M. and Torring, J. 1993. *Free Radical Research Communication* 19: 297–301.

Finch-Savage, W. E., Hendry, G. A. F., Atherton, N. M. 1994. *Proceedings of the Royal Society of Edinburgh* 102B: 257–260.

Halliwell, B. and Gutteridge, J. M. C. 1989. *Free Radicals in Biology and Medicine*, 2nd Edn. Oxford: Clarendon Press.

Hendry, G. A. F. 1986. *New Phytologist* 102: 239–247.

Hendry, G. A. F. 1993. *Seed Science Research* 3: 141–153

Hendry, G. A. F. 1994. *Proceedings of the Royal Society of Edinburgh* 102B: 155–166.

Hendry, G. A. F.and Crawford, R. M. M. 1994. *Proceedings of the Royal Society of Edinburgh* 102B: 1–10.

Hendry, G. A F., Finch-Savage, W. E., Thorpe, P. C., Atherton, N. M., Buckland, S. M., Nilsson, K. A., Seel, W. A. 1992. *New Phytologist* 122: 273–279.

Khan, M. M., Hendry, G. A. F., Atherton, N. M. and Vertucci, C. W. 1996. (This volume).

Leprince, O., Atherton, N., Deltour, R. and Hendry, G. A. F. 1994. *Plant Physiology* 104: 1333–1339.

Leprince, O., Vertucci, C.W., Hendry, G.A.F. and Atherton, N.M. 1995. *Physiologia Plantarum* 94: 233–240.

Lesham, Y. Y. 1988. *Free Radicals in Biology and Medicine* 5: 39–42.

Merzylak, M. N. and Hendry, G. A. F. 1994. *Proceedings of the Royal Society of Edinburgh* 102B: 459–472.

Ohlrogge, J. B. and Kernan, T. P. 1982. *Plant Physiology* 70: 791–796.

Osborne, D. J. 1980. In: *Senescence in Plants*. Boca Raton, Florida: CRC Press.

Parish, D. J. and Leopold, A. C. 1978 *Plant Physiology* 61: 365–368.

Priestley, D. A. 1986. *Seed Aging*. Ithaca: Cornell University Press.

Roberts, E. H. 1972. In: *Viability of Seeds* pp 253–306 (ed. E. H. Roberts). London: Chapman and Hall.

Sun, W. Q. and Leopold, A. C. 1995. *Physiologia Plantarum* 94: 94–104.

Wilson, D. O. Jr. and McDonald, M. B. Jr. 1986. *Seed Science and Technology* 14: 296–300.

74. Changes in Superoxide Dismutase, Catalase and Glutathione Reductase Activities in Sunflower Seeds during Accelerated Ageing and Subsequent Priming

C. BAILLY[1], A. BENAMAR[1], F. CORBINEAU[2] and D. CÔME[2]

[1]LRPV, Dépt Science et Technologie des Semences, 16 bd Lavoisier, 49045 Angers cédex 01; [2]Physiologie Végétale Appliquée, Université Pierre et Marie Curie, Tour 53, 1er étage, 4 place Jussieu, 75252 Paris cédex 05, France

Abstract

Sunflower (*Helianthus annuus* L.) seeds progressively lost their germination ability during accelerated ageing. Seed deterioration was associated with an increase in malondialdehyde (MDA) and conjugated diene contents, suggesting that lipid peroxidation was involved in loss of seed viability. Accelerated ageing also resulted in a decrease in the activity of the detoxifying enzymes superoxide dismutase (SOD), catalase (CAT) and glutathione reductase (GR), SOD activity being the least affected.

The initial germination ability of aged seeds was progressively restored by a subsequent osmopriming with a polyethylene glycol-6000 solution (–2 MPa). This treatment resulted in a decrease in lipid peroxidation, estimated by MDA and conjugated diene contents. It was also associated with a progressive increase in SOD, CAT and GR activities, which reached levels similar to those found in unaged seeds. Our results clearly demonstrate that sunflower seed deterioration during accelerated ageing is closely related to a decrease in the activities of detoxifying enzymes and to an increase in lipid peroxidation. The reversibility of these biochemical events during osmopriming suggests that the cell detoxifying system, by preventing accumulation of toxic forms of oxygen, might play a key role in seed resistance to deterioration.

Introduction

Seeds progressively lose their viability during prolonged storage (Roberts, 1972; Priestley, 1986). Among the biochemical processes involved in seed deterioration, oxidative injuries and lipid peroxidations have often been considered as major events (Priestley, 1986; Hendry, 1993). However, the involvement of lipid peroxidation in seed ageing seems to be highly related to the species concerned, since some studies (Harman and Mattick, 1976; Stewart and Bewley, 1980) have demonstrated such a relationship whereas others (Priestley and Leopold, 1979; Kalpana and Madhava Rao, 1994) have not.

Higher plants have evolved various protective mechanisms against oxidative injuries. These mechanisms include free radical- and peroxide-scavenging enzymes, such as superoxide dismutase (SOD), catalase (CAT) and glutathione reductase (GR), and antioxidants as for example ascorbate and a-tocopherol.

R.H. Ellis, M. Black, A.J. Murdoch, T.D. Hong (eds.), Basic and Applied Aspects of Seed Biology, pp. 665–672.
© *1997 Kluwer Academic Publishers, Dordrecht. Printed in Great Britain.*

SOD dismutates the superoxide radical $O_2^{\cdot-}$, thus preventing the formation of hydroxyl radicals involved in lipid peroxidation (Gutteridge and Halliwell, 1990). Hydrogen peroxide may be removed by CAT (Fridovich, 1986) or through the ascorbate–glutathione cycle which involves GR (Smith *et al.*, 1989). Moreover, GR controls the level of glutathione which is known to be a potent scavenger of toxic forms of oxygen (Alsher, 1989).

In other respects, priming is known to improve seed germination and vigour (Heydecker and Coolbear, 1977; Bray, 1995). Many biochemical changes have been associated with the beneficial effects of priming, but they seem to be species specific (Bray, 1995).

The aims of the present work were (i) to study the effects of an accelerated ageing treatment on sunflower seed viability in relation to oxidative processes, and (ii) to determine whether a subsequent priming treatment may restore the initial germination ability of aged sunflower seeds and whether this reinvigoration is associated with the reversibility of the deleterious biochemical events occurring during ageing.

Materials and Methods

Plant Material

Experiments were carried out with seeds (achenes) of sunflower (*Helianthus annuus* L., cv Briosol) harvested in 1993 and obtained from Lesgourgues & Co. (Peyrehorade, South-West of France). Seeds were stored at 20°C (70% RH) for at least 9 months before the experiments started; they were therefore non-dormant (Corbineau *et al.*, 1990).

Accelerated Ageing and Priming Treatments

Accelerated ageing was performed by exposing the seeds to 45°C in tightly closed boxes with 100% RH. Priming treatment was carried out by placing aged seeds (5 d of accelerated ageing) on a layer of cotton wool moistened with a polyethylene glycol-6000 solution (–2 MPa) at 15°C.

For each treatment, water uptake was determined by weighing daily 3 samples of 10 seeds. Seed moisture content was calculated on a dry weight basis. Dry weight was obtained by oven drying at 105°C for 3 d.

Germination Tests

Whole seeds (achenes) were germinated in darkness at 25°C or 15°C in 10-cm-diameter Petri dishes (25 seeds per dish, 8 replicates), on a layer of cotton wool moistened with distilled water. A seed was considered as germinated when the radicle had pierced the envelopes (seed coat + pericarp). Germination counts were made daily for 7 to 10 d.

Lipid Peroxidation

Lipid peroxidation was estimated by spectrophotometric measurements of malondialdehyde (MDA) and conjugated diene contents of naked seeds (seeds without pericarp) as described respectively by Heath and Parker (1968) and by Gidrol *et al.* (1989). Extracts were performed with 2 seeds for MDA determination and 10 seeds for conjugated diene measurements. Results presented are the means of the values obtained with 4 different extracts and are expressed as % of MDA and conjugated diene contents of unaged seeds.

Enzyme Extraction and Assays

The pericarp of achenes subjected to various durations of accelerated ageing or of priming was removed, and 2 g of naked seeds were ground in a mortar and homogenized with 0.4 g polyvinylpyrrolidone in 20 ml of 0.1 M phosphate buffer (pH 7.8) containing 2 mM dithiothreitol, 0.1 mM EDTA and 1.25 mM PEG-4000. The homogenate was centrifuged at $16\,000g$ for 15 min. The resulting supernatant was filtered through Miracloth, desalted on PD10 column (Pharmacia) and used for enzyme assays. All steps of the extraction procedure were carried out at 1–4°C.

Superoxide dismutase (SOD, EC 1.15.11) activity was estimated according to Giannopolitis and Ries (1977), by measuring the ability of the enzyme extract to inhibit the photochemical reduction of nitroblue tetrazolium.

Catalase (CAT, EC 1.11.1.6) activity was measured according to Clairbone (1985). It was estimated by the decrease of absorbance of H_2O_2 at 240 nm.

Glutathione reductase (GR, EC 1.6.4.2) activity was determined according to Esterbauer and Grill (1978), by following the rate of NADPH oxidation at 340 nm.

SOD, CAT and GR activities of each extract were measured 3 times, and the results presented correspond to the means of the values obtained with 3 different extracts. Activities are expressed as % of the activities measured in control unaged seeds.

Results

Effects of Accelerated Ageing

Effects on Germination and Moisture Content of Seeds

Figure 1 shows the effects of accelerated ageing on the germination of seeds at 25°C (optimal temperature for sunflower seed germination) and on their moisture content. All unaged seeds germinated within 1 d. Accelerated ageing resulted in loss of the germination ability, and this deleterious effect increased with extending period of ageing. Almost all seeds were dead after 7 d of ageing.

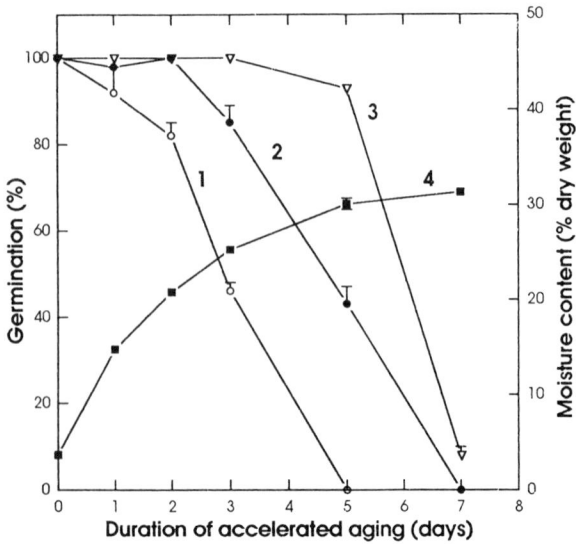

Figure 1. Effects of duration of accelerated ageing on the subsequent germination percentages obtained after 1 d (1), 2 d (2) and 7 d (3) at 25°C, and on seed moisture content (4). Means of 8 (germination percentages) or 3 (moisture content) replicates. Vertical bars correspond to SD. Where no bars are shown the spread of SD is less than the size of the symbols

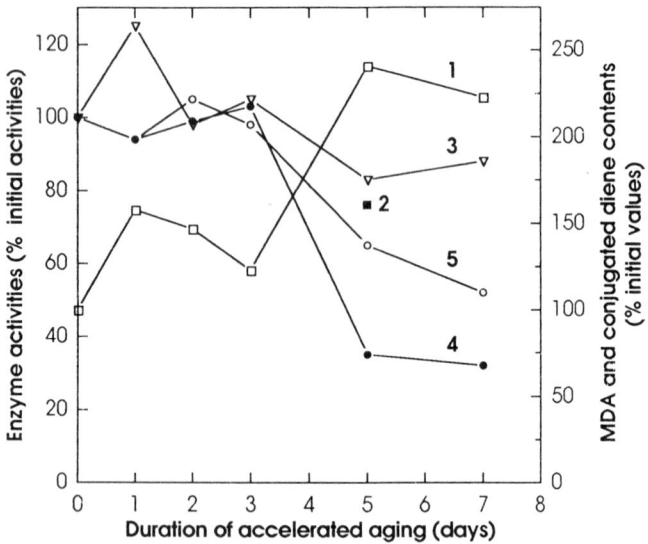

Figure 2. Effects of duration of accelerated ageing on MDA (1) and conjugated diene (2) contents of naked seeds, and on SOD (3), CAT (4) and GR (5) activities. Conjugated diene content was measured only in unaged seeds and in seeds aged for 5 days

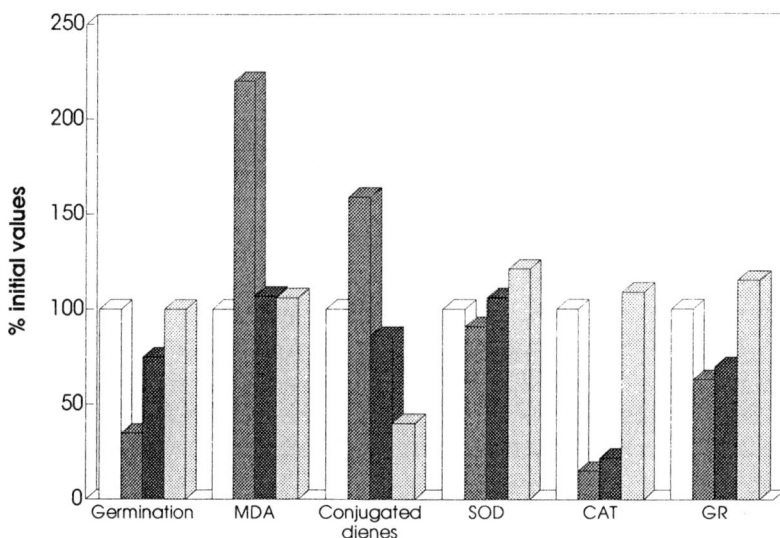

, control unaged seeds; ▨ , 5 days of ageing; ▨ , 5 days of ageing followed by 3 days of priming; ▨ , 5 days of ageing followed by 7 days of priming.

Figure 3. Effects of 5 days of accelerated ageing, followed or not by 3 or 7 d of priming, on the germination percentages obtained after 3 days at 15°C, MDA and conjugated diene contents, and SOD, CAT and GR activities. All results are expressed as percentages of values obtained with unaged seeds

Loss of seed viability was associated with an increase in seed moisture content which was 3.7–4.0% (dry weight basis) before ageing and reached 30–31% after 5 to 7 d of ageing.

Effects on Lipid Peroxidation and SOD, CAT and GR Activities

The effects of accelerated ageing on MDA and conjugated diene contents, and on SOD, CAT and GR activities are shown in Figure 2. MDA content increased during the ageing treatment, particularly by the fifth day. Conjugated diene content also reached a high value (160% of the value obtained with unaged seeds) after 5 d of ageing. These increases in MDA and conjugated diene contents revealed lipid peroxidation. SOD activity was not markedly affected by the ageing treatment but, when performed longer than 3 d, accelerated ageing resulted in a strong decrease in CAT activity and, to a lesser extent, in GR activity. Preliminary results (not presented) have shown that the decrease in

GR activity was associated with a significant decrease in glutathione content.

Effects of Subsequent Priming

The various effects of 3 d and 5 d of priming of seeds previously aged for 5 d are summarized in Figure 3.

Effects on Germination of Aged Seeds

After 5 d of accelerated ageing only about 30% of seeds germinated within 3 d at 15°C (a suboptimal temperature for germination of sunflower seeds) as against 100% for unaged seeds. However, all of them germinated within 7–8 d (data not shown) and were then fully viable but of low vigour. Subsequent priming restored the germination ability of these aged seeds, and the longer the treatment the faster the germination; after 7 d of priming aged seeds germinated as well at 15°C as unaged ones.

Effects on Lipid Peroxidation and SOD, CAT and GR Activities

Priming resulted in a marked decrease in lipid peroxidation, as estimated by MDA and conjugated diene contents. MDA content of aged seeds rapidly reached a level similar to that found in unaged seeds and then remained constant, whereas conjugated diene content continuously decreased during priming and became lower than in unaged seeds. SOD activity remained at a high level. CAT and GR activities increased and reached similar levels as in control unaged seeds by the seventh day. The recovery of GR activity was associated with an increase in glutathione content (data not shown).

Discussion and Conclusion

Accelerated ageing performed at 45°C and 100% RH resulted in a decrease in germination rate and then in loss of viability of sunflower seeds. This deleterious effect of accelerated ageing was associated with an increase in seed moisture content.

The increases in MDA and conjugated diene contents suggest that lipid peroxidation is involved in the loss of sunflower seed viability induced by accelerated ageing. Similar results have been obtained by Gidrol *et al.* (1989) and Gay (1991) with the same species.

The activities of the detoxifying enzymes SOD, CAT and GR were affected to different degrees by the ageing treatment. SOD activity remained quite high, thus suggesting that this enzyme is probably not involved in sunflower seed deterioration during ageing, as already demonstrated in soybean (Stewart and Bewley, 1980) and peanut (Sung and Jeng, 1994) seeds. CAT and GR were much more sensitive to the ageing treatment. They exhibited a marked decrease in their activities after 3 d of treatment. The fall in CAT and GR activities was associated with the increase in MDA content, suggesting that the lack of these enzyme activities led to peroxide accumulation and to lipid peroxidation.

Sunflower seeds artificially aged for 5 d progressively recovered their initial germination ability when primed with a PEG solution. A priming treatment seems therefore to be efficient in repairing the damage caused by ageing. Priming has already been demonstrated to improve germination of aged seeds such as tomato seeds (De Vos *et al.*, 1994). The reinvigoration of aged seeds might be related to the repair processes of DNA, RNA and membranes that have been shown to occur during priming treatments (Bray, 1995).

Recovery of the germination ability of aged seeds during priming was associated with a decrease in MDA and conjugated diene contents, indicating a fall in lipid peroxidation processes which was probably linked to the recovery of detoxifying enzyme (SOD, CAT and GR) activities occurring also during the priming treatment. The priming treatment therefore led to full restoration of the cell detoxifying mechanisms which were strongly altered during ageing. The straight relationship between these reversible biochemical events and the germinability of seeds strongly suggests involvement of oxidative processes in sunflower seed deterioration. CAT, by preventing the accumulation of hydrogen peroxide, and GR, probably through the formation of the antioxidant compound glutathione, might play a major role by controlling the level of lipid peroxidation associated with the loss of seed viability.

[handwritten annotation: Changes but not necessarily fatal ones? seeds reimbibed for germination.]

References

Alsher, R.G. 1989. *Physiologia Plantarum* 77: 457–464.
Bray, C.M. 1995. In: *Seed Development and Germination*, pp.767–789 (eds J. Kigel and G. Galili). New York, Basel, Hong Kong: Marcel Dekker.
Clairbone, A. 1985. In: *Handbook of Methods for Oxygen Radical Research*, pp. 283–284 (ed. R.A. Greenwald). Boca Raton: CRC Press.
Corbineau, F., Bagniol, S. and Côme, D. 1990. *Israel Journal of Botany* 39: 313–325.
De Vos, C.H.R., Kraak, H.L. and Bino, R.J. 1994. *Physiologia Plantarum* 92: 131–139.
Esterbauer, H. and Grill, D. 1978. *Plant Physiology* 61: 119–121.
Fridovich, I. 1986. *Archives of Biochemistry and Biophysics* 147: 1–11.
Gay, C. 1991. Ph. D. Thesis. Université Pierre et Marie Curie, Paris, France.
Giannopolitis, C.N. and Ries, S.K. 1977. *Plant Physiology* 59: 309–314.
Gidrol, X., Serghini, H., Noubhani, A., Mocquot, B. and Mazliak, P. 1989. *Physiologia Plantarum* 76: 591–597.
Gutteridge, J.M.C. and Halliwell, B. 1990. *Trends in Biochemical Science* 15: 129–135.
Harman, G.E. and Mattick, L.R. 1976. *Nature* 260: 323–324.
Heath, R.L. and Parker, L. 1968. *Archives of Biochemistry and Biophysics* 125: 189–198.
Hendry, G.A.F. 1993. *Seed Science Research* 3: 141–153.
Heydecker, W. and Coolbear, P. 1977. *Seed Science and Technology* 3: 353–425.
Kalpana, R. and Madhava Rao, K.V. 1994. *Seed Science and Technology* 22: 253–260.
Priestley, D.A. 1986. *Seed Aging. Implications of Seed Storage and Persistence in the Soil*, pp. 304. Ithaca: Cornell University Press.
Priestley, D.A. and Leopold, A.C. 1979. *Plant Physiology* 63: 726–729.

Roberts, E.H. 1972. *Viability of Seeds.* London: Chapman & Hall.
Smith, I.K., Vierheller, T.L. and Thorne, C. 1989. *Physiologia Plantarum* 77: 449–456.
Stewart, R.R.C. and Bewley, J.D. 1980. *Plant Physiology* 65: 245–248.
Sung, J.M. and Jeng, T.L. 1994. *Physiologia Plantarum* 91: 51–55.

75. The Response of Hydrated Recalcitrant Seeds to Long-Term Storage

N.W. PAMMENTER, N. MOTETE and P. BERJAK

Department of Biology, University of Natal: Durban, Private Bag X10, Dalbridge, 4014 South Africa

Abstract

Recalcitrant seeds will ultimately die if stored in the hydrated state. This could be because newly-shed recalcitrant seeds are metabolically active and undergo germinative metabolism in storage. This in turn could generate a requirement for additional water, which could impose a long-term water stress if water is not supplied. Increases in hydrated storage lifespan might be achieved by reducing metabolic rate and thus the rate of post-shedding development. This paper presents some data from storage experiments with seeds of *Avicennia marina* designed to test this hypothesis. To reduce germinative metabolism naked seeds were coated with alginate in which ABA had been incorporated. Coating the seeds with alginate or alginate with ABA extended the lifespan by up to a factor of three to four. Germination rate, respiration, consumption of storage reserves and ultra-structural data indicated that a reduction of the rate of post-shedding development may have contributed to, but could not fully explain the enhanced storage lifespan of coated seeds. No evidence of water stress in bulk tissue could be detected. It was concluded that dehydration of root primordia might have contributed to the early viability loss of uncoated seeds. Alginate coating may have reduced the rate of dehydration of the primordia, contributing to the extended storage lifespan of coated seeds.

Abbreviations: NC, seeds stored with the pericarp removed only (No Coat); ALG, seeds stored with the pericarp replaced by an alginate coating; ALG+ABA, seeds stored with the pericarp replaced by an alginate coating in which ABA had been incorporated; ER, endoplasmic reticulum.

Introduction

Recalcitrant seeds are characterized by their sensitivity to desiccation. Such seeds also ultimately lose viability if maintained in the hydrated state, even if microbial contamination is controlled. Upon reviewing the limited available data on storage lifespan of recalcitrant seeds, Pammenter *et al.* (1994) presented two hypotheses concerning hydrated storage lifespan: (i) storage lifespan was inversely related to the rate of post-shedding germinative metabolism, and (ii) germinative processes in storage generated a requirement for additional water, and if this requirement was not met, a mild but prolonged water stress would ensue that would ultimately prove fatal. The importance of the degree of development to the desiccation sensitivity of recalcitrant seeds, both prior

R.H. Ellis, M. Black, A.J. Murdoch, T.D. Hong (eds.), Basic and Applied Aspects of Seed Biology, pp. 673–687.
© 1997 Kluwer Academic Publishers, Dordrecht. Printed in Great Britain.

(Hong and Ellis, 1990; Finch-Savage, 1992; Tompsett and Pritchard, 1993), and subsequent (Farrant *et al.*, 1986) to shedding has been well documented. It would not be surprising if developmental stage was also important in storage physiology. This paper describes some experiments designed to test the hypotheses proposed by Pammenter *et al.* (1994).

Assessment of the developmental status of stored seeds is not straight-forward. It could involve factors such as speed of subsequent germination, metabolic rate, storage reserve consumption or ultrastructural characteristics such as appearance of mitochondria, endomembrane development or extent of vacuolation. Similarly, it is generally not a simple matter to manipulate the storage lifespan of recalcitrant seeds. Many species are sensitive to low (but above freezing) temperatures, and controlling water content at levels high enough to prevent desiccation damage is virtually impossible. Seeds of *Avicennia marina* (Forssk.) Vierh. have a characteristic that may permit manipulation of the post-shedding development rate. The fruit of this species has a pericarp that is sloughed shortly after shedding if the seeds are moistened or placed in a humid atmosphere. The pericarp is the only tissue in the fruit with significant amounts of ABA (Farrant *et al.*, 1993), and it has been suggested that this ABA prevents precocious germination. The pericarp is also heavily infected by fungi (Mycock and Berjak, 1990) and removal of this tissue is necessary for any long-term storage. It could be suggested that replacement of the pericarp by a coating, such as alginate, containing ABA would inhibit post-shedding germinative development and increase storage lifespan.

In the experiments described here the pericarps were removed from fruits of *A. marina* and the seeds were stored in a saturated atmosphere with no coats, with an alginate coating, or with an alginate coating into which ABA had been incorporated. Viability was assessed in relation to presumed measures of post-shedding development and tissue water relations were investigated to assess the possibility of the development of water stress.

Materials and Methods

Seeds were collected locally and immediately transported to the laboratory. Pericarps were removed by briefly soaking the seeds in water. Some of the seeds were coated with a potassium alginate paste (pH 6, viscosity 28 000 cps; Kelp Products (Pty) Ltd, Simonstown, South Africa) diluted 6:4 (v/v) with distilled water, and immersed in a 100 mM solution of $CaCl_2$ for ten minutes to facilitate alginate polymerisation. For some of the seeds sufficient ABA was incorporated in the alginate to give a quantity of ABA applied to the seed equivalent to that naturally occurring in the pericarp (approximately 250 ng per seed). Coated seeds were left on the laboratory bench overnight to permit the alginate to dry partially. If the alginate was too wet germination ensued immediately. Seeds were stored in layers no more than three seeds deep on grids over saturated paper towelling in sealed buckets at room temperature.

For all assays, except respiration, the alginate coating was removed from the seeds prior to measurements. Water content (ten replicates) was measured on individual axes and cotyledons by drying at 80°C to constant weight. Water content was expressed on a dry weight basis. For germination assays 20 seeds per test were germinated in wet vermiculite in a greenhouse (25°C day, 18°C night, natural lighting). Germination and establishment was assessed twice daily for 40 days (d). Respiration was measured as CO_2 production by individual seeds (five replicates) in an open gas exchange system. Soluble sugars were measured on freeze dried samples (four replicates) by the phenol/sulphuric acid method (Buysse and Merckx, 1993). For electrolyte leakage isolated axes (ten replicates) were soaked in distilled water and electrical conductance was measured over a 6 h period. Leakage rate was taken as the slope of the linear portion of the time course and was expressed on a fresh weight basis. Water potential was measured on the distal portion of the hypocotyl or on cotyledon discs (four replicates of each) using thermocouple psychrometry. Osmotic potential was measured on the same tissue pieces after freezing and thawing. Turgor was calculated as the difference. For ultrastructural studies excised root primordia were prepared by conventional methods. Cell size and the contribution of various subcellular structures in cross section were quantified using a point grid overlay. The significance of changes in measured parameters over time, and differences between treatments, were assessed using linear regression with the appropriate significance tests for multiple values of Y (replicates) for each value of X (Sokal and Rohlf, 1981).

Results

The final germination achieved by seeds stored for varying times is shown in Figure 1. The germinability of non-coated seeds declined after only five days in storage and by 19 d seedling establishment was zero (establishment data not shown). Coating seeds with ALG or ALG+ABA increased storage lifespan considerably, with ALG+ABA seeds being fully germinable at 80 d storage and showing full seedling establishment after 60 d (establishment data not shown). It was not anticipated that seeds coated with alginate without ABA would have such an extended lifespan and stored material was depleted prior to the loss of viability. At the final harvest (52 d) both germination and seedling establishment were high. Storage lifespan of the ALG seeds could not be established, but it is unlikely that they would have survived longer than those treated with ALG+ABA. There was considerable variation in the rate of electrolyte leakage from axes, particularly during the early stages of storage (Fig. 2), but there were no trends with storage time and no differences between treatments. Electrolyte leakage did not appear to be a good indicator of viability under the experimental conditions used here.

The lag phase of germination decreased until such time as it was zero (germination occurred in storage). This was delayed in the alginate coated seeds

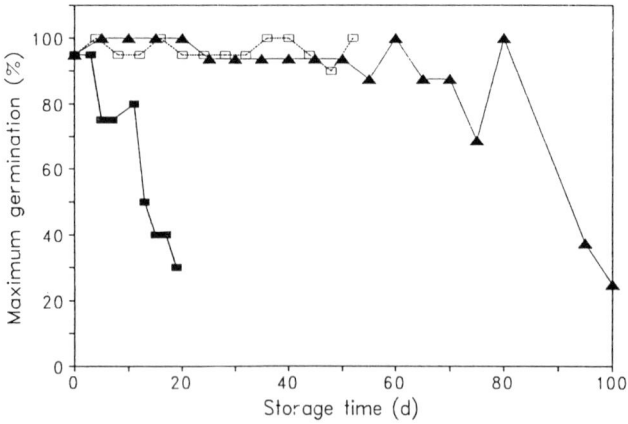

Figure 1. Germination of seeds of *A. marina* stored under hydrated conditions. ■, NC; □, ALG; ▲, ALG+ABA

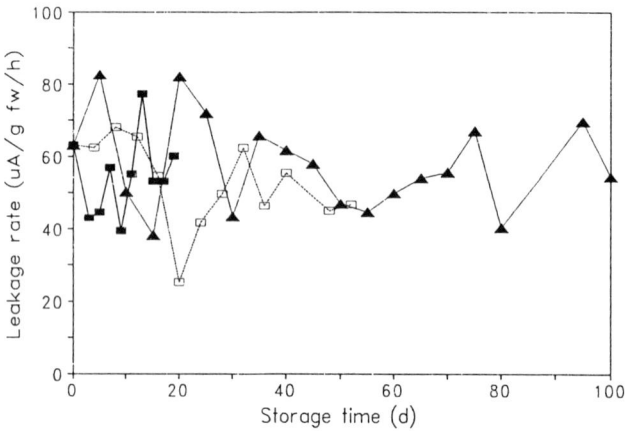

Figure 2. Rate of electrolyte leakage from seeds of *A. marina* stored under hydrated conditions. ■, NC; □, ALG; ▲, ALG+ABA

Table 1. Time taken for stored seeds to show germination in storage (i.e. germination lag phase reduced to zero)

NC	ALG	ALG+ABA
7	12	20

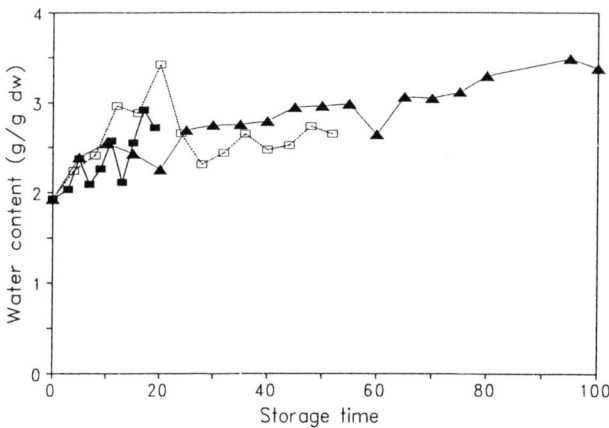

Figure 3. Water content of cotyledons of seeds of *A. marina* stored under hydrated conditions. ■, NC; □, ALG; ▲, ALG+ABA

relative to the non-coated seeds and further delayed in the ALG+ABA seeds (Table 1). The increased time prior to germination in storage is consistent with the hypothesis that rate of post-shedding development was reduced by the alginate coating, and that this effect was more pronounced in those seeds exposed to ABA.

The water content of the axes did not change during storage (data not shown) but that of the cotyledons increased by as much as 75% over the full duration of the experiment (Fig. 3; the lack of statistical significance of the ALG data is probably a consequence of the considerable variation among seeds, rather than the lack of a trend). This increase in water content suggests that the cotyledons were absorbing water from the saturated atmosphere. The difference in lifespan between the treatments is unlikely to be related to this phenomenon as the rate at which water content increased did not differ among the species.

Respiration rate showed considerable variation, particularly in the non-coated seeds and during the early stages of storage (Fig. 4). However, there were no trends with time and no differences among treatments. A depression of metabolic activity, as assessed by whole seed respiration rate, did not appear to be involved in the extension of storage lifespan of the coated seeds. The continued respiration in storage brought about a considerable depletion of cotyledon soluble sugars, to 40% of that initially present (Fig. 5). In *A. marina* soluble sugars constitute the major reserve material (Farrant *et al.*, 1992). Although the decline in soluble sugars in the ALG+ABA treated seeds appears to be slower, the difference is not statistically significant. It is unlikely that viability loss in stored seeds was brought about by a depletion of reserves: non-coated seeds showed 50% loss of viability at soluble sugar levels at which the coated seeds were fully viable.

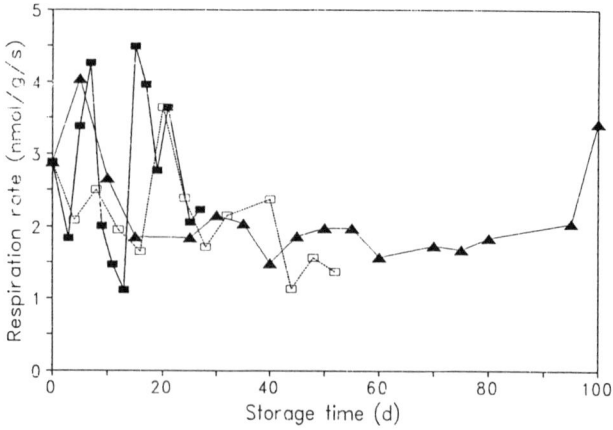

Figure 4. Respiration rate of whole seeds of *A. marina* stored under hydrated conditions. ■, NC; □, ALG; ▲, ALG+ABA

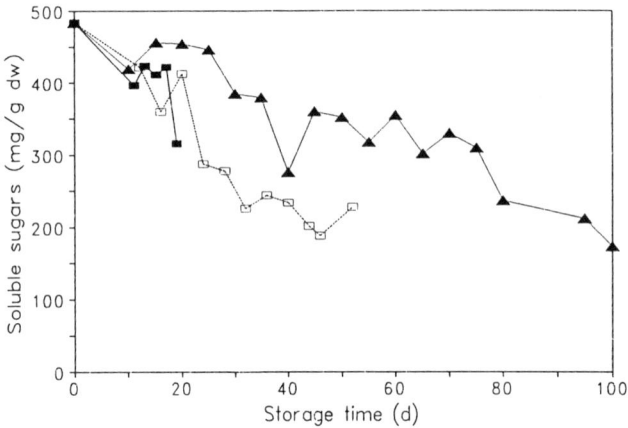

Figure 5. Soluble sugar content of seeds of *A. marina* stored under hydrated conditions. ■, NC; □, ALG; ▲, ALG+ABA

The consumption of a considerable portion of the reserves during storage was accompanied by a decrease in the dry weight of the cotyledons (Fig. 6) with up to 50% of dry weight being lost over the course of the experiment. There were no significant differences in the rate at which this dry matter loss occurred. The putative absorption of water by the cotyledons is an artefact of the method of expression of water content. The increase in water content of the cotyledons was actually a consequence of the decrease in dry weight. The absolute water content

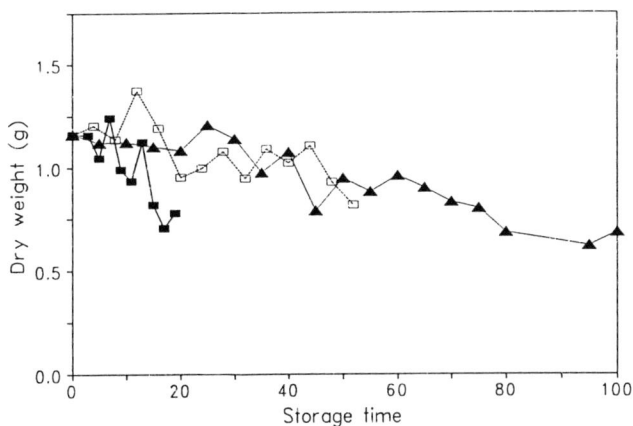

Figure 6. Dry weight of cotyledons of seeds of *A. marina* stored under hydrated conditions. ■, NC; □, ALG; ▲ ALG+ABA

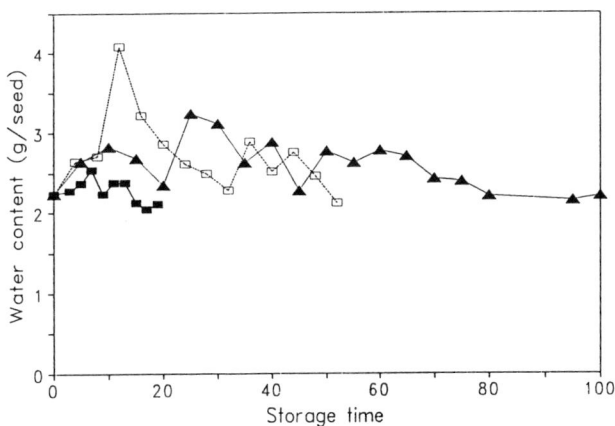

Figure 7. Absolute water content per seed of *A. marina* seeds stored under hydrated conditions. ■, NC; □, ALG; ▲, ALG+ABA

per seed did not change with storage time and there were no differences among the treatments (Fig. 7).

The water potential of the bulk seed tissue showed a significant increase in all treatments during storage. The data for the cotyledons are shown in Figure 8; axes showed essentially similar trends and the data are not shown. Osmotic potential similarly showed a significant increase with increasing storage time, such that turgor remained positive and did not change during storage (Fig. 9).

The ultrastructural changes occurring during storage in the cells of the root

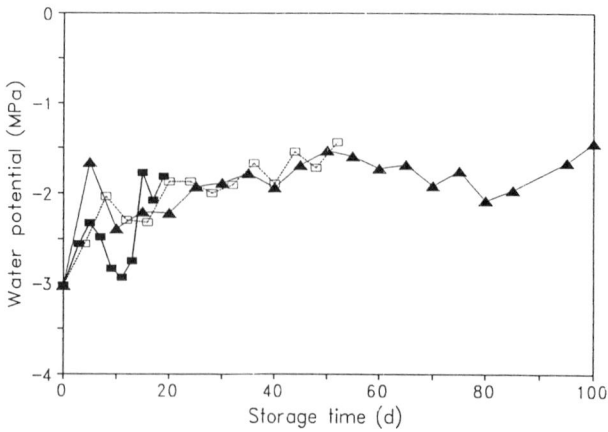

Figure 8. Water potential of cotyledons of seeds of *A. marina* stored under hydrated conditions. ■, NC; □, ALG; ▲, ALG+ABA

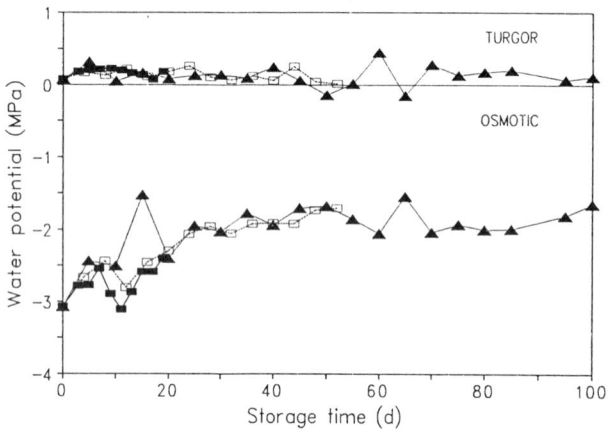

Figure 9. Osmotic and turgor potentials of seeds of *A. marina* stored under hydrated conditions. ■, NC; □, ALG; ▲, ALG+ABA

primordia are illustrated in Figures 10–15. These studies were undertaken on a different seed batch the previous season. Although not identical, survival data were similar in that non-coated seeds had lost viability within 23 d, ALG treated seeds were still fully viable at 30 d, and ALG+ABA seeds were beginning to lose vigour after 70 d in storage. After 15 d storage the ultrastructural status of the meristematic cells of NC material was similar to fresh material (Fig. 10A) while cortical cells showed vacuolar fusion (Fig. 10B) and some showed extensive

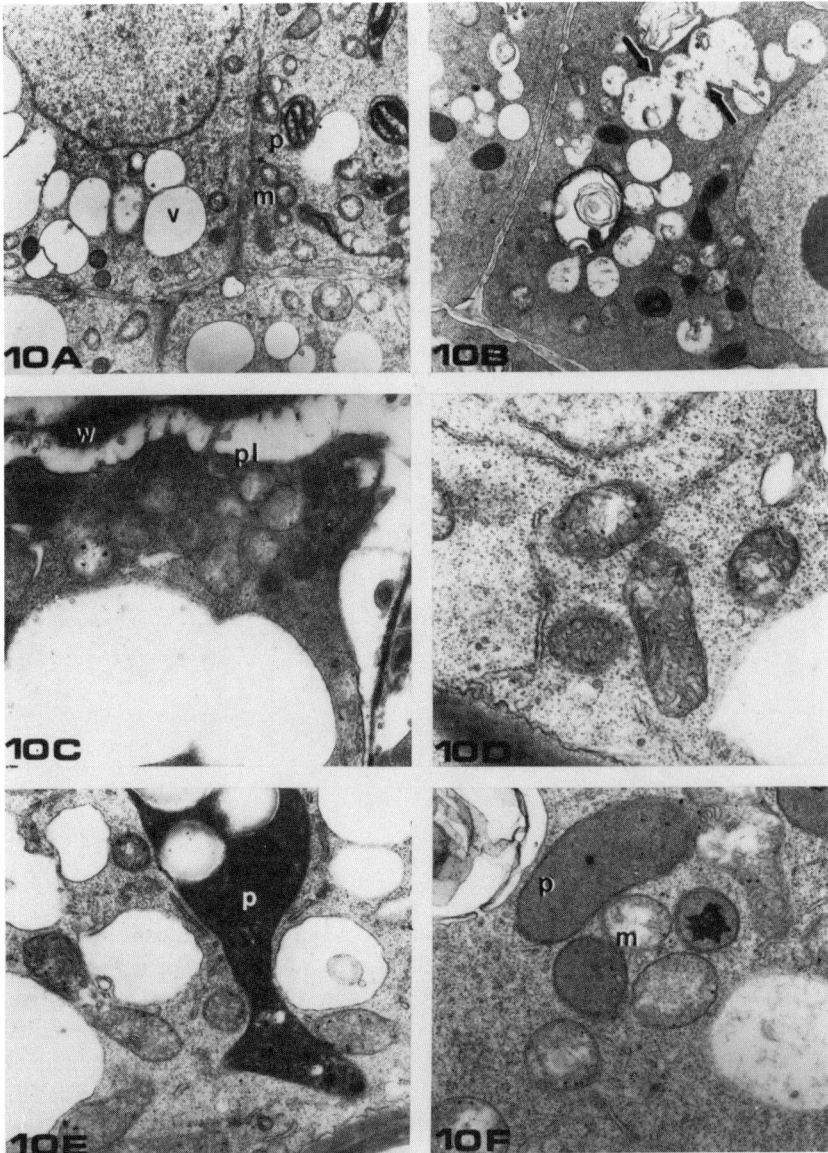

Figures 10A–C. Ultrastructural details of root primordia of NC seeds of *A. marina* after 15 d hydrated storage. A, meristematic cells; B and C, cortical cells. Fusion of vacuoles arrowed in 10C. Abbreviations for this and other micrograph plates: G, Golgi body; m, mitochondrion; p, plastid; pl, plasmalemma; rER, rough endoplasmic reticulum; v, vacuole; w, wall.

Figures 10D–F. Cortical cells of ALG (D and E) and ALG+ABA (F) after 15 d hydrated storage

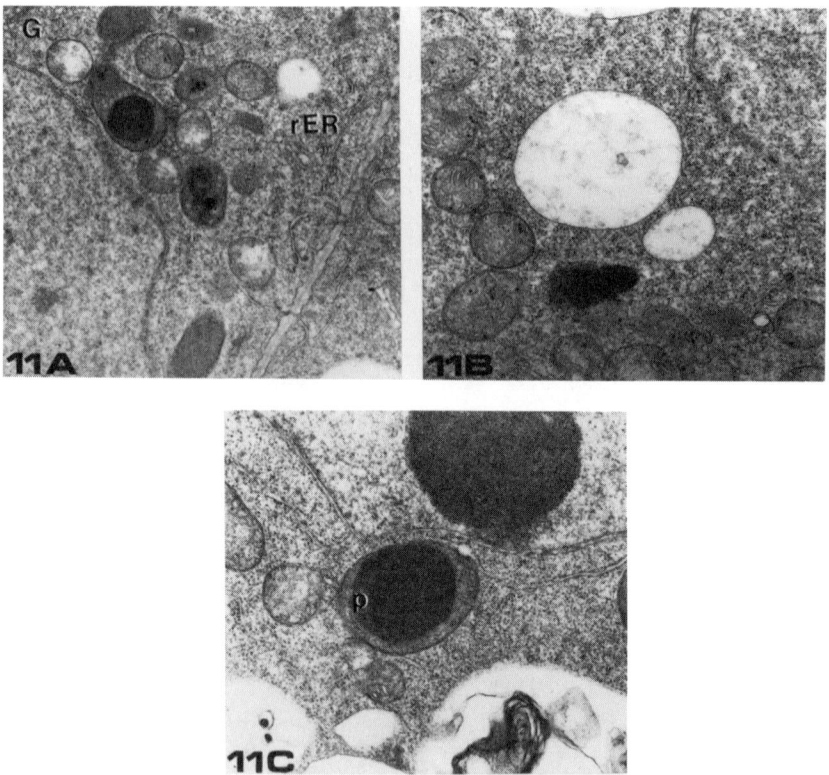

Figures 11A–C. Cortical cells of material stored for 23 d. A and B, ALG seeds; C, ALG+ABA seeds

deterioration (Fig. 10C). In cortical cells of ALG treated seeds mitochondria were well developed (Fig. 10D) and plastids had accumulated dense material (Fig. 10E). In the cells of ALG+ABA seeds mitochondria and plastids were quiescent in appearance (Fig. 10F). By 23 d of storage all NC seeds were dead and ultrastructure is not illustrated. ALG treated seeds showed development of Golgi bodies and rER (Fig. 11A) and active mitochondria and polysomes (Fig. 11B). At this stage the ALG+ABA material maintained a relative quiescent state (Fig. 11C). After 30 d storage the cells of ALG treated seeds maintained an appearance indicative of intense activity (Fig. 12A), although some aberrations of mitochondrial inner membranes were apparent. ALG+ABA treated material began to show signs of increased vacuolation by this stage (Fig. 12B). After 39 d storage Golgi bodies and elongated rER profiles were evident (Fig. 13) and at 49 d the active appearance of the cells was maintained (Fig. 14A) and extensive vacuolation had occurred (Fig. 14B). By 70 d of storage the cells of the ALG+ABA treated material showed well developed Golgi bodies as well as

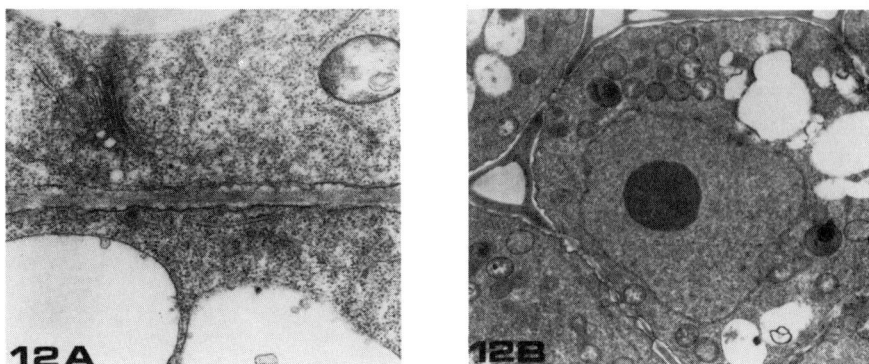

Figures 12A and B. Cortical cells of material stored for 30 d. A, ALG; B, ALG+ABA

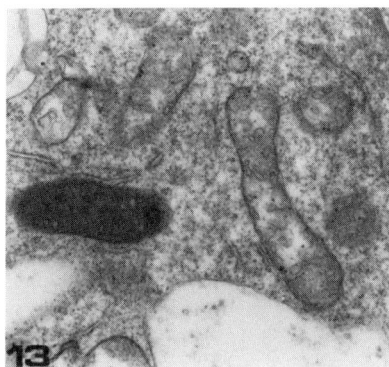

Figure 13. Cortical cells of ALG material stored for 39 d

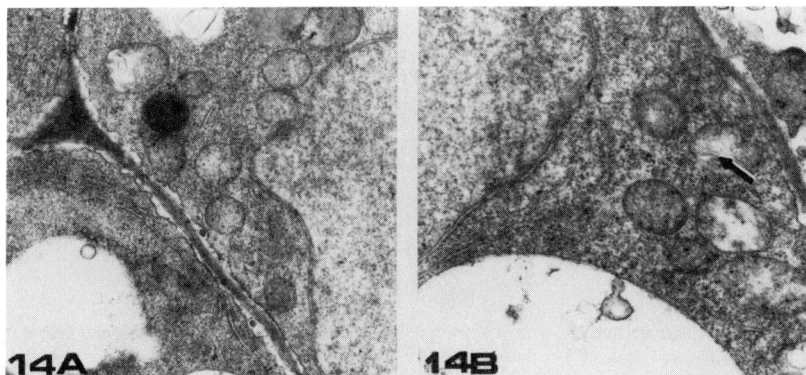

Figures 14A and B. ALG+ABA material stored for 49 d. A, meristematic cells; B, cortical cells, incipient deterioration of mitochondrion arrowed

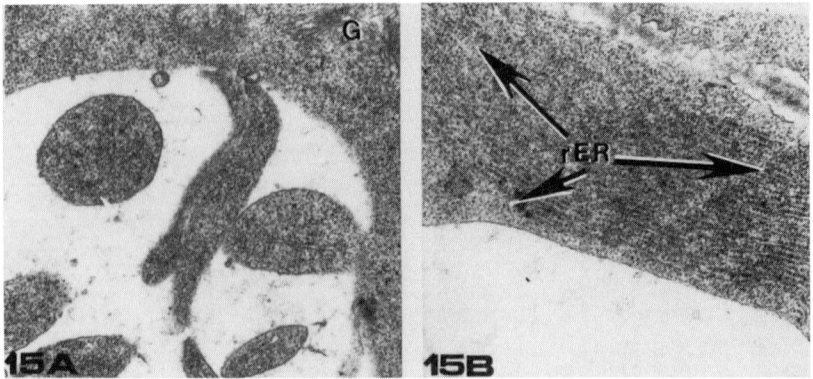

Figures 15A and B. Meristematic cells of ALG+ABA material stored for 70 d

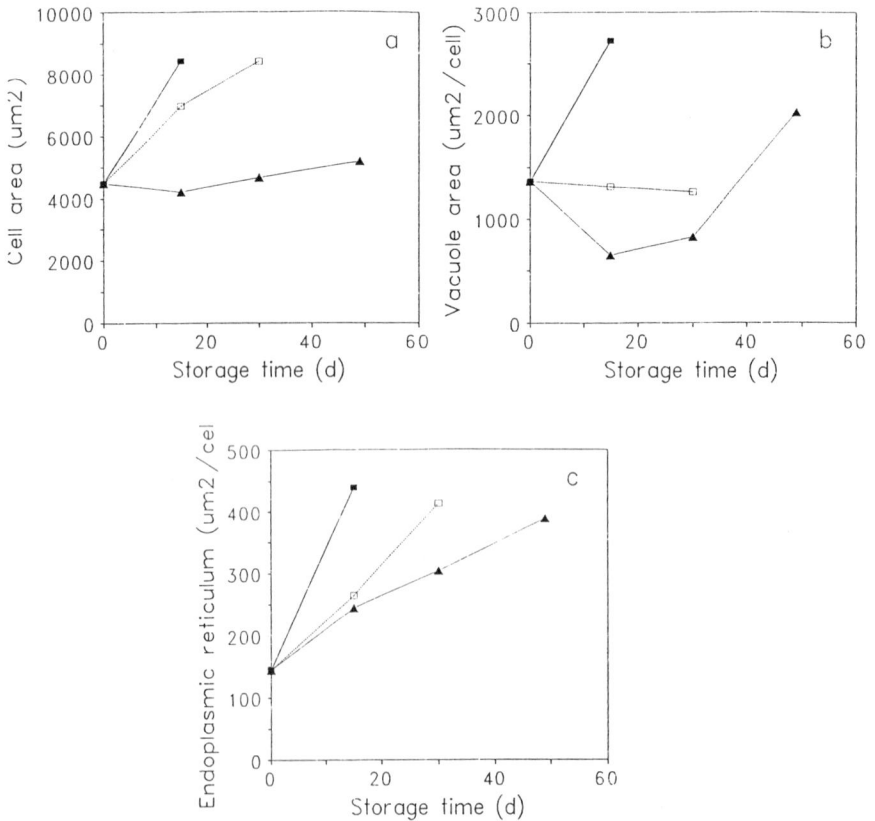

Figures 16a–c. Quantitative changes in a, meristematic cell area; b, area of vacuole per cell; c, area of ER per cell

intense vacuolation, including apparent autophagic activity (Fig. 15A), and remarkable development of the rER (Fig. 15B). Quantitative changes in meristematic cell size, and vacuolation and ER development in cortical cells is shown in Figure 16. All the ultrastructural changes are consistent with the concept of germinative metabolism occurring in storage, with the rate of this development being in the order NC > ALG > ALG+ABA.

Discussion

This experiment was conducted to test two hypotheses:

i) The hydrated storage lifespan of recalcitrant seeds is inversely related to post-shedding development rate. A corollary of this is that reducing the rate would lead to an extended storage lifespan.

ii) During storage a developmentally imposed water stress develops, with the corollary that this water stress would develop more slowly in seeds where post-shedding development rate had been reduced.

An evaluation of the first hypothesis requires that the method of assessing the rate of post-shedding development be considered. Two methods were used here; speed of germination after storage, and general level of metabolism. The speed of germination was measured by the germination lag, or alternatively, the time taken for the lag to reduce to zero (i.e. time to first germination in storage). By this criterion both alginate coatings reduced the rate of post-shedding development, with the ABA treatment having a greater effect. The ultrastructural data support the concept of post-shedding development varying among the treatments. The changes were similar to those described previously for short-term stored seeds of *A. marina* (Berjak *et al.*, 1984; Farrant *et al.*, 1986), which have been interpreted as germinative metabolism occurring in storage, regressing to deterioration with continued storage. These changes occurred most rapidly in the NC material, followed by the ALG treatment, with the changes slowest in the ALG+ABA seeds. The treatments had a marked effect on the storage lifespan of *A. marina*. Unfortunately, the storage lifespan of ALG treated seeds was not determined, and it is not possible to show whether the difference in post-shedding developmental rate between the two coating treatments resulted in different storage lifespans. On the basis of speed of germination and ultrastructural evidence, the first hypothesis cannot be rejected. The general level of metabolism, as assessed by respiration rate and depletion of storage reserves, did not differ among the treatments. On the basis of this evidence the first hypothesis cannot be supported. Alternatively, the metabolic activity of the whole seed may not be a good indicator of rate of post-shedding development.

No evidence could be found of a water stress developing during storage. On the contrary, water potentials increased rather than declined, although turgor did not change. The low water potentials of the fresh seeds (around –3 MPa) are necessary because *A. marina* is a mangrove and the seeds are shed into a saline habitat (osmotic potential of sea water approximately –2.5 MPa). These low water potentials are generated in part by the presence of considerable quantities of soluble sugars as the major reserve. The increase in water and osmotic potentials during storage was probably a consequence of the decline in soluble reserves. The water potential data do not support the second hypothesis.

Although there are indications that the extended storage lifespan of coated seeds was related to a decrease in the rate of post-shedding development, it is felt that this is not the only contributing factor. The difference in time to first germination in storage between the NC and ALG+ABA seeds was only 13 d, yet the difference in storage lifespan was approximately 70 d. Although germination and ultrastructural evidence suggested that post-shedding development was faster in ALG than in ALG+ABA seeds, the ALG seeds did not lose viability earlier than the ALG+ABA seeds. In an earlier study of Mumford and Brett (1982), application of ABA to cacao seeds failed to curtail germination during hydrated storage. Yu *et al.* (1993) extended of the storage lifespan of litchi seeds by pelleting in alginate, and attributed the results to prevention of desiccation. This explanation would appear not to be applicable to the *A. marina* seeds as bulk water contents did not decline. In fact, all the studies on bulk seed tissue failed to show differences between the treatments. Furthermore, electrolyte leakage did not correspond to survival data, suggesting that the bulk of the tissue was not damaged, despite the failure of the seeds to germinate. This raises the question as to which tissue is susceptible to damage and is essential for germination to occur. In *A. marina* seeds the root primordia at the distal end of the hypocotyl are near the surface, covered by only a few cell layers with little cuticularization. It is suggested that the non-coated seeds lost viability because the root primordia suffered desiccation damage under the storage conditions used here. Although the bulk of the tissue was healthy, germination could not occur because the root primordia were non-functional. Loss of water from, and declines in activity of such a restricted tissue would not be apparent in the measurements made on bulk tissue. The extended lifespan of the coated seeds may have been due in part to a slowing of post-shedding development, but it is suggested that a more important effect was that the alginate coating reduced the rate at which the root primordia lost water and suffered desiccation damage. If this suggestion is correct, it emphasizes the necessity of considering the effects of stress at the microlevel, rather than aT the level of the whole seed. Unfortunately it is difficult to perform biochemical and physiological studies of such minute tissue pieces.

References

Berjak, P., Dini, M. and Pammenter, N.W. 1984. *Seed Science and Technology* 12: 365–384.
Buysse, J. and Merckx, R. 1993. *Journal of Experimental Botany* 44: 1627–1629.

Farrant, J.M., Berjak, P., Cutting, J.G.M. and Pammenter, N.W. 1993. *Seed Science Research* 3: 55–63.

Farrant, J.M., Pammenter, N.W. and Berjak P. 1986. *Physiologia Plantarum* 67: 291–298.

Farrant, J.M., Pammenter, N.W. and Berjak P. 1992. *Annals of Botany* 70: 75–86.

Finch-Savage, W.E. 1992. *Seed Science Research* 2: 17–22.

Hong, T.D. and Ellis, R.H. 1990. *New Phytologist* 116: 589–596.

Mumford, P.M. and Brett, A.C. 1982. *Tropical Agriculture* 59: 306–310.

Mycock, D.J. and Berjak, P. 1990. *Phytophylactica* 22: 843–846.

Pammenter, N.W., Berjak, P., Farrant, J.M., Smith, M.T. and Ross, G. 1994. *Seed Science Research* 4: 187–191.

Sokal, R.R. and Rohlf, F.J. 1981. *Biometry. The Principles and Practice of Statistics in Biological Research*, Second Edition. San Francisco: W.H. Freeman and Company.

Tompsett, P.B. and Pritchard, H.W. 1993. *Annals of Botany* 71: 107–116.

Yu, X.-P., Jian, F., Sheng, S.-J., Sun, C.-K., Xu, X.-Y. and Zheng, J. 1993. *Fourth International Workshop on Seeds. Basic and Applied Aspects of Seed Biology*, pp 863–865 (eds D. Côme and F. Corbineau). Paris: ASFIS.

76. Progress in the Understanding and Manipulation of Desiccation-Sensitive (Recalcitrant) Seeds

P. BERJAK and N.W. PAMMENTER
Department of Biology, University of Natal, Durban, Private Bag X10, Dalbridge, 4014 South Africa

Abstract

Seeds of many species behave in a 'non-orthodox' manner and have been variously categorized as sub-orthodox, intermediate or recalcitrant. Those of some species have been recorded as showing anomalous behaviour in their being placed in more than one of these categories. Individual seed species may, in fact, be assigned to some point on a behavioural continuum subtended by the extremes of orthodoxy and recalcitrance. The nature of seed recalcitrance is discussed in terms of two components, desiccation sensitivity and post-harvest behaviour. Both components differ among species, and seeds of any one species may show variability within or between seasons. Developmental stage and the tissue water status influence both these components and show considerable variation among species and within individual species inter- and intra-seasonally. This variability may, to some extent, contribute to supposedly anomalous behaviour. Storage lifespan is also affected by the seed-associated mycoflora. Practical problems associated with short-term storage of hydrated recalcitrant seeds, and the possibility of cryopreservation for long-term storage are also important issues.

Abbreviations: a_w, water activity; $g\,g^{-1}$, g water per g dry matter; ψ_w, water potential; PVS, post-vascular separation; RH, relative humidity; SOD, superoxide dismutase.

Categorization of Seed Types

Since the early 1970s and until fairly recently, seeds showing disparate storage behaviour have been categorized as being either orthodox or recalcitrant (Roberts, 1973). These categories, however, take cognizance of only extremes of post-harvest behaviour based principally on whether the seeds of particular species are desiccation tolerant or are sensitive to dehydration. Orthodox seeds are understood as those which develop desiccation tolerance during their formation, undergo substantial drying as the final developmental phase, may be further dehydrated after they are shed and will survive in this state for a considerable time. On the other hand, although several recalcitrant seed species have been recorded as undergoing a measure of dehydration during their development (see e.g. reviews by Finch-Savage, 1995; Vertucci and Farrant, 1995), they do not undergo maturation drying *per se*, and are shed at relatively high water contents. Recalcitrant seeds are desiccation-sensitive both before and after shedding and have very limited post-harvest lifespans, even in the hydrated condition.

R.H. Ellis, M. Black, A.J. Murdoch, T.D. Hong (eds.), Basic and Applied Aspects of Seed Biology, pp. 689–703.
© *1997 Kluwer Academic Publishers, Dordrecht. Printed in Great Britain.*

It has, however, become apparent that even among desiccation-sensitive seeds there are degrees of recalcitrant behaviour, with some being highly intolerant of even slight water loss and others withstanding a considerable degree of dehydration without any apparently deleterious effects. This led to the proposal that the category of recalcitrant seeds should be viewed as comprizing a continuum of species, the behaviour of which varies from extreme dehydration sensitivity to tolerance of the loss of a substantial proportion of the water present at shedding (Farrant *et al.*, 1988). Those authors considered that other properties, particularly the hydrated storage lifespan and possible chilling-sensitivity, should be taken into account when ranking individual seed species within the continuum. Since then, Ellis *et al.* (1990; 1991a,b) have defined as a third category of seed-type, those that will withstand desiccation down to relatively low moisture contents (*c.* 6–12% [0.06–0.14 g g^{-1}]), but are chilling-sensitive in the dehydrated state. Such seeds are said to show intermediate storage behaviour.

The three categories are, however, open-ended and it is frequently difficult to place a seed species quite unequivocally within a particular grouping. Certain seeds, e.g. those of wild rice, *Zizania palustris*, show differential responses to dehydration, depending on the temperature during drying (Kovach and Bradford, 1992; Berjak *et al.*, 1994; Vertucci *et al.*, 1994; 1995) which confers properties other than those traditionally considered to characterize recalcitrance. The seeds of neem, *Azadirachta indica*, provide a more extreme example of anomalous behaviour. Those of Asian provenance have been categorized as being 'more or less recalcitrant', while those of African origin (Burkina Faso) were described as orthodox (Gamene *et al.*, 1994). However, a recent study revealed neem seeds originating in Africa (from a coastal provenance in Kenya) to lose viability gradually as they were equilibrated to ambient RH. Furthermore, the newly-shed seeds were chilling sensitive in the hydrated state, showed a highly differentiated ultrastructure and underwent germination-associated changes in storage under high RH conditions (Berjak *et al.*, 1995a). These features are all consistent with recalcitrant seed behaviour.

Such considerations have led to the comment that, 'recalcitrance is not an all-or-nothing situation' (Berjak and Pammenter, 1994). Those authors have suggested that far more specific information than merely the desiccation and chilling responses of seeds be used to assign particular species to the categories, intermediate or recalcitrant - and to positions on the continuum between these two categories. The characteristics that have been proposed include: the natural habitat of the species; the time taken to reach 50% of the original viability (V_{50}) when hydrated seeds are stored at a specified temperature; mean water content of separated embryonic axes and storage tissues at V_{50} under specified drying conditions; the water activity (a_w) corresponding to the water content at V_{50}; chilling sensitivity of the hydrated seeds; chilling sensitivity at the lowest water content commensurate with viability retention (i.e. intermediate behaviour); the responses to dehydration at various specified temperatures; and whether or not the seeds show dormancy. Berjak and Pammenter (1994) have suggested that provision of this information will enable individual seed species to be integrated into a continuum of behaviour from intermediate to highly recalcitrant. As an extrapolation of this proposal, it is likely that a continuum of seed behaviour from intermediate to the most desiccation tolerant of orthodox seeds can also be demonstrated. Developmental characteristics in relation to the degree of desiccation tolerated, as discussed by Finch-Savage (1995) and Vertucci and Farrant (1995), are likely to emerge as important to the 'fine-tuning' of seed categorization.

Recalcitrant Seeds

Collectively, recalcitrant seeds remain metabolic throughout development to shedding and continue to accumulate dry mass. Although some reduction in water content is a characteristic of most temperate as well as tropical species (as documented by Nautiyal and Purohit, 1985; Nkang and Chandler, 1986; Ray and Sharma, 1987; Dodd *et al.*, 1989; Hong and Ellis, 1990; Dickie *et al.*, 1991; Berjak *et al.*, 1992; 1993a; Tompsett, 1992; Tompsett and Pritchard, 1993; Fu *et al.*, 1994; Finch-Savage, 1995), the significant decline in water content and metabolic shut-down associated with maturation drying of orthodox seeds does not occur. On the contrary, recalcitrant seeds will either entrain germinative metabolism at the shedding water content or, under certain conditions (Berjak *et al.*, 1989; Finch-Savage *et al.*, 1993), will continue with pre-germination development. Recalcitrant seeds are characterized by desiccation sensitivity, although there is a developmental stage at which most species are relatively at their most desiccation tolerant. This appears to coincide with the lowest water content attained (Finch-Savage, 1995). Nevertheless, usually recalcitrant seeds are sufficiently hydrated at shedding that germination commences without any additional water (e.g. Berjak *et al.*, 1989; Farrant *et al.*, 1988; 1993a; Finch-Savage, 1995). It is assumed that the desiccation sensitivity of recalcitrant seeds is intimately associated with their persistent state of metabolic activity. A further trait of recalcitrant seed species is that all will, sooner or later, die under storage conditions that do not permit any significant degree of water loss (Pammenter *et al.*, 1994).

Evolutionary Considerations

While there is much ongoing work linking recalcitrant seed behaviour with certain developmental, physiological and biophysical phenomena, evolutionary and ecological considerations have not enjoyed similar attention. The work of von Teichman and van Wyk (1994) makes a notable exception, and deserves close consideration. Those authors have linked the phenomenon of recalcitrance with bitegmic, crassinucellate ovules that have nuclear endosperm development, in 45 families of dicotyledonous plants.

Von Teichman and van Wyk (1994) are of the opinion that pachychalazy, large-seededness and recalcitrance may be regarded as ancestral traits among the dicotyledons. Those authors make the point that the integuments and/or chalaza of ovules of recalcitrant species are extensively vascularized and that in many the chalaza is massive (the pachychalazal condition) and surrounds the embryo sac to varying degrees. The seeds of most recalcitrant species, whether gymnosperms or angiosperms, are large (e.g. Chin *et al.*, 1984; Tompsett, 1992). Large seededness and extensive vascularization are considered to be traits of more primitive families, while seeds of more advanced taxa are generally small with very little or no vascularization (von Teichman and van Wyk, 1994). According to those authors most pachychalazal seeds have a hypostase (comprised of thin-walled cells containing polyphenolics) adjacent to the chalaza. Translocation of nutrients is suggested to be greatly facilitated by the extensive vascularization of the pachychalazal seed (von Teichman and van Wyk, 1994). Recalcitrant seeds continue to accumulate dry weight up to the time they are shed (Finch-Savage, 1995); thus the postvascular separation phase (PVS phase [Vertucci and Farrant, 1995]) characterizing orthodox seed development must be assumed to be abbreviated to a state of virtual non-existence. Concomitantly though, seed water content declines somewhat. This could be as a consequence of water loss, or the accumulation of dry matter at a greater rate than the acquisition of water. In either case, there will be a decline in turgor and/or a greater proportion of cell water will be interacting with cellular components, leading to a decline in water potential (ψ_w). In this regard, the amount of non-freezable (structure-associated) water in embryonic axes of *Landolphia kirkii* increases as the seeds mature (Berjak *et al.*, 1992). The hypostase tissue has been reported to become secondarily lignified and probably cuticularized in some mature recalcitrant seeds (von Teichman and van Wyk, 1994) which, together with the other seed coverings, could retard water loss, so contributing to a slow dehydration rate of the seed before and after shedding.

Most recalcitrant seeds considered by von Teichman and van Wyk (1994) are exalbuminous, leading those authors to suggest that transfer of the storage function from endosperm to embryo might have been an early development in large, recalcitrant seeds (although they cite Takhtajan [1991] as considering this feature to represent an evolutionary advance in angiosperms generally). Transfer of the nutrients either to the cotyledons (the usual situation) or their accumulation in the axis which consequently becomes hypertrophied in some species (Corbineau and Côme, 1986; Berjak, 1995) facilitates their immediate availability for the rapid germination which characterizes many recalcitrant seed species. The nature of the stored reserves may also have a bearing on post-shedding seed behaviour. Starch constitutes the major stored reserve of many large, exalbuminous, recalcitrant seed species (von Teichman and van Wyk, 1994), while the highly-recalcitrant mangrove species, *Avicennia marina* stores mainly soluble sugars (Farrant *et al.*, 1992; 1993a). von Teichman and van Wyk (1994), making the observation that intermediate seeds may contain predominantly other sorts of reserves (e.g. lipid in *Coffea arabica*) suggest that various degrees of intermediate–recalcitrant behaviour (Farrant *et al.*, 1988; Berjak *et al.*, 1989) may be related to the type, site and amount of stored reserves.

There has been much discussion about the factors which might confer desiccation tolerance in orthodox seeds, with particular attention being paid recently to the dehydrin-like or LEA proteins (e.g. Close *et al.*, 1989; Dure *et al.*, 1989; Kermode, 1990; Blackman *et al.*, 1991; 1992; Dure, 1993). While such proteins are apparently absent from the tropical species, *Avicennia marina*, they have been recorded as accumulating during development in the seeds of several temperate, recalcitrant species (Finch-Savage *et al.*, 1994a; Gee *et al.*, 1994) which are desiccation sensitive. While the expression of LEAs alone is considered inadequate to confer desiccation tolerance (e.g. Blackman *et al.*, 1991), their presence is thought to be implicated in facilitation of the dehydrated state (e.g. Vertucci and Farrant, 1995). In view of the fact that such proteins do occur in desiccation-sensitive seeds, it might be argued that their expression is, in fact, representative of the ancestral seed condition and that their rôle in conferring desiccation tolerance was actuated only when co-components conferring this property developed. Alternatively, as LEA accumulation accelerates rapidly around the point of maximum mass accumulation in orthodox seeds (e.g. Galau *et al.*, 1991), it is possible that this accumulation phase is not reached in recalcitrant seeds, as their dry mass continues to increase while they remain attached to the parent plant (Finch-Savage, 1995).

Variability

One of the greatest problems occurring when working with recalcitrant seeds, is the degree of variability among different harvests of the same species, and even from seed-to-seed within a single harvest (Berjak *et al.*, 1989; Finch-Savage and Blake, 1994). There are also obviously marked differences among species in characteristics such as seed size, structure and composition, degree of embryo development and water content at shedding, relative desiccation sensitivity and how readily water is lost. Natural provenance of individual species and the possibility of chilling sensitivity confer additional differences among seeds that have desiccation sensitivity as their common property. It is these differences among species that underlie the varying degrees of recalcitrance (Farrant *et al.*, 1988; Berjak *et al.*, 1989; Berjak and Pammenter, 1994).

Most recalcitrant species, with the exception of a few from the tropics (e.g. *Avicennia marina* [Farrant *et al.*, 1993a]; *Symphonia globulifera* [Corbineau and Côme, 1986]; *Barringtonia racemosa* [Berjak, 1995]), have seeds to which the embryonic axis contributes only an insignificant proportion of the total volume. Seemingly invariably, axis water content is higher than that of the cotyledons; however, considerable variation in axis water content has been recorded both intra- and inter-seasonally (Berjak *et al.*, 1989; Berjak *et al.*, 1993a; 1995b; Finch-Savage and Blake, 1994; Finch-Savage, 1995). In a detailed study on *Quercus robur*, Finch-Savage and Blake (1994) reported that differing growth patterns in cotyledons and axes occurred in different years, resulting in the production of seeds of markedly different size. Not only were differences in seed water content recorded among the different harvests (Finch-Savage and Blake, 1994), but axis water contents also varied from year to year (Finch-Savage, 1995). For *Q. robur*, Finch-Savage and Blake (1994) also reported that seed shed late in the season had lower water contents than those shed earlier which, together with the observation that ψ_w does decline as the seeds mature, suggests that shedding occurs at different maturity stages, even within a single species.

Studies on *Zizania* spp. illustrate the complexity arising from variability in recalcitrant seeds (Vertucci *et al.*, 1994; 1995). These show considerable variation in maturity status at harvest which can be related to the relative desiccation sensitivity (Vertucci *et al.*, 1994; 1995). However, those authors have suggested that despite the variation in critical water content above which survival will occur at different maturity states and temperatures, these water contents correspond to a common single water activity (a_w *c.* 0.90). Vertucci *et al.* (1995) have consequently proposed a model to predict the optimum storage conditions for seeds of *Zizania* spp. that takes into account a_w in relation to water content and temperature.

Developmental Aspects

Irrespective of the relative state of maturity at shedding, recalcitrant seeds are hydrated and metabolically active, and will continue development. In the case of the highly recalcitrant tropical species, *A. marina*, the seeds become fully germinable after about 83% of the time from anthesis to natural shedding. From this stage on, if the ABA-containing pericarp is removed, then germination occurs (Farrant *et al.*, 1993b). In other cases, the ability to germinate is acquired earlier, while reserve deposition is still occurring (Nautiyal and Purohit, 1985; Hong and Ellis, 1990; Finch-Savage, 1992; Tompsett and Pritchard, 1993; Fu *et al.*, 1994), but in no case is the onset of germinability dependent on any water loss (Finch-Savage, 1995). Indeed, with premature, but germinable seeds of *A. marina*, the germination lag is equivalent to the period between the age at which the seeds were harvested, and the age at natural shedding (Farrant *et al.*, 1993b).

Progression towards germination of post-harvest, recalcitrant seeds at water contents that permit this to occur, appears to be the rule among recalcitrant seeds, although the time taken for visible manifestation of this varies among species (Berjak *et al.*, 1989; 1992; 1993a; Farrant *et al.*, 1989; Finch-Savage, 1995). This period may be very protracted in seeds that are shed in a less-than-mature condition (Berjak *et al.*, 1989; Farrant *et al.*, 1989). While this may be a characteristic of the seeds of some species (e.g. *Scadoxus membranaceous* [Farrant *et al.*, 1989]), in others the pre-germination lag may be inconsistent, occurring as a consequence of intra- or inter-seasonal variability in the relative state of development at shedding.

Finch-Savage (1992; 1995) has suggested in view of the many similarities between the development of orthodox and recalcitrant seeds, that the explanation of recalcitrance may reside in a truncated development. That is, that recalcitrant seeds may not enter the developmental phase (of orthodox seeds) during which tolerance to extreme desiccation develops, as natural shedding occurs before this, when dry mass is still accumulating. In their analysis, Vertucci and Farrant (1995) propose that recalcitrance may either be a consequence of early interruption of development, or of a curtailed PVS stage.

With the exception of *A. marina*, all recalcitrant seeds so far studied become increasingly desiccation tolerant during development (Hong and Ellis, 1990; Finch-Savage, 1992; Berjak *et al.*, 1993a,b; Tompsett and Pritchard, 1993), although none ever achieves a truly tolerant state. Without exception, all species become increasingly desiccation sensitive with the onset of germinative metabolism (Farrant *et al.*, 1988; Berjak *et al.*, 1989; 1992; 1993a; Vertucci and Farrant, 1995), which may be initiated a short while before shedding (Finch-Savage, 1995) in those seeds that become less tolerant at this stage (e.g. *Litchi chinensis* and *Clausena lansium* [Fu *et al.*, 1994]). Generally also, an increasing resistance to water loss accompanies increasing desiccation tolerance (Finch-Savage, 1995). Interestingly though, although germinating seeds become increasingly desiccation sensitive, drying rate has been found to be relatively slow (Berjak *et al.*, 1992; 1993a) in the two species for which this has been assessed. Finch-Savage (1995) ascribes the wide range of desiccation tolerance among recalcitrant seed species at shedding, to the varying degree to which this property is expressed during development. Consequently, that author suggests that acquisition of desiccation tolerance should be viewed as a quantitative phenomenon.

Desiccation Sensitivity

While the deleterious effects of dehydration may be recognizable by a syndrome of symptoms at various levels, including cell collapse, damage to intracellular membranes and the cytoskeleton, and biochemical and biophysical abnormalities, the underlying deficiencies that result in desiccation-related degeneration remain unresolved. In order to understand these phenomena, desiccation tolerance and sensitivity should be studied in tandem. When events conferring tolerance are unequivocally resolved, then it should be possible to identify differences that might contribute to our understanding of desiccation sensitivity.

Primary phenomena characterizing the onset of maturation drying in orthodox seeds include intracellular de-differentiation and a decline in metabolic activity (e.g. Berjak *et al.*, 1984), which culminate in the desiccated, quiescent condition. Similar events do not occur in recalcitrant seeds, although a measure of dehydration (e.g. Finch-Savage, 1995) and limited organelle de-differentiation (Farrant *et al.*, 1992) may take place in most species. This, coupled with the fact that active metabolism does not cease at any stage in recalcitrant seeds, is likely to be one of the factors contributing to their desiccation sensitivity. Metabolism is likely to become deranged and membranes and other subcellular components damaged, as water is lost (see below). Another aspect of sub-cellular organization that might contribute to desiccation tolerance – or its lack – is the degree of vacuolation. In many orthodox seeds, effective vacuolar volume is reduced either through shrinkage or as proteins accumulate within these compartments. Generally other, extravacuolar storage products also accumulate, effectively buffering the cells against the consequences of mechanical stresses accompanying dehydration (e.g. Vertucci and Farrant, 1995). While the extent to which such intra-cellular phenomena occur in recalcitrant seeds varies, there is not the dramatic reduction in vacuolar and cytoplasmic volume that characterizes maturation of orthodox seed tissues. This, perhaps coupled with a possible relative rigidity of the cell walls, might contribute to the damaging effects of the mechanical stresses imposed by dehydration.

The production and implication of dehydrins or LEA proteins has been discussed in the context of evolutionary considerations of seed recalcitrance, above. While there is now unequivocal evidence that dehydrins do occur in a spectrum of recalcitrant seed species, it may be the absence or insufficiency of *particular* proteins that contributes to the inability of recalcitrant seeds to tolerate any significant degree of dehydration (Finch-Savage *et al.*, 1994a; Finch-Savage, 1995). Nevertheless, the dehydrins that are present could confer a measure of desiccation tolerance, commensurate with the reduction in water content that accompanies development in many recalcitrant seed species (Finch-Savage *et al.*, 1994a). However, it has been concluded that even for orthodox seeds, the mere presence of dehydrins does not confer complete desiccation tolerance (e.g. Blackman *et al.*, 1991), but the quantity of such proteins and/or their activity in combination with other protectants may be important (Blackman *et al.*, 1992).

Convincing evidence has accumulated that the presence of sucrose and certain oligosaccharides in favourable mass ratios (Horbowicz and Obendorf, 1994), contributes to, rather than accounts for, desiccation tolerance in dry, orthodox seeds (Leopold *et al.*, 1994). Sucrose, in combination with oligosaccharides promotes glass formation, in which a high-viscosity, amorphous, non-equilibrium condition is achieved at suitable water contents. The glassy state is held to protect macromolecules against denaturation and perhaps prevent the liquid crystalline gel phase transition in membranes (e.g. Leopold *et al.*, 1994). There is apparently no lack of sucrose and raffinose or stachyose in most recalcitrant seed tissues examined (e.g. Berjak *et al.*, 1989; Farrant *et al.*, 1993a; Finch-Savage *et al.*, 1993; Finch-Savage and Blake, 1994). However, the mass ratios may be unfavourable for the formation of anything but the most transient glasses. Work on *Quercus rubra* has shown that a glassy state could be achieved in recalcitrant seed tissues (Sun *et al.*, 1994), even though oligosaccharides were not detectable in this species. This glass was, however, not conducive to the maintenance of membranes in the liquid crystalline phase in the cotyledons of the desiccation-sensitive *Q. rubra*. It should also be noted that glass transitions at ambient temperatures are achieved only at water contents which are lethal to recalcitrant seeds (e.g. Leopold *et al.*, 1994; Sun *et al.*, 1994) unless perhaps the axes have been subjected to flash drying (Farrant *et al.*, 1993a). In the case of flash-dried material, freezing is the only way viability can be retained (see below). This attests to the transient nature of such glasses or their inability to facilitate tolerance of the desiccated state under ambient or even refrigerated conditions.

In their recent review, Vertucci and Farrant (1995) have considered the dimension of intracellular water in considerable detail, linking desiccation-sensitivity and -tolerance with metabolic events and aspects of seed development. Those authors have proposed critical water levels to exist at which discrete changes in metabolic activity occur. In particular, they have identified the following: above a ψ_w of -1.5 MPa (>0.7 g g^{-1} [dilute solution]), cell division and germinative metabolism will proceed (hydration level 5). At hydration level 4, (down to ψ_w -3 MPa [0.45 g g^{-1}]) damage will not occur in most mature seeds (indeed, protein synthesis and repair are suggested to be possible), but cell division (Myers *et al.*, 1992), and therefore germination, are precluded. However, hydration at this level presumably permits ongoing development, as has been described for recalcitrant seeds of *Quercus robur*, the axes of which will continue to accumulate dry weight at the expense of the cotyledons (Finch-Savage *et al.*, 1992). On the other hand, immature embryos and seedlings and highly-recalcitrant seeds, such as those of *Avicennia marina*, are proposed to die at this hydration level. In hydration level 3 (which extends down to ψ_w -11 MPa (0.25 g g^{-1}), lipid bilayer phase transitions will be initiated and all recalcitrant seeds will die, while intermediate seeds are proposed to lose viability rapidly below this hydration level. Considering the water characteristic of each level to have distinct properties and functions, the loss of water in a particular level eliminates the functions it provides (Vertucci and Farrant, 1995). Those authors propose that tissues which are not damaged when a particular type of water is removed, have the mechanisms required to avoid or tolerate the resultant stress. Thus the removal of concentrated solution or capillary water (Type 4) will kill the highly-vacuolated seeds of *Avicennia marina*, which are poised for germination, while dehydration that removes water forming bridges over macromolecules or hydrophobic moieties (Type 3) results in the loss of viability of other recalcitrant seeds. It is likely that at this hydration level, restricted, but deranged, metabolism will occur and free radicals may be generated (see below). It is interesting that, as long as the developmental stage permits it, the more rapidly axes from recalcitrant seeds can be dried, the greater is the water loss that they will tolerate, which is the basis of flash drying (Berjak *et al.*, 1990). There is, however, a lower limit to the degree of dehydration commensurate with viability retention that can be achieved by flash drying. This has been proposed to be the level of non-freezable (Type 2) water (Pammenter *et al.*, 1991; 1993). Flash drying may limit the uncontrolled consequences of deranged metabolism (e.g. Berjak *et al.*, 1989; 1993; Pammenter *et al.*, 1991; Pritchard, 1991), which might occur with time in slowly-dried seeds, particularly at hydration level 3 (Vertucci and Farrant, 1995). Upon dehydration, DNA in desiccation-sensitive material loses integrity (Osborne and Boubriak, 1994) and lipid phase transitions occur (reviewed by Vertucci and Farrant, 1995). While the latter may be reversible, protein structure may be irretrievably damaged (Sowa *et al.*, 1991) and/or these macromolecules may be lost from the bilayer.

Free radicals appear to play a major rôle in deteriorative processes in seeds generally, and dehydration-stressed recalcitrant seeds provide no exception (Hendry *et al.*, 1992; Hendry, 1993; Finch-Savage *et al.*, 1994b). The occurrence of peroxidized membrane lipids resulting in structural changes, inactivation of membrane-associated proteins and changes in membrane permeability are indications of membrane deterioration having its origin in the uncontrolled consequences of free radical generation (Hendry, 1993). Dehydration of active recalcitrant seeds is bound to generate metabolic stresses which might well lead to uncontrolled free radical generation and activity in a situation where normal functioning of antioxidants and enzymatic free radical scavengers is impaired (Senaratna and McKersie, 1986; Smith and Berjak, 1995). Leprince *et al.* (1990), working with desiccation-sensitive maize seedlings (which could be considered to be akin to recalcitrant seeds), showed not only that an organic free radical accumulated and lipid peroxidation significantly increased during dehydration, but also that the activity of superoxide dismutase (SOD) was severely affected. Uncontrolled generation of free radicals occurring in several dehydration-stressed, recalcitrant seed species has been reported by Finch-Savage *et al.* (1994b). Those authors showed that lipid peroxidation and accumulation of a stable free radical occurred within axes of *Castanea sativa*, *Aesculus hippocastanum* and *Quercus robur* during dehydration, in advance of viability loss. For *Q. robur*, free-radical induced damage has been shown to be initiated in axes dried below 0.89 g g^{-1} (Hendry *et al.*, 1992). It is interesting that axes from other recalcitrant seed species lose viability at similar, or even higher, water contents when dehydrated slowly (e.g. Berjak *et al.*, 1984; Pammenter *et al.*, 1991).

Generally the axes of recalcitrant seeds are at a considerably higher water content than are the cotyledons (Berjak *et al.*, 1989) and Finch-Savage *et al.* (1993) have demonstrated that for *Q. robur* there is a higher proportion of matrix-bound water in the cotyledons. According to those authors, this may underlie the greater desiccation sensitivity of the cotyledons relative to the axes. Despite this, for *Q. robur* no significant increase in peroxidation was found to occur in the cotyledons as a consequence of dehydration, whereas a rapid accumulation of free radicals was measured in the axes (Finch-Savage *et al.*, 1993). Those authors suggested that this may be a consequence of the different protective mechanisms that operate: in the axes, the activity of protective enzymes and the content of the anti-oxidant, α-tocopherol, were found to diminish while increasingly high SOD and glutathione reductase activity occurred in the cotyledonary tissue. Thus, at least in the axes of recalcitrant seeds, there is convincing evidence that uncontrolled free radical generation –

probably as a consequence of deranged metabolism – accompanies dehydration. Free radical generation may well be the primary event that underlies a variety of lesions in the axes of dehydration-stressed, recalcitrant seeds. For example, formation of gel-phase domains associated with lateral phase separation in membranes with the consequent displacement of membrane proteins, could well be a consequence of uncontrolled free radical activity (Senaratna and McKersie, 1986; McKersie *et al.*, 1988). Such damage could be the cause of viability loss that occurs in slowly-dried, recalcitrant seeds which are still at relatively high water contents and also in flash-dried axes of some species (Pammenter *et al.*, 1991; 1993). In this regard, the success of flash drying in lowering water content close to the level of non-freezable water without axis viability loss, has been ascribed to the rate at which dehydration is achieved. Nevertheless, Pammenter *et al.* (1993) have indicated that for some species marked desiccation damage becomes apparent at considerably higher water contents, despite rapid water removal.

Chilling Sensitivity

A recent study on hydrated seeds of *Azadirachta indica* indicated that, in response to chilling, a decline in viability was accompanied by ultrastructural degeneration: mitochondria and plastids in axis cells lost internal organization and vacuoles generally collapsed. Areas of advanced degradation also occurred, comprizing cells in which the plasmalemma was discontinuous and vesiculated (Berjak *et al.*, 1995a). Under the circumstances that recalcitrant seeds are metabolic and may have entrained germinative metabolism, rather than seek parallels with imbibitional chilling injury, the deleterious effects of cold may be equated with damage sustained by germinating seeds or seedlings. These effects include decreased respiratory rate which has been suggested to be a consequence of cold-lability of particular TCA cycle enzymes (Herner, 1990). This could well be bound up with the regression of the mitochondrial inner membrane system (Berjak *et al.*, 1995a). Two key enzymes of glycolysis, phosphofructokinase and pyruvate kinase, have also been identified as being cold labile (Guy, 1990). The vacuolar collapse reported for cold-exposed *A. indica* cells (Berjak *et al.*, 1995a) might have been a consequence of dismantling of the cytoskeleton in response to chilling (Raison and Orr, 1990). This, in turn, would affect glycolysis, in view of the structural association between key glycolytic enzymes and actin microfilaments (Masters, 1984). Maintenance of intracellular spatial organization (in which water plays an integral rôle) includes the existence of multi-enzyme particles (Hrazdina and Jensen, 1992), such as those of glycolysis (Masters, 1984). If some key enzymes of both the glycolytic pathway and the TCA cycle become impaired, then out-of-phase metabolism must be likely (Lyons and Breidenbach, 1990). As a consequence, free radical activity might escalate to proportions where considerable damage could accumulate, if enzymatic and anti-oxidant scavenging systems operate inefficiently (Senaratna and McKersie, 1986; McKersie *et al.*, 1988; Smith and Berjak, 1995). As a result, membrane lipids would be adversely affected.

In the opinion of Murata and Nishida (1990), membrane lipids might well be a primary intracellular site of chilling injury, and Raison and Orr (1990) have suggested the phase change from the liquid crystalline to the gel state to be likely. In recent work on tomato fruit, Sharom *et al.* (1994) have demonstrated that phase changes in cellular membranes are, in fact, induced by chilling. Even if the phase change were reversible, it is likely that membrane proteins would have been displaced, as discussed above (see Desiccation sensitivity).

Conservation of Recalcitrant Germplasm

Depending on the species, hydrated, intact recalcitrant seeds can be stored only for periods from days to months (e.g. Chin and Roberts, 1980; Berjak *et al.*, 1989; Berjak *et al.*, 1995b). Storage lifespan may, however, be prolonged by various manipulations (our unpublished observations) but the effective extension of longevity is still not useful for long-term conservation of the germplasm. One of the major difficulties even in the short-term, is that the high RH conditions necessary to prolong storage lifespan of the seeds, are also conducive to the proliferation of micro-organisms, especially as chilling is precluded in many instances (Berjak, 1995). As the vigour of wet-stored recalcitrant seeds declines as a result of inherent changes (Pammenter *et al.*, 1994), it has been suggested that natural defence mechanisms (e.g. the elicitation of phytoalexins) fail, thus facilitating fungal invasion of the debilitated seed tissues (Berjak, 1995). Short of *in situ* conservation, and minimal-growth storage of seedlings or *in vitro* cultures, the only option for preservation of the germplasm, and thus the biological diversity of species with recalcitrant seeds, is cryostorage (Berjak *et al.*, 1995b).

Cryostorage of recalcitrant germplasm involves the maintenance of zygotic embryonic axes, explants of various kinds, or somatic embryoids in liquid nitrogen (–196°C) or liquid nitrogen vapour (at *c.*–150°C). At such temperatures, low energy levels should preclude molecular movement and thus reactions, although events such as free radical generation and macromolecule damage by ionising radiation cannot be eliminated (Grout, 1990). While it may seem enigmatic to suggest that freezing chilling-sensitive material will succeed, there are distinct differences between chilling and freezing that allow this. It is essential, in the first instance, that the material is rapidly dehydrated to a range of water contents that obviates both dehydration damage and the lethal injury that results from ice crystal formation. This can be achieved by a combination of flash drying (or other, relatively rapid means of lowering the water content) and very rapid freezing (Wesley-Smith *et al.*, 1992; Berjak *et al.*, 1995b). Additionally, the specimen to be frozen must be as small as possible and, when necessary, cryoprotectants or other appropriate pre-treatments are required. Using this approach, we have obtained acceptable to 100% survival of zygotic axes from a variety of recalcitrant seed species, as well as of somatic embryoids (Berjak *et al.*, 1995b; Mycock *et al.*, this Volume). Other investigators too, have reported acceptable to good survival in cryostorage of zygotic and somatic material of various species (reviewed by Berjak *et al.*, 1995b).

However, a detailed protocol that is generally applicable, especially for zygotic axes of a broad variety of species, remains unattainable at present (our unpublished observations). This is probably because of the essential variability of the axes from recalcitrant seeds which, as discussed above, is presumed to be the outcome of differences in developmental status, water content and biochemical variation from one seed species to the next.

References

Berjak, P. 1995. In: *Improving the Handling and Storage of Tropical Tree Species with Recalcitrant and Intermediate Characteristics*, IPGRI/DFSC Workshop, Humlebaek, Denmark: In press.

Berjak, P. and Pammenter, N.W. 1994. *Seed Science Research* 4: 263–264.

Berjak, P., Dini, M. and Pammenter, N.W. 1984. *Seed Science and Technology* 12: 365–384.

Berjak, P., Farrant, J.M. and Pammenter, N.W. 1989. In: *Recent Advances in the Development and Germination of Seeds*, pp. 89–108 (ed. R.B. Taylorson). New York: Plenum Press.

Berjak, P., Pammenter, N.W. and Vertucci, C.W. 1992. *Planta* 186: 249–261.

Berjak, P., Vertucci, C.W. and Pammenter, N.W. 1993a. *Seed Science Research* 3: 155–166.

Berjak, P., Farrant, J.M., Mycock, D.J. and Pammenter, N.W. 1990. *Seed Science and Technology* 18: 297–310.

Berjak, P., Bradford, K.J., Kovach, D.A. and Pammenter, N.W. 1994. *Seed Science Research* 4: 111–121.

Berjak, P., Farrant, J.M., Pammenter, N.W., Vertucci, C.W. and Wesley-Smith, J. 1993b. In: *Fourth International Workshop on Seeds. Basic and Applied Aspects of Seed Biology*, pp. 705–714 (eds D. Côme and F. Corbineau). Paris: ASFIS.

Berjak, P., Campbell, G.K., Farrant, J.M., Omondi-Oloo, W. and Pammenter, N.W. 1995a. *Seed Science and Technology* 23: 779–792.

Berjak, P., Mycock, D.J., Wesley-Smith, J., Dumet, D. and Watt, M.P. 1995b. In: *International Workshop on* in vitro *Conservation of Plant Genetic Resources*, Kuala Lumpur, Malaysia: In press.

Blackman, S.A., Obendorf, R.L. and Leopold, A.C. 1992. *Plant Physiology* 100: 225–230.

Blackman, S.A., Wettlaufer, S.H., Obendorf, R.L. and Leopold, A.C. 1991. *Plant Physiology* 96: 868–874.

Chin, H.F., Hor, Y.L. and Mohd Lassim, M.B. 1984. *Seed Science and Technology* 12: 428–436.

Chin, H.F. and Roberts, E.H. 1980. *Recalcitrant Crop Seeds*. Kuala Lumpur: Tropical Press.

Close, T.J., Kortt, A.A. and Chandler, P.M. 1989. *Plant Molecular Biology* 13: 95–108.

Corbineau, F. and Côme, D. 1986. *Seed Science and Technology* 14: 585–591.

Dickie, J.B., May, K., Morris, S.V.A. and Titley, S.E. 1991. *Seed Science Research* 1: 149–162.

Dodd, M.C., van Staden, J. and Smith, M.T. 1989. *Annals of Botany* 64: 297–310.

Dure, L.S. III. 1993. In: *Control of Plant Gene Expression*, pp. 325–335 (ed D.P.S. Verma). Boca Raton, FL, U.S.A.: CRC Press.

Dure, L.S. III, Crouch, M., Harada, J., Ho, T-H.D., Mundy, J., Quatrano, R., Tamas, T. and Sung, Z.R. 1989. *Plant Molecular Biology* 12: 475–486.

Ellis, R.H., Hong, T.D. and Roberts, E.H. 1990. *Journal of Experimental Botany* 41: 1167–1174.

Ellis, R.H., Hong, T.D. and Roberts, E.H. 1991a. *Seed Science Research* 1: 69–72.

Ellis, R.H., Hong, T.D., Roberts, E.H. and Soetisna, U. 1991b. *Seed Science Research* 1: 99–104.

Farrant, J.M., Pammenter, N.W. and Berjak, P. 1988. *Seed Science and Technology* 16: 155–166.

Farrant, J.M., Pammenter, N.W. and Berjak, P. 1992. *Annals of Botany* 70: 75–86.
Farrant, J.M., Pammenter, N.W. and Berjak, P. 1993a. *Seed Science Research* 3: 1–13.
Farrant, J.M., Berjak, P. and Pammenter, N.W. 1993b. *Annals of Botany* 71: 405–410.
Finch-Savage, W.E. 1992. *Seed Science Research* 2: 17–22.
Finch-Savage, W.E. 1995. In: *Improving the Handling and Storage of Tropical Tree Species with Recalcitrant and Intermediate Characteristics*, IPGRI/DFSC Workshop, Humlebaek, Denmark: In press.
Finch-Savage, W.E. and Blake, P.S. 1994. *Seed Science Research* 4: 127–133.
Finch-Savage, W.E., Pramanik, S.K. and Bewley, J.D. 1994a. *Planta* 193: 478–485.
Finch-Savage, W.E., Hendry, G.A.F. and Atherton, N.M. 1994b. *Proceedings of the Royal Society of Edinburgh* 102B: 257–260.
Finch-Savage, W.E., Clay, H.A., Blake, P.S. and Browning, G. 1992. *Journal of Experimental Botany* 43: 671–679.
Finch-Savage, W.E., Grange, R.I., Hendry, G.A.F. and Atherton, N.M. 1993. In: *Fourth International Workshop on Seeds. Basic and Applied Aspects of Seed Biology*, pp. 723–730 (eds D. Côme and F. Corbineau). Paris: ASFIS.
Fu, J-R., Jin, J.P., Peng, Y.F. and Xia, Q.H. 1994. *Seed Science Research* 4: 257–261.
Galau, G.A., Jakobsen, K.S. and Hughes, D.W. 1991. *Physiologia Plantarum* 81: 280–288.
Gamene, C.S., Kraak, H.L. and van Pijlen, J.G. (1994). In: *Proceedings of the International Workshop on Desiccation Tolerance and Sensitivity of Seeds and Vegetative Plant Tissues*, Kruger National Park, South Africa.
Gee, O.H., Probert, R.J. and Coomber, S.A. 1994. *Seed Science Research* 4: 135–141.
Grout, W.W. 1990. In: *Plant Tissue Culture. Applications and Limitations*, pp. 394–411 (ed. S.S. Bhojwani). Amsterdam: Elsevier.
Guy, C.L. 1990. *Annual Review of Plant Physiology* 41: 187–223.
Hendry, G.A.F. 1993. *Seed Science Research* 3: 141–153.
Hendry, G.A.F., Finch-Savage, W.E., Thorpe, P.C., Atherton, N.M., Buckland, S.M., Nilsson, K.A. and Seel, W.A. 1992. *New Phytologist* 122: 273–279.
Herner, R.C. 1990. In: *Chilling Injury of Horticultural Crops*, pp. 51–69 (ed. C.Y. Wang). Boca Raton, FL, U.S.A.: CRC Press Inc.
Hong, T.D. and Ellis, R.H. 1990. *New Phytologist* 116: 589–596.
Horbowicz, M. and Obendorf, R.L. 1994. *Seed Science Research* 4: 385–405.
Hrazdina, G. and Jensen, R.A. 1992. *Annual Review of Plant Physiology* 43: 241–267.
Kermode, A.R. 1990. *Critical Reviews in Plant Sciences* 9: 155–195.
Kovach, D.A. and Bradford, K.J. 1992. *Journal of Experimental Botany* 43: 747–757.
Leprince, O., Deltour, R., Thorpe, P.C., Atherton, N.M. and Hendry, G.A.F. 1990. *New Phytologist* 116: 573–580.
Leopold, A.C., Sun, W.Q. and Bernal-Lugo, I. 1994. *Seed Science Research* 4: 267–274.
Lyons, J.M. and Breidenbach, R.W. 1990. In: *Chilling Injury of Horticultural Crops*, pp. 223–233 (ed. C.Y. Wang). Boca Raton, FL, U.S.A.: CRC Press Inc.
Masters, C. 1984. *Journal of Cell Biology* 99: 222s–225s.
Myers, P.N., Setter, T.L., Madison, J.T. and Thompson, J.F. 1992. *Plant Physiology* 99: 1051–1056.
McKersie, B.D., Senaratna, T., Walker, M.A., Kendall, E.J. and Hetherington, R.P. 1988. In: *Senescence and Aging in Plants*, pp. 442–464 (eds D. Noodén and A.C. Leopold). San Diego: Academic Press.
Murata, N. and Nishida, I. 1990. In: *Chilling Injury of Horticultural Crops*, pp. 181–199 (ed. C.Y. Wang). Boca Raton, FL, U.S.A.: CRC Press Inc.
Nautiyal, A.R. and Purohit, A.N. 1985. *Seed Science and Technology* 13, 59–68.
Nkang, A. and Chandler, G. 1986. *Journal of Plant Physiology* 126: 243–256.
Osborne, D.J. and Boubriak, I.I. 1994. *Seed Science Research* 4: 175–185.
Pammenter, N.W., Vertucci, C.W. and Berjak, P. 1991. *Plant Physiology* 96: 1093–1098.
Pammenter, N.W., Vertucci, C.W. and Berjak, P. 1993. In: *Fourth International Workshop on Seeds. Basic and Applied Aspects of Seed Biology*, pp. 867–872 (eds D. Côme and F. Corbineau). Paris: ASFIS.

Pammenter, N.W., Berjak, P., Farrant, J.M., Smith, M.T. and Ross, G. 1994. *Seed Science Research* 4: 187–191.

Pritchard, H.W. 1991. *Annals of Botany* 67: 43–49.

Raison, J.K. and Orr, G.R. 1990. In: *Chilling Injury of Horticultural Crops*, pp. 145–164 (ed. C.Y. Wang). Boca Raton, FL, U.S.A.: CRC Press Inc.

Ray, P.K. and Sharma, S.B. 1987. *Scientia Horticulturae* 33: 213–221.

Roberts, E.H. 1973. *Seed Science and Technology* 1: 499–514.

Senaratna, T. and McKersie, B.D. 1986. In: *Membranes, Metabolism and Dry Organisms*, pp. 85–101 (ed. A.C. Leopold). Ithaca, London: Comstock.

Sharom, M., Willemot, C. and Thompson, J.E. 1994. *Plant Physiology* 105: 305–308.

Smith, M.T. and Berjak, P. 1995. In: *Seed Development and Germination*, pp. 701–746 (eds J. Kigel and G. Galili). New York: Marcel Dekker Inc.

Sowa, S., Vertucci, C.W., Crane, J., Pammenter, N.W. and Berjak, P. 1991. *Agronomy Abstracts*, p. 170.

Sun, W.Q., Irving, T.C. and Leopold, A.C. 1994. *Physiologia Plantarum* 90: 621–628.

Takhtajan, A. 1991. *Evolutionary Trends in Flowering Plants*. New York: Columbia University Press.

Tompsett, P.B. 1992. *Seed Science and Technology* 20: 251–267.

Tompsett, P.B. and Pritchard, H.W. 1993. *Annals of Botany* 71: 107–116.

Vertucci, C.W. and Farrant, J.M. 1995. In: *Seed Development and Germination*, pp. 237–271 (eds J. Kigel and G. Galili). New York: Marcel Dekker Inc.

Vertucci, C.W., Crane, J., Porter, R.A. and Oelke, E.A. 1994. *Seed Science Research* 4: 211–224.

Vertucci, C.W., Crane, J., Porter, R.A. and Oelke, E.A. 1995. *Seed Science Research* 5: 31–40.

von Teichman, I. and van Wyk, A.E. 1994. *Seed Science Research* 4: 225–239.

Wesley-Smith, J., Vertucci, C.W., Berjak, P., Pammenter, N.W. and Crane, J. 1992. *Journal of Plant Physiology* 140: 596–604.

77. Heat Stable Proteins and Desiccation Tolerance in Recalcitrant and Orthodox Seeds

J.R. FU, X.Q. YANG, X.C. JIANG, J.X. HE and S.Q. SONG

School of Life Sciences, Zhongshan University, Guangzhou 510275, People's Republic of China

Abstract

During development of peanut seeds, a set of low molecular weight polypeptides was identified most of which were heat-stable. For wampee seeds, only one band of heat-stable proteins could be checked out by SDS-PAGE. It had a molecular weight of 20 kDa and constituted 1% of the total proteins. By Western blot analysis, wampee seed axes can express dehydrin-like proteins (at 67–88 days (d) after anthesis) but the level was very low. Stachyose and sucrose increased and maltose declined with the acquisition of desiccation tolerance during embryogenesis of peanut seed. Following the development of wampee seed, glucose and maltose decreased gradually, while fructose, sucrose and stachyose increased markedly. It may be assumed that the heat-stable and low molecular weight polypeptides synergize with soluble sugars in the acquisition of desiccation tolerance in peanut seed though not in wampee seed.

Introduction

Recalcitrant seeds cannot tolerate desiccation (Roberts 1973). *Clausena lansium* (wampee) seed is recalcitrant (Fu *et al.*, 1989a,b; Hoffmann and Steiner, 1989). Fu *et al.* (1994) reported that desiccation tolerance of wampee seeds increased to a maximum value and then decreased during seed development and maturation. In contrast, desiccation tolerance of peanut axes (an orthodox seed) increased during seed development and maturation.

As for orthodox seeds, a lot of data shows that LEA proteins and soluble sugars are associated with the development of desiccation tolerance (Caffrey *et al.*, 1988; Crowe *et al.*, 1988; 1992). However, desiccation tolerance in recalcitrant seeds may be different from that in orthodox seeds. Recent data have shown that desiccation tolerance was acquired gradually during maturation of developing seeds (Leprince *et al.*, 1990; Sun and Leopold 1993; Fu *et al.*, 1994). There were no LEA proteins in *Avicennia marina* (Farrant *et al.*, 1992) but Finch-Savage (1994) stated that proteins which have homology with dehydrins have been identified immunologically in mature seeds of five desiccation sensitive tree species. The relationship between desiccation sensitivity and LEA proteins (or dehydrins) of recalcitrant seeds is uncertain.

The aim of this work is to test the correlation between the levels of heat-stable proteins and desiccation tolerance in wampee, a recalcitrant seed, and peanut, an orthodox seed. The status of oligosaccharides, which have been recognized as protective agents, is also assessed during seed development.

R.H. Ellis, M. Black, A.J. Murdoch, T.D. Hong (eds.), Basic and Applied Aspects of Seed Biology, pp. 705–713.
© *1997 Kluwer Academic Publishers, Dordrecht. Printed in Great Britain.*

Materials and Methods

Plant Material

The experimental materials were seeds of wampee (*Clausena lansium* (Lour.) Skeels cultivar 'JiXin') and peanut (*Arachis hypogaea* L. cultivar 'Yue-you 116'). Wampee seeds were collected at weekly intervals from mid-maturation to the full ripened stage (43–85 DAA, days after anthesis) and peanut seeds were collected at 10 d intervals from 25 DAP (days after pegging) to 65 DAP.

Germination Test

Seeds were surface sterilized with 1.5% NaOCl for 15 min, and embryonic axes were excised and incubated in MS+B_5+3% sucrose (peanut) or WPM (wampee) medium with 2% sucrose (McCown and Lloyd, 1981) at 25°C for seven days. Germination was defined as an increase of more than 50% in the length of the axis.

Conductivity Tests

Isolated axes were washed with distilled water to remove surface electrolytes and air-dried to desired water contents. The axes were then incubated in water and electrolyte leakage was measured after 12 h. Electrolyte leakage was expressed as percentage of the total electrical conductivity after axes were killed by boiling in water for 10 min.

Heat Stable Protein Extraction

Freshly dissected axes frozen in liquid nitrogen were ground to a powder with sand in a cold pestle and mortar. For peanut embryonic axes, n-hexane (1:10, w: v) was used to remove fat at –20°C overnight. 400 µl cold extraction buffer was added to 30–40 mg defatted powder. The powder was then homogenized in ice cold extraction with 50 mM Tris-HCl (pH 8.0), containing 500 mM NaCl, 10 mM B-ME, 5 µg.ml leucipeptin, 1 mM PMSF (dissolved in isopropanol).

The resulting homogenate was centrifuged at 16 000*g* at 4°C for 10 min. To obtain heat stable proteins, the supernatant was boiled for 10 min then kept on ice for 15 min and then centrifuged as above. Protein (50 µg) was subject to electrophoresis on 10–18% polyacrylamide–SDS and stained with Coomassie brilliant blue. Protein concentration in the resulting supernatant was estimated by the method of Bradford (1976). The level of heat stable protein was expressed as the ratio of heat stable to soluble protein.

According to the method of Yamada *et al.* (1979), storage proteins were isolated from cotyledons of peanut seeds and subjected to repeated isoelectric precipitation to isolate arachin, conarachin and 2S albumin. 2S albumin was purified by sephadex G-75 column chromatography. Their heat stable properties were assayed as above.

Determination of Protein Synthesis

Wampee seeds were collected at different developmental stages. [^{3}H]leucine (10 μCi/ml) was used as tracer and embryonic axes were incubated at 28°C for 4 h. The sample was assayed with a liquid scintillation counter (PACKARD 2000CA, TRI-CARB).

Western Blotting

After phenol extraction and precipitation with ammonium acetate in methanol, samples were dried under vacuum and dissolved in the electrophoresic loading buffer. After fractionating with 10% SDS-PAGE (Laemmli, 1970), proteins were electrotransferred on to nitrocellulose filters. The filters were blocked for 2 h with nonfat milk and then allowed to immunoreact with antiserum against pea dehydrin. The signals were revealed by using horseradish peroxidase-conjugated goat secondary antibodies and the colour developed with the substrate diaminobenzidine tetrahydrochloride (Sambrook *et al.,* 1989).

Saccharides Assayed by HPLC

85% alcohol preheated to 75°C was added to 300 mg dried defatted powder (as above), and shaken for 40 min at 72–75°C. The slurry was incubated at 75°C for 30 min. The supernatants were dried in the vortex evaporator (XZ-803) at 60°C. The dried extract, after dissolution in 1.5 ml double-distilled H$_2$O, was filtered through a DEAE-cellulose column to remove the pigments. The aqueous elution (5 ml) was collected and analyzed by HPLC (Shimadzu LC-6AC-R3A) to determine the types and amounts of saccharides.

Results

Heat Stable Proteins and Desiccation Tolerance in Peanut Seed

Desiccation tolerance was acquired at 40–55 DAP and reached a maximum at physiological maturity (65 DAP). As seeds developed, the germination percentage increased, but desiccation tolerance developed somewhat later.

During development, heat stable proteins increased conspicuously. 25- and 35-DAP embryonic axes with low desiccation tolerance contained less heat stable proteins (15.1% and 30.6%, respectively), 40-DAP embryonic axes had a higher content of heat stable proteins (42–44%), and heat stable proteins at 60- and 65-DAP embryonic axes were the highest. When artificial drying was given to 65-DAP embryonic axes, the amount of heat stable protein increased (Fig. 1).

Embryonic axes of peanut seeds at 25 DAP and 35 DAP having low germination percentage did not endure flash-drying and gave a low vigour index, whereas the slow-drying raised the germination percentage and vigour index (data not shown).

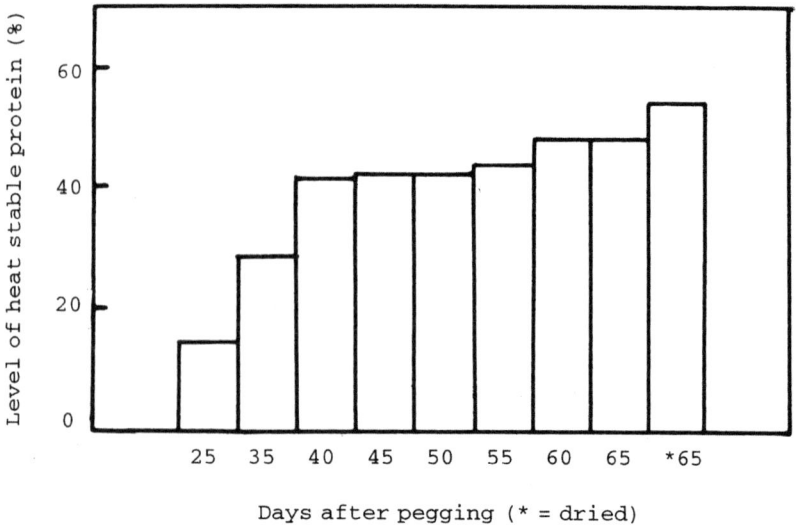

Figure 1. Changes in the relative amounts of heat stable proteins extracted from developing peanut axes. The levels of heat stable protein were expressed as a ratio of heat stable to soluble protein

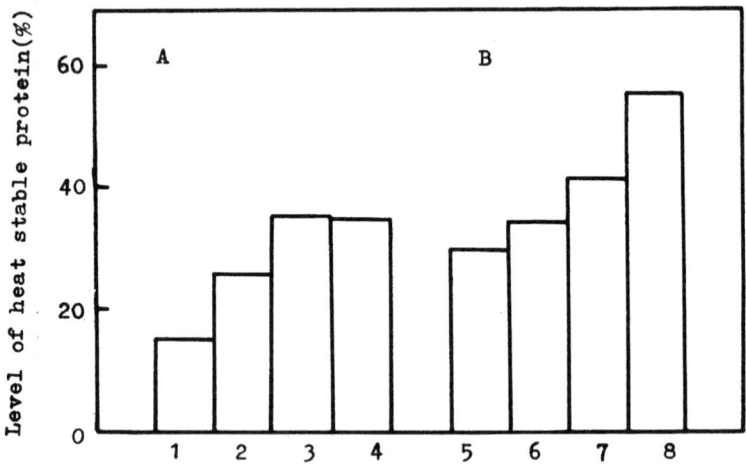

Figure 2. Changes in level of heat stable proteins in peanut axes. A: 25-DAP axes cultured in a row on a series of MS+B5 media in which sucrose contents were 3% (as control) (1), 8% (2), 14% (3) and 21% (4). The samples were harvested after 48 h incubation. B: 35-DAP axes subjected to slow-drying for 0 d (5), 3 d (6), 5 d (7) and 10 d (8)

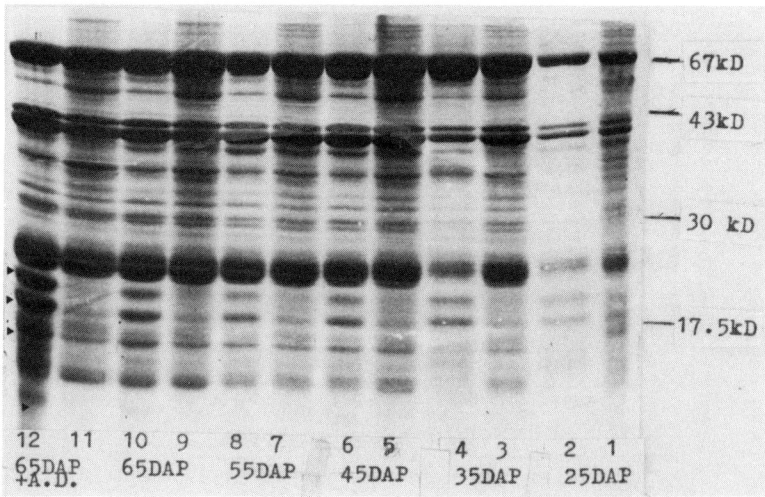

Figure 3. SDS-PAGE analysis of heat stable proteins (lanes 2, 4, 6, 8, 10, 12) and soluble proteins (lanes 1, 3, 5, 7, 9, 11) extracted from peanut axes. Equal amounts of protein from 25- to 65-DAP were loaded per track, heat stability of 2S albumin is indicated (by arrow)

The amount of heat stable proteins increased rapidly 35 DAP in embryonic axes which had been subjected to slow-drying for 3–10 days (Fig. 2B). The addition of sucrose as an osmoticum also increased the amount of heat stable proteins (Fig. 2A). At the same time, desiccation tolerance of peanut seeds also increased (data not shown).

The soluble proteins extracted by buffer with high salt (2 M NaCl) were mainly storage proteins including arachin, conarachin and 2S albumin. Some storage proteins, including some sub-units of arachin and conarachin, particularly 2S albumin (15.5 kDa, 17 kDa, 18 kDa), were heat stable. The soluble proteins and the heat stable proteins were assayed by electrophoresis as shown in Fig. 3.

Heat Stable Proteins and Desiccation Tolerance in Wampee Seed

Wampee embryonic axes of different developmental stages were subjected to flash-drying. The electrolyte leakage of all the samples which were maintained at 32% moisture content (w.b.) was measured. Embryonic axes at the physiological maturity (74 DAA) had the lowest leakage rate (Fig. 4). It was shown that the membrane of embryonic axes at physiological maturity was unimpaired and more tolerant to desiccation, but those of immature and overripe ones was less tolerant.

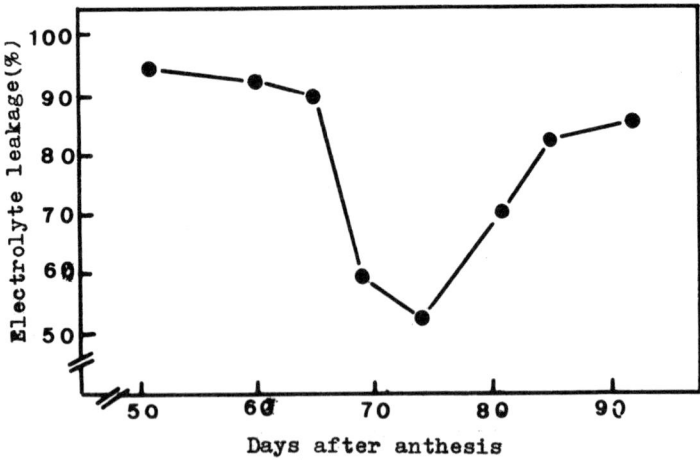

Figure 4. Electrolyte leakage of wampee seed axes in different developmental stages after predrying to 32% moisture content

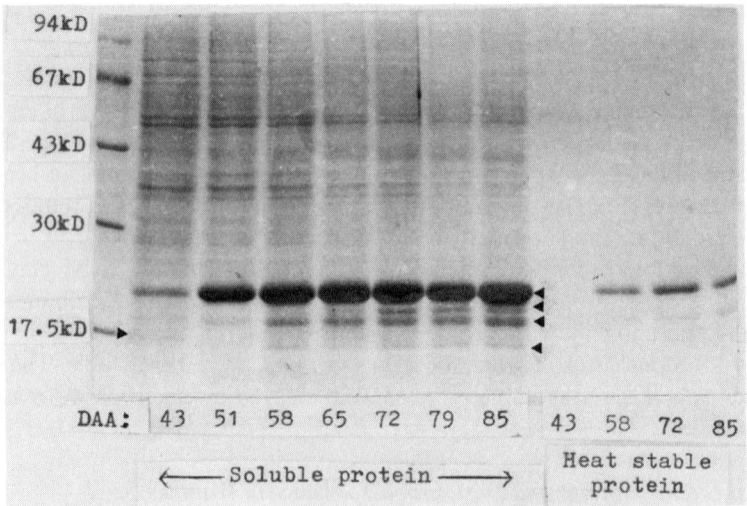

Figure 5. SDS-PAGE analysis of the soluble and heat stable proteins in wampee cotyledons at different developmental stages. Equal amounts of soluble proteins were loaded or used to extract heat stable proteins for each track before electrophoresis. Numbers at the left of the photograph indicate Kilodaltons of markers. Numbers beneath the photograph indicate days after anthesis. Arrows indicate those polypeptides which have obvious changes during development (Left, 17.5 kDa; right, from upper: 20, 19, 18, 14 kDa, respectively)

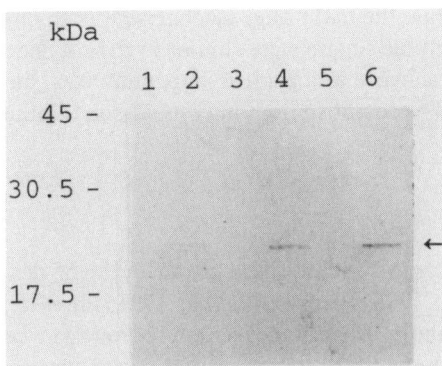

Figure 6. Expression of dehydrin in the developing wampee seed, 1, 3, 5 representing expression of dehydrin in cotyledons at 46, 67 and 88 DAA, respectively; 2, 4, 6 representing expression of dehydrin in embryonic axes at 60, 67 and 88 DAA, respectively

The dry weight of wampee seeds increased rapidly from 50 to 70 DAA (Fu *et al.*, 1994), but protein synthesis rate decreased in later stages of development (data not shown).

There were no heat stable proteins in 43-DAA embryonic axes. In 51-DAA axes, a 20 kDa protein appeared and became more prominent as development proceeded. Among the band of 20 kDa proteins, only part was heat stable. By SDS-PAGE, heat stable proteins began to appear from 58 DAA. The electrophoretic pattern in embryonic axes and in cotyledons were the same (Fig. 5). It constituted 1% of the content of total proteins in seed.

Expression of Dehydrin Protein

Soluble proteins were isolated from cotyledons and axes of developing wampee seeds and the expression of LEA protein (dehydrin) was investigated by Western blot analysis. The results indicated that in the whole range of development investigated (60, 67 and 88 DAA), a 22 kDa dehydrin-like protein was detected in wampee axes using pea dehydrin antiserum (Fig. 6); however, no immuno-signal was observed from cotyledon samples. This result suggested that wampee seed can only express dehydrin-like proteins in axes. It appeared that dehydrin started to synthesize at 60 DAA and small amounts accumulated as the seed matured.

Saccharides and Desiccation Tolerance

As to 25- and 35-DAP axes of peanut seed, the main soluble sugars were sucrose and maltose, there being no oligosaccharides present. When seeds matured (65 DAP axes) and had acquired desiccation tolerance they contained a high content of sucrose, while the main oligosaccharide was stachyose. The contents and constituents of soluble sugars were similar in the cotyledons.

There was much stachyose and sucrose in peanut axes, but in wampee axes, stachyose and sucrose were not so high in comparison to other soluble sugars.

Discussion

Germination percentage is usually used as an index of desiccation tolerance (Koster and Leopold, 1988; Blackman *et al.*, 1991; Fu *et al.*, 1994). However, sometimes the germination percentage is inappropriate as when immature seeds are unable to germinate. In this paper, both germination percentage and electrolyte leakage measurements were used to determine desiccation tolerance. A key role in the acquisition of desiccation tolerance of developing seeds is played by changes in cellular membranes (Le Page-Degivry and Garelbo, 1991). The ability to maintain membrane integrity and stability upon drying is essential for seed axes in desiccation tolerance. Sun and Leopold (1993) introduced the electrolyte leakage method to show the relation of membrane integrity to desiccation tolerance. The threshold of desiccation tolerance can be quantitatively determined by this method. They found that electrolyte leakage offered a quantitative measure for desiccation tolerance.

Blackman *et al.* (1991) stated that the presence of heat stable protein was related to desiccation tolerance. In our experiments, the increase of heat stable protein accompanied the development of peanut and wampee seeds and was correlated with the onset of desiccation tolerance. As for wampee axes, though desiccation tolerance increased more or less during development (Fu *et al.*, 1994), the tolerance to drying was still low. In peanut axes, there were many heat stable proteins including a lot of storage proteins which may have enabled them to tolerate drying. Blackman *et al.* (1991) considered that the heat stable storage proteins may have some protective function. In peanut axes, 2S albumins are the interesting ones but whether they play a protective role in drying must be investigated further.

There was in wampee seeds an expression of dehydrin-like proteins, increasing in the axes from 60 to 88 DAA. However, there was no simple correlation between the presence of dehydrin-like proteins and the ability of seeds to survive desiccation. This result showed that the dehydrin-like protein is unlikely to be implicated in the desiccation tolerance of wampee seeds. Blackman *et al.* (1991) concluded that dehydrin alone was not sufficient to induce desiccation tolerance. Gee *et al.* (1994) also reported that 20 kDa dehydrins were detected both in desiccation-tolerant and desiccation-intolerant seeds. The level of dehydrins in wampee axes was probably also too low to prevent desiccation damage.

The large amounts of stachyose and sucrose in peanut compared to wampee axes, may also be a factor influencing desiccation tolerance.

In conclusion, desiccation tolerance is acquired during seed development and maturation and may be related to heat stable proteins and sugars, but other factors should also be considered.

Acknowledgements

The project was supported by the National Natural Science Foundation of China and the Natural Science Foundation of Guangdong Province. We also thank Dr. P.M. Chandler, CSIRO, Canberra for kindly donating dehydrin antibody.

References

Blackman, S.A.,Wettlaufer, S.H., Obendorf, R.L. and Leopold, A.C. 1991. *Plant Physiology* 96: 868–874.
Blackman, S.A., Obendorf, R.L. and Leopold, A.C. 1992. *Plant Physiology* 100: 225–230.
Bradford, M.M. 1976. *Analytical Biochemistry* 72: 248–252.
Caffrey, M., Fonseca, V. and Leopold, A.C. 1988. *Plant Physiology* 86: 754–758.
Crowe, J.H., Crowe, L.M. and Chapman, D. 1984. *Science* 223: 701–703.
Crowe, J.H., Crowe, L.M., Carpenter, J.F., Rudolph, A.S., Aurell Wistrom, C., Spargo, B.J. and Anchordoguy, T.J. 1988. *Biochimica et Biophysica Acta* 947: 367–384.
Crowe, J.H., Hoekstra, F.A. and Crowe, L.M. 1992. *Annual Review of Physiology* 54: 579–599.
Farrant, J.M., Berjak, P. and Pammenter, N.W. 1992.*Plant Growth Regulation* 11: 257–265.
Farrant, J.M., Pammenter, N.W. and Berjak, P. 1993. *Seed Science Research* 3: 1–13.
Finch-Savage, W.E. 1992. *Seed Science Research* 2: 17–22.
Finch-Savage, W.E. 1994. *Abstracts of International Workshop on Desiccation Tolerance and Sensitivity of Seeds and Vegetative Plant Tissues*, pp. 20, South Africa.
Fu, J.R., Zhang, B.Z., Wang, X.F. and Qiao, Y.Z. 1989a. In: *22nd International Seed Testing Congress*, Abstracts, p.9, Zurich, ISTA Secretariat.
Fu, J.R., Zhang, B.Z., Wang, X.F., Qiao, Y.Z. and Huang, X.L. 1989b. In: *International Symposium of Horticultural Germplasm, Cultivated, Wild, Part I. Fruit Trees*, pp. 121–125. Beijing: International Academic Publishers.
Fu, J.R., Jin, J.P., Peng, Y.F. and Xia, Q.H. 1994. *Seed Science Research* 4: 257–261.
Gee, O.H., Probert, R.J. and Coomber, S.A. 1994. *Seed Science Research* 4: 135–141.
Hofmann, P. and Steiner, A.M. 1989. *Landwirtschaftliche Forschung* 42: 310–323.
Koster, K.L. and Leopold, A.C. 1988. *Plant Physiology* 88: 829–832.
Laemmli, U.K. 1970. *Nature* 227: 680–685.
Le Page-Degivry, M.T. and Garello, G. 1991. *Seed Science Research* 1: 221–227.
Leprince, O., Bronchart, R. and Deltour, R. 1990. *Plant Cell Environment* 13: 539–546.
McCown, B.H. and Lloyd, G. 1981. *Horticulture* 16: 453–459
Roberts, E.H. 1973. *Seed Science and Technology* 1: 499–514.
Sambrook, J., Fritsch, E.F. and Manniatis, T. 1989. *A Laboratory Manual*, Cold Spring Harbor Laboratory Press.
Sun, W.Q. and Leopold, A.C. 1993. *Physiologia Plantarum* 87: 403–409.
Yamada, T., Aibara, S. and Morita, Y. 1979. *Agricultural Biological Chemistry* 43: 2563–2568.

78. Cellular and Metabolic Events Associated with Dehydration of Recalcitrant *Araucaria angustifolia* Embryos

F. CORBINEAU, L. SALMEN ESPINDOLA, D. VINEL and D. CÔME

Physiologie Végétale Appliquée, Université Pierre et Marie Curie, Tour 53, 1er étage, 4 place Jussieu, 75252 Paris cédex 05, France

Abstract

The aim of the present work was to investigate the sequence of some cellular and metabolic events occurring in the embryonic axes, which might be related to the loss of viability of recalcitrant *Araucaria angustifolia* embryos during dehydration in the open air at 25°C and 55% relative humidity. The decreases in the ability for protein synthesis and in the capacity to convert 1-aminocyclopropane 1-carboxylic acid to ethylene, which were observed respectively at 0.5 and 1.5 h of dehydration, were very early indicators of deterioration. A high increase in leakage of electrolytes, which indicated a deterioration of cell membrane properties, was observed by the third–fourth h of desiccation. ATP content and energy charge also rapidly decreased during dehydration. However, energy charge cannot be a good marker of damage, since reimbibition of embryos restored its value close to that measured in non-dehydrated axes.

Introduction

Seed dehydration, which results in reduced metabolism, is the normal terminal event in the development of many seeds (Bewley and Black, 1994). Such seeds that can be stored in the dry state are called orthodox (Roberts, 1973). Seeds of several species have been termed recalcitrant (Roberts, 1973) because, as opposed to orthodox seeds, they are high in moisture content and cannot withstand intensive desiccation. This applies to various large-seeded hardwoods (e.g. *Castanea, Quercus, Juglans*) and numerous important tropical and sub-tropical trees (King and Roberts, 1979; Chin and Roberts, 1980).

Unfortunately, while there is considerable information on biochemical injury during loss of viability of orthodox seeds (Priestley, 1986), only few data concern the metabolic damage associated with dehydration injury of recalcitrant seeds. The aim of the present work was to precise the cellular and metabolic consequences of desiccation in *Araucaria angustifolia* embryos, which are typically recalcitrant (Salmen Espindola *et al.*, 1994), and to determine the possible sequence of these events occurring in the embryonic axis.

R.H. Ellis, M. Black, A.J. Murdoch, T.D. Hong (eds.), Basic and Applied Aspects of Seed Biology, pp. 715–721.
© *1997 Kluwer Academic Publishers, Dordrecht. Printed in Great Britain.*

Materials and Methods

Plant Material and Dehydration Method

Experiments were carried out with embryos isolated from freshly harvested *Araucaria angustifolia* seeds collected in the southern part of Brazil. Embryo viability was estimated by the germination percentages obtained after 7 days (d) at 25°C as described by Salmen Espindola *et al.* (1994).

In order to study the effects of dehydration on embryo viability and on metabolic damage induced in the embryonic axis by desiccation, isolated embryos were placed for various periods in the open air at 25°C and 55% relative humidity. Moisture content of whole embryos or embryonic axes (15 replicates) was calculated on a dry weight basis. Dry weight was obtained by oven drying the embryos or the embryonic axes at 105°C for 3 d.

Electrolyte Leakage Measurements

Solute leakage was determined by placing 4 embryonic axes in 10 ml distilled water at 25°C and measuring the conductivity of the medium with a K 220 CONSORT conductimeter after 2 h of soaking. Results are expressed as percentages of the total leakage from axes boiled for 5 min in water. They correspond to the means of 4 measurements $\pm$ SD.

Measurement of ACC Conversion to Ethylene

The conversion of 1-aminocyclopropane 1-carboxylic acid (ACC) to ethylene was measured by placing 3 embryonic axes in tightly closed 15-ml flasks containing 0.5 ml ACC solution (1 mM) in water. After 24 h incubation at 30°C, a 1-ml gas sample was taken from each flask and injected into a gas chromatograph (type 330, Girdel-France) equipped with a flame ionization detector and an activated alumina column for ethylene determination. Results are the means of 3 measurements $\pm$ SD, and are expressed as the percentages of ethylene produced by axes from freshly isolated (non-dehydrated) embryos (4.5 $\pm$ 2.1 nl per h and per axis).

Adenosine Phosphate Assays

Adenosine phosphates were extracted from one isolated axis according to Olempska-Beer and Bautz Freeze (1984). ATP, ADP and AMP contents of the extracts were measured using the bioluminescence method (Strehler and Totter, 1952) with a pico-ATP biophotometer. ADP and AMP were transformed into ATP as described by Saglio *et al.* (1979). The results obtained are expressed in nmol per g dry matter and are the means of 5–8 measurements $\pm$ SD.

The energy charge was calculated by the ratio (ATP+0.5 ADP)/(ATP+ADP +AMP) defined by Atkinson (1968).

Measurement of [³⁵S] Methionine Incorporation in Total Protein

Four embryonic axes were sterilized with 1% calcium hypochlorite for 10 min. After washing in sterilized water, they were incubated in 200 μl 0.6 kBq μl⁻¹ [³⁵S]methionine (Amersham, UK) for 2 h. At the end of the *in vivo* labelling period, organs were washed with sterilized water and their proteins were extracted as described by Salmen Espindola *et al.* (1994). Methionine incorporation into total protein was expressed as percentage of the [³⁵S]methionine uptake by the embryonic axes. Results presented correspond to the means of 3 measurements ±SD and to the percentages of methionine incorporation by axes from freshly isolated (non-dehydrated) embryos (21.9±3.7 % uptake).

Results

Sensitivity of Embryos to Desiccation

The mean moisture content of freshly isolated (non-dehydrated) embryos was about 120% (dry weight basis). It decreased rapidly during desiccation, which resulted in the loss of embryo viability (Fig. 1). Around 50% of the embryos were dead when their moisture content had fallen by about 60–65%. The critical moisture content at which viability was completely lost was around 25–30%.

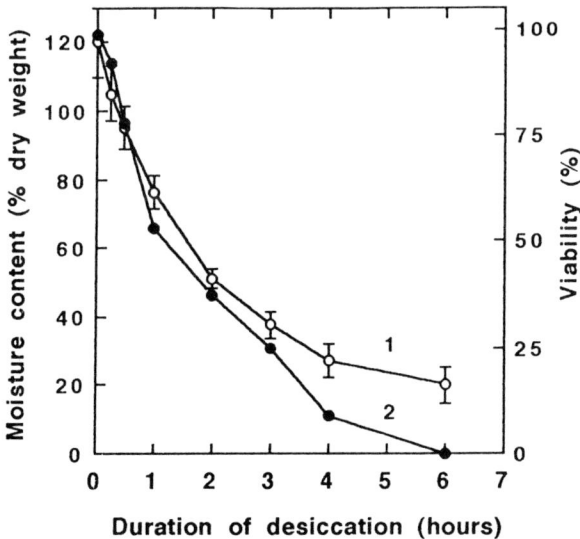

Figure 1. Effects of duration of desiccation on moisture content (1) and viability (2) of embryos. Means of 3 replicates ±SD (moisture content) and of 2 replicates (viability)

Electrolyte Leakage

Electrolyte leakage from embryonic axes increased progressively with decreasing moisture content (Fig. 2, curve 1). However, it was not a good indication of germination ability of embryos since it significantly increased when the moisture content fell to about 40%, i.e. when about 70% of the embryos had become unable to germinate (cf Fig. 1).

Conversion of ACC to Ethylene

Desiccation of embryos was also associated with a decrease in the ability of the embryonic axes to convert ACC to ethylene (Fig. 2, curve 2). This decrease occurred when the moisture content of axes reached less than 70%, i.e. after only 1 h of dehydration. Below 50% moisture content, the ACC-dependent ethylene production was almost nil.

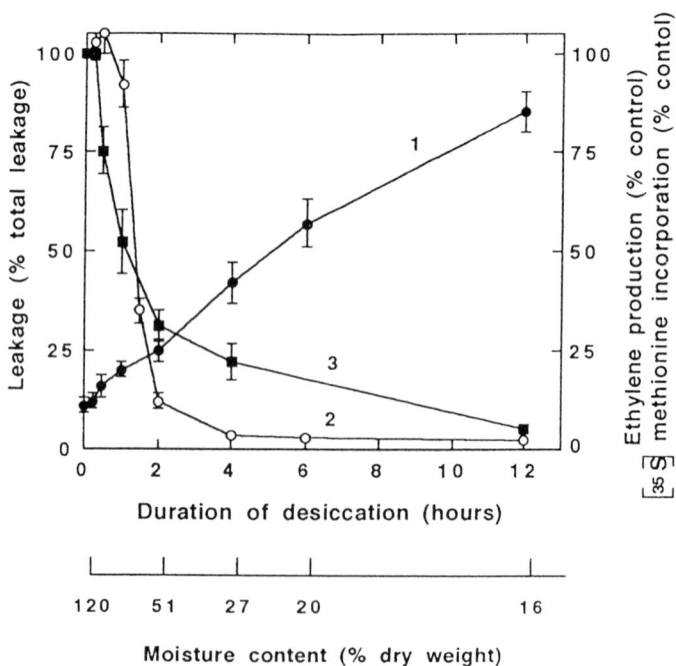

Figure 2. Effects of duration of desiccation of embryos and of corresponding moisture content of embryonic axes on electrolyte leakage (1), conversion of ACC to ethylene (2) and [^{35}S]methionine incorporation in total protein (3) by embryonic axes. Means of 3 (ethylene production and [^{35}S]methionine incorporation) or 4 (leakage) replicates ±SD

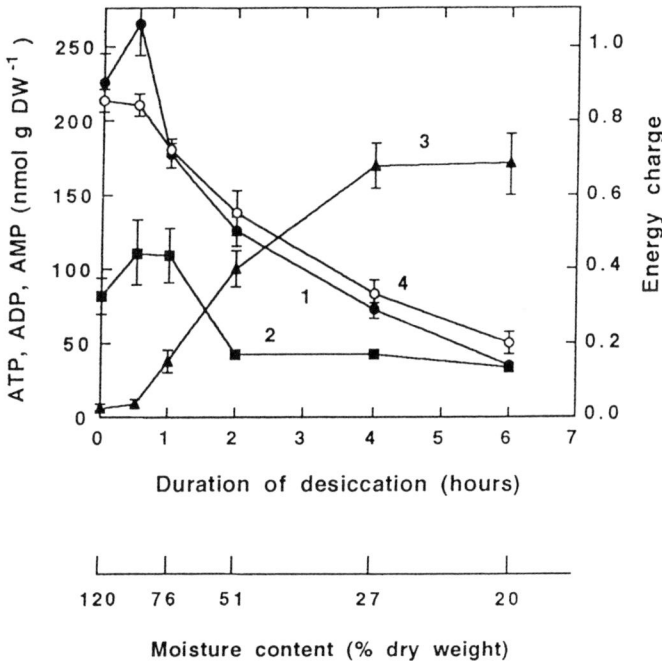

Figure 3. Effects of duration of desiccation of embryos and of corresponding moisture content of embryonic axes on ATP (1), ADP (2) and AMP (3) contents of embryonic axes, and on energy charge (4). Means of 5 to 8 measurements $\pm$ SD

Incorporation of $[^{35}S]$ Methionine in Total Protein

Dehydration resulted in a decrease in incorporation of $[^{35}S]$methionine in total protein in the embryonic axes (Fig. 2, curve 3), although the methionine uptake was not strongly reduced (data not shown). A small decrease in moisture content was sufficient to inhibit protein synthesis. The incorporation of methionine in proteins was reduced by about 25% and 75% when the moisture content fell to about 95% and 40%, respectively.

Energy Metabolism

The adenylate pool (ATP+ADP+AMP) of the embryonic axes was significantly reduced only by the sixth day of desiccation (data not shown), i.e. when all embryos had become unable to germinate (cf Fig. 1). But dehydration induced a clear decrease in ATP and ADP levels and an increase in AMP content (Fig. 3). These changes in adenine nucleotide levels resulted in a decline in the energy charge (Fig. 4). These changes in energy metabolism were noticeable by the first hour of desiccation.

Discussion and Conclusion

The moisture content at which viability of *Araucaria angustifolia* embryos is completely lost is about 25–30%. It is similar to that reported by Corbineau and Côme (1986, 1988) for *Mangifera indica, Symphonia globulifera* and *Hopea odorata* seeds and by Fu *et al.* (1990) for *Litchi chinensis* and *Euphorbia longan* seeds.

Embryos of freshly harvested *Araucaria angustifolia* seeds are metabolically active, though cytological observations by Salmen Espindola *et al.* (1994) have shown that most of the nuclei of the meristem zone of the radicle are quiescent. These embryos are characterized by high protein synthesis and energy charge. Dehydration results in decreases in various metabolic activities, among which loss of the ability to incorporate methionine in proteins is one of the earliest indicators of cell deterioration. Desiccation also induces an increase in electrolyte leakage and a decrease in the ability to convert ACC into ethylene. These results are consistent with the concept that cell membranes are progressively damaged by dehydration. Similar increase in solute leakage during desiccation was observed in recalcitrant seeds of *Quercus robur* by Poulsen and Eriksen (1992), silver maple and areca palm by Becwar *et al.* (1982), and *Landolphia kirkii* by Pammenter *et al.* (1991). Decrease in the ability to oxidize ACC into ethylene is also a good indication of membrane injury since the *in vivo* activity of ACC oxidase is known to depend on membrane integrity (Odawara *et al.*, 1977). Our results show that the decline in ACC oxidase activity is an earlier indicator of membrane deterioration than the increase in electrolyte leakage. As in orthodox seeds (Bewley, 1979), decreases in respiratory activity (Salmen Espindola *et al.*, 1994) and in ATP level and energy charge are associated with desiccation of *Araucaria angustifolia* embryos. However, reimbibition of embryos restores the energy charge to a value close to that measured in non-dehydrated embryos (data not shown). As the oxygen uptake (Salmen Espindola *et al.*, 1994), the energy charge cannot therefore be considered as a marker of damage induced by desiccation.

Our results show that a sequence of irreversible cellular and metabolic damage is associated with desiccation of *Araucaria angustifolia* embryonic axes, but it is difficult to know whether it is a cause or consequence of the loss of viability. Oxidative processes and free radical accumulation will be the subject of further investigations, since they are usually supposed to be involved in cell deterioration during dehydration (Bewley, 1979; Leprince *et al.*, 1990; Hendry *et al.*, 1992). In this point, study of the possibility of repairing injury upon rehydration through free radical- and peroxide-scavenging enzyme activities and/or antioxidant compounds might be required.

References

Atkinson, D.E. 1968. *Biochemistry* 7: 4030–4034.

Becwar, M.R., Stanwood, P.C. and Roos, E.E. 1982. *Plant Physiology* 69: 1132–1135.

Bewley, J.D. 1979. *Annual Review of Plant Physiology* 30: 195–238.

Bewley, J.D. and Black, M. 1994. *Seeds. Physiology of Development and Germination,* Second Edition, pp. 445. New York, London: Plenum Press.

Chin, H.F. and Roberts, E.H. 1980. *Recalcitrant Crop Seeds,* pp. 152. Kuala Lumpur: Tropical Press SDN.

Corbineau, F. and Côme, D. 1986. *Seed Science and Technology* 14: 585–591.

Corbineau, F. and Côme, D. 1988. *Seed Science and Technology* 16: 97–103.

Fu, J.R., Zhang, B.Z., Wang, X.F., Qiao, Y.Z. and Huang, X.L. 1990. *Seed Science and Technology* 18: 743–754.

Hendry, G.A.F., Finch-Savage, W.E., Thorpe, P.C., Atherton, N.M., Buckland, S., Nilson, K.A. and Seel, W.E. 1992. *New Phytologist* 122: 273–279.

King, M.W. and Roberts, E.H. 1979. *The storage of recalcitrant seeds. Achievements and possible approaches,* pp. 96. Rome: International Board for Plant Genetic Resources.

Leprince, O., Deltour, R., Thorpe, P.C., Atherton, N.M. and Hendry, G.A.F. 1990. *New Phytologist* 116: 573–580.

Odawara, S.A., Watanabe, H. and Imaseki, H. 1977. *Plant Physiology* 18: 569–575.

Olempska-Beer, Z. and Bautz Freeze, E. 1984. *Analytical Biochemistry* 140: 236–245.

Pammenter, N.W., Vertucci, C. and Berjak, P. 1991. *Plant Physiology* 96: 1093–1098.

Poulsen, K.M. and Eriksen, E.N. 1992. *Seed Science Research* 2: 215–221.

Priestley, D.A. 1986. *Seed aging. Implications for seed storage and persistance in the soil,* pp. 304. Ithaca, New York: Cornell University Press.

Roberts, E.H. 1973. *Seed Science and Technology* 1: 499–514.

Saglio, P.H., Daniels, M.J. and Pradet, A. 1979. *Journal of General Microbiology* 110: 13–20.

Salmen Espindola, L., Noin, M., Corbineau, F. and Côme, D. 1994. *Seed Science Research* 4: 193–201.

Strehler, B.L. and Totter, J.R. 1952. *Archives of Biochemistry and Biophysics* 40: 28–40.

79. Freezing Stress, Osmotic Strain, and their Viscoelastic Coupling

R.J. WILLIAMS, H.T. MERYMAN, M.St.J. DOUGLAS and P.M. MEHL

Transfusion and Cryobiology Research Program, Naval Medical Research Institute, 8901 Wisconsin Avenue, Bethesda, MD 20889-5607, USA

Abstract

This paper attempts to explain why, against seemingly overwhelming odds, certain organisms can tolerate freezing to low temperatures, why some attempts to imitate this in nontolerant organisms have been partially successful, and why narrowly constrained rates of drying, cooling and heating are critical to that success. Below the ice point, temperature lowering reduces the vapour pressure of water, and the semipermeability of membranes imposes a large mechanical or 'osmotic' stress on cells which undergo a deformation, or strain, borne principally by the cytoskeleton. Nonetheless, even lethal stresses can have their effects deferred for minutes or even hours, providing a 'window' for cryopreservation. We propose that because the connection between stress and strain is not elastic but viscoelastic, the strain is spread over time and its intensity diluted. Beyond limits of time and intensity, relaxation in the cytoskeleton will become irreversible. We offer an Avrami model in which an Arrhenius expression models the temperature effect and elastic moduli supply an activation energy, to provide a rational basis for the development of cryopreservation techniques.

Introduction

Our ancestors have, over the last two to three billion years, provided us with a homeostatic planet. They have buried the toxic salts and the excess carbon. They have produced a diversity so vast, complex and generally benign that we as biologists can wander across it for all of our careers without discovering that it has edges. But it does have edges beyond which the world is not benign; the edge which concerns us here is reached when the temperature falls below the ice point and the laws of biochemistry give way to the laws of physics. Applying physical laws to biological function may be difficult, but there are no known exceptions (Thompson, 1942). Unlike chemistry, physics has not evolved and cryobiologists who ignore or deny its laws do so to their detriment (Muldrew and McGann, 1994).

Freezing Stress

It must be emphasized that the stress and strain referred to in the title are not metaphors, but used the way that physicists use them, a deforming force and a deformation, respectively. As the growth of extracellular ice increases solution concentration, intracellular water can leave the cell but its volume cannot be

R.H. Ellis, M. Black, A.J. Murdoch, T.D. Hong (eds.), Basic and Applied Aspects of Seed Biology, pp. 723–735.
© *1997 Kluwer Academic Publishers, Dordrecht. Printed in Great Britain.*

made up by other materials because of membrane semipermeability with the result that osmotic work is done upon cells. In 1877, the botanist, Pfeffer, first showed this by measuring the hydrostatic head required to equilibrate sucrose solutions and pure water across a semipermeable barrier (Smith, 1962) From Pfeffer's data, van't Hoff demonstrated that the solutes follow Boyle's law for gases (Hammel and Scholander, 1976) and if this is taken literally, the potential forces are extreme. Temperature reduction of one degree in frozen solutions has the same effect on vapour pressure as adding about half a mole of solute. In terms of chemical potential, this is about 10 atm (1 MPa) per degree; or, in terms of osmotic or hydrostatic potential across a semipermeable membrane, a column of water over 100 m high per degree. Experience suggests that these huge forces are not actually realized, since many seeds and plants and some animals can be frozen, or even dried, if they have become properly adapted. The changes necessary for adaptation are to a degree understood but not necessarily accessible to the researcher attempting to confer artificial tolerance in recalcitrant seeds or their embryos.

Cryopreservation confers two effects of low temperature: (1) the Arrhenius effect, that temperature lowering reduces exponentially the rates of catabolic reactions; and (2) vitrification or glass formation, that low temperatures so increase the viscosity of solutions that their chemical potentials cannot be achieved in time periods of practical significance. Unfortunately, most active living things are such dilute solutions that the benefits of vitrification accrue only at cryogenic temperatures and thus cryopreserved materials must be brought across broad cooling and warming regimes where ice growth and the resulting osmotic forces will prevail. Indeed, between the ice point and the glass transition temperature, temperature lowering increases, rather than decreases the rate of injury (Takahashi, 1977); this is the principle freezing lesion and failure to address it is the source of essentially all failures to cryopreserve.

The obvious way to prevent ice damage is to prevent ice formation. In order to suppress ice formation completely or at least to restrict it to tolerated levels at cooling rates slow enough to be near equilibrium, it would be necessary to replace over half the extra- and intracellular water with a liquid which does not freeze, such as glycerol, dimethyl sulphoxide or special vitrification cocktails (Fahy *et al.*, 1991). This has only worked for red cells (Meryman and Hornblower, 1971) and a few other types of cells (Takahashi *et al.*, 1986) or small groups of cells (Rall and Fahy, 1984) but has not yet found use with complex tissues or organs (Fahy *et al.*, 1990). Alternatively, one could cool rapidly (Luyet, 1937; Meryman and Kafig, 1954), but with few exceptions heat transfer limitations make this practicable only for specimens of dimensions suitable for electron microscopy. A middle road has been found and exploited for essentially all practical cryopreservation procedures now in use in which low 'non-toxic' concentrations of a cryoprotectant such as glycerol or dimethyl sulphoxide allow partial survival over relatively narrow ranges of cooling and heating rates. The reasons for this remain unclear, but an emerging hypothesis will be developed below.

Over thirty years ago, when we were first attempting to develop practical methods for freezing whole units of red cells, the two predominant views of freezing injury were: dehydration (Luyet and Gehenio, 1939); and electrolyte toxicity (Lovelock, 1954). Both involved changes in chemical potential to values below or above some threshold. To our dismay, our experiments disallowed both. Solute (and solvent) concentration at the point of injury varied widely in different types of cells; injury resulted when volume reduction through water loss had reached a point of no return at about 1/3 of the normal water content (Meryman, 1968; Williams and Meryman, 1965; Williams, 1970). This observation, inauspiciously dubbed 'minimum volume', has been confirmed since in a number of plant and animal models, though the value of the limit varies. At that time, we had no plausible model for how a mechanical stress, as opposed to a chemical potential ('toxicity') could be injurious. We did, however, have a practical solution to our immediate problem: replace intracellular water with materials that didn't freeze or otherwise prevented cell volume reduction. Meryman's *gedanken* experiment was to 'fill the cell with rocks'; for human red cells, glycerol sufficed (Meryman and Hornblower, 1971).

'Rocks', which must be soluble enough to suppress crystallization, abound in the world of drying- or cold-tolerant organisms. In orthodox seeds, the mature embryo cells may be (from the viewpoint of the physical chemist) supersaturated aqueous solutions of at least 80% (w/w) sugars admixed with proteins, which through some magic do not crystallize, and which are at or near their glass transition temperatures under ambient conditions (Williams and Leopold, 1989; Leopold *et al.*, 1992; Koster, 1991). Bark cells in cold-hardened poplar trees, though dilute below saturation, use a version of the same manoeuvre (Hirsh *et al.*, 1985). They contain perhaps a 20% sugar solution, but during freezing, intracellular water distills to extracellular ice, shrinking the cells. This raises the intracellular concentration to 70%; cell shrinkage, hindered by a stiff cell wall, is still within the minimum volume limits, and this concentrated solution vitrifies at $-28°C$, a temperature the trees otherwise tolerate. Both of these systems are remarkably stable and will suffice for example.

Osmotic Strain

Measuring the volume of cells in the presence of ice can be nearly unachieveable. Since the days of Kylin (1917) it has been routine to test cold tolerance by exposing cells at temperatures above freezing to solutions whose freezing point depression is known, as the injuries produced are remarkably similar. While plant cells plasmolyze very differently (Siminovitch *et al.*, 1967) because the air–water interfaces are replaced by water–water interfaces, the cells are clearly visible, separate from the cell wall and form shapes which are easily analysed according to Archimedes' rules (Höfler, 1917), simplifying the stress–strain analysis. In every type of cell we have examined (Williams, 1981), hyperosmotic stress, Π, produced a reduction in cell volume, V, which followed the Boyle–van't Hoff equation:

$$(V-b)/(V_o-b) = \Pi_o/\Pi$$

toward the osmotically insensitive volume, b, at concentrations near isotonic, but deviated sharply near the point of injury, remaining larger than expected. By far the most dramatic discrepancy was seen in cortical parenchyma of the cold hardened dogwood, *Cornus florida* L. (Williams and Williams, 1976). These cells withstood freezing to $-26°C$, hyperosmotic exposure to $CaCl_2$ solutions with freezing points as low as $-23°C$, but did not plasmolyze in solutions whose freezing point depression exceeded $-4°C$. This observation was confirmed in a few freezing experiments.

So long as this deviation persists, the process is reversible, and cells survive release of the osmotic stress. However, at the point of injury, the volume discrepancy disappears and the $\Pi-V$ signature returns to or near the expected value. At this point reduction in osmolality is accompanied by cell injury and volumes below those seen during compression (Williams and Hope, 1981). Because these curves so resemble the classic stress–strain curves of mechanical analysis, up to and beyond the point of plastic deformation or 'yield point', we have adopted that interpretation. When these experiments were first done in the early 1970's, we were still convinced, on the basis of monolayer data, that the membrane was the source of the resistance, a notion eventually dispelled by Prof. John McGrath (personal communication, 1983)). More recent evidence has led us to believe that an irreversible or plastic deformation of the cytoskeleton is the principle lesion in osmotic stress injury and that the loss of bilayer is secondary. The cytoskeleton after freezing is crumpled (Morisset *et al.*, 1993) while the lipids remain in bilayers even though separated from their cytoskeletal elements (MAPs).

Viscoelasticity

Recalcitrant seeds or embryos can be defined as lacking adequate rocks and thus susceptible to osmotic stress. In addition to protective additives, successful attempts to cryopreserve them have generally involved rather rapid drying and cooling steps whose rates were crucial to success. Our cumulative experience has led us to suggest that there are constraints on the time during which an embryo can withstand a specific level of stress. We will develop this element of time because we hypothesize that it provides the 'window of opportunity' which permits cryopreservation, and that it is a consequence of viscoelasticity.

Viscoelasticity is an elasticity which changes with time, exhibiting the properties of both a fluid and a solid. While intuitively simple, the mathematics required is difficult and measurements of viscoelastic properties of biological materials have been few. Some measurements of the elastic properties of biological materials can be considered definitive (Twombly *et al.*, 1994; Discher *et al.*, 1994) but in most the duration was insufficient to allow a clear distinction between a true elasticity and a viscoelastic behavior. The report by Gusta *et al.* (1983) of the depression of low temperature exotherms (homogeneous nucleation) to below −50°C in hardwoods continually exposed to low temperatures and Levitt's (1986) examination of the reestablishment of turgor in stressed and isolated cabbage leaves are unquestionable examples of a sizeable viscoelastic component in cell walls. More recently, the viscoelasticity of cytoskeleton has been preliminarily but directly explored by atomic force microscopy (Putnam *et al.*, 1994).

Viscoelastic moduli are defined in terms of a frequency, often using versions of the Eyring equation for a thermodynamic overview (Starkweather, 1990; 1991; 1993), or of the free volume and related models (Perez, 1994) or percolation models (Hunt, 1992) at the atomic or molecular level. As with all footnotes to Boltzmann, these leave one disquieted, longing for the way science used to be, but the approach has revolutionized the plastics industry by allowing simple predictions and extrapolations of curing and molding rates; because their example has proved generally so valuable and instructive, we have adopted it. While we will study the rate at which a process can proceed, our interest is the persistence of living states very far from equilibrium. Thus, we use the reciprocal of frequency, the time constant, in our analysis.

The standard automobile suspension serves as a useful metaphor for viscoelasticity. It contains an elastic spring in parallel with a viscous shock absorber whose combined action is to dilute bumps in the road by spreading the shock out over time: the stress has been deflected to reduce the intensity of the strain. Energy is conserved and the piper ultimately paid but, in the limit, the system fails suddenly. To substantiate this model, we will show that the freezing stress produces a measurable osmotic strain which eventually forces a plastic or viscoelastic deformation of the sort commonly seen in materials testing, and that it is the irreversibility of this deformation which results in cell injury. Finally, we will model this injury, using standard equations from materials science, and explore how they might contribute to a better understanding of the requirements for cryopreservation.

A simple paradigm for stress failure in a composite involves having the stress supply the 'activation energy' in an Arrhenius model. This model has worked in a simple qualitative way, but a deeper examination of mechanism requires an array of ancillary assumptions (Meakin, 1991). A dozen years ago, Williams and Takahashi (1983) adapted this to model kinetic data of osmotic stress induced cytolysis in unfertilized sea urchin eggs, and found that it simply didn't work. Hirsh eventually realized that these data could be fit to a set of standard models of either viscoelasticity or crystallization and, for its simplicity, chose the Johnson–Mehl–Avrami equation (Williams *et al.*, 1993a,b):

$$\ln N_o/N = (t/\tau)^n$$

where $N/N_o =$ the ratio of surviving cells, $t =$ time, $\tau =$ the time constant and $n =$ the Avrami constant which reflects the number of dimensions in which the change is taking place.

The time constant is treated as an Arrhenius function of temperature, in which the osmotic stress supplies part of the activation energy:

$$\ln \tau = \ln A + \Delta H_o / kT - m \ln P/P_o$$

where A is a pre-exponential constant, ΔH_o is an intrinsic activation enthalpy, k is Boltzmann's constant, T is the absolute temperature and m is a modulus, or efficiency, relating the strain of cytolysis to the osmotic stress and expressed as a ratio of vapor pressure of the osmolyte to isotonic, P/P_o.

What follows is a materials science perspective on the osmotic stress/strain relationships during drying or freezing of living cells and some preliminary measurements on a very simple type of cell, the human red cell. We report two experiments. The first quantifies the stress–strain relationship and the second the resultant viscoelastic relaxation (= death). Living tissues are the most complex composites ever developed, containing as viscoelastic elements plant cell walls, organelles, cytoskeleton and membrane, often quite differentiated. Our assumption is that we can learn much from those who have wrestled with fibreglass or carbon fibre composites and turned them into products ranging from tennis racquets to Stealth aircraft.

Methods and Materials

Human erythrocytes were obtained by conventional blood bank procedures. Cells were used within 5 d of collection. The plasma was removed by centrifugation 2.5 min at 1000g and the red cells resuspended in an excess of isotonic sodium chloride containing phosphate buffer at pH 7.2.

To determine cell stress–strain relationships, we measured cell volume as a function of osmotic stress. A dilution technique was used, based upon the fact that cells suspended in a hyperosmotic solution lose water as their volume is reduced. The resulting dilution of the suspending solution reflects the magnitude of the change in volume. The cell suspension was centrifuged and the osmolality of the supernatant measured in a freezing point osmometer (Advanced Instruments). The specific gravity of the packed cells was determined by weighing 10 ml in a volumetric flask. Two ml aliquots of the packed cells were placed in weighed test tubes and the net weight of cells determined with an accuracy of ± 10 mg. A series of independent measurements of the true haematocrit using the isotope dilution technique was used to determine the relative volume of cells and extracellular salt solution. Four ml of a series of 27 hyperosmotic salt solutions was then added. The tubes were then again weighed to determine the net weight of the added solution. The suspension was centrifuged and the osmolality of the supernatant determined, using the relationship:

$$\% \text{ isotonic volume} = 100 - (H-C-E)/AB \times 100$$

where: $H = (CD+EF)/G$, A = initial haematocrit of packed cells corrected for trapped plasma, B = initial volume of packed cells, C = volume of packed cell supernatant, D = osmolality of packed cell supernatant, E = volume of hypertonic diluent, F = osmolality of diluent, G = osmolality of supernatant after mixing and H = total extracellular volume after mixing.

The intracellular osmolality of the cell was determined by haemolyzing an aliquot of the hypertonic cell suspension and measuring the osmolality of the mixture. The apparent internal osmolality of the cells could be calculated by:

$$\Pi_c = (V_t \Pi_t - V_s \Pi_s) V_c$$

where: Π_c = intracellular osmolality, V_t = total sample volume, Π_t = osmolality of whole sample (haemolyzed), V_s = extracellular volume, Π_s = extracellular osmolality and V_c = volume of cells. The reproducibility of these measurements by freezing point depression left much to be desired. To spot-check the values, cells were centrifuged at $14000g$, carefully blotted and measured in a vapor pressure osmometer. Using [131I]albumin, we estimated that the extracellular space was 3.5% of packed cell volume, and calculated the intracellular osmolality as:

$$\Pi_c = 0.965\Pi_t + 0.035\Pi_s$$

To determine the viscoelastic strain we measured the time constants for cell survival. Ten microliters of a 1% cell suspension in hyperosmotic NaCl solution were placed between coverslips and examined on a thermostatted stage (Linkam) held at temperatures from 5°C to 45°C on an interference microscope (Zeiss/Jena). The microscope objective was a $40 \times$, N.A. 0.5 Cassegrainian long working distance objective and about 20 cells in the field of view could be examined on a video monitor at approximately 3000 diameters magnification. The lytic process was recorded on a time lapse videotape recorder and analyzed later. The data presented in this paper indicate duration of stress in seconds to the beginning of visible injury to individual cells. The regression analysis values for n and τ are from cytolytic data, and for ΔH_0 and m from regression analysis of isopiestic or isothermal sets of values of τ, respectively. For this preliminary account, 34 experiments were completed and analysed before the work was interrupted by equipment failure and relocation of the Laboratory.

Results

Osmotic Stress and Strain

Sixteen complete osmometric experiments were conducted, with four replicate measurements made on all samples. Replicate measurements did not differ by more than 0.5%, making these among the most precise volumetric measurements ever reported. Individual units showed variations from this general curve, but these were highly repeatable and characteristic of the particular sample. In Figure 1, a curve representing 4 determinations of a single unit has been constructed. The regression line projects to a b intercept of 44.3%. Assuming a nonaqueous cell volume of 30% of isotonic volume, this implies a discrepancy of 14%, consistent with the estimates of other investigators, generally based on data from less hypertonic conditions. It is notable that at higher osmolalities the curves deviate considerably from a smooth curve. The latter effect is heightened by the compression of the ordinate at high osmolalities, but the deviations are significant and because of the close agreement of multiple measurements cannot be dismissed as experimental error.

The observed differences between intra- and extracellular osmolality measured on cells in hyperosmotic suspension are shown in Figure 1. These experiments rule out the possibility that the disparity is the result of some 'osmotic coefficient', a correction for the difference between observed and expected volume at a given osmolality: these data compare osmolality directly.

These data, assuuming a constant surface area during stress, allow us to calculate a Hooke's Law elastic modulus:

$$(\Delta \Pi_e - \Delta \Pi_o) / \Pi_e = [E V_o / \Pi_o (V_o - b)] \Delta V / V_o$$

where E is the elastic modulus, and the subscripts e and o are the stressed and initial values, respectively, in milliosmolar units. The value is 124.64 ± 4.67 (SEM) (300 to 400 kPa).

Viscoelastic Strain

The practicable length of experiment using these methods ranged from about 10 min to one working day. Thus, the accessible combination of temperature and osmotic stress in this hyperexponential process was along a rather narrow cusp.

Three steps can be distinguished (Fig. 2). First the cell loses volume as the result of osmotic stress. Next, it loses apparent surface area, becoming a spherical vesicle. Finally, after some period from a few seconds to many minutes, it loses semipermeability and its haemoglobin content. In addition to a slow Fick's Law diffusion which reduces the transmembrane osmotic potential, there appear to be two JMA processes proceeding simultaneously whose relative contributions to cell loss depend upon the specifics of temperature and osmotic stress, with the fast process favoured by higher temperatures and higher levels of osmotic stress. The slow process has an Avrami exponent of one or less and reflects an ordinary Arrhenius mechanism. The fast process has an Avrami exponent of approximately two, and is illustrated in Figure 2.

Figure 3 illustrates the kinetics of the rapid JMA process in a sample exposed to 1800 mosm NaCl and held at 45°C. The time constant surprisingly approaches half an hour. Though this curve is continuous, there is an initial period of 8 min during which only one cell undergoes lysis.

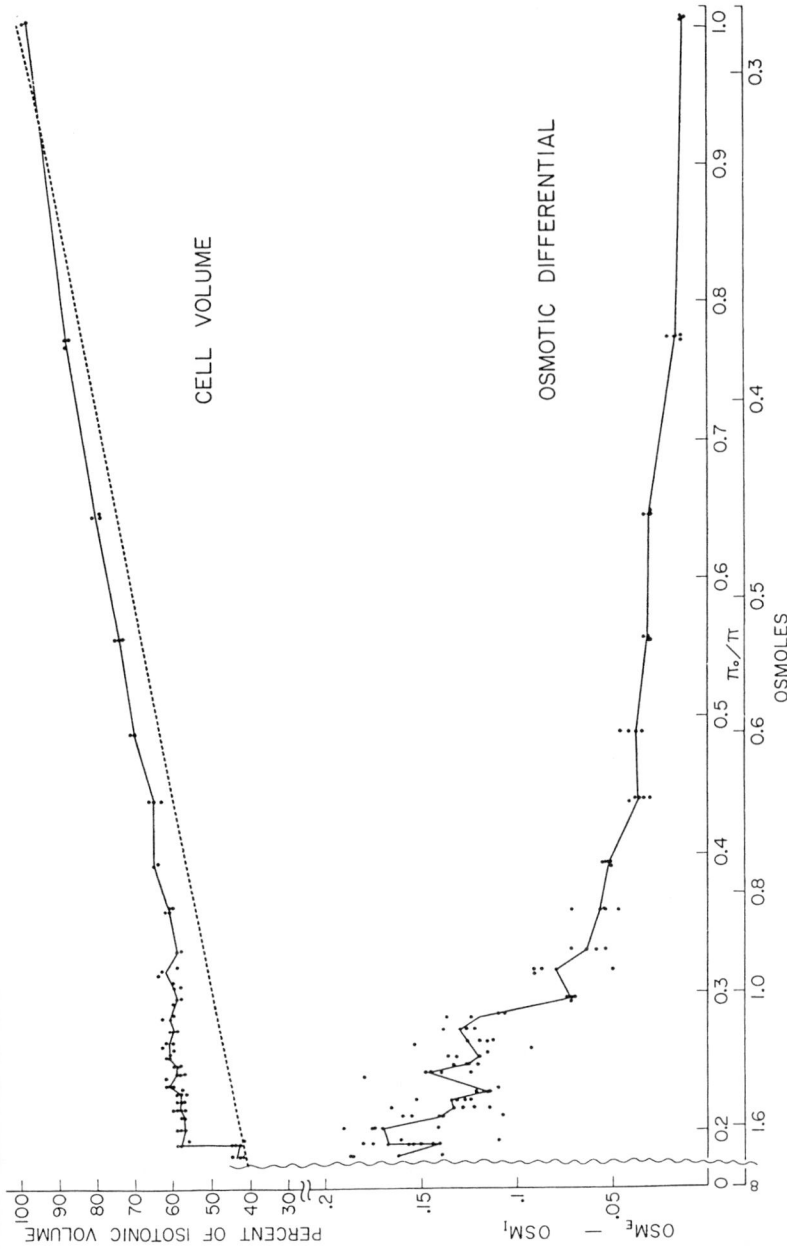

Figure 1. The volume of red cells as a function of hyperosmotic stress. Top. Cells remain progressively larger than predicted at high concentrations. Bottom. This volume discrepancy is reflected in an osmotic potential difference between extra- and intracellular solutions

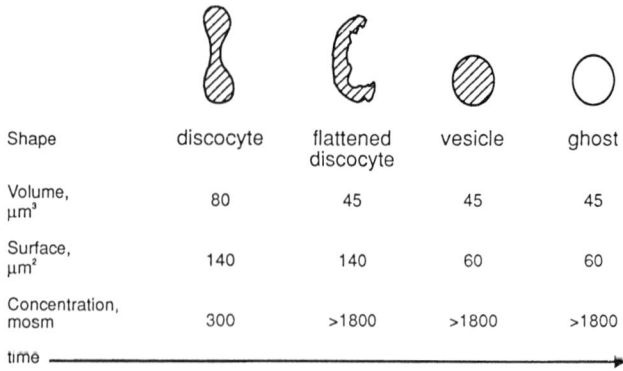

Shape	discocyte	flattened discocyte	vesicle	ghost
Volume, μm^3	80	45	45	45
Surface, μm^2	140	140	60	60
Concentration, mosm	300	>1800	>1800	>1800
time				

Figure 2. The progress of lysis in hyperosmotically stressed red cells

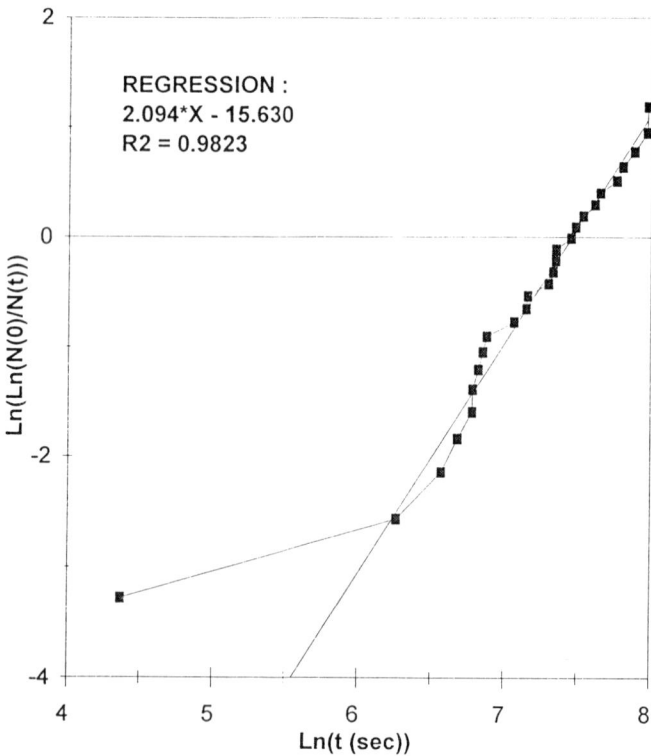

REGRESSION :
2.094*X - 15.630
R2 = 0.9823

Figure 3. The progress of lysis in a population of red cells held at 45°C in 1800 mosm NaCl

Table 1. Time constants, τ, Arrhenius constants, ΔH_0, and visoelastic moduli, m, as a function of temperature and intensity of osmotic stress

Temp, °C	Osmotic stress, mosm					m
	1800	2200	2400	2600	2800	
45	1632±95	1080±11	546±164			0.51
35		2145±992	2898±265			0.45
25	8417	8075±1300	6280±673			0.13
15			58 247	4831		3.73
5				23 643		
ΔH_0	−38.6	−47.1	−72.1	−91.4		

Table 1 summarizes the results of all experiments. Values of τ are given for each combination of temperature and osmotic stress. Values for the ΔH_o are given at the bottom of each column and values for m at the right of each row. The Arrhenius constants are unexceptional, with a 'Q_{10}' of 2 to 4. The elastic constants reflecting the coupling between stress and strain are difficult to interpret, but with the exception of the last (based on only two measurements) are within the constraints of $0 < m < 1$ and thus can be considered efficiencies.

Discussion

The red cell volumetric experiments are the only ones we are aware of which show directly the magnitude of the osmotic discrepancy in animal cells. This adds the cytoskeleton to the plant or microbial cell wall known to resist osmotic stress. The modulus, 3–4 atmospheres, is over a thousand times too large to be accounted for by the lipid bilayer (McGrath, unpublished) and it must therefore be the result of mechanical resistance in the cytoskeleton, and possibly from dissolved protein as well.

The viscoelastic relaxation data support earlier work on osmotic lysis in sea urchin eggs and in red cells undergoing 'thermal shock' (Williams *et al.*, 1993a, b). Because the loss of intracellular soluble protein, haemoglobin, occurs subsequent to the collapse of the cell, loss of membrane semipermeability, traditionally cited as the cause of cytolysis, instead appears to be the result of it. And because $n \sim 2$, we propose that the fast lytic process in the current isothermal experiments represents a two-dimensional stress failure in the cytoskeleton. The red cell cytoskeleton is a network of filamentous protein, anchored to proteins which penetrate the bilayer, with interlocking hexagonal and pentagonal rings – an elaboration of the geodesic dome. Entropic forces cause it to be reflected back upon itself rather like a drying fishnet. This folding is stabilized by the presence of specific crosslinking interactions (Vertessy and Steck, 1989). The bilayer is stretched across this framework, and on this microscopic scale where surface forces prevail, more resembles a coat of drying paint than a tent supported on poles.

It is easy to conceive how this structure, stretched to its limits by hypo-osmotic exposure, might rupture like a balloon, a process called 'fragility'. Less simple is conceiving how compressive forces, again essentially isotropic, can cause destruction. We do know that cytoskeletal elements can be extremely stiff (Williams, 1991) and that their disassembly can enhance cryosurvival (Morisset *et al.*, 1993).

The activation energies implied by the time constants were unexceptional for biological results. Nor are the trends in ΔH or m unexpected. One potential complicating factor not taken into account is a first order phase change near +12°C which has been inferred from a variety of experiments on many different types of cells but never satisfactorily identified. We suspect that this, too, may be attributable to the cytoskeleton and not the bilayer. Attempts to attribute the failure to the hyperosmotic stress give one value for m which is outside the acceptable range of 0 to 1, but it is derived from two experiments only. It is the long latency seen in Figure 3 which gives hope to cryobiologists: it is the 'window of opportunity' through which we must pass before the time runs out.

At this early stage, the structural details are less important than the testable model. If we can specify a relationship between temperature, intensity of stress, time and injury in a specific cell or tissue, it should be possible to find an optimal path to cryopreservation. While even this optimal path may not suffice in specific cases, we should be heartened by the degree of success already achieved.

References

Discher, D.E., Mohandas, N. and Evans, E.A. 1994. *Science* 266: 1032–1035.
Fahy, G.M., Khirabadi, B.S. and Mehl, P.M. 1991. *Cryobiology* 28: 511–512.
Fahy, G.M., Sauer, J. and Williams, R.J. 1990. *Cryobiology* 27: 492–510.
Gusta, L.V., Tyler, N.L. and Chen, T.H-H. 1983. *Plant Physiology* 72: 122–128.
Hammel, H.T. and Scholander, P.F. 1976. *Osmosis and Tensile Solvent.* Berlin: Springer-Verlag. 133 pp.
Hirsh, A.G., Williams, R.J. and Meryman, H.T. 1985. *Plant Physiology* 79: 41–56.
Höfler, K. 1917. *Berichte des deutsches botanisches Gesellschaft* 35: 710–726.
Hunt, A. 1992. *Solid State Communications* 84: 701–704.
Koster, K.L. 1991. *Plant Physiology* 96: 302–304.
Kylin, 1917. *Berichte des deutsches botanisches Gesellschaft* 35: 370–384.
Leopold, A.C., Bruni, F. and R.J. Williams. 1992. In: *Water and Life* pp.161–169 (eds. Somero *et al.*) Berlin: Springer-Verlag.
Levitt, J. 1986. *Plant Physiology* 82: 147–153.
Lovelock, J.M. 1954. *Biochemical Journal* 56: 265–270.
Luyet, B.J. 1937. *Biodynamica* 1(29): 1–14.
Luyet, B.J. and Gehenio, P.M. 1939. *Life and Death at Low Temperatures.* Normandy, MO: Biodynamica.
Meakin, P. 1991. *Science* 252: 226–234.

Meryman, H.T. 1968 *Nature* 218: 333–336.

Meryman, H.T. and Hornblower, M. 1971. *Bibliographia Hematologica* 38 (II): 244–249, Basel: Karger.

Meryman, H.T. and Kafig, E. 1954. *Proceedings of the Society for Experimental Biology and Medicine* 90: 587–589.

Morisset, C., Gazeau, C., Hansz, J. and Dereuddre, J. 1993. *Protoplasma* 173: 35–47.

Muldrew, K. and McGann, L.E. 1994. *Biophysical Journal* 66: 532–541.

Perez, J. 1994. *Journal of Food Engineering* 22: 89–114.

Putnam, C.A.J., van der Werf, K.O., de Grooth, B.G., van Hulst, N.F. and Greve, J. 1994. *Biophysical Journal* 67: 1749–1753.

Rall, W.F. and Fahy, G.M. 1984. *Nature* 313: 573–575.

Siminovitch, D., Gfeller, F. and Rheaume, B. 1967. In: *Cellular Injury and Resistance in Freezing Organisms*. Sapporo, Institute for Low Temperature Science, pp. 93–118.

Smith, H.W. 1962. *Circulation* 26: 987–1012.

Starkweather, H.W., Jr. 1990. *Macromolecules* 23: 328–332.

Starkweather, H.W., Jr. 1991. *Polymer* 32: 2443–2448.

Starkweather, H.W., Jr. 1993. *Macromolecules* 26: 4805–4808.

Takahashi, T.A. 1977. *Contributions from the Institute of Low Temperature Science* B19: 1–48.

Takahashi, T.A., A.G. Hirsh, E.F. Erbe, J.B. Bross, R.L. Steere and R.J. Williams. 1986. *Cryobiology* 23: 103–115.

Thompson, D.W. 1942. *On Growth and Form.* (1961 Abridged edition, ed. J.T. Bonner) University Press, Cambridge, 346 pp.

Twombly, B., Cassell, B. and Miller, A.T. 1994. *Proceedings of the Twenty-Third North American Thermal Analysis Society Conference* 23: 288–293.

Vertessy, B.G. and Steck, T.L. 1989. *Biophysical Journal* 55: 255–262.

Williams, J.M. and Williams, R.J. 1976. *Plant Physiology* 58: 243–247.

Williams, R.C., Jr. 1991. *Biophysical Journal* 59: 59a.

Williams, R.J. 1970. *Comparative Biochemistry and Physiology* 35: 145–161.

Williams, R.J. 1981. In: *Analysis and Improvement of Plant Cold Hardiness* pp 89–115, (eds. C.R. Olien and M.N. Smith,) CRC Press, Boca Raton, FL.

Williams, R.J. and Hope, H.J. 1981. *Cryobiology* 18: 146–154.

Williams, R.J. and Leopold, A.C. 1989. *Plant Physiology* 89: 977–981.

Williams, R.J. and Meryman, H.T. 1965. *Cryobiology* 1: 317–323.

Williams, R.J. and Takahashi, T.A. 1983. *Cryobiology* 20: 716–717.

Williams, R.J., Takahashi, T.A. and Hirsh, A.G. 1993a. *Thermochimica Acta* 203: 493–501.

Williams, R.J., Hirsh, A.G. Meryman, H.T. and Takahashi, T.A. 1993b. *Journal of Thermal Analysis* 40: 857–862.

80. Characterization of Intracellular Glasses in Bean Axes with Relevance to Storage Stability

O. LEPRINCE[1] and C.W. VERTUCCI[2]

[1]Brassica and Oilseed Research Department, John Innes Centre, Norwich NR4 7UH, UK; [2]USDA-ARS, National Seed Storage Laboratory, Fort Collins CO 80521, USA

Abstract

Although evidence indicates that water in seed tissues exhibits glass properties, the physiological role of glasses in storage stability is still conjectural. We believe that this shortcoming is due to our lack of understanding the implications of the glassy state. Using differential scanning calorimetry we studied the thermal behaviours of glass transitions in axes of bean at temperatures between $-120°C$ and $+120°C$. Three types of thermal behaviours associated with the glass transition were observed in tissues with water contents between 0.03 and 0.45 g H_2O/g DW. The appearance, the temperature and the amount of energy released during these transitions were dependent on the tissue water content. Below 0.03 and above 0.45 g H_2O/g DW, no glass transitions were observed. Because water content influences the glass–liquid transition behaviour, a Tg/water content curve is not entirely satisfactory to describe the glassy state of water in dry seeds. We present evidence that the non-equilibrium nature of glasses, the thermal history during glass formation and the complexity of molecular interactions between the glass-forming solutes and water are important parameters to characterize the intracellular glass properties. These parameters are also relevant with seed storage stability.

Abbreviations: DSC, differential scanning calorimetry; g/g, g H_2O/g DW; Tg, glass–liquid transition temperature; WC, water content.

Introduction

Research into seed storage has long established the crucial role of the physical state and properties of water upon which rely the nature and kinetics of degradative reactions in seed tissues. The physical state of water has been characterized by the motional and thermal properties of intracellular water using two distinct thermodynamic theory standpoints. Degradative reactions are viewed to be controlled by the chemical potential of water or its 'availability' for chemical reactions which can be described by equilibrium thermodynamics (Vertucci, 1992; Vertucci and Roos, 1993; Nelson and Labuzza, 1994). Alternatively the state of water has been considered by invoking the concept of glasses in which slow molecular diffusion due to the extremely high viscosity renders chemical reactions improbable (Slade and Levine, 1991; Williams *et al.*, 1993; Leopold *et al.*, 1994). In order to assess the glassy state of water, one must refer

R.H. Ellis, M. Black, A.J. Murdoch, T.D. Hong (eds.), Basic and Applied Aspects of Seed Biology, pp. 737–746.
© *1997 Kluwer Academic Publishers, Dordrecht. Printed in Great Britain.*

to the non-equilibrium thermodynamics principles.

Despite the experimental evidence that water in dry seeds exhibits some of the properties representative of a glassy state (Williams, 1994), the evidence for a physiological role of intracellular glasses is still conjectural. Data on the implication of glasses in desiccation tolerance are contradictory (Bruni and Leopold, 1992; Sun *et al.*, 1994; Williams and Leopold, 1995). The glass model is not entirely satisfactory to link the water properties with storage stability because the Tg/WC combination does not predict the experimental evidence that drying to extremely low WC results in increased ageing rates (Vertucci and Roos, 1993). In addition, in soyabean and maize at very low WC, there is a poor correspondence in the relationships between WC and experimental Tg values and between WC and maximum storage temperature values derived from the viability equation (Sun and Leopold, 1994). Clearly the applicability of the glass principles to the storage technology of orthodox seeds requires further investigation.

We believe that most of the glass principles which have been extensively studied in the food, polymer and materials sciences have not sufficiently permeated into seed storage science. The objective of this paper is to identify beyond the glass transition temperature (Tg) additional aspects characteristic of intracellular glasses which may give better insight into the nature of glasses in dry seeds. Using DSC as an experimental tool and theoretical principles on glasses as guidelines, we have investigated the thermal behaviour of glass transitions in axes of bean with water contents between 0 and 1 g/g and temperatures between −120 and +120°C.

The Relevance of the Non-Equilibrium Nature of the Glasses in Seed Storage Stability

A glass at temperatures below Tg is in a non-equilibrium state. This concept is of fundamental importance when studying the effects of time and external conditions (e.g. pressure, temperature, humidity) on glasses (Tant and Wilkes, 1981; Jewell and Shelby, 1990; Slade and Levine, 1991). The concept of non-equilibrium originates from the way a glass is formed. Vitrification consists of rapidly bringing the liquid system out of reach of its equilibrium state by inducing a rapid increase in viscosity. Because of the high viscosity (i.e. around 10^{12} to 10^{14} Pa.s), the glass is inhibited from rapidly reaching its equilibrium state. Nevertheless there is still a 'driving force' for the glass to achieve equilibrium. For that reason the glass is described as kinetically stable. With respect to storage stability, the kinetic stability is an important factor because it controls changes in the structure and properties of a glass over long periods of time. Understanding the kinetic stability of intracellular glasses in seeds is needed before applying glass principles to storage technology.

The non-equilibrium behaviour of glasses is reflected in the thermal behaviour of the glass–liquid transition (Tant and Wilkes, 1981). In order to gain insights into the non-equilibrium nature of glasses, a calorimetric analysis of thermal events considered as glass transitions was performed in bean axes at various WC. Thermal events that can be attributed to glass–liquid transitions were observed at decreasing WC from 0.45 to 0.03 g/g. The glass–liquid transitions exhibited three different types of thermal behaviours which depend on temperature and sample WC (Fig. 1). Type 1 and 2 appeared as endothermic shifts in the heat flow which were diagnosed as second-order-like transitions and were observed in samples with WC between 0.45 and 0.06 g/g. The appearance of a devitrification event (Td) in samples containing 0.45 and 0.3 g/g distinguishes type 1 from type 2 (Fig. 1). Type 3 appeared as an endothermic peak superimposed on a second-order-like transition and was found in samples with WC between 0.09 and 0.03 g/g and at temperatures between 18 and 70°C. Similar observations have been reported for various organic and inorganic glasses where the glass–liquid transition behaviours varied according to the initial WC of the glass (Jewell and Shelby, 1990; Angell *et al.*, 1994). From a kinetic point of view, the glass–liquid transition is understood as a distribution of molecular rearrangements of different types, kinetics and energy (so-called relaxation processes) (Tant and Wilkes, 1981; Angell *et al.*, 1994; Perez, 1994). It follows that the differences in glass behaviours observed in bean axes may result from differences in the kinetics and/or types of molecular rearrangements, these being dependent on the sample WC. Therefore, over long periods of time, one may suspect that the intracellular glasses will move towards equilibrium following different routes or thermodynamical paths (1, 2, 3) which depends on the WC. We suspect that the different routes to achieve equilibrium may also induce different kinetics of degradative reactions during long-term storage. So far the relationship between temperature and the state of water in seeds of various species has been represented by a Tg/WC curve (Leopold *et al.*, 1994). The Tg/WC curve delimits the temperature/WC boundaries of the non-equilibrium *vs* equilibrium states and therefore may determine the critical values of temperature and water content for optimum conditions of storage. A similar curve was established from our observations in bean axes (Fig. 2, solid line). However, one can see that the Tg/WC curve *per se* does not adequately reflect the differences in thermal behaviours of the glass transition. Thus additional parameters are needed to take them into account. In Figure 2, we overcame this problem by using different symbols for each type of glass transitions and by adding a time factor as third dimension to the Tg/WC relationship.

The Relevance of the Thermal History during the Glass Formation

Rapidly cooling the cytoplasmic liquid below its melting temperature is not the physiological route for glass formation in maturing seeds but it is the most instructive to understand the importance of the thermal history on glass properties. During vitrification and rise in viscosity, the structure of the liquid system becomes 'frozen' in a particular structural configuration. During this

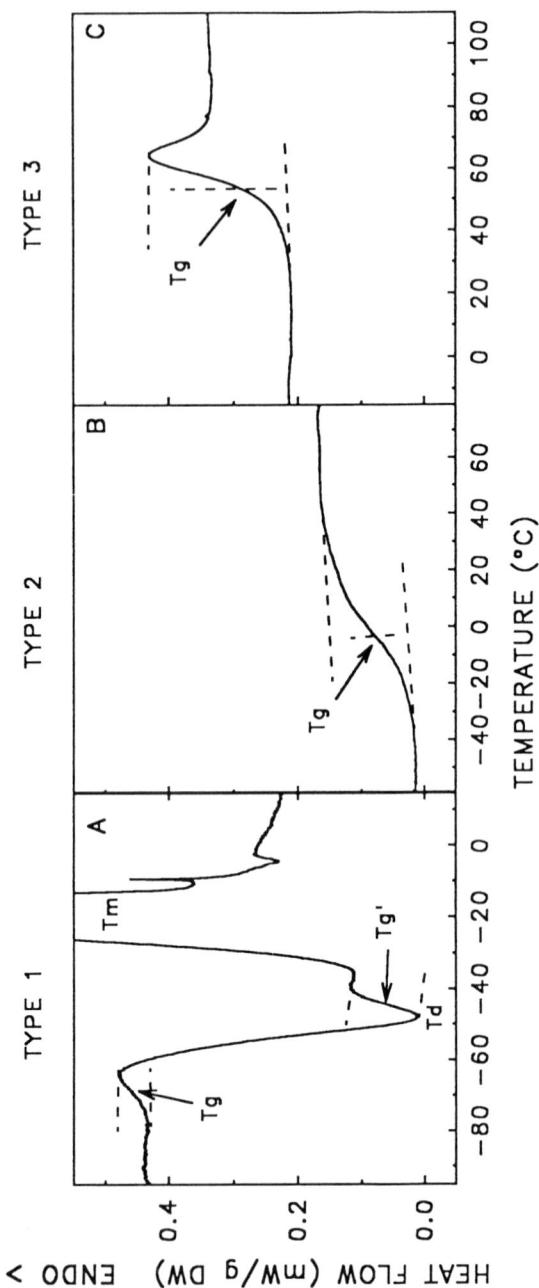

Figure 1. DSC heating thermograms of the glass–liquid transitions representative of the three different thermal behaviours (Types 1, 2, 3) in bean embryonic axes at 0.39 (A), 0.12 (B) and 0.05 g H$_2$O/g DW (C), respectively. Thermal transitions were measured at least 30°C below Tg using a Perkin-Elmer DSC-7 interfaced with a thermal analysis data station. Cooling and heating rates were set at 20°C/min and 10°C/min respectively. The arrows indicate the onset of the glass–liquid transition (Tg). The dashed lines correspond to the interpretation of the theoretical baselines of equal heat capacity which were used to determine Tg. Thermograms are normalized with respect of the sample dry weight and corrected for the curvature. Td, devitrification temperature, Tm temperature of the ice melting peak

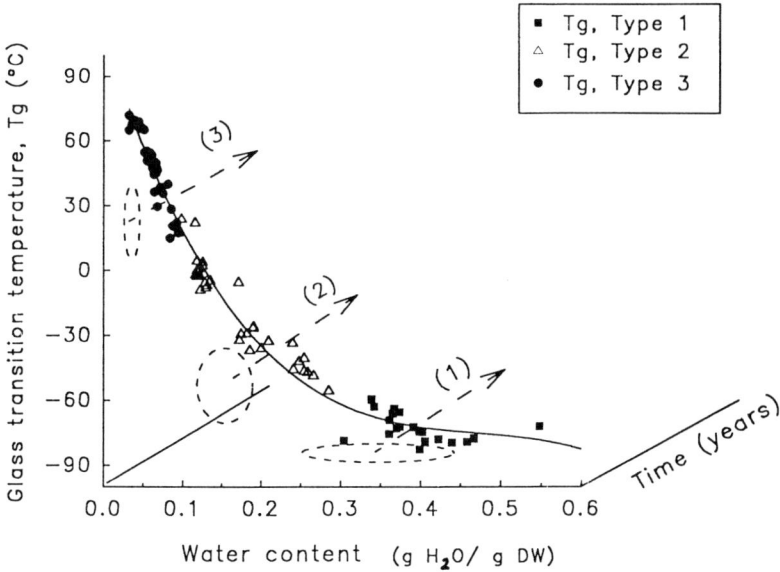

Figure 2. The relationship between the glass–liquid transition temperature and water content in bean embryonic axes. The different symbols reflect the different types of thermal behaviours (type 1, ■; 2, △ ; 3, ●). Time as a third dimension has been added to emphasize the effect of time on the Tg/WC relationship and the non-equilibrium nature of glasses. The three arrows depict the three different thermodynamic paths followed by intracellular glasses in order to achieve equilibrium over long periods of time

process the system has also been prevented from exploring other configurations which may have been of lower energy (Angell *et al.*, 1994; Perez, 1994). The glass theory predicts that a 'frozen' disordered structure can adopt a certain number of different configurations (or structural degree of freedom) which depend on the thermal history during glass formation (Tant and Wilkes, 1981; Angell *et al.*, 1994). Therefore, can intracellular glasses in seeds adopt several configurations depending on their thermal history? To answer this question, we studied the effects of a brief exposure to 100°C followed by a rapid cooling on the thermal behaviour described above. The glass transitions were recorded during the first heating scan to 100°C then during the second heating scan after the samples were rapidly cooled to the desired temperature (Fig. 3, inserts). At WC >0.1 g/ g, the glass transition behaviours were similar in the first and second scans, suggesting that at these WCs, the ability to form a glass is resistant to the heating treatment. Between 0.1 and 0.04 g/g, the endothermic peak at Tg disappeared and a second-order-like transition was observed instead. In addition, we observed that the high temperature manipulation significantly depressed the WC/Tg relationship (Fig. 3C). It is unlikely that the sample WC

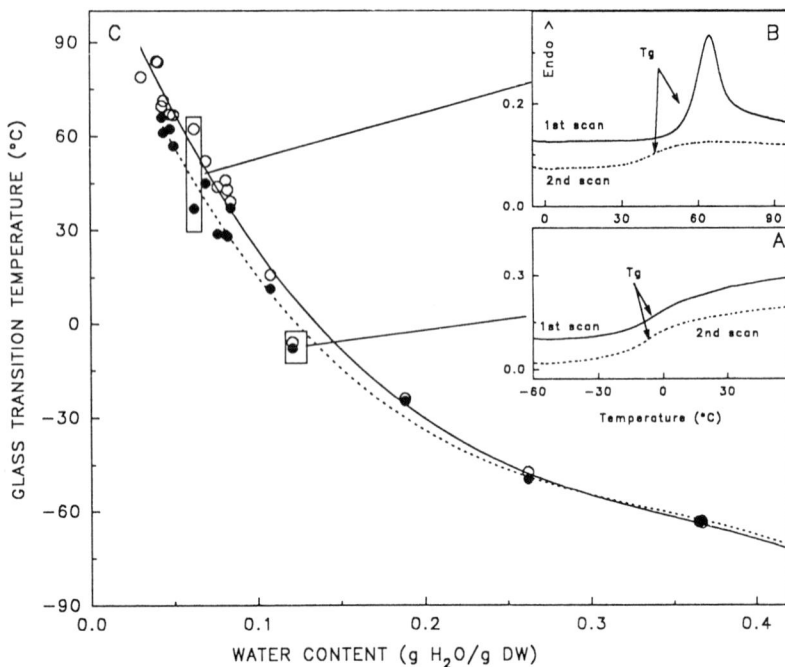

Figure 3. A and B (Inserts). DSC heating thermograms describing the effect of a 100°C exposure on the glass–liquid transition. The first scan (solid line) was obtained following the same conditions as for Figure 1. The heating scan was stopped at 100°C and samples were rapidly cooled to the desired temperature then rescanned at 10°C/min (dashed line) (A = 0.13 and B = 0.06 g H$_2$O/g DW). C. Relationship between water content and glass–liquid transition temperature occurring during the first heating (○) to 100°C and the immediate second heating (●)

or the concentration of the glass-forming solutes have been altered by the 100°C exposure. This indicates that in dry seeds the thermal history influences the glass structure and transition to a liquid state. From a physiological point of view, thermal history would correspond both to physiological and environmental conditions which lead to the formation of the intracellular glasses during maturation drying of the embryo and to postharvest conditions such as priming. Thus it is reasonable to assume that seed storage stability will depend on the thermal history during intracellular glass formation. As concluded above, caution must be taken when the Tg/WC relationship is used to determine the optimum temperature and WC for long-term storage.

The Absence of a Glassy State at Very Low Water Content

Below 0.03 g/g glass transitions were no longer observed between −100 and 120°C, suggesting an inherent instability of the intracellular glassy matrices. Loss of glassy structures at extremely low water content has also been reported in some food systems (Noel *et al.*, 1990). The absence of a glass transition below 0.03 g/g supports the view that the physical state of water below the Tg/WC curve (Fig. 2) is complex and cannot be explained solely by assessing the presence of a glassy state. Whether there is a link between the decline in longevity observed in seeds of several species stored at extremely low WC (Vertucci and Roos, 1993) and the absence of glass transition deserves ample investigation.

The Complexity of Interactions Between Water and the Glass-Forming Solutes

The amorphous characteristic is another important parameter defining a glassy state. Dry amorphous materials usually exhibit a single Tg value (e.g. for cellulose 227°C (Hancock and Zografi, 1994), 50°C for sucrose (Slade and Levine, 1991)). When another component is added to the amorphous structure, Tg can be lowered or increased according to the nature and concentration of the additive. This effect is called respectively 'plasticizing' and 'antiplasticizing effect' (Slade and Levine, 1991; Williams *et al.*, 1993). Like most dry amorphous solids, dry seeds can absorb certain amounts of water according to the RH of the environment. This results in a decrease in Tg, which pattern closely corresponds to the Tg/WC shown in Figure 2. Based on this evidence, several authors have proposed that water acts as a plasticizer in dry seeds (Leopold *et al.*, 1994; Williams and Leopold, 1995) in a similar way as in low-moisture food systems (Slade and Levine, 1991). However, the theory on plasticizing effects also implicates that there is a total and 'perfect' mixing of the glass-forming solutes and water (i.e. water acts as a neutral diluent), making the intracellular glass a perfectly homogeneous, amorphous liquid (Slade and Levine, 1991; Hancock and Zografi, 1994). In addition, according to the plasticizing theory, Tg should not be greatly influenced by the potential interactions between water (the plasticizer) and the glass components. Intuitively, we can rule out the possibility that glasses in seeds are made from an uniform, single homogeneous liquid. Instead it is likely to be a complex mixture of sugars, proteins, amino acids and salts whose concentrations vary among organelles. Besides, intracellular glasses are constrained by interfaces such as membranes. Thus it is reasonable to speculate that complex interactions must occur within the glassy matrix and with its environment.

We wished to further establish whether water actually acts as a plasticizer in aqueous glasses or whether the specific interactions described above influence Tg. If water is proven to be an effective plasticizer, then the Tg/WC combinations could be rapidly predicted using a single value of Tg of dry seeds and

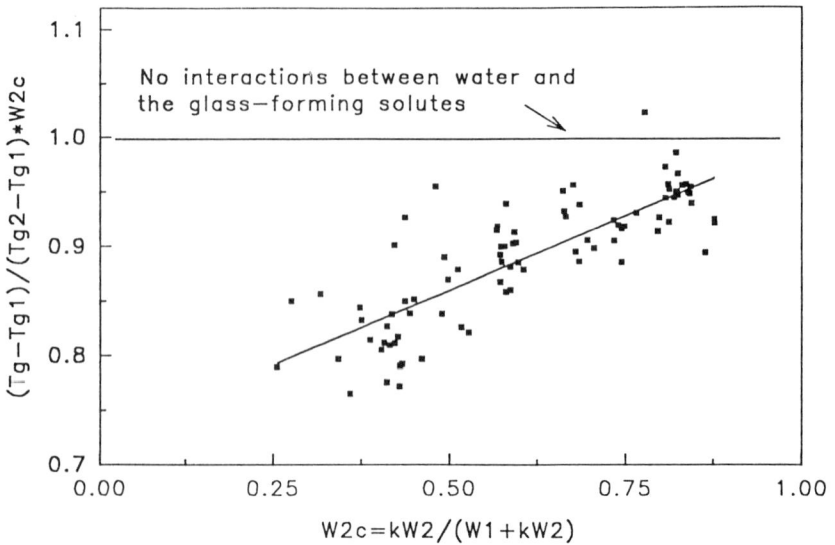

Figure 4. A study of the influence of the glass components on the expression Tg using the Schneider equation (see text) and a Schneider plot (Schneider, 1989). The Schneider plot tests whether water acts as a plasticizer in intracellular glasses in which case the potential interactions between water and the glass components do not influence Tg. If so, the result of the Schneider equation should give a horizontal line of slope = 0 and y-intercept = 1. In contrast the Schneider plot obtained from our experimental data (■) shows a significant deviation from the horizontal line, indicating that water should not be considered a plasticizer in aqueous, intracellular glasses

without resorting to extensive and time-consuming determinations of a range of Tg values at various WC. Such a prediction can be achieved by different models developed in the food, polymer and material sciences. These models relate Tg to the phase composition of mixed amorphous systems and describe the plasticizing effect (see Schneider, 1989; Slade and Levine, 1991; Hancock and Zografi, 1994 and Leprince and Vertucci, 1995 for references and theoretical background.). To test the plasticizing effect of water in bean axes, we applied the Schneider model (Equations 1 and 2) to our data (Schneider, 1989).

$$W_{2c}[(Tg-Tg_1)/(Tg_2-Tg_1)] = (1+K_1)-(1+K_2)W_{2c}+K_2W_{2c}^2 \qquad \text{(Eq. 1)}$$

where $W_{2c} = kW_2/(W_1+kW_2)$, Tg is the predicted glass–liquid transition temperature of the mixture; Tg_1 and Tg_2, the Tg values of the glass-forming solutes and pure water, respectively. W_1 and W_2 are the weight fractions of the glass-forming solutes and water respectively; k is an experimentally determined coefficient (Schneider, 1989). $Tg_2 = 134$ K and W_2 corresponds to the sample WC (on a fresh weight basis). W_1, Tg_1 and k were calculated from our Tg/WC curve (Fig. 2) using fitting experiments (data not shown). The values for Tg_1 and k were 415 K and 11.5 respectively. K_1 and K_2 are specific coefficients which describe the specific interactions between the glass components and water and account for the fact that these components are not perfectly soluble. If K_1 and $K_2 = 0$, the glass components do not interact and are completely dissolved in the system. In consequence water can be described as a plasticizing agent. The above equation is then reduced to a simpler form:

$$W_{2c} [(Tg-Tg_1)/(Tg_2-Tg_1)] = 1 \qquad \text{(Eq. 2)}$$

It follows that a plot of W_{2c} values obtained from equation 2 (so-called Schneider plot) gives a horizontal line (Fig. 4). In the case where K_1 and/or K_2 are not nil, the Schneider plot exhibits positive or negative deviations from the horizontal line. The result of the Schneider plot using our observations shows that our experimental data are scattered away from the horizontal line, indicating that various interactions within water and the glass-forming solutes affect the glass transition behaviour. It follows that water does not act as a plasticizer to form a homogenous amorphous structure in seed tissues.

Conclusions

Our data demonstrate why a Tg/WC relationship is not entirely satisfactory to link the different water states in seed tissues with storage stability. Here we have identified four factors which need to be taken into account to characterize the water state in terms of 'intracellular glasses'. These factors are the non-equilibrium nature of glasses which is dependent on the sample WC, the thermal history prior to glass formation, the absence of a glass–liquid transition at very low WC and the complexity of the intracellular structures involved in the glassy matrix.

Reference

Angell, C.A., Bressel, R.D., Green, J.L., Kanno, H., Oguni, M., Sare, E.J. 1994. *Journal of Food Engineering* 22: 115–142.
Bruni, F. and Leopold, A.C. 1992. *Biophysical Journal* 63: 63–672.
Hancock, B.C. and Zografi, G. 1994. *Pharmaceutical Research* 11: 471–477.
Jewell, J.M. and Shelby, J.E. 1990. *Journal of Applied Physics* 68: 6159–6169.
Leopold, A.C., Sun, W.Q. and Bernal-Lugo, I. 1994. *Seed Science Research* 4: 267–274.
Leprince, O. and Vertucci, C.W. 1995. *Plant Physiology* 109: 1471–1481.
Nelson, K.A. and Labuza, T.P. 1994. *Journal of Food Engineering* 22: 271–289.
Noel, T.R., Ring, S.G. and Whittam, M.A. 1990. *Trends in Food Science and Technology* September: 62–67.
Perez, J. 1994. *Journal of Food Engineering* 22: 89–114.
Schneider, R.A. 1989. *Polymer* 30: 771–779.

Slade, L. and Levine, H. 1991. In: *Water Relationships in Food*, pp. 29–101 (eds H. Levine and L. Slade). New York: Plenum Press.

Sun, W.Q. and Leopold, A.C. 1994. *Annals of Botany* 74: 601–604.

Sun, W.Q., Irving, T.C. and Leopold, A.C. 1994. *Physiologia Plantarum* 90: 621–628.

Tant, M.R. and Wilkes, G.L. 1981. *Polymer Engineering and Science* 21: 874–895.

Vertucci, C.W. 1992. In: *Basic and Applied Aspects of Seed Biology*, pp. 739–746 (eds D. Côme and C. Corbineau). Paris: AFSIS.

Vertucci, C.W. and Roos, E.E. 1993. *Seed Science Research* 3: 201–213.

Williams, R.J. 1994. *Annals of Botany* 74: 525–530.

Williams, R.J. and Leopold, A.C. 1995. *Seed Science Research* 5: 117–120.

Williams, R.J., Hirsh, A.G., Takahashi, T.A. and Meryman, H.T. 1993. *Japanese Journal of Freeze Drying* 39: 3–12.

81. DNA Extraction from Dry Seeds for RAPD Analyses

M.B. McDONALD, L.J. ELLIOT and P.M. SWEENEY

Department of Horticulture and Crop Science, The Ohio State University, Columbus, OH 43210, USA

Abstract

Random amplified polymorphic DNA (RAPD) has emerged from biotechnologi-
cal research as a practical genetic identity determinant useful in plant breeding,
seed production, and seed testing programmes. This report describes a DNA
extraction procedure from dry seeds of maize (*Zea mays* L.), cotton (*Gossypium
hirsutum* L.), soyabean (*Glycine max* [L.] Merr.), wheat (*Triticum aestivum* L.) and
red clover (*Trifolium pratense* L.) that can be successfully employed in subsequent
RAPD amplification of DNA fragments. The technique also extracted DNA from
dry peanut (*Arachis hypogaea* L.) seeds, but no amplification fragments were
observed following RAPD amplification. This procedure is inexpensive, simple,
fast, avoids the use of growing plant tissue, and is applicable to a number of crops.

Introduction

Electrophoresis of seed proteins has proven to be highly versatile in varietal
identification testing of seeds and this technology has been incorporated into the
Rules of the International Seed Testing Association and described by the
Association of Official Seed Analysts (1991). Most electrophoretic systems
employ either starch or polyacrylamide gels as the preferred media in which
protein separations based on molecular size and charge density are made.
Another approach is the use of isoelectrically focused gels that accomplish
protein separations based on their position within a pH gradient, the protein
ultimately coming to equilibrium at a pH where the molecule is no longer
charged (isoelectric point or pI). All of these systems possess advantages and
disadvantages from a commercial perspective (Table 1). Principal among these
is the ability to evaluate sufficient numbers of seeds at the lowest cost. Starch gel
electrophoresis systems have been described that allow adequate numbers of
seeds (usually 80) to be run at one time (Stuber *et al.*, 1988). More importantly,
starch gels can be moulded that permit up to five slices of the parent gel, each gel
slice being evaluated for different enzyme staining patterns. This system enables
the equivalent of five electrophoretic evaluations of the same seed samples,
thereby significantly reducing costs and analysis time. Starch gel electrophoresis
of corn seed proteins provides reproducible and standardized results within and
among seed testing laboratories (McDonald, 1990).

Other approaches have been identified to make electrophoresis more efficient
and cost effective. One technique is to miniaturize and computerize the

R.H. Ellis, M. Black, A.J. Murdoch, T.D. Hong (eds.), Basic and Applied Aspects of Seed Biology, pp. 747–753.
© *1997 Kluwer Academic Publishers, Dordrecht. Printed in Great Britain.*

Table 1. Advantages and disadvantages of polyacrylamide, starch, and IEF (isoelectric focusing electrophoresis) for varietal identification of seeds

Polyacrylamide	Starch	IEF
Advantages		
Technical system well defined	Technical system well defined	Technical system well defined
Excellent band resolution	Inexpensive	Uses charge (pI) rather than charge density and size of proteins
	Gels can be sliced	Gels can be blotted
		Short running time (1.5 h)
Disadvantages		
Expensive	Standardization of gels	Expensive
Cannot slice gel	Long running time (5–6 h)	Cannot slice gel
Potentially toxic	Poor band resolution	

electrophoresis process. This results in less time being committed to electrophoresis, staining and destaining of gels. Such an approach has been developed, evaluated and found commercial application (McDonald and Drake, 1990). Another strategy is to utilize non-denaturing isolectric focussing gels that cannot be sliced and blot the proteins from the parent gel onto a nitrocellulose transfer membrane. This blotting technique is another approach to obtaining multiple enzyme staining patterns from one electrophoretic run and provides greater resolution of banding patterns than obtained on a starch gel (McDonald, 1991).

Still, these techniques fail to differentiate a number of varieties in some crops. Greater sensitivity in genotypic identification is desired. In 1990, a new genetic assay called random amplified polymorphic DNA (RAPD) was reported (Welsh and McClelland, 1990; Williams et al., 1990). This technique uses a single arbitrarily chosen oligonucleotide primer that hybridises to the genomic DNA template at two different sites, one on each strand of the complementary DNA. Under appropriate temperature alternations, a thermostable DNA polymerase is able to synthesize discrete DNA products (usually 200 to 2000 base pairs long) that can be resolved on an agarose gel following electrophoresis (Fig. 1). Each primer has the capability to consistently direct amplification of several unique DNA fragments in the genome. Some amplified fragments or patterns of fragments may be unique to a genotype and hence useful in varietal identification.

The use of RAPDs in genetic purity testing possesses several important advantages for seed technologists compared to other systems such as protein electrophoresis. First, RAPDs have the capability to provide greater potential discrimination of varieties since the nucleotide composition of a gene is being

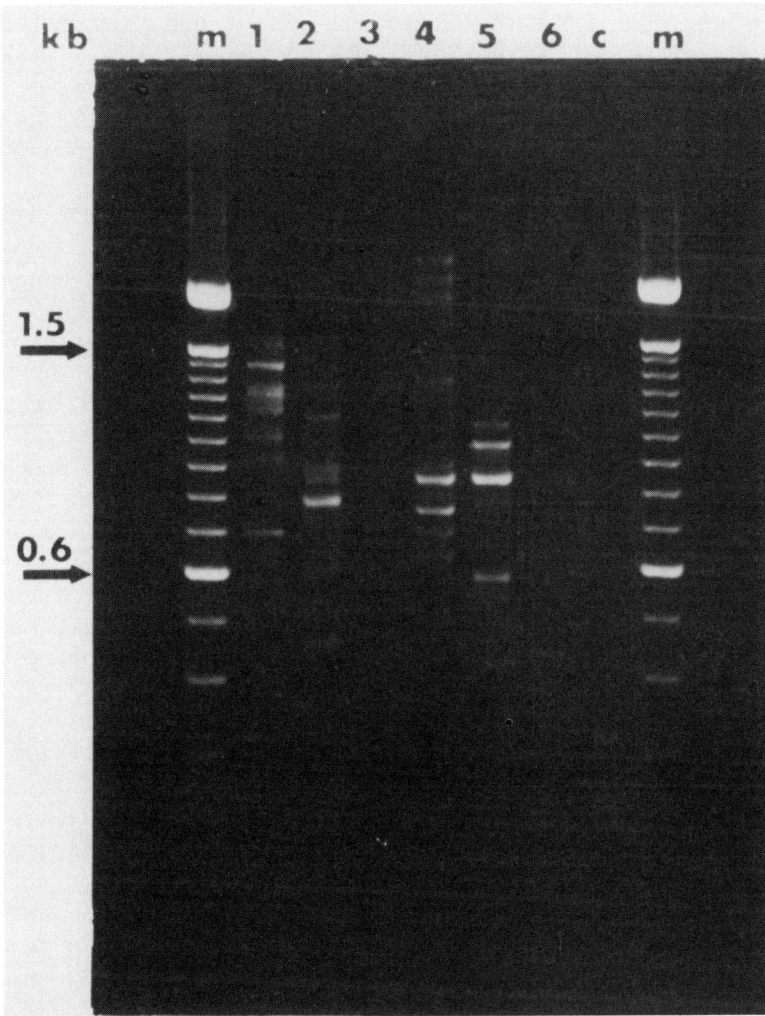

Figure 1. RAPD products derived from dry seeds using primer 262. From left to right: m = molecular weight marker (BRL 100 bp ladder); lanes 1 to 6 are red clover, wheat, peanut, soyabean, cotton, and maize, respectively; c = control; m = molecular weight marker

directly determined rather than the product of a gene such as an enzyme. Second, the versatility of RAPDs is greater than protein electrophoresis. Over 700 primers are available for screening in the former compared to approximately 20 enzyme systems in the latter. Third, RAPDs do not pose the potential human health and environmental disposal issues associated with radioisotopes used in most other DNA technologies such as restriction fragment length polymorphism (RFLP) assays. Fourth, RAPDs require the same general equipment and technical expertise as protein electrophoresis. However, the purchase of a DNA thermocycler ($6000 US) is recommended. Fifth, the cost of and time to complete a RAPD analysis is equivalent to current protein electrophoresis protocols ($1.00 US per lane, 24 h).

These significant advantages justify the examination of RAPDs for genetic purity analyses in seed testing. While the technique remains relatively new, many advances in protocol are being made. For example, the DNA template concentration, magnesium concentration, cycling temperatures and times, and base composition of the primers affect the success of RAPD analysis (Tingey *et al.*, 1992; Fritsch *et al.*, 1993). Many reports investigating DNA extraction protocols for specific primer-driven polymerase chain reaction (PCR) technology have been published (Tai and Tanksley, 1990; Berthomieu and Meyer, 1991; Edwards *et al.*, 1991; Langridge *et al.*, 1991; Luo *et al.*, 1992). Most of these use DNA extractions that involve a number of laborious and expensive steps, require access to an ultracentrifuge, and some take up to a day for sample preparation. In all of these reports, some portion of the growing plant is required as the source of DNA. This is a major limitation in a seed testing laboratory because the time required to grow tissue from seed (at least 7 d) is unacceptable and makes the incorporation of this technology into routine seed testing unlikely.

From a seed testing perspective, the preferred part of the plant to use in a RAPD analysis would be the dry seed. Few studies have critically examined DNA extraction procedures for RAPD analyses. We are aware of only one report that used seed tissue in a RAPD study (Benito *et al.*, 1993). That study reported RAPD amplification products from the endosperm of closely related barley (*Hordeum vulgare* L.) and rye (*Secale cereale* L.), but used an unpurified DNA supernatant in the amplification step and it remains unclear whether unpurified DNA extracted from whole seeds would prove reliable for varietal identification. Protocols using simple DNA extraction procedures specific primer-driven PCR (Edwards, 1991) and random primer-driven PCR (Oard and Dronavalli, 1992) have been reported. These protocols are not complicated and offer the advantage of a DNA precipitation step that eliminates unwanted contaminants from a concentrated seed extract. The Edwards *et al.* (1991) protocol is faster and has been successfully used to extract DNA from fine fescue (*Festuca longifolia, F. rubra* L. subsp. *rubra,* and *F. rubra* L. subsp. *commutata*) leaves which was then used in RAPD analyses (Sweeney *et al.,* 1993). The objective of this study was to evaluate the Edwards *et al.* (1991) DNA extraction protocol from dry seeds as an inexpensive procedure for use in varietal identification of a number of crops.

Materials and Methods

DNA Extraction

Five seeds each of maize, cotton, soyabean, peanut, wheat, and red clover were crushed between two paper germination towels with a hammer and 20 mg of the crushed seed tissue placed in a centrifuge tube followed by the addition of 100 µl of extraction buffer (200 mM Tris-HCl [pH 7.5], 288 mM NaCl, 25 mM ethylenediamine tetraacetic acid [EDTA], and 0.5% sodium dodecyl sulphate). After further maceration of the seed tissue with a pipette tip, 900 µl of extraction buffer were added to make a total of 1 ml and the centrifuge tube and its contents vortexed for 30 s and then centrifuged for 2 min at 10 144g. After centrifugation, the supernatant was retained and centrifuged again for 4 min at 10 144g. Absolute isopropanol (900 µl) was added to the remaining supernatant and allowed to precipitate the DNA for 2 min. The DNA was pelleted by centrifugation for 7 min at 10 144g and the pellet dried in a vacuum desiccator for 15 min.

Following drying, the DNA pellet was resuspended in 150 µl TE (10 mM Tris-HCl [pH 7.5], 1 mM EDTA) for 2 min at room temperature. The DNA extract was quantified in a TKO 100 Dedicated Minifluorometer (Hoefer Scientific, San Francisco, CA) by adding 2 µl of the DNA resuspension to a 2 ml solution consisting of 10 mM Tris, 200 mM NaCl, 1 mM EDTA (pH 7.4), and 0.1 µg/ml Hoechst dye 33258 (Polysciences, Warrington, PA). Hoescht dye is highly specific for DNA and its fluorescence is substantially increased when it binds to DNA (Perkin-Elmer, 1986). This allows an estimate of extracted DNA based on fluorescence intensity. The extracted DNA was then diluted with the TE buffer to 2 µg/ml for amplification.

Amplification

The isolated DNA was denatured for 5 min at 96°C and immediately cooled in ice for 3 min. After this time, 2 ng of the denatured DNA was added to a reaction tube containing 24 µl of the following: 20 mM Tris-HCl (pH 8.4), 10 mM KCl, 3 mM MgCl$_2$, 0.2 mg/ml gelatin, 0.1 µM primer 262 (Biotechnology Laboratory, University of British Columbia, Manitoba, BC), 60 µM each dATP, dCTP, dGTP, dTTP, and 1.5 units TAQ DNA polymerase (Perkin Elmer, Fairfield, OH). A control containing all the reactants excluding extracted seed DNA was also subjected to amplification. Primer 262 (CGCCCCAGT) was selected because it has a high proportion (80%) of G+C nucleotides and a previous study (Fritsch *et al.*, 1993) indicated that it generated numerous amplification fragments in a number of diverse plant species.

A drop of mineral oil was added to the 25 µl reaction mixture to minimize evaporation from the reaction tubes. The reaction tube and mixture were placed in a DNA Thermal Cycler 480 (Perkin Elmer, Fairfield, OH) programmed for the following temperature cycle: 94°C 1 min, 40°C 1 min, 72°C 1 min. This cycle was repeated 39 times.

Electrophoresis

A 1.5% agarose gel containing 89 mM Tris (pH 8.0), 89 mM H_2BO_3, 2 mM EDTA, and 1.34 mM ethidium bromide was used to assay the amplification products. Ethidium bromide specifically binds to DNA and this complex fluoresces in the presence of UV light (Lepecq and Paoletti, 1967). Two µl of a loading and tracking dye (30% glycerol and 0.25% bromophenol blue in distilled water) were added to the amplified reaction mixture and 10 µl of the amplified sample loaded onto the gel and electrophoresis conducted at a constant 3.24 V/cm for 2 h. Following electrophoresis, the amplified DNA products were viewed and photographed using a Transluminator Model 3-3100 (Fotodyne, Hartland, WI).

Results and Discussion

Use of the DNA extraction protocol described in this report successfully extracted DNA from dry seeds of the six crops examined. The DNA concentrations (in parentheses) for each crop were maize (29 µg/ml), cotton (77 µg/ml), soyabean (68 µg/ml), peanut (32 µg/ml), wheat (37 µg/ml), and red clover (25 µg/ml). Most crops displayed multiple fragments under the electrophoretic assay when primer 262 was used in the amplification reaction (Fig. 1). No bands were observed in the control lane. In addition, no amplified products were found for peanut seeds and only one was detected for maize seeds. In the former case, the high oil content of peanut seeds may have adversely interfered with the amplification process. In this situation, while the extraction protocol was successful in isolating DNA from peanut seeds, a defatting step prior to extraction may be necessary to eliminate contaminating oils. In the latter case, the single amplified DNA product for maize seeds might be attributed to the selection of the 262 primer which failed to produce multiple fragments in the species. Subsequent tests with other primers might reveal a greater diversity in banding patterns.

These results demonstrate that the quality and quantity of DNA extracted from dry seeds using the modified Edwards *et al.* (1991) protocol is sufficient for use in RAPD analysis. These data indicate that this DNA extraction procedure for RAPDs is useful for a majority of crops although it is not presently universally adaptable (peanut is one exception). The potential of RAPDs for improving the quality of plant breeding, seed production, and seed testing programmes warrants further study and the DNA extraction protocol from dry seeds described here appears readily adaptable to most routine, standardized seed testing laboratory situations.

Acknowledgements

Salaries and research support were provided by state and Federal funds appropriated to the Ohio Agricultural Research and Development Centre, Ohio State University.

References

Association of Official Seed Analysts. 1991. *Cultivar Purity Testing Handbook. Contribution Number 33*. 78 pp. Ames: Association of Official Seed Analysts.

Benito, C., Figueiras, A. M., Zaragoza, C., Gallego, F. G. and de la Peña, A. 1993. *Plant Molecular Biology* 21: 181–183.

Berthomieu, R. and Meyer, C. 1991. *Plant Molecular Biology* 17: 555–557.

Edwards, K., Johnstone, C. and Thomson, C. 1991. *Nucleic Acids Research* 19: 1349.

Fritsch, P., Hanson, M. A., Spore, C. D., Pack, P. E. and Riesebert, L. H. 1993. *Plant Molecular Biology Reporter* 11: 10–20.

Langridge, U., Schwall, M. and Langridge, P. 1991. *Nucleic Acids Research* 19: 6954.

Lepecq, J. B. and Paoletti, C. 1967. *Journal of Molecular Biology* 27: 87–106.

Luo, G., Hepburn, A. G. and Widholm, J. M. 1992. *Plant Molecular Biology Reporter* 10: 319–323.

McDonald, M. B. 1990. *Proceedings Annual American Seed Trade Association Corn and Sorghum Research Conference* 45: 43–53.

McDonald, M. B. 1991. *Seed Science and Technology* 19: 33–40.

McDonald, M. B. and Drake, D. M. 1990. *Seed Science and Technology* 18: 89–96.

Oard, J. H. and Dronavalli, S. 1992. *Plant Molecular Biology Reporter* 10: 236–241.

Perkin-Elmer. 1986. *Biotechnology Technical Report*. 4 pp. Norwalk: Perkin Elmer.

Stuber, C. W., Wendel, J. F., Goodman, M. M. and Smith, J. S. C. 1988. *North Carolina Agricultural Research Service, North Carolina State University Technical Bulletin* 286: 1–87.

Sweeney, P. M., Danneberger, T. K. and Kamalay, J. C. 1993. *International Turfgrass Society Research Journal* 7: 768–774.

Tai, T. H. and Tanksley, S. D. 1990. *Plant Molecular Biology Reporter* 8: 297–303.

Tingey, S. V., Rafalski, J. A. and Williams, J-. G. K. 1992. *Proceedings of the Symposium on Applied RAPD Techniques, Plant Breeding*, pp. 3–8, Crop Science Society of America, Madison, WI.

Welsh, J. and McClelland, M. 1990. *Nucleic Acids Research* 18: 7213–7218.

Williams, J. G. K., Kubelik, A. R., Livak, K. J., Rafalski, J. A. and Tingey, S. V. 1990. *Nucleic Acids Research* 18: 6531–6535.

82. Some Aspects of the Biological Control of Seed Storage Fungi

C. CALISTRU[1], M. McLEAN[2] and P. BERJAK[1]

[1]Plant Cell Biology Research Unit, Department of Biology, University of Natal, Private Bag X10, Dalbridge 4014; [2]Department of Physiology, Faculty of Medicine, University of Natal, P.O. Box 17039, Congella 4013, South Africa

Abstract

Under storage conditions of ambient temperature and relative humidity in South Africa, Fusarium moniliforme and Aspergillus flavus proliferate. Both fungal species are generally toxigenic under local storage conditions. The rural community chooses visibly mouldy maize for beer-making: such infected maize often contains high levels of mycotoxins which may be associated with prevalence of oesophageal cancer in certain areas. The aim of this investigation is to ascertain the feasibility of using Trichoderma species as a biocontrol agent against seed-associated pathogenic aspergilli and fusaria.

Introduction

The fungi associated with developing plants and seeds in the field and stored seeds generally belong to the genera Penicillium, Aspergillus and Fusarium. Fusarium moniliforme is ubiquitous in newly-harvested maize, persisting for variable periods in storage, while Aspergillus flavus may represent the final group of species in the succession of aspergilli after grain storage under high temperature and/or high humidity (McLean and Berjak, 1987). Many of these fungi are producing toxigenic secondary metabolites (mycotoxins) under South African storage conditions. In many rural commodities, hand-harvested maize is manually sorted into two lots: apparently healthy seed, used for general consumption, and visibly mouldy maize used for animal feed and beer-making. The fungal contaminated maize frequently contains high levels of mycotoxins, which may account for the prevalence of oesophageal cancer in certain areas (Rheeder et al., 1992). Since pathogenic fungi may be present within the tissues of stored seeds (Mycock et al., 1992), these contaminants will not be eradicated by external fungicide treatment. It has been reported previously that some isolates of Trichoderma species may be used as potential biocontrol agents of other more pathogenic fungal species (Whipps and McQuilken, 1993). The aim of the present investigation is to ascertain whether certain strains and/or species of Trichoderma may be considered as biological control agents against potentially toxigenic strains of seed-associated Aspergillus and Fusarium.

R.H. Ellis, M. Black, A.J. Murdoch, T.D. Hong (eds.), Basic and Applied Aspects of Seed Biology, pp. 755–762.
© 1997 Kluwer Academic Publishers, Dordrecht. Printed in Great Britain.

Materials and Methods

Fungal Isolates and Growth Conditions

The two strains of *A. flavus* and *F. moniliforme* used in this study were isolated from the local maize seeds. The *Trichoderma* species (three *T. harzianum* strains – T1, T2, T3 and three strains of *T. viride* – T4, T5 and T6) were supplied by the Plant Protection Institute, Pretoria. The other three strains of *Trichoderma* spp. (T7, T8 and T9) were supplied by the Department of Physiology, University of Natal, Durban. All fungal isolates have been periodically reinoculated and grown on Potato Dextrose Agar (PDA) medium at 25°C.

Dual Cultures of Trichoderma Species and the Pathogenic Fungi

Individually, nine strains of *Trichoderma* species (*T. harzianum* and *T. viride*) were challenged by one strain of *F. moniliforme* and one of *A. flavus*. The fungi were inoculated at opposite edges of the PDA plate and incubated for 10 days (d) at 25°C. Six replicates were used. Plates were inspected after three, five and seven days, and the diameter of fungal colonies measured.

Scanning Electron Microscopy (SEM) Observation of Hyphal Interaction Areas

Mycelial samples from the contact regions of dual cultures were sampled six days after inoculation. The plugs were fixed overnight in 2.5% gluteraldehyde, postfixed for 1 h in 0.5% osmium tetroxide, and dehydrated out in a graded alcohol series. Following critical point drying, the specimens were coated with gold in a Polaron Sputter Coater and viewed with an Hitachi S520 SEM.

Assay of Antifungal Activities of Culture Filtrates of Trichoderma Strains

The technique of this experiment was adapted from the method of Jackson *et al.* (1991). The effects of different carbon (glucose, maltose and sucrose) and nitrogen sources (L-alanine and KNO_3) in a basal medium (Czapek-Dox agar) on the possible production of inhibitory products by four aggressive *Trichoderma* strains were investigated. Each strain of *Trichoderma* was inoculated onto 30 ml of medium (carbon/nitrogen source in the basal medium) and incubated for 10 d in constant light at 25°C. Two mililiters of sterile culture filtrates of each fungal strain were added to 20 ml of PDA. PDA plates containing no culture filtrate served as controls. Agar discs with actively growing colonies of *Fusarium* and *Aspergillus* were placed in the centre of agar plates. Radial colony growth was measured for six replicates per study after 3 and 6 d of incubation.

Test for Antifungal Properties of Volatiles Released From Trichoderma Isolates

The technique of Dick and Hutchinson (1966) was used to test the production of volatiles by *Trichoderma* species. Plugs from actively growing margins of the most aggressive *Trichoderma* colonies were each placed in the center of PDA plates. Plates inoculated with plugs of *Aspergillus* and *Fusarium* in the same manner were incubated for 4 d previously, then inverted over the plates with *Trichoderma* plugs. A 0.45 μm pore sized membrane was placed between the two dishes. Control plates were the pathogenic fungi inverted over uninoculated agar dishes. The radial colony diameter was assessed after 6 d for six replicates incubated at 25°C.

Investigation of Extracellular Enzyme Production by Trichoderma *Isolates*

Complex carbon sources (protein, pectin, starch) media were prepared according to Hankin and Anagnostakis (1975) and the lipid agar was adapted from the method of El Azzabi *et al.* (1981). Additionally, a cellulose agar and a chitin agar were prepared in the same way. The ability to produce extracellular lipase, protease, amylase, cellulase and chitinase were investigated according to Gahan (1984). The colony diameter of three replicates was measured 10 and 15 d after inoculation.

Results and Discussion

Macroscopic Observations of Fungal Growth in Dual Culture

The study of fungal growth in dual cultures revealed that four strains of *Trichoderma* species (*T. harzianum* T1, T2 and *T. viride* T5, T6) had an inhibitory effect on the growth of both *F. moniliforme* and *A. flavus*. The first contact between both *Trichoderma* strains and either of the two pathogenic fungi was observed within 5 d after inoculation. Following 7 d of growth, intense colonisation of the agar by *Trichoderma* had occurred, with the mycelium of each pathogen appearing almost encircled by the aggressive *Trichoderma* colony (Table 1). In the plates with dual cultures of *Fusarium* and *Trichoderma*, an inhibition zone around the pathogen was observed, without there being close contact of the hyphae of the two fungi. This may possibly indicate the liberation of diffusable inhibitory substances by *Trichoderma* strains. In contrast, in the *Aspergillus–Trichoderma* combination, hyphae of aggressive *T. harzianum* or *T. viride* overrun the *A. flavus* colony.

Scanning Electron Microscopy Observation of Hyphal Interaction Areas

Mycelial samples collected from the interaction region between aggressive and non-aggressive *Trichoderma* and the pathogenic fungi were viewed with SEM. A parallel study of *Trichoderma-Fusarium* and *Trichoderma–Aspergillus* cultures revealed a different interaction patern. When *Fusarium* was co-cultured against the aggressive *Trichoderma* spp., the hyphae of the two fungi did not make intimate contact, as was the case for *Aspergillus–Trichoderma* combination.

Table 1. Colony diameter of Trichoderma spp. and either A. flavus or F. moniliforme in dual cultures

Fungal species	Diameter of fungal colony (mm)[a]		
	Day 3	Day 5	Day 7
F. moniliforme	27.1 ±1.37	35.5 ±1.28	35.8 ±0.98
T. harzianum (T1)	32.4 ±3.52	72 ±2.52	78.8 ±1.16
F. moniliforme	20.3 ±1.21	32.1 ±2.31	34 ±1.26
T. harzianum (T2)	39.8 ±1.29	42.6 ±2.8	50.5 ±1.81
F. moniliforme	14.3 ±1.36	24 ±2.28	36 ±4.81
T. harzianum (T3)	6.7 ±0.41	11.8 ±3.97	18.2 ±4.91
F. moniliforme	18.1 ±0.98	31.1 ±0.98	46 ±2.82
T. viride (T4)	2.5 ±2.24	10 ±2.28	25.8 ±2.31
F. moniliforme	25.6 ±2.3	40.6 ±3.61	44.1 ±2.31
T. viride (T5)	25.3 ±2.50	21.5 ±1.37	57.5 ±1.51
F. moniliforme	18.5 ±1.37	29.5 ±0.49	38 ±1.0
T. viride (T6)	32.8 ±2.56	55.9 ±6	67.8 ±2.4
F. moniliforme	18.5 ±1	32.1 ±0.75	44 ±1
Trichoderma (T7)	6.8 ±0.4	15.6 ±2.58	21.1 ±1.32
F. moniliforme	23.8 ±1.1	40.3 ±2.5	49.6 ±4.45
Trichoderma (T8)	18.6 ±2.33	27.3 ±3.2	38.8 ±6.4
F. moniliforme	21.5 ±0.49	37 ±2.44	44 ±1.26
Trichoderma (T9)	8.41 ±1.42	12.5 ±0.49	23.6 ±4.17
A. flavus	23 ±0.89	32.6 ±0.40	47.3 ±3
T. harzianum (T1)	28 ±0.89	45 ±0	58.8 ±1.16
A. flavus	22.3 ±0.87	31.1 ±0.49	45.8 ±0.98
T. harzianum (T2)	35.9 ±3	55.3 ±0.51	70.5 ±1
A. flavus	20.3 ±0.81	26.1 ±0.68	42 ±1
T. harzianum (T3)	13.5 ±1.42	16.9 ±2.1	38.1 ±2.2
A. flavus	21.4 ±0.9	29 ±1	50.1 ±2.63
T. viride (T4)	20.8 ±0.51	34 ±0	41.1 ±1.49
A. flavus	22.5 ±0.54	30.5 ±0.49	47.8 ±3.1
T. viride (T5)	15.5 ±2.72	34.58 ±2.8	68 ±3.2
A. flavus	19.5 ±2.97	31.2 ±1.4	49.5 ±1.87
T. viride (T6)	21.9 ±0.2	36 ±0	68 ±0
A. flavus	21.9 ±1.56	33 ±0	49 ±3.8
Trichoderma (T7)	17 ±1.1	32.8 ±1.5	32.8 ±1.9
A. flavus	21.5 ±0.4	30.5 ±0.4	45.6 ±2.25
Trichoderma (T8)	19.2 ±4	30.4 ±0.8	47.1 ±1.32
A. flavus	32.1 ±3.37	31.5 ±1.9	45.6 ±2.22
Trichoderma (T9)	16 ±1.2	23 ±0.8	34.4 ±1.35

[a]Values represent the means ±standard deviations of six replicates

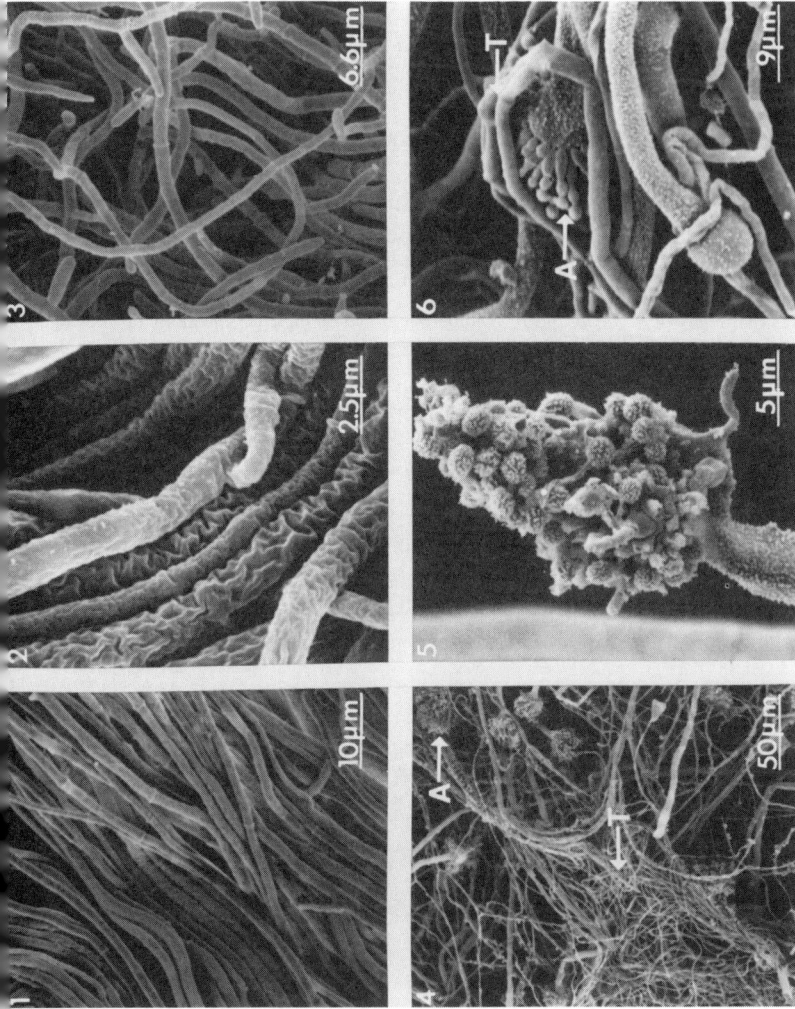

Figures 1 and 2. Portion of a colony of *F. moniliforme* co-cultured with *T. harzianum*; *Figure 3.* Aerial hyphae, which were predominant when *Fusarium* was co-cultured with *Trichoderma* spp.; *Figure 4.* Hyphal interaction between *A. flavus* and aggressive *Trichoderma*; *Figure 5.* Aberrant microhead of *A. flavus*; *Figure 6.* The interaction region between *A. flavus* and an aggressive *Trichoderma*. No hyphal penetration was observed

Hyphae from the growth-restricted colonies of *Fusarium* co-cultured with an aggressive strain of *T. harzianum* were generally wrinkled, prostate and adpressed (Figs. 1 and 2). In contrast, when *Fusarium* was co-cultured with a non-aggressive strain of *Trichoderma*, aerial hyphae predominated (Fig. 3), resulting in a fluffy, well-developed colony, similar to control colonies. When *A. flavus* was cultured with an aggressive *Trichoderma* strain (Fig. 4), numerous aberrant microheads were observed in the aspergilla colony (Fig. 5). Additionally, the inhibitory effect on *Aspergillus* hyphae (Fig. 6) did not result from hyphal penetration of *A. flavus* by *Trichoderma* [i.e. parasitism, which has been reported in the literature (Elad *et al.*, 1983)].

Inhibitory Effects of Cultures Filtrates of Trichoderma Strains

As measured by colony diameter (Table 2), culture filtrates of *Trichoderma* spp. had little or no effect on colony diameter of either *F. moniliforme* or *A. flavus*. Of the three carbon sources, growth of both fungal pathogens was greatest on the

Table 2. Colony diameters of *A. flavus* and *F. moniliforme* grown on different media incorporating different carbon and nitrogen sources, and liquid culture extracts of *T. harzianum* and *T. viride* strains

		Inhibition of fungal growth[a]			
		F. moniliforme		Pathogen *A. flavus*	
Medium	Antagonist	Control	Day 6	Control	Day 6
Maltose	*T. harzianum* (T1)	62.2 ±0.33	60.2 ±1.47	54.7 ±0.52	54.7 ±0.88
	T. harzianum (T2)	62.2 ±0.33	60.9 ±2	54.7 ±0.52	57.41 ±0.96
	T. viride (T5)	62.2 ±0.33	64.9 ±1.27	54.7 ±0.52	49.8 ±0.73
	T. viride (T6)	62.2 ±0.33	59.9 ±1.59	54.7 ±0.52	59.9 ±0.69
Sucrose	*T. harzianum* (T1)	58.2 ±0.46	59.9 ±0.94	57.9 ±0.67	64.55 ±0.95
	T. harzianum (T2)	58.2 ±0.46	57.9 ±1.46	57.9 ±0.67	60 ±0.63
	T. viride (T5)	58.2 ±0.46	60.2 ±0.56	57.9 ±0.67	59.9 ±0.69
	T. viride (T6)	58.2 ±0.46	58 ±0.72	57.9 ±0.67	61.7 ±2.18
Glucose	*T. harzianum* (T1)	61.1 ±0.78	57.3 ±0.62	58.9 ±0.19	60.8 ±0.54
	T. harzianum (T2)	61.1 ±0.78	58.2 ±0.36	58.9 ±0.19	63.96 ±0.67
	T. viride (T5)	61.1 ±0.78	57.6 ±0.79	58.9 ±0.19	61 ±0.66
	T. viride (T6)	61.1 ±0.78	60.3 ±0.75	58.9 ±0.19	66 ±0.78
L-alanine	*T. harzianum* (T1)	64.5 ±56	61 ±0	54.1 ±0.51	59.4 ±1.21
	T. harzianum (T2)	64.5 ±0.48	64.8 ±0.83	54.1 ±0.51	51.8 ±1.65
	T. viride (T5)	64.5 ±0.48	63 ±1.55	54.1 ±0.51	59.4 ±0.65
	T. viride (T6)	64.5 ±0.48	62.8 ±0.42	54.1 ±0.51	59.4 ±0.53
KNO₃	No growth	No growth	No growth	No growth	No growth

[a]Values represent the mean ± standard deviations of six replicates

Table 3. Colony diameter of pathogenic fungi grown on agar plates inverted over *Trichoderma* plates

Pathogen	Control	T. harzianum (T1)	T. harzianum (T2)	T. viride (T5)	T. viride (T6)
		Hyphal growth (mm)[a] on plates in combination with:			
A. flavus	74.6 ±0.51	74 ±0.7	71.9 ±0.66	67.3 ±0.87	71.5 ±0.8
F. moniliforme	74 ±0.8	74.1 ±0.51	75.3 ±0.4	57.2 ±0.97	69.9 ±0.8

[a]Values represent means ± standard deviations of six replicates

medium incorporating maltose. On this carbon source, a single strain of *T. viride* (T5) marginally inhibited the radial growth of *A. flavus*. Additionally, filtrates of this *T. viride* strain, when incorporated into a glucose medium, slightly inhibited the growth of *F. moniliforme*. All fungi had an optimal growth on the medium incorporating the organic nitrogen source (L-alanine), in comparison with the medium containing inorganic nitrogen source (KNO_3), where no growth was recorded.

The explanation for these negligible inhibitory effects may be two-fold: first, *Trichoderma* spp. may not have produced substances that were inhibitory to the other fungal species, and secondly, if such factors were indeed produced, the 1:10 dilution may have resulted in a concentration too low to effect any inhibitory response. Jackson *et al.* (1991), using the same dilution (1:10) of culture filtrates of several isolates of *T. viride*, found no inhibition of growth of three pathogens. *Gliocladium virens* and *T. pseudokoningii* were, however, effective in inhibiting the radial growth of the pathogens. Where it is at this stage not possible to demonstrate any statistical differences of the effect of culture filtrates of *Trichoderma* strains on the pathogenic fungal growth, it would appear that *T. viride* (T5) is possibly the most antagonistic towards the two pathogens.

Possible Volatiles Produced by Trichoderma *Isolates*

In general, colony growth of *A. flavus* and *F. moniliforme* was minimally inhibited or showed no response when exposed to cultures of *Trichoderma* isolates. The exception was, however, in the instance of *T. viride* (T5), where colony diameter of both pathogens was reduced. In comparison with the controls (Table 3), this inhibition of colony diameter was 10% for *A. flavus*, and 23% for *F. moniliforme*.

These results would suggest, on the whole, that volatile compounds do not play an important role in any inhibitory effects observed in colony growth and

Table 4. The growth of four *Trichoderma* stains on different osmotic potentials, measured as a diameter (mm) in agar-plate culture

| | Fungal species[a] | | | | | | | |
| Medium | T. harzianum (T1) | | T. harzianum (T2) | | T. viride (T5) | | T. viride (T6) | |
	10	15	10	15	10	15	10	15
Lipid 5% NaCl	22.6 ±0.52	28.2 ±0.64	22.4 ±0.64	29.6 ±0.70	50.3 ±0.65	58.7 ±0.79	3.7 ±0.43	4.5 ±0.36
Lipid 10% NaCl	–	–	5.8 ±0.28	5.8 ±0.28	24.6 ±0.81	30 ±0.2	–	–
Protein 5% NaCl	17.8 ±2.23	27.9 ±0.3	25.7 ±0.4	30.9 ±1	40 ±0.15	55.7 ±1.15	24.4 ±0.87	35.5 ±0.51
Protein 10% NaCl	–	–	1.6 ±0.57	1.6 ±0.57	7.6 ±0.69	19.3 ±0.57	1 ±0	1 ±0
Starch 5% NaCl	43.9 ±1	50.9 ±0.83	15.6 ±0.57	20.8 ±0.23	6.7 ±0.46	8 ±0	3 ±0.1	3 ±0.1
Starch 10% NaCl	–	–	–	–	3.8 ±0.23	6.8 ±0.2	2 ±0	2 ±0
PGA 5% NaCl	1 ±0	3 ±0.2	5.9 ±0.11	7.1 ±0.32	8.1 ±0.96	15.4 ±0.51	2 ±0	3.4 ±0.52
PGA 10% NaCl	–	–	–	–	14.3 ±0.61	21.6 ±0.57	–	–
Cellulose 5% NaCl	19.6 ±0.52	25.5 ±0.77	26.6 ±1.16	32.6 ±1.18	25.6 ±0.69	35.9 ±0.36	20.9 ±0.17	25.9 ±1
Cellulose 10% NaCl	–	–	2.6 ±0.57	2.6 ±0.57	14.2 ±0.66	22.2 ±0.64	–	–
Chitin	Full[b]	Full	Full	Full	Full	Full	Full	Full

[a]Values represent the mean ±standard deviations of three replicates

[b]Full: full plate, in excess of 88 mm

Table 5. Extracellular enzyme production of *Trichoderma* spp.

Extracellular enzyme	T. harzianum		T. viride	
	T1	T2	T5	T6
Amylase	+[a]	+	+	+
Cellulase	+	+	+	+
Pectinase	+	−[b]	+	−
Lipase	+	+	+	−
Protease	−	−	−	−
Chitinase	−	−	−	−

[a]Enzyme detected

[b]Enzyme not detected

morphology of the pathogens, with the possible exception of *T. viride* (T5). In the instance of *F. moniliforme*, the observed inhibition zone, where SEM demonstrated no hyphal contact between *T. viride* and the pathogen, may have resulted partly from the production of volatiles by the *Trichoderma* strain.

Investigation of Extracellular Enzyme Production by Trichoderma *Isolates*

The ability of four strains of *Trichoderma* spp. to produce extracellular enzymes was investigated using solid media. The different concentrations of salt incorporated in basal media had an effect on fungal growth (Table 4). With increasing osmotic pressure, there was a general trend for a decrease in fungal growth. *T. viride* (T5) appeared the least affected, growing prolifically on all media.

When stained appropriately to demonstrate depletion of the nutrient source (hence demonstrating extracellular enzyme production), all *Trichoderma* strains exhibited amylase and cellulase activity. *T. viride* (T5) was, on the whole, the most prolific enzyme producer (Table 5).

Conclusions

In summary, evidence has been presented in this study that four strains of *Trichoderma* spp. were capable of curtailing the growth of *A. flavus* and *F. moniliforme*, two fungal species important in effecting plant/seed deterioration. It has been reported that the inhibitory action of aggressive *Trichoderma* strains may arise from mycoparasitism, competition for nutrients and/or antibiosis (Whipps and McQuilken, 1993). Based on the scanning electron microscopy studies of the present investigation, it would appear that mycoparasitism plays no role in the observed inhibitory response. The production of extracellular enzymes and volatile compounds by most *Trichoderma* strains, notably *T. viride* (T5), would suggest that competition and antibiosis may be the mechanisms operating in the *Trichoderma*–pathogen combinations considered in the present investigations.

References

Dick, C.M. and Hutchinson, S.A. 1966. *Nature, London* 211: 868.

El Azzabi, T.S., Clarke, J.H. and Hill, S.T. 1981. *Journal of the Science of Food and Agriciculture* 32: 493–497.

Elad, Y., Chet, I., Boyle, P. and Henis Y. 1983. *Phytopathology* 73: 85–88.

Gahan, P.B. 1984. *Plant Histochemistry and Cytochemistry*, pp. 118–121. Academic Press.

Hankin, L.H. and Anagnostakis, S.L. 1975. *Mycologia* 67: 597–607.

Jackson, A.M., Whipps, J.M., Lynch, J.M. and Bazin, M.J. 1991. *Biocontrol Science and Technology* 1: 43–51.

McLean, M. and Berjak, P. 1987. *Seed Science and Technology* 15: 831–850.

Mycock, D.J., Rijkenberg, F.H.J. and Berjak, P. 1992. *Seed Science and Technology* 20: 1–13.

Rheeder, J.P., Marasas, W.F.O., Thiel, P.G., Shephard, G.S. and van Schalkwyk, D.J. 1992. *Phytopathology* 82: 353–357.

Whipps, J.M. and McQuilken, M.P. 1993. *Exploitation of Microorganisms* (ed. D.G. Jones). London: Chapman and Hall.

83. Preservation of *Quercus robur* L. Embryonic Axes in Liquid Nitrogen

P. CHMIELARZ

Institute of Dendrology, Polish Academy of Sciences, 62–035 Kórnik, Poland

Abstract

Seeds of pedunculate oak show 'recalcitrant' seed storage behaviour and cannot be stored for more than three winters without loss of germinability. The aim of the experiments was to find out whether it is possible to protect excised oak embryonic axes from injuries caused by partial desiccation and freezing in liquid nitrogen (LN2).

Embryonic axes were sterilized, cryoprotected, coated with calcium alginate, slowly dehydrated to 24% moisture content (fresh weight basis) and finally frozen, first slowly to –20°C and afterwards rapidly to –196°C. After rewarming (at 40°C) and 3 weeks of *in vitro* culture survival of the embryonic axes was assessed. After exposure of embryonic axes to LN2 (–196°C) the following results were obtained: 4% of embryonic axes with growing shoot, 8% with growing root, 0% with growing shoot and root, 78% of axes producing callus, 6% of greening embryonic axes without shoot and root growth, so the total survival was 96%. Regeneration of plants from LN2-treated embryonic axes is, however, impossible because so few surviving embryonic axes had a growing shoot and none had growing shoots and roots.

Introduction

The acorns of pedunculate oak (*Quercus robur* L.) are classified as recalcitrant. Recalcitrant seeds are not able to tolerate desiccation (Roberts, 1973). Farrant *et al.* (1988) claimed that no successful method has been found for long-term storage of recalcitrant seeds. According to Roberts *et al.* (1984) the most promising method of germplasm conservation for recalcitrant seeds could be storage in liquid nitrogen (LN2). Bajaj (1985) also suggested that the germplasm of recalcitrant seeds could possibly be conserved through cryopreservation of their excised embryos. However, to survive at very low temperatures seeds must be dried prior to freezing. So far, recalcitrant seeds of very few species survived storage in liquid nitrogen. Probably the desiccation of embryonic axes instead of whole embryos or seeds should be investigated in detail prior to work on cryopreservation. Pence (1990) tested cryostorage of embryonic axes of several large-seeded temperate trees after desiccation.

Poulsen (1992) attempted to cryopreserve embryonic axes of *Quercus robur* but no axes survived exposure to –20°C.

In the experiments reported here we have tried to find out whether it is

R.H. Ellis, M. Black, A.J. Murdoch, T.D. Hong (eds.), Basic and Applied Aspects of Seed Biology, pp. 765–769.
© *1997 Kluwer Academic Publishers, Dordrecht. Printed in Great Britain.*

possible to protect pedunculate oak embryonic axes from injuries, caused by partial desiccation and freezing in LN2 for the purpose of long-term storage of the genetic diversity of oaks.

Materials and Methods

Acorns were collected in mid-west Poland from the ground during shedding in mid-October. Seeds used for experiments were stored for two weeks in a non-sealed bag at 3°C at a moisture content of 43% (fresh weight basis). The excised embryonic axes of acorns were disinfected with 0.1% mercuric chloride for 2.5 min and rinsed thoroughly in aseptic distilled water four times for one min. The moisture content of acorns was determined in three replicates for acorns (7 acorns each) and embryonic axes (30 axes in each replicate) (oven method, 105°C/48 h for acorns, 105°C/24 h for embryonic axes).

After sterilization and rinsing in sterile water, embryonic axes were treated by the following cryoprotectants: 0.5 M solution of sucrose 20 h (5°C), followed by 0.75 M sucrose 1 h (20°C), 1 M sucrose 1 h (20°C) and finally 1 M glycerol 1 h (20°C).

Embryonic axes destined for cryopreservation were then coated with calcium alginate. For this purpose they were suspended in a liquid calcium-free medium, supplemented with 3% sodium alginate, macro- and micronutrients and sucrose 20 g/l. This mixture was then dropped (inside one drop was only one embryonic axis) into a liquid medium containing calcium chloride, in order to form calcium alginate beads.

After half an hour of shaking on a rotary shaker ($\sim$130 rpm) in calcium chloride, beads of about 3 mm in diameter and containing one embryonic axis were then surface dehydrated by 1 h of drying (20°C) under air flow in a laminar cabinet (preliminary phase of desiccation). After that desiccation of beads was continued, still in sterile conditions, in tightly sealed Petri dishes above silica gel (7 cm^3 of silica gel/one Petri dish/30 embryonic axes) 20 h (main phase of desiccation) until the moisture content of embryonic axes was reduced to 24%. After the latter phase of desiccation the moisture content of beads and embryonic axes was determined using the oven method (105°C/24 h), separately for embryonic axes and the coatings alone (6 replicates with 15 axes or coatings).

Before freezing the coated embryonic axes were placed in sterile plastic cryovials (1.8 ml). They were frozen initially under controlled conditions to 0°C (stage I), at a cooling rate of 2°C/min followed by freezing to –20°C at a cooling rate 1°C/min (stage II) and finally in the stage III – by ultra rapid cooling from –20°C to –196°C through direct immersion of samples in LN2 for 24 h.

After 24 h of storage in LN2 embryonic axes were thawed by rapid rewarming in a water bath at 40°C for 5 min. Cryovials were surface sterilized by immersion in 70% ethanol.

Table 1. Percentage of *Q. robur* embryonic axes surviving (growth *in vitro*) before freezing and after freezing, after 3 weeks of *in vitro* culture

Criteria of growth *in vitro*	Embryonic axes not stored at −196°C (control)	Embryonic axes after 24 h of storage at −196°C	
	Growth[a] (%)	Survival[b] (%)	Recovering[c] (shoot growth, shoot and root growth, % with reference to control) (%)
Shoot	28 a	4 b	14
Root	13 c	8 d	–
Shoot and root	40 e	0 f	0
Callus	0 g	78 h	–
Greening of embryonic axes	19 i	6 i	–
Sum	100	96	14

[a,b]Means with the same letter not significantly different with Duncan's test (*p* = 0.05)

[c]Explants recorded as 'recovering' were those with shoot elongation or shoots and root elongation with a normal pattern of development. Recovering explants plus those which formed callus, plus embryonic axes with only root growth and only greening were classified as 'surviving'

Coated embryonic axes were cultured *in vitro* for 24 h. After that the coatings were removed and the axes were transferred in to a fresh medium of the same composition. All embryonic axes were cultured separately in single glass tubes. Survival of embryonic axes was assessed after 3 weeks of their *in vitro* culture (3 replicates, 30 embryonic axes in each). Survival of embryonic axes was determined after each stage of the experiment i.e. after sterilization (control), after cryoprotection, after desiccation and finally after storage in LN2. Criteria of survival of individual embryonic axes were: shoot growth, root growth, shoot and root growth, callus growth and greening of embryonic axes without shoot and root growth. The following elements were used for the *in vitro* culture medium: macronutrients – Quoirin and Lepoivre (1977), micronutrients – Murashige and Skoog (1962), vitamins, 1 mg/l BAP (6-benzylaminopurine), 0.25 mg/l zeatin, sucrose 30 g/l, agar 6 g/l; the pH was adjusted to 5.8 prior to autoclaving. Cultures were incubated at 26 ± 2°C in 16/8 h light/darkness photoperiod.

Results and Discussion

Adding up all percent values of survival (all criteria of growth of embryonic axes) we have obtained 96% (Table 1) of survival of embryonic axes after their exposure to LN2.

SHOOT GROWTH

SHOOT & ROOT GROWTH

ROOT GROWTH

CALLUS GROWTH

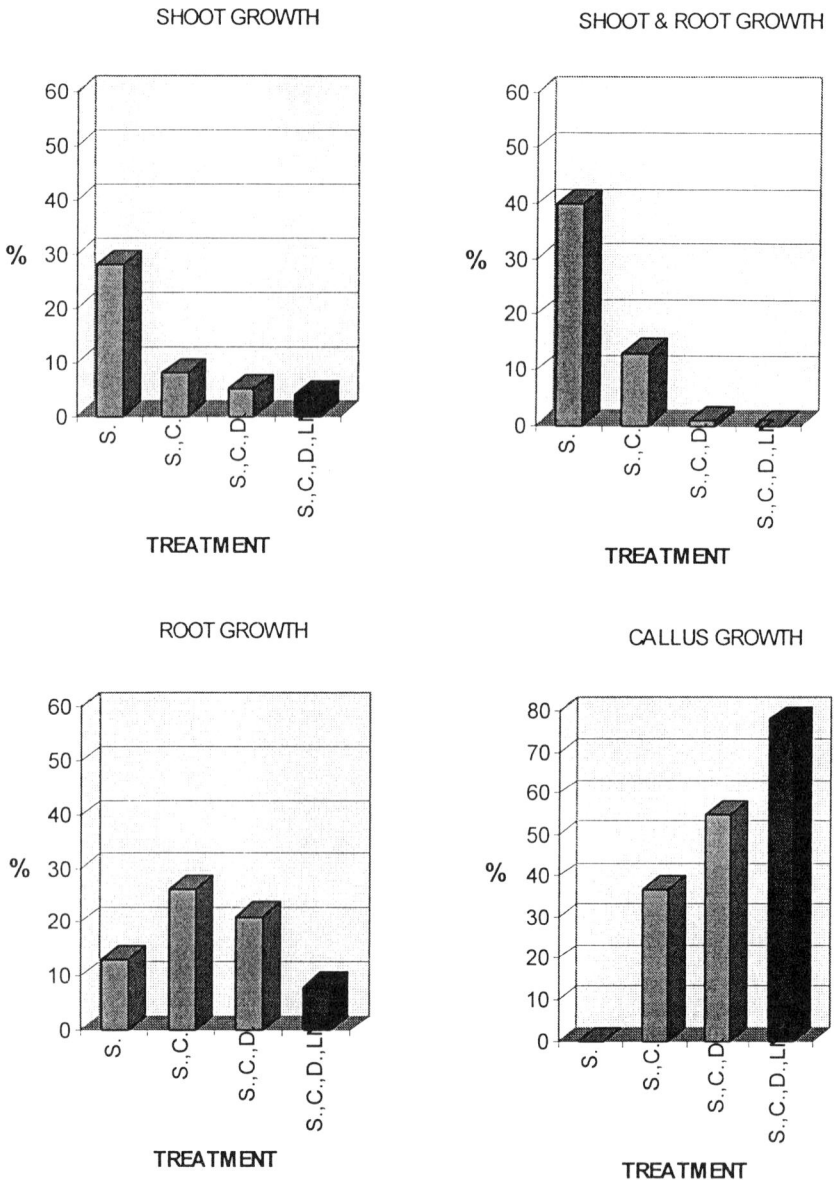

Figure 1. Quercus robur L. Survival (growth on culture) of embryonic axes after their exposure to liquid nitrogen (–196°C/24 h), rewarmed at 40°C. Growth of embryonic axes *in vitro* after 3 weeks of culture. Before freezing embryonic axes were dehydrated to 24% of moisture content (fresh weight basis)

S, embryonic axes after sterilization; SC, embryonic axes after sterilization and cryoprotection; SCD, embryonic axes after sterilization, cryoprotection, desiccation; SCDLN, embryonic axes after sterilization, cryoprotection, desiccation and storage in liquid nitrogen

Regeneration of plants from explants ('recovering') is still not possible because of the low number of embryonic axes with growing shoots or shoots and roots (Table 1 and Fig. 1). We have observed that partial desiccation of embryonic axes always increased *in vitro* the number of axes producing roots, in comparison with non-dessicated embryonic axes (Fig. 1b, root growth).

In the case when it is still not possible to preserve genetic diversity of oaks in gene banks for a long period of time, investigations should be continued to increase the number of normally growing *in vitro* (shoot only or shoot and root) embryonic axes surviving after exposure to LN2.

References

Bajaj, Y.P.S. 1985. In: *Cryopreservation of plant cells and organs.* (ed. K.K. Kartha) pp. 228–242, CRC Press, Florida.

Farrant, J.M., Pammenter, N.W. and Berjak, P.1988. *Seed Scence and Technology* 16: 155–166.

Murashige, T. and Skoog, E. 1962. *Physiology Plantarum* 15: 473–497.

Pence, V.C. 1990. *Cryobiology* 27: 212–218.

Poulsen, K.M. 1992. *Cryo-Letters* 13: 75–82.

Quoirin, M. and Lepoivre, P. 1997. *Acta Horticultura* 78: 437–442.

Roberts, E.H. 1973. *Seed Science and Technology* 1: 449–514.

Roberts, E.H., King, M.W. and Ellis, R.H. 1984. In: *Conservation and Evaluation.* (eds. J.H.W. Holden and J.T. Williams), pp. 38–52. London.

84. Desiccation Tolerance and Cryopreservation of Embryonic Axes of Recalcitrant Species

D. DUMET and P. BERJAK

Centre for Indigenous Plant Use Research, Department of Biology, University of Natal, Durban, Private Bag X10, Dalbridge, 4014 South Africa

Abstract

Because recalcitrant seeds are generally large and show very curtailed longevity even if stored hydrated, cryopreservation of their embryonic axes seems to be the most obvious way to create a gene bank. The aim of this investigation was to determine the minimal water content tolerated by embryos of five recalcitrant species and their subsequent survival after cryopreservation. As sucrose is known to improve desiccation- and cryopreservation-tolerance of many plant tissues, experiments were performed with both non-pre-treated and sucrose pre-treated embryonic axes. Embryonic axes of *Camellia sinensis* and *Azadirachta indica* tolerated dehydration to very low water contents, 0.14 and 0.23 g water per g of dry weight (g.g^{-1}), respectively. In such dry states most of the axes survived cryopreservation. Below 0.29, 0.42 and 0.34 g.g^{-1}, no survival after desiccation was recorded for *Trichilia dregeana, Artocarpus heterophyllus* and *Landolphia kirkii* embryos, respectively. Survival at lower water contents, 0.16 and 0.21 g.g^{-1}, respectively for *Trichilia dregeana* and *Artocarpus heterophyllus* could be obtained after a high sucrose pre-treatment. In contrast, viability was lost below 0.42 g.g^{-1} in the case of pre-treated axes of *Landolphia kirkii*. Concerning the last three species, survival after cryopreservation was recorded only for sucrose pre-treated, desiccated axes of *Trichilia dregeana*.

Introduction

Seeds have been classified into two categories, orthodox or recalcitrant, depending on their desiccation tolerance (Roberts, 1973). When mature, orthodox seeds have low water contents (20% or lower on a fresh weight basis [FWB]) and can withstand further desiccation down to 1 to 5% (FWB) without losing viability (Roberts and King, 1980). In such dry states, seed metabolism is minimal or even ceases and the lower the storage temperature the longer they can be stored (Roberts and Ellis, 1977). Mature recalcitrant seeds are character-ized by high water contents at shedding (e.g. up to 67%, in the case of *Avicennia marina* [Berjak *et al.*, 1984]), and a high sensitivity to desiccation (Roberts and Ellis, 1977). Moreover, they are often chilling sensitive (Chin and Roberts, 1980). Metabolism of recalcitrant seeds remains active, their life span is limited and their long-term storage is not possible. Seeds of a few species which were initially classified as recalcitrant have recently been shown to tolerate desiccation to between 5 and 12% (Ellis *et al.*, 1990; 1991). However, as opposed to

R.H. Ellis, M. Black, A.J. Murdoch, T.D. Hong (eds.), Basic and Applied Aspects of Seed Biology, pp. 771–776.
© *1997 Kluwer Academic Publishers, Dordrecht. Printed in Great Britain.*

orthodox types, such seeds are sensitive to low temperatures when desiccated. This category of seeds has been called intermediate (Ellis *et al.*, 1990).

Many indigenous and cultivated species in South Africa produce recalcitrant seeds. Consequently, their long term storage is not straightforward and cryostorage appears to offer the only means for conservation of the germplasm, i.e. storage at very low temperature, generally in liquid nitrogen (–196°C). At this temperature, cell division and all metabolic processes cease so that tissues can, in theory, be stored for several thousand years (Ashwood-Smith and Friedman, 1979). However, due to their high water contents, recalcitrant seeds must be subjected to specific treatments before cryopreservation in order to avoid the formation of lethal ice crystals. In practice, the smaller the tissue volume, the easier can it be conditioned for cryopreservation. However, because recalcitrant seeds are often large, cryopreservation of their embryonic axes appears to be the only way to create a gene bank from seed sources.

During the past few years, various investigations have shown that zygotic embryonic axes of recalcitrant (e.g. coconut) or intermediate (e.g. coffee, oil palm) seeds (Engelmann *et al.*, 1995) and from various large-seeded temperate species (Pence, 1990) can be successfully cryopreserved after partial desiccation. Similar processes have been used for cryopreservation of somatic embryos (Anandarajah and McKersie, 1990; Dumet *et al.*, 1993). Somatic embryos can be likened, in some respects, to zygotic embryos of recalcitrant or intermediate species as their water content remains high throughout their development. Moreover, a high sucrose pre-treatment has been reported to be of benefit in promoting desiccation and/or freezing tolerance of zygotic and somatic embryos (Monier and Leddet, 1978; Engelmann, 1986; Anandarajah and McKersie, 1990; Assy-Bah and Engelmann, 1992; Dumet *et al.*, 1993).

The aim of the present investigation was to determine the minimal water content tolerated by embryonic axes of five different recalcitrant species (*Camellia sinensis* [L.] O. Kuntze, *Azadirachta indica* A. Juss., *Trichilia dregeana* Sond., *Artocarpus heterophyllus* Lamk. and *Landolphia kirkii* Dyer), with or without a high sucrose pre-treatment, and the subsequent survival of the axes after cryopreservation.

Materials and Methods

Embryonic axes were excised from seeds removed from newly-harvested fruit. They were then surface sterilized in a 1% solution of commercial sodium hypochlorite for 20 min, followed by three rinses in sterile, distilled water. After a rapid surface drying on filter paper, some axes were directly placed on culture media as controls, while others were first pre-treated on a high sucrose medium (0.75 M) or directly desiccated with silica gel. Desiccation was performed in 50 ml air-tight glass vials containing 30 g of silica gel except for *A. indica* where desiccation was carried out in cryotubes containing 0.5 to 1 g of silica gel (5 axes per glass vial or cryotube). After various desiccation periods (from 0 to 4 h) some of the axes were transferred to culture medium while others were frozen. Cryopreservation was performed in cryotubes that were directly immersed in liquid nitrogen for at least 1 h. For thawing, cryotubes were plunged in a 40°C water bath for 2 min. The basic culture medium consisted on half-strength Murashige and Skoog (1962) salts and organics, 0.14 M sucrose and 0.8% agar. In the case of *L. kirkii* embryonic axes, 0.6 µM kinetin, 1.00 µM naphthalene-acetic acid, 1.4 µM gibberellic acid ind 4 g.l^{-1} charcoal were added to the culture medium. Cultures were maintained at 27°C with a 16 h photoperiod. Determination of the water content, expressed in g water per g dry weight, was performed after desiccation of the samples at 80°C for two days (d).

Results

Depending on the species, embryo initial water contents varied between 2.33 and 4.02 g water per g dry weight (g.g^{-1}) (Table 1). After a high sucrose pre-treatment, these values dropped to between 1.88 and 2.24 g.g^{-1} (Table 1).

Table 1. Initial water content of embryonic axes and those achieved after high sucrose pre-treatment

	Not sucrose pre-treated	Sucrose pre-treated
C. sinensis	3.46 ± 0.93	1.88 ± 0.4
A. indica	2.33 ± 0.49	Not tested
T. dregeana	4.02 ± 0.43	2.03 ± 0.32
A. heterophyllus	3.38 ± 0.69	2.24 ± 0.61
L. kirkii	2.93 ± 0.82	2.12 ± 0.61

Axes of *C. sinensis* and *A. indica* could be desiccated to very low water contents (0.14 and 0.23 g.g^{-1}, respectively) without losing viability (Table 2). In comparison, embryonic axes of *T. dregeana*, *A. heterophyllus* and *L. kirkii* appeared more sensitive to a water loss, only some surviving when desiccated to moisture contents of 0.29, 0.42 and 0.34 g.g^{-1}, respectively (Table 2). Whatever the species, surviving desiccated axes showed similar development to their non-desiccated controls. Sucrose pre-treatment improved desiccation tolerance of *T. dregeana* and *A. heterophyllus* axes, facilitating lowering of their water content to 0.16 and 0.21 g.g^{-1}, respectively (vs 0.29 and 0.42 g.g^{-1} when not pre-treated) (Table 2). However, while embryo viability remained high in case of *T. dregeana* (10/15), only 1 out of 5 axes of *A. heterophyllus* survived. Sucrose pre-treatment had a negative effect on *C. sinensis* and *L. kirkii* axes. For *C. sinensis*, 80% of the pre-treated axes survived to a water content of 0.18 g.g^{-1}

(vs 100% survival at 0.14 g.g^{-1} for non pre-treated axes) while for *L. kirkii* axes viability was lost below 0.42 g.g^{-1} (vs 0.34 g.g^{-1} for non pre-treated axes) (Table 2). Moreover, after high sucrose pre-treatment, whether followed by a desiccation period or not, embryonic axes of *C. sinenis*, *A. heterophyllus* and *L. kirkii* showed abnormal growth (becoming swollen rather than elongating). This phenomenon was not observed in the case of *T. dregeana* embryonic axes where a high sucrose pre-treatment seemed to promote radicle elongation (data not shown).

Very high survival rates after cryopreservation were recorded for desiccated embryonic axes of *C. sinensis* and *A. indica* (90 and 100%, respectively). A high sucrose pre-treatment was essential to obtain a limited survival of 28% after cryopreservation of *T. dregeana* embryonic axes. Desiccated and cryopreserved axes of *C. sinensis* and *A. indica* showed similar development to their unfrozen controls. That was not the case with cryopreserved embryonic axes of *T. dregeana* which showed only poor growth characterized by the development of greened cell clumps, presumably the fore-runners of callus. None of the embryonic axes of *A. heterophyllus* or *L. kirkii* survived the temperature of liquid nitrogen under the present experimental conditions (Table 2).

Table 2. Lowest water content, g water per g dry weight (g.g^{-1}), commensurate with viability retention (LWCV), survival at LWCV and survival after cryopreservation (LN) at LWCV of embryonic axes with or without a high sucrose pre-treatment

	Not pre-treated			Pre-treated		
	LWCV (g.g^{-1})	Survival at LWCV	Survival after LN at LWCV	LWCV (g.g^{-1})	Survival at LWCV	Survival after LN at LWCV
C. sinensis	0.14±0.02	10/10	9/10	0.18±0.08	4/5	2/5
A. indica	0.23±0.06	10/10	10/10	Not tested	Not tested	Not tested
T. dregeana	0.29±0.01	7/20	0/20	0.16±0.04	10/15	7/25
A. heterophyllus	0.42±0.01	1/5	0/5	0.21±0.05	1/5	0/5
L. kirkii	0.34±0.07	4/5	0/5	0.42±0.04	2/4	0/5

Discussion

Despite being categorized in the same seed-type group, i.e. recalcitrant, desiccation tolerance of isolated embryonic axes of these five species is very different, as is their amenability to cryopreservation.

Two out of the five (*C. sinensis* and *A. indica*) species possess highly desiccation-tolerant axes. These can be dehydrated to such low water contents that subsequent freezing does not affect their viability. It is interesting that while intact seeds of *A. indica* can be rapidly desiccated to very low water content (4.5% FWB, data not shown) without losing viability, seeds of *C. sinensis* are recalcitrant in the true sense (Berjak *et al.*, 1989). While *A. indica* seeds from certain provenances have been categorized as orthodox (Bellefontaine and Audinet, 1993), slow drying (over 4 months) of seeds from this batch was accompanied by 80% loss of viability. Additionally, these *A. indica* seeds are chilling-sensitive in the hydrated state (Berjak *et al.*, 1995). Their categorization as orthodox/intermediate/recalcitrant is thus anomalous.

For two other species, *T. dregeana* and *A. heterophyllus*, axis desiccation-tolerance can be improved by a high sucrose pre-treatment. However, the minimum water content achieved commensurate with viability retention might not be low enough to avoid the formation of lethal ice crystals, resulting in the destruction of whole axes (*A. heterophyllus*) or parts thereof (*T. dregeana*). In this last case, the proportion of surviving cells appears to be insufficient to ensure normal axis development. In this experiment, only one recovery medium was used for each species. Attention should be paid to the recovery conditions since slight modifications have been shown to enhance considerably the recovery rates of different somatic and zygotic embryos (Engelmann *et al.*, 1995). For axes of *L. kirkii*, a high sucrose pre-treatment further decreased desiccation tolerance. This may be due to the fact that water content decreases slowly when embryos are placed on a high sucrose medium. Berjak and co-workers (1989) have previously shown for this species that axes survive very rapid drying to substantially lower water contents than do axes from slowly dried seeds.

For all species, except *T. dregeana*, a high sucrose pre-treatment led to abnormal growth of embryos. Consequently, for these desiccation and sucrose-sensitive species, an alternative stategy of cryopreservation must be developed. This may require the use of non-sucrose cryoprotectants and very fast freezing rates, in order to cryopreserve axes at higher water contents.

It is interesting to note that axes of *A. heterophyllus* and *L. kirkii*, which show the highest sensitivity to desiccation, derived from seeds rich in latex. It is possible that during desiccation the latex crystallises and cannot re-dissolve during re-imbibition of the axes. This phenomenon could be responsible for damage by mechanical means (crystal formation). Some survival after cryopreservation of partially mature, desiccated embryos of *A. heterophyllus* has been previously achieved (e.g. by Chandel *et al.*, 1995). At that developmental stage, axis latex content might be lower than when mature (as they were in the present experiment). Furthermore, less mature axes may be more desiccation tolerant, which would facilitate survival to lower water contents and hence successful cryopreservation.

Further experiments are necessary in order to elucidate the differences in desiccation and cryopreservation tolerance of the species presently investigated. This includes DSC (differential scanning calorimetry) analysis of axes to reveal the thermodynamic properties of the tissue water. Such studies will reveal if the proportion of free : bound water is similar from one species to another, and if survival of cryopreservation can be correlated with free water loss. The occurrence of specific sugar combinations and proportions, and of specific proteins has been correlated with desiccation tolerance (Koster and Leopold, 1988; Blackman *et al.*, 1991; 1992; Kovach *et al.*, 1992). It would be interesting to ascertain if the difference in desiccation tolerance of these five species is related to the presence or absence of such factors. Besides the absence of such characteristics, other attributes may interfere with the capacity of a tissue to survive very low water content. The nature and content of certain secondary metabolites might be one of them.

References

Abdelnour-Esquivel, A., Villalobos, V. and Engelmann, F. 1992. *Cryo-Letters* 13: 297–302.

Anandarajah,K. and McKersie, B.D. 1990. *Plant Cell Reports* 9: 451–455.

Ashwood-Smith, M.S. and Friedman, G.B. 1979. *Cryobiology* 16: 132–140.

Assy-Bah, B. and Engelmann, F. 1992. *Cryo-Letters* 13: 117–126.

Bellefontaine, R. and Audinet, M. 1993. *IUFRO Symposium, Ouagadougou, Burkina Faso.* Bachuys Publishers, The Netherlands.

Berjak, P., Dini, M. and Pammenter, N.W. 1984. *Seed Science and Technology* 12: 365–384.

Berjak, P., Farrant, J.M. and Pammenter, N.W. 1989. In: *Recent Advances in the Development and Germination of Seeds,* pp 89–108 (ed. R.B. Taylorson). NewYork, London: Plenum Press.

Berjak, P., Campbell, G.K., Farrant, J.M., Omondi-Oloo, W. and Pammenter, N.W. 1995. *Seed Science and Technology* 23: 779–792.

Blackman, S.A., Wettlaufer, S.H., Obendorf, R.L. and Leopold, A.C 1991. *Plant Physiology* 96: 868–874.

Blackman, S.A., Obendorf, R.L. and Leopold, A.C. 1992. *Plant Physiology* 100: 225–230.

Chandel, K.P.S., Chaudhury, R. and Radhamani, J. 1994. *IBGRI-NBPGR Report.*

Chin, H.F.and Roberts, E.H. 1980. In: *Recalcitrant Crop Seeds.* Kuala Lumpur, Malaysia: Tropical Press. SDN.BDH.

Dumet, D., Engelmann, F., Chabrillange, N. and Duval, Y. 1993. *Plant Cell Reports* 12: 352–355.

Ellis, R.H., Hong, T.D. and Roberts, E.H. 1990. *Journal of Experimental Botany* 41: 1167–1174.

Ellis, R.H., Hong, T.D., Roberts, E.H. and Soetisna, U. 1991. *Seed Science Research* 1: 99–104.

Engelmann, F. (1986). *Thèse de Doctorat,* Université Paris 6, 228pp.

Engelmann, F., Dumet, D., Chabrillange, N., Abdelnour-Esquivel, A., Assy-Bah, B., Dereuddre, J. and Duval, Y. 1995. *Plant Genetic Resources Newsletter* 103: 27–31.

Koster, K.L. and Leopold, A.C. 1988. *Plant Physiology* 88: 829–832.

Kovach, D.A., Still, D.W. and Bradford, K.J. 1992. *Proceedings of the 4th International Workshop on Seeds, Basic and Applied Aspects of Seeds Biology, Vol 3.* Angers, France, pp 851–856.

Monier, M. and Leddet, C. 1978. *Comptes Rendus de l'Academie des Sciences de Paris,* 287: 615–618.

Murashige, T. and Skoog, F. 1962. *Physiologia Plantarum* 15: 473–497.

Normah, M.N., Chin, H.F. and Hor, Y.L. 1986. *Pertanika* 9: 299–303.

Pence, V.C. 1990. *Cryobiology* 27: 212–218.

Roberts, E.H. 1973. *Seed Science and Technology,* 1: 499–514.

Roberts, E.H. and Ellis, R.H. 1977. *Nature* (London), 268: 431–432.

Roberts, E.H. and King, M.W. 1980. In: *Recalcitrant Crop Seeds.* (eds H.F. Chin and E.H. Roberts). Kuala Lumpur, Malaysia: Tropical Press.

85. Studies of Hot Water Treatments for Curtailing Seed-Associated Mycoflora

D.P. ERDEY[1], D.J. MYCOCK[2] and P. BERJAK[1]

[1]*Plant Cell Biology Research Unit, Department of Biology, University of Natal, Durban, Private Bag X10, Dalbridge, 4014;* [2]*Department of Botany, University of the Witwatersrand, P.O. Wits, Johannesburg, 2050, South Africa*

Abstract

The effects of immersion of maize (*Zea mays*) seed in water at 55, 57 and 60°C, for varying times were assessed in terms of fungal status, seed germination and seedling establishment. These assessments were conducted immediately after treatment, after re-dehydration for 2 days (d) in a cool air stream, and following a 1 month storage period under either cold (4°C) or ambient (25°C) conditions (RH 33% and 91%, respectively). The level of internal contamination, represented almost entirely by *Fusarium moniliforme* Sheldon, declined significantly when assessed immediately after the treatment, the efficacy of which increased with increasing temperature and duration of treatment. Although there was an associated decline in germination when treated for longer than 15 and 5 min at 55 and 57°C, respectively, seedling emergence was unaffected by a treatment duration of 5 min. Subsequent storage for 1 month, even under conditions that would normally encourage fungal proliferation (i.e. 25°C and 91% RH), did not further alter the initial effect of the treatment, for short durations, on infestation levels and seedling establishment. However, seed vigour was somewhat reduced. Therefore, while treatment at 60°C was too extreme, that at 55 and 57°C shows promise in producing a crop less prone to fusarial pathogenesis.

Abbreviations: mc, moisture content; %WMB, percentage wet mass basis; RH, relative humidity.

Introduction

Large quantities of stored grain are lost annually to invading pests and deterioration by fungi is particularly prevalent in the tropics and sub-tropics. Seed fungal proliferation is marked by a succession of species, starting with the field fungi which are succeeded by the storage fungi, the result of which ultimately reduces seed quality and viability (Christensen and Kaufmann, 1974). In addition, these fungi also produce toxins that pose a serious health hazard to the consumer (Shotwell, 1991). It is therefore desirable that methods to minimize/eradicate the infection within seeds, without adversely affecting seed quality, be developed. One such potential treatment involves the immersion of seeds in hot water, the efficacy of which appears to be dependent on both the temperature and duration of treatment (Daniels, 1983; Agarwal and Sinclair,

R.H. Ellis, M. Black, A.J. Murdoch, T.D. Hong (eds.), Basic and Applied Aspects of Seed Biology, pp. 777–785.
© *1997 Kluwer Academic Publishers, Dordrecht. Printed in Great Britain.*

1987; Berjak *et al.*, 1992). In the present study, the effect of immersion at 55, 57 and 60°C for periods of 5 to 60 min on the levels of internal fungal infestation, particularly by *Fusarium moniliforme* Sheldon, and germinability of maize (*Zea mays*) seeds was determined. *F. moniliforme* is of great concern in the southern African region as it is extremely toxigenic (Shotwell, 1991) and is also responsible for ear-rot in maize (Rheeder *et al.*, 1993). Elimination of this, and other seed-infecting fungi, by hot water treatments would therefore be greatly advantageous, particularly amongst Third World subsistence communities, as the technique is both inexpensive and easily applied.

Materials and Methods

Seeds

Caryopses (seeds) of a yellow maize variety (*Zea mays* var. PAN 6480), harvested in 1993, were obtained from the Pannar Seed Company, Greytown, Kwazulu-Natal, South Africa and maintained in hermetic storage at 4°C for 2 years until used.

Hot Water Treatment

Seeds were soaked in sterile, distilled water for 4 h at ambient temperature, after which they were transferred to boiling tubes (12 seeds per tube) containing 20 ml of pre-heated (either at 55, 57 or 60°C) sterile distilled water. The tubes were maintained in a water bath at the appropriate temperature from 5 to 60 min each. Following the duration of treatment at the appropriate temperature, the seeds were removed from the tubes and seed temperature was gradually reduced to ambient by continuous washing with sterile distilled water at room temperature.

The effect of the treatment on seed internal fungal status, moisture content and germination was assessed either immediately, after re-dehydration in a cool air stream for 2 d and after subsequent storage for 1 month. These parameters were also assessed for control seeds and seeds that were imbibed for 4 h.

Internal Infection

Seeds were surface sterilized (Mycock *et al.*, 1988) for 20 min and internal fungal status was assessed for 100 half seeds each time (except control seeds, n = 300), as previously described by Mycock *et al.* (1988).

Moisture Content

The moisture contents (mc) of both the embryo (scutellum and embryonic axis) and endosperm (pericarp, aleurone layer and endosperm) were determined gravimetrically on 25 individual seeds (control, 75 seeds) per sample. Results are expressed on a percentage wet-mass basis (%WMB). Statistical tests were performed using the ANOVA multivariate function of Statgraphics at the 0.05 confidence level.

Germination

Following surface sterilization (Mycock *et al.*, 1988) for 20 min, 100 seeds (control, n = 300) were set to germinate at 25°C in sterile Petri dishes for 5 d. The seeds were then planted in vermiculite and grown under greenhouse conditions for a further 2 weeks. The percentage seeds germinated was therefore determined after 5 d, and the percentage seedlings established after a total of 18 d, from the start of imbibition.

Storage

Following re-dehydration, seeds from each treatment (including the control material) were dusted with Benlate (benzimidazole 500 mg/kg) and separated into 2 sub-samples. Each sub-sample was placed in a sterile vial over a saturated solution of either $MgCl_2$ or KNO_3 (equivalent to 33% and 91% RH) (Vertucci and Roos, 1993) and stored for 1 month at 4°C and 25°C, respectively. After storage, seeds were first rinsed with sterile distilled water prior to seed fungal status and germination assessments.

Results and Discussion

The seeds used in this study exhibited relatively low levels of fungal infection at the start, with unimbibed (control) seeds being only 20% infected (Table 1). For this, and any of the treatments that yielded fungal infection, the predominant genus isolated was *Fusarium* spp., represented almost entirely by *F. moniliforme* Sheldon. Other pathogens, when isolated, included *Diploida* sp., *Cladosporium* sp., *Aspergillus* spp. and *Penicillium* spp., but their occurrences were erratic and low (results not shown). Although described as a fungus associated primarily with newly harvested seeds (Christensen and Kaufmann, 1974), the persistence of *F. moniliforme* in seed stored hermetically at low temperatures, as was the case in the present study, was not unexpected (McLean and Berjak, 1987; Berjak *et al.*, 1992). Recent reports suggest this fungus to be an important deterrent to invasion by other seed-infecting fungi, including other *Fusarium* spp. (van Wyk *et al.*, 1988; Rheeder *et al.*, 1990), *Diploida* spp. (Rheeder *et al.*, 1990) and *Aspergillus flavus* (Zummo and Scott, 1992).

Both seed infection levels and seed germination were unaffected by the 4 h imbibition and imbibition/re-dehydration treatments (Tables 1 and 2), when compared with control material. However, the number of seedlings established after 18 d was less than that originally germinated for seeds imbibed for 4 h

(Tables 1 and 2). This decline in seedling establishment could possibly be attributed to post-emergence damping off (Agarwal and Sinclair, 1987). Imbibing seeds, either for germination or pre-soaking treatments, may stimulate pathogen growth (Agarwal and Sinclair, 1987) and, in fact, actively growing mycelia could be seen in some cases. Although seedling emergence has been shown to be unaffected if seeds are infected with *F. moniliforme* (Rheeder *et al.*, 1990), other *Fusarium* spp. (Nakagawa and Yamaguchi, 1989; Rheeder *et al.*, 1990) and *Diploida* spp. (Rheeder *et al.*, 1990) have been reported to be aggressive pathogens. Seeds imbibed for 4 h (Table 1) as well as those subsequently re-dehydrated (Table 2) were characterized by mcs significantly higher than that of the control seeds, possibly making them more prone to post-emergence damping off.

Table 1. Effect of imbibtion for 4 h at room temperature, followed by hot water treatment at 55, 57 and 60°C on total internal infection levels, percentage germination and water uptake of maize seeds, compared with that of unimbibed (control) seeds

	Duration of water treatment (min)	Total internal infection (%)	Moisture content (%WMB)[b]		Germination (%)[a] (d = 5)	Seedling emergence (%)[a] (d = 18)
			Embryo	Endosperm		
Unimbibed	0	20	9.22 a	11.88 A	98	97
Imbibed	0	20	28.98 b	22.60 B	99	93
Imbibed 4 h;	5	10	31.48 c	23.96 C	97	94
hot water	10	3	32.40 c	24.09 C	94	96
treated (55°C)	15	3	32.79 c	24.07 C	98	99
	30	2	35.36 d	25.93 DE	89	91
	45	0	36.90 de	26.49 DEF	87	94
	60	0	39.95 ef	27.98 GH	74	92
Imbibed 4 h;	5	4	31.59 c	23.59 C	97	98
hot water	10	3	32.04 c	24.36 C	89	95
treated (57°C)	15	0	32.62 c	24.11 C	89	95
	30	0	36.50 de	25.91 DE	83	94
	45	0	37.15 de	26.64 EF	60	87
	60	0	40.20 f	27.02 F	49	72
Imbibed 4 h;	5	4	32.54 c	23.59 C	78	95
hot water	10	2	32.24 c	24.47 C	65	90
treated (60°C)	15	0	32.29 c	24.16 C	66	73
	30	0	36.02 d	25.59 D	13	27
	45	0	38.18 ef	27.05 FG	9	12
	60	0	41.31 f	28.52 H	1	1

[a]Represents the mean of 100 seeds (except control, n = 300)

[b]Represents the mean of 25 seeds (except control, n = 75). Values with different letters (embryo = lower case, endosperm = upper case) are significantly different at the 0.05 confidence level

Immersing seeds in water from 5 to 60 min at 55, 57 and 60°C, respectively, significantly decreased the level of internal infection, the efficacy of which increased with increasing temperature and duration of treatment (Table 1). Compared with unimbibed (control) material, seeds immersed in water for 15 min at 55°C exhibited an 85% reduction in contamination, while those treated at 57 and 60°C for the same duration appeared completely uninfected (Table 1). The efficacy of the hot water treatment in reducing seed infection after only 5 min may be explained in terms of the peripheral location of *Fusarium* spp., particularly *F. moniliforme*, within the pericarp (Russell and Berjak, 1983; Agarwal and Sinclair, 1987) and pedicel (Bacon *et al.*, 1992) of maize seeds. Some members of the genus *Aspergillus* have also been found to be peripherally located (Mycock and Berjak, 1992). Both the embryo and endosperm showed a slight, though significant, increase in water content at all temperatures after only

Table 2. Effect of re-dehydration on total internal infection levels, percentage germination and water loss of pre-imbibed and hot water treated maize seeds, compared with that of unimbibed (control) seeds

	Duration of water treatment (min)	Total internal infection (%)	Moisture content (%WMB)[b]		Germination (%)[a] (d = 5)	Seedling emergence (%)[a] (d = 18)
			Embryo	Endosperm		
Unimbibed	0	20	9.22 ghi	11.88 C	98	97
Imbibed 4 h; re-dehydrated	0	20	10.08 l	13.40 L	99	91
Imbibed 4 h; hot water treated (55°C); re-dehydrated	5	8	9.75 jk	12.24 EF	97	98
	10	5	9.09 fgh	12.04 D	98	98
	15	5	9.43 hij	12.59 H	94	98
	30	3	9.52 hij	12.76 IJ	90	97
	45	0	9.57 ij	12.96 K	86	95
	60	0	9.27 ghi	12.32 FG	60	91
Imbibed 4 h; hot water treated (57°C); re-dehydrated	5	7	8.35 bc	12.33 FG	96	97
	10	0	8.89 efg	12.40 G	81	96
	15	0	8.90 efg	11.99 CD	93	90
	30	0	8.79 cde	12.00 CD	84	91
	45	0	8.46 bcd	12.07 DE	66	76
	60	0	9.19 fghi	12.91 JK	44	65
Imbibed 4 h; hot water treated (60°C); re-dehydrated	5	4	7.77 a	11.17 A	89	97
	10	0	7.78 a	11.27 A	69	91
	15	0	9.28 ghi	12.36 FG	57	84
	30	0	9.75 jk	12.06 HI	12	25
	45	0	9.40 hij	12.61 HI	1	1
	60	0	8.20 ab	11.72 B	0	0

[a]Represents the mean of 100 seeds (except control, n = 300)

[b]Represents the mean of 25 seeds (except control, n = 75). Values with different letters (embryo = lower case, endosperm = upper case) are significantly different at the 0.05 confidence level

Table 3. Effect of storage for 1 month at 4°C (33% RH) on total internal infection levels, percentage germination and moisture content of re-dehydrated maize seeds that were previously pre-imbibed and hot water treated, compared with that of unimbibed (control) seeds

Duration of water treatment (min)	Total internal infection (%)	Moisture content (%WMB)[b]		Germination (%)[a] (d = 5)	Seedling emergence (%)[a] (d = 18)
		Embryo	Endosperm		
Unimbibed; stored at 4°C (33% RH)	0	12	9.15 ghi	11.80 DEF	97	98
Imbibed 4 h; re-dehydrated; stored at 4°C (33% RH)	0	19	9.17 fghi	12.39 KLM	92	97
Imbibed 4 h; hot water treated (55°C); re-dehydrated; stored at 4°C (33% RH)	5	10	8.87 efgh	11.72 CDEF	84	93
	10	2	8.44 bcde	11.85 EFG	81	94
	15	5	9.35 hi	12.55 LM	90	99
	30	3	9.46 i	12.33 JKL	77	96
	45	0	8.79 defg	12.60 M	70	96
	60	0	8.81 defg	12.13 HIJ	64	81
Imbibed 4 h; hot water treated (57°C); re-dehydrated; stored at 4°C (33% RH)	5	2	8.87 efgh	12.20 IJK	71	95
	10	1	9.02 fghi	12.10 HI	74	100
	15	3	7.98 ab	11.53 C	79	95
	30	0	8.00 ab	11.68 CDE	56	82
	45	0	8.36 bcd	11.52 C	34	71
	60	0	8.88 efgh	12.50 LM	35	59
Imbibed 4 h; hot water treated (60°C); re-dehydrated; stored at 4°C (33% RH)	5	1	7.76 a	11.00 A	70	98
	10	0	8.00 ab	11.28 B	53	85
	15	0	8.28 bc	11.94 FGH	54	81
	30	0	8.37 bcd	11.62 CD	16	30
	45	0	9.17 fghi	12.06 GHI	1	2
	60	0	7.67 a	11.65 CDE	0	0

[a]Represents the mean of 100 seeds (except control, n = 300)
[b]Represents the mean of 25 seeds (except control, n = 75). Values with different letters (embryo = lower case, endosperm = upper case) are significantly different at the 0.05 confidence level

5 min treatment, compared with seeds imbibed for 4 h (Table 1). Seed germination was unaffected up to a 15 min treatment period at 55°C and 5 min at 57°C, while treatment at 60°C appeared to affect germination adversely at all durations tested (Table 1). As in the case of the unimbibed and 4 h imbibed seeds (see earlier), seeds treated at 55°C for 5 and 10 min, were also prone to post-emergence damping off (yielding 94 and 96% seedling establishment after 18 d, respectively).

Both *Fusarium* spp. (Agarwal and Sinclair, 1987; Bacon *et al.*, 1992) and *Aspergillus* spp. (Mycock and Berjak, 1992) have also been located within the seed embryonic tissue. Contaminants located within these tissues may be systemically transmitted to, and, as a consequence, could affect the growth of developing seedlings (Agarwal and Sinclair, 1987). Thus it would appear that at 55°C, a treatment duration of 15 min was required to reduce embryo-associated seed infection, while at 57°C, a 5 min treatment was sufficient. In fact, treatment at 55°C for 15 min exhibited superior starting material and improved seedling establishment, as did 57°C for 5 min (Table 1). These results are in agreement with those of Daniels (1983) who found that reducing seed infection by hot water treatment significantly reduced the levels of subsequent seedling infection.

Re-dehydration following the treatment appeared to enhance the reduction of internal infection levels as seeds immersed at 57 and 60°C showed 0% infection

Table 4. Effect of storage for 1 month at 25°C (91% RH) on total internal infection levels, percentage germination and moisture content of re-dehydrated maize seeds that were previously pre-imbibed and hot water treated, compared with that of unimbibed (control) seeds

	Duration of water treatment (min)	Total internal infection (%)	Moisture content (%WMB)[b]		Germination (%)[a] (d = 5)	Seedling emergence (%)[a] (d = 18)
			Embryo	Endosperm		
Unimbibed; stored at 25°C (91% RH)	0	11.3	10.98 de	12.70 B	82	100
Imbibed 4 h; re-dehydrated; stored at 25°C (91% RH)	0	19	11.28 efg	13.72 IJK	89	98
Imbibed 4 h;	5	7	10.94 cde	13.26 F	86	99
hot water	10	2	11.64 fgh	13.39 FG	72	96
treated (55°C);	15	1	10.41 bc	13.21 EF	80	96
re-dehydrated;	30	2	10.91 bcde	13.48 GH	55	85
stored at 25°C	45	0	10.59 bcd	13.61 HIJ	54	84
(91% RH)	60	0	11.41 efg	13.61 HIJ	53	78
Imbibed 4 h;	5	3	11.82 ghi	13.53 GH	71	99
hot water	10	1	11.74 ghi	13.65 HIJ	54	88
treated (57°C);	15	0	11.17 ef	13.61 HI	67	94
re-dehydrated;	30	0	11.36 efg	13.51 GH	47	75
stored at 25°C	45	0	11.68 fghi	13.80 JK	19	55
(91% RH)	60	0	12.19 i	13.85 K	12	39
Imbibed 4 h;	5	1	9.46 a	12.35 A	67	88
hot water	10	0	10.40 b	12.63 B	44	75
treated (60°C);	15	0	12.04 hi	13.59 HI	30	54
re-dehydrated;	30	0	11.41 efg	13.03 DE	4	9
stored at 25°C	45	0	11.01 de	12.91 CD	0	0
(91% RH)	60	0	9.60 a	12.80 BC	0	0

[a]Represents the mean of 100 seeds (except control, n = 300)

[b]Represents the mean of 25 seeds (except control, n = 75). Values with different letters (embryo = lower case, endosperm = upper case) are significantly different at the 0.05 confidence level

after a treatment duration of only 10 min (Table 2). In addition, most treatments were characterized by improved germination and seedling emergence (Table 2), compared with the same treatments assessed immediately after the hot water treatment (Table 1). The seeds, in most cases, were re-dehydrated to levels lower than that of the control, following treatment (Table 2). The activity of fungi within seeds is dependent on the water content of the seed tissues, with dry seeds yielding lower levels of infection (Christensen and Kaufmann, 1974). Soaking followed by re-dehydration, when not too extreme (as may be the case in this study), has been shown to improve germination (Nath *et al.*, 1990; 1991). Thus either one, or a combination of both of these factors could help explain the reduced infection and enhanced germination of the re-dehydrated seeds. Also, this could explain why, upon re-dehydration, those seeds treated for 5 and 10 min at 55°C did not show signs of the suggested post-emergence damping off (compare Tables 2 and 1).

It is well established that the storage of seeds under cold, dry conditions allows for improved longevity compared with humid, warm conditions (Justice and Bass, 1978). As expected, therefore, seed germination declined for both 4 h imbibed and unimbibed seeds after a storage period of 1 month at 25°C and 91% RH (compare Tables 1 and 4), when compared with seeds at the start of the investigation (Table 1). However, a similar trend was evident for seeds stored at 4°C and 33% RH (Table 3), although not as marked.

In order to determine the effect of internally located seed mycoflora only, seeds were dusted with Benlate prior to storage under either regime. This fungicide has been reported to reduce seed infestation but with associated reductions in seed germination (Champawat, 1990; Aveling *et al.*, 1993). However, although germination was initially reduced, seedling emergence was enhanced for both the unimbibed and imbibed treatments following storage (Tables 3 and 4). At the same time, seed infection was reduced (Tables 3 and 4), suggesting that Benlate dusting could be partially effective in reducing surface fungal contamination during storage, ultimately resulting in superior field emergence.

As previously reported (Berjak *et al.*, 1992), cold storage contained the levels of fungal infestation in seeds previously treated with hot water, to a minimum (Table 3). This, however, was also apparent when the treated seeds were stored at 25°C and 91% RH (Table 4), conditions that would normally encourage fungal growth (Christensen and Kaufmann, 1974). The effect of the hot water treatment, with the possible synergistic influence of the surface fungicide, on reducing seed infestation levels appears therefore to be via elimination, rather than fungistasis. Seed germination was reduced for both storage regimes tested (Tables 3 and 4), when compared with seeds tested immediately after the hot water treatment (Table 1). The establishment of seedlings from seeds from the following treatments: 55°C – 15 min, 57°C – 10 min and 60°C – 5 min, was similar to that of control seeds, following cold storage (Table 3). The same was true for seeds treated at 55 and 57°C, but for a treatment duration of 5 min only, following storage at ambient conditions (Table 4).

In conclusion, therefore, the immersion of seeds in water at 55 and 57°C, for a relatively short time shows great promise as both a pre-sowing and pre-storage seed disinfection treatment, without greatly affecting seed/seedling quality. This treatment also has the advantages of both simplicity and negligible cost, and could therefore be of tremendous benefit to subsistence farming.

References

Agarwal, V.K. and Sinclair, J.B. 1987. *Principles of Seed Pathology*, vol.1, pp. 17–76. Boca Raton, Florida: CRC Press.

Aveling, T.A.S., Snyman, H.G. and Naude, S.P. 1993. *Plant Disease* 77: 1009–1011.

Bacon, C.W., Bennett, R.M., Hinton, D.M. and Voss, K.A. 1992. *Plant Disease* 76: 144–148.

Berjak, P., Whittaker, A. and Mycock, D.J. 1992. *South African Journal of Science* 88: 346–349.

Champawat, R.S. 1990. *Journal of Phytological Research* 3: 133–136.

Christensen, C.M. and Kaufmann, H.H. 1974. In: *Storage of Cereal Grains and their Products*, pp. 158–192 (ed. C.M. Christensen). Minnesota: American Association of Cereal Chemists.

Daniels, B.A. 1983. *Plant Disease* 67: 609–611.

Justice, O.L. and Bass, L.N. 1978. *Principles and Practices of Seed Storage*, pp.26–56. Washington, D.C.: Science and Education Federal Research.

McLean, M. and Berjak, P. 1987. *Seed Science and Technology* 15: 831–850.

Mycock, D.J., Lloyd, H.L. and Berjak, P. 1988. *Seed Science and Technology* 16: 647–653.

Mycock, D.J. and Berjak, P. 1992. *South African Journal of Botany* 58: 139–144.

Nakagawa, A. and Yamaguchi, T. 1989. *JARQ* 23: 94–99.

Nath, S., Coolbear, P., Hampton, J.G. and Cornford, C.A. 1990. *Proceedings Agronomy Society of New Zealand* 20: 51–57.

Nath, S., Coolbear, P. and Hampton, J.G. 1991. *Crop Science* 31: 822–826.

Rheeder, J.P., Marasas, W.F.O., and Van Wyk, P.S. 1990. *Phytopathology* 80: 131–134.

Rheeder, J.P., Marasas, W.F.O., and Van Schalkwyk, D.J. 1993. *Phytophylactica* 25: 43–48.

Russell, G.H. and Berjak, P. 1983. *Seed Science and Technology* 11: 441–448.

Shotwell, O.L. 1991. In: *Mycotoxins and Animal Foods*, pp. 325–340 (eds J.E. Smith and R.S. Henderson). London: CRC Press.

Van Wyk, P.S., Scholtz, D.J. and Marasas, W.F.O. 1988. *Plant and Soil* 107: 251–257.

Vertucci, C.W. and Roos, E.E. 1993. *Seed Science Research* 3: 201–213.

Zummo, N. and Scott, G.E. 1992. *Plant Disease* 76: 771–773.

86. Behaviour of Membranes and Proteins during Natural Seed Ageing

E.A. GOLOVINA, W.F. WOLKERS and F.A. HOEKSTRA

Department of Plant Physiology, Wageningen Agricultural University, Arboretumlaan 4, 6703 BD Wageningen, The Netherlands

Abstract

Membrane integrity and protein secondary structure were studied in seeds of different plant species directly after harvest and after long-term storage in a seed bank. Electron paramagnetic resonance spin label studies revealed extensive membrane damage in the aged, nonviable seeds. In contrast, Fourier self-deconvolved infrared spectra of the proteins in these embryos did not reveal changes in relative peak height and band position of the different protein secondary structures with ageing. Extended β-sheet structures typical of protein denaturation were also not observed. We conclude that in spite of the loss of membrane integrity with seed ageing the protein secondary structure in desiccation tolerant (orthodox) dry seeds is maintained during several decades of open storage.

Introduction

During maturation and subsequent desiccation orthodox seeds lose their water but retain their viability. Stabilization of membranes and proteins is considered crucial for acquisition of desiccation tolerance (Crowe *et al.*, 1987; 1992). For the long term survival of seeds in dry storage the stabilized membranes and proteins are favourably embedded in a stable glassy matrix (Williams and Leopold, 1989; Sun and Leopold, 1993).

Loss of seed viability with ageing is usually linked with the loss of membrane integrity (Priestley, 1986; Bewley and Black, 1994). There is also some evidence that proteins change with ageing (Crocker and Barton, 1953; Roberts, 1972; Priestley, 1986). Their extractability decreases with ageing, which was suggested to be caused by denaturation or disulphide bridge formation (reviewed in Priestley, 1986). However, there is no direct evidence of possible changes in protein secondary structure (denaturation) with ageing. Moreover, it is also not clear whether protein denaturation precedes or follows membrane disruption.

To study membrane integrity in seeds an electron paramagnetic resonance (EPR) spin label technique has been developed (Smirnov *et al.*, 1992; Golovina and Tikhonov, 1994). The evaluation of membrane integrity by this method is based on the differential permeabilities of cell membranes for neutral spin probe molecules, and charged paramagnetic broadening agents. Extensive membrane disruption was detected by this technique in wheat embryos during long-term

R.H. Ellis, M. Black, A.J. Murdoch, T.D. Hong (eds.), Basic and Applied Aspects of Seed Biology, pp. 787–796.
© *1997 Kluwer Academic Publishers, Dordrecht. Printed in Great Britain.*

storage (Golovina and Tikhonov, 1994).

Fourier transform infrared (FTIR) spectroscopy has been used to study secondary structure of dehydrated proteins *in situ* (Wolkers and Hoekstra, 1995; 1996). Each type of secondary structure causes a characteristic set of IR-absorption bands in the amide-I region of the spectrum, which mainly arise from the C = O stretching vibrations (Susi *et al.*, 1967). These bands are very sensitive to changes in the nature of the hydrogen bonds arising from the different types of secondary structure. Information on desiccation-induced protein denaturation has been obtained this way (Prestrelski *et al.*, 1993). A recent FTIR study on desiccation tolerant pollen has shown that in spite of the loss of viability during accelerated ageing, protein secondary structure was conserved (Wolkers and Hoekstra, 1995). However, the mechanism of viability loss during accelerated ageing may be different from that during natural ageing because of the different conditions that prevail over a much longer period of time.

Here we report on membrane integrity and protein secondary structure in seed embryos of different plant species after long-term natural ageing, as studied by an EPR spin probe technique and FTIR spectroscopy.

Materials and Methods

Plant Material

Seeds kept in open storage (15–20°C, 30% RH) were from the seed collection of the CPRO/DLO, Wageningen, The Netherlands, and initially had high germinative capacities (>85%). They were harvested 20–30 years ago, whereas control seeds from various sources were harvested in 1994. Seeds of onion (*Allium cepa*) cv. Revro (1969) and cv. Dinaro (1994), radish (*Raphanus sativus*) cv. Kabouter (1969) and cv. Foxyred (1994), white cabbage (*Brassica napus*) cv. R.O. Cross (1969) and cv. Bejo (1994), melon (*Cucumis melo*) cv. Resistant (1965) and cv. Ardor (1994), and paprika (*Capsicum annuum*) cv. Hot lips (1975) and cv. Balaton (1994) were used for this study. All control seeds germinated better than 88%, whereas the aged seeds did not germinate at all (Table 1).

EPR Measurements

The EPR spectra were obtained at room temperature with a Bruker model ESP-300E spectrometer. Microwave power was 2.02 mW and the modulation amplitude was 1 G. Under this condition saturation and overmodulation of EPR spectra were excluded. The water soluble, stable nitroxide radical Tempone (4-oxo-2,2,6,6-tetramethyl-1-piperidinyloxy) from Sigma (St Louis, Missouri, USA) was used as the spin probe. Potassium ferricyanide was used as the broadening agent.

Seeds were imbibed in water for 6 h, and embryos were isolated and

Table 1. Partition parameter R calculated from EPR spectra for embryos and cotyledons of viable seeds and aged, nonviable seeds. The percentage of viable cells in the dead seeds was calculated from this parameter R, assuming 100% viable cells in the seeds harvested in 1994

Species	Year of harvest	Germination (%)	Partition parameter R and % of viable cells	
			Embryo axis	Cotyledon
Cabbage	1994	>90	5.1 (100%)	1.0 (100%)
Cabbage	1969	0	0.0 (0%)	0.0 (0%)
Melon	1994	>95	1.1 (100%)	1.3 (100%)
Melon	1965	0	0.3 (27%)	0.5 (39%)
Onion	1994	>90	2.5 (100%)	n.d.*
Onion	1969	0	0.0 (0%)	n.d.*
Radish	1994	>90	7.3 (100%)	1.9 (100%)
Radish	1969	0	2.6 (36%)	1.2 (63%)
Paprika	1994	>88	1.4 (100%)	1.0 (100%)
Paprika	1975	0	0.5 (35%)	0.5 (50%)

*n.d., not determined

incubated in a fully broadened solution of Tempone (1 mM Tempone + 120 mM $K_3Fe(CN)_6$) for 15 min. One or several embryos (depending on size) were then loaded into a capillary (2 mm diameter) together with 2 µL solution and placed in the spectrometer cavity for spectrum recording. Four spectra were accumulated to improve the signal/noise ratio.

Principle of Nitroxide Spin Probe Technique

A typical EPR spectrum from an aqueous 1 mM Tempone solution is a triplet with narrow equidistant lines with an isotropic hyperfine splitting constant of about 17 G (Griffith *et al.*, 1974). Because of the amphiphilic character of Tempone molecules they also dissolve in apolar organic solvents. Due to polarity effects the distance between peaks in the EPR spectrum then decreases to 15 G. An aqueous solution of Tempone may be broadened by potassium ferricyanide via spin–spin interaction (Eaton and Eaton, 1978). The broadening leads to the apparent disappearance of the EPR spectrum of Tempone.

When viable cells are placed in this broadened solution, the signal of Tempone will appear again, because only the spin probe molecules can penetrate the cells and the charged $Fe(CN)_6^{-3}$ ions cannot, which excludes the broadening inside (Smirnov *et al.*, 1992). Membranes of viable cells are permeable to Tempone molecules because of their amphiphilic character and relatively small molecular size, with penetration times of several min. Only the signal from

Tempone molecules outside of the cells will be broadened. Thus, the triplet signal is exclusively derived from Tempone molecules inside the cells.

Figure 1 shows a typical EPR spectrum of Tempone localized inside embryonic cells of viable onion seeds. In the high-field region of this EPR spectrum (right side of the spectrum) splitting into two peaks can be observed. This splitting can be explained by the partitioning of the amphiphilic probe into lipid bodies (designated L) and into the aqueous phase of the cytoplasm (designated A). Spectra from Tempone in the lipid phase and the aqueous phase are easily resolved in high field lines because of the combined effect of the polarity differences from both the isotropic hyperfine splitting and the g-values (Shimshick and McConnel, 1973).

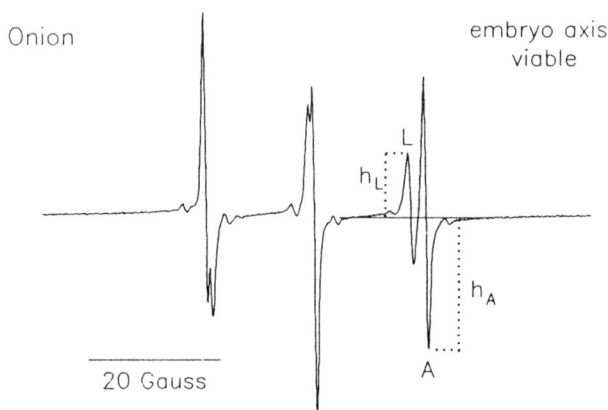

Figure 1. EPR spectra of Tempone in embryo cells of viable onion seeds. The parameter R is calculated as the ratio (h_A/h_L) of the peak heights to baseline of the hydrophilic component (A) and the hydrophobic component (L)

In the case of loss of membrane integrity ferricyanide ions can penetrate the cell and broaden the signal of Tempone molecules localized in the aqueous cytoplasm. In this case only the hydrophobic component of the EPR spectrum will be left because of the inability of the charged ferricyanide ions to penetrate the lipid bodies. The intensity of this hydrophobic signal can be used as a measure of the total amount of cells in the sample, whereas the intensity of the polar component represents the amount of cells with intact membranes. The ratio R between the heights of the polar and hydrophobic peaks (h_A/h_L) to the baseline allows for a quantitative assessment of the cellular membrane integrity of the sample.

IR-spectroscopy

FTIR spectra were recorded on a Perkin-Elmer 1725 spectrometer equipped with a liquid nitrogen-cooled mercury/cadmium/telluride detector and a Perkin-Elmer microscope interfaced to a personal computer as described by Wolkers and Hoekstra (1995). Embryo axes and cotyledons were isolated and transverse sections were made for use in the FTIR microscope. One slice of appropriate thickness was sandwiched between two diamond windows and mounted in a brass temperature-controlled cell. After purging the optical bench for at least 1 h with dry, CO_2-free air (Balston; Maidstone, Kent, England) at a flow rate of 25 L.min^{-1}, spectra were recorded at room temperature. Embryo slices of more than 2 seeds were investigated for every case, and typical spectra are shown in the figures. The acquisition parameters were: 4 cm^{-1} resolution, 512 co-added interferograms, 2 cm.s^{-1} moving mirror speed, 3500–900 cm^{-1} wavenumber range, and triangle apodization function. The time needed for acquisition and processing of a spectrum was 4.5 min.

Spectral analysis and display were carried out using the Infrared Data Manager Analytical Software, version 3.5 (Perkin-Elmer). The spectral region between 1800 and 1500 cm^{-1} was selected. This region contains the amide-I and amide-II absorption bands of the protein backbones. Deconvolved spectra were calculated, using the interactive Perkin-Elmer routine for Fourier self-deconvolution. The parameters for the Fourier self-deconvolution procedure were: smooth factor 15.0 and a width factor of 30.0 cm^{-1}. Deconvolved spectra were normalized.

Results and Discussion

Membrane Integrity as Studied by EPR Spin Probe Technique

Figures 2 and 3 depict EPR spectra of Tempone in embryos and cotyledons from both recently harvested viable seeds and nonviable seeds that were aged for a few decades in open storage at 20°C. In the nonviable cabbage seeds from the 1969 harvest (Fig. 2), the peak representing the hydrophilic component was completely absent, both in the embryo and the cotyledons, which is interpreted to mean that all the cells had disrupted membranes. Radish seeds from 1969 were also nonviable, but the hydrophilic component was still visible, albeit at a much lower amplitude than in the viable 1994 seeds.

In Table 1 the partition parameter R is given for the seeds of Figures 2 and 3 and for some other seeds. From this parameter the proportion of viable cells in the aged seeds was calculated assuming that cells in the non-aged seeds were 100% viable (Table 1). This calculation is considered legitimate on the premise that the $Fe(CN)_6^{-3}$ ions can enter leaky cells in quantities sufficient for full broadening of EPR spectra of Tempone localized in the aqueous phase of the cytoplasm. Intermediate leakage would have caused increased peak widths of

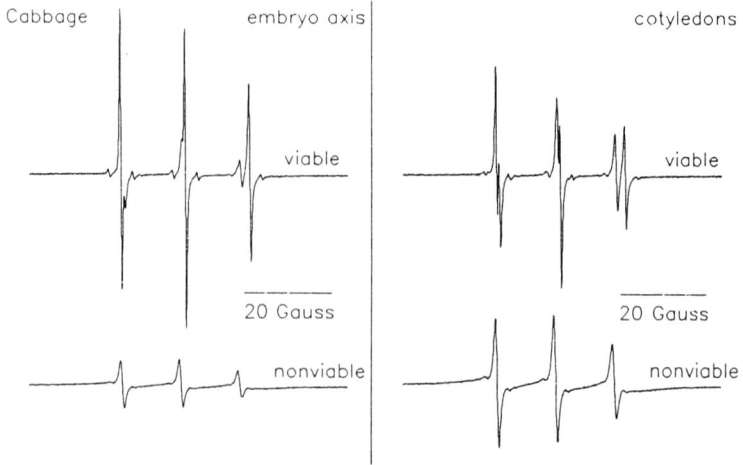

Figure 2. EPR spectra of Tempone in embryo and cotyledon cells of viable and nonviable cabbage seeds

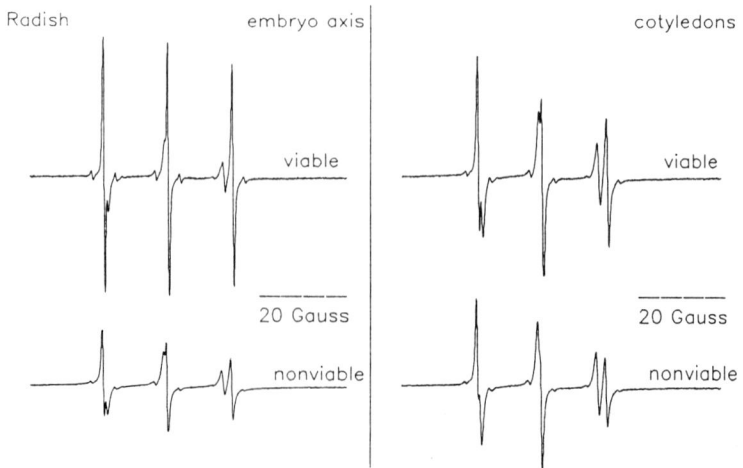

Figure 3. EPR spectra of Tempone in embryo and cotyledon cells of viable and nonviable radish seeds

the hydrophilic component in the spectra without apparent disappearance of the signal, a phenomenon that we did not observe. On account of the reduction of the R-values in the nonviable embryo axes of radish, melon and paprika, we conclude that the embryos still contained a certain amount of viable cells, not exceeding approximately 36%. A similar trend was observed in the cotyledons of the aged seeds studied, but in this case the percentage of viable cells was sometimes higher (up to 63%). Apparently, viability is determined by the proportion of surviving axis cells rather than cotyledonous cells.

The data presented in Table 1 show that there is no fixed value for the partition parameter in viable seeds. This can be attributed to the varying oil contents. More oil leads to a lower value of the parameter R. Cotyledons of the oily cruciferae seeds, for example, have lower R values than the seed axes (Table 1). Differences between seed species and tissues can thus be explained. The stage of germination also determines the value of the parameter R. When cells enlarge, the total intracellular volume which is inaccessible to ferricyanide ions will increase, leading to an increase of R as well. But since a study of these differences was not the objective of this work, we did not investigate it further.

The presence of a considerable amount of cells with intact membranes in aged seeds as shown for radish, melon and paprika apparently is not sufficient for germination. However, such intact cells from nonviable aged seeds have been shown to resume growth *in vitro* (Innocenti *et al.*, 1983). It is also likely that the survival of certain cells is not essential for germination, as has been demonstrated by vital staining with tetrazolium salts (rules of ISTA). However, necrosis of even a few meristematic cells of the embryo axis may result in viability losses, since their division is essential for seedling establishment. In conclusion, the EPR spin probe technique as described above, gives direct evidence of membrane disruption during seed ageing.

Protein Secondary Structure as Studied by FTIR Spectroscopy

Fourier deconvolved IR-spectra of non-aged and aged embryos of five different seed species in the wavenumber region where proteins absorb are shown in Figure 4. It can be seen that at least three bands occur in the amide-I region, located around 1638, 1655–1658 and 1684 cm^{-1}, representing different types of protein secondary structure. The bands around 1745 cm^{-1} are due to the ester bonds of the lipids. Amplitudes of this peak varied and depended on how much oil remained after sample preparation. Amide bands have previously been assigned to α-helical structure (1657 cm^{-1}), random coil structure (1652 cm^{-1}), turn structure (1637 and 1680 cm^{-1}) and β-sheet structures (1637 and 1690 cm^{-1}) (Wolkers and Hoekstra, 1995 and references therein). The relative contribution of α-helical structures was high in all the seed species studied. Despite the long storage period, IR-spectra of embryos from aged seeds were virtually identical to the non-aged seeds, and only slight differences could be observed between such spectra in the case of cabbage and radish embryos. This slight change mainly concerned an increase in the aged embryos of the turn

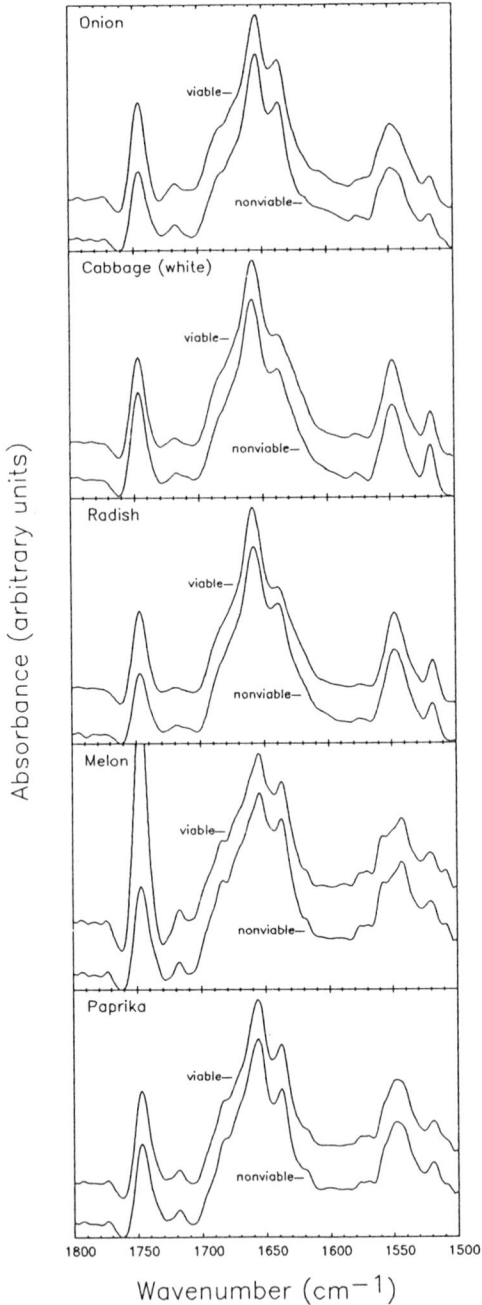

Figure 4. Fourier self-deconvolved IR absorption spectra of the amide regions of embryo slices from dry viable seeds and nonviable seeds that were aged in open storage (20°C/30% RH) for a few decades. The viable seeds were harvested in 1994. The aged seeds were from 1969 in the case of onion, white cabbage and radish and from 1965 and 1975 for melon and paprika, respectively

structure band around 1638 cm^{-1} relative to the α-helix band around 1658 cm^{-1}. Protein denaturation generally involves the formation of irreversible protein aggregates of the extended β-sheet type, which leads to an absorption band around 1627 cm^{-1}. Such bands were not detected in naturally aged seeds. This indicates that protein secondary structure is conserved during long-term storage inspite of the loss of viability, and that protein denaturation does not occur. The extreme stability of protein structures in dry seeds has been confirmed earlier by their excellent heat stability (Golovina *et al.*, 1996) and is in agreement with the extreme heat tolerance of dry seeds (Barton, 1961).

There is evidence that powerful oxidative processes occur in ageing seeds (De Vos *et al.*, 1994, and references therein). This should also lead to protein denaturation. However, because proteins are immobilized in a solid glassy matrix in dry seeds, their secondary structures may be conserved. When this glassy matrix disappears upon rehydration, one would expect protein denaturation proportional to the age of the seed. Oxidizing agents then become active. To test if this is the case, aged and non-aged radish seeds, which had the most pronounced differences between spectra, were imbibed for 6 h to allow the proteins to rearrange and then were dried (spectra not shown). In this case, possible differences in protein secondary structure between viable and nonviable embryos would be amplified. However, no changes were observed, either between the viable and nonviable embryos, or between spectra before and after the 6 h hydration/drying treatment.

Conclusions

This study on changes in membrane integrity and protein secondary structure in seeds after long term natural ageing shows that proteins are conserved whereas cellular membranes are disrupted. Apparently cellular membranes in dry seeds are more sensitive to the ageing processes during storage than proteins. We suggest that the relatively high amount of α-helical structures, which characterizes proteins of orthodox seeds is associated with long term stability of these proteins in situ.

References

Barton, L.V. 1961. *Seed Preservation and Longevity*, First Edition. New York: Interscience Publishers.

Bewley, J.D. and Black, M. 1994 *Seeds, Physiology of Development and Germination*, pp. 445. New York, London: Plenum Press.

Crocker, W. and Barton, L.V. 1953. *Physiology of Seeds*, pp. 36–41. Waltham: Chronica Botanica Company.

Crowe, J.H., Crowe, L.M., Carpenter, J.F. and Aurell Wistrom, C. 1987. *Biochemical Journal* 242: 1–10.

Crowe, J.H., Hoekstra, F.A. and Crowe, L.M. 1992. *Annual Review of Physiology* 54: 570–599.

De Vos, C.H.R., Kraak, H.L. and Bino, R.J. 1994. *Physiologia Plantarum* 92: 131–139.

Eaton, S.S. and Eaton, G.R. 1978. *Coordination Chemistry Reviews* 26: 207–262.

Golovina, E.A. and Tikhonov, A.N. 1994. *Biochimica et Biophysica Acta* 1190: 385–392.

Golovina, E.A., Wolkers, W.F. and Hoekstra, F.A. 1996. *Comparative Biochemistry and Physiology* (in press)

Griffith. O.H., Dehlinger, P.J. and Van, S.P. 1974. *Journal of Membrane Biology* 15: 159–192.

Innocenti, A.M., Bitonti, M.B. and Bennici, A. 1983. *Caryologia* 36: 83–87.

Prestrelski, S.J., Tedeschi, N., Arakawa, T. and Carpenter, J.F. 1993. *Biophysical Journal* 65: 661–671.

Priestley, D.A. 1986. *Seed aging: Implications for Seed Storage and Persistence in Soil*, pp. 304. Ithaca, London: Comstock Publ. Assoc.

Roberts, E.H. 1972. *Viability of Seeds*. pp. 448. London: Chapman and Hall Ltd.

Shimshick, E.J. and McConnel, H.M. 1973. *Biochemistry* 12: 2351–2360.

Smirnov, A.I., Golovina, E.A., Yakimchenko, O.E., Aksyonov, S.I. and Lebedev, Ya.S. 1992. *Journal of Plant Physiology* 140: 447–452.

Sun, W.Q. and Leopold, A.C. 1993. *Physiologia Plantarum* 89: 767–774.

Susi, H., Timasheff, S.N. and Stevens, L. 1967. *The Journal of Biological Chemistry* 242: 5460–5466.

Williams, R.J. and Leopold, A.C. 1989. *Plant Physiology* 89: 977–981.

Wolkers, W.F. and Hoekstra, F.A. 1995. *Plant Physiology* 109: 907–915.

Wolkers, W.F. and Hoekstra, F.A. 1996. *Comparative Biochemistry and Physiology* (in press).

87. Effect of Cryopreservation on Seed Germination of Different Leguminosae Species

E. GONZÁLEZ-BENITO[1], J.M. PITA[2], C. PÉREZ[1] and
F. PÉREZ-GARCIA[2]

[1]Departamento de Biología Vegetal, Escuela Técnica Superior de Ingenieros Agrónomos;
[2]Departamento de Biología Vegetal, Escuela Universitaria de Ingeniería Técnica Agrícola, Universidad Politécnica de Madrid, Ciudad Universitaria, 28040 Madrid, Spain

Abstract

The effect of cryopreservation on seed germination and vigour has been studied in lucerne (*Medicago sativa* cv. Capitata), red clover (*Trifolium pratense*) and bean (*Phaseolus vulgaris* cv. Strike). Non-desiccated (8–10% moisture content) and desiccated seeds (4–5% moisture content) were cryopreserved. Samples were frozen by direct immersion in liquid nitrogen for 1 or 30 days (d). Thawing took place at room temperature. Seed germination was carried out under three different temperature regimes (15°C, 25°C and 25°C day/15°C night), always with 16 h day/8 h night photoperiod. In lucerne, cryopreservation for 30 d increased seed germination significantly for the three incubation regimes studied. However, incubation temperature interacted significantly with desiccation and cryopreservation in clover and bean, respectively. There were no significant differences among treatments (desiccation × cryopreservation) for clover at 15°C and 25/15°C. Desiccation significantly decreased bean germination in the three regimes studied. In these two species, there was a decrease in hypocotyl length due to cryopreservation and/or desiccation when germination took place at 25°C. In clover and bean, incubation temperature could have had some influence in increasing differences among desiccation and cryopreservation treatments.

Introduction

Seeds from many species have been reported to survive liquid nitrogen exposure (see review in Stanwood, 1985). Several legume species are included in that review; rapid cooling was used in most cases and seeds had relatively low moisture content (6–14%). Seeds of wild legume species have also been cryopreserved successfully (González-Benito *et al.*, 1994). However, seedling vigour has not been so extensively studied (Stanwood and Roos, 1979; Pritchard *et al.*, 1988). Seed vigour can predict field performance more accurately than the standard germination test (Kim *et al.*, 1994). Indices related to seedling growth have been included in vigour studies of several species, e.g. plumule length (Perry, 1987; Kim *et al.*, 1994), fresh weight (Roos and Manalo, 1971) or root length (Stanwood and Roos, 1979). In the present work, the effects of cryopreservation, desiccation, and their interaction with incubation temperature on seed germination and vigour (hypocotyl length) have been studied in three Leguminosae species: lucerne, red clover and bean.

R.H. Ellis, M. Black, A.J. Murdoch, T.D. Hong (eds.), Basic and Applied Aspects of Seed Biology, pp. 797–802.
© *1997 Kluwer Academic Publishers, Dordrecht. Printed in Great Britain.*

Materials and Methods

Three Leguminosae species were studied: lucerne (*Medicago sativa* cv. Capitata), red clover (*Trifolium pratense*) and bean (*Phaseolus vulgaris* cv. Strike).

Seeds were desiccated in a chamber with silica gel for 40 d. Moisture content was determined on a fresh weight basis after drying in an oven at 105°C for 24 h, using 3 replicates.

Desiccated and non-desiccated seeds were included in cryovials or, for bean seeds, wrapped in aluminium foil, plunged into liquid nitrogen (LN) and kept there for 1 or 30 d. Thawing took place at room temperature. Frozen and non-frozen seeds were set to germinate at three different temperature regimes (15°C, 25°C and 25°C day/15°C night) with 18-h day/6-h night photoperiod and $30 \, \mu mol \, s^{-1} \, m^{-2}$ light intensity. Clover and lucerne seeds were sown in 7 cm diameter Petri dishes (4 replicates × 25 seeds) with two sheets of filter paper moistened with 3 ml distilled water, which was replaced periodically. For bean seeds, 9 cm diameter Petri dishes were used (5 replicates × 20 seeds each) and filter paper was moistened with 6 ml distilled water.

Germination (scored as radicle emergence) percentages recorded at 8 d after sowing were subjected to analysis of variance after arcsin transformation. Seedling vigour was measured by hypocotyl length 5 d after sowing, except for bean seeds incubated at 15°C which were recorded on the seventh d. Statistical analyses were carried out using SAS software.

Results

After desiccation seed moisture content was reduced from 8–10% to 4–5% in the three species (Table 1). Responses to LN exposure and desiccation, in terms of germination percentages and hypocotyl length, differed among species. Hypocotyl length data were analysed independently for each temperature.

In lucerne, there was no significant effect of desiccation and temperature on germination; however, an overall increase in germination for seeds stored in LN for 30 d vs. control seeds was observed ($p < 0.01$) (Table 1). The responses of hypocotyl length did not follow a clear pattern: at 25°C and 25/15°C there was a significant effect of the LN storage period, with best results in non-frozen seeds and seeds stored for 30 d in LN, respectively (Table 2), and no effect of desiccation. At 15°C, there was a significant interaction between LN exposure period and desiccation.

There was a significant interaction of incubation temperature with desiccation and cryopreservation on clover and bean seed germination, respectively. Therefore, statistical analyses were carried out by comparing means within each incubation regime.

Clover seed germination was not affected by treatments when incubation temperature was 15°C; in the other two regimes there was a significant interaction between LN and desiccation (Table 1). At 25°C for non-desiccated

Table 1. Effect of preservation in liquid nitrogen (LN) for 0 d (control), 1 d or 30 d and desiccation on seed germination at three different incubation temperatures (means ±SE)

Species	Desiccation[a]	Moisture content (%)	Germination (%) Incubation temperature								
			15°C			25°C			25°C day/15°C night		
			Control	1 d LN	30 d LN	Control	1 d LN	30 d LN	Control	1 d LN	30 d LN
Lucerne	−	7.9±0.1	99±1	100±0	99±1	99±1	99±1	100±0	96±2	99±1	100±0
	+	4.2±0.0	97±2	98±1	100±0	97±1	98±1	100±0	98±2	100±0	99±2
Clover	−	8.7±0.0	96±2	94±2	97±1	88±2	76±3	91±2	93±2	89±2	90±3
	+	4.6±0.1	92±1	97±2	95±3	89±2	94±2	95±4	77±4	93±3	89±2
Bean	−	10.6±0.0	94±3	81±3	100±0	99±1	94±2	100±0	95±1	100±0	99±1
	+	4.6±0.0	74±12	61±6	71±2	99±1	86±6	83±19	92±3	92±4	92±7

[a] −, non-desiccated seeds; +, desiccated seeds for 40 d in a chamber with silica gel

Table 2. Effect of preservation in liquid nitrogen (LN) for 0 d (control), 1 d or 30 d and desiccation on seedling hypocotyl length at three different incubation temperatures (means ± SE)

| Species | Desiccation[a] | Hypocotyl length (mm) Incubation temperature | | | | | | | | |
| | | 15°C | | | 25°C | | | 25°C day/15°C night | | |
		Control	1 d LN	30 d LN	Control	1 d LN	30 d LN	Control	1 d LN	30 d LN
Lucerne	−	13±0	12±0	10±1	17±1	13±1	13±0	12±0	12±0	14±1
	+	8±1	13±0	7±2	14±1	12±0	15±2	13±0	10±0	13±1
Clover	−	3±0	3±0	4±0	13±1	9±1	11±1	6±0	7±0	8±1
	+	3±0	6±1	3±1	10±1	10±0	9±1	7±1	8±0	7±1
Bean	−	6±0	9±1	9±0	20±1	16±1	12±1	7±1	12±0	10±1
	+	3±1	3±0	2±1	8±1	11±1	12±1	4±0	6±1	4±1

[a] −, non-desiccated seeds; +, desiccated seeds for 40 d in a chamber with silica gel

seeds, those which were cryopreserved for one d showed lower germination percentage than the other two treatments (control and 30 d in LN) ($p < 0.05$). On the other hand, there were no differences in the germination of desiccated seeds. At 25/15°C there was no significant effect among treatments. Seedling vigour after desiccation and cryopreservation treatments varied also depending on the incubation regimes (Table 2).

In bean, desiccation decreased germination especially at 15°C ($p < 0.001$), with less significance at 25°C and 25/15°C ($p < 0.05$) (Table 1). Similar results were observed for hypocotyl length (Table 2); in this case, however, with high significance level for all temperature regimes ($p < 0.001$). It should be noted that, in this species, 13% of seeds showed their cotyledons detached after LN storage, although only apparently intact seeds were sown.

Discussion

So far, the effect of incubation temperature on germination after desiccation and/or cryopreservation has not been widely studied. Germination tests are usually carried out as recommended by ISTA (International Seed Testing Association, 1985). For wild species, where less information is available, the best incubation temperature for germination tests is less well established before cryopreservation (Iriondo *et al.*, 1992). In our study it has been observed that, in some species like lucerne, incubation temperature did not have an effect on germination responses after different cryopreservation and desiccation treatments. However, this was not the case for clover and bean, where incubation temperature interacted with cryopreservation and desiccation in their effect on germination.

The damage caused by desiccation in bean seeds has been reported in other legume species such as *Pisum sativum* (Ellis *et al.*, 1990). However, as those authors demonstrated, damage was caused by imbibition, as it could be avoided by prehumidification. In the present study, desiccation damage was more evident at the lowest incubation temperature tested (15°C).

The hypocotyl length data recorded in the present study, as a means of vigour measurement, have not provided consistent results, especially for lucerne and clover. Only a negative effect of desiccation on bean hypocotyl length was observed. Other vigour tests should be explored to evaluate more accurately the effect of cryopreservation and desiccation, i.e. epicotyl and seedling dry weight (Roos and Manalo, 1971) or abnormal seedling percentage (Boyce, 1987).

Cracking of embryo tissues has been observed in other Leguminosae seeds following liquid nitrogen exposure (Sakai and Noshiro, 1975; Pritchard *et al.*, 1988). In the latter work, *Trifolium arvense* embryo damage was observed with scanning electron microscope after four cycles of freezing in LN and rewarming. In the present study embryo cracking was only evident in the species with large seeds, i.e. bean, where damage was shown as total detachment of cotyledons.

This fact should be taken into account in terms of the number of seeds being actually cryopreserved.

Acknowledgements

This work was supported by the CICYT project AMB 93-0092.

References

Boyce, K.G. 1987. *Seed Science and Technology* 15 (supplement): 466–468.

Ellis, R.H., Hong, T.D. and Roberts, E.H. 1990. *Seed Science and Technology* 18: 131–137.

González-Benito, M.E., Caze-Filho, J. and Pérez, C. 1994. *Plant Varieties and Seeds* 7: 23–27.

International Seed Testing Association. 1985. *Seed Science and Technology* 13: 299–319.

Iriondo, J.M., Pérez, C. and Pérez-García, F. 1992. *Seed Science and Technology* 20: 165–171.

Kim, S.H., Choe, Z.R., Kang, J.H., Copeland, L.O. and Elias, S.G. 1994. *Seed Science and Technology* 22: 59–68.

Perry, D.A. 1987. In: *Handbook of Vigour Test Methods,* pp. 10–20. Zurich: The International Seed Testing Association.

Pritchard, H.W., Manger, K.R. and Prendergast, F.G. 1988. *Annals of Botany* 62: 1–11.

Roos, E.E. and Manalo, J.R. 1971. *HortScience* 6: 347–348.

Sakai, A. and Noshiro, M. 1975. In: *Crop Genetic Resources for Today and Tomorrow,* pp. 317–326 (eds. O.H. Frankel and J.C.Hawkes). Cambridge: Cambridge University Press.

Stanwood, P.C. 1985. In: *Cryopreservation of Plant Cells and Organs,* pp. 199–226 (ed. K.K. Kartha). Boca Raton: CRC Press.

Stanwood, P.C. and Roos, E.E. 1979. *HortScience* 14: 628–630.

88. Changes of Carbohydrate Contents During Natural and Accelerated Ageing of Some Vegetable Seeds

M. HORBOWICZ

Research Institute of Vegetable Crops, 96-100 Skierniewice, Konstytucji 3 Maja 1/3, Poland

Abstract

The aim of the studies was to determine if changes of mono- and oligosaccharides correlate with deterioration following accelerated and natural ageing of watercress (*Lepidium sativum* L.) and white cabbage (*Brassica oleracea* L. var. *capitata*) seeds. Seeds were aged at 75% RH and 45°C for 0 to 14 days (d) before analyses of carbohydrate composition and germination tests. Sucrose, raffinose, stachyose and *myo*-inositol levels did not change during ageing of either species. Glucose was positively correlated with germination and shoot length in aged cabbage seeds. During ageing of watercress seeds glucose was relatively stable. Watercress seeds were more resistant to accelerated ageing and contained four times more raffinose than cabbage seeds. In aged seeds glucose was converted to an unknown derivative which from gas chromatography–mass spectrometry was probably gluconic acid or gluconic acid lactone.

Introduction

Seeds store reserve materials such as lipids, proteins and carbohydrates (Bewley and Black, 1985). These reserves are utilized during germination, and some of them, specially oligosaccharides, may preserve seed viability during dry storage (Blackman *et al.*, 1992). Soluble oligosaccharides accumulate rapidly during seed maturation (Horbowicz and Obendorf, 1994). They play an important role in membrane and protein stabilization (Crowe *et al.*, 1987; Leprince *et al.*, 1993). According to Wettlaufer and Leopold (1991) during seed ageing the mono-sugars, mainly glucose, can convert to products of Amadori and Maillard reactions. Glucose is a very reactive carbohydrate because it contains an easily oxidized aldehyde group. The aim of this paper is to determine if glucose and oligosaccharide levels correlate with deterioration during rapid ageing of white cabbage and watercress seeds.

Materials and Methods

Seeds of white cabbage (*Brassica oleracea* L. var. *capitata*), and watercress (*Lepidium sativum* L.) were aged for two weeks at 45°C, 75% RH (over a saturated sodium chloride solution). Carbohydrate composition was analysed

R.H. Ellis, M. Black, A.J. Murdoch, T.D. Hong (eds.), Basic and Applied Aspects of Seed Biology, pp. 803–808.
© *1997 Kluwer Academic Publishers, Dordrecht. Printed in Great Britain.*

before and after 1, 5, 9, and 14 d of ageing. Three replicates of ten seeds were used for carbohydrate determinations. In watercress, seeds were analysed with the seed coat, while in white cabbage the seed coat was removed. Glucose, sucrose, raffinose, stachyose, *myo*-inositol and an unknown carbohydrate derivative were determined as TMS-derivatives using gas chromatography (Horbowicz *et al.*, 1980). Mass spectra of the unknown carbohydrate were determined using an HP5890 gas chromatograph equipped with a mass selective detector – MSD HP5971A and a 12 m × 0.2 mm (ID) capillary column coated with HP-1 methyl silicone stationary phase (0.33 μm film thickness). Additionally the carbohydrate composition in 10-year old (naturally aged) seeds of both species was measured. These seeds were non-germinable.

Percentage germination and shoot lengths were measured on seeds that were randomly removed from the seed population undergoing accelerated ageing. One replicate contained 25 seeds. Seeds were germinated on wet germination paper in the dark for 7 d at 20–22°C.

All analyses were performed in three replicates, and data were subject to analysis of variance, using PC program for one-factor analysis of variance (Student's *t*-test, $\alpha = 0.05$ or $p = 0.05$). Correlation coefficients were also determined.

Results and Discussion

Sucrose was the major soluble carbohydrate in seeds of watercress and white cabbage (Table 1) being about three times higher than stachyose. Sucrose, stachyose, raffinose and *myo*-inositol contents did not change in cabbage and watercress seeds aged at 45°C and 75% RH. According to Sun and Leopold (1993), during accelerated ageing of soyabean seed the sucrose, stachyose and raffinose levels did not change significantly. Sucrose, stachyose and *myo*-inositol levels were very similar in seeds of both species, while raffinose content was four times higher in watercress than in cabbage seeds. Relatively large amounts of glucose were detected in non-aged cabbage and watercress seeds (Fig. 1). Changes in glucose during ageing were different in cabbage in comparison to watercress. In cabbage seeds ageing caused a quick decline of glucose while in watercress during the first 5 d of ageing the glucose level increased, and then slightly decreased over the next 9 d. The final amounts of glucose in aged and non-aged watercress seeds were nearly the same, while in cabbage there was four times less glucose in non-viable aged seeds than before ageing. This difference between the two vegetables can be explained by the sampling technique. The cabbage was analyzed without the seed coat whereas whole seeds of watercress were analyzed as whole seeds, because of difficulties in removing the seed coat.

Seeds of cabbage and watercress contain an unknown carbohydrate, the content of which increased during seed ageing (Fig. 1). According to GC–MS data that unknown substance could be a gluconic acid lactone, or another glucose derivative, or a mixture. Mass spectra of the unknown substance had

Table 1. Changes of sucrose, raffinose, stachyose and *myo*-inositol content during ageing of watercress and cabbage seeds (mg g^{-1} dry weight)

Ageing time	Sucrose W[a]	C[b]	Raffinose W	C	Stachyose W	C	*myo*-Inositol W	C
Non-aged	31.25	28.89	11.18	2.51	11.29	11.27	0.24	0.39
1 day	30.88	29.30	11.66	2.59	12.01	10.17	0.24	0.40
5 days	33.37	30.10	12.93	3.22	10.14	11.69	0.32	0.48
9 days	32.43	33.02	11.60	3.54	10.03	10.08	0.34	0.43
14 days	29.86	30.65	11.07	3.52	9.56	9.78	0.38	0.32
10 years	33.08	29.39	12.40	1.93	7.86	6.01	0.34	0.22
LSD, $\alpha = 0.05$	ns	ns	ns	0.60	2.05	1.92	ns	ns

[a]Watercress seeds; [b]Cabbage seeds; ns, non-significant differences

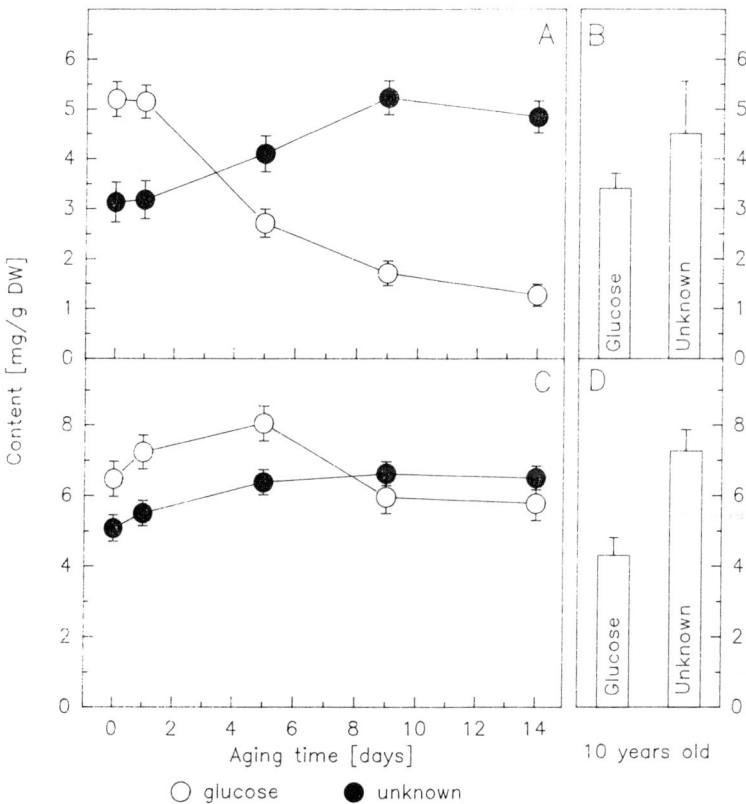

Figure 1. Content of glucose and unknown carbohydrate in 14 d accelerated aged and 10-year-old seeds of cabbage (A,B) and watercress (C,D). Vertical bars represent standard deviations of the mean at $p = 0.05$

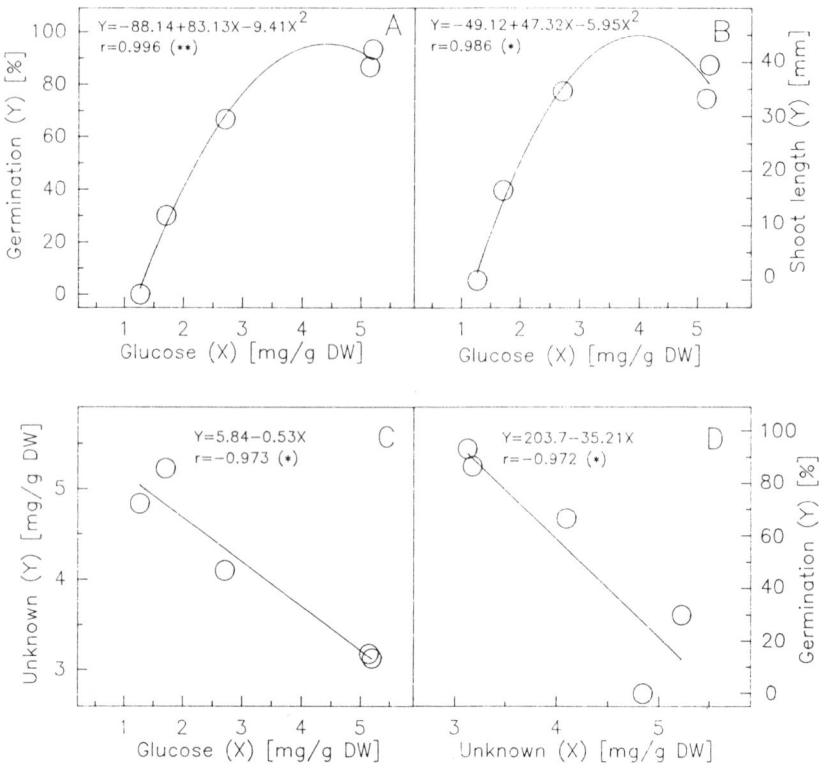

Figure 2. Relationships between glucose and germination (A), shoot length (B), unknown carbohydrate (C), and between unknown carbohydrate and germination (D). Vertical bars represent standard deviation (SD) of the mean at $p = 0.05$

the following main M/Z signals: 73, 103, 147, 217, 249, 307, 334, 363, 451 and 466. Mass spectra of the TMS derivative of gluconic acid lactone contained in the computer library had the following main M/Z signals: 73, 103, 147, 204, 217, 243, 305, 333, 361, 451 and 466. The last 466 signal is the so-called molecular ion of TMS-glucolactone and was present in seed samples of both vegetables investigated.

Sucrose, raffinose, and *myo*-inositol contents in 10-year old cabbage and watercress seeds were the same as in non-aged seeds (Table 1). The stachyose level was significantly lower in old non-viable seeds of both species in comparison to non-aged ones. The glucose contents in old seeds of both species were lower than in non-aged seeds (Fig. 1). In old cabbage seeds the glucose level was similar to that after 5 or 9 d of accelerated ageing. The seeds after 5 and 9 d of ageing treatment had 66.7% and 30.0% of germination, while 10-year

Table 2. Germination percent and shoot length of non-aged and aged watercress and cabbage seeds

Ageing time (days)	Germination (%)		Shoot length (mm)	
	Watercress	Cabbage	Watercress	Cabbage
Non-aged	100.0	93.3	39.5	39.5
1	100.0	86.7	33.2	33.3
5	86.7	66.7	19.9	34.7
9	53.3	30.0	8.6	16.5
14	10.0	0.0	1.7	0.0
LSD, $\alpha = 0.05$	10.5	11.5	3.3	9.8

old seeds were all non-germinable. The glucose level in old watercress seeds was lower than in seeds not aged and after 14 d of accelerated ageing. The levels of the unknown carbohydrate in old cabbage and watercress seeds and after 14 d of accelerated ageing were quite similar, and were about 50% higher than in non-aged, fully germinable seeds (Fig. 1). The unknown carbohydrate or its derivative could be a marker of watercress and cabbage quality. However, this presumption requires further studies using seeds of another species, first of all from the *Cruciferae* family. Moreover the unknown substance should be isolated, purified and fully identified using NMR and other techniques.

There is a significant correlation between glucose and the unknown carbohydrate in cabbage seeds during seed ageing (Fig. 2A). When the glucose content dropped the unknown substance increased. It appears that glucose may be converted to the unknown substance during ageing. Other significant correlations were found between glucose and germination percentage, the unknown carbohydrate level and germination percentage, and glucose and shoot length of aged cabbage seeds (Figs. 2B, C and D). Glucose, germination percentage and shoot length of cabbage seeds decreased during the accelerated ageing but such correlations were not found in watercress seeds.

Watercress seeds survived conditions of accelerated ageing more easily than cabbage seeds (Table 2). Germination percentages of watercress were higher after 1, 5, 9 and 14 d of ageing treatment. Better survival of accelerated ageing of watercress seeds may be associated with four times higher content of raffinose in comparison to cabbage seeds (Tables 1 and 2).

Acknowledgement

The author wishes to thank The Kosciuszko Foundation for support of the studies.

References

Bewley, J.D. and Black, M. 1985. *Seeds. Physiology of Development and Germination.* New York and London: Plenum Press.

Blackman, S., Obendorf, R.L. and Leopold, A.C. 1992. *Plant Physiology* 100: 225–230.

Crowe, J.H., Crowe, L.M., Carpenter, J.F. and Aurell Wistrom, C. 1987. *Biochemical Journal* 242: 1–10.

Horbowicz, M., Czapski, J. and Bakowski, J. 1980. *Acta Alimentaria Polonica* 6: 227–236.

Horbowicz, M. and Obendorf, R.L. 1994. *Seed Science Research* 4: 385–405.

Leprince,O., Hendry, G.A.F. and McKersie, B.D. 1993. *Seed Science Research* 3: 231–246.

Sun, W.Q. and Leopold, A.C. 1993. *Physiologia Plantarum* 89: 767–774.

Wettlaufer, S.C. and Leopold, A.C. 1991. *Plant Physiology* 97: 165–169.

89. Light Promotes Free Radical Accumulation in Ageing Soyabean Seeds

M.M. KHAN[1], G.A.F. HENDRY[1], N.M. ATHERTON[2] and
C. WALTERS VERTUCCI[3]

[1]NERC Unit of Comparative Plant Ecology, Department of Animal and Plant Sciences; [2]Department of
Chemistry, The University of Sheffield, Sheffield S10 2TN, UK; [3]USDA – ARS National Seed Storage
Laboratory 1111 S. Mason St., Fort Collins, CO 80521–4500 USA

Abstract

The role played by oxygen, as a free radical, in ageing, death and decay is well
established in animal systems. In plants, including seeds, its role is less certain –
plants are richly endowed with anti-oxidants (e.g. vitamins C and E). As a model
system for studying the role of oxidative damage in ageing seeds, a study was
undertaken to evaluate the effect of accelerated ageing on germinability, hypocotyl
development and free radical processes. Accelerated ageing was achieved by
incubating seed over H_2SO_4 at 35°C (1% RH) in the light (90 $\mu mol\, m^{-2}\, s^{-1}$) or in
the dark for 7, 14, 35, 56 and 69 days (d). The results show that accelerated ageing
under darkness and light caused significantly different rates of germination and
hypocotyl extension. Damage in the form of enhanced lipid peroxidation and
organic free radical accumulation was observed and largely confined to the testa.
This was significantly greater in seeds aged under illumination. Both lipid
peroxidation and free radical accumulation was significantly correlated with the
decline in germinability and hypocotyl expansion in rapidly aged seeds, particularly in seeds exposed to light. The strong correlation between loss of germinability
and activity of free radical-linked processes in aged seeds strengthens the evidence
that the loss of viability is closely linked to oxidative damage.

Introduction

Seeds deteriorate during prolonged storage, but the rate of deterioration varies
greatly among species (Priestley, 1986; Roberts, 1986). While the exact cause of
loss of seed viability is still not well-defined, many studies have suggested
damage to the membrane as one causative factor (e.g. Parrish and Leopold,
1978; Pearce and Abdel Samad, 1980; Bewley, 1986). In the presence of oxygen,
ageing of seed is associated with peroxidation of polyunsaturated fatty acids
(Stewart and Bewley, 1980; Wilson and McDonald, 1986; Hendry et al., 1992;
Hendry, 1993). Lipid peroxidation and associated membrane damage may be
important in the deterioration of stored seeds and reduced longevity of seeds
under natural conditions.

It is known from vegetative tissue that light can induce the production of free
radicals which, in turn, can lead to the destruction of macromolecules (Levitt,
1980; Foyer et al., 1994; Cakmak et al., 1995), but the mechanism by which

R.H. Ellis, M. Black, A.J. Murdoch, T.D. Hong (eds.), Basic and Applied Aspects of Seed Biology, pp. 809–815.
© 1997 Kluwer Academic Publishers, Dordrecht. Printed in Great Britain.

light-induced oxidative damage can arise in seeds is not clear. A study was undertaken to investigate the germinability, hypocotyl expansion, accumulation of stable free radicals and lipid peroxidation of soyabean seeds, aged rapidly over 10 weeks in the light and in the dark.

Material and Methods

Storage and Ageing Treatment

Seeds of soyabean (*Glycine max* L. Merr. cv. Williams 82), were aged over H_2SO_4 at 35°C (1% RH) in the light (90 µmol m^{-2} s^{-1}) or in the dark. Samples were removed from the ageing treatment at 7, 14, 35, 56 and 69 d.

Germination Treatment

Seeds were pre-hydrated for 16 h and then rolled in Whatman 1 filter paper in contact with a reservoir of water suffucent to keep the seeds moist. After 96 h at 25°C in the dark, the percentage germination and hypocotyl expansion were recorded.

Lipid Peroxidation Product Estimation

Lipid peroxidation was determined as the concentration of thiobarbituric acid-reactive substances, equated with malonyealdehyde (MDA), and quantified from the second derivative spectrum with standards prepared from 1,1,3,3-tetra-ethoxypropane as described by Hendry *et al.* (1993).

Electron Paramagnetic Resonance (EPR) Response

Electron paramagnetic resonance (EPR) spectra were recorded on a Bruker ER200D spectrometer (Leprince *et al.*, 1990). Free radical concentrations were estimated by the amplitude of the second derivative spectrum corrected for instrument gain and expressed on an unimbibed seed weight basis.

Statistical Analysis

Statistical data throughout were expressed as the mean values ($\pm$SE) of minimum replication of five samples or three for cotyledon or single samples of up-to twenty testa in the case of EPR responses.

Results

Seed Ageing and Viability

Seed material aged rapidly in the light and dark over 69 days showed a decrease in germination to 80% in light-treated and 91% in dark-treated seeds (Fig. 1). There was a 2-fold decline in hypocotyl length over the 10 weeks (Fig. 2).

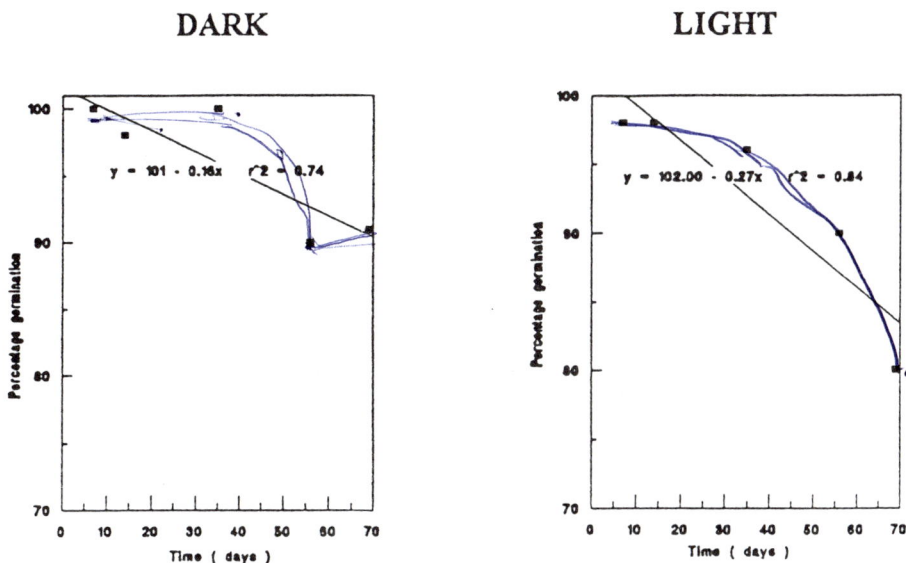

Figure 1. Percentage germination of soyabean seeds aged under dark or light

Lipid Peroxidation in Cotyledon and Testa

Lipid peroxidation increased with time in both testa and cotyledon in both light- and dark-treated seeds (Fig. 3). The highest increase (2-fold) occurred in the testa of light-treated tissue and was strongly correlated with length of ageing treatment ($r^2 = 0.97$).

Free-radical Processes in Cotyledon and Testa

In absolute terms the amplitude of the EPR signal was nearly twice as great in light-treated as in dark-treated testa (Figs. 4 and 5). There was also a strong correlation between the increase in EPR response in the testa and the decline in germination in light-treated seeds ($r^2 = 0.94$). There was no significant difference in EPR response in the cotyledons of seeds aged rapidly under light or dark.

DARK   LIGHT

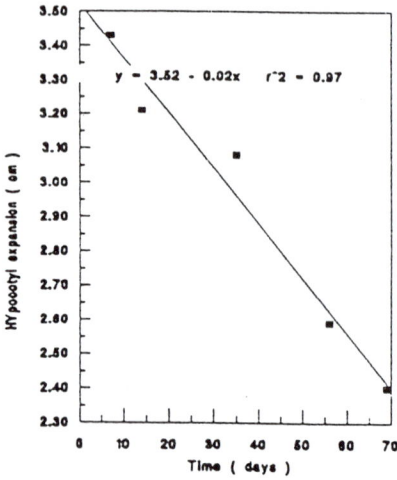 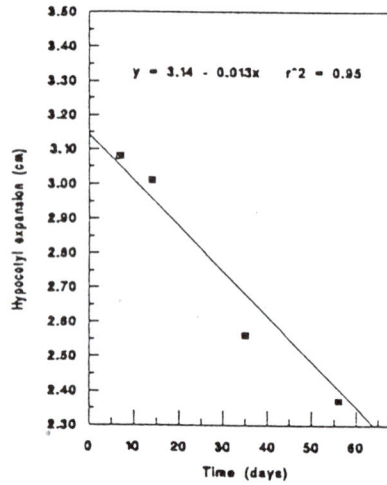

Figure 2. Effect of accelerated ageing on hypocotyl expansion of soyabean seeds aged under dark or light

DARK   LIGHT

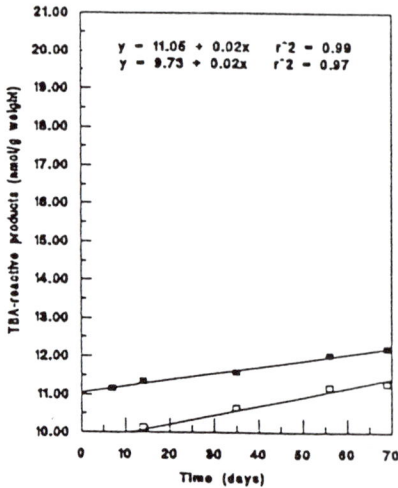 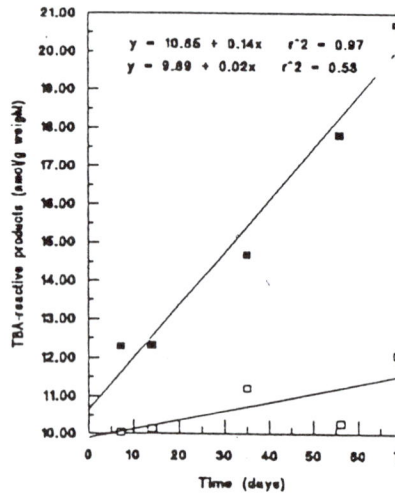

Figure 3. Lipid peroxidation (as concentration of thiobarbituric acid reactive products/g wt in testa (closed symbols) and cotyledons (open symbols) aged under dark or light

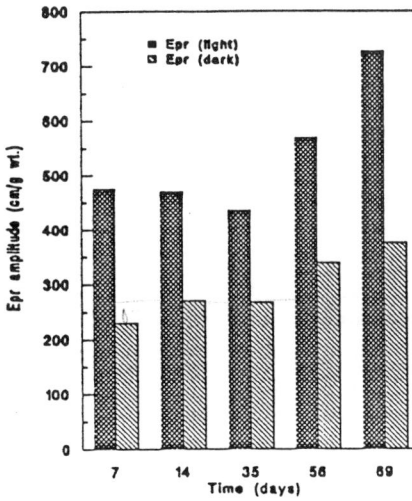

Figure 4. Free-radical build-up: EPR response, measured as the amplitude of the signal at constant instrument settings/g wt of seed testa aged under dark or light

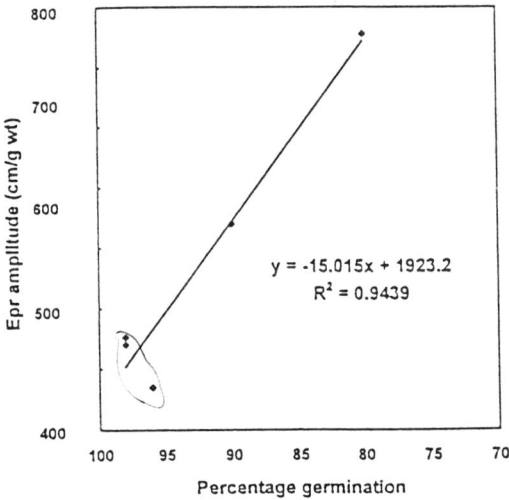

Figure 5. Relationship between EPR in testa and percentage germination in soyabean seeds aged under dark or light

Discussion

The results described here demonstrate that both lipid peroxidation and free radical accumulation were significantly correlated with the decline in germinability and hypocotyl expansion in rapidly aged soyabean seeds. In absolute values lipid peroxidation and free-radical build-up were significantly greater in seeds aged under illumination. The location for lipid peroxidation and free-radical build-up was the testa (rather than the cotyledon), an unexpected observation given the high concentration of lipids in the cotyledon. Similarly, the most pronounced evidence of stable free-radical accumulation was in the testa, again in seeds under illumination.

The strong correlation between loss of viability and activity of free radical-linked processes in rapidly aged seeds reported here strengthens the evidence that loss of viability is closely linked to the ravages of oxidative damage. The finding that most of the damage was located in the testa raises questions about the function of the testa during seed storage and the molecular connection, if any, between testa and embryo. If the testa had been removed at the start of the rapid ageing treatments, would such seeds have retained maximum viability?

The generation of free radicals and the associated peroxidation of lipids, is substantially enhanced on illumination, raising further questions on the nature of the light receptor and the possible significance of pigmented as opposed to translucent testa. These are questions which we are currently addressing.

Acknowledgements

M.M. Khan acknowledges support from the Ministry of Education, Govt. of Pakistan.

References

Bewley, J. D. 1986. In: *Physiology of Seed Deterioration*. pp 27–45 (eds. M. B. McDonald Jr. and C. J. Nelson). Madison, WI: Crop Science Society of America Inc.

Cakmak, I., Atli, M., Kaya, R., Evliya, H. and Marschner, H. 1995. *Journal of Plant Physiology* 146: 355–360.

Foyer, C. H., Lelandais, M. and Kunert, K. J. 1994. *Physiologia Plantarum* 92: 696–717.

Hendry, G. A. F. 1993. *Seed Science Research* 3: 273–279.

Hendry, G. A. F., Finch-Savage, W. E., Thorpe, P. C., Atherton, N. M., Bukland, S. M., Nilsson., K. A. and Seel, W. E. 1992. *The New Phytologist* 122: 273–279.

Hendry, G. A. F., Thorpe, P. C. and Merzlyak, M. N. 1993. In: *Methods in Comparative Plant Ecology*, pp 154–156 (eds. G. A. F. Hendry and J. P. Grime). London: Chapman & Hall.

Leprince, O., Deltour, R., Thorpe, P. C., Atherton, N. M. and Hendry, G. A. F. 1990. *The New Phytologist* 116: 573–580.

Levitt, J. 1980. *Responses of Plants to Environmental Stresses, V2. Water, Radiation, Salt and other Stresses*. NY and London: Academic Press.

Parrish, D. J. and Leopold, A. C. 1978. *Plant Physiology* 61: 365–368.

Pearce, R. S. and Abdel Samad, I. M. 1980. *Journal of Experimental Botany* 31: 1283–1290.
Priestley, D. A. 1986. In: *Seed Aging,* pp. 125–195 (ed. D. A. Priestley). New York, NY: Comstock Publishing Associates.
Roberts, E. H. 1986. In: *Physiology of Seed Deterioration*, pp. 101–123 (eds. M. B. McDonald, Jr. and C. J. Nelson). Madison WI: Crop Science Society of America, Inc.
Stewart, R. R. C. and Bewley, J. D. 1980. *Plant Physiology* 65: 245–246.
Wilson, D. O. and McDonald, M. B. 1986. *Seed Science and Technology* 14: 269–300.

90. Cryopreservation of Seeds and Embryonic Axes of Several *Citrus* Species

M.N. NORMAH and M.N. SITI DEWI SERIMALA

Department of Botany, Faculty of Life Sciences, Universiti Kebangsaan Malaysia, 43600 UKM, Bangi, Selangor, Malaysia

Abstract

Citrus aurantifolia seeds can be successfully cryopreserved after desiccating them to a moisture content of 12.93% (50% viability) while seeds of *C. halimii* gave only 25% viability after cryopreservation with a moisture content of 9.5%. Seeds of *C. hystrix* however, are sensitive to desiccation as they failed to germinate when the moisture content was reduced to 27%, and thus did not survive cryopreservation. The embryonic axes of the three *Citrus* species gave higher percentage of survival after cryopreservation. 100% survival was obtained with *C. aurantifolia* and *C. halimii* embryonic axes with moisture contents of 9–11% and 16.6% respectively. With *C. hystrix* axes, the highest survival obtained was 60% at a moisture content of 11.04%. Cryopreservation methods i.e. encapsulation–dehydration, slow freezing and vitrification were further employed for the embryonic axes of *C. hystrix*; however, there was no improvement in the survival percentage obtained.

Introduction

Cryopreservation is an important tool in conservation of plant germplasm. Successful cryopreservation of *Citrus* has been restricted to a few reports, where whole seeds (Mumford and Grout, 1979), ovules (Bajaj, 1984), somatic embryos (Marin and Duran-Vila, 1988), zygotic embryos (Radhamani and Chandel, 1992) or nucellar cells (Kobayashi *et al.*, 1990) have been used.

South-East Asia is known to be the centre of diversity and distribution of *Citrus* especially in the subfamily of Aurantioideae (family Rutaceae) (Jones and Ghani, 1987). *Citrus aurantifolia* (Christm. & Panzer) Swingle is commonly known as lime, sour lime or common lime. The fruit is used in nearly every home in South-East Asia, mainly to flavour food, but also to prepare drinks and for a variety of medicinal applications. The juice of *Citrus hystrix* DC. is used for seasoning and as insecticides. The leaves are commonly used to season food. *Citrus halimii* B. C. Stone is an endemic species to Malaysia and has been noted to be at the edge of extinction. These cultivated, semi cultivated or wild *Citrus* need to be conserved before they are extinct or being replaced with new cultivars. In the present study, an attempt was made to desiccate the seeds and excised embryonic axes of *C. aurantifolia*, *C. hystrix* and *C. halimii* as a means to long-term conservation of germplasm.

R.H. Ellis, M. Black, A.J. Murdoch, T.D. Hong (eds.), Basic and Applied Aspects of Seed Biology, pp. 817–823.
© *1997 Kluwer Academic Publishers, Dordrecht. Printed in Great Britain.*

Materials and Methods

Plant Material

Freshly harvested fruits of *Citrus aurantifolia, C. hystrix* and *C. halimii* were collected from Pontian, Johor; Kepala Batas, Kedah and Fairlie Estate (Boh Plantation) respectively.

Culture Medium

A basal MS (Murashige and Skoog, 1962) medium supplemented with 30 g l^{-1} sucrose, 7 g l^{-1} Difco agar and 0.3 mg l^{-1} BAP (benzyladenine purine) was used. The pH was adjusted to 5.7 before autoclaving.

Establishment of Cultures

Seeds were surface sterilized with absolute ethanol for 2 min followed by 20% commercial Clorox with a few drops of Tween 20 for 5 min. The seeds were then rinsed with sterile distilled water 3–5 times. Embryonic axes were excised aseptically from these seeds by removing the testa and separating the axes from the cotyledons with a scalpel blade. Axes were then cultured on the described medium and allowed to grow. All cultures were maintained at 26 ± 1°C under 8 h photoperiod with light intensity of 25 μmol m^{-2} s^{-1}. *In vitro* grown seedlings were subcultured onto MS medium supplemented with 0.5 mg l^{-1} NAA to induce better growth of shoots and roots. These seedlings were later thoroughly washed with sterile distilled water to remove traces of medium. They were then planted into small pots containing a mixture of sand, soil and organic material (3:2:1). The potted plants were covered with perforated plastic and were kept in a shade house at 26–30°C with light intensity of 22–30 μmol m^{-2} s^{-1}.

Cryopreservation Methods

Desiccation

Seeds of *C. aurantifolia, C. hystrix* and *C. halimii* were desiccated by placing batches of 80 seeds at ambient temperature 26 ± 2°C for different periods of time. Excised embryonic axes were desiccated in batches of 90 for 0, 1, 2, 3 and 4 h in the sterile air flow of a laminar flow cabinet. At the end of each desiccation period, 20 seeds/30 axes were used for determining moisture content using a low constant temperature oven method (103 ± 2°C for 16 h) and the moisture was calculated on a fresh weight basis. 30 of the remaining seeds/axes were germinated/cultured as controls for the respective desiccation period.

Another group of 30 seeds/embryonic axes were wrapped in aluminium foil envelopes and immersed directly in liquid nitrogen (–196°C) for 16 h. The embryonic axes were then removed from the cryo-tank, thawed in a water bath ($40 \pm 2°C$) and tested for germination. The experiment was repeated twice.

Seeds which produce a morphologically normal seedling are considered viable. Embryonic axes were recorded as surviving when a fully developed seedling (with shoot and root) was obtained.

For *C. hystrix*, the following cryopreservation methods were further employed for embryonic axes.

Encapsulation–Dehydration

After excision, the embryonic axes were encased in alginate beads (3% low viscosity alginic acid with 0.75 M sucrose) for 72-h pretreatment in liquid MS medium with 0.1 M, 0.5 M or 1.0 M sucrose. Following pretreatment, beads were separated out on sterile petri dishes and air dried in the laminar flow hood for 0, 1, 2 or 4 h, placed in cryotubes and plunged into liquid nitrogen. Each treatment combination consisted of ten embryonic axes and was repeated twice. Thawing was carried out by rewarming the samples at room temperature ($25 \pm 1°C$) for 5–10 min. Encapsulated embryonic axes were then planted on recovery medium.

Vitrification

Embryonic axes were precultured on MS medium supplemented with 5% DMSO and 5% glucose; 1.2 M sorbitol; 1.0 M sorbitol or 0.6 M sorbitol. PVS 2 cryoprotectant (30% glycerol, 15% ethylene glycol and 15% DMSO in liquid MS medium with 0.4 M sucrose) was added to cryotubes on ice and embryonic axes were added and stirred. After 20 min cryotubes were submerged in liquid nitrogen. Samples were rewarmed in a water bath ($40 \pm 2°C$), rinsed in liquid MS medium with 1.2 M sucrose and cultured on recovery medium. Each treatment consisted of ten embryonic axes and was repeated twice.

Slow Freezing

Embryonic axes were acclimatized for 2 days (d) at room temperature ($25 \pm 1°C$) on MS medium with 5% DMSO and additional Gelrite ($0.3 \ gl^{-1}$) in the medium. The samples were then transferred to 0.25 ml liquid MS medium in 2 ml cryotubes and 1 ml of the cryoprotectant PGD (10% each polyethylene glycol, glucose and DMSO in MS liquid medium) was added over 30 min. A 30 min equilibration at 4°C was followed by cooling at 0.3°C/min to –35°C and plunging in liquid nitrogen. Samples were thawed for two min in $40 \pm 1°C$ water then rinsed in liquid MS medium and cultured on recovery medium. Ten embryonic axes were used for each treatment and was repeated twice.

Results and Discussion

Table 1 shows the percentage viability of *Citrus* seeds following cryopreservation at various moisture contents. The viability of *C. aurantifolia* seeds was not much affected when whole seeds were desiccated from about 50% moisture to 12% moisture. After 4 h of desiccation, at a moisture content of 12.93%, *C. aurantifolia* seeds survived cryopreservation (50% germination). A further reduction in moisture content reduced the seed viability after cryopreservation. A similar result (53% germination was obtained when *Coffea liberica* seeds were cryopreserved at a moisture content of 16.7% (Normah and Vengadasalam, 1992). Seeds of *C. hystrix* lost moisture when desiccated for 6 h. The loss in moisture was greater for the first 2 h of desiccation. The viability of fresh seeds was considerably low and significantly lost when whole seeds were desiccated to about 27% moisture content. No viability was obtained after cryopreservation. The viability of *C. halimii* seeds was reduced significantly when whole seeds were desiccated from about 60% moisture to 9.5% moisture. However, at a moisture content of 9.46% the seeds gave 25% viability after cryopreservation.

Table 1. Percentage viability of *Citrus* seeds following cryopreservation at various initial moisture contents

Species	Desiccation (h)	Moisture content (%)	Viability of non-cryopreserved seeds (%)	Viability of cryopreserved seeds (%)
C. aurantifolia	0	49.68[a]	88 [a]	0
	1	28.65[b]	86 [ab]	0 [e]
	2	21.34[c]	85 [ab]	0 [e]
	3	20.50[c]	83 [ab]	13 [d]
	4	12.93[d]	82 [ab]	50 [a]
	5	12.27[d]	76 [ab]	38 [b]
	6	11.99[d]	73 [b]	28 [c]
C. hystrix	0	60.66[a]	50 [a]	0
	1	43.77[b]	45 [a]	0
	2	27.02[c]	0 [b]	0
	3	22.12[c]	0 [b]	0
	4	18.35[c]	0 [b]	0
	5	15.88[c]	0 [b]	0
	6	14.95	0 [b]	0
C. halimii	0	61.28	91.67	0.00[a]
	3	29.84	58.33	0.00[a]
	6	9.46	16.67	25.00[b]

For each species, values in the same column having the same superscript are not significantly different at $p = 0.05$ based on Duncan Multiple Range test

Seeds of *C. aurantifolia* and *C. halimii* survived considerably greater desiccation

Table 2. Percentage survival of *Citrus* embryonic axes following cryopreservation at various initial moisture contents

Species	Desiccation (h)	Moisture content (%)	Survival of non-cryopreserved axes (%)	Survival of cryopreserved axes (%)
C. aurantifolia	0	15.08[a]	93.33[a]	100.00[a]
	1	9.02[a]	100.00[a]	93.33[b]
	2	11.66[a]	100.00[a]	100.00[a]
	3	5.50[b]	96.67[a]	100.00[a]
	4	3.91[b]	96.67[d]	96.67[a]
C. hystrix	0	39.44[a]	100[a]	0[e]
	1	15.31[b]	70[b]	50[c]
	2	11.04[b]	65[c]	60[a]
	3	9.45[c]	65[c]	55[b]
	4	5.89[c]	30[d]	30[d]
C. halimii	0	61.71[a]	100.00[a]	0.00[d]
	1	16.62[b]	100.00[a]	100.00[a]
	2	8.90[c]	77.78[ab]	66.67[b]
	3	6.73[c]	44.44[bc]	38.89[c]
	4	5.92[c]	38.89[c]	16.67[cd]

For each species, values in the same column having the same superscript are not significantly different at $p = 0.05$ based on Duncan Multiple Range test

than that typically observed in recalcitrant seeds, which was in agreement with reports on *Coffea arabica* (Ellis *et al.*, 1990). Mumford and Grout (1979) had shown that *C. limon* seeds could successfully be cryopreserved at a moisture content of 5.4%.

Embryonic axes of *C. aurantifolia* survived cryopreservation when their moisture content was between 3.9–15.1% (Table 2). The survival rate was high (93–100%) at these moisture levels. Freshly excised embryonic axes of *C. hystrix* had moisture content of approximately 39%. The moisture content of axes decreased rapidly within the first hour of desiccation but the decrease was more gradual thereafter. Embryonic axes with no desiccation showed 100% viability when cultured *in vitro*. However, with increasing duration of desiccation, there was a decrease in viability down to 30% at 4 h. Following cryopreservation, none of the axes with moisture content 39% (0 h desiccation) were viable. 50–60 % survival was obtained when the cryopreserved axes contained 9.5–11% moisture. The best percentage viability (60%) was obtained when the axes were cryopreserved at a moisture content of approximately 11%. Moisture content below this significantly decreased the percentage viability. With *C. halimii*, 100% survival was obtained when embryonic axes were desiccated to 16.6% moisture content. Axes with lower moisture content (5.92–8.9%) survived at a lower percentage (16.7–66.7%). Surviving axes of the three *Citrus* species studied, after cryopreservation swelled and turned green within one week of culture. Compared to the control, the development of the root and shoot of the cryopreserved axes was slightly slower. However, both types of seedlings were found to be morphologically uniform.

The results of the present study indicate that the excised embryonic axes of *C. aurantifolia, C. halimii, C. hystrix* are markedly tolerant to desiccation. The axes of *C. aurantifolia* were tolerant at a wider range of moisture content (3–15%) whereas the axes of *C. halimii* and *C. hystrix* were most tolerant at 16.6% and 11.04% moisture content, respectively. Radhamani and Chandel (1992) had shown that 14% moisture content was optimal for trifoliate orange axes, to successfully tolerate the low freezing temperature of –196°C.

C. aurantifolia embryonic axes showed an orthodox seed behaviour when they survived liquid nitrogen storage at a low moisture content of 3.9%. A similar result was reported for *Vigna* embryonic axes (Normah and Vengadasalam, 1992). *C. hystrix* and *C. halimii* axes however, showed a characteristic of recalcitrant seeds where survival of embryonic axes is usually greatest when the moisture content is between 14–20% (Normah *et al.,* 1986; Krishnapillay, 1989; Radhamani and Chandel, 1992).

Survival of *C. hystrix* axes with other cryopreservation methods was much lower than obtained with the desiccation method. The best percentage survival using encapsulation–dehydration, vitrification and slow freezing methods was 38%, 5% and 20%, respectively. The results of this study indicated the importance of screening several techniques in order to obtain a suitable or an appropriate method of conservation.

This study shows that by manipulating the moisture content for *C. aurantifolia* and *C. halimii,* both the whole seeds and excised embryos can successfully be cryopreserved. For *C. hystrix,* only the embryonic axes can be cryopreserved. The desiccated embryonic axes of the three *Citrus* species do not lose viability after rapid cooling, and storage at the temperature of liquid nitrogen (–196°C), thus indicating the possibility of using the desiccation of axes method for long-term conservation of these species.

Acknowledgements

This work was supported by the Malaysian National IRPA grant.

References

Bajaj, Y. P. S. 1984. *Current Science* 53: 1215–1216.
Ellis, R. H., Hong, T. D. and Roberts, E. H. 1990. *Journal of Experimental Botany* 41: 1167–1174.
Jones, D. T. and Ghani, F. D. 1987. *Malaysian Applied Biology* 16: 139–144.

Kobayashi, S., Sakai, A. and Ohyama, I. 1990. *Plant Cell, Tissue Organ Culture* 23: 18–20.
Krishnapillay, B. 1989. Ph.D. Thesis. Universiti Pertanian Malaysia, Serdang, Malaysia.
Marin, M. L. and Duran-Vila, N. 1988. *Plant Cell, Tissue Organ Culture* 14: 51–57.
Mumford, P. M. and Grout, B. W. W. 1979. *Seed Science and Technology* 7: 401–410.
Murashige, T. and Skoog, F. 1962. *Physiologia Plantarum* 15: 473–497.
Normah, M. N. and Vengadasalam, M. 1992. *Cryo-Letters* 13: 199–208.
Normah, M. N., Chin, H. F. and Hor, Y. L. 1986. *Pertanika* 9: 299–303.
Radhamani, J. and Chandel, K. P. S. 1992. *Plant Cell Reports* 11: 372–374.

Current Plant Science and Biotechnology in Agriculture

1. H.J. Evans, P.J. Bottomley and W.E. Newton (eds.): *Nitrogen Fixation Research Progress.* Proceedings of the 6th International Symposium on Nitrogen Fixation (Corvallis, Oregon, 1985). 1985   ISBN 90-247-3255-7

2. R.H. Zimmerman, R.J. Griesbach, F.A. Hammerschlag and R.H. Lawson (eds.): *Tissue Culture as a Plant Production System for Horticultural Crops.* Proceedings of a Conference (Beltsville, Maryland, 1985). 1986   ISBN 90-247-3378-2

3. D.P.S. Verma and N. Brisson (eds.): *Molecular Genetics of Plant-microbe Interactions.* Proceedings of the 3rd International Symposium on this subject (Montréal, Québec, 1986). 1987   ISBN 90-247-3426-6

4. E.L. Civerolo, A. Collmer, R.E. Davis and A.G. Gillaspie (eds.): *Plant Pathogenic Bacteria.* Proceedings of the 6th International Conference on this subject (College Park, Maryland, 1985). 1987   ISBN 90-247-3476-2

5. R.J. Summerfield (ed.): *World Crops: Cool Season Food Legumes.* A Global Perspective of the Problems and Prospects for Crop Improvement in Pea, Lentil, Faba Bean and Chickpea. Proceedings of the International Food Legume Research Conference (Spokane, Washington, 1986). 1988   ISBN 90-247-3641-2

6. P. Gepts (ed.): *Genetic Resources of* Phaseolus *Beans.* Their Maintenance, Domestication, Evolution, and Utilization. 1988   ISBN 90-247-3685-4

7. K.J. Puite, J.J.M. Dons, H.J. Huizing, A.J. Kool, M. Koorneef and F.A. Krens (eds.): *Progress in Plant Protoplast Research.* Proceedings of the 7th International Protoplast Symposium (Wageningen, The Netherlands, 1987). 1988   ISBN 90-247-3688-9

8. R.S. Sangwan and B.S. Sangwan-Norreel (eds.): *The Impact of Biotechnology in Agriculture.* Proceedings of the International Conference The Meeting Point between Fundamental and Applied in vitro Culture Research (Amiens, France, 1989). 1990.
                                                                   ISBN 0-7923-0741-0

9. H.J.J. Nijkamp, L.H.W. van der Plas and J. van Aartrijk (eds.): *Progress in Plant Cellular and Molecular Biology.* Proceedings of the 8th International Congress on Plant Tissue and Cell Culture (Amsterdam, The Netherlands, 1990). 1990
                                                                   ISBN 0-7923-0873-5

10. H. Hennecke and D.P.S. Verma (eds.): *Advances in Molecular Genetics of Plant–Microbe Interactions.* Volume 1. 1991   ISBN 0-7923-1082-9

11. J. Harding, F. Singh and J.N.M. Mol (eds.): *Genetics and Breeding of Ornamental Species.* 1991   ISBN 0-7923-1094-2

12. J. Prakash and R.L.M. Pierik (eds.): *Horticulture – New Technologies and Applications.* Proceedings of the International Seminar on New Frontiers in Horticulture (Bangalore, India, 1990). 1991   ISBN 0-7923-1279-1

13. C.M. Karssen, L.C. van Loon and D. Vreugdenhil (eds.): *Progress in Plant Growth Regulation.* Proceedings of the 14th International Conference on Plant Growth Substances (Amsterdam, The Netherlands, 1991). 1992   ISBN 0-7923-1617-7

14. E.W. Nester and D.P.S. Verma (eds.): *Advances in Molecular Genetics of Plant–Microbe Interactions.* Volume 2. 1993   ISBN 0-7923-2045-X

15. C.B. You, Z.L. Chen and Y. Ding (eds.): *Biotechnology in Agriculture.* Proceedings of the First Asia-Pacific Conference on Agricultural Biotechnology (Beijing, China, 1992). 1993   ISBN 0-7923-2168-5

Current Plant Science and Biotechnology in Agriculture

KLUWER ACADEMIC PUBLISHERS – DORDRECHT / BOSTON / LONDON